# Ethnopharmacology and Drug Discovery for COVID-19: Anti-SARS-CoV-2 Agents from Herbal Medicines and Natural Products

Jen-Tsung Chen
Editor

# Ethnopharmacology and Drug Discovery for COVID-19: Anti-SARS-CoV-2 Agents from Herbal Medicines and Natural Products

Springer

*Editor*
Jen-Tsung Chen
Department of Life Sciences
National University of Kaohsiung
Kaohsiung, Taiwan

ISBN 978-981-99-3663-2 ISBN 978-981-99-3664-9 (eBook)
https://doi.org/10.1007/978-981-99-3664-9



This Springer imprint is published by the registered company Springer Nature Singapore Pte Ltd.
The registered company address is: 152 Beach Road, #21-01/04 Gateway East, Singapore 189721, Singapore

# Preface

In recent years, the global outbreak of the coronavirus disease 2019 (COVID-19) has caused a severe public health crisis and consequently significant economic losses worldwide. COVID-19 is an RNA virus disease caused by the infection of severe acute respiratory syndrome coronavirus 2 (SARS-CoV-2), which tends to mutate into new strains, therefore posing an ongoing threat to human health. In this scenario, the timely development and subsequent application of drugs and vaccines for managing COVID-19 are extremely challenging. Fortunately, traditional and herbal medicines are well known and are recognized as beneficial and therapeutic agents for various health conditions as well as for the prevention and treatment of coronavirus disease. These medicines have been used for thousands of years across different areas of the world to treat various diseases, and are now evaluated for their roles in managing COVID-19 based on the research system of ethnopharmacology. Bioactive compounds derived from certain traditional or herbal medicines have shown antiviral, anti-inflammatory, and antioxidative properties, suggesting their potential effectiveness against SARS-CoV-2.

This book first presents a complete overview of COVID-19, followed by in-depth analyses of relevant literature and a comprehensive discussion and perspectives on promising anti-SARS-CoV-2 agents based on ethnopharmacology, which covers Chinese medicine, traditional medicines of India and Africa, Turkish folk medicine, essential oils, some well-known medicinal plants, and so on. Additionally, individual chapters extensively explore and refine the roles of black cumin and Gan Cao in combating SARS-CoV-2, focusing on their possible mechanisms of action targeting viral proteins.

In this book, the current status of therapeutic strategies against COVID-19 was summarized and the authors concluded a perspective of potential options for the future. For drug discovery, phytochemicals from medical plants have shown potential as immune enhancers against SARS-CoV-2. Some well-established traditional herbal formulations have exhibited promising antivirus activity, making them potential candidates for managing COVID-19 and its post-illness phase. Nutraceuticals derived from dietary plants, spices, and fruits have been suggested to possess the ability to inhibit coronavirus activity, potentially preventing future infections or severe illness by emerging variants. Using computer-aided approaches including molecular docking and network pharmacological tools, researchers have gained insights into the possible molecular mechanisms of anti-SARS-CoV-2 activity, and these techniques can predict candidate compounds from medicinal plants that can be explored for their potential in combating COVID-19. In the context of post-illness recovery from COVID-19, bioactive compounds from selected medicinal plants have shown therapeutic potential for neurological disorders. Additionally, ethnopharmacological interventions are proposed to reduce kidney damage and enhance kidney function.

This book provides a summary of the current knowledge in the field of ethnopharmacology for the management of coronavirus diseases. It presents in-depth studies that have proven the anti-SARS-CoV-2 activity of bioactive constituents found in a number of traditional and herbal medicines, including several well-established formulations. All chapters were organized by experts with reliable publications in this field and based on scientific findings from literature and therefore making this book an ideal reference for students, teachers, and researchers in a range of related fields involving life sciences, drug discovery, natural products, bioactive compounds, ethnopharmacology, systems pharmacology, bioinformatics, traditional herbal medicines, and viral diseases. As the editor, I'm grateful to all the contributors for their valuable chapters, and the instruction and help from the editorial office of the publisher are much appreciated.

Kaohsiung, Taiwan Jen-Tsung Chen

# Contents

# Chapter 1
# COVID-19: An Overview of Virology, Mutations, Pathology, Epidemiology, Diagnosis, Preventions, and Treatments

**Priya Chaudhary, Safaet Alam, Nazim Uddin Emon, Devendra Singh, Pracheta Janmeda, Anca Oana Docea, Daniela Calina, and Javad Sharifi-Rad**

Priya Chaudhary and Safaet Alam contributed equally with all other contributors.

P. Chaudhary · P. Janmeda (✉)
Department of Bioscience and Biotechnology, Banasthali Vidyapith, Banasthali, Rajasthan, India
e-mail: pracheta@banasthali.in

S. Alam
Drug and Toxins Research Division, BCSIR Laboratories Rajshahi, Bangladesh Council of Scientific and Industrial Research, Rajshahi, Bangladesh

N. U. Emon (✉)
Department of Chemistry and Biochemistry, CEMB Program, University of Arkansas, Fayetteville, United States of America

D. Singh
Department of Chemistry, Mohanlal Sukhadia University, Udaipur, Rajasthan, India
e-mail: dsingh@mlsu.ac.in

A. O. Docea
Department of Toxicology, University of Medicine and Pharmacy of Craiova, Craiova, Romania

D. Calina (✉)
Department of Clinical Pharmacy, University of Medicine and Pharmacy of Craiova, Craiova, Romania

J. Sharifi-Rad (✉)
Facultad de Medicina, Universidad del Azuay, Cuenca, Ecuador



J.-T. Chen (ed.), *Ethnopharmacology and Drug Discovery for COVID-19: Anti-SARS-CoV-2 Agents from Herbal Medicines and Natural Products*,
https://doi.org/10.1007/978-981-99-3664-9_1

## 1.1 Introduction

Coronaviruses (CoVs) relate to a big family of viruses that are known to cause mild to moderate types of respiratory tract infections in human beings (Docea et al. 2020; Rajib et al. 2021). In the twenty-first century, there have been three notable outbreaks of coronaviruses originating from animal sources that raised concerns about global transmission (Udriştoiu et al. 2021a). After the outburst of coronavirus disease 2019 (COVID-19) caused by severe acute respiratory syndrome coronavirus-2 (SARS-CoV-2), the question of its evolution via mutation comes into view. Viral mutation can increase in frequency due to rapid epidemiology, random genetic drift, and natural selection. Because of these forces, it is very difficult to identify if a pandemic changes the result of an infection or the single-point mutation (Grubaugh et al. 2020; Rajib et al. 2021). CoVs belong to four different genera (δ, γ, β, and α) and the major CoVs responsible for infection of the respiratory tract in human beings are human CoV-229E, CoV-OC43 (Palestino et al. 2020). Severe acute respiratory syndrome (SARS), a respiratory disease, initially originated in Foschan, China, in November 2002 with about 8098 cases and thereafter spread to 29 different nations with a mortality rate of 9.6%, whereas Middle East respiratory syndrome (MERS) was recovered from the sample of lung organ of a 60-year-old dead patient in Saudi Arabia in 2012 (Fehr et al. 2017). Both the syndromes are beta coronaviruses that belong to lineages B and C and are considered a spreader of zoonotic diseases. These types of diseases are listed as a top priority by the World Health Organization (WHO) as they are posing harm to the security of the public's well-being on a global scale (Azhar et al. 2019; To et al. 2013).

Later on, in December 2019, some of the local healthcare authorities reported some studies regarding the causes of pneumonia of unknown origin in patients that were epidemiologically linked to a wet animal and seafood wholesale market in Wuhan, China (Hossain et al. 2021; Udriştoiu et al. 2021b).

After the isolation and deep genomic sequencing of the virus from the samples of the lower respiratory tract, it was determined that the virus was a new beta coronavirus whose title was eventually mentioned as SARS-CoV-2 by the International Committee on Taxonomy of Viruses (Yuen et al. 2020). The genomic sequence of the SARS-CoV-2 was submitted to WHO on Jan 12, 2020, and this initiated research in different parts of the sphere to find out a specific means of polymerase chain reaction (PCR) based test for the analysis of SARS-CoV-2 (Amodio et al. 2020). The SARS-CoV-2 belongs to group 2B and has shown about 70% of similarity with the genomic sequence of earlier SARS-CoV (Yang et al. 2020). Many inclusive factors such as smoking, drinking, and the ABO blood group have been found to contribute to the risk of various infectious diseases in a number of studies (Fan et al. 2020). With the help of laboratory tests, in which there is a reduction in the absolute value of lymphocytes, it was determined that SARS-CoV-2 primly acts on T-lymphocyte specifically the same as reported by SARS-CoV (Chen and Guo 2020). The infection of SARS-CoV-2 is found to be milder in comparison to MERS and SARS-CoVs in terms of transmissibility, mortality rate, and severity which increase the chances of cases remaining undetected (Tsatsakis et al. 2020a). Therefore, WHO recommends that people wash their hands, cover their faces with a mask, avoid consumption of undercooked animals, and people experiencing any symptoms of cough, cold, fever, or problem in breathing, are advised to visit hospitals for their medical assessment. Some vaccines and antiviral agents such as lopinavir/ritonavir and Remdesivir have shown promising effects in the MERS-CoV and SARS-CoV-infected animal models. Therefore, these drugs are being used to treat SARS-CoV-2 (Mahase 2020). This review aims to highlight the important information available for coronaviruses that need to be understood for their control and cure.

## 1.2 Review Methodology

The literature review is based on the articles and research papers published in English and indexed on different authentic medical and non-medical databases, i.e., Pubmed/MEDLINE, Web of Science, Wiley Online Library, Sciencedirect, Scopus, NCBI, Google Scholar, World Health Organization, Clinical Trials Arena, Worldometers.info, Centre for Health Security.

The next MeSH terms were used for searching: "Antiviral Agents/therapeutic use," "Betacoronavirus,' "COVID-19," "COVID-19 Testing," "COVID-19 Vaccines," "China," "Clinical Laboratory Techniques/methods," "Coronavirus Infections/diagnosis," "Coronavirus Infections/epidemiology," "Coronavirus Infections/physiopathology," "Coronavirus Infections/prevention & control," "Coronavirus Infections/therapy," "Genome, Viral," "Humans," "Pandemics/prevention & control," "Pneumonia, Viral/epidemiology," "Pneumonia, Viral/physiopathology," "Pneumonia, Viral/prevention & control," "Pneumonia, Viral/therapy," "Public Health," "SARS-CoV-2,″ "Viral Vaccines."

Results of the different database searches are reviewed to collect the matter for this article. Finally, 73 out of 252 articles about SARS-CoV, MERS-CoV, and COVID-19 were selected from the aforementioned databases to complete the literature review.

## 1.3 Emergence and Evolution of SARS-CoV-2

### *1.3.1 Known Coronaviruses*

A total of seven coronaviruses are known to date that is capable to infect a human being. Among them, HKU1, OC43, NL63, and 229E are known for the mildest infection of the upper respiratory tract, others are SARS-CoV and MERS-CoV which are known for typical pneumonia and the seventh one is SARS-CoV-2. Some of these coronaviruses are circulating in the animals such as cats, pigs, bats, and camels which jump to human beings (known as spillover events) and result in life-threatening ailments (Bassetti et al. 2020).

### 1.3.2 *Origin of Coronavirus*

According to the latest studies, bats and snakes are considered the main reservoir of SARS-CoV-2 as shown in Fig. 1.1 but according to WHO, the exact origin of SARS-CoV-2 is still unidentified and according to Wang et al. (2020a) the mortality rate of SARS-CoV-2 is about 2.84% which is lower as compared to and MERS-CoV (34.4%) and SARS-CoV (9.6%) mortality rates (Gralinski and Menachery 2020).

### 1.3.3 *Evolution of SARS-COV-2*

Through the phylogenetic analysis of about 103 genomes with the genome of SARS-CoV-2, it was determined that the two main lineages co-occurred are S-type (ancestral type) L-type (obtained after single-nucleotide polymorphism) but the majority of the cases of Wuhan was of L-type, worldwide S-type along with coinfection with both the types (reported in a USA patient) is more notable. There is no latest proof that the immunity in the control of one lineage will come up with protection against another one but this suggested that a patient cured of one lineage can go through the infection with another lineage from a different strain. Therefore, there is an urgent need for the discovery of a vaccine that must focus on the

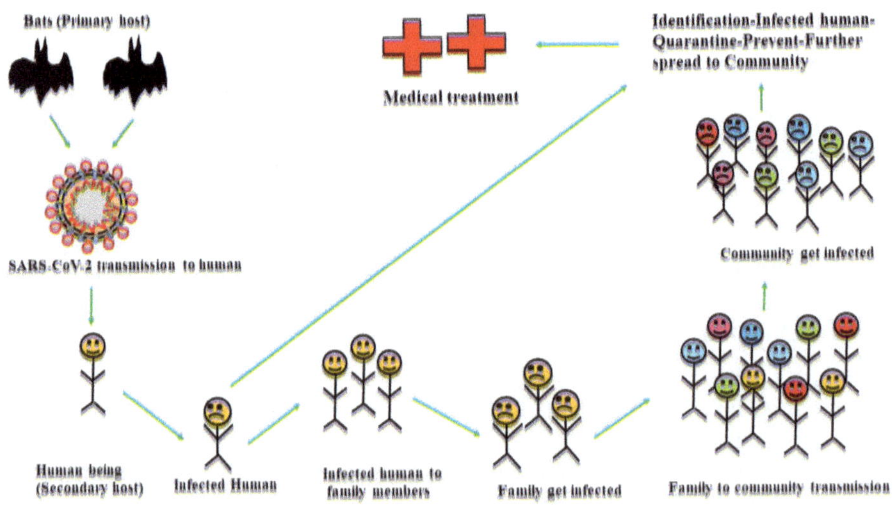


**Fig. 1.1** The possible origin and transmission pathways of coronaviruses

conserved region in both types of lineages. Additionally, a discrete vaccine for lineage must also be fabricated (Kakodkar et al. 2020).

### *1.3.4 Virology and Genome of SARS-CoV-2*

Coronaviruses (CoVs) are positive sense, enveloped, single-stranded ribonucleic acid (RNA) viruses of 26.4–31.7 kb of the genome and are greatly diverse (Hui et al. 1920; Verma et al. 2020). In the taxonomy of the virus, CoV is related to the Nidovirales order, Ortho Coronavirinae subfamily, and Coronaviridae family. Based on cryo-electron microscopy and tomography, coronavirus is spherical having ~125 nm diameter. The most notable characteristic of coronavirus is the spikes of club shape, protruding from the exterior of the virion which makes it look like a solar corona, resembling their name coronavirus (Fehr et al. 2015) as shown in Fig. 1.2.

The arrangement of the genome is generally identical in all CoVs. They contain two untranslated regions (UTRs) with coding genes, accessory ORFs, and transcription regulatory motifs (TRFs).

Four types of structural proteins are reported from the coronaviruses: spike (S) protein, nucleocapsid (N), envelope (E), and membrane (M), all of which are encoded by the 3′ end of the virus genome. The S protein encodes for the spikes of CoVs and is cleaved into S1 and S2 domains (Receptor binding and cell membrane fusion domain). The M and E genes are well sustained and code for two small transmembrane proteins in all CoVs. The N protein is a structural protein, that encodes for the packaging of the virus. Like the other CoVs, the SARS-CoV-2 also has many small ORFs (ORF 10, 9, 13, and 14) at the downstream side of N-gene but the functionality of these small ORFs and N-gene of SARS-CoV-2 is not investigated yet (Biswas et al. 2020).

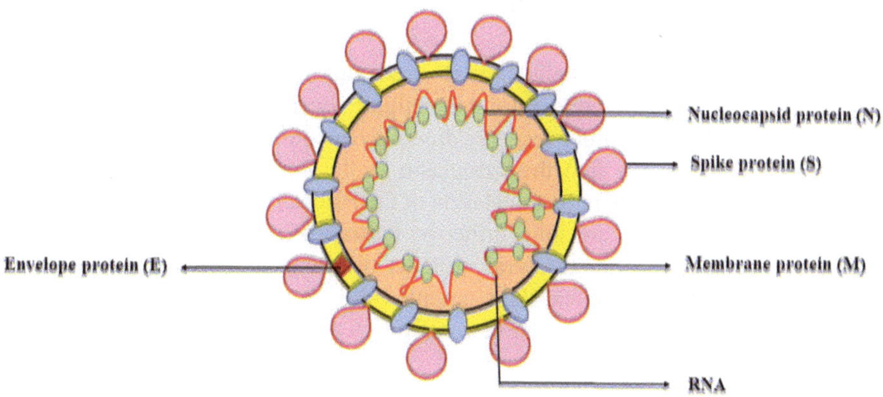


**Fig. 1.2** The structure of SARS-CoV-2

### 1.3.5 *Possible Mutations in SARS-CoV-2*

Researchers observed various point mutations in the viral genome within different geographic regions when the SARS-CoV-2 is transmitted out of Asia. Three-point mutations in positions 23,403, 14,408, and 3036 were reported in Europe while North America reported having a mutation in positions 18,060, 17,857, and 17,746 but these mutations have not been recognized in Asia. The occurrence and number of point mutations recognized in Asia are raising over time (Zhang et al. 2018).

1. *D614G mutation:* in comparison to human immunodeficiency virus (HIV), SARS-CoV-2 is changing at a very slow rate with its rate of transmission. Korber et al. (Kang et al. 2006) identify a mutation in the samples of many individuals with COVID-19. It is at the 614th position of the spike protein; the aspartate (D) amino acid was replaced by glycine (G). Virologists named this mutation D614G mutation. It was determined that the spread of G614 is more rapid as it is more infectious in comparison to D614. But to fill in the gaps that how D614G affects the pandemic, more in vivo studies are needed (Grubaugh et al. 2020).
2. *NSP6 and ORF10 mutation:* The worldwide analysis of SARS-CoV-2 genomes revealed the presence of two types of mutation, i.e., the NSP6 and the ORF10. Mutation in NSP6 could induce some alterations in the expression of the virus's relationship with its host, specifically autophagic lysosomal machinery. Based on previous studies related to SARS-CoV, it was determined that the deletion of 29 nucleotide segments disrupts the region of ORF9 with ORF10 and 11 as well. No clinical significance related to this deletion is known as these deleted segments co-exist with the non-deleted variant within the same specimen. Alterations in such types of viral regions should be frequently studied as they modify the pathogenicity of SARS-CoV-2 (Benvenuto et al. 2020).

## 1.4 Life Cycle and Physiopathology of Coronaviruses

Coronaviruses target the host cells through endocytosis. The SARS or SARS-CoV-2 and MERS-CoV protein cohere to the cellular receptor angiotensin-converting enzyme 2 (ACE2) and cellular receptor dipeptidyl-peptidase 4 (DPP4) that induces the release of viral RNA into the cytoplasm of the host cells (Islam et al. 2020; Tsatsakis et al. 2020b). Then the translation of ORF1ab and ORF1a yields pp1ab and pp1a polyproteins, which are split further by proteases enzyme and are encoded by ORF1a to produce 16 non-structural proteins resulting in the formation of RNA replicase-transcriptase complex. This complex directs the fabrication of RNAs of negative sense [(−) RNA] through the replication and transcription process. In the

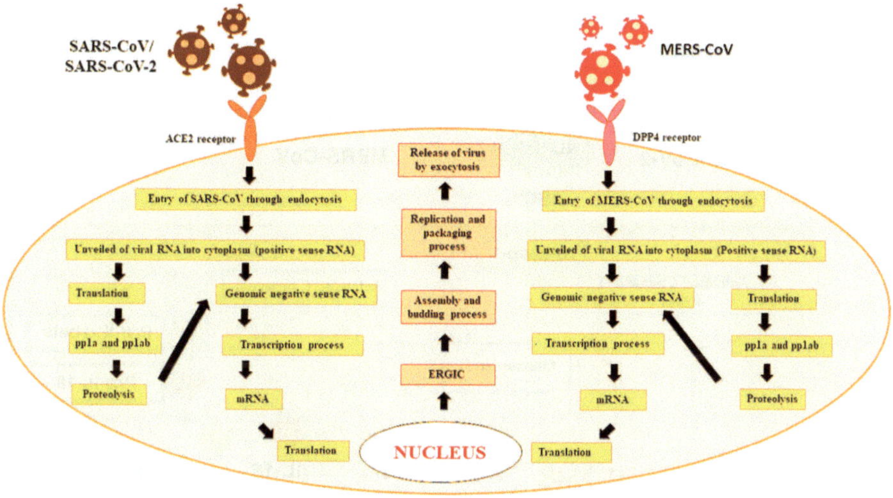


**Fig. 1.3** Release of virions through the exocytosis process from the infected host cell

course of replication, negative sense RNA copies are used as templates for the production of full-length positive sense RNA and during the transcription process, a subset of 7–9 sub-genomic RNAs is produced through the transcription of discontinuous type. Although various open reading frames (ORFs) may be present on the different sub-genomic mRNAs only the first ORF (that to the 5′ end) is being translated (Islam et al. 2021). Followed by budding into the lumen of the ERGIC (endoplasmic reticulum (ER)-Golgi intermediate compartment), nucleocapsids of the virus get assembled, and after that virions release through the exocytosis process from the infected host cell (Gathara 2020) as shown in Fig. 1.3.

Recent evidence showed the increased level of IL-1beta in the serum of SARS-CoV-2-infected individuals, which indicates cell pyroptosis. Therefore, this suggests that the activity of cell pyroptosis is somehow related to the SARS-CoV-2 pathogenesis (Yang 2020). In the Caspase-1-dependent classical pathway, when a pathogen enters a cell, a well-conserved molecular structure (pathogen-associated molecular pattern; PAMPs) present on the top of the pathogen will be identified by the NOD-like receptors protein 3 (NLRP3) on the membrane of the cell. This is a tie to the pro-caspase-1 with the help of adaptor protein ASC (Apoptosis-associated speck-like protein containing a caspase recruitment domain) of the cell to devise a multiprotein complex required for the activation of caspase-1. Activated caspase-1 takes part in innate immunity, and switches on the inflammatory cytokines of pro-inflammatory such as pro-IL-18 and pro-IL-1β in various chief inflammatory diseases, supporting the production of IL-18 and IL-1β to set an inflammatory response in cell pyroptosis. In non-classical (caspase-4, 5, 11 dependent) pathway,

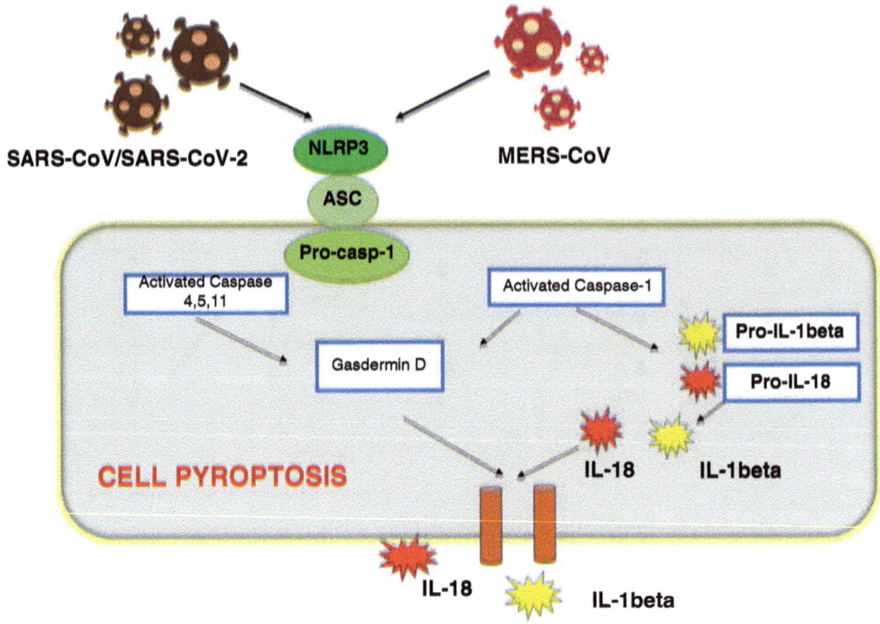


**Fig. 1.4** Cell pyroptosis signaling of MERS-CoV/ SARS-CoV/SARS-CoV-2

Pro-caspase-1, NLRP3, and adapter ASC involve together in the formation of the NLRP3 inflammasome, which is needed for the inflammation process. Unlike apoptosis, pyroptosis majorly includes the activation of caspases (4, 5, 11) and inflammation, which incises and polymerizes the members of the Gasdermin family (GSDMD). The cleavage product of the N-terminal of GSDMD could result in the release of cellular content, inflammation, swelling, and lysis to retard the replication of invaded pathogens. Molecules are named the damage-associated molecular patterns (DAMPs), also called alarm factors released by the damaged cells. Similar to the PAMPs, the DAMPs can also be identified by the NLRP3 which results in pro-inflammatory responses and pyroptosis.

Pyroptosis is reported as a quicker process and it is also followed by the production of a great quantity of pro-inflammatory factors in comparison to apoptosis which is a phagocytic, orderly shrinking, and neat process. As a result, pyroptotic cells of both pathways showed the characteristic of decreased viability, cell number, and cell death (Jiang et al. 2020) as shown in Fig. 1.4.

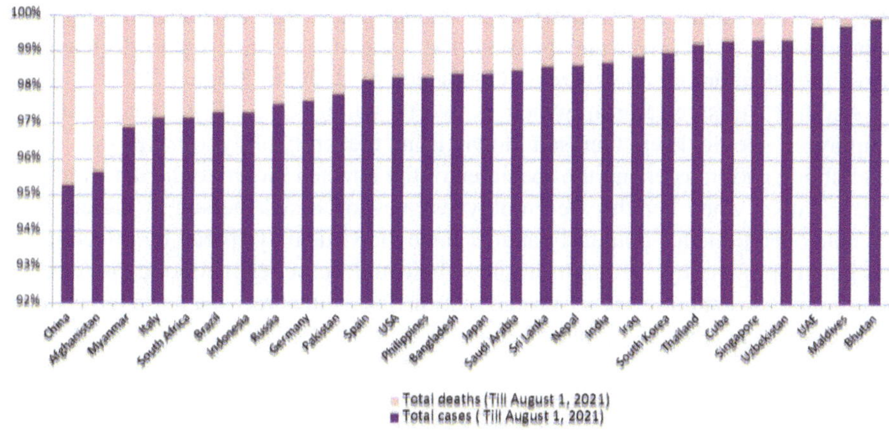


**Fig. 1.5** Pink & violet bars together indicating the total number of infected cases and pink bar solely indicating the mortality rate of SARS-CoV-2 worldwide

## 1.5 COVID-19 Epidemiology, Diagnosis, and Clinical Symptoms

### *1.5.1 Epidemiology*

#### 1.5.1.1 Statistic of COVID-19 Cases with Their Mortality Rate Across the Globe

As of now, the number of coronavirus cases is dropping all over the world, but this number is continuing to peak up along with new mortality cases in some parts of the world (Ashraf et al. 2021) as shown in Fig. 1.5 till the first August 2021.

#### 1.5.1.2 Preventions

Lots of specific treatments and vaccines for SARS-CoV-2 are developed or in the development phase (Hernández et al. 2021; Kostoff et al. 2021). Treatment is concerned and grounded on the clinical state of the patients. As a common safeguard, the public visiting barns, markets, farms, or some other locations where other animals and dromedary camels are available, should follow some general measures of hygiene, which include wearing of mask properly, washing hands regularly before and after touching these animals, keep yourself away from the ill animals and be informed all the time. Products acquired from animals that are properly processed through pasteurization or cooking are considered safe for consumption, but care should also be taken to avoid the chances of cross-contamination with uncooked foods. Immune-compromised individuals should avoid the intake of undercooked meat and the products procured from the camel (World Health Organization 2019;

Ramadan and Shaib 2019). Individuals should put an end to spreading any misconceptions and incorrect details about the SARS-CoV-2 and try to decrease the panic prevalent among the citizens (Goumenou et al. 2020; Singhal 2020).

Another measure/caution which is worth taking care of is that healthcare employees should keep themself down from the probability of exposure while assembling and transferring the specimen of the infected victim to the lab. A goggle is mandatory when health care employees question a sick person at the clinic or execute operations for doubtful patients since the virus may spread by infecting the conjunctiva of the eye through droplets. It would be better if robotics is used for questioning or handling the suspected victim. This will surely decrease the chances of exposure to well-being-related helpers (Wang et al. 2020a).

#### 1.5.1.3 Control Measures by China and Other Countries

On 31 December 2019, the WHO was notified by Chinese Health officials which had induced health officials in Taiwan, Macau, and Hong Kong to increase their border supervision. This also created a sense of fear and concern regarding the emergence of a serious and novel threat to a living being (Hui and Zumla 2019). Bus services for long-distance routes have been banned in China and watch drones have been used across the country to instruct civilians regarding safety measures if they are not taking them (Ayittey et al. 2020). Additionally, China had rapidly created hospitals and assigned them to exclusively handle patients with SARS-CoV-2, closing them off from other disease treatments (Brueck et al. 2020). However, some countries such as Hong Kong, India, Philippines, Malaysia, Papua New Guinea, North Korea, the United Kingdom, Singapore, the USA, and Japan have banned visas policy to the citizen of Hubei or China temporarily to prevent the spread of the virus to their countries (Ayittey et al. 2020). Various countries such as the USA, Australia, Japan, India, Canada, New Zealand, Malaysia, Bangladesh, Egypt, Turkey, South Korea, and some EU countries have evacuated their citizen from China. All these countries kept their citizens in detention for 14 days, which was reported as the incubation phase of SARS-CoV-2 (Ayittey et al. 2020). Across the globe, many nations (Italy, France, UK, USA, Spain, Germany, India, and many more) have enforced limitations on the motion of their citizens having strict lockdowns in their countries with specific guidelines. They have also imposed several fines on persons who avoid the rules. To break the chain of coronavirus, many nations have ramped up their testing process so that they can find out the infected cases at right time and quarantine them as soon as possible to prevent a further rise in the cases.

**Table 1.1** Laboratory and clinical tests for SARS-CoV-2

| Diagnosis | SARS-CoV-2 |
|---|---|
| Molecular diagnosis | RT-PCR<br>RNA extraction real-time RT-PCR with SARS-CoV-2-specific probes and primers tests were performed in biosafety level III services (Li et al. 2020) |
| Hematological parameters | (25%) leukopenia<br>(63%) lymphopenia<br>(5%) thrombocytopenia<br>(73%) higher level of lactate dehydrogenase prothrombin<br>IP-10 (interferon gamma-induced protein 10)<br>D-dimer<br>MIP1A (macrophage inflammatory protein alpha), creatinine kinase, C-reactive protein<br>TNF-α (tumor necrosis factor-alpha)<br>MCP1 (monocyte chemotactic protein 1)<br>GCSF (granulocyte colony-stimulating factor)<br>IL-7,10,1 along with deduction in CD8 and CD4 lymphocytes<br>(Bassetti et al. 2020; Hassan et al. 2020; Huang et al. 2020; Rajnik et al. 2021; Wang et al. 2020a; Yuan et al. 2020) |
| Radiographic features | 100% chest abnormalities<br>(bilateral lobular, subsegmental areas of consolidation, and bilateral ground-glass opacity) |
| Clinical features | For mild cases: Muscle pain, headache, sore throat, mild fever, dry cough, malaise, congestion.<br>For moderate cases: Tachypnea, cough, and breath shortness.<br>For severe cases: Respiratory distress, tachypnea, dyspnea, septic shock, severe pneumonia, acute respiratory sepsis;<br>For critical cases: multiorgan dysfunction, cardiac injury, septic shock, respiratory failure anemia (Bassetti et al. 2020; Cascella et al. 2021; Hassan et al. 2020; Wang et al. 2020a) |
| Serological tests | The detection of SARS-CoV-2: Various serological assays:<br>Neutralization<br>Protein microarray<br>Chemiluminescence assay (CLIA)<br>Western blot (WB)<br>Enzyme-linked immunosorbent assay (ELISA)<br>Immunofluorescence assay (IFA) (Yan et al. 2020) |
| P-BEST | A novel pooling-based SARS-CoV-2 testing method is easy to use and increases the capacity of testing, used in diagnostic laboratories (Shental et al. 2020). |

### *1.5.2 Laboratory and Clinical Tests*

Diagnosis of COVID-19 involves the combination of one or more techniques that are listed in Table 1.1 and can be done in a biosafety level III facility (Azhar et al. 2019; Bai et al. 2020; Neagu et al. 2021; Richardson et al. 2004).

### 1.5.3 Clinical Features

#### 1.5.3.1 Risk Factors Related to SARS-COV-2

Smoking

There is some evidence that smokers are more prone to bacterial and viral infections, as compared to non-smokers (Jain et al. 2021). The WHO has reported that the hand-to-mouth action and sharing of tobacco through water pipes are the common factors that raise the chances of smokers getting infected with SARS-CoV-2 (Grundy et al. 2020).

Drinking

Awareness about the use of alcohol-based hand sanitizers to prevent the spread of infection may arise an erroneous rumor on social media sites that consuming alcohol might be a solution against SARS-CoV-2 (Calina et al. 2021). Furthermore, based on meta-analysis and systematic review, Simou et al. (Simou et al. 2018) reported that the consumption of alcohol raises the pneumonic condition in individuals. Therefore, there is a direct correlation between the consumption of alcohol and the amount of ACER2 in the human body (Mallet et al. 2020).

Blood Group

Previous studies have established a correlation between blood group with cancer, cardiovascular, and infectious diseases. The blood group is not correlated with the risk of death and intubation in patients with SARS-CoV-2. According to Latz et al. (2020), blood type O has a low correlation to testing positive, blood types $Rh^+$, AB, and B are more correlated with positive testing, and blood type A does not correlate with a positive test. Zietz et al. (2020) reported that blood type A sufferers are at more risk of hospitalization, while blood type O has a very low risk, which indicated that the ABO blood type could be utilized as a biomarker to test the severity of novel coronavirus infection.

Clinical Symptoms

Based on the diagnosed cases, the median days from the appearance of the first symptom to the death of the individual were found to be 14 days in SARS-CoV-2 that are similar to the median days of MERS-CoV but in the case of SARS-CoV, it was 3.8 days to 17.4 days. It was found that the progression of the disease was faster in older people with a median day of 5.7 to 11.5 in comparison to younger people. Therefore, more attention should be given to older people who have the possibility of being targeted by a novel coronavirus (Bai et al. 2020).

According to clinical data published in January 2020, SARS-CoV-2 infected patients show many symptoms which include myalgia, fatigue, cough, and fever. All the infected patients reported pneumonia. About 63% of patients were investigated with lymphopenia (63%) as compared to leucopenia which was found only in 25% of the patients (Hui et al. 2020).

On other hand, the level of many inflammatory cytokines such as TNF-alpha, IFN-gamma, MCP-1, IL-10, IL-9, IL-8, IL-7, and IL-1beta was also found to be higher in the plasma of infected individuals as compared to the healthy ones. These characteristics of infection can be associated with pyroptosis of the cell (Yang 2020).

## 1.6 Potential Treatments

Unlike some recently discovered vaccines, no established therapeutics have been still established (Calina et al. 2020b) Thus, to tackle the pandemic, researchers and other related authorities are focusing on the repurposing of existing drugs. Among repurposed drugs, antiviral drugs: Remdesivir, favipiravir, lopinavir/ritonavir; antiparasitic drugs: hydroxychloroquine, chloroquine, ivermectin; antibiotic drugs: azithromycin, doxycycline, and other drugs such as ACEIs, ARBs, and rivaroxaban are most common which have initially demonstrated anti-COVID-19 actions employing multiple mechanisms of actions (Figs. 1.6, 1.7, 1.8, and 1.9) (Alam et al. 2021; Kostoff et al. 2020).

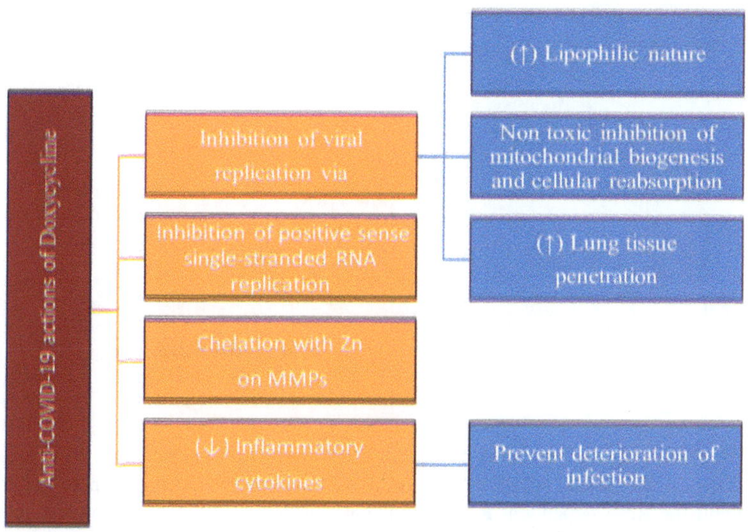


**Fig. 1.6** Doxycycline's anti-SARS-CoV-2 activities encompass viral replication suppression and positive sense single-stranded RNA replication inhibition and chelation with Zn on MMPs and down-regulating inflammatory cytokines

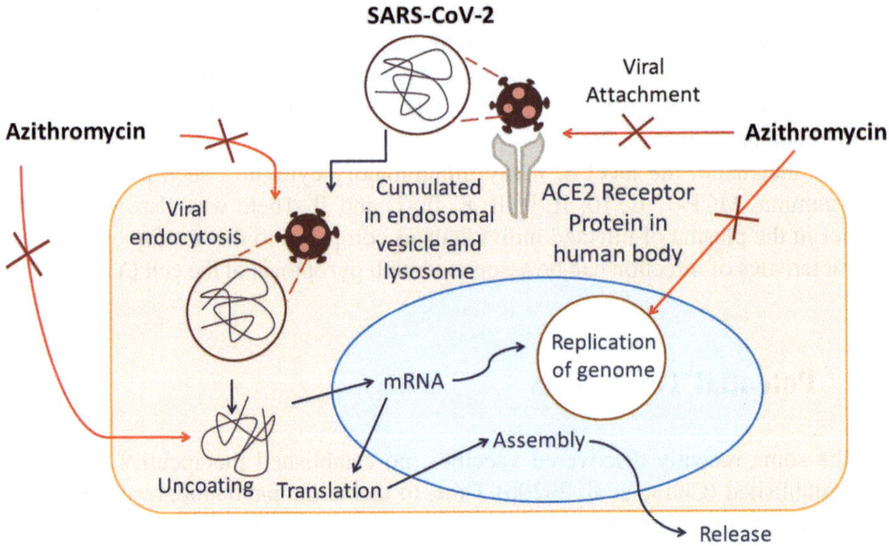


**Fig. 1.7** Azithromycin exhibits suppression of SARS-CoV-2's uncoating, modification of the host's antiviral responses mediated by interferon (IFN) pathway, and interference with the binding interaction of SARS-CoV-2 spike protein and host receptor ACE2 protein, all of which lead to a decrement in viral replication and viral endocytosis

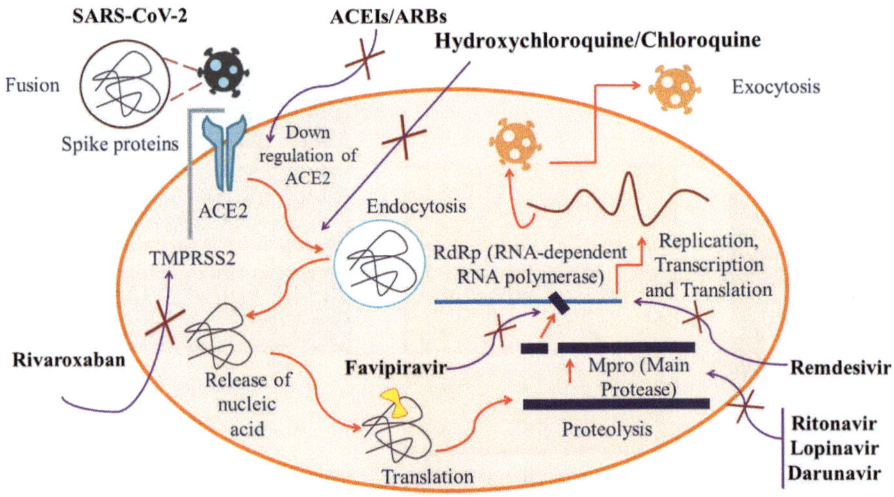


**Fig. 1.8** Down-regulation of ACE2 by ACEIs/ARBs. Favipiravir and Lopinavir/Ritonavir, respectively, inhibit the RNA-dependent RNA Polymerase (RdRp) and the main protease (Mpro). By elevating the cell pH using Hydroxychloroquine/Chloroquine, viral fusion onto the cell membrane and endocytosis can be prevented. Rivaroxaban inhibits the transmembrane protease serine 2 (TMPRSS2) enzyme which may prevent SARS-CoV-2 from infecting the host by preventing the fragmentation of spike protein into the S1 and S2 subunits

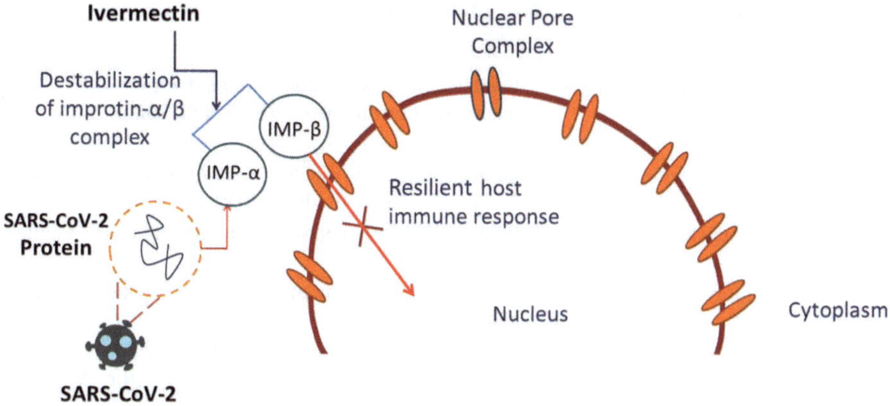


**Fig. 1.9** Ivermectin can exert anti-SARS-CoV-2 activity by impeding nuclear transport, which is caused by the instability of the IMP−/α/β complex followed by disruption of RNA viruses

On the contrary, after getting the initial approval few drugs have been deprecated by WHO, i.e., Remdesivir, lopinavir-ritonavir, hydroxychloroquine, and ivermectin, and FDA monotherapy and/or co-therapy of chloroquine and hydroxychloroquine (Sarkar et al. 2020; Torequl Islam et al. 2020). Different research groups in Hong Kong, Australia, Singapore, the USA, Germany, India, Canada, the UK, and China have reinforced the discovery of a particular vaccine for the treatment of novel coronavirus (Calina et al. 2020a, c; Huang et al. 2020). The possible treatments available for the control of the epidemic are listed in Table 1.2.

**Table 1.2** Possible treatments for SARS-CoV-2 infection

| Possible treatments for SARS-CoV-2 | Functions |
|---|---|
| Support oxygen therapy with high-flow nasal oxygen (HFNO) | Used in case of respiratory failure, decreases the requirement for intubation (Setti et al. 2020) |
| Allopathic treatments | |
| Corticosteroid (Dexamethasone) | Decreases the death rate by one-third among SARS-CoV-2 infected patients (Ledford 2020) |
| Antiviral (Remdesivir GS5734) | Inhibits RNA polymerase in a number of viruses (Gordon et al. 2020) |
| Antiviral (Arbidol) | Reduces the death rate of SARS-CoV-2 patients (Wang et al. 2020b) |
| Antiviral drug (Hydroxychloroquine) | Reduces the viral load (Gautret et al. 2020) |
| Ivermectin | Approved by FDA for the control of SARS-CoV-2 replication in vitro (Kramy 2020) |
| Anti-flu drug (Oseltamivir) | Used to treat SARS-CoV-2 sufferers (Chen et al. 2020) |
| Anticoagulant (Enoxaparin) | In case of known thrombosis or thrombophilia, an anti-coagulant is required (Kollias et al. 2020) |
| Inflammation inhibitors (Tocilizumab) | Anti-IL-6 receptor antibody (Berry and Fontana 2020) |
| Inflammation inhibitors (Sarilumab) | Anti-IL-6 receptor antibody (Berry and Fontana 2020) |
| Inflammation inhibitors (Anakinra) | Recombinant IL-1 receptor antagonists are utilized in the case of autoinflammatory disorders (Cavalli et al. 2020) |
| Inflammation inhibitors (Acalabrutinib) | Selective Bruton tyrosine inhibitors regulate the signaling and activation of macrophages (Roschewski et al. 2020) |
| Traditional treatments | |
| Qingfeipaidu decoction | Controlled the symptoms of SARS-CoV-2 (Ren et al. 2020) |
| Lianhuaqingwen | Protect control of SARS-CoV-2 (Runfeng et al. 2020) |
| Shuanghuanglian oral liquid | Carried out the inhibition of novel coronavirus (Wang et al. 2020b) |
| Ethnomedicinal plants | Bioactive compounds with antiviral activity (Verma et al. 2021) |
| The auxiliary blood purification system | |
| Extracorporeal blood purification technology | Eliminate the storm of cytokine, remove the inflammatory factors, correct the imbalance of electrolytes, and maintain the balance of acid-base to control the capacity load of patients in an effective way (Ke et al. 2020) |
| Yoga therapy | |
| Bhastrika pranayama | Free the lungs from unnecessary microbes and air and drains the excessive mucous out from the nasal mucosa (Chaudhary and Janmeda 2020) |
| AnulomVilom pranayama | Removes congestion, maintains a free flow of energy, boosts up the immune system, and helps to overcome the imbalance of the autonomous nervous system (Chaudhary and Janmeda 2020) |
| Kapalabhati | Clears respiratory blockages, improves mental function (Chaudhary and Janmeda 2020) |

(continued)

**Table 1.2** (continued)

| Possible treatments for SARS-CoV-2 | Functions |
|---|---|
| Meditation | Maintain a healthy mind, improve immunity, and micro-circulation and reduce the level of stress in the body (Reddy et al. 2021) |
| Vaccines | |
| Inactivated vaccine | CoronaVac by Sinovac Biotech, China, and another by Bharat Biotech, India which targets nucleus, envelope, matrix, and spike proteins developed by Sinovac and under phase III trial (Krammer 2020) |
| Live attenuated vaccine | Developed by the Serum Institute of India and Codagenix; this vaccine was attenuated with the help of codon de-optimization and under the pre-clinical stage (Krammer 2020) |
| RNA vaccine | RNA vaccines developed by the Chinese Liberation Army and Imperial College London are under phase I; by Arcturus and CureVac under phase I/II trial, by Moderna and Pfizer under phase III (Ewer et al. 2021; Jackson et al. 2020) |
| Viral vectors vaccine | ChAdOx1nCoV-19 by CanSino and Janssen; another by Gamaleya Research Institute is under phase III and by ReiThera is under phase I (Krammer 2020) |

## 1.7 Conclusions

COVID-19 is one of the most catastrophic pandemics of the last century. As the COVID-19 outbreak spreads across the world, public awareness of the disease has grown, with more information available now than only a few years ago. More investigation is still required especially for public health and clinical assessments to effectively control and cure this viral infection. With an alarming rise in infected sufferers though with a ray of hope, globally many vaccine developments are under trial and some of them are already approved by many nations for use. Mass awareness should be created about the virus since it is highly transmissible. Besides, awareness regarding the usage of repurposing drugs against SARS-CoV-2 should also be raised as it can cause serious drug resistance cases and can cost us a lot in near future. To recapitulate, the future evolution of this disease is still not clear, thus the public should be very cautious in dealing with the virus since it is highly transmissible.

**Disclosure Statement** No potential conflict of interest was reported by the authors.

**Funding** This work is not supported by any funding agency or institution.

**Ethical Approval** Not applicable.

**Consent to Publication** Not applicable.

**Author Contributions** All authors made a significant contribution to the work reported, whether that is in the conception, study design, execution, acquisition of data, analysis, and interpretation, or in all these areas; that is, revising or critically reviewing the article; giving final approval of the version to be published; agreeing on the journal to which the article has been submitted; and confirming to be accountable for all aspects of the work. All authors have read and agreed to the published version of the manuscript.

## References

Alam S, Kamal TB, Sarker MMR, Zhou J-R, Rahman SA, Mohamed IN (2021) Therapeutic effectiveness and safety of repurposing drugs for the treatment of COVID-19: position standing in 2021. Front Pharmacol 12:659577

Amodio E, Vitale F, Cimino L, Casuccio A, Tramuto F (2020) Outbreak of novel coronavirus (SARS-CoV-2): first evidences from international scientific literature and pending questions healthcare. Healthcare (Basel) 8:51

Ashraf O, Virani A, Cheema T (2021) COVID-19: an update on the epidemiological, clinical, preventive, and therapeutic management of 2019 novel coronavirus disease. Crit Care Nurs Q 44:128

Ayittey FK, Dzuvor C, Ayittey MK, Chiwero NB, Habib A (2020) Updates on Wuhan 2019 novel coronavirus epidemic. J Med Virol 92:403

Azhar EI, Hui DS, Memish ZA, Drosten C, Zumla A (2019) The middle east respiratory syndrome (MERS). Infect Dis Clin 33:891–905

Bai Y, Yao L, Wei T, Tian F, Jin D-Y, Chen L, Wang M (2020) Presumed asymptomatic carrier transmission of COVID-19. JAMA 323:1406–1407

Bassetti M, Vena A, Giacobbe DR (2020) The novel Chinese coronavirus (2019-nCoV) infections: challenges for fighting the storm. Wiley Online Library

Benvenuto D et al (2020) Evolutionary analysis of SARS-CoV-2: how mutation of Non-Structural Protein 6 (NSP6) could affect viral autophagy. J Infect 81:e24–e27

Berry SK, Fontana RJ (2020) Potential treatments for SARS-CoV-2 infection. Clin Liver Dis 15: 181

Biswas A, Bhattacharjee U, Chakrabarti AK, Tewari DN, Banu H, Dutta S (2020) Emergence of Novel Coronavirus and COVID-19: whether to stay or die out? Crit Rev Microbiol 46:182–193

Brueck H, Miller AM, Feder S (2020) China took at least 12 strict measures to control the coronavirus. They could work for the US, but would likely be impossible to implement. Business Insider

Calina D et al (2020a) Towards effective COVID-19 vaccines: updates, perspectives and challenges (review). Int J Mol Med 46:3–16. https://doi.org/10.3892/ijmm.2020.4596

Calina D et al (2020b) COVID-19 vaccines: ethical framework concerning human challenge studies. Daru 28:807–812. https://doi.org/10.1007/s40199-020-00371-8

Calina D et al (2020c) Recent advances, approaches and challenges in targeting pathways for potential COVID-19 vaccines development. Immunol Res 68:315–324. https://doi.org/10.1007/s12026-020-09154-4

Calina D et al (2021) COVID-19 pandemic and alcohol consumption: impacts and interconnections. Toxicol Rep 8:529–535. https://doi.org/10.1016/j.toxrep.2021.03.005

Cascella M, Rajnik M, Aleem A, Dulebohn S, Di Napoli R (2021) Features, evaluation, and treatment of coronavirus (COVID-19). StatPearls

Cavalli G et al (2020) Interleukin-1 blockade with high-dose anakinra in patients with COVID-19, acute respiratory distress syndrome, and hyperinflammation: a retrospective cohort study. Lancet Rheumatol 2:e325–e331

Chaudhary P, Janmeda P (2020) Spreadable virus of the 21st centuary: COVID-19. Environ Conserv J 21:13–18

Chen YLQ, Guo D (2020) Emerging coronaviruses: genome structure, replication, parthenogenesis. J Virol 92:418–423

Chen N et al (2020) Epidemiological and clinical characteristics of 99 cases of 2019 novel coronavirus pneumonia in Wuhan, China: a descriptive study. Lancet 395:507–513

Docea AO et al (2020) A new threat from an old enemy: re-emergence of coronavirus (review). Int J Mol Med 45:1631–1643. https://doi.org/10.3892/ijmm.2020.4555

Ewer KJ et al (2021) T cell and antibody responses induced by a single dose of ChAdOx1 nCoV-19 (AZD1222) vaccine in a phase 1/2 clinical trial. Nat Med 27:270–278

Fan Q, Zhang W, Li B, Li D-J, Zhang J, Zhao F (2020) Association between ABO blood group system and COVID-19 susceptibility in Wuhan. Front Cell Infect Microbiol 10:404

Fehr AR, HJ PSM, Bickerton E, Britton P. An overview of their replication and pathogenesis; Section 2 genomic organization. Methods Mol Biol Springer 2015;1282:1–23.

Fehr AR, Channappanavar R, Perlman S (2017) Middle East respiratory syndrome: emergence of a pathogenic human coronavirus. Annu Rev Med 68:387–399

Gathara P (2020) Kenya is turning a public health crisis into a law-and-order one. Washington Post

Gautret P, Jc L, Parola P, Hoang V, Meddeb L, Mailhe M, Doudier B, Courjon J, Giordanengo V, Vieira V et al (2020) Hydroxychloroquine and azithromycin as a treatment of cOVId-19: results of an open-label non-randomized clinical trial. Int J Antimicrob Agents 56:105949

Gordon CJ, Tchesnokov EP, Feng JY, Porter DP, Götte M (2020) The antiviral compound Remdesivir potently inhibits RNA-dependent RNA polymerase from Middle East respiratory syndrome coronavirus. J Biol Chem 295:4773–4779

Goumenou M et al (2020) COVID-19 in Northern Italy: an integrative overview of factors possibly influencing the sharp increase of the outbreak (review). Mol Med Rep 22:20–32. https://doi.org/10.3892/mmr.2020.11079

Gralinski L, Menachery V (2020) Return of the coronavirus: 2019-nCoV. Viruses 12:135

Grubaugh ND, Hanage WP, Rasmussen AL (2020) Making sense of mutation: what D614G means for the COVID-19 pandemic remains unclear. Cell 182:794–795

Grundy EJ, Suddek T, Filippidis FT, Majeed A, Coronini-Cronberg S (2020) Smoking, SARS-CoV-2 and COVID-19: a review of reviews considering implications for public health policy and practice. Tob Induc Dis 18:58

Hassan SA, Sheikh FN, Jamal S, Ezeh JK, Akhtar A (2020) Coronavirus (COVID-19): a review of clinical features, diagnosis, and treatment. Cureus 12:e7355

Hernández AF, Calina D, Poulas K, Docea AO, Tsatsakis AM (2021) Safety of COVID-19 vaccines administered in the EU: should we be concerned? Toxicol Rep 8:871–879. https://doi.org/10.1016/j.toxrep.2021.04.003

Hossain R et al (2021) In silico screening of natural products as potential inhibitors of SARS-CoV-2 using molecular docking simulation. Chin J Integr Med 28:249. https://doi.org/10.1007/s11655-021-3504-5

Huang C et al (2020) Clinical features of patients infected with 2019 novel coronavirus in Wuhan, China. Lancet 395:497–506

Hui DS, Zumla A (2019) Severe acute respiratory syndrome: historical, epidemiologic, and clinical features. Infect Dis Clin 33:869–889

Hui DS et al (1920) The continuing epidemic threat of novel coronaviruses to global health-the latest novel coronavirus outbreak in Wuhang, China. Int J Infect Dis 91:264–266

Hui DS et al (2020) The continuing 2019-nCoV epidemic threat of novel coronaviruses to global health—the latest 2019 novel coronavirus outbreak in Wuhan, China. Int J Infect Dis 91:264–266

Islam MT, Salehi B, Karampelas O, Sharifi-Rad J, Docea AO, Martorell M, Calina D (2020) High skin melanin content, vitamin D deficiency and immunity: potential interference for severity of COVID-19. Farmacia 68:970–983. https://doi.org/10.31925/farmacia.2020.6.3

Islam MT et al (2021) The role of HMGB1 in the immune response to SARS-COV-2 infection: from pathogenesis towards a new potential therapeutic target. Farmacia 69:621–634. https://doi.org/10.31925/farmacia.2021.4.1

Jackson LA et al (2020) An mRNA vaccine against SARS-CoV-2—preliminary report. N Engl J Med 383(20):1920–1931

Jain D et al (2021) Tobacco smoking and liver *cancer* risk: potential avenues for carcinogenesis. J Oncol 2021:5905357. https://doi.org/10.1155/2021/5905357

Jiang M, Qi L, Li L, Li Y (2020) The caspase-3/GSDME signal pathway as a switch between apoptosis and pyroptosis in cancer. Cell Death Discov 6:1–11

Kakodkar P, Kaka N, Baig M (2020) A comprehensive literature review on the clinical presentation, and management of the pandemic coronavirus disease 2019 (COVID-19). Cureus 12(4):e7560

Kang S-W, Rane NS, Kim SJ, Garrison JL, Taunton J, Hegde RS (2006) Substrate-specific translocational attenuation during ER stress defines a pre-emptive quality control pathway. Cell 127:999–1013

Ke C, Wang Y, Zeng X, Yang C, Hu Z (2020) 2019 Novel coronavirus disease (COVID-19) in hemodialysis patients: a report of two cases. Clin Biochem 81:9–12

Kollias A, Kyriakoulis KG, Dimakakos E, Poulakou G, Stergiou GS, Syrigos K (2020) Thromboembolic risk and anticoagulant therapy in COVID-19 patients: emerging evidence and call for action. Br J Haematol 189:846–847

Kostoff RN et al (2020) Vaccine- and natural infection-induced mechanisms that could modulate vaccine safety. Toxicol Rep 7:1448–1458. https://doi.org/10.1016/j.toxrep.2020.10.016

Kostoff RN, Calina D, Kanduc D, Briggs MB, Vlachoyiannopoulos P, Svistunov AA, Tsatsakis A (2021) Why are we vaccinating children against COVID-19? Toxicol Rep 8:1665–1684. https://doi.org/10.1016/j.toxrep.2021.08.010

Krammer F (2020) SARS-CoV-2 vaccines in development. Nature 586:516–527

Kramy P (2020) Terapi potensial terkini corona virus disease 2019 (COVID19). Cermin Dunia Kedokteran 47:297–298

Latz CA et al (2020) Blood type and outcomes in patients with COVID-19. Ann Hematol 99:2113–2118

Ledford H (2020) Coronavirus breakthrough: dexamethasone is first drug shown to save lives. Nature 582:469–469

Li J, Wang X, Chen J, Cai Y, Deng A, Yang M (2020) Association between ABO blood groups and risk of SARS-CoV-2 pneumonia. Br J Haematol 190:24–27

Mahase E (2020) China coronavirus: what do we know so far? BMJ 368:m308

Mallet J, Dubertret C, Le Strat Y (2020) Addictions in the COVID-19 era: current evidence, future perspectives a comprehensive review. Prog Neuropsychopharmacol Biol Psychiatry 106:110070

Neagu M et al (2021) Back to basics in COVID-19: antigens and antibodies-completing the puzzle. J Cell Mol Med 25:4523–4533. https://doi.org/10.1111/jcmm.16462

Palestino G, García-Silva I, González-Ortega O, Rosales-Mendoza S (2020) Can nanotechnology help in the fight against COVID-19? Expert Rev Anti-Infect Ther 18:849–864

Rajib H et al (2021) Amentoflavone, new hope against SARS-CoV-2: an outlook through its scientific records and an in silico study. Pharmacogn Res 13:149

Rajnik M, Cascella M, Cuomo A, Dulebohn SC, Di Napoli R (2021) Features, evaluation, and treatment of coronavirus (COVID-19). Uniformed Services University Of The Health Sciences

Ramadan N, Shaib H (2019) Middle East respiratory syndrome coronavirus (MERS-CoV): a review. Germs 9:35

Reddy RK, Charles WN, Sklavounos A, Dutt A, Seed PT, Khajuria A (2021) The effect of smoking on COVID-19 severity: a systematic review and meta-analysis. J Med Virol 93:1045–1056
Ren J-L, Zhang A-H, Wang X-J (2020) Traditional Chinese medicine for COVID-19 treatment. Pharmacol Res 155:104743
Richardson SE, Tellier R, Mahony J (2004) The laboratory diagnosis of severe acute respiratory syndrome: emerging laboratory tests for an emerging pathogen. Clin Biochem Rev 25:133
Roschewski M et al (2020) Inhibition of Bruton tyrosine kinase in patients with severe COVID-19. Science Immunol 5(48):eabd0110
Runfeng L et al (2020) Lianhuaqingwen exerts anti-viral and anti-inflammatory activity against novel coronavirus (SARS-CoV-2). Pharmacol Res 156:104761
Sarkar C et al (2020) Potential therapeutic options for COVID-19: current status, challenges, and future perspectives. Front Pharmacol 11:572870. https://doi.org/10.3389/fphar.2020.572870
Setti L et al (2020) Airborne transmission route of COVID-19: why 2 meters/6 feet of inter-personal distance could not be enough. Multidisciplinary Digital Publishing Institute
Shental N et al (2020) Efficient high-throughput SARS-CoV-2 testing to detect asymptomatic carriers. Sci Adv 6:eabc5961
Simou E, Britton J, Leonardi-Bee J (2018) Alcohol and the risk of pneumonia: a systematic review and meta-analysis. BMJ Open 8:e022344
Singhal T (2020) Review on COVID19 disease so far. Indian J Pediatr 87:281–286
To KK, Hung IF, Chan JF, Yuen K-Y (2013) From SARS coronavirus to novel animal and human coronaviruses. J Thorac Dis 5:S103
Torequl Islam M et al (2020) A perspective on emerging therapeutic interventions for COVID-19 front. Public Health 8:281. https://doi.org/10.3389/fpubh.2020.00281
Tsatsakis A et al (2020a) SARS-CoV-2 pathophysiology and its clinical implications: an integrative overview of the pharmacotherapeutic management of COVID-19. Food Chem Toxicol 146: 111769. https://doi.org/10.1016/j.fct.2020.111769
Tsatsakis A et al (2020b) COVID-19, an opportunity to reevaluate the correlation between long-term effects of anthropogenic pollutants on viral epidemic/pandemic events and prevalence. Food Chem Toxicol 141:111418. https://doi.org/10.1016/j.fct.2020.111418
Udriştoiu AL et al (2021a) COVID-19 and artificial intelligence: an approach to forecast the severity of diagnosis. Life 11:1281
Udriştoiu AL et al (2021b) COVID-19 and artificial intelligence: an approach to forecast the severity of diagnosis. Life (Basel) 11:1281. https://doi.org/10.3390/life11111281
Verma D et al (2020) Canthin-6-One 9-O-Beta-Glucopyranoside: an inhibitor of SARS-CoV-2 (COVID 19) proteases PLpro and Mpro/3CLpro
Verma D et al (2021) Potential inhibitors of SARS-CoV-2 (COVID 19) proteases PLpro and Mpro/3CLpro: molecular docking and simulation studies of three pertinent medicinal plant natural components. Curr Res Pharmacol Drug Discov 2:100038
Wang W, Tang J, Wei F (2020a) Updated understanding of the outbreak of 2019 novel coronavirus (2019-nCoV) in Wuhan, China. J Med Virol 92:441–447
Wang Z, Yang B, Li Q, Wen L, Zhang R (2020b) Clinical features of 69 cases with coronavirus disease 2019 in Wuhan, China. Clin Infect Dis 71:769–777
World Health Organization (2019) Infection prevention and control during health care for probable or confirmed cases of Middle East respiratory syndrome coronavirus (MERS-CoV) infection: interim guidance: updated October 2019. World Health Organization
Yan Y, Chang L, Wang L (2020) Laboratory testing of SARS-CoV, MERS-CoV, and SARS-CoV-2 (2019-nCoV): current status, challenges, and countermeasures. Rev Med Virol 30:e2106
Yang M (2020) Cell pyroptosis, a potential pathogenic mechanism of 2019-nCoV infection. Available at SSRN 3527420
Yang X et al (2020) Clinical course and outcomes of critically ill patients with SARS-CoV-2 pneumonia in Wuhan, China: a single-centered, retrospective, observational study. Lancet Respir Med 8:475–481
Yuan X et al (2020) Changes of hematological and immunological parameters in COVID-19 patients. Int J Hematol 112:553–559

Yuen K-S, Ye Z-W, Fung S-Y, Chan C-P, Jin D-Y (2020) SARS-CoV-2 and COVID-19: the most important research questions. Cell Biosci 10:1–5
Zhang C et al (2018) Gut microbiota dysbiosis in male patients with chronic traumatic complete spinal cord injury. J Transl Med 16:1–16
Zietz M, Zucker J, Tatonetti NP (2020) Testing the association between blood type and COVID-19 infection, intubation, and death. MedRxiv

# Chapter 2
# The Recent Development of Therapeutic Strategies Against COVID-19

**Hai-Long Zhang**

## 2.1 Introduction

Coronavirus disease 2019 (COVID-19) is a contagious disease, which is caused by severe acute respiratory syndrome coronavirus-2 (SARS-CoV-2). COVID-19 can cause death and some symptoms, which usually include fever, dry cough, tiredness, sore throat, etc., were found vary between variants (Williams et al. 2019; Azer 2020). The capacity of infection also differs between variants; for example, the Delta variant has been found to cause more infections due to its faster spread than its previous variants. It had led to a resurgence of COVID-19 in the United States, India, and many other countries. Globally, there have been about 515 million confirmed cases of COVID-19 on 9 May 2022 (World Health Organization n.d.-a). It was reported approximately 81 million confirmed cases of COVID-19 on 9 May 2022 in the United States (World Health Organization n.d.-b).

SARS-CoV-2 has caused severe damage to public health, also reducing international activity and consequently weakening the global economy. Importantly, the virus has a high tendency to develop new variants; therefore, it is significant to develop an effective therapeutic strategy against its fast-changing characteristic. The scientists must ask a question about what kind of therapeutic strategies should be used in combating COVID-19. Indeed, the factors of time, efficacy, and safety are the most important considerations in combating COVID-19. Surely, it is very critical to choose a therapeutic strategy against COVID-19 by considering the coordination of these factors. This chapter provides new insights into a therapeutic strategy against COVID-19 by patent evidence and proposes a novel therapeutic strategy against COVID-19 for the future.

H.-L. Zhang (✉)
Central International Intellectual Property (Baotou) Co., Ltd, Baotou, China

 
J.-T. Chen (ed.), *Ethnopharmacology and Drug Discovery for COVID-19: Anti-SARS-CoV-2 Agents from Herbal Medicines and Natural Products*,
https://doi.org/10.1007/978-981-99-3664-9_2

To know about the current therapeutic strategy against COVID-19, a granted patent analysis was conducted, including granted patent documents of China, the United States, Russia, Japan, and Korea. The keywords of TA "COVID-19" or "SARS-CoV-2" were used to search and identify relevant patent documents. All granted patents from December 2019 to May 2022 were retrieved.

## 2.2 Patent Filing Profile

There are about 163 patent documents in China, 142 patent documents in Russia, and 172 patent documents in the USA, as shown in Fig. 2.1. China, the USA, and Russia play a leading role in the number of patents related to COVID-19 (Figs. 2.1 and 2.2). Japan and Korea has fewer number of patients compared to China, the USA, and Russia. China, the USA, and Russia have lots of granted patents, as shown in Table 2.1, which lists the recently granted patents from the three countries. One reason is that the patents of these three countries have the fastest examination procedure for filing patents. On the other hand, this revealed the three countries may have some better policies or regulations for patents, and research and development when compared to others.

## 2.3 Patent Distribution

Federalnoe Gosudarstvennoe from Russia had the strongest patent position surrounding COVID-19, which filed 21 granted patents. The Academy of Military Medical Sciences from China and Federalnoe Byudzhetnoe Uchrezhdenie from

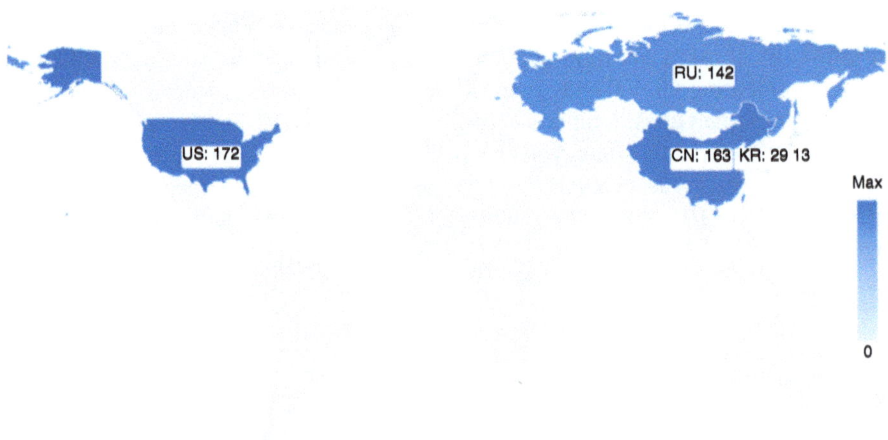


**Fig. 2.1** Patent filing profile trend from the five countries

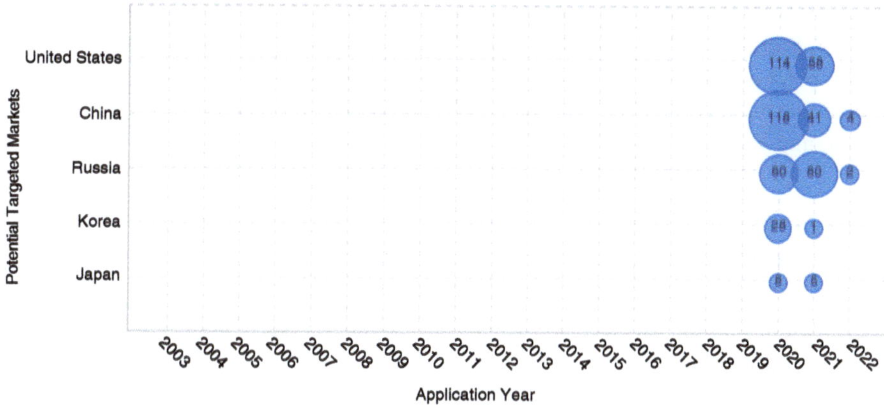


**Fig. 2.2** Patent filing profile trends by different years

**Table 2.1** Recent granted patents related to COVID-19

| Publication number | Application date | Publication date |
|---|---|---|
| RU2769220C1 | 2022-02-10 | 2022-03-29 |
| CN114150004B | 2022-02-09 | 2022-04-22 |
| CN114150005B | 2022-02-09 | 2022-04-22 |
| CN114107225B | 2022-01-25 | 2022-04-19 |
| CN114085930B | 2022-01-21 | 2022-04-12 |
| RU2766285C1 | 2022-01-13 | 2022-03-11 |
| RU2769224C1 | 2021-12-30 | 2022-03-29 |
| RU2765729C1 | 2021-12-29 | 2022-02-02 |
| CN113981152B | 2021-12-28 | 2022-03-25 |
| RU2769828C1 | 2021-12-28 | 2022-04-06 |
| RU2765731C1 | 2021-12-22 | 2022-02-02 |
| RU2769223C1 | 2021-12-22 | 2022-03-29 |
| CN113913561B | 2021-12-15 | 2022-03-04 |
| CN113913406B | 2021-12-14 | 2022-04-08 |
| RU2767624C1 | 2021-12-08 | 2022-03-18 |
| CN113881812B | 2021-12-03 | 2022-02-15 |
| CN113862286B | 2021-12-03 | 2022-03-04 |
| CN113862285B | 2021-12-03 | 2022-03-04 |
| RU2761904C1 | 2021-11-26 | 2021-12-13 |
| RU2770357C1 | 2021-10-28 | 2022-04-15 |
| RU2764954C1 | 2021-10-11 | 2022-01-24 |
| RU2764024C1 | 2021-09-30 | 2022-01-12 |
| RU2764015C1 | 2021-09-30 | 2022-01-12 |
| RU2764022C1 | 2021-09-30 | 2022-01-12 |
| RU2764020C1 | 2021-09-30 | 2022-01-12 |
| RU2764023C1 | 2021-09-30 | 2022-01-12 |

Russia were also prolific assignees in patent filing, which filed a total of 22 granted patents. Figure 2.3 shows the number of patents of the top ten applicants who filed over 6 patents. Around 16.6% of the patents were assigned to the top ten applicants, and approximately 33.1%, 31.4%, and 27% of the patents were assigned to the USA, China, and Russia, respectively.

## 2.4 Technology Areas

As shown in Fig. 2.4, there were five “peaks” in the 3D landscape. Peak 1 represents the patent documents related to traditional Chinese medicines and chemical drugs. Peak 2 and peak 3 represent the patent documents related to vaccines and antibodies. Peak 4 and peak 5 cover the patent documents related to detecting methods, physical protection, and medicinal devices.

There are four therapeutic or prevention strategies against COVID-19, which are based on traditional Chinese medicines, chemical drugs, vaccines, and antibodies. The patents related to traditional Chinese medicines and chemical drugs were clustered in peak 1, as shown in Fig. 2.4.

Peak 1 covered the technologies of traditional Chinese medicines and chemical drugs, e.g., *Centipeda minima* (L.) in preparation application for the treatment of COVID-19 and transmucosal dosage forms of Remdesivir.

Peak 2 focused on the technological area of vaccines for COVID-19, e.g., inactive vaccines and their application, mRNA vaccines and their application, etc. This peak appeared in the 3D landscape a few months after the outbreak of the pandemic.

Peak 3 concerned with the antibodies of COVID-19 and antibody application in the clinic, e.g., antibodies of SARS-CoV-2 and methods thereof. In Fig. 2.4, the 3D landscape shows that the timing of peak 3 (vaccine events) appeared later than peak 2 (antibody events). The number of patents on antibodies was the same as the number of patents on vaccines according to the size of the two peaks.

Peak 4 focused on the technology in assays and detecting methods of SARS-CoV-2, e.g., assays for the detection of SARS-CoV-2. The number of patents related to the detecting method of SARS-CoV-2 was a large proportion of all granted patent documents.

Peak 5 was engaged in physical protection and medicinal devices, e.g., face shield, and the medical devices.

Notably, as shown in Fig. 2.4, peak 3 was slim in the 3D landscape when compared with other peaks, and it indicated that patents related to antibodies showed more similarity. In contrast to peak 3, peak 5 and peak 1 showed broad peaks in the 3D landscape, which indicated less similarity among the patents.

The patents related to traditional Chinese medicines were only filed by applicants from China. Chinese applicants filed patents covering traditional Chinese medicines, chemical drugs, vaccines, and antibodies. The applicants from the USA and Russia filed patents that only cover chemical drugs, vaccines, and antibodies.

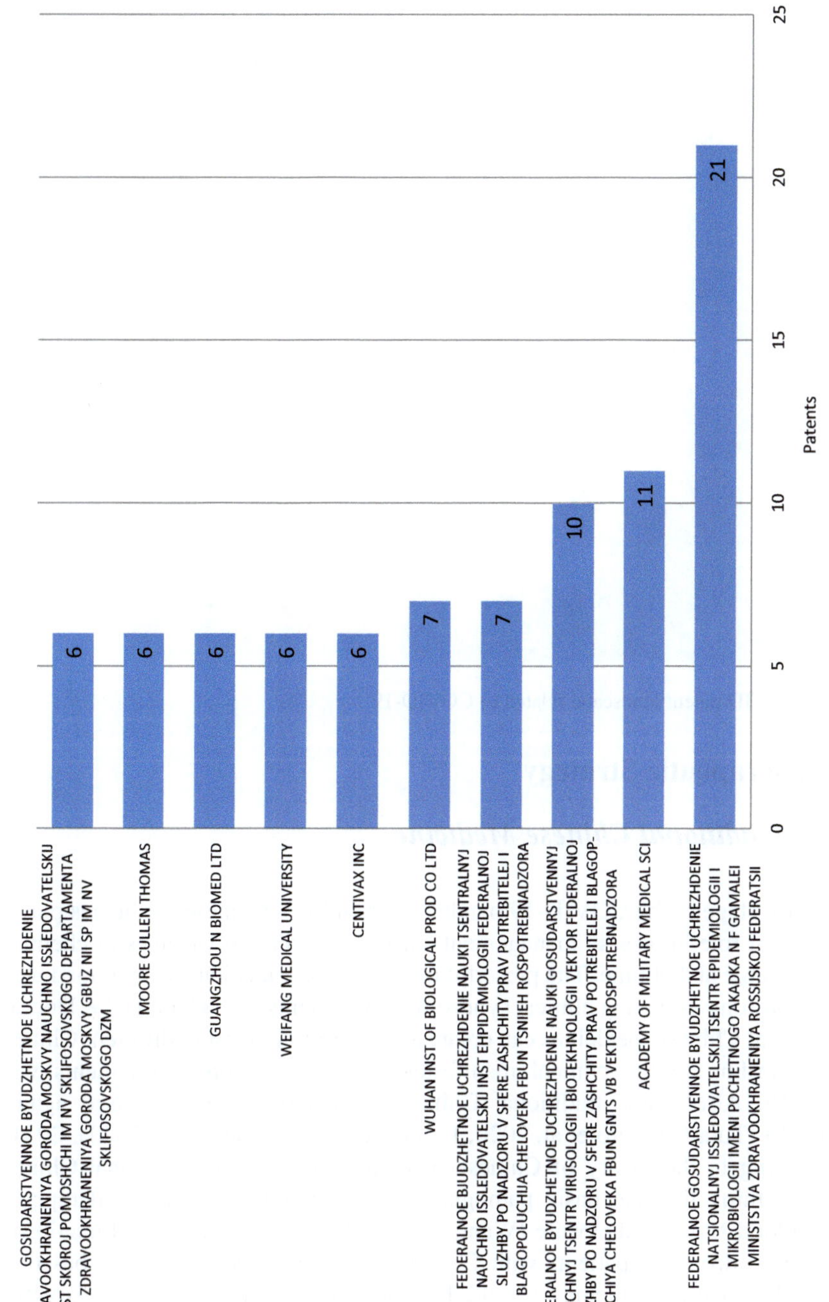


**Fig. 2.3** The top ten patent assignees related to COVID-19

**Fig. 2.4** The 3D patent landscape related to COVID-19

## 2.5 Therapeutic Strategy

### *2.5.1 Traditional Chinese Medicine*

In the early stage of the COVID-19 pandemic, traditional Chinese medicines played an important role in disease management since there were no vaccines or chemical drugs available at that time. The prescriptions of Chinese medicines are immediately available as long as having clinical syndromes for patients with COVID-19. Doctors of traditional Chinese medicines can create a prescription by using different Chinese herbs according to the clinical manifestations and syndrome of patients with COVID-19. Also, some traditional Chinese medicines have been used against COVID-19 (Tuta-Quintero et al. 2020; Li and Xu 2021; Wang et al. 2020a, 2021).

According to the theory of Chinese medicines, doctors of traditional Chinese medicines think that the basic pathogenesis of COVID-19 is due to dampness, poison, blood stasis, and closure (Wang et al. 2020b). Based on clinical manifestations and syndrome of patients with COVID-19, the therapeutic method based on traditional Chinese medicines should be concentrated on aspects of dispelling dampness of the body, dispelling heat and damp toxins of the body, and clearing

**Table 2.2** Traditional Chinese medicines against COVID-19

| Trademark/generic name | Developing company | Approved/EUA date |
|---|---|---|
| Qingfei Paidu decoction | NA | NA |
| Lianhuaqingwen capsule | Yiling pharmaceutical | Mar 2020 |

dampness in the triple warmer of the body, as well as strengthening essential Qi of the body (Luo 2020).

Based on our knowledge, traditional Chinese medicines have been involved in the management of COVID-19 in China in 2019. One of the typical prescriptions of traditional Chinese medicines is Qingfei Paidu decoction, as shown in Table 2.2.

Qingfei Paidu decoction comprises 21 Chinese herbs, as follows: Ephedrae Herba with 9 g, Glycyrrhizae Radix with 6 g, Armeniacae Semen Amarum with 9 g, Poria with 15 g, Bupleuri Radix with 16 g, Gypsum Raw with 15–30 g, Cinnamomi Ramulus with 9 g, Alismatis Rhizoma with 9 g, Umbellatus Polyporus with 9 g, Atractylodis Macrocephalae Rhizoma with 9 g, Scutellariae Radix with 6 g, Pinelliae Preparata Rhizome with 9 g, Zingiberis Recens Rhizoma with 9 g, Asteris Radix with 9 g, Farfarae Flos with 9 g, Belamcandae Rhizoma with 9 g, Dioscoreae Rhizoma with 12 g, Aurantii Immaturus Fructus with 6 g, and Pogostemonis Herba with 9 g, Citri Reticulatae Pericarpium with 6 g, Asari Herba with 6 g (National Administration of Traditional Chinese Medicine n.d.).

The mechanism study of Qingfei Paidu decoction showed that the composition of this prescription comprised 292 chemical compounds and 214 targets related to these compounds, and the key targets included AKT serine/threonine kinase 1 (AKT1) target, IL6 interleukin 6 (IL6), mitogen-activated protein kinase 8 (MAPK8) target, mitogen-activated protein kinase 1 (MAPK1) target as well as Jun proto-oncogene target (National Administration of Traditional Chinese Medicine n.d.). In general, Chinese herbal medicines play their role in the body through the approach of multi-components delivery, multi-target action, and multi-pathway engagements.

Wu et al. treated 98 cases of COVID-9 by using Qingfei Paidu Decoction in China. With the 3-day treatment of Qingfei Paidu Decoction, the total effect was 84.22%, of which the recovery rate was 21.14%. After 6 days of treatment, the total effect was 90.15%, of which the recovery rate was 31.34%. With 9 days of treatment, the total effect reached 92.09%, meanwhile, the recovery rate was 41.13% (Wang et al. 2020a). Some side effects, including nausea and vomiting, dizziness, and rash, were found, and the total percentage of side effects was about 7%. Specifically, four of them showed nausea and vomiting, two of them showed dizziness, and one of them showed a rash.

Another clinical study showed that 98 cases of COVID-9 were treated by using Qingfei Paidu; the total effect was about 88%. The side effects of Qingfei Paidu treatment were about 5.3%. There were 2 patients with vomiting, 1 patient with nausea, and 1 patient with a rash. However, the side effects were mild and disappeared without other treatment (Hu et al. 2020).

Basically, according to the theory of traditional Chinese medicines, Qingfei Paidu decoction may play a role in anti-SARS-CoV-2 activity by their capacity on dispelling dampness and eliminating heat and turbidity, as well as nourish the lung and spleen, detoxify, etc.

Notably, the National Health Commission of China proposed to officially use Qingfei Paidu for the treatment of COVID-19. Therefore, the Qingfei Paidu decoction as a general prescription of traditional Chinese Medicines was used for the treatment of COVID-19.

Another popular traditional Chinese medicine, namely Lianhuaqingwen capsule, has been used in the management of COVID-19. Lianhuaqingwen capsule comprises over 13 Chinese herbs, including Honey-Fried Mahuangephedrae Herba, Bitter Apricot, Isatidis Radix, Dryopteridis Crassirhizomatis Rhizoma, Houttuyniae Herba, Pogostemonis Herba, Rhei Radix Et Rhizoma, Rhodiolae Crenulatae Radix Et Rhizoma, Glycyrrhizae Radix Et Rhizoma, Menthol, Forsythiae Fructus, Lonicerae Japonicae Flos, and Gypsum Fibrosum (Yiling Pharmaceutical n.d.).

Lv et al. conducted a clinical study by a collection of 63 patients with COVID-19 who received therapy from Lianhua Qingwen Granules. The clinical results showed that the rate of symptom disappearance in fever was 86.7%, and the rate of symptom disappearance in cough and weakness was 55.6% and 82.5%, respectively. The rate of symptom disappearance in short breath was 68.2%. There are no other side effects in the treatment group (Lv et al. 2020).

Compared with Qingfei Paidu decoction, Lianhuaqingwen has a broader clinical efficacy, which also could be used for the treatment of influenza, and patients with the symptoms of fever, cough, headache, etc. Therefore, Lianhuaqingwen demonstrates lower therapeutic specificity for clinical manifestations and syndrome of COVID-19 compared to Qingfei Paidu decoction. Traditional Chinese medicine treatment could be used for patients with COVID-19, and the earlier treatment is a benefit for patients with COVID-19.

It is a significant feature that traditional Chinese medicines were involved in COVID-19 in China. Traditional Chinese medicines have a unique feature for the therapy of patients with COVID-19 compared to chemical drugs and vaccines, e.g., immediately availability and flexibility. Both chemical drugs and vaccines need time to be developed and studied. Nevertheless, traditional Chinese Medicines do not need to long-time to be developed.

In China, there are two parallel medical systems for combating diseases, which are traditional Chinese medicines and Western medicines. Both these two medical systems have their advantages and disadvantages in the treatment of diseases. In the past years, traditional Chinese medicines have benefited Chinese people in combating COVID-19. A competitive advantage is that the appearance of a therapeutic method of traditional Chinese medicines against COVID-19 is much earlier than chemical drugs, antibody as well as vaccines. Theoretically, it takes about many years to develop a vaccine, which is much later than the clinical use of traditional Chinese medicines.

### 2.5.2 Chemical Drugs

It appears that chemical drugs are not a very good option for the treatment of COVID-19 at present. The time to develop chemical drugs is longer than that for vaccines, and the time of developing an effective chemical drug is quite difficult to predict. Although there were some patents related to chemical drugs at the early stage of emerging of COVID-19, most of these chemical drugs were developed for a new medical application or repurposed drugs against COVID-19 by using known chemical compounds or drugs. Usually, the efficiency of these chemical drugs is quite low. An example is that Remdesivir is used for the therapy of patients with COVID-19. It is a much more difficult and time-consuming process to develop effective chemical drugs against COVID-19.

Some chemical drugs have been approved by the U.S. Food and Drug Administration (FDA), e.g., Nirmatrelvir and Ritonavir. The U.S. FDA has issued an emergency use authorization for Pfizer's Paxlovid, including Nirmatrelvir and Ritonavir, for the treatment of COVID-19 in adults on Dec 2021 (The U.S. Food and Drug Administration n.d.-a). Nirmatrelvir and Ritonavir could reduce the risk of death or hospitalization by about 89% compared to placebo in patients with COVID-19 (Anon n.d.). However, the safety and effectiveness of Paxlovid for the treatment of COVID-19 need time to be further evaluated.

The U.S. FDA approved an emergency use authorization for remdesivir in the treatment of patients with COVID-19 on October 2020 (The U.S. Food and Drug Administration n.d.-b). Recently, molnupiravir was approved as an emergency use authorization by the U.S. FDA for the treatment of patients with COVID-19 on December 2021 (The U.S. Food and Drug Administration n.d.-c). Molnupiravir is provided by prescription only and should be used for patients with mild to moderate COVID-19, and be initiated as soon as possible after being confirmed as COVID-19. Recently approved or emergency use authorization chemical drugs are listed in Table 2.3.

### 2.5.3 Vaccines

Vaccines are administered to the body, which stimulates the production of antibodies and provides immunity against specific viruses. The vaccine approach is the most effective therapeutic strategy against COVID-19. About 1 year after the emergence of COVID-19, its vaccines were first available. Several vaccine products are

**Table 2.3** Chemical drug against COVID-19

| Trademark/generic name | Developing company | Approved/EUA date |
|---|---|---|
| Paxlovid | Pfizer | Dec 2021 |
| Remdesivir | Gilead Sciences | Oct 2020 |
| Molnupiravir | Merck | Dec 2021 |

available and show good safety and effectiveness (Voysey et al. 2021; Dagan et al. 2021; Emary et al. 2021; Al Kaabi et al. 2021; Logunov et al. 2021). Since vaccines are a globally mainstream approach to combating COVID-19, many vaccine products have been approved by the U.S. FDA and the National Medical Products Administration to use against COVID-19, as listed in Table 2.4.

The U.S. FDA announced the first approval of the COVID-19 vaccine (Comirnaty) developed by Pfizer-BioNTech (The U.S. Food and Drug Administration n.d.-d) on August 2021. Comirnaty was developed by using mRNA technology, which is indicated for active immunization to prevent COVID-19.

The FDA approved the second vaccine of COVID-19 called Spikevax on January 2022 (The U.S. Food and Drug Administration n.d.-e). The Spikevax vaccine was developed by Moderna and developed using mRNA technology, which is indicated for active immunization to prevent COVID-19 for people 18 years and older. The FDA announced the third approval of EUA for the COVID-19 vaccine called Janssen COVID-19 on Feb 2021 (The U.S. Food and Drug Administration n.d.-f). The emergency use authorization indicated that the Janssen COVID-19 vaccine was allowed to prevent COVID-19 for patients 18 years and older. The Janssen COVID-19 vaccine is developed by using a specific virus, which is called adenovirus type 26.

Currently, available vaccines against COVID-19 are from several different countries, and different countries may use different technologies to develop vaccines. For example, many companies in China develop COVID-19 vaccine products by using an inactive approach (Al Kaabi et al. 2021). Many companies in the USA develop vaccine products by using the mRNA technology approach (Dagan et al. 2021; The U.S. Food and Drug Administration n.d.-d). However, the inactive vaccines demonstrated more safety than the mRNA vaccines (Al Kaabi et al. 2021). At this stage, the vaccine plays a leading role and is a dominating approach in combating the global COVID-19 pandemic. The recently approved or emergency use authorization vaccines are listed in Table 2.4.

**Table 2.4** Vaccine products against COVID-19

| Trademark/generic name | Developing company | Approved/EUA date |
|---|---|---|
| BBIBP-CorV | Sinopharm | Aug 2021 |
| CoronaVac | Sinovac | Jun 2020 |
| Convidecia | Cansinobio | Feb 2021 |
| Covilo | Wuhan Institute of Biological Products, Sinopharm | Jul 2021 |
| Kconvac | Biokangtai | May 2021 |
| Recombinant COVID-19 vaccine | Anhui Zhifei Longcom biopharmaceutical | Mar 2021 |
| Comirnaty | Pfizer-Biontech | Aug 2021 |
| Spikevax | Moderna | Jan 2022 |
| Janssen COVID-19 vaccine | Janssen Biotech | Feb 2021 |

### 2.5.4 Antibodies

Antibodies are also an effective therapeutic approach with high specificity. Currently, antibodies are mainly used in an emergency for patients with mild-to-moderate COVID-19, especially for patients who are at high risk of or severe COVID-19. Clinical use of antibodies against COVID-19 is not as extensive as vaccines. So far, several antibodies have been approved (or an emergency use authorization) by the U.S. FDA, as listed in Table 2.5.

The U.S. FDA approved an emergency use authorization for Bamlanivimab and Etesevimab for patients with COVID-19 on February 2021 (The U.S. Food and Drug Administration n.d.-g). This combination antibody is particularly suitable for patients with high-risk or severe COVID-19 (Food and Drug Administration 2021). Bamlanivimab and Etesevimab should be administered together. The combination use of Bamlanivimab and Etesevimab has revealed good clinical effectiveness in people with moderate COVID-19 (Weinreich et al. 2021; Dougan et al. 2021). The treatment of antibodies should be started as soon as possible, as long as the patients are confirmed as positive for COVID-19, which will be a benefit for people with COVID-19.

Casirivimab and Imdevimab, administered together, were approved as emergency use authorization by the FDA for the treatment of COVID-19 in November 2020 (The U.S. Food and Drug Administration n.d.-h). Casirivimab and Imdevimab were developed by Regeneron Pharmaceutical. Casirivimab and Imdevimab should be administered together by intravenous infusion, which is suitable for the treatment of patients with mild to moderate or high-risk COVID-19.

The FDA approved an emergency use authorization for the monoclonal antibody of Sotrovimab for the treatment of COVID-19 on May 2021 (The U.S. Food and Drug Administration n.d.-i). Sotrovimab was developed by GlaxoSmithKline LLC. Sotrovimab is suitable for the treatment of patients with high-risk or severe COVID-19.

Bebtelovimab was approved as an emergency use authorization by the FDA for the treatment of COVID-19 patients on February 2022 (The U.S. Food and Drug Administration n.d.-j). Bebtelovimab was developed by Eli Lilly and Company. Bebtelovimab is suitable for the treatment of COVID-19 in patients who are at high risk of or severe COVID-19. Particularly, it is suitable for the treatment of patients

**Table 2.5** Antibodies against COVID-19

| Trademark/generic name | Developing company | Approved/EUA date |
|---|---|---|
| Sotrovimab | GlaxoSmithKline | May 2021 |
| Casirivimab | Regeneron pharmaceutical | Nov 2020 |
| Imdevimab | Regeneron pharmaceutical | Nov 2020 |
| Bebtelovimab | Eli Lilly | Feb 2022 |
| Bamlanivimab | Eli Lilly | Feb 2021 |
| Etesevimab | Eli Lilly | Feb 2021 |

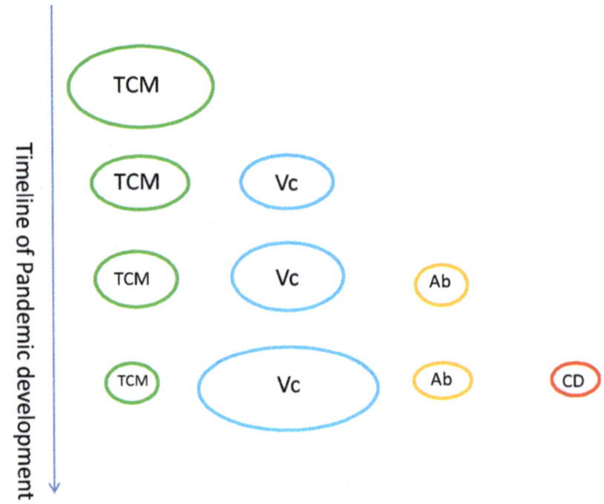


**Fig. 2.5** Current option of therapeutic strategy. *TCM* traditional Chinese medicines, *VC* vaccines, *AB* antibodies, *CD* chemical drugs

confirmed as Omicron variants of COVID-19. The recently approved or emergency use authorization antibodies are listed in Table 2.5.

In general, most antibodies were approved for emergency use authorization by the U.S. FDA. The antibodies are suitable for the treatment of mild to moderate patients confirmed as COVID-19, as well as COVID-19 patients who are at high risk or progress to severe. Usually, they should be administered together if two antibodies were used as a package.

In general, there are four therapeutic strategies, including traditional Chinese medicines, chemical drugs, vaccines, and antibodies. Each therapeutic strategy has its advantages and disadvantages in clinical use. Thus, the option of therapeutic strategies according to the progression of COVID-19 is a skillful piece of work. As shown in Fig. 2.5, the author depicts a therapeutic strategy pattern, which shows the option of therapeutic strategies against COVID-19. Each circle indicates therapeutic proportion as the development of the pandemic of COVID-19. The larger circle indicates the larger proportion of use in the whole therapeutic strategy. At the beginning of the emergence of COVID-19, traditional Chinese medicines are the best option for therapeutic strategy against COVID-19 since it is unavailable for vaccines, chemical drugs, or antibodies. About 1 year later, with the development of vaccines, vaccines will be a major trend in therapeutic methods against COVID-19. Antibodies are also available like vaccines, but based on consideration in therapeutic strategy, the clinical use of antibodies is not as extensive as vaccine products. The chemical drug and antibody approaches play a minor role in the whole therapeutic strategies against COVID-19. Hence, at this stage, the overall therapeutic strategy pattern against COVID-19 could be presented as a model as shown in Fig. 2.5.

However, to face potential new variants of coronavirus in the future, the pattern of therapeutic strategies may need some modification. As shown in Fig. 2.6, at the beginning of the emergence of a novel coronavirus, traditional Chinese medicines are suggested to be one of the best options for earlier therapy against future novel coronavirus, since it is immediately available, flexible, safe, and 1000-year practical

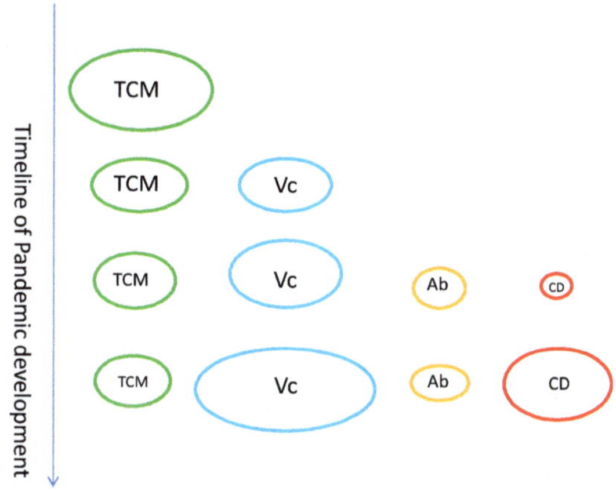


**Fig. 2.6** Future option of therapeutic strategy. *TCM* traditional Chinese medicines, *VC* vaccines, *AB* antibodies, *CD* chemical drugs

experience. About 1 year later, vaccines are a major therapeutic method in therapeutic strategies. The difference is that chemical drugs will play a more important role in therapeutic methods against a possible novel coronavirus because new technological development will make it possible to find an effective target, screen the most active compounds, and shorten the time of R&D of chemical drugs. In my opinion, with the rapid development, the potential role of traditional Chinese medicines in the management of COVID-19 variants in the future may become more important. Thus, the overall therapeutic strategy pattern will be presented in a modified model as shown in Fig. 2.6.

The chapter depicts a novel therapeutic strategy pattern against COVID-19. The novel therapeutic strategy pattern will be a benefit against COVID-19. Scientific application and options of the therapeutic method according to therapeutic strategy patterns can save time and improve efficiency and safety in the treatment of patients with COVID-19. The therapeutic strategy pattern will help humans against COVID-19 effectively.

## 2.6 Conclusion

The chapter provides an opinion on the categories of four therapeutic strategies for the treatment of COVID-19. Each strategy uses a central treatment by traditional Chinese medicines, chemical drugs, vaccines, and antibodies. Traditional Chinese medicines are the fastest available therapeutic approach against COVID-19 and have shown good efficacy and safety in the treatment of patients infected with SARS-CoV-2. The vaccine approach is the most reliable therapeutic approach against COVID-19 and its development and application took about a year after the outbreak of COVID-19. Developing chemical drugs against COVID-19 is a time-consuming approach with a high failure risk. The therapeutic strategy that is based on the use of antibodies has the nature of high specificity. It could be used for the treatment of mild to moderate COVID-19 patients, or those patients who are at high risk of or severe COVID-19. The development time of antibodies is even longer than vaccines. Overall, this chapter summarizes the current status of therapies against COVID-19 and suggests a potential strategy against new variants.

## References


Al Kaabi N, Zhang Y, Xia S, Yang Y, Al Qahtani MM, Abdulrazzaq N, Al Nusair M, Hassany M, Jawad JS, Abdalla J et al (2021) Effect of 2 inactivated SARS-CoV-2 vaccines on symptomatic COVID-19 infection in adults: a randomized clinical trial. JAMA 326:35

Anon (n.d.) Pfizer's Novel COVID-19 Oral Antiviral Treatment Candidate Reduced Risk of Hospitalization or Death by 89% in Interim Analysis of Phase 2/3 EPIC-HR Study. https://www.pfizer.com/news/press-release/press-release-detail/pfizers-novel-covid-19-oral-antiviral-treatment-candidate

Azer SA (2020) COVID-19: pathophysiology, diagnosis, complications and investigational therapeutics. New Microbes New Infect 37:100738

Dagan N, Barda N, Kepten E, Miron O, Perchik S, Katz MA, Hernán MA, Lipsitch M, Reis B, Balicer RD (2021) BNT162b2 mRNA Covid-19 vaccine in a nationwide mass vaccination setting. N Engl J Med 384:1412

Dougan M, Nirula A, Azizad M, Mocherla B, Gottlieb RL, Chen P, Hebert C, Perry R, Boscia J, Heller B et al (2021) Bamlanivimab plus Etesevimab in mild or moderate Covid-19. N Engl J Med 385:1382

Emary KRW, Golubchik T, Aley PK, Ariani CV, Angus B, Bibi S, Blane B, Bonsall D, Cicconi P, Charlton S et al (2021) Efficacy of ChAdOx1 nCoV-19 (AZD1222) vaccine against SARS-CoV-2 variant of concern 202012/01 (B.1.1.7): an exploratory analysis of a randomised controlled trial. Lancet 397:1351–1362; https://linkinghub.elsevier.com/retrieve/pii/S0140673621006280

Food and Drug Administration. Fact sheet for healthcare providers: emergency use authorization (EUA) of Bamlanivimab and Etesevimab. No Title. 2021. https://www.fda.gov/media/145802/download

Hu GM, He CX, Sun QL, Wan BB, Li YB et al (2020) Preliminary study on the clinical efficacy of "Qingfei detox granule" in the treatment of COVID-19. Tianjin J Tradit Chin Med 37:999–1004

Li J-G, Xu H (2021) Chinese medicine in fighting against Covid-19: role and inspiration. Chin J Integr Med 27:3–6

Logunov DY, Dolzhikova IV, Shcheblyakov DV, Tukhvatulin AI, Zubkova OV, Dzharullaeva AS, Kovyrshina AV, Lubenets NL, Grousova DM, Erokhova AS et al (2021) Safety and efficacy of an rAd26 and rAd5 vector-based heterologous prime-boost COVID-19 vaccine: an interim analysis of a randomised controlled phase 3 trial in Russia. Lancet 397:671–681

Luo YFCJT (2020) "Three-in-one system" treatment of novel coronavirus pneumonia. Fujian J Tradit Chin Med 51:12–14

Lv RB, Wang WJ et al (2020) Clinical observation on Lianhua Qingwen granules combined with western medicine conventional therapy in the treatment of 63 suspected cases of COVID-2019. J Tradit Chinese Med 61:655–659

National Administration of Traditional Chinese Medicine (n.d.). http://yzs.satcm.gov.cn/zhengcewenjian/2020-02-07/12876.html

The U.S. Food and Drug Administration (n.d.-a). https://www.fda.gov/news-events/press-announcements/coronavirus-covid-19-update-fda-authorizes-first-oral-antiviral-treatment-covid-19

The U.S. Food and Drug Administration (n.d.-b). https://www.fda.gov/drugs/news-events-human-drugs/fdas-approval-veklury-remdesivir-treatment-covid-19-science-safety-and-effectiveness

The U.S. Food and Drug Administration (n.d.-c). https://www.fda.gov/news-events/press-announcements/coronavirus-covid-19-update-fda-authorizes-additional-oral-antiviral-treatment-covid-19-certain

The U.S. Food and Drug Administration. (n.d.-d) No Title. https://www.fda.gov/news-events/press-announcements/fda-approves-first-covid-19-vaccine

The U.S. Food and Drug Administration (n.d.-e). https://www.fda.gov/emergency-preparedness-and-response/coronavirus-disease-2019-covid-19/spikevax-and-moderna-covid-19-vaccine

The U.S. Food and Drug Administration (n.d.-f). https://www.fda.gov/news-events/press-announcements/fda-issues-emergency-use-authorization-third-covid-19-vaccine

The U.S. Food and Drug Administration (n.d.-g). https://www.fda.gov/news-events/press-announcements/coronavirus-covid-19-update-fda-authorizes-monoclonal-antibodies-treatment-covid-19-0

The U.S. Food and Drug Administration (n.d.-h). https://www.fda.gov/news-events/press-announcements/coronavirus-covid-19-update-fda-authorizes-monoclonal-antibodies-treatment-covid-19

The U.S. Food and Drug Administration (n.d.-i). https://www.fda.gov/news-events/press-announcements/coronavirus-covid-19-update-fda-authorizes-additional-monoclonal-antibody-treatment-covid-19

The U.S. Food and Drug Administration (n.d.-j). https://www.fda.gov/news-events/press-announcements/coronavirus-covid-19-update-fda-authorizes-new-monoclonal-antibody-treatment-covid-19-retains

Tuta-Quintero EA, Suárez-Ramirez V, Pimentel J (2020) Efficacy and safety of traditional Chinese medicine in COVID-19: a scoping review. Rev Int Acupunt 14:132–150

Voysey M, Clemens SAC, Madhi SA, Weckx LY, Folegatti PM, Aley PK, Angus B, Baillie VL, Barnabas SL, Bhorat QE et al (2021) Safety and efficacy of the ChAdOx1 nCoV-19 vaccine (AZD1222) against SARS-CoV-2: an interim analysis of four randomised controlled trials in Brazil, South Africa, and the UK. Lancet 397:99

Wang RQ, Yang SJ, Xie CG, Shen QL, Li MQ, LX. (2020a) Clinical efficacy of qingfei detox decoction in the treatment of COVID-19. Med Pharm Clin 36:13–18

Wang YG, Qi WS, Ma JJ, Ruan LG, Lu YR, LXC. (2020b) Clinical features of novel coronavirus (2019-ncov) pneumonia and its treatment based on syndrome differentiation. J Tradit Chin Med 61:281–285

Wang WY, Xie Y, Zhou H, Liu L (2021) Contribution of traditional Chinese medicine to the treatment of COVID-19. Phytomedicine 85:153279

Weinreich DM, Sivapalasingam S, Norton T, Ali S, Gao H, Bhore R, Xiao J, Hooper AT, Hamilton JD, Musser BJ et al (2021) REGEN-COV antibody combination and outcomes in outpatients with Covid-19. N Engl J Med 384:238

Williams VR, Pajak D, Salt N, Leis JA (2019) 2448. Clinical presentation and outcomes of long-term care residents with coronavirus respiratory infection: a retrospective cohort study. Open Forum Infect Dis 6:S846–S846

World Health Organization (n.d.-a). https://covid19.who.int/

World Health Organization (n.d.-b). https://covid19.who.int/region/amro/country/us

Yiling Pharmaceutical (n.d.). http://www.yiling.cn/contents/75/4904.html

[illegible]

# Chapter 3
# Plant Immunoenhancers: Promising Ethnopharmacological Candidates for Anti-SARS-CoV-2 Activity

**Salar Hafez Ghoran, Fatemeh Taktaz, and Seyed Abdulmajid Ayatollahi**

## Abbreviations

| | |
|---|---|
| $^{18}$F-FDG | $^{18}$F-fluoro-deoxy-D-glucose |
| $^{18}$F-FDG-PET | $^{18}$F-fluoro-deoxy-D-glucose-positron emission tomography |
| $3CL^{pro}$ | 3 chymotrypsin-like protease |
| 5-LOX | 5-Lipoxygenase |
| ACE-2 | Angiotensin-converting enzyme-2 |
| ADMET | Adsorption, Desorption, Metabolism, Excretion, and Toxicity |
| ARDS | Acute respiratory distress syndrome |
| AREs | Antioxidant-response elements |
| AS-CDs | Carbon dots derived from *Allium sativum* |
| BAL | Bronchoalveolar lavage |
| BRD4 | Bromodomain-containing protein 4 |
| CAT | Catalase |

S. Hafez Ghoran (✉)
H.E.J. Research Institute of Chemistry, International Center for Chemical and Biological Sciences, University of Karachi, Karachi, Pakistan

Phytocehmistry Research Center, Shahid Beheshti University of Medical Sciences, Tehran, Iran
e-mail: S.Hafezghoran@iccs.edu

F. Taktaz
Phytocehmistry Research Center, Shahid Beheshti University of Medical Sciences, Tehran, Iran

Department of Advanced Medical and Surgical Sciences, University of Campania, "Luigi Vanvitelli", Naples, Italy
e-mail: F.taktaz@gmail.com

S. A. Ayatollahi (✉)
Phytocehmistry Research Center, Shahid Beheshti University of Medical Sciences, Tehran, Iran

Department of Pharmacognosy, School of Pharmacy, Shahid Beheshti University of Medical Sciences, Tehran, Iran
e-mail: A.ayatollahi@sbmu.ac.ir

 
J.-T. Chen (ed.), *Ethnopharmacology and Drug Discovery for COVID-19: Anti-SARS-CoV-2 Agents from Herbal Medicines and Natural Products*,
https://doi.org/10.1007/978-981-99-3664-9_3

| | |
|---|---|
| CCL | Chemokine ligand |
| COPD | Chronic obstructive pulmonary disease |
| CRP | C-reactive protein |
| DENV-2 | Dengue virus |
| DPPH | 1,1′-diphenyl-2-picrylhydrazyl radical |
| DTH | Delayed-type hypersensitivity |
| ECs | Endothelial cells |
| ED | Endothelial dysfunction |
| EGCG | Epigallocatechin-3-gallate |
| ELISA | Enzyme-linked immunosorbent assay |
| G-CSF | Granulocyte-colony stimulating factor |
| GM-CSF | Granulocyte-macrophage colony-stimulating factor |
| GSH-Px | Glutathione-peroxidase |
| HAV | Hepatitis A virus |
| HTVS | High throughput screening |
| IFN-$\gamma$ | Interferon-$\gamma$ |
| iNOS | Inducible nitric oxide synthase |
| LPS | Lipopolysaccharide |
| MAPK | Mitogen-activated protein kinase |
| MCP-1 | Monocyte chemoattractant protein 1 |
| MD | Molecular dynamics |
| MDA | Malondialdehyde |
| MERS-CoV | Middle East respiratory syndrome coronavirus |
| MIP-1$\alpha$ | Macrophage inflammatory protein-1$\alpha$ |
| MPO | Myeloperoxidase |
| $M^{pro}$ | Main protease |
| NF-$\kappa$B | Nuclear factor-$\kappa$B |
| NLRP3 | NLR family pyrin domain containing 3 |
| NO | Nitric oxide |
| NQO1 | NAD(P)H quinone dehydrogenase 1 |
| PGE2 | Prostaglandin E2 |
| $PL^{pro}$ | Papain like protease |
| PMNs | Polymorphonuclear cells |
| RBD | Receptor-binding domain |
| RdRp | RNA-dependent RNA polymerase |
| ROS | Reactive oxygen species |
| SARS-CoV-2 | Severe acute respiratory syndrome coronavirus type 2 |
| SI | Selectivity Index |
| SOC | Standard of care |
| SOD | Superoxide dismutase |
| SP | Standard precision |
| $SpO_2$ | Peripheral oxygen saturation |
| TAOC | Total antioxidant capacity |
| TCM | Traditional Chines medicine |

| | |
|---|---|
| TLR2 | Toll-like receptor 2 |
| TNF-α | Tumor necrosis factor α |
| UHPLC | Ultra-high-performance liquid chromatography |
| vWF | von Willebrand factor |
| WBC | White blood cells |

## 3.1 Introduction

An undesirable outbreak of a kind of infectious respiratory disease, coronavirus disease (COVID) was quickly initiated in Wuhan, China, becoming a worldwide pandemic, which took the lives of millions of lives and caused economic setbacks. Scientists realized that there is a genetic similarity between severe acute respiratory syndrome coronavirus type 2 (SARS-CoV-2) and the previously emerged coronaviruses, such as SARS-CoV and MERS-CoV (Middle East respiratory syndrome coronavirus) (Hafez Ghoran et al. 2021a). It is well studied that in COVID-19 pathogenesis, various mechanisms, including extensive virus replication, massive necrosis and apoptosis in lung cells, oxidative stress, the high response of inflammation mediators, overexpression of pro-inflammatory factors, cytokine storm, ACE2 downregulation, and the impairment of renin-angiotensin system are involved (Jafarzadeh et al. 2021). Using traditional medicine and looking at previous experiences with viral infections and drug discovery development, certain medicinal plants and herbal preparations have been prescribed to treat SARS-CoV disease. Therefore, increasing daily mortality and morbidity rate during the SARA-CoV-2 pandemic called researchers to pay more attention to nature-derived secondary metabolites, which have proven to be an effective treatment to tackle this devastating pandemic (Hafez Ghoran et al. 2021a). So far, for human SARS-CoV diseases, four types have been characterized: β-HCoV (OC43), α-HCoV (229E), CoV-NL63, and CoV-HKU1. Besides targeting the before-mentioned HCoV viruses, researchers figured that inhibiting a key enzyme in the SARS-CoV-2 host receptor, angiotensin-converting enzyme 2 (ACE2), could be a promising way to stop the virus activity (Zhang et al. 2020a; Geller et al. 2012). From the other point of view, pathological results of SARS-CoV-2 positive individuals displayed a cytokine storm of around 20% in the patient's plasma, which leads to destructive and uncontrolled inflammation. This phenomenon on the molecular scale is because of the overproduction of various cytokines, including interleukins (IL; i.e., IL-6, IL-7, IL-8, IL-9, IL-1β, IL-17), interferon-γ (IFN-γ), tumor necrosis factor-alpha (TNF-α), granulocyte-macrophage colony-stimulating factor (GM-CSF), and granulocyte-colony stimulating factor (G-CSF) as well as chemokine ligands (CCL) viz. CCL2, CCL3, and CCL4. The high concentration of these destroying factors has been recorded in the patient's plasma bearing SARS-CoV-2 and even in the dying stage (Xu et al. 2020). In addition, consideration of anti-inflammation and immune-nanomedicines approaches (Xu et al. 2020; Bonam et al. 2021) leads to feeling better and reduces the time of hospital discharge. From an ethnopharmacology point of view, finding promising drugs from nature by getting clues from traditional medicine has been well-recognized for reducing infection and viral adhesion. These herbal formulas, including bioactive secondary metabolites, could block the interaction between host cells and viruses by inhibiting receptor-mediated recognition. Finally yet importantly, increasing the general immune system through various types of herbal products, nutraceuticals, and dietary supplements could be the alternative way to stop the mortality caused by SARS-CoV-2 (Bonam et al. 2020; Hensel et al. 2020). Therefore, the main goal is to look for new anti-viral drugs because viral diseases, such as SARS, MERS, and Ebola, are the main threat to humans. The use of medicinal plants is embedded in everyday life because these are easily available, patient-friendly with high curative effects and do not have toxicity and side effects when compared to synthetic drugs. Several researchers have applied plant-derived natural compounds showing great potency to combat this phenomenon. For instance, the anti-viral assessment of some traditional medicinal plants along with their extracts, such as *Allium sativum* L., *Artemisia annua* L., *Anthemis hyaline* DC., *Camellia sinensis* (L.) Kuntze, *Citrus sinensis* L. Osbeck, *Echinacea purpurea* (L.) Moench, *Nigella sativa* L., *Zinziber officinalis*, *Hypericum perforatum* L., *Scutellaria baicalensis* Georgi., *Rheum officinale* Baill., *Juniperus formosana* Hayata., *Euphorbia neriifolia* L., *Glycyrrhiza glabra* L., and *Lycoris radiata* Herb. showed their potency against SARS-CoV in vitro. By the way, for a further pre-clinical and clinical survey, these plants have been recommended (Hafez Ghoran et al. 2021a; Hensel et al. 2020). Overall taken, in this chapter, the author(s) aims to provide current knowledge on some medicinal plants of high ethnopharmacological interests worldwide as potential antiviral and anti-inflammatory sources. These plant species have been prescribed as medicinal plants in folk medicine around the world. Therefore, various plant-derived natural metabolites have salient potential bioactivity for controlling SARA-CoV-2 by enhancing the immune system.

## 3.2 Reactive Oxygen Species (ROS) and Inflammation Promotion

An important signaling process in the progression of inflammation is oxidative stress caused by reactive oxygen species (ROS). These are oxygen-containing free radicals, such as superoxide anion ($O_2^{-\bullet}$; unstable), hydroxyl radical ($OH^{\bullet}$; unstable), hydrogen peroxide ($H_2O_2$; stable), and nitric oxide ($NO^{\bullet}$). On this occasion, chronic inflammation can develop by the action of free radicals that damage healthy cells and infections and injuries will have the potency to trigger the immune response. Macrophages, known as immune cells, produce free radicals when they fight off invading microorganisms (Yahfoufi et al. 2018). Normally, a small portion of ROS is produced by cells and is necessary for maintaining homeostasis and function in cells. In this mechanism, physiological imbalances increase the production of inflammatory factors, cytokines, and chemokines, leading to oxidized proteins, lipids, and DNA and even altering their function (Ansari et al. 2020). The inflammatory response to a viral infection that damages pulmonary cells may also be aggravated by ROS generated by the unchecked flow of inflammatory mediators. Moreover, this condition brings about the desquamation and degradation of alveolar epithelial cells, production of the hyaline membrane, and pulmonary edema (Tian et al. 2020). To more extent, inflammation is induced by ROS through NF-κB activation and elevated transcription of genes encoding cytokines. $H_2O_2$ can also inhibit NF-κB activation by tyrosine phosphorylation of the nuclear factor-enhancing kappa light chains of activated B cells (IκB) (Pérez-Torres et al. 2021). These physiological events translocate the active NF-κB to the nucleus, where it induces the overexpression of many harmful biomolecules, such as IL-1β, IL-6, and TNF-α, all involved in the inflammatory process (Sul and Ra 2021). The current knowledge manifested that inflammation is an immune response to various infections or damages that can lead to many detrimental and complex diseases. With the growing body of knowledge on inflammatory-related factors, inflammasomes particularly nucleotide-binding oligomerization domain (NOD)-like receptor-containing pyrin domain 3 (NLRP3), have been well studied. Several preclinical studies indicate that antioxidant components can suppress NLPR3 inflammasome activation and scavenge ROS (Sho and Xu 2019). Another important signal molecule for inflammation in the body is NO, produced by acting inducible nitric oxide synthase (iNOS), a carrier of gas-to-cell intracellular information transmission. During an inflammatory response, the iNOS expression will increase, resulting in a high NO release (Wang et al. 2017). However, in oxidative stress conditions, a redox-sensitive transcription factor, nuclear factor erythroid 2-related factor 2 (Nrf2), regulates the oxidant/antioxidant system, besides the regulating antioxidant-related components by cytoprotective genes, phase 2 detoxifying genes (i.e., superoxide dismutase; SOD and NAD(P)H quinone dehydrogenase 1; NQO1), and stress genes (i.e., heme oxygenase 1; HO-1). Indeed, Nrf2 regulates the antioxidant-response elements (AREs). Most importantly, Nrf2/ARE activation acts as a clue to show the protection of body tissues against inflammatory mediators and oxidative stress-associated diseases (Khan et al. 2017). By the way, the progression of diseases happens by inflammation and oxidative stress, two interlocked routes, and jointly interfere with the therapy through multiple signaling pathways (Fig. 3.1).

## 3.3 Cross-Talk between Inflammation and Viral Infection

Viral infection triggers a reverse immune response referring to a difference in the immune response of individuals to pathogens based on their genetic diversity. Viral infection also promotes ROS production through leukocytes. In addition to ROS overproduction and antioxidant depletion, viral replication is also increased along with viral-associated complications. Lung, kidneys, intestines, and heart, which are rich in ACE2, are primarily affected by an impaired defense system causing massive

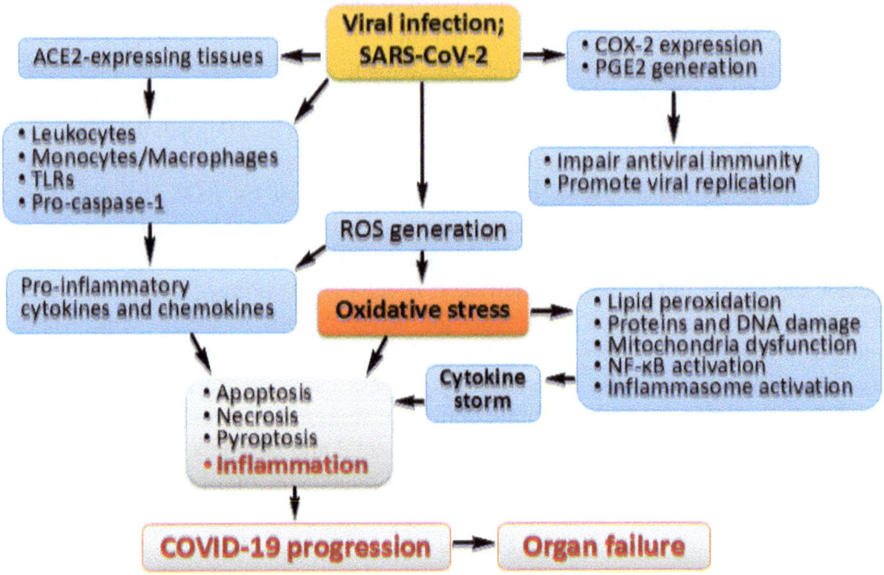


**Fig. 3.1** Metabolic pathogenesis of SARS-CoV-2 infection and its consequences on oxidative stress and inflammation

tissue damage. In the lungs, damaged cells bring innate inflammation mediated by pro-inflammatory factors, such as macrophages and granulocytes. However, in severe respiratory dysfunction, an uncontrolled immune response with the corresponding lung inflammation leads to damaged lung tissues, impaired lung function, and reduced lung capacity (Bose et al. 2021). In a viral infection, a well-regulated innate immunity is the first line of defense. However, during severe COVID-19 conditions, hyperinflammation (cytokine storm) results in acute respiratory distress syndrome (ARDS) and represses immunity (Jafarzadeh et al. 2021). Another line of defense includes toll-like-receptor (TLR), which stimulates the NF-κB pathway, leading monocyte cells to produce pro-inflammatory mediators, viz. IL-6, IL-1β, and TNF-α. Furthermore, by recruiting other leukocytes and directly regulating the antiviral pathways, TLR can control viral infection. In the case of COVID-19, the high levels of oxidative stress by inflammatory factors; on the other hand, the reduction of IFN-α and IFN-β makes it worse (Wang et al. 2020). As previously discussed, chronic viral infections can induce inflammation and immunosuppression, resulting in a dysfunctional immune system. In acute viral infections, type I interferons (IFN-Is) produce inflammation; however, IFN-Is also promote chronicity and limit immunity. Noted that chronic inflammation mediated by IFN-I does not promote viral replication alone but results in progressive immune dysfunction and disease (Snell et al. 2017). Most significantly, in order to check the extracellular and intracellular environment for infection signs, the body's innate immune system recruits the pattern recognition receptors. The activation of pattern recognition receptors by virus nucleic acids induces the expression and secretion of type I interferons (IFN-Is), which are key mediators for antiviral immunity (Stok et al. 2020). In the case of acute COVID-19, the highest concentrations of pro-inflammatory cytokines and chemokines have been observed in the patient's plasma (Mehta et al. 2020). Viral infections frequently trigger severe inflammatory responses that may function protectively or destructively in the innate immune system. Several virus-related diseases, including MERS-CoV, SARS-CoV, and SARS-CoV-2, which deregulate immune and inflammatory responses, bring about organ failure (Li et al. 2021). In general, viral infection not only decreases the antioxidant status but also increases the production of harmful oxidants and ROS, which is related directly to the enhancement of viral replication and viral-related diseases. Concerning SARS-CoV-2 infection, the Nrf2 pathway, which acts as a protective agent against oxidative stress, is suppressed, while NF-κB is activated, increasing inflammation and oxidative damage (Nema et al. 2021). As reported in SARS-CoV patients, both viral infection and viral replication can lead to high levels of virus-associated pyroptosis and vascular leakage in airway epithelial cells. Pyroptosis occurs as part of programmed cell death in cytopathic viruses, which is an intense form of inflammation because of high levels of IL-1β secretion (Chen et al. 2010). During viral infection, NLRP3 inflammasome induces caspase 1, ultimately leading to the cleavage of gasdermin-D (GSDMD), an important cellular membrane protein that leads to cell lysis (Man et al. 2017). Other innate immune cells are neutrophils and leukocytes. In addition to clearing pathogens and debris via phagocytosis, neutrophils are also involved in the inflammatory response (Rosales 2020). There are a lot of attention given to the endothelial dysfunction (ED) disease causing by viral infection like COVID-19. Due to activation and dysfunction of endothelial cells, previously discussed, pro-inflammatory cytokines and chemokines (i.e., monocyte chemoattractant protein-1; MCP-1), von Willebrand factor (vWF) antigen, vWF activity, and factor VIII are increased. As a result, associated with SARS-CoV-2 infection, high concentrations of severe phase reactants, such as C-reactive protein (CRP), IL-6, and D-dimer, have been reported (Zhang et al. 2020b). When tissue is exposed to viral infection, leukocytes influx in excess and activate tissue-resident leukocytes. There are various ways in which clinical inflammation can relieve or aggravate the disease, may not have any effect on the disease, or maybe both alleviate and aggravate the disease at different times, depending on the patient (Janeway 1953). Concerning SARS-CoV-2 and its complicated process, understanding the in-depth process and evaluating antiviral and anti-inflammatory components required more efforts to attenuate acute COVID-19 (Fig. 3.1).

## 3.4 Immunobooster Ability of Plants through Antioxidant and Anti-inflammatory Activity

In recent years, the role of medicinal plants has been increasingly emphasized and, in addition to their use as either traditional drugs or sources of bioactive compounds, they are "supportive tools" to enhance the immunity system. In this framework, a combination of classical and modern ethnopharmacology used as health support and alternative strategies has been wieldy encouraged by several types of research due to the benefits these herbal preparations can bring to human health, as well as mitigating health problems. However, the complexity of plant materials and their phytomolecules should be carefully considered, since in some cases, they not only are not immunoenhancer but also toxic agents. In fact, Mother Nature does not provide standardized medicinal plants, and in-depth studies on herbal preparations and their extracts are required in several research areas including, toxicology, pharmacology, clinical biochemistry, etc. (Das 2021). The brilliant examples of plant immunoenhancers belong to the following plant families: Alliaceae (*Allium* sp.), Caryophyllaceae (*Syzygium* sp.), Hypericaceae (*Hypericum* sp.), Lamiaceae (*Salvia* sp. and *Mentha* sp.), Leguminosae (*Curcuma* sp. and *Glycyrrhiza* sp.), Menispermaceae (*Tinospora* sp.), Ranunculaceae (*Nigella* sp.), Rutaceae (*Citrus* sp.), Zingiberaceae (*Zingiber* sp.), etc. The plants have been consumed conventionally in Arabic, Ayurvedic, Chinese, Tibetan, Unani-Tibb, and various folkloric systems of ethnomedicine to establish strong immunity in the human body and avoid viral infection (Hafez Ghoran et al. 2021a). Overall, there are still an array of plants that play as immune boosters and anti-SARS-CoV agents, some of which are mentioned in Table 3.1.

### *3.4.1* Glycyrrhiza glabra *L.*

Liquorice (*Glycyrrhiza glabra* L.; Leguminosae/Fabaceae) has been used traditionally for thousands of years. Being a perennial plant, it has been used not only as a nutraceutical plant in food products but also for medicinal purposes, including antiallergic and anti-inflammatory properties. In folk medicine, liquorice roots have been widely consumed for asthma, cough, colds, and even chronic obstructive pulmonary disease (COPD) (Bhat et al. 2012). Moreover, some conducted studies in China recommended using liquorice (*Glycyrrhiza* sp.) for deactivating COVID-19 (Wang and Yang 2021). To activate the body's immune system, the liquorice extract can maturate and differentiate the lymphocytes and macrophages (Ng et al. 2021). Liquorice (*Glycyrrhiza* sp.) contains various bioactive metabolites, of which glycyrrhizin (**1**) and glycyrrhetinic acid (Fig. 3.2), belonging to the saponin triterpenoids, are the main active natural products showing potential inhibition of cortisol metabolism because of their steroid-like structures. Flavonoids especially isoflavonoids and chalcones are the other phytochemicals that have been reported from liquorice (Bhat et al. 2012). In terms of antioxidant activity, the liquorice extract, despite scavenging DPPH$^{•}$ radicals (1,1-diphenyl-2-picrylhydrazyl radicals), exhibited significant protective effects towards liposomal membrane lipid peroxidation and liberation of blood ROS. The liquorice's antioxidant activity is because of isoflavonoid derivatives, coumarins, and chalcones (Račková et al. 2007). The plant extract also increases the enzymatic antioxidant status, including catalase (CAT), superoxide dismutase (SOD), glutathione-peroxidase (GSH-Px), and total

**Table 3.1** Some traditional medicinal plants for increasing the body's general immunity and anti-coronavirus activity

| Plant name | Family | Common name | Part of use | Anti-inflammatory and immunoenhance mechanisms | Anti-SARS-CoV activity | Reference |
|---|---|---|---|---|---|---|
| *Alnus japonica* (Thunb.) Steud | Betulaceae | Japanese Alder | Bark | Inhibition of the production of lipopolysaccharide (LPS)-induced pro-inflammatory cytokines, including IL-6 and IL-1β in RAW264.7 macrophage cells | Inhibitory activity against SARS-CoV-$PL^{pro}$ | Park et al. (2012), Lee et al. (2017) |
| *Angelica keiskei* Ito | Apiaceae | Ashitaba | Roots | Inhibition of endothelin-1 gene expression through the suppression of NF-κB | Dose-dependently, inhibitory activity against SARS-CoV-$3CL^{pro}$ and SARS-CoV-$PL^{pro}$ | Sugii et al. (2005), Mahmud et al. (2021) |
| *Azadiracta indica* A. Juss | Meliaceae | Neem | Leaves and flowers | Stimulation of pro-inflammatory cytokines, including IFN-γ, IL-1, and TNF-α | In silico inhibitory activity against SARS-CoV-2 $PL^{pro}$ and SARS-CoV-$M^{pro}$ | Das (2021) |
| *Brassica oleracea* L. | Brassicaceae | Broccoli | Aerial parts | Increase of TNF-α production by LPS-stimulated macrophages | Dose-dependently, inhibitory activity against SARS-CoV-2-Wuhan-Hu-1 virus-related cell death in non-human primate Vero C1008 cells and HCoV-OC43 | Das (2021), Ordonez et al. (2022) |
| *Bupleurum* sp. | Apiaceae | Chai Hu | Roots | Inhibition of NF-κB | Inhibitory activity against HcoV-229E attachment and penetration at concentrations of 0.25–25 μmol/L | Lei et al. (2020), Cheng et al. (2006) |
| *Camellia sinensis* | Theaceae | Tea | Leaves and flowers | Inhibition of IL-1β, TNF-α, and NO production in LPS-induced liver inflammation | Inhibitory activity against SARS-CoV-$3CL^{pro}$ activity and suppression of SARS-CoV replication | Chen et al. (2005, 2012) |

(continued)

**Table 3.1** (continued)

| Plant name | Family | Common name | Part of use | Anti-inflammatory and immunoenhance mechanisms | Anti-SARS-CoV activity | Reference |
|---|---|---|---|---|---|---|
| *Cibotium barometz* L. | Cibotiaceae | Golden chicken fern | Rhizomes | Dose-dependently, suppression of NO production and IL-6 secretion, in IFN-γ/LPS-induced mouse peritoneal macrophage cells | Inhibitory activity against SARS-CoV replication at concentrations of 25 and 200 mg/mL | Wen et al. (2011), Lee et al. (2010) |
| *Cullen corylifolium* (L.) Medicus | Fabaceae | Buguzhi | Whole plant | Dose-dependently, suppression of PGE2 and NO production in LPS-induced microglia; Downregulation of COX-2 (protein and mRNA) and iNOS in LPS-induced microglia | Dose-dependently, inhibitory activity against SARS-CoV-$PL^{pro}$ | Kim et al. (2018), Raman et al. (2022) |
| *Curcuma longa* L. | Zingiberaceae | Turmeric | Rhizomes | Reduction of NO production in RAW264.7 macrophages along with T helper cell activation. | Inhibitory activity against S protein/SARS-CoV-2 viral entry | Das (2021), Merarchi et al. (2021) |
| *Emblica officinalis* Gaertn | Phyllanthaceae | Gooseberry | Fruits | Modulating oxidative stress, inflammatory mediators, and cytokine storm, including IL-6, TNF-α, and COX-2 secretion | Inhibitory activity against SARS-CoV-2 replication | Ratheesh et al. (2022) |
| *Galla chinensis* | Anacardiaceae | Chinese gall | Nutgall | Inhibition of secretion of inflammatory cytokines and enhancing the expression of anti-inflammatory factors | Inhibition of binding with SARS-CoV spike protein | Sun et al. (2018), Yi et al. (2004) |
| *Glycyrrhiza glabra* L. | Leguminosae | Liquorice | Roots | Increase of NF-κB and IL-10 production | Inhibitory activity against SARS-CoV replication | Das (2021) |
| *Houttuynia cordata* Thunb. | Sauruaceae | Chameleon plant | Aerial parts | Inhibition of LPS-induced PGE2 and COX-2 production in mouse peritoneal macrophages | Inhibitory activities against SARS-CoV $3CL^{pro}$ and SARS-CoV RNA-dependent RNA polymerase (RdRp) | Lau et al. (2008), Li et al. (2011) |

| | | | | | | |
|---|---|---|---|---|---|---|
| *Isatis indigotica* fort. | Brassicaceae | Dyer's woad | Roots | Inhibition of influenza A virus, viral RNA-stimulated pro-inflammatory cytokines, and IFNs via inactivation of p38 MAPK and NF-κB pathways | Inhibitory activity against SARS-CoV $3CL^{pro}$ | Liang et al. (2020) Lin et al. (2005) |
| *Moringa oleifera* lam. | Moringaceae | Drum stick | Aerial parts | Increase of production of IL-2 and antibody-dependent cellular cytotoxicity (ADCC) | Inhibitory activity against SARS-CoV-2 | Das (2021) |
| *Ocimum sanctum* L. | Labiatae/ Lamiaceae | Tulsi | Leaves | Stimulation of cytokines, i.e. IL-4, IFN-γ, T helper cells, and increase of the phagocytic activity | Inhibitory activity against SARS-CoV-2 $M^{pro}$ using molecular dynamic stimulation | Das (2021), Saravanan et al. (2022) |
| *Picrorhiza kurroa* Royale ex Benth | Scrophulariaceae | Kutki | Seeds | Modulating oxidative stress, inflammatory mediators, and cytokine storm, including IL-6, TNF-α, and COX-2 secretion | Inhibitory activity against SARS-CoV-2 replication | Ratheesh et al. (2022) |
| *Piper nigrum* L. | Piperaceae | Black pepper | Seeds | Increase of the phagocytic index and proliferation of macrophage cells | Inhibitory activity against SARS-CoV-2 $M^{pro}$ | Majdalawieh and Carr (2010), Gutierrez-Villagomez et al. (2020) |
| *Scrophularia scorodonia L.* | Scrophulariaceae | Figwort | Aerial parts | Dose-dependently, inhibitory activity against LPS-induced TNF-α, PGE2, and NO production | Inhibitory activity against HcoV-229E attachment and penetration at concentrations of 0.25–25 μmol/L | Cheng et al. (2006), Dıaz et al. (2004) |
| *Scutellaria baicalensis* Georgi | Labiatae/ Lamiaceae | Chinese skullcap | Aerial parts | Downregulation of NO, iNOs, COX-2, PGE2, NF-κB, and IκBα<br>Downregulation of pro-inflammatory cytokines, including TNF-α, IL-2, IL-6, IL-12, and IL-1β | Inhibitory activity against SARS-CoV-2 replication and SARS-CoV-2 $3CL^{pro}$. Inhibitory activity against SARS-CoV helicase (nsP13) | Liu et al. (2021), Kim et al. (2009), Yu et al. (2012) |

(continued)

**Table 3.1** (continued)

| Plant name | Family | Common name | Part of use | Anti-inflammatory and immunoenhance mechanisms | Anti-SARS-CoV activity | Reference |
|---|---|---|---|---|---|---|
| *Tinospora cordifolia* (Willd.) Miers | Menispermaceae | Gurjo, Guduchi | Leaves | Modulating oxidative stress, inflammatory mediators, and cytokine storm, including IL-6, TNF-α, and COX-2 secretion | Inhibitory activity against SARS-CoV-2 replication | Ratheesh et al. (2022) |
| *Tribulus terrestris* L. | Zygophyllaceae | Bindi | Leaves | Reduction of expression of NO, PGE2, and COX-2 in LPS-induces RAW264.7 macrophage cells<br>Downregulation of inflammatory mediators, including IL-6, IL-10, and TNF-α | Dose-dependently, inhibitory activity against SARS-CoV-$PL^{pro}$ | Song et al. (2014), Ko et al. (2015) |
| *Torreya nucifera* (L.) Siebold & Zucc | Taxaceae | Japanese nutmeg-yew | Leaves | Inhibitory activity against NO and PGE2 production;<br>Dose-dependently, blocking the production of iNOS mRNA, COX-2, and TNF-α;<br>Reduction of NF-κB | Inhibitory activity against SARS-CoV $3CL^{pro}$ | Kim et al. (2016), Ryu et al. (2010) |
| *Toona sinensis* (A. Juss.) Reom | Meliaceae | Cedrela sinensis | Leaves | Inhibitory activities against COX-1 and COX-2 | Inhibitory activity against SARS-CoV viral entry and SARS-CoV replication | Chen et al. (2008), Hu et al. (2016) |
| *Withania somnifera* (L.) Dunal | Solanaceae | Ashwagandha | Roots | Increase of phagocytic activity of peritoneal macrophages | Inhibitory activity against SARS-CoV-2 RdRp;<br>Inhibitory activity against human trans-membrane serine protease-2 (TMPRSS2) protein binding potency using computational analysis | Das (2021), Prajapati et al. (2022) |

| | | | | | | |
|---|---|---|---|---|---|---|
| *Zingiber officinale* Roscoe | Zingiberaceae | Ginger | Rhizomes | Increase of production of splenocyte proliferation, and a decrease of inflammatory mediators, including IL-6, IL-1β, and TNF-α in activated macrophages | Inhibitory activity against SARS-CoV-2 $3CL^{pro}$ | Das (2021), Zubair et al. (2021) |

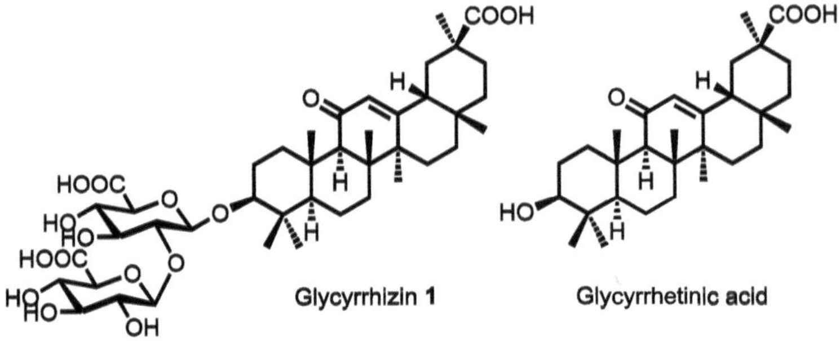


**Fig. 3.2** Structures of glycyrrhizin (**1**) and glycyrrhetinic acid reported from *Glycyrrhiza glabra* L.

antioxidant capacity (TAOC), as well as decreases the carbachol-induced concentration in lung tissue (Ng et al. 2021). In in vivo model of asthma induced by ovalbumin (OVA), **1** showed the alleviation of allergic asthma through the increase of IFN-γ levels, a decrease of inflammatory mediators (i.e., IL-4 and IL-5), and the levels of eosinophil in bronchoalveolar lavage (BAL), reduction of OVA-specific IgE, and upregulation of total IgG2a in serum. In addition to alleviation of lung inflammation induced by carrageenan, compound **1** attenuated mucus production in mice models. In the mice model on lung tissue, the liquorice root extract containing flavonoid compounds modulated lipopolysaccharide (LPS)-induced pulmonary inflammation via inhibition of responsible cells (i.e., lymphocytes, macrophages, and neutrophils) and suppression of IL-1β and TNF-α expression (Pastorino et al. 2018). Noted that the liquorice ethanolic extract, besides scavenging the ROS, potentially reduces the levels of inflammatory markers, such as interleukins, TNF-α, NF-κB, IFN-γ, and PI3K, and increases the increment of serum LgA, LgG, and LgM levels together with spleen lymphocytes proliferation (Ng et al. 2021). Interestingly, for compound **1,** a broad spectrum of antiviral activities against various kinds of viruses has been reported (Hafez Ghoran et al. 2021a), so that compound **1** strongly inhibited the replication of SARS-CoV in Vero cells when tested against two coronavirus strains (i.e., FFM-1 and FFM-2) isolated from SARS positive patients, with selectivity index (SI = 67). It is reported that at the beginning of SARS replication, **1** interrupted the adsorption and viral entry (Cinatl et al. 2003). Van de Sand et al. revealed that treatment of **1** potentially inhibited SARS-CoV-2 main protease ($M^{pro}$), blocking the replication process. Therefore, because of these positive effects of glycyrrhizin, consuming tea prepared from black liquorice roots was recommended for SARS-CoV-2 patients (van de Sand et al. 2021). Taken together, the inhibition of SARS-CoV-2 main proteases (i.e., 3C-like protease; $3CL^{pro}$, papain-like protease; $PL^{pro}$, and main protease; $M^{pro}$) and S proteins have been reported for liquorice extract (Ng et al. 2021).

### *3.4.2* Citrus aurantium *L.*

*Citrus* species is the most substantial and medicinal fruit of Rutaceae plants, which is freshly consumed in folk medicine or as an additive in food supplements. Some biological activities, including antioxidant, anti-inflammatory, antimicrobial, cancer prevention, and degenerative diseases, have been reported for *Citrus* species, which have a high quantity of bioactive metabolites such as polyphenols, phenolics, flavonoids, essential oils, and vitamins. Using the blooms of *Citrus aurantium* L. (known as bitter orange) are deeply rooted in Iranian traditional medicine because of heart tonic and antidepressant properties (Hafez Ghoran et al. 2020). Besides using a stimulant and appetite suppressant agent, the plant is also consumed in traditional Chinese medicine for treating cardiovascular problems, indigestion, nausea, and cancer (Popoola et al. 2022). Reverse phase-HPLC analyses of *C. aurantium* flowers indicated the presence of phenolics and flavonoids such as gallic acid (**2**), pyrogallol (**3**), syringic acid (**4**), caffeic acid (**5**), quercetin (**6**), naringin (**7**), and rutin (**8**) (Fig. 3.3) as the bioactive compounds. Besides worthwhile contributions in antioxidant activity, *Citrus* polyphenols and crude polysaccharides showed significant anti-inflammatory and immunoenhacment activity on LPS-treated RAW264.7 murine macrophage cells via suppression of mitogen-activated protein kinase (MAPK) and NF-κB processes (Kang et al. 2011), reduction of nitric oxide generation, stimulation of IL-6 and TNF-α secretion, and promotion of IL-6, IL-1β, TNF-α, and iNOS mRNA expression. LC-MS analysis of crude polyphenol extract of *C. aurantium* blossoms represented **7**, **8**, eriocitrin (**9**), neoeriocitrin (**10**), rhoifolin (**11**), hesperidin (**12**), neohesperidin (**13**), veronicastroside (**14**), and hesperetin (**15**) (Fig. 3.3) as the main phytoconstituents (Shen et al. 2017). Local people in China, India, and Africa are consuming the decoction of *C. aurantium* blossoms to treat viral infections, especially COVID-19, and stimulate the body's immunity. The pathways were blocking the entry and replication of SARS-CoV-2 and even suppressing the expression of pro-inflammatory markers leading to an anti-inflammation property (Popoola et al. 2022). Clinically, *C. aurantium* together with two *Citrus* sp. (*C. reticulata* and *C. grandis*) boosts digestive health and relieves the cough in SARS-CoV-2 patients. Compounds **7** and **15**, as the main phenolic compounds in *Citrus* plants, potently prevent the cytokine storm (expression of IL-6, IL-1β, COX-2, and iNOS). Thus, flavonoids in *Citrus* fruits can be anti-SARS-CoV-2 agents and/or can increase the host immunity through anti-inflammatory activities (Liu et al. 2022).

### *3.4.3* Hypericum perforatum *L.*

St. John's wort (*Hypericum perforatum* L.; Hypericaceae) belongs to Europe, western Asia, and northern Africa, and has therapeutic implications in minor anxiety and depressive episodes. Various bioactive constituents such as hyperforin (**16**; a

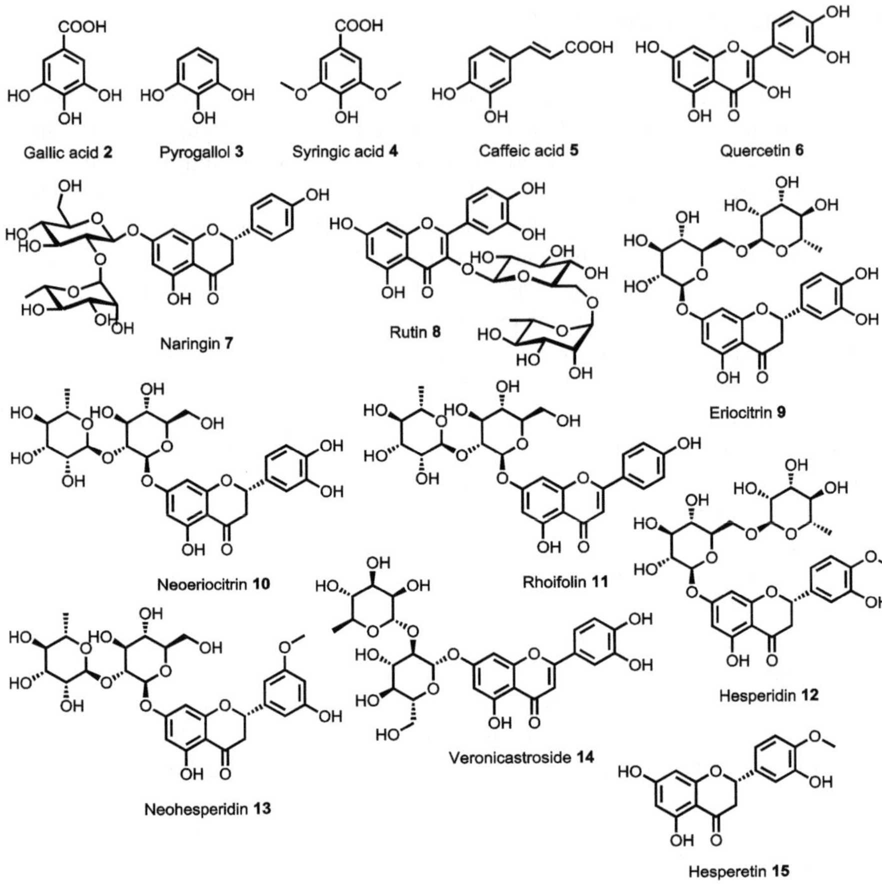


**Fig. 3.3** Some bioactive phytochemicals, **2**–**15**, reported from *Citrus aurantium* L.

prenylated phloroglucinol derivative), hypericin (**17**; a naphthodianthrone derivative) (Fig. 3.4), flavonoids, flavonoid glycosides, bioflavonoids, and tannins are responsible for diverse therapeutic effects. Based on the Iranian ethnopharmacological documents, in the northern part of Iran, the flowering branches of the plant are used for relaxing the nervous system, improving the immune system, refreshment of respiratory and uterus systems, astringent, anti-HIV and anticancer activities (Bahmani et al. 2018). Most *H. perforatum* fractions were active against DPPH$^{\bullet}$ and superoxide radicals, nitric oxide production, and lipid peroxidation. These results indicated that caffeoylquinic acids, flavonoids, and phenolic acids, including **6**, **8** (Fig. 3.3), hyperoside (**18**), quercitrin (**19**), 3,8'-biapigenin (**20**), amentoflavone (**21**) (Fig. 3.4) dealt with antioxidant activity, while phloroglucinols and naphtodianthrones did not show significant effects (Orčić et al. 2011). Owing to the presence of these bioactive components, the plant displays high antioxidant activity, which decreases, in a concentration-

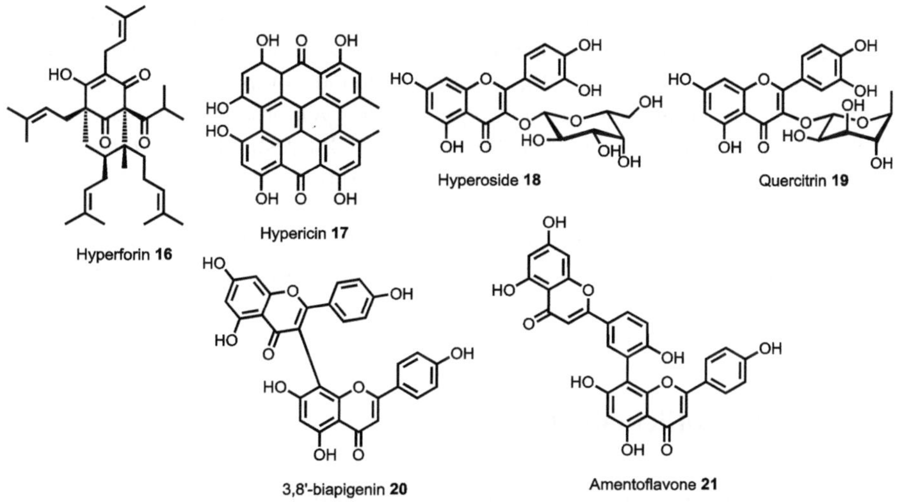


**Fig. 3.4** Some bioactive phytochemicals, **16–21**, reported from *Hypericum perforatum* L.

dependent manner, the peroxidative and chlorinating activity of human leukocyte myeloperoxidase (MPO). This phenomenon further causes the anti-inflammatory activity of *H. perforatum* (Pabuçcuoğlu et al. 2003). Moreover, having been stimulated with *N*-formyl-methionyl-leucyl-phenylalanine, plant-derived metabolites like hyperguinone and hyperforin strongly reduced oxygen production by polymorphonuclear cells (PMNs) (Heilmann et al. 2003). Not only does *H. perforatum* have immunomodulatory activity but it also has a potent antiviral activity and because of this, the plant is a candidate against COVID-19. The extract of *H. perforatum* and hyperforin (**16**) strongly inhibited the pro-inflammatory mediators and β-cell lines isolated from rat and human pancreatic islets (Novelli et al. 2014). The same authors evaluated the mechanism of action and proposed there is a simultaneous blockage of the multiple phosphorylation steps by *H. perforatum* extract together with compound **16**, which are involved in a cytokine storm, including IL-1β, TNF-α, IFN-γ, NF-κB, MAPK, and JAK/STAT pathways. Furthermore, the plant preparations deactivated the transcription of apoptotic and inflammatory genes, such as COX-2, iNOS, CXCL9, and CXCL10 chemokines, as well as the production of ICAM-1 adhesion molecules. Most significantly, despite the activation of some kinases (i.e., JNK, ERK1/2, IKK, and Akt), *H. perforatum* extract and compound **16** inhibited DNA binding and phosphorylation of NF-κB and STAT-1 (Novelli et al. 2016). Associated with the protective effects of *H. perforatum* on various animal models against chronic inflammation, Menegazzi et al. included that the plant extract is able to attenuate lung inflammation caused by carrageenan in mice through deactivation of NF-κB and STAT-3, inhibition of IL-1β and TNF-α secretion, downregulation of ICAM-1, infiltration of lung neutrophils, and nitration of cellular proteins (Menegazzi et al. 2006). Emerged by the multitasking behavior of *H. perforatum* and its major component, hyperforin (**16**), and showing deserved anti-inflammatory activity, Masiello et al. proposed the plant as the promising candidate to either inhibit or block the drawbacks of cytokine storm in COVID-19 patients. Administrated the extract of *H. perforatum* and **16**, researchers found compatible results with the selected antiviral drugs, such as chloroquine/hydroxychloroquine, human immunoglobulins, and LMW heparin (Masiello et al. 2020). Interestingly, the *H. perforatum* fractions rich in flavonoid and catechin showed various antiviral activities. Using molecular docking study, the phytoconstituents of *H. perforatum*, for example, hypericin (**17**) and isohypericin, showed effective anti-SARS-CoV-2 activity, where high binding affinity was observed for inhibition of SARS-CoV-2 $M^{pro}$ (Yalçın et al. 2021).

### *3.4.4* Salvia officinalis *L.*

Sage (*Salvia officinalis* L.; Lamiaceae/Labiatae family) is a multipurpose culinary herb and medicinal plant native to the northern Mediterranean area and widely distributed over the hillsides and shores of southern Europe. Despite variously reported bioactivities such as antimicrobial, antiviral, antioxidant, anti-inflammatory, cytotoxic, hypoglycemic, and astringent, the plant has an effective immunomodulatory activity (Capek and Matulova 2013). The *Salvia* sp. extract contains the most potent antioxidant and immunoenhancer metabolites, including polysaccharides, terpenoids (i.e., carnosol **22**, carnosic acid **23**, and ursolic/oleanolic acid **24**/**25**), phenylpropanoids (i.e., caffeic acid **5** and rosmarinic acid **26**), flavonoids (i.e., genkwanin **27** and cirsimaritin **28**) (Fig. 3.5), and other phenolic acids. In traditional medicine, the decoction and infusion of the plant and its preparations are used to treat stomach disorders, mouth and throat mucus, diarrhea, periodontitis, gingivitis, and gingival bleeding, as well as abscesses (Hafez Ghoran et al. 2021b, 2022a). Rahman et al. showed that daily supplementation of *S. officinalis* leaves

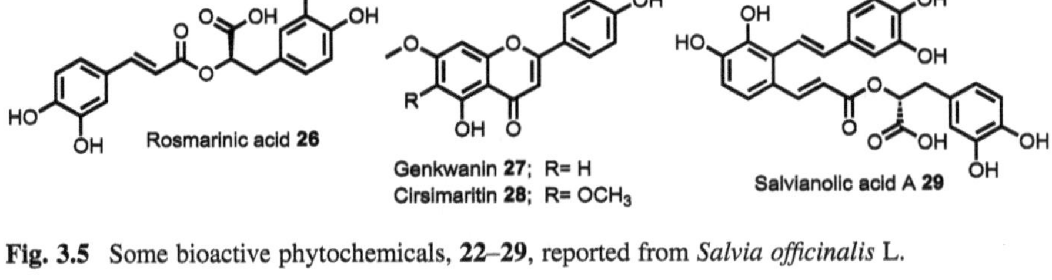


**Fig. 3.5** Some bioactive phytochemicals, **22**–**29**, reported from *Salvia officinalis* L.

could efficiently enhance the antioxidant-immune capacity along with resistance against gram-positive bacteria *Aeromonas sobria* infection in *Cyprinus carpio*. At concentrations of 4 and 8 $g.Kg^{-1}$, not only the function of the digestive system was improved by increasing the activity of intestinal enzymes (lipase and amylase), but also the acting of enzymatic antioxidants (SOD and GSH-Px) along with the immunity effects (nitric oxide, anti-protease, and immunoglobulin M) significantly improved. Moreover, both dietary concentrations enhanced the splenic genes expression (*SOD*, *TLR-2*, *IL-6*, *IL-10*, and *IL-1β*), the intestinal genes expression (*Slc26a6*, *PepT1* and/or *Slc15a1*), and muscular genes expression (*SOD* and *IGF-1*) whereas these doses resulted in downregulation of *MSTN* (Rahman et al. 2022). In another research on female Wistar albino rats, *S. officinalis* exhibited anti-inflammation activity in the lung, liver, kidney, and hemolysate versus induced by LPS and oxidative stress via enhancing the enzymatic antioxidant activity of CAT, SOD, and GSH-Px, and regulating molecules in the inflammation pathway. The mechanism of action of *S. officinalis* was detected by the uptake of $^{18}$F-fluoro-deoxy-D-glucose ($^{18}$F-FDG) in hepatic and pulmonary cells and determining the inflammation status using scanning of $^{18}$F-fluoro-deoxy-D-glucose positron emission tomography ($^{18}$FDG-PET). Results indicated that the plant extract attenuates the levels of malondialdehyde (MDA), NO, TNF-α, and NF-κB (Kolac et al. 2017). Moreover, the ethanolic extract of *S. officinalis*, obtained by the soxhlet method, showed better antioxidant activity and decreased the secretion of IL-6 and TNF-α, rather than hydroethanolic and aqueous extracts. The bioactivity probably is because of the presence of compounds **22**, **23**, and **26**. The plant extract had higher to equal anti-inflammatory activity than reference drugs such as diclofenac, salicylic acid, and celecoxib (Vieira et al. 2020). Apart from these activities, the plant extract showed effective antiviral activity against various viral infections. According to Loizzo et al., the essential oil of Lebanon *S. officinalis*, containing the highest 1,8-cineole (43.6%), significantly inhibited SARS-CoV replication (Loizzo et al. 2008). In Unani traditional medicine (or Greek medicine), which is based on natural formulations, it is believed that a combined plant extract with ethanol provides a hand sanitizer that has an appropriate efficiency against COVID-19 (Azad et al. 2020). Using a molecular docking study, salvianolic acid A (**29**) (Fig. 3.5) obtained from *S. officinalis* showed good binding affinity to inhibition of SARS-CoV-2 $M^{pro}$ (Ibrahim et al. 2020).

### *3.4.5* Mentha x piperita *L.*

Peppermint (*Mentha* × *piperita* L.; Lamiaceae/Labiatae family) is used as a most common herbal tea or tisanes. Despite using the leaves in food flavoring applications, the plant is traditionally consumed for treating throat inflammation, oral mucosa, cold, fever, digestive problems, antimicrobial and antiviral activities, as well as detoxifying the respiratory system. Moreover, peppermint tea/infusion, brewed from plant leaves or essential oils, is prescribed for respiratory-related disorders, including anti-congestive and expectorants. Evidence-based studies

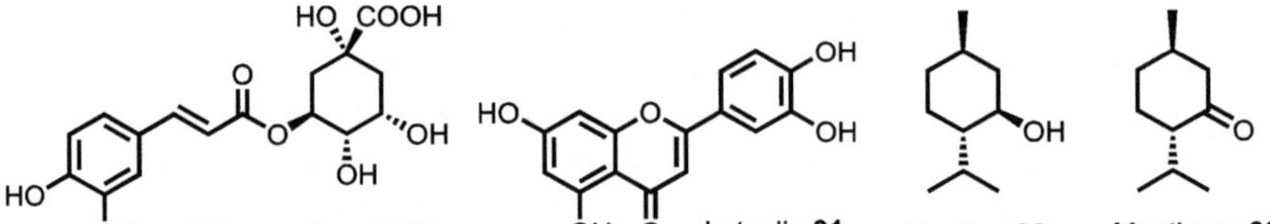

Chlorogenic acid **30** Luteolin **31** Menthol **32** Menthone **33**

**Fig. 3.6** Some bioactive phytochemicals, **30**–**33**, reported from *Mentha* × *piperita* L.

revealed various biological activities such as antioxidant, antiallergenic, antibacterial, anticancer, antidiabetic, antiviral, anti-inflammatory, larvicidal, radio-protective properties, and genotoxicity. The plant leaves are rich in phenolics (i.e. caffeic acid **5**, rosmarinic acid **26**, and chlorogenic acid **30**), flavonoids (i.e., eriocitrin **9**, hesperidin **12**, and luteolin **31**), lignans, and essential oils (i.e. menthol **32** and menthone **33**) (Fig. 3.6) (Mahendran and Rahman 2020). Both products of *M. x piperita* (i.e., essential oil and ethanol extract) represented an effective anti-inflammation in LPS-induced RAW264.7 cells by reducing NO release and suppressing IL-6, TNF-α, and prostaglandin E2 (PGE2) secretion. It is proposed that these processes are linked to virus infection. For instance, respiratory syncytial virus (RSV) brings about the induction of TNF-α secretion. Therefore, the plant could increase innate immunity by interfering with viral entry. Meanwhile, peppermint essential oils decrease the inflammation and lipid peroxidation caused by oxidative stress (Li et al. 2017). In another research, using LPS-stimulated RAW264.7 cells, Cho et al. evaluated the anti-inflammatory properties of the combination of *M.* × *piperita*, *Chrysanthemum zawadskii*, and *Glycyrrhiza glabra* (MCG). MCG mixture reduced the iNOS, IFN-β, and COX-2 expression (inflammatory markers) and IL-6, IL-1β, and TNF-α secretion (pro-inflammatory cytokines). Meanwhile, the MCG mixture suppressed the LPS-stimulated phosphorylation of STAT1, AKT, Iκb, and NF-κB, inhibiting the translocation of NF-κB into the nucleus (Cho et al. 2021). Furthermore, the combination of white tea and *M.* × *piperita* synergistically decreased the above-mentioned pro-inflammatory cytokines. An in-depth study on the mechanism revealed that the herbal preparations inhibited p-IκB-α and MAPK phosphorylation. Therefore, the anti-inflammatory activity was suggested through NF-κB deactivation and MAPK inhibition (Xia et al. 2021). Since there is no data regarding the anti-SARS-CoV-2 activity of *M.* × *piperita* (peppermint), Aldwihi et al. reported that before and during COVID-19, using peppermint-containing preparations like infusion or decoction effectively reduces the period of hospitalization. Based on a clinical study in Saudi Arabia, compared with 22% of hospitalized patients, ~78% of non-hospitalized patients consumed *M.* × *piperita*. Hence, by dietary consumption of peppermint, the time of hospitalization decrease during infection with COVID-19 (Aldwihi et al. 2021). Another adjuvant approach for counteracting SARS-CoV-2 is using aromatherapy and inhalation of essential oils. Based on a randomized clinical trial on peripheral oxygen saturation ($SpO_2$) in COVID-19 patients, the nebulizer effects of rose oil and peppermint oil were compared. Peppermint oil showed better results and significantly increased the mean $SpO_2$ compared to rose oil and the control group. As a consequence, the need for ventilation machines decreased, leading to getting more satisfaction from patients (Ommi et al. 2020).

### *3.4.6* Allium sativum *L.*

Garlic (*Allium sativum* L.) and onion (*Allium cepa* L.), belonging to the Alliaceae family, are the most researched and best medicinal herbs in traditional medicine for most human health-related diseases. Despite the pungent smell, which is because of organosulphur-containing compounds (i.e. diallyl trisulfide **34**, alliin **35**, and allicin **36**) (Fig. 3.7), *Allium* plants are composed of vitamins (i.e., A, B, C, and E), carbohydrates (i.e., glucose and sucrose), minerals (i.e., phosphor, potassium, and selenium), non-volatile organosulphur-containing compounds (i.e., *S*-allylcystein **37**) (Fig. 3.7), other secondary metabolites, including flavonoids (i.e., quercetin **6**), triterpenoids, steroidal saponins, and nitrogenous compounds, which are in charge of various biological and/or pharmacological activities (Hafez Ghoran et al. 2021c). In traditional medicine of the Far East, *Allium* plants are used because of their preventive properties in cardiovascular diseases, regulation of blood pressure, lowering of blood sugar and cholesterol levels, increase of the immune system, and have antibacterial, antiviral, antifungal, antiparasitic, antitumoral, and antioxidant activities (Ayaz and Alpsoy 2007). Both plants, garlic and onion, have been able to provide better conditions for COVID-19 patients since they possess immunomodulatory, anti-inflammatory, antioxidant, antiviral, antihypertensive, antidiabetic, antimutagenic, and anticarcinogenic activities. According to one in silico study, garlic's metabolites can bond with the amino acids in the active site of SARS-CoV-2 $M^{pro}$ through hydrogen bonds (Aldwihi et al. 2021). Clinical evidence reported that the levels of T helper cells in COVID-19 patients are low, and the inflammation is getting worst. However, *A. sativum* supplementation downregulated IL-6, TNF-α, leptin, leptin receptor, and PPAR-γ levels; on the other hand, it significantly upregulated cytotoxic T cells, T helper cells, and natural killers (NK) (Khubber et al. 2020). In another research, Kalkal et al. suggested that carbon dots derived from *A. sativum* (AS-CDs) could be potential theranostic agents and salient candidates for antiviral and bioimaging purposes. Concerning SARS-CoV-2 patients, AS-CDs through nebulization might downregulate the production of NF-κB and MAPKinase, and detoxify ROS species, which leads to the rejuvenation

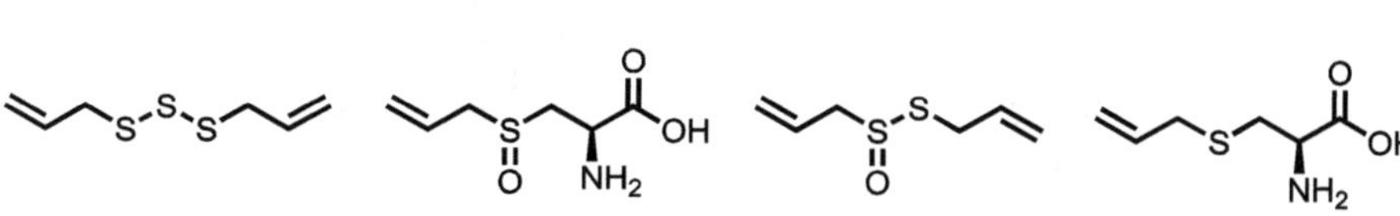


Diallyl trisulfide **34** Alliin **35** Allicin **36** *S*-Allylcysteine **37**

**Fig. 3.7** Some bioactive phytochemicals, **34**–**37**, reported from *Allium sativum* L.

of the body's immunity (Kalkal et al. 2021). Keiss et al. reported that the garlic powder extract and diallyldisulfide modulate inflammatory cytokines expression in LPS-stimulated human blood through NF-κB deactivation and a significant decrease of IL-1β and TNF-α production (Keiss et al. 2003). Because of the modulation of cytokine storm, immunoglobulin production, phagocytosis, and activation of macrophages, the *Allium* plants are good candidates to promote the innate immunity system in COVID-19 patients (Donma and Donma 2020). Moreover, *A. sativum* was on the list of Peruvian herbal medicine to treat COVID-19 and respiratory symptoms during the pandemic. It is well documented that the plant inhibited SARS-CoV-2 replication (Villena-Tejada et al. 2021).

### *3.4.7* Zingiber officinalis *Roscoe*

Ginger (*Zingiber officinale* Roscoe.; Zingiberaceae family) is deeply rooted in ethnopharmacology worldwide. In Iranian, Japanese, Chinese, and Ayurvedic traditional medicine, ginger is consumed for various cancer types, pains, vomiting, cough, and cold symptoms, sore throat, arthritis, indigestion, hypertension, rheumatism, respiratory and infectious diseases. In Malaysia, local people use the plant to treat various immune and infectious diseases (Khodaie and Sadeghpoor 2015). Several studies reported that the ginger extracts with its bioactive constituents, such as vallinoids (i.e. gingerols **38**, shogaols **39**, and paradols **40**) (Fig. 3.8), curcuminoids, flavonoids, and volatile oils, viz. zingerone **41** (Fig. 3.8), zingerol, zingiberene, possess antioxidant, antimicrobial, anti-inflammatory, anticancer, and immunomodulatory activities (Harun and Mohamad 2022). Some animal studies on lung inflammation reported that the *Z. officinale* extract and its bioactive components decline the production of inflammatory markers (i.e., IL-6, IL-1β, and TNF-α) and oxidative stress through decreased MDA and MPO levels, as well as total oxidant content. The plant also prevents lung tissue denaturation and DNA oxidation and

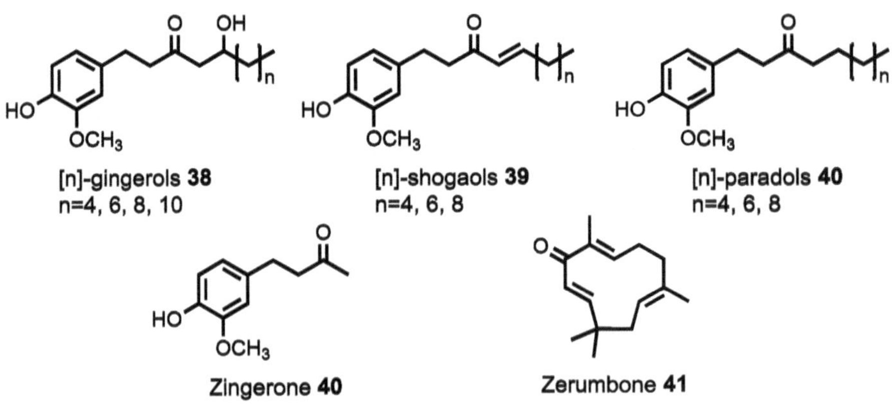


**Fig. 3.8** Some bioactive phytochemicals, **38–41**, reported from *Zingiber officinale* Roscoe

increases the endogenous antioxidant components (Cifci et al. 2018). 6-Shogaol derived from *Z. officinale* extract plays as an antioxidant component that suppresses ROS, iNOS, and COX-2 production and upregulates the levels of antioxidants (i.e., GSH, Nrf2, quinone-1, and hemeoxygenase-1). In a model of cisplatin-mediated toxicity, zingerone **41**, another ginger metabolite, effectively refines the enzymatic antioxidant (SOD, CAT, and GPx) activities and even enhances GSH production. Meanwhile, compound **41** reduces IL-6, IL-1β, TNF-α, NF-κB, iNOS, and COX-2 expression. Overall, ginger and its secondary metabolites decline the oxidative factors and stimulate the oxidative stress-attenuating proteins. The plant has the capability of managing the health condition of COVID-19 patients (Jafarzadeh et al. 2021). In a study on LPS-induced human colonic epithelial cell lines, *Z. officinale* extract with 6-gingerol and 6-shogaol significantly downregulated IL-6, IL-8, PGE2, and iNOS expression, as well as deactivated NF-κB. Noteworthy to mention that both compounds also avoided the production of COX-2 and PGE2 in vitro in microglia and colonic epithelial cells, which were induced by LPS (Kim et al. 2017). *Z. officinale* extract also downregulated the synthesis of prostaglandins and leukotrienes through enzymatic inactivation, including 5-lipoxygenase (5-LOX), COX-1, and COX-2 (van Breemen et al. 2011) proposing the possibility of reduction of hyper-inflammation in COVID-19 patients. Moreover, in LPS-induced human macrophage cells, zerumbone (**42**) (Fig. 3.8), a sesquiterpenoid compound from *Z. officinale*, suppressed the expression of IL-6, TLR2, TLR4, COX-2, and myeloid differentiation factor 88 (MyD88) as well as NF-κB expression (Kim and Yun 2019). In terms of anti-SARS-CoV-2 activity, computational studies showed that 6-, 8-, and 10-gingerols potently inhibit the SARS-CoV-2 $PL^{pro}$. Above all, 6-gingerol showed potential binding affinity to the proteins dealing with virus replication, such as SARS-CoV-2 $M^{pro}$, SARS-CoV-2 $3CL^{pro}$, and cathepsin K (Oso et al. 2022). Another molecular docking study reported that the ginger metabolites (i.e., gingerol, shogaol, geraniol, zingerone, zingiberene, and zingiberenol) not only attach with the SARS-CoV-2 $M^{pro}$ catalytic domain but also destroy the S protein-ACE2 interaction (Ahkam et al. 2020). To more extent, based on a randomized-controlled study in Iran using combination therapy by *Z. officinale* and Echinacea, the clinical symptoms such as breath shortness, coughing, and muscular pain of those none-hospitalized COVID-19 patients were more alleviated than those COVID-19 outpatients used the standard protocol using hydroxychloroquine alone (Mesri et al. 2021). Another randomized-controlled trial was carried out on ARDS patients who were fed for 21 days with an enteral diet containing the rich ginger extract. Compared to the control group, besides decreasing the secretion of IL-1, IL-6, and TNF-α in serum, better oxygenation was observed; however, the time of using electronic ventilation was reduced. Nevertheless, organ failure, barotrauma, and mortality rates happened in patients treated with ginger the same as in the control group. As patients with pulmonary complications have symptoms (i.e., ARDS, fibrosis of lung tissue, sepsis, and pneumonia) similar to COVID-19 signs, ginger and its secondary metabolites play a substantial role in COVID-19 management (Thota et al. 2020).

### *3.4.8* Syzygium aromaticum *L.*

Clove (*Syzygium aromaticum* L.; family Myrtaceae) is a precious spice composed of dried flower buds, originally from Maluku islands in Indonesia, which is used for food preservation by local people. The worldwide use of clove in folk medicine is due to various bioactive phytochemicals, including hydrocarbons, phenylpropanoids (i.e., eugenol **42** and eugenyl acetate **43**), terpenoids (i.e., β-caryophyllene **44**), steroids (i.e., stigmasterol **45**), flavonoids (i.e., kaempferol **46**), tannins (i.e., bicornin **47**), and phenolic compounds (i.e., ellagic acid **48**) (Fig. 3.9) (El-Saber Batiha et al. 2020). In folk medicine, the essential oil of *S. aromaticum* is used as a painkiller in dental care and in curing tooth infections and toothache. The plant oil consisting of eugenol, eugenol acetate, and thymol shows high antioxidant activity and reduces oxidative stress (Nam and Kim 2013). However, it is also administered for burns and wounds. In Ayurvedic and Chinese traditional medicine, the plant is used not only as a warming and stimulating agent but also for treating respiratory disorders, anti-inflammatory, antiviral, and immunostimulatory activities. Nowadays, *S. aromaticum* with a combination of other medicinal plants, such as *Eucalyptus globulus*, *Cymbopogon citratus*, and *Zingiber officinale* is prescribed by traditional healers to prevent and control SRAS-CoV-2-associated diseases (Vicidomini et al. 2021). Moreover, in the mice model, the essential oil of *S. aromaticum* enhanced the total white blood cells (WBC) and increased the delayed-type hypersensitivity (DTH) response. The plant oil also improved the humor- and cell-mediated immune responses, which is recommended for COVID-19 patients to improve their health (Carrasco et al. 2009). Using enzyme-linked immunosorbent assay (ELISA),

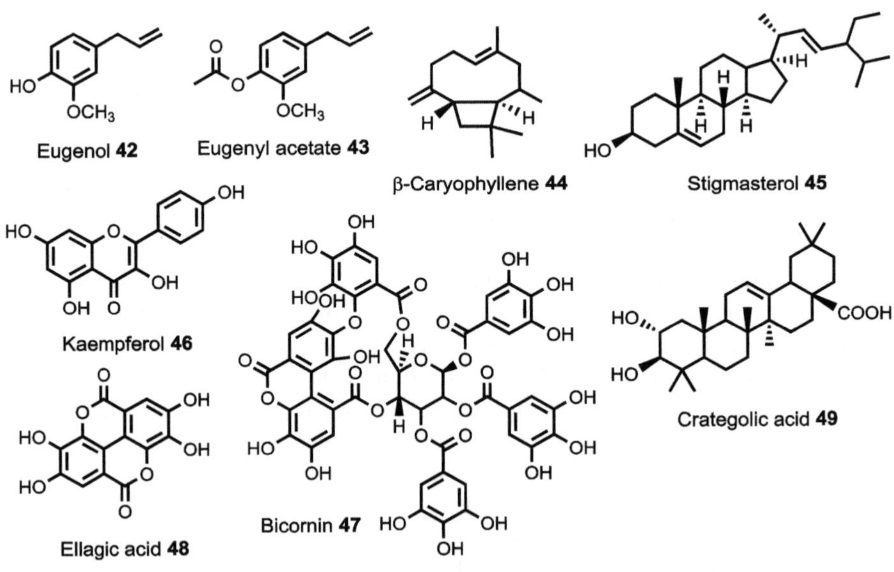


**Fig. 3.9** Some bioactive phytochemicals, **42–49**, reported from *Syzygium aromaticum* L.

Bachiega et al. concluded that *S. aromaticum* extract and its main compound, eugenol **42**, possess immunomodulatory/anti-inflammatory activity through activation of macrophages to secret the pro-inflammatory cytokines. *S. aromaticum* extract, at 100 μg/mL, significantly reduced IL-6, IL-10, and IL-1β production, while eugenol **42** inhibited the IL-10 and IL-6 production at 50–100 μg/mL (Bachiega et al. 2012). The same behavior was observed by the water-soluble part of the hydroalcoholic extract of *S. aromaticum* and eugenol **42**, which inhibited IL-6 and IL-1β secretion in BALB/mice using the ELISA method (Rodrigues et al. 2009). Regarding anti-COVID-19 activity, the *S. aromaticum* water and ethanol extracts containing 34 identified phenolic compounds, such as flavonoids, flavonoid glycosides, and tannins, repressed the binding of SARS-CoV-2 spike protein to ACE2 and inhibited the ACE2 activity in a dose-dependent manner. Interestingly, besides potent antiradical activity against $DPPH^{\bullet}$ and $ABTS^{\bullet+}$ radicals for water extract, its potency was also better than ethanol extract in ACE2 inhibitory activity. On the other hand, by possessing the highest total phenolic content (TPS), the ethanolic extract of *S. aromaticum* relatively detoxified the unstable $OH^{\bullet}$ radicals. Considering these results, clove (*S. aromaticum*) has a good potency to decrease the infection of SARS-CoV-2 and the development of COVID-19 (Li et al. 2022). A molecular docking study on *S. aromaticum*-derived compounds showed that crategolic acid **49** (Fig. 3.9), an oleanane-type triterpenoid, has the best binding affinity to SARS-CoV-2 $M^{pro}$ through the protein-ligand interaction, which is because of amino acids and hydrogen bonds (Yunus 2021).

## *3.4.9* Curcuma longa *L.*

Turmeric (*Curcuma longa* L.; Leguminosae/Fabaceae family), native to the south of Asia, is commonly utilized as a spice or dye for culinary purposes with a long history of traditional uses. Various classes of natural products, including essential oils, curcuminoids (i.e., curcumin **50**, demethoxycurcumin **51**, and bisdemethoxycurcumin **52**) (Fig. 3.10), phenylpropanoids, flavonoids, flavonoid glycosides, and phenolic compounds, have been reported from the *C. longa*, however, the yellow color and pharmacological/biological effects of turmeric are largely attributed to the polyphenols like curcuminoid and sesquiterpenoid compounds (Hafez Ghoran et al. 2022b; de Oliveira Filho et al. 2021). In traditional medicine,

**Fig. 3.10** Some bioactive phytochemicals, **50–52**, reported from *Curcuma longa* L.

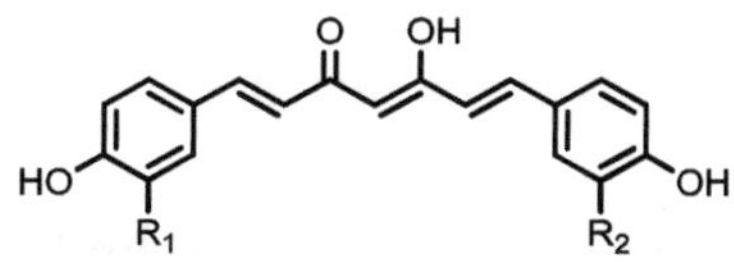


Curcumin **50**; $R_1 = R_2 = OCH_3$
Demethoxycurcumin **51**; $R_1 = OCH_3$, $R_2 = H$
Bisdemethoxycurcumin **52**; $R_1 = R_2 = H$

the *C. longa* rhizomes, composed of ~70% of curcumin, are used for jaundice, rheumatism, hepatic disorders, indigestive, carminative, blood purifier, wound healing, stomachic, and alleviation of chronic inflammation. The plant is also used for boosting immune system response and reduction of the pain of inflammation in several diseases (Hamidpour et al. 2015). Memarzia et al., in their review, represented that *C. longa* and its main constituent, curcumin (~70%), can reduce the MDA and NO levels. In contrast, they enhanced the thiol, SOD, and CAT in oxidative stress conditions. Most significantly, both products improved the functionality of immunoglobulin E (IgE), IL-4, IL-17, IFN-γ levels, and T helper cells (Th1 and Th2), as well as transforming growth factor beta (Memarzia et al. 2021). The same authors, furthermore, delineated that *C. longa* and curcumin **50** relieve the tracheal smooth muscle showing their bronchodilatory effects in chronic obstructive pulmonary disorders. In experimental animal models, *C. longa* and compound **50** treatment alleviated most respiratory dysfunctions through antioxidant, anti-inflammatory, and immunomodulatory mechanisms (Memarzia et al. 2022). Looking at the mechanism of action of curcumin against SARS-CoV-2 activity, it is revealed that curcumin interacted with spike protein or ACE2 protein in the COVID-19-stimulated signal transduction pathway. Several transcription factors (i.e., Nrf2, NF-κB, Wnt/β-catenin, STAT-3, and p38/MAPK), which deal with SARS-CoV-2 infection, were also suppressed by treatment of **50**. Moreover, compound **50** decreased virus-associated inflammation by suppressing IL-6, IL-8, IL-10, TNF-α, TNF-β, and COX-2 expression in patients suffering from COVID-19 (Subhan et al. 2020). Thimmulappa et al. concluded that **50** inhibits SARS-CoV-2 activity through the following mechanisms: interacting with membrane proteins of the virus, interference in virus entry, suppression of virus proteases, and stimulation of host immune-boosting reactions. Therefore, compound **50** showed protective behavior on ARDS and lethal patients by NF-κB deactivation and downregulation of IL-6, inflammasome, and HMGB1 pathways, suggesting curcumin (**50**) as an ideal prophylactic therapeutic for COVID-19 (Thimmulappa et al. 2021). It is worth mentioning that a combination of curcumin and bromelain, a cysteine protease obtained from the pineapple stem, showed immunomodulatory effects leading to interference in the vital steps of SARS-CoV-2 pathophysiology. As mentioned before, these nutraceuticals potentially reduced the secretion of pro-inflammatory mediators and inhibited the transcription factors. Based on in silico studies, both bromelain and curcumin prevented SARS-CoV-2 entry into the cells, while curcumin also inhibited viral replication (Kritis et al. 2020).

### *3.4.10* Nigella sativa *L.*

Black cumin (*Nigella sativa* L.; Ranunculaceae family) is another immunoenhancer herbal medicine traditionally consumed for asthma, cough, bronchitis, chest congestion, acute headache, fever, back pain, dizziness, and inflammation worldwide. Pharmacologically and experimentally, black cumin (the seeds of *N. sativa*) has

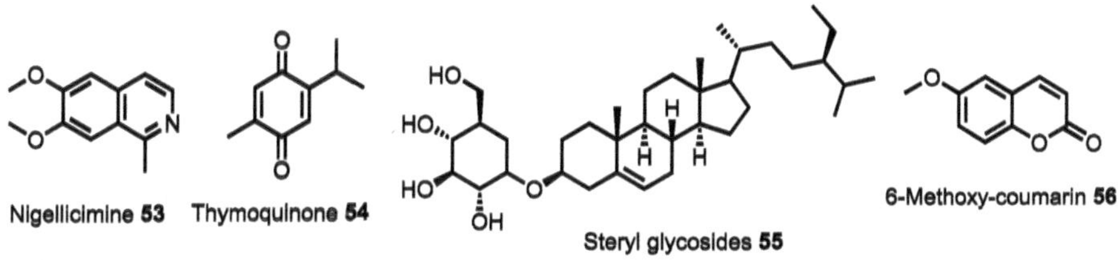


**Fig. 3.11** Some bioactive phytochemicals, **53**–**56**, reported from *Nigella sativa* L.

shown various biological properties, such as antibacterial, antioxidant, anti-inflammatory, immunomodulatory, antidiabetic, anti-tumor, hepatoprotective, analgesic, gastroprotective, bronchodilator, and antihypertensive effects. *N. sativa* contains various phytochemicals such as alkaloids (i.e., nigellicimine **53**), terpenoids (i.e., thymoquinone **54**), saponins (i.e., steryl glycosides **55**), coumarins (i.e., 6-methoxy-coumarin **56**) (Fig. 3.11), flavonoids, and phenolic acids (Ahmad et al. 2021). One of the best attitudes of *N. sative* is an immunomodulatory activity through recruiting T helper cells and NK cells. As well as improving the function of NK cells, *N. sative* proteins significantly increased the levels of T cells (i.e., CD4 and CD8) (Haq et al. 1999). According to a clinical trial on 48 healthy young volunteers, the effects of three doses of 0.5, 1.0, and 2.0 g capsule-containing *N. sativa* powder on the body's immunity and general health were in favor of the group who received a 1.0 g dose. The total increase of lymphocyte, $CD^{3+}$, and $CD^{4+}$ counts was observed in 1.0 g administration showing the plant's immunopotentiation ability by enhancement of T helper cells (Salem et al. 2021). In another research on allergic asthma induced in the murine model, the *N. sativa* fixed oil reduced the production of inflammatory cells (i.e., IL-2, IL-10, and IL-12), IFN-γ, IgG1, and IgG2a (Abbas et al. 2005). Despite the before-mentioned potential properties, *N. sativa* also exhibits antiviral activity and prevents getting ARDS, the principal cause of SARS-CoV-2 mortality. Therefore, *N. sativa* possessing anti-inflammatory and immunomodulatory could be an adjuvant medicinal plant for patients suffering from SARS-CoV-2 infection. (Islam et al. 2021). Considering several clinical trials, inventions, and patent literature, Imran et al. reported that *N. sativa*, *N. sativa* oil, and its phytochemicals have all potency to prevent, manage, and treat COVID-19 infection among high-risk patients (Imran et al. 2022). According to the computational studies, among *N. sativa* secondary metabolites, only α-hederin, nigelledine, hederagenin, thymohydroquinone, and thymoquinone, are potential SARS-CoV-2 inhibitors, which might inhibit SARS-CoV-2 replication (Koshak and Koshak 2020).

### *3.4.11* Tinospora cordifolia *(Willd.) Miers*

Several Himalayan herbal medicines are showing high anti-COVID-19 activity, for example, *Tinospora cordifolia* (Willd.) Miers (Menispermaceae family). *T. cordifolia* is a native and well-recognized medicinal plant in India and Nepal and is used for hay fever, motility, diabetes, high cholesterol, stomach upset, gout, and itchy skin infections caused by mites (scabies) (Choudhary et al. 2013). The plant contains alkaloids (i.e., tetrahydropalmatine **57**), phenyl propanoide glycosides (i.e., cordifolioside A **58**), terpenoids (i.e., tinocordiside **59**, tinosponone **60**, and cordiofolioside B **61**) (Fig. 3.12), polysaccharides, and steroids, which are responsible for the immunomodulatory activity, thereby demonstrating the diverse biological/pharmacological properties of the plant (Yates et al. 2022). In recent years, the significant immune booster properties of *T. cordifolia* have been well documented through fighting the acute and chronic inflammation and inducing autoimmune-like hepatitis or unmasking an underlying autoimmune chronic liver disease (Nagral et al. 2021). Currently, millions of people in the Himalayan region (Nepal, India, and Bhutan) are using the decoction of *T. cordifolia* for COVID-19 treatment; however, not sufficient scientific reports have been found to support its use (Singh et al. 2021). In a clinical study, an Ayurvedic herbal formulation containing *T. cordifolia* and *Piper longum* was added to the standard of care protocol, and the results were compared with the standard protocol. As a

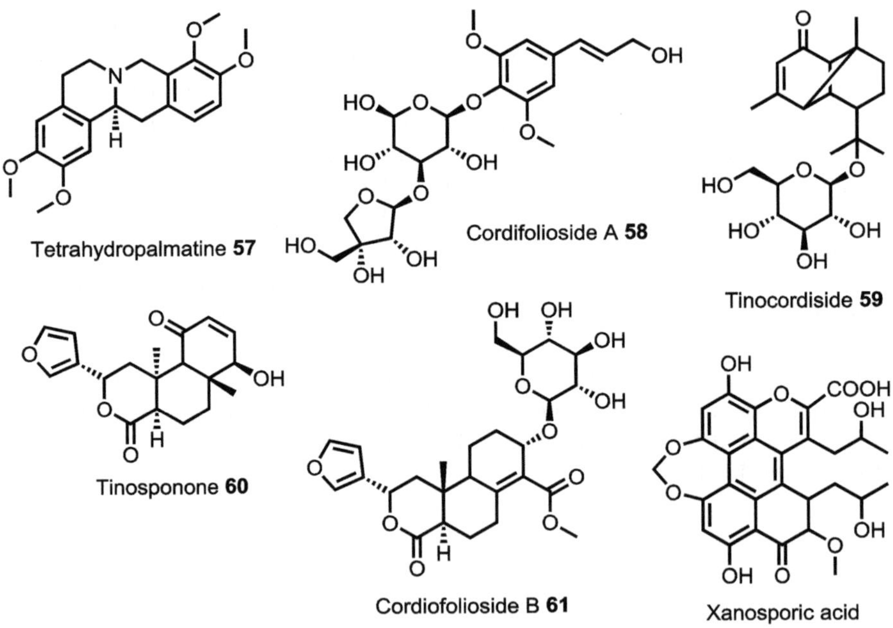


**Fig. 3.12** Some bioactive phytochemicals, **57**–**61**, reported from *Tinospora cordifolia* (Willd.) Miers

consequence, the time of recovery and hospital stay decreased. In post-discharge patients, the general feeling was improved by the Ayurveda add-on formulation even 3 months later (Kataria et al. 2022). Using the humanized Zebrafish model, a reliable in vivo system to evaluate viral pathologies, the inhibition of SARS-CoV-2 activity was studied by a tri-herbal medicine, Coronil, composed of *T. cordifolia* (Willd.) Miers, *Withania somnifora* (L.) Dunal, and *Ocimum sanctum* L. Besides inhibition of SARS-CoV-2 spike protein by Coronil, the secretion of pro-inflammatory cytokines (i.e., IL-6, IL-1β, and TNF-α) was attenuated, and the activity of TNF-α stimulated NF-κB/AP-1 transcription was reduced, at human equivalent doses of 12 and 58 μg/kg. Ultra-high-performance liquid chromatography (UHPLC) analysis revealed rosmarinic acid, betulinic acid, ursolic acid, withanoside IV-V, withaferine A, withanone, cordifolioside A, mangoflorine, and palmatine, as the major phytoconstituents in Coronil, which are responsible for immunomodulatory properties (Balkrishna et al. 2020).

There are several computational studies on *T. cordifolia* and its bioactive compounds. According to a molecular docking study, Balkkrishna et al. concluded that tinocordiside (**59**) interrupts the electrostatic bonds between host ACE2 and the virus spike protein receptor-binding domain (RBD). As a consequence, tinocordiside attenuates or blocks the entry of SARS-CoV-2 into the host cells (Balkrishna et al. 2021). Among 11 phytoconstituents of *T. cordifolia*, only tinosponone (**60**), cordiofolioside B (**61**), xanosporic acid (Fig. 3.12), tembetarine, and berberine showed significant docking scores when they studied in silico against SARS-CoV-2 $3CL^{pro}$ complex. Further docking study along with ADMET (Adsorption, Desorption, Metabolism, Excretion, and Toxicity) evaluation revealed that tinosponone (**60**) was the best SARS-CoV-2 $3CL^{pro}$ inhibitor (the highest binding affinity of − 7.7 kcal/mol) (Krupanidhi et al. 2021). In another molecular dynamics (MD) simulation study, Shree et al. analyzed the anti-SARS-CoV-2 $M^{pro}$ activity of 28 *T. cordifolia*-derived phytoconstituents. Computational results revealed that tinocordiside has a binding energy of −8.1 Kcal/mol, indicating that it could block the SARS-CoV-2 $M^{pro}$. (Shree et al. 2022). Furthermore, from 31 phytochemicals reported from *T. cordifolia*, only amritoside recorded good docking scores to SARS-CoV-2 $M^{pro}$, calculated by three different methods, including high throughput screening, standard precision, and blind docking using Autodock Vina (−5.63, − 7.61, and − 7.2 Kcal/mol, respectively). ADMET, Drug-likeness, and prediction of bioactivity score of amritoside validated its ability to be a candidate for COVID-19 $M^{pro}$ inhibitor (Murugesan et al. 2021).

## 3.5 Phenolic and Polyphenol Compounds against SARS-CoV

### 3.5.1 *Flavonoids and Flavonoid Glycosides*

Flavonoids ($C_6$-$C_3$-$C_6$) are placed in the group of phenolic compounds, which are responsible for the yellow color in plants, and are well-known for their antioxidant activity. Indeed, flavonoids are classified into the following subclasses: flavanones, flavonols, flavanols, flavones, isoflavones, chalcones, anthocyanins, and auranes that could be found in high abundance in food resources and fruits. Besides antioxidant activity, these compounds are in charge of efficient anti-inflammatory, immunomodulatory, antibacterial, anti-viral, and anticancer properties. Most significantly, flavonoids can inhibit the SARS-CoV-2 activity by attaching to the spike protein and consequently interfere with the ACE2 receptors and blockage of entry into the host cells. This phenomenon has been approved using several computational studies and bioinformatics surveys indicating quercetin, quercitrin, luteolin, kaempferol, naringenin, fisetin, theaflavin, and epigallocatechin-3-gallate (EGCG) are some potential SARS-CoV-2 inhibitors (Liskova et al. 2021). The importance of natural flavonoids and flavonoid glycosides in to fight against SARS-CoV has been reviewed and discussed in different mechanisms of action. Quercetin (**6**), naringin (**7**), rhoifolin (**11**), hesperidin (**12**), neohesperidin (**13**), hesperetin (**15**) (Fig. 3.3), luteolin (**31**) (Fig. 3.6), kaempferol (**46**) (Fig. 3.9), baicalein (**62**), scutellarin (**63**), myricetin (**64**), scutellarein (**65**), naringenin (**66**), nobiletin (**67**), herbacetin (**68**), morin (**69**), pectolinarin (**70**), and quercetin-3-*O*-β-D-galactoside (**71**) (Fig. 3.13) are some of the potential flavonoids that are appropriate for drug discovery and development of HCoV-229E inhibitors (Hafez Ghoran et al. 2021a). Generally, flavonoids modulate the inflammation caused by COVID-19 and increase immune responses. These compounds bring about overexpression of pro-inflammatory mediators, systematic immune response attenuation, and consequently, reduction of hyper-inflammation in COVID-19 patients (Roshanravan et al. 2020). An animal model on obese mice showed that apigenin (**72**) (Fig. 3.13) significantly ameliorated the plasma levels of IL-6, TNF-α, and IFN-γ (Jung et al. 2016). In addition, flavonoids such as hesperetin (**15**) (Fig. 3.3) and chrysin (**73**) (Fig. 3.13) regulate the function of immune cells not only by increasing the concentrations of T helper lymphocytic cells and NK cells but also by improving macrophage functions through nitric oxide release and modulation of liposomal activity (Sassi et al. 2017). In another research, the beneficial effects of *Citrus* flavonoids like hesperidin (**12**) (Fig. 3.3) were highlighted against SARS-CoV-2 activity because of their antioxidant, anti-inflammatory, and antiviral abilities (Bellavite and Donzelli 2020). In an animal model of mice co-treated with LPS, Compound **12** could also target anti-inflammatory responses leading to IL-33 and TNF-α reduction (Al-Rikabi et al. 2020). As discussed, significant affinity to suppress pro-inflammatory cytokines production during COVID-19 is achieved by flavonoid treatment. In silico studies recorded a high binding score to the helicase, ACE2 protease sites, and the

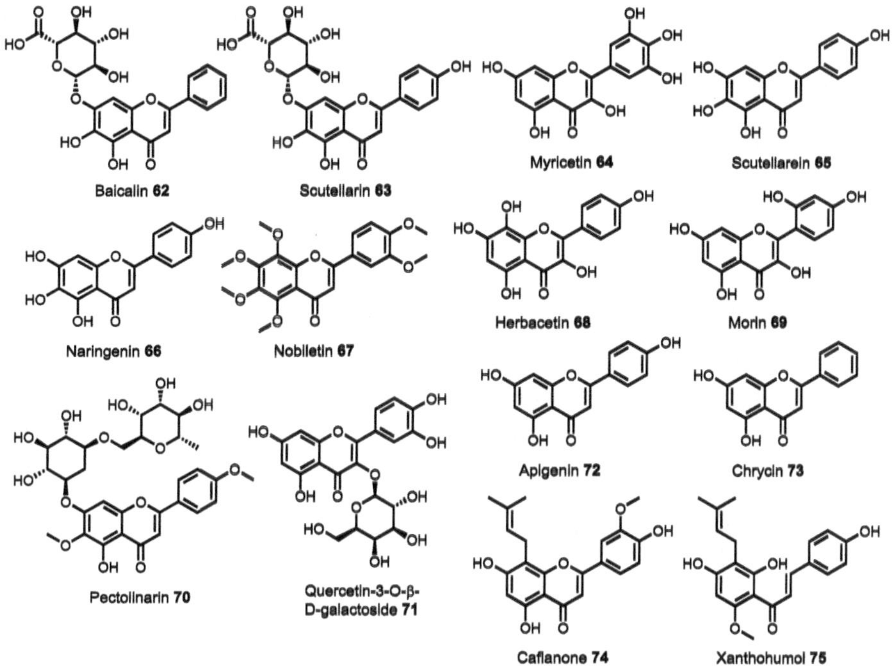


**Fig. 3.13** Some bioactive flavonoids and flavonoid glycoside metabolites, **62–75**, against SARS-CoV-2 infection

spike protein. In this respect, caflanone (**74**) (Fig. 3.13) effectively repressed the cytokines secretion, including IL-6, IL-8, IL-1β, TNF-α, and macrophage inflammatory protein-1α (MIP-1α) (Ngwa et al. 2020). It is worth mentioning that some active flavonoids approved in traditional Chines medicine (TCM), i.e., naringenin, kaempferol, quercetin, baicalein, luteolin, and wogonin, can manage the COVID-19 process through anti-inflammatory and immunomodulatory properties, regulation of pro-inflammatory markers, blocking viral adsorption, and inhibition of viral replication, resulting in protection of the targeted organs. Using network pharmacology for the characterization of potential phytomolecules and regulate the IL-6, detected at all stages of COVID-19 infection, even the time of dying, this ability does not restrict to quercetin **6**, rutin **8** (Fig. 3.3), and luteolin **31** (Fig. 3.6); however, other flavonoids can prevent and rehabilitate the immune system in COVID-19 patients (Niu et al. 2021). The same behavior was found in those TCM formulas consisting of quercetin, which causes ACE2 downregulation in COVID-19 patients (Niu et al. 2020). Flavonoids potentially target ACE2-associated pathways in COVID-19. Through COVID-19 infection, AngII converts to Ang1–7 resulting in AngII inflammatory function. Pro-inflammatory responses are happened by increasing AngII production and angiotensin II receptor type 1 (AT1R) activation. AngII and AT1R activate NF-κB and MAPK signaling pathways, leading to IL-1, IL-6, IL-1β, and TNF-α overproduction. Indeed, in COVID-19 patients, increased AngII levels in circulatory and plasma and further lung damage have been detected (Mahmudpour et al. 2020). In one animal model on hypertensive rats, the AngII levels were alleviated by hesperidin treatment (Wunpathe et al. 2018), and in the same way, kaempferol inhibited the inflammation and oxidative stress in cardiac fibroblast cells through the reduction of AngII-stimulated collagen accumulation (Du et al. 2019). In in vivo study on rats, Choucry et al. reported the extract of *Crateva nurvala* Buch. Hamm, which is rich in flavonoids, stimulates Nrf2 and reduces the secretion of pro-inflammatory mediators (i.e., IL-6, TNF-α, and NF-κB) (Choucry et al. 2018). The same behavior was observed by the flavonoid content of *Apios Americana* Medikus, decreasing the secretion of inflammatory cytokines and activating the Nrf2-KEAP1 pathway in LPS-induced macrophage cells (Chu et al. 2019). Interestingly, by in vitro and in vivo models, the effects of xanthohumol (**75**) (Fig. 3.13), belonging to the subgroup of flavonoids, viz. chalcone class, against oxidative stress and inflammatory damage were evaluated. Compound **75** significantly ameliorated LPS-stimulated acute lung injury through AMPK/GSK3β-Nrf2 signaling pathway (Lv et al. 2017). Considering the valuable anti-SARS-CoV-2 points of flavonoids into account, these naturally occurring compounds, i.e., EGCG, luteolin, nobiletin, apigenin, scutellarin, myricetin, and baicalin, could also modulate other SARS-CoV-2-associated pathways, including activation of NLRP3 inflammasome, Nrf2, TLRs, and bromodomain-containing protein 4 (BRD4) (Liskova et al. 2021).

### *3.5.2 Other Phenolic Compounds*

Ma et al. investigated the molecular mechanism of aloin (**76**) (Fig. 3.14), also known as barbaloin; the active anthraquinone glycoside found in *Aloe vera* juice, on LPS-induced inflammatory response. The level of LPS-induced iNOS expression was reduced by aloin treatment, and IL-6, IL-1β, TNF-α, and NO production was decreased dose-dependently. Meanwhile, aloin (**76**) not only suppressed LPS-induced JAL1-ATAT1/3 activation and STAT1/3 nuclear translocation but also inhibited LPS-induced ROS production. Indeed, aloin (**76**) modulated LPS-stimulated inflammation by blocking ROS-mediate activation in the JAK1-ATAT1/3 signaling pathway, thereby blocking the nuclear translocation of STAT1/M (Ma et al. 2018). One oral mouth rinse containing some bioactive compounds, such as aloin isoforms A & B, hexetidine, chlorhexidine, triclosan, methyl salicylate, menthol, eucalyptol, sodium fluoride, and povidone, was in vitro studied to inhibit the main enzymes in SARS-CoV-2 (i.e., $PL^{pro}$ and $3CL^{pro}$). Interestingly, aloin isoforms A & B selectively showed inhibitory activity against $PL^{pro}$ proteolytic activity with $IC_{50}$ values of 13.16 and 16.08 μM, respectively. A molecular docking study represented a hydrogen bond between Tyr268 and aloin isoforms, acting as a curtail factor for their $PL^{pro}$ proteolytic activity. On the other hand, with the aid of molecular dynamic studies, both aloin isoforms were found to inhibit the activity of $PL^{pro}$ deubiquitination (Lewis et al. 2022).

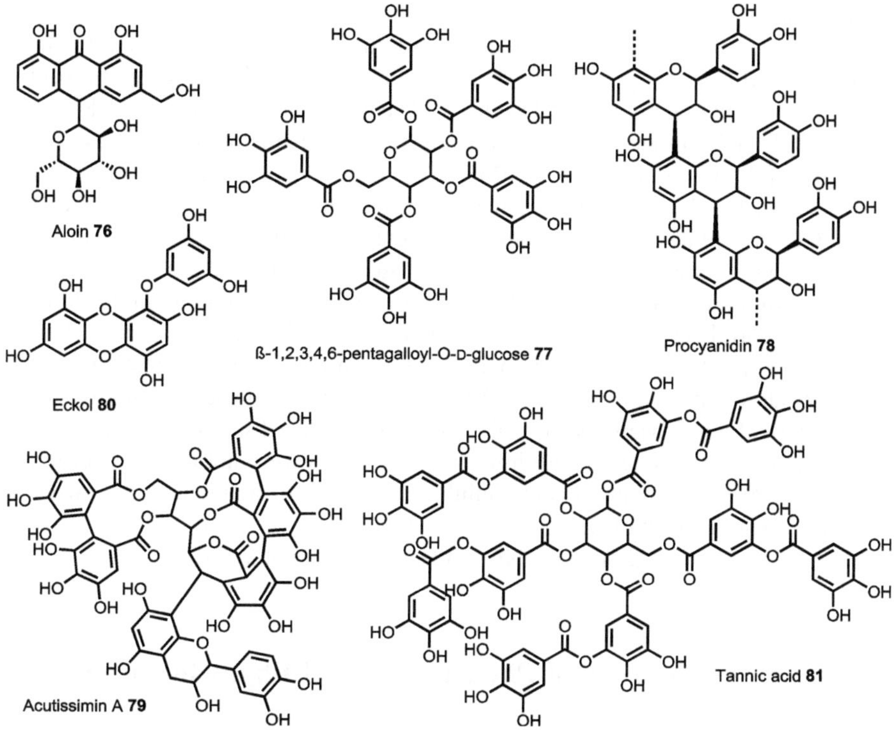


**Fig. 3.14** Some bioactive phenolic compounds, **76–81**, against SARS-CoV-2 infection

Another group of phenolic compounds is tannins, which are classified into various subclasses, such as hydrolysable tannins (i.e., β-1,2,3,4,6-pentagalloyl-*O*-D-glucose **77**), condensed tannins (i.e., procyanidin **78**), complex tannins (i.e., acutissimin A **79**), and phlorotannins (i.e., eckol **80**) (Fig. 3.14), representing potential antioxidant, antiviral, antimicrobial, anti-inflammatory, and anticancer activities. These compounds are found in fruits, seeds, wood, the bark of trees, various type of wild plants, and even in brown algae. The use of tannins in the leather industry dates back centuries ago, and in traditional medicine, they attenuate infections and diseases. Other applications of high tannin-containing plants in folk medicine refer to anti-diabetic, wound healing, cardiovascular protection, and anti-diarrhea properties (Pizzi 2021). A high abundance of inflammatory mediators and reduction of T helper lymphocytic cells (i.e. CD4+ and CD8+) have been reported in the serum of acute COVID-19 patients (Tang et al. 2020). Interestingly, using tannins along with other polyphenols has been in clinical trials on COVID-19 patients. For example, combined oral tannins derived from chestnut and quebracho with B12 vitamin decreased the TNF-α and MIP-1α levels in hospitalized COVID-19 patients, resulting in diarrhea and pneumonia symptoms reduction (Pisarevsky et al. 2021). In one animal study, the tannins from persimmon (scientific name: *Diospyros virginiana*) were pre-administrated by oral gavage to the Syrian hamsters.

Results showed that the plant extract reduced the severity of infection, SARS-CoV-2 transmission, and pneumonia and decreased the inflammation caused by IL-6, TNF-α, and IFN-γ overproduction (Furukawa et al. 2021). Many scientific studies have reported that tannic acid (**81**) (Fig. 3.14) can potentially inhibit SARS-CoV-2 $M^{pro}/3CL^{pro}$ ($IC_{50}$ range = 1 μM - 13.4 mM) and transmembrane protease serine 2 (TMPRSS2; $IC_{50}$ range = 2.31 μM - 50.0 μM), a vital protease located at cell surface related to the viral entry regulated by its linking to the ACE2. SARS-CoV-2 $M^{pro}/3CL^{pro}$ were also inhibited, besides the combination of tannic acid with polyphenols, with other tannin compounds, such as punicalagin and dimeric proanthocyanidins (Wang et al. 2022). Systematic inflammation in COVID-19 patients is correlated with gut dysbiosis, which is antibiotic-independent. Milano et al., in their randomized placebo-controlled trial on 124 patients, administrated oral tannin to survey the symptoms of COVID-19, cytokine response, and gut dysbiosis. Interestingly, having been treated with an oral tannin-based supplement, the systematic inflammation was decreased; however, the results were not clinically efficient. Moreover, MIP-1α was significantly inhibited with tannin supplementation where it correlated negatively with Bifid bacterium abundance; on the other hand, it positively correlated with IL-1β and TNF-α release (Molino et al. 2022).

## 3.6 Synergic Activity of Phenolic-Derived Phytochemicals

An array of experimental studies underlines the effectiveness of combination therapy of natural products to represent the synergic ability against the concerned health disorder. According to the study of Zhao et al. on high fat-stimulated inflammation in rats, combined quercetin with resveratrol treatments at doses of 240 and 120 mg/kg/day, respectively, considerably modulate the levels of IL-6 and TNF-α and the circulating inflammatory marker, such as MCP-1, (Zhao et al. 2017).

As discussed in Sect. 3.4.9, curcumin has efficient anti-SARS-CoV-2 activities by inhibiting infection pathways. In terms of enhancement of bioavailability and efficiency, different kinds of curcumin-loaded nanocarriers were developed and investigated. For instance, a curcumin nanosystem (Sinacurcumin®, a soft gel capsule type composed of 40 mg curcuminoids in nanomicelles) was evaluated not for antiviral activity but for suppression of inflammation in COVID-19 patients. Sinacurcumin® potentially attenuates the release of inflammatory markers, resulting in better condition and refreshment in patients with COVID-19 (Valizadeh et al. 2020). To this respect, Dourado et al. suggested two approaches for inhibition of SARS-CoV-2 activity by curcumin, including (**a**) the bioavailability enhancement leads to antiviral activity through delivery by nanoparticles and (**b**) the synergic behavior of curcumin and nanoparticles themselves increase a dual mechanism of action (Dourado et al. 2021). Another interesting phenolic compound is quercetin showing high potency in immunomodulatory, antioxidant, anti-inflammatory, and antiviral activities. Many scientific reports have highlighted the anti-viral properties of quercetin in synergism with zinc, vitamins C, D, and E, as well as other polyphenol compounds. Therefore, the synergistic behavior of quercetin could validate its efficiency of being an anti-SARS-CoV-2 drug because of well-documented data on prevention, mitigation, and therapy against COVID-19 (Mani et al. 2020). In another study, the *Houttuynia cordata* Thunb. ethyl acetate extract consisting of quercetin, quercitrin, and rutin was in vitro evaluated against dengue virus (DENV-2) and murine coronavirus infection. The plant extract showed anti-coronavirus activity ($IC_{50}$ = 0.98 mg/mL) and inhibited the dengue virus ($IC_{50}$ = 7.50 mg/mL). Quercetin alone inhibited both viruses, while quercetrin only inhibited DENV-2, and rutin did not exhibit any virucidal activities. Interestingly, the combination of quercetin with quercetrin increased the anti-DENV-2 activity. Further in vivo assessment showed that the plant extract had no chronic oral toxicity when the C57BL/6 mice were fed in doses >2000 mg/kg, and histologically, the main organs remained normal (Chiow et al. 2016). As previously discussed, ACE2 and TMPRSS2 are related to the viral entry to the host cells. The higher the TMPRSS2 expression in cells, the more vulnerable to coronavirus. Therefore, by either inhibition or downregulation of TMPRSS2 activity in human cells, it is expected to be a kind of therapeutic approach toward SARS-CoV-2 infection. On this occasion, kaempferol, belonging to flavonol compounds, inactivated the transcription of TMPRSS2 at about 49.14% and 79.48% at concentrations of 5 and 15 mM, respectively. Moreover, expression of TMPRSS2 was significantly reduced when the enzyme was treated with the standardized flavonoids' complex consisting of luteolin, quercetin, and kaempferol showing considerable synergic activity in lower concentrations (Fuzimoto and Isidoro 2020). Most recently, combined resveratrol and pterostilbene showed inhibitory activity toward the replication of the influenza virus and SARS-CoV-2 infection. De Angelis et al. examined A5+ products consisting of polyphenols, other micronutrients, and polydatin, a resveratrol glycoside found in grape juices, against both viral infections. Compared with the polydatin treatment alone, A5+ not only demonstrated a better efficiency in inhibiting both viruses but also considerably reduced the viral proteins expression and virus release showing the possibility of blocking the replication of the virus process. Moreover, in influenza virus-infected cells, A5+ treatment potently decreased the IL-6 production, which is proposed to be a promising anti-inflammatory agent during the infection. Hence, the synergic activity in the A5+ mixture enhances the antiviral activity rather than individual compounds (De Angelis et al. 2021).

## 3.7 Conclusions

- Ethnopharmacology and medicinal plants can be effective and adjuvant in halting SARS-CoV-2 replication.
- Traditional medicine potentially offers some benefits: reduction of inflammation, enhancement of innate immunity of the body, and improving the health condition of post-hospitalized patients.

- *Glycyrrhiza* sp., *Citrus* sp., *Hypericum* sp., *Salvia* sp., *Mentha* sp., *Allium* sp., *Zingiber* sp., *Syzygium* sp., *Curcuma* sp., *Nigella* sp., and *Tinospora* sp., are some examples of medicinal plants used in folk medicine worldwide to prevent ARDS and treat COVID-19 infection.
- In COVID-19, the increased cytokine storm and circulatory and lung inflammatory markers were observed. Therefore, attenuating these detrimental factors by plant-derived phytochemicals could be promising agents in SARS-CoV-2 therapy.
- In SARS-CoV-2 infection, ethnopharmacology and medicinal plants are recruited to attenuate ROS/oxidative stress and consequently reduce pro-inflammatory cytokines and chemokines.
- Chronic virus infections promote ROS production through leukocytes, monocytes, and macrophages, along with antioxidant depletion and viral replication.
- Phenolic compounds like flavonoids and flavonoid glycosides derived from medicinal plants show anti-inflammatory and immunomodulatory abilities through blocking NF-κB activation and suppression of IL-6, IL-1β, and TNF-α secretion.
- Preclinical studies showed that antioxidants suppress the activation of the NLPR3 inflammasome and scavenge ROS.
- The synergistic activity of phenolic compounds provides promising sights for the treatment of SARS-CoV-2 infection.
- Hypothetically, the effects of natural compounds and plant immunoenhancers on SARS-CoV-2 infection require corroboration in further preclinical, epidemiological, and clinical studies.

## References

Abbas AT, Abdel-Aziz MM, Zalata KR, Tel-D AA-G (2005) Effect of dexamethasone and *Nigella sativa* on peripheral blood eosinophil count, IgG1 and IgG2a, cytokine profiles and lung inflammation in murine model of allergic asthma. Egypt J Immunol 12(1):95–102

Ahkam AH, Hermanto FE, Alamsyah A, Aliyyah IH, Fatchiyah F (2020) Virtual prediction of antiviral potential of ginger (*Zingiber officinale*) bioactive compounds against spike and $M^{Pro}$ of SARS-CoV2 protein. Berkala Penelitian Hayati J Biol Res 25(2):52–57

Ahmad MF, Ahmad FA, Ashraf SA, Saad HH, Wahab S, Khan MI et al (2021) An updated knowledge of Black seed (*Nigella sativa* Linn.): review of phytochemical constituents and pharmacological properties. J Herb Med 25:100404

Aldwihi LA, Khan SI, Alamri FF, AlRuthia Y, Alqahtani F, Fantoukh OI et al (2021) Patients' behavior regarding dietary or herbal supplements before and during COVID-19 in Saudi Arabia. Int J Environ Res Public Health 18(10):5086

Al-Rikabi R, Al-Shmgani H, Dewir YH, El-Hendawy S (2020) In vivo and in vitro evaluation of the protective effects of hesperidin in lipopolysaccharide-induced inflammation and cytotoxicity of cell. Molecules 25(3):478

Ansari MY, Ahmad N, Haqqi TM (2020) Oxidative stress and inflammation in osteoarthritis pathogenesis: role of polyphenols. Biomed Pharmacother 129:110452

Ayaz E, Alpsoy HC (2007) Garlic (*Allium sativum*) and traditional medicine. Turkiye parazitolojii dergisi 31(2):145–149

Azad MM, Al Mahmud A, Islam MS, Gouhar AI (2020) Common sage (*Salvia officinalis*) antiviral role: potentiality of a Unani hand sanitizer in COVID-19 (corona virus) second wave control. Asian J Med Biol Res 6(4):611–617

Bachiega TF, de Sousa JPB, Bastos JK, Sforcin JM (2012) Clove and eugenol in noncytotoxic concentrations exert immunomodulatory/anti-inflammatory action on cytokine production by murine macrophages. J Pharm Pharmacol 64(4):610–616

Bahmani M, Khaksarian M, Rafieian-Kopaei M, Abbasi N (2018) Overview of the therapeutic effects of *Origanum vulgare* and *Hypericum perforatum* based on Iran's ethnopharmacological documents. J Clin Diagn Res 12(7)

Balkrishna A, Solleti SK, Verma S, Varshney A (2020) Application of humanized zebrafish model in the suppression of SARS-CoV-2 spike protein induced pathology by tri-herbal medicine coronil via cytokine modulation. Molecules 25(21):5091

Balkrishna A, Pokhrel S, Varshney A (2021) Tinocordiside from *Tinospora cordifolia* (Giloy) may curb SARS-CoV-2 contagion by disrupting the electrostatic interactions between host ACE2 and viral S-protein receptor binding domain. Comb Chem High Throughput Screen 24(10): 1795–1802

Bellavite P, Donzelli A (2020) Hesperidin and SARS-CoV-2: new light on the healthy function of citrus fruits. Antioxidants 9(8):742

Bhat H, Sampath P, Pai R, Bollor R, Baliga M, Fayad R. Indian medicinal plants as immunomodulators: scientific validation of the ethnomedicinal beliefs. Bioactive food as dietary interventions for arthritis and related inflammatory diseases: bioactive food in chronic disease states. 2012:215

Bonam SR, Kaveri SV, Sakuntabhai A, Gilardin L, Bayry J (2020) Adjunct immunotherapies for the management of severely ill COVID-19 patients. Cell Rep Med 1(2):100016

Bonam SR, Kotla NG, Bohara RA, Rochev Y, Webster TJ, Bayry J (2021) Potential immunonanomedicine strategies to fight COVID-19 like pulmonary infections. Nano Today 36:101051

Bose P, Sunita P, Pattanayak SP (2021) Molecular insights into the crosstalk between immune inflammation nexus and SARS-CoV-2 virus. Curr Microbiol 78(11):3813–3828

Capek P, Matulova M (2013) An arabino (glucurono) xylan isolated from immunomodulatory active hemicellulose fraction of *Salvia officinalis* L. Int J Biol Macromol 59:396–401

Carrasco FR, Schmidt G, Romero AL, Sartoretto JL, Caparroz-Assef SM, Bersani-Amado CA et al (2009) Immunomodulatory activity of *Zingiber officinale* Roscoe, *Salvia officinalis* L. and *Syzygium aromaticum* L. essential oils: evidence for humor-and cell-mediated responses. J Pharm Pharmacol 61(7):961–967

Chen C-N, Lin CP, Huang K-K, Chen W-C, Hsieh H-P, Liang P-H et al (2005) Inhibition of SARS-CoV 3C-like protease activity by theaflavin-3, 3′-digallate (TF3). Evid Based Complement Alternat Med 2(2):209–215

Chen C-J, Michaelis M, Hsu H-K, Tsai C-C, Yang KD, Wu Y-C et al (2008) *Toona sinensis* Roem tender leaf extract inhibits SARS coronavirus replication. J Ethnopharmacol 120(1):108–111

Chen S et al (2010) Decision-making optimization of TMT: a simulated annealing algorithm analysis. J Serv Manag 3:363–368

Chen B-T, Li W-X, He R-R, Li Y-F, Tsoi B, Zhai Y-J et al (2012) Anti-inflammatory effects of a polyphenols-rich extract from tea (*Camellia sinensis*) flowers in acute and chronic mice models. Oxidative Med Cell Longev 2012:537923

Cheng PW, Ng LT, Chiang LC, Lin CC (2006) Antiviral effects of saikosaponins on human coronavirus 229E in vitro. Clin Exp Pharmacol Physiol 33(7):612–616

Chiow K, Phoon M, Putti T, Tan BK, Chow VT (2016) Evaluation of antiviral activities of *Houttuynia cordata* Thunb. extract, quercetin, quercetrin and cinanserin on murine coronavirus and dengue virus infection. Asian Pac J Trop Med 9(1):1–7

Cho BO, Shin JY, Kang HJ, Park JH, Hao S, Wang F et al (2021) Anti-inflammatory effect of *Chrysanthemum zawadskii*, peppermint, *Glycyrrhiza glabra* herbal mixture in lipopolysaccharide-stimulated RAW264. 7 macrophages. Mol Med Rep 24(1):1–8

Choucry MA, Khalil MN, El Awdan SA (2018) Protective action of *Crateva nurvala* Buch. Ham extracts against renal ischaemia reperfusion injury in rats via antioxidant and anti-inflammatory activities. J Ethnopharmacol 214:47–57

Choudhary N, Siddiqui M, Azmat S, Khatoon S (2013) *Tinospora cordifolia*: ethnobotany, phytopharmacology and phytochemistry aspects. Int J Pharm Sci Res 4(3):891

Chu Q, Yu X, Jia R, Wang Y, Zhang Y, Zhang S et al (2019) Flavonoids from apios americana medikus leaves protect RAW264. 7 cells against inflammation via inhibition of MAPKs, Akt-mTOR pathways, and Nfr2 activation. Oxidative Med Cell Longev 2019:1563024

Cifci A, Tayman C, Yakut HI, Halil H, Cakir E, Cakir U et al (2018) Ginger (*Zingiber officinale*) prevents severe damage to the lungs due to hyperoxia and inflammation. Turkish J Med Sci 48(4):892–900

Cinatl J, Morgenstern B, Bauer G, Chandra P, Rabenau H, Doerr H (2003) Glycyrrhizin, an active component of liquorice roots, and replication of SARS-associated coronavirus. Lancet 361(9374):2045–2046

Das K (2021) Herbal plants as immunity modulators against COVID-19: a primary preventive measure during home quarantine. J Herb Med 32:100501

De Angelis M, Della-Morte D, Buttinelli G, Di Martino A, Pacifici F, Checconi P et al (2021) Protective role of combined polyphenols and micronutrients against influenza a virus and SARS-CoV-2 infection in vitro. Biomedicines 9(11):1721

de Oliveira Filho JG, de Almeida MJ, Sousa TL, dos Santos DC, Egea MB (2021) Bioactive compounds of turmeric (*Curcuma longa* L.). In: Bioactive compounds in underutilized vegetables and legumes, pp 297–318

Dıaz AM, Abad MJ, Fernández L, Silván AM, De Santos J, Bermejo P (2004) Phenylpropanoid glycosides from *Scrophularia scorodonia*: in vitro anti-inflammatory activity. Life Sci 74(20): 2515–2526

Donma MM, Donma O (2020) The effects of *Allium sativum* on immunity within the scope of COVID-19 infection. Med Hypotheses 144:109934

Dourado D, Freire DT, Pereira DT, Amaral-Machado L, Alencar ÉN, de Barros ALB et al (2021) Will curcumin nanosystems be the next promising antiviral alternatives in COVID-19 treatment trials? Biomed Pharmacother 139:111578

Du Y, Han J, Zhang H, Xu J, Jiang L, Ge W (2019) Kaempferol prevents against ang II-induced cardiac remodeling through attenuating ang II-induced inflammation and oxidative stress. J Cardiovasc Pharmacol 74(4):326

El-Saber Batiha G, Alkazmi LM, Wasef LG, Beshbishy AM, Nadwa EH, Rashwan EK (2020) *Syzygium aromaticum* L. (Myrtaceae): traditional uses, bioactive chemical constituents, pharmacological and toxicological activities. Biomol Ther 10(2):202

Furukawa R, Kitabatake M, Ouji-Sageshima N, Suzuki Y, Nakano A, Matsumura Y et al (2021) Persimmon-derived tannin has antiviral effects and reduces the severity of infection and transmission of SARS-CoV-2 in a Syrian hamster model. Sci Rep 11(1):1–11

Fuzimoto AD, Isidoro C (2020) The antiviral and coronavirus-host protein pathways inhibiting properties of herbs and natural compounds-Additional weapons in the fight against the COVID-19 pandemic? J Tradit Complement Med 10(4):405–419

Geller C, Varbanov M, Duval RE (2012) Human coronaviruses: insights into environmental resistance and its influence on the development of new antiseptic strategies. Viruses 4(11): 3044–3068

Gutierrez-Villagomez JM, Campos-García T, Molina-Torres J, López MG, Vázquez-Martínez J (2020) Alkamides and piperamides as potential antivirals against the severe acute respiratory syndrome coronavirus 2 (SARS-CoV-2). J Phys Chem Lett 11(19):8008–8016

Hafez Ghoran S, Fadaei Dashti M, Maroofi A, Shafiee M, Zare-Hoseinabadi A, Behzad F et al (2020) Biosynthesis of zinc ferrite nanoparticles using polyphenol-rich extract of *Citrus aurantium* flowers. Nanomed Res J 5(1):20–28

Hafez Ghoran S, El-Shazly M, Sekeroglu N, Kijjoa A (2021a) Natural products from medicinal plants with anti-human coronavirus activities. Molecules 26(6):1754

Hafez Ghoran S, Firuzi O, Asadollahi M, Stuppner H, Alilou M, Jassbi AR (2021b) Dammarane-type triterpenoid saponins from *Salvia russellii* Benth. Phytochemistry 184:112653
Hafez Ghoran S, Rahimi H, Kazemi A, Scognamiglio M, Naderian M, Iraji A et al (2021c) *Allium hooshidaryae* (Alliaceae); Chemical compositions, biological and ethnomedicine uses. J Ethnopharmacol 274:113918
Hafez Ghoran S, Taktaz F, Mozafari AA, Tunçtürk M, Sekeroglu N, Kijjoa A (2022a) Uncommon terpenoids from *Salvia* species: chemistry, biosynthesis and biological activities. Molecules 27(3):1128
Hafez Ghoran S, Calcaterra A, Abbasi M, Taktaz F, Nieselt K, Babaei E (2022b) Curcumin-based nanoformulations: a promising adjuvant towards cancer treatment. Molecules 27(16):5236
Hamidpour R, Hamidpour S, Hamidpour M, Sohraby M, Hamidpour M (2015) Turmeric (*Curcuma longa*): from a variety of traditional medicinal applications to its novel roles as active antioxidant, anti-inflammatory, anti-cancer, and anti-diabetes. Int J Pharmacol Phytochem Ethnomed 1(1):37–45
Haq A, Lobo PI, Al-Tufail M, Rama NR, Al-Sedairy ST (1999) Immunomodulatory effect of *Nigella sativa* proteins fractionated by ion exchange chromatography. Int J Immunopharmacol 21(4):283–295
Harun NH, Mohamad MF (2022) The immunomodulatory effects of *Zingiber officinale* (Ginger): a systematic review. Res J Pharm Technol 15(8):3776–3781
Heilmann J, Winkelmann K, Sticher O (2003) Studies on the antioxidative activity of phloroglucinol derivatives isolated from *Hypericum* species. Planta Med 69(03):202–206
Hensel A, Bauer R, Heinrich M, Spiegler V, Kayser O, Hempel G et al (2020) Challenges at the time of COVID-19: opportunities and innovations in antivirals from nature. Planta Med 86(10): 659–664
Hu J, Song Y, Mao X, Wang Z-J, Zhao Q-J (2016) Limonoids isolated from *Toona sinensis* and their radical scavenging, anti-inflammatory and cytotoxic activities. J Funct Foods 20:1–9
Ibrahim MA, Abdelrahman AH, Hussien TA, Badr EA, Mohamed TA, El-Seedi HR et al (2020) In silico drug discovery of major metabolites from spices as SARS-CoV-2 main protease inhibitors. Comput Biol Med 126:104046
Imran M, Khan SA, Alshammari MK, Alkhaldi SM, Alshammari FN, Kamal M et al (2022) *Nigella sativa* L. and COVID-19: a glance at the anti-COVID-19 chemical constituents, clinical trials, inventions, and patent literature. Molecules 27(9):2750
Islam MN, Hossain KS, Sarker PP, Ferdous J, Hannan MA, Rahman MM et al (2021) Revisiting pharmacological potentials of *Nigella sativa* seed: a promising option for COVID-19 prevention and cure. Phytother Res 35(3):1329–1344
Jafarzadeh A, Jafarzadeh S, Nemati M (2021) Therapeutic potential of ginger against COVID-19: is there enough evidence? J Tradit Chin Med Sci 8(4):267–279
Janeway CA (1953) Agammaglobulinemia. Trans Assoc Am Phys 66:200–202
Jung UJ, Cho Y-Y, Choi M-S (2016) Apigenin ameliorates dyslipidemia, hepatic steatosis and insulin resistance by modulating metabolic and transcriptional profiles in the liver of high-fat diet-induced obese mice. Nutrients 8(5):305
Kalkal A, Allawadhi P, Pradhan R, Khurana A, Bharani KK, Packirisamy G (2021) *Allium sativum* derived carbon dots as a potential theranostic agent to combat the COVID-19 crisis. Sensors Int 2:100102
Kang SR, Park KI, Park HS, Lee DH, Kim JA, Nagappan A et al (2011) Anti-inflammatory effect of flavonoids isolated from Korea Citrus aurantium L. on lipopolysaccharide-induced mouse macrophage RAW 264.7 cells by blocking of nuclear factor-kappa B (NF-κB) and mitogen-activated protein kinase (MAPK) signalling pathways. Food Chem 129(4):1721–1728
Kataria S, Sharma P, Ram JP, Deswal V, Singh M, Rana R et al (2022) A pilot clinical study of an add-on Ayurvedic formulation containing *Tinospora cordifolia* and *Piper longum* in mild to moderate COVID-19. J Ayurveda Integr Med 13(2):100454

Keiss H-P, Dirsch VM, Hartung T, Haffner T, Trueman L, Auger J et al (2003) Garlic (*Allium sativum* L.) modulates cytokine expression in lipopolysaccharide-activated human blood thereby inhibiting NF-κB activity. J Nutr 133(7):2171–2175
Khan NM, Haseeb A, Ansari MY, Devarapalli P, Haynie S, Haqqi TM (2017) Wogonin, a plant derived small molecule, exerts potent anti-inflammatory and chondroprotective effects through the activation of ROS/ERK/Nrf2 signaling pathways in human Osteoarthritis chondrocytes. Free Radic Biol Med 106:288–301
Khodaie L, Sadeghpoor O (2015) Ginger from ancient times to the new outlook. Jundishapur J Nat Pharm Prod 10(1):e18402
Khubber S, Hashemifesharaki R, Mohammadi M, Gharibzahedi SMT (2020) Garlic (*Allium sativum* L.): a potential unique therapeutic food rich in organosulfur and flavonoid compounds to fight with COVID-19. Nutr J 19(1):1–3
Kim M-J, Yun J-M (2019) Molecular mechanism of the protective effect of zerumbone on lipopolysaccharide-induced inflammation of THP-1 cell-derived macrophages. J Med Food 22(1):62–73
Kim EH, Shim B, Kang S, Jeong G, Lee J-s, Yu Y-B et al (2009) Anti-inflammatory effects of *Scutellaria baicalensis* extract via suppression of immune modulators and MAP kinase signaling molecules. J Ethnopharmacol 126(2):320–331
Kim SH, Park JG, Hong YD, Kim E, Baik K-S, Yoon DH et al (2016) Src/Syk/IRAK1-targeted anti-inflammatory action of *Torreya nucifera* butanol fraction in lipopolysaccharide-activated RAW264. 7 cells. J Ethnopharmacol 188:167–176
Kim Y, Kim DM, Kim JY (2017) Ginger extract suppresses inflammatory response and maintains barrier function in human colonic epithelial Caco-2 cells exposed to inflammatory mediators. J Food Sci 82(5):1264–1270
Kim DH, Li H, Han YE, Jeong JH, Lee HJ, Ryu J-H (2018) Modulation of inducible nitric oxide synthase expression in LPS-Stimulated BV-2 microglia by prenylated chalcones from *Cullen corylifolium* (L.) Medik. through inhibition of I-κBα degradation. Molecules 23(1):109
Ko HJ, Ahn EK, Oh JS (2015) N-*trans*-ρ-caffeoyl tyramine isolated from *Tribulus terrestris* exerts anti-inflammatory effects in lipopolysaccharide-stimulated RAW 264.7 cells. Int J Mol Med 36(4):1042–1048
Kolac UK, Ustuner MC, Tekin N, Ustuner D, Colak E, Entok E (2017) The anti-inflammatory and antioxidant effects of *Salvia officinalis* on lipopolysaccharide-induced inflammation in rats. J Med Food 20(12):1193–1200
Koshak AE, Koshak EA (2020) *Nigella sativa* L. as a potential phytotherapy for coronavirus disease 2019: a mini-review of in silico studies. Curr Ther Res 93:100602
Kritis P, Karampela I, Kokoris S, Dalamaga M (2020) The combination of bromelain and curcumin as an immune-boosting nutraceutical in the prevention of severe COVID-19. Metabol Open 8: 100066
Krupanidhi S, Abraham Peele K, Venkateswarulu T, Ayyagari VS, Nazneen Bobby M, John Babu D et al (2021) Screening of phytochemical compounds of *Tinospora cordifolia* for their inhibitory activity on SARS-CoV-2: an in silico study. J Biomol Struct Dyn 39(15):5799–5803
Lau K-M, Lee K-M, Koon C-M, Cheung CS-F, Lau C-P, Ho H-M et al (2008) Immunomodulatory and anti-SARS activities of *Houttuynia cordata*. J Ethnopharmacol 118(1):79–85
Lee J-Y, Ko S-H, Lee Y-J, Lee S-Y, Park H-J, Shin T-Y et al (2010) Anti-inflammatory effect of MeOH extract of *Cibotium barometz* in IFN-γ and LPS-stimulated mouse peritoneal macrophage. Korean J Pharmacogn 41(2):108–114
Lee M, Shim S-Y, Sung SH (2017) Triterpenoids isolated from *Alnus japonica* inhibited LPS-induced inflammatory mediators in HT-29 cells and RAW264. 7 cells. Biol Pharm Bull 40(9):1544–1550
Lei Z, Zou G, Gao Y, Yao Y, Peng C, Shu J et al (2020) A new triterpenoid and a new flavonoid glycoside isolated from *Bupleurum marginatum* and their anti-inflammatory activity. Nat Prod Res 34(24):3492–3498

Lewis DS, Ho J, Wills S, Kawall A, Sharma A, Chavada K et al (2022) Aloin isoforms (A and B) selectively inhibits proteolytic and deubiquitinating activity of papain like protease (PLpro) of SARS-CoV-2 in vitro. Sci Rep 12(1):1–11

Li W, Zhou P, Zhang Y, He L (2011) *Houttuynia cordata*, a novel and selective COX-2 inhibitor with anti-inflammatory activity. J Ethnopharmacol 133(2):922–927

Li Y, Liu Y, Ma A, Bao Y, Wang M, Sun Z (2017) In vitro antiviral, anti-inflammatory, and antioxidant activities of the ethanol extract of *Mentha piperita* L. Food Sci Biotechnol 26(6): 1675–1683

Li F, Li J, Wang P-H, Yang N, Huang J, Ou J et al (2021) SARS-CoV-2 spike promotes inflammation and apoptosis through autophagy by ROS-suppressed PI3K/AKT/mTOR signaling. Biochim Biophys Acta Mol Basis Dis 1867(12):166260

Li Y, Liu Z, Zeng M, El Kadiri A, Huang J, Kim A et al (2022) Chemical compositions of clove (*Syzygium aromaticum* (L.) Merr. & L.) extracts and their potentials in suppressing SARS-CoV-2 spike protein–ACE2 binding, inhibiting ACE2, and scavenging free radicals. J Agric Food Chem 70(45):14403–14413. https://doi.org/10.1021/acs.jafc.2c06300

Liang X, Huang Y, Pan X, Hao Y, Chen X, Jiang H et al (2020) Erucic acid from Isatis indigotica Fort. suppresses influenza A virus replication and inflammation in vitro and in vivo through modulation of NF-κB and p38 MAPK pathway. J Pharmaceut Anal 10(2):130–146

Lin C-W, Tsai F-J, Tsai C-H, Lai C-C, Wan L, Ho T-Y et al (2005) Anti-SARS coronavirus 3C-like protease effects of *Isatis indigotica* root and plant-derived phenolic compounds. Antivir Res 68(1):36–42

Liskova A, Samec M, Koklesova L, Samuel SM, Zhai K, Al-Ishaq RK et al (2021) Flavonoids against the SARS-CoV-2 induced inflammatory storm. Biomed Pharmacother 138:111430

Liu H, Ye F, Sun Q, Liang H, Li C, Li S et al (2021) *Scutellaria baicalensis* extract and baicalein inhibit replication of SARS-CoV-2 and its 3C-like protease in vitro. J Enzyme Inhib Med Chem 36(1):497–503

Liu W, Zheng W, Cheng L, Li M, Huang J, Bao S et al (2022) *Citrus* fruits are rich in flavonoids for immunoregulation and potential targeting ACE2. Nat Prod Bioprospect 12(1):1–10

Loizzo MR, Saab AM, Tundis R, Statti GA, Menichini F, Lampronti I et al (2008) Phytochemical analysis and in vitro antiviral activities of the essential oils of seven Lebanon species. Chem Biodivers 5(3):461–470

Lv H, Liu Q, Wen Z, Feng H, Deng X, Ci X (2017) Xanthohumol ameliorates lipopolysaccharide (LPS)-induced acute lung injury via induction of AMPK/GSK3β-Nrf2 signal axis. Redox Biol 12:311–324

Ma Y, Tang T, Sheng L, Wang Z, Tao H, Zhang Q et al (2018) Aloin suppresses lipopolysaccharide-induced inflammation by inhibiting JAK1-STAT1/3 activation and ROS production in RAW264. 7 cells. Int J Mol Med 42(4):1925–1934

Mahendran G, Rahman LU (2020) Ethnomedicinal, phytochemical and pharmacological updates on Peppermint (*Mentha× piperita* L.)—A review. Phytother Res 34(9):2088–2139

Mahmud S, Uddin MAR, Paul GK, Shimu MSS, Islam S, Rahman E et al (2021) Virtual screening and molecular dynamics simulation study of plant-derived compounds to identify potential inhibitors of main protease from SARS-CoV-2. Brief Bioinform 22(2):1402–1414

Mahmudpour M, Roozbeh J, Keshavarz M, Farrokhi S, Nabipour I (2020) COVID-19 cytokine storm: the anger of inflammation. Cytokine 133:155151

Majdalawieh AF, Carr RI (2010) In vitro investigation of the potential immunomodulatory and anti-cancer activities of black pepper (*Piper nigrum*) and cardamom (*Elettaria cardamomum*). J Med Food 13(2):371–381

Man SM, Karki R, Kanneganti TD (2017) Molecular mechanisms and functions of pyroptosis, inflammatory caspases and inflammasomes in infectious diseases. Immunol Rev 277(1):61–75

Mani JS, Johnson JB, Steel JC, Broszczak DA, Neilsen PM, Walsh KB et al (2020) Natural product-derived phytochemicals as potential agents against coronaviruses: a review. Virus Res 284: 197989

Masiello P, Novelli M, Beffy P, Menegazzi M (2020) Can *Hypericum perforatum* (SJW) prevent cytokine storm in COVID-19 patients? Phytother Res 34(7):1471–1473

Mehta P, McAuley DF, Brown M, Sanchez E, Tattersall RS, Manson JJ, HLH Across Speciality Collaboration, UK (2020) COVID-19: consider cytokine storm syndromes and immunosuppression. Lancet (London, England) 395(10229):1033

Memarzia A, Khazdair MR, Behrouz S, Gholamnezhad Z, Jafarnezhad M, Saadat S et al (2021) Experimental and clinical reports on anti-inflammatory, antioxidant, and immunomodulatory effects of *Curcuma longa* and curcumin, an updated and comprehensive review. Biofactors 47(3):311–350

Memarzia A, Saadat S, Behrouz S, Boskabady MH (2022) Curcuma longa and curcumin affect respiratory and allergic disorders, experimental and clinical evidence: a comprehensive and updated review. Biofactors 48(3):521–551

Menegazzi M, Di Paola R, Mazzon E, Muià C, Genovese T, Crisafulli C et al (2006) *Hypericum perforatum* attenuates the development of carrageenan-induced lung injury in mice. Free Radic Biol Med 40(5):740–753

Merarchi M, Dudha N, Das BC, Garg M (2021) Natural products and phytochemicals as potential anti-SARS-CoV-2 drugs. Phytother Res 35(10):5384–5396

Mesri M, Saber SSE, Godazi M, Shirdel AR, Montazer R, Koohestani HR et al (2021) The effects of combination of *Zingiber officinale* and Echinacea on alleviation of clinical symptoms and hospitalization rate of suspected COVID-19 outpatients: a randomized controlled trial. J Complement Integr Med 18(4):775–781

Molino S, Pisarevsky A, Badu S, Wu Q, Mingorance FL, Vega P et al (2022) Randomized placebo-controlled trial of oral tannin supplementation on COVID-19 symptoms, gut dysbiosis and cytokine response. J Funct Foods 99:105356

Murugesan S, Kottekad S, Crasta I, Sreevathsan S, Usharani D, Perumal MK et al (2021) Targeting COVID-19 (SARS-CoV-2) main protease through active phytocompounds of ayurvedic medicinal plants–*Emblica officinalis* (Amla), *Phyllanthus niruri* Linn. (Bhumi Amla) and *Tinospora cordifolia* (Giloy)–A molecular docking and simulation study. Comput Biol Med 136:104683

Nagral A, Adhyaru K, Rudra OS, Gharat A, Bhandare S (2021) Herbal immune booster-induced liver injury in the COVID-19 pandemic-a case series. J Clin Exp Hepatol 11(6):732–738

Nam H, Kim M-M (2013) Eugenol with antioxidant activity inhibits MMP-9 related to metastasis in human fibrosarcoma cells. Food Chem Toxicol 55:106–112

Nema NK, Mamdapur GMN, Sarojam S, Khamborkar SD, Sajan LC, Sabu S et al (2021) Preventive Medicinal Plants and their Phytoconstituents against SARS-CoV-2/COVID-19. Pharmacogn Rev 13(4)

Ng SL, Khaw K-Y, Ong YS, Goh HP, Kifli N, Teh SP et al (2021) Licorice: a potential herb in overcoming SARS-CoV-2 infections. J Evid Based Integr Med 26:2515690X21996662

Ngwa W, Kumar R, Thompson D, Lyerly W, Moore R, Reid T-E et al (2020) Potential of flavonoid-inspired phytomedicines against COVID-19. Molecules 25(11):2707

Niu W, Wu F, Cui H, Cao W, Chao Y, Wu Z et al (2020) Network pharmacology analysis to identify phytochemicals in traditional Chinese medicines that may regulate ACE2 for the treatment of COVID-19. Evid Based Complement Alternat Med 2020:7493281

Niu W-h, Wu F, Cao W-y, Wu Z-g, Chao Y-C, Peng F et al (2021) Network pharmacology for the identification of phytochemicals in traditional Chinese medicine for COVID-19 that may regulate interleukin-6. Biosci Rep 41(1)

Novelli M, Beffy P, Menegazzi M, De Tata V, Martino L, Sgarbossa A et al (2014) St. John's wort extract and hyperforin protect rat and human pancreatic islets against cytokine toxicity. Acta Diabetol 51(1):113–121

Novelli M, Menegazzi M, Beffy P, Porozov S, Gregorelli A, Giacopelli D et al (2016) St. John's wort extract and hyperforin inhibit multiple phosphorylation steps of cytokine signaling and prevent inflammatory and apoptotic gene induction in pancreatic β cells. Int J Biochem Cell Biol 81:92–104

Ommi D, Saeedi N, Hajizadeh N, Mohseni G, Teymourian H (2020) Complementary effects of *Mentha piperita* (Peppermint) and *Rosa damascene* extract (Rose oil) on SpO2 in patients with COVID-19: a Randomized Clinical Trial. Herb Med J 5(3):119–126

Orčić DZ, Mimica-Dukić NM, Francišković MM, Petrović SS, Jovin EĐ (2011) Antioxidant activity relationship of phenolic compounds in *Hypericum perforatum* L. Chem Cent J 5(1):1–8

Ordonez AA, Bullen CK, Villabona-Rueda AF, Thompson EA, Turner ML, Merino VF et al (2022) Sulforaphane exhibits antiviral activity against pandemic SARS-CoV-2 and seasonal HCoV-OC43 coronaviruses in vitro and in mice. Commun Biol 5(1):1–11

Oso BJ, Adeoye AO, Olaoye IF (2022) Pharmacoinformatics and hypothetical studies on allicin, curcumin, and gingerol as potential candidates against COVID-19-associated proteases. J Biomol Struct Dyn 40(1):389–400

Pabuçcuoğlu A, Konyalıoğlu S, Baş M, Meral GE (2003) The in vitro effects of *Hypericum* species on human leukocyte myeloperoxidase activity. J Ethnopharmacol 87(1):89–92

Park J-Y, Jeong HJ, Kim JH, Kim YM, Park S-J, Kim D et al (2012) Diarylheptanoids from *Alnus japonica* inhibit papain-like protease of severe acute respiratory syndrome coronavirus. Biol Pharm Bull 35(11):2036–2042

Pastorino G, Cornara L, Soares S, Rodrigues F, Oliveira MBP (2018) Liquorice (*Glycyrrhiza glabra*): a phytochemical and pharmacological review. Phytother Res 32(12):2323–2339

Pérez-Torres I, Castrejón-Téllez V, Soto ME, Rubio-Ruiz ME, Manzano-Pech L, Guarner-Lans V (2021) Oxidative stress, plant natural antioxidants, and obesity. Int J Mol Sci 22(4):1786

Pisarevsky AA, Mingorance FL, Vega P, Stefanolo JP, Repetti JA, Ludueña G et al (2021) Fr578 Oral tannins reduce proinflammatory cytokines associated with diarrhea and pneumonia in hospitalized COVID-19 patients. Gastroenterology 160(6):S-371

Pizzi A (2021) Tannins medical/pharmacological and related applications: a critical review. Sustain Chem Pharm 22:100481

Popoola TD, Segun PA, Ekuadzi E, Dickson RA, Awotona OR, Nahar L et al (2022) West African medicinal plants and their constituent compounds as treatments for viral infections, including SARS-CoV-2/COVID-19. DARU. J Pharm Sci:1–20

Prajapati KS, Singh AK, Kushwaha PP, Shuaib M, Maurya SK, Gupta S et al (2022) *Withania somnifera* phytochemicals possess SARS-CoV-2 RdRp and human TMPRSS2 protein binding potential. Vegetos:1–20

Račková L, Jančinová V, Petríková M, Drábiková K, Nosáľ R, Štefek M et al (2007) Mechanism of anti-inflammatory action of liquorice extract and glycyrrhizin. Nat Prod Res 21(14):1234–1241

Rahman ANA, Van Doan H, Elsheshtawy HM, Dawood A, Salem SM, Sheraiba NI et al (2022) Dietary *Salvia officinalis* leaves enhances antioxidant-immune-capacity, resistance to *Aeromonas sobria* challenge, and growth of *Cyprinus carpio*. Fish Shellfish Immunol 127: 340–348

Raman K, Rajagopal K, Islam F, Dhawan M, Mitra S, Aparna B et al (2022) Role of natural products towards the SARS-CoV-2: a critical review. Ann Med Surg 80:104062

Ratheesh M, Sunil S, Sheethal S, Jose SP, Sandya S, Ghosh OSN et al (2022) Anti-inflammatory and anti-COVID-19 effect of a novel polyherbal formulation (Imusil) via modulating oxidative stress, inflammatory mediators and cytokine storm. Inflammopharmacology 30(1):173–184

Rodrigues T, Fernandes A Jr, Sousa J, Bastos J, Sforcin J (2009) In vitro and in vivo effects of clove on pro-inflammatory cytokines production by macrophages. Nat Prod Res 23(4):319–326

Rosales C (2020) Neutrophils at the crossroads of innate and adaptive immunity. J Leukoc Biol 108(1):377–396

Roshanravan N, Seif F, Ostadrahimi A, Pouraghaei M, Ghaffari S (2020) Targeting cytokine storm to manage patients with COVID-19: a mini-review. Arch Med Res 51(7):608–612

Ryu YB, Jeong HJ, Kim JH, Kim YM, Park J-Y, Kim D et al (2010) Biflavonoids from *Torreya nucifera* displaying SARS-CoV 3CLpro inhibition. Bioorg Med Chem 18(22):7940–7947

Salem A, Bamosa A, Alam M, Alshuraim S, Alyalak H, Alagga A et al (2021) Effect of *Nigella sativa* on general health and immune system in young healthy volunteers; a randomized, placebo-controlled, double-blinded clinical trial. F1000Research 10:1199

Saravanan KM, Zhang H, Senthil R, Vijayakumar KK, Sounderrajan V, Wei Y et al (2022) Structural basis for the inhibition of SARS-CoV2 main protease by Indian medicinal plant-derived antiviral compounds. J Biomol Struct Dyn 40(5):1970–1978

Sassi A, Bzéouich IM, Mustapha N, Maatouk M, Ghedira K, Chekir-Ghedira L (2017) Immunomodulatory potential of hesperetin and chrysin through the cellular and humoral response. Eur J Pharmacol 812:91–96

Shen C-Y, Yang L, Jiang J-G, Zheng C-Y, Zhu W (2017) Immune enhancement effects and extraction optimization of polysaccharides from *Citrus aurantium* L. var. *amara* Engl. Food Funct 8(2):796–807

Sho T, Xu J (2019) Role and mechanism of ROS scavengers in alleviating NLRP3-mediated inflammation. Biotechnol Appl Biochem 66(1):4–13

Shree P, Mishra P, Selvaraj C, Singh SK, Chaube R, Garg N et al (2022) Targeting COVID-19 (SARS-CoV-2) main protease through active phytochemicals of ayurvedic medicinal plants–*Withania somnifera* (Ashwagandha), *Tinospora cordifolia* (Giloy) and *Ocimum sanctum* (Tulsi)–a molecular docking study. J Biomol Struct Dyn 40(1):190–203

Singh RS, Singh A, Kaur H, Batra G, Sarma P, Kaur H et al (2021) Promising traditional Indian medicinal plants for the management of novel Coronavirus disease: a systematic review. Phytother Res 35(8):4456–4484

Snell LM, McGaha TL, Brooks DG (2017) Type I interferon in chronic virus infection and cancer. Trends Immunol 38(8):542–557

Song YH, Kim DW, Curtis-Long MJ, Yuk HJ, Wang Y, Zhuang N et al (2014) Papain-like protease ($PL^{pro}$) inhibitory effects of cinnamic amides from *Tribulus terrestris* fruits. Biol Pharm Bull 37(6):1021–1028

Stok JE, Vega Quiroz ME, van der Veen AG (2020) Self RNA sensing by RIG-I–like receptors in viral infection and sterile inflammation. J Immunol 205(4):883–891

Subhan F, Khalil AAK, Zeeshan M, Haider A, Tauseef I, Haleem SK et al (2020) Curcumin: from ancient spice to modern anti-viral drug in COVID-19 pandemic. Life Sci 1(Suppl):69–73

Sugii M, Ohkita M, Taniguchi M, Baba K, Kawai Y, Tahara C et al (2005) Xanthoangelol D isolated from the roots of *Angelica keiskei* inhibits endothelin-1 production through the suppression of nuclear factor-κB. Biol Pharm Bull 28(4):607–610

Sul O-J, Ra SW (2021) Quercetin prevents LPS-induced oxidative stress and inflammation by modulating NOX2/ROS/NF-kB in lung epithelial cells. Molecules 26(22):6949

Sun K, Song X, Jia R, Yin Z, Zou Y, Li L et al (2018) Evaluation of analgesic and anti-inflammatory activities of water extract of *Galla chinensis* in vivo models. Evid Based Complement Alternat Med 2018:6784032

Tang Y, Liu J, Zhang D, Xu Z, Ji J, Wen C (2020) Cytokine storm in COVID-19: the current evidence and treatment strategies. Front Immunol 11:1708

Thimmulappa RK, Mudnakudu-Nagaraju KK, Shivamallu C, Subramaniam KT, Radhakrishnan A, Bhojraj S et al (2021) Antiviral and immunomodulatory activity of curcumin: a case for prophylactic therapy for COVID-19. Heliyon 7(2):e06350

Thota SM, Balan V, Sivaramakrishnan V (2020) Natural products as home-based prophylactic and symptom management agents in the setting of COVID-19. Phytother Res 34(12):3148–3167

Tian S, Hu W, Niu L, Liu H, Xu H, Xiao S-Y (2020) Pulmonary pathology of early phase SARS-COV-2 pneumonia. J Thorac Oncol 15:700–704

Valizadeh H, Abdolmohammadi-Vahid S, Danshina S, Gencer MZ, Ammari A, Sadeghi A et al (2020) Nano-curcumin therapy, a promising method in modulating inflammatory cytokines in COVID-19 patients. Int Immunopharmacol 89:107088

van Breemen RB, Tao Y, Li W (2011) Cyclooxygenase-2 inhibitors in ginger (*Zingiber officinale*). Fitoterapia 82(1):38–43

van de Sand L, Bormann M, Alt M, Schipper L, Heilingloh CS, Steinmann E et al (2021) Glycyrrhizin effectively inhibits SARS-CoV-2 replication by inhibiting the viral main protease. Viruses 13(4):609

Vicidomini C, Roviello V, Roviello GN (2021) Molecular basis of the therapeutical potential of clove (*Syzygium aromaticum* L.) and clues to its anti-COVID-19 utility. Molecules 26(7):1880

Vieira SF, Ferreira H, Neves NM (2020) Antioxidant and anti-inflammatory activities of cytocompatible *Salvia officinalis* extracts: a comparison between traditional and Soxhlet extraction. Antioxidants 9(11):1157

Villena-Tejada M, Vera-Ferchau I, Cardona-Rivero A, Zamalloa-Cornejo R, Quispe-Florez M, Frisancho-Triveño Z et al (2021) Use of medicinal plants for COVID-19 prevention and respiratory symptom treatment during the pandemic in Cusco, Peru: a cross-sectional survey. PLoS One 16(9):e0257165

Wang Z, Yang L (2021) Chinese herbal medicine: fighting SARS-CoV-2 infection on all fronts. J Ethnopharmacol 270:113869

Wang Q, Jiang H, Li Y, Chen W, Li H, Peng K et al (2017) Targeting NF-kB signaling with polymeric hybrid micelles that co-deliver siRNA and dexamethasone for arthritis therapy. Biomaterials 122:10–22

Wang J, Li Q, Yin Y, Zhang Y, Cao Y, Lin X et al (2020) Excessive neutrophils and neutrophil extracellular traps in COVID-19. Front Immunol 11:2063

Wang S-C, Chou I-W, Hung M-C (2022) Natural tannins as anti-SARS-CoV-2 compounds. Int J Biol Sci 18(12):4669

Wen C-C, Shyur L-F, Jan J-T, Liang P-H, Kuo C-J, Arulselvan P et al (2011) Traditional Chinese medicine herbal extracts of *Cibotium barometz*, *Gentiana scabra*, *Dioscorea batatas*, *Cassia tora*, and *Taxillus chinensis* inhibit SARS-CoV replication. J Tradit Complement Med 1(1): 41–50

Wunpathe C, Potue P, Maneesai P, Bunbupha S, Prachaney P, Kukongviriyapan U et al (2018) Hesperidin suppresses renin-angiotensin system mediated NOX2 over-expression and sympathoexcitation in 2 K-1C hypertensive rats. Am J Chin Med 46(04):751–767

Xia X, Lin Z, Shao K, Wang X, Xu J, Zhai H et al (2021) Combination of white tea and peppermint demonstrated synergistic antibacterial and anti-inflammatory activities. J Sci Food Agric 101(6): 2500–2510

Xu Z, Shi L, Wang Y, Zhang J, Huang L, Zhang C et al (2020) Pathological findings of COVID-19 associated with acute respiratory distress syndrome. Lancet Respir Med 8(4):420–422

Yahfoufi N, Alsadi N, Jambi M, Matar C (2018) The immunomodulatory and anti-inflammatory role of polyphenols. Nutrients 10(11):1618

Yalçın S, Yalçınkaya S, Ercan F (2021) Determination of potential drug candidate molecules of the *Hypericum perforatum* for COVID-19 treatment. Curr Pharmacol Rep 7(2):42–48

Yates CR, Bruno EJ, Yates ME (2022) Tinospora Cordifolia: a review of its immunomodulatory properties. J Diet Suppl 19(2):271–285

Yi L, Li Z, Yuan K, Qu X, Chen J, Wang G et al (2004) Small molecules blocking the entry of severe acute respiratory syndrome coronavirus into host cells. J Virol 78(20):11334–11339

Yu M-S, Lee J, Lee JM, Kim Y, Chin Y-W, Jee J-G et al (2012) Identification of myricetin and scutellarein as novel chemical inhibitors of the SARS coronavirus helicase, nsP13. Bioorg Med Chem Lett 22(12):4049–4054

Yunus G (2021) Herbal compounds from *Syzygium aromaticum* and *Cassia acutifolia* as a shield against SARS-CoV-2 Mpro: a molecular docking approach. Biointerface Res Appl Chem 11: 14853–14865

Zhang H, Penninger JM, Li Y, Zhong N, Slutsky AS (2020a) Angiotensin-converting enzyme 2 (ACE2) as a SARS-CoV-2 receptor: molecular mechanisms and potential therapeutic target. Intensive Care Med 46(4):586–590

Zhang J, Tecson KM, McCullough PA (2020b) Endothelial dysfunction contributes to COVID-19-associated vascular inflammation and coagulopathy. Rev Cardiovasc Med 21(3):315–319

Zhao L, Cen F, Tian F, Li M-J, Zhang Q, Shen HY et al (2017) Combination treatment with quercetin and resveratrol attenuates high fat diet-induced obesity and associated inflammation in rats via the AMPKα1/SIRT1 signaling pathway. Exp Ther Med 14(6):5942–5948. https://doi.org/10.3892/etm.2017.5331

Zubair MS, Maulana S, Widodo A, Pitopang R, Arba M, Hariono M (2021) GC-MS, LC-MS/MS, Docking and molecular dynamics approaches to identify potential SARS-CoV-2 3-chymotrypsin-like protease inhibitors from *Zingiber officinale* Roscoe. Molecules 26(17):5230

# Chapter 4
# Herbal Formulations in Fighting Against the SARS-CoV-2 Infection

**Jothi Dheivasikamani Abidharini, Biju Reji Souparnika, James Elizabeth, Ganesh Vishalini, Sidhic Nihala, Velayathaprabhu Shanmugam, Rengasamy Lakshminarayanan Rengarajan, Natchiappan Senthilkumar, Boro Arthi, and Arumugam Vijaya Anand**

## Abbreviations

| | |
|---|---|
| ACE2 | Angiotensin-converting enzyme 2 |
| Bp | Base pair |
| E | Envelope |
| ER | Endoplasmic reticulum |
| $HSYF^{pro}$ | Hanshiyi Formula |
| IFN-β | interferon- beta |
| JGF | Jing Guan Fang |
| KK | Kabasura kudineer |
| Kbp | kilo base pair |
| $M^{pro}$ | Main protease |
| M | Membrane |

J. D. Abidharini · B. R. Souparnika · J. Elizabeth · G. Vishalini · S. Nihala · B. Arthi (✉) · A. V. Anand (✉)
Department of Human Genetics and Molecular Biology, Bharathiar University, Coimbatore, Tamil Nadu, India
e-mail: avahgmb@buc.edu.in

V. Shanmugam
Department of Biotechnology, Bharathiar University, Coimbatore, Tamil Nadu, India

R. L. Rengarajan
Rashtriya Uchchatar Shiksha Abhiyan (RUSA), Madurai Kamaraj University, Madurai, Tamil Nadu, India

N. Senthilkumar
Chemistry and Bioprospecting Division, Institute of Forest Genetics and Tree Breeding (IFGTB-ICFRE), Coimbatore, Tamil Nadu, India

 
J.-T. Chen (ed.), *Ethnopharmacology and Drug Discovery for COVID-19: Anti-SARS-CoV-2 Agents from Herbal Medicines and Natural Products*,
https://doi.org/10.1007/978-981-99-3664-9_4

| | |
|---|---|
| N | Nucleocapsid |
| $PL^{pro}$ | Papain-like protease |
| RBD | Receptor binding domain |
| SARS-CoV-2 | Severe Acute Respiratory Syndrome Coronavirus 2 |
| S | Spike |
| TCM | Traditional Chinese medicine |
| TMPRSS2 | Transmembrane serine protease 2 |
| TXYF | Tongxie Yaofang |
| WHO | World Health Organization |

## 4.1 Introduction

Herbs are culinary-based plants that are been used to flavor foods and also used as medicines for a long time. These plants are found to have preventative properties against diseases such as diabetes, different heart disorders, and cancer. These plants are rich in antioxidant components. Medicines are derived from these herbs and are used for primary health care in developing countries and are used with the general belief that these drugs are without any side effects, cheap, and easily available (Pal and Shukla 2003; Nargis Begum et al. 2009; Vijayakumar et al. 2015). These are called formulations that are the extracted solution from plants. These formulations are prepared from one or more processed herbs in a particular specific quantity that contains the entire beneficial component for diagnosing and treating the disorders in humans and animals. These herbal formulas can be prepared by extraction, distillation, expression, fractionation, purification, fermentation, and powder formation and are obtained from both fresh and dried plants, fungi, lichens, and algae are involved in forming these formulations (Manikandan and Anand 2016; Vijayakumar et al. 2019; Abubakar and Haque 2020).

These herbal formulations are used as traditional medicines that are used for alleviating various ailments. Traditional medicine is the therapeutic practice of physicians using indigenous medicine. According to World Health Organization (WHO), traditional medicine consists of therapeutic practices that are existing before the development of modern medicines and are still in use to this date (Pal and Shukla 2003; Vijaya Anand et al. 2020). Herbs are been used in traditional medicines for a long time in different regions of the world, their use in the prevention of different diseases and disorders and their treatment, and their valuable pharmaceutical properties have attracted the attention of the world. The WHO has launched a WHO Traditional Medicine Strategy 2014–2023, for corroborating medicinal plant use in management and healthcare services. Using the medicinal plant as an alternative treatment for diseases and maintaining health is an expanding trend (Govindharajan et al. 2020; Teh et al. 2021).

Traditional medicines are been practiced in regions such as China, India, Japan, Korea, and other parts of Asia and also in countries belonging to Africa and Europe.

Indian traditional medicine Ayurveda is the oldest traditional medicinal practice in the world (Arumugam et al. 2020; Saravana Prabha et al. 2020; Anand et al. 2021; Kuchi Bhotla et al. 2021a, b; Chandra Manivannan et al. 2022). It is the ancient Vedic knowledge of healing science that is still been practiced. Ayurveda is known as the "Mother of all healing" and the philosophy behind Ayurveda is preventing unwanted suffering and living a long and healthy life. The history of this traditional medicine can be traced back to the pre-Vedic periods between 4000 BC and 1500 BC. In Ayurveda there are two types of herbal formulation one is purely herbal preparation or Kasthoushadhies and the other is herbiomineral metallic preparation also known as Rasaushadhies. The formulation in Ayurveda is based on two principles using a single drug or using more than one which is also known as polyherbal formulation (Parasuraman et al. 2014).

There are different ways of preparation of herbal formulas and ways of their administration. The herbal formulas are prepared in the form of decoctions, powders, pills, alcohols, and tinctures. A decoction is the dominant form of herbal formula preparation (Dharmananda 1997). It is the common form of administering herbal medicine and is taken orally. It is used for acute conditions; can also be used as a herbal bath for the affected region or the whole body (Yang 2010). It is mainly prepared from the hard parts of the herbs such as roots, bark, and seeds, by heating the required quantity of herbs in water for about 30 minutes (Nagalingam 2017). A dried decoction is another form of herbal preparation, it was first developed in Japan by drying the liquid in the decoction to form a syrup form (Nafiu et al. 2017). Powders and pills are other forms for the preparation of crude herbs, prepared by drying the plants to a sufficient state when they can be crushed to fine particles. This form of herbal formulation can have advantages over liquid formulations.

The powdered herbs are mixed with binder components like honey or malt syrup which is formed into tables or pills which upon ingestion directly can be dissolved to the power form in the stomach (Dharmananda 1997). Tinctures are prepared by soaking different parts of the herbal plants such as leaves, berries, bark, and roots in solvents like alcohol or vinegar. These are concentrated extracts of the herbs. The tinctures can be prepared from fresh plant parts, drying causes a change in their chemistry and may lose their efficiency (Fougère and Wynn 2007). These preparations are an ideal form of use as medicine because of their longer shelf-life and also their mixing is easy (Wynn and Fougère 2007).

## 4.2 SARS-CoV-2

Coronavirus disease 2019 (COVID-19) is caused by the severe acute respiratory syndrome coronavirus 2 (SARS-CoV-2), which was identified by WHO as a pandemic on 11 March 2020. It was first identified in the respiratory tract of a patient with pneumonia in Wuhan, Hubei, China. It is an enveloped positive non-segmented RNA virus that belongs to the genus of β-coronavirus which is the variant of previously identified SARS-CoV and MERS-CoV that caused the pulmonary failure

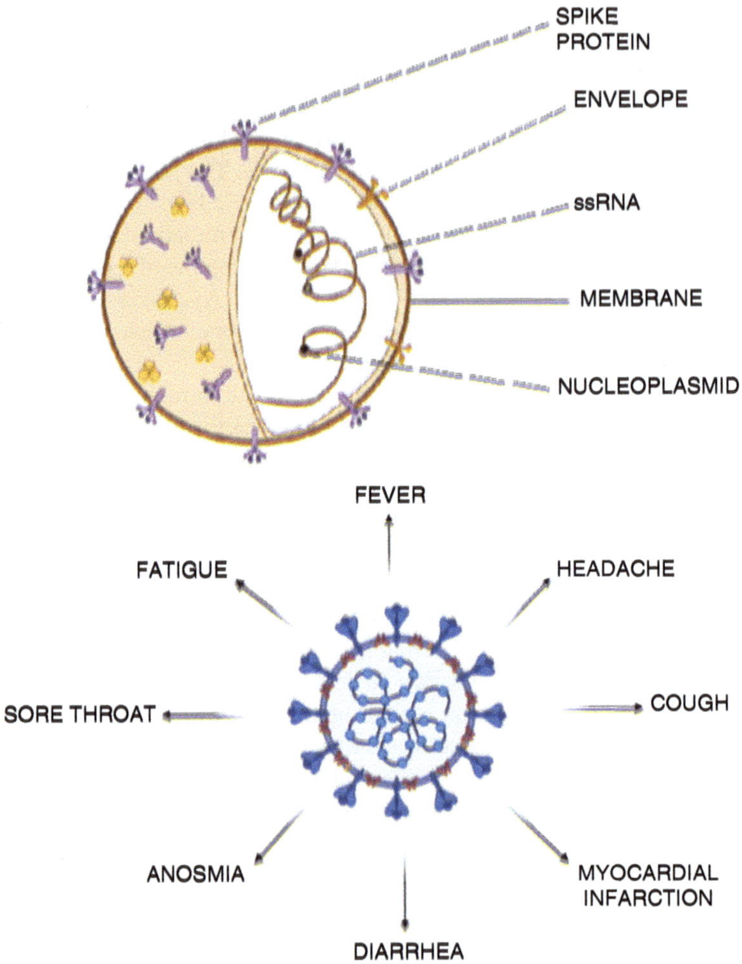


**Fig. 4.1** SARS-CoV-2 and different symptoms caused by infection due to COVID-19

and infection in the respiratory tract (Astuti 2020). This virus is a member of the order Nidovariles of the family Coronaviridae.

The total genome content of this virus is 29,891 base pairs (bp) approximately 30 kilobase pairs (kbp) with a GC content of 38%. This virus has four structural proteins spike protein (S), membrane (M), envelope (E), and nucleocapsid (N) (Fig. 4.1) (Dhama et al. 2020). It has 16 nonstructural and 5–8 accessory proteins along with the structural proteins. This virus enters the host cells through the respiratory epithelium where the angiotensin-converting enzyme 2 (ACE2) receptors are expressed. The viral spike protein attaches with the host protein transmembrane serine protease 2 (TMPRSS2) which facilitates the virus entry and also a replication of the viral genome (Cascella et al. 2022). The cause of the severity of these diseases is divided into three mild, severe, and critical. The manifestation of COVID-19 is fever (Fig. 4.1) and dry cough which are the mild symptoms along with pneumonia which is the common cold symptom (Shanmugam et al. 2020; Kuchi Bhotla et al. 2020; Kuchi Bhotla et al. 2021a, b; Pushparaj et al. 2022; Meyyazhagan et al. 2022).

Other than respiratory complications, other complications such as myocardial injury (Fig. 4.1), arrhythmia, and neurological complications such as myalgia, hypogeusia, hyposmia, and stroke are reported in some cases along with digestive symptoms and liver injury. In critical cases, patients progress to ARDS septic shock and coagulation dysfunction which finally lead to multiple organ failure and death (Wang et al. 2020; Alimohamadi et al. 2021). SARS-CoV-2 attacks the mucus-producing goblet cells and ciliated cells as indicated by the topology of expression of host entry factors by tropism studies in both ex vivo and in vitro, other than this endothelial cells are also found to be susceptible to this virus, the binding of the S protein with the host cells relies on the protease enzymes in the host cells (Pizzato et al. 2022).

## 4.3 SARS-CoV-2 Proteins

### *4.3.1 Spike Protein (S)*

S protein is one of the important viral proteins, found in different variations in the different variants of the coronavirus mainly in the sequence of the protein and differences in domains of the protein. This protein has two domains S1 and S2, S1 domain binds to the receptors, which are composed of two sub-domain an N-terminal domain and the C-terminal sub-domain or also known as the receptor binding domain (RBD). The S1 domain of the S protein is found to be less preserved between SARS-CoV and SARS-CoV-2, but the S2 fusion domain preserved nearly 90% of the sequence between the two variants (Hatmal et al. 2020). The S protein of SARS-CoV-2 is about 1273 amino acids long but in the SARS-CoV, the protein total length is about 1255, both the domains of this protein fold and are related to having specific functions. This protein is highly glycosylated at the N glycosylation sites, in some sites O- glycosylation is also found in this protein of SARS-CoV-2 (Xia 2021). The S proteins bind to ACE2, a cellular receptor which after the formation of the complex translocated into the cellular endosomes, where this protein is cleaved by cathepsin L also called endosomal acid proteases to activate the activity of fusion of this protein (Du et al. 2009). The SARS-CoV-2, S protein forms a loop that is rich in Arginine residues and is exposed but this loop is not found in SARS-CoV and other related coronavirus variants such as MERS-CoV, OC43, and HKU1 (Hoffmann et al. 2020). Mutation at the N501Y position in the RBD of the S protein exhibited a stronger association rate of the protein with ACE2 receptors also the association was faster but the dissociation compared to the association was slower, which caused a higher rate of transmission of this SARS-CoV-2 variant (Tian et al. 2021).

### 4.3.2 Membrane Protein (M)

M protein is one of the main structural proteins of SARS-CoV-2. It binds to the other structural protein; binding of this protein will stabilize the binding of the other structural protein to RNA. This M protein incorporates the S proteins which affect the attachment of the host cells with the virus and also its entry (Thomas 2020). The interaction of the M protein with the S protein retains it in the endoplasmic reticulum (ER)-Golgi intermediate compartment, which also stabilizes the structure of the M protein. It has a role in B-cell response and has higher immunogenicity for T-cell response, than other nonstructural proteins (Alharbi and Alrefaei 2021). It is the most abundant structural protein and is a mushroom-shaped dimer, consisting of two transmembrane domain-swapped three-helix bundles and two intravirion domains, it further forms assembles of oligomers and is similar to the ion channel ORF3a in the SARS-CoV-2 (Zhang et al. 2022). M protein can suppress the expression of interferon-beta (IFN-β) and other interferon-stimulated genes that are induced by the genes such as RIG-I, MDA4, IKK ϵ, and TBK1 and also to inhibits the dimerization of TBK1 and phosphorylation of IRF3 (Sui et al. 2021).

### 4.3.3 Nucleocapsid (N) Protein

The N protein is an important structural protein antigen of SARS-CoV-2. It has roles in the packaging of the RNA molecule and also has a role in releasing the viral particles. It enters along with the viral RNA and facilitates the replication of RNA and the processing of the assembly of the viral particles (Zeng et al. 2020). The viral genome is encapsulated in the N protein and thereby protects the viral genome from the host intracellular environment. It is produced high in the infected cells and also enhances the transcription of RNA and is one of the essential factors for the replication of the viral genome (Savastano et al. 2020).

The N protein of SARS-CoV is homologous to the nucleocapsid of SARS-CoV-2, it was found to spatiotemporally regulate the localization of the protein and also its function, and in further bioinformatic analysis it was found that this protein can regulate the condensation of protein with the RNA or key proteins in the host in vivo (Cascarina and Ross 2020). The C-terminal of this protein is minimally homologous to that of the protein of the human coronavirus and is highly immunogenic (Dobaño et al. 2021). In the HEK293T and Calu-3 cells the N protein is found to interact with 160 cellular proteins, these proteins have roles in ribosome biogenesis, RNA-associated processes, ribonucleoprotein complex biogenesis, ribosomal large and small subunit biogenesis, RNA binding, and catalyze the translation and process of transcription (Zheng et al. 2020).

### 4.3.4 Envelope (E) Protein

The E protein is one of the less studied proteins among all the structural proteins in SARS-CoV-2. A higher amino acid variation at positions 55, 56, and 69, altered the binding of the E protein with the PALS1 protein which is a tight junction-associated protein that may have a role in the pathogenesis of COVID-19 (Rahman et al. 2021). This protein is an integral membrane protein and has a role in coronavirus maturation, assembly, and virulence it has a PDZ binding motif in its C terminus (Toto et al. 2020). This protein affects the assembly of the virus and budding, it is pentamer viroprotein, which is responsible for the transfer of ion particles in the direction from the lumen to the cytosol. Palmitoylation of this pentamer E protein has a great role in its stability, as the palmitoyl group is lost the porin size is reduced and may collapse too (Sun et al. 2021). The E protein produced by the wild-type SARS-CoV-2 can be translocated inside the cellular organelles like the ER-Golgi intermediate, in which the pH in the organelle increases the expression of this protein (Cabrera-Garcia et al. 2021).

### 4.3.5 Main Protease ($M^{pro}$)

Main protease ($M^{pro}$) also known as $3CL^{pro}$ is encoded by the Nsp5 gene and is one of the important proteins in the SARS-CoV-2 variant and is found to have a role in gene expression and replication in coronavirus (Bzówka et al. 2020). The viral replication is carried out by two proteins ppa and ppab. These two proteins are synthesized as an overlapping polyprotein by the replicase gene of the SARS-CoV-2 variant, $M^{pro}$ processes this polyprotein into two functional polypeptides by proteolytically cleaving the overlapping protein (Singh and Mishra 2021). It cleaves the polypeptides after the glutamine residue which is a unique character of a protease as no other proteases in humans has the same cleavage specificity as this protease enzyme. The active site of this protein is composed of four sites S1, S1, S2, and S4 (Qiao et al. 2021). This protein is considered one of the targeted candidates for drug discovery and is one of the best-studied targets. SARS-CoV-2 variant, $M^{pro}$ has some similarities with the betacoronavirus SARS-CoV and MERS-CoV which too caused serious epidemics (Sheik Amamuddy et al. 2020).

### 4.3.6 Papain-Like Protease ($PL^{pro}$)

Papain-like protease ($PL^{pro}$) protein is an important coronavirus enzyme that is required for the processing of the polyproteins into a functional complex of replicase and also enables its spread (Shin et al. 2020). This protein is secreted by the gene Nsp3, this protease enzyme cleaves the Nsp1, Nsp2, and Nsp3 (Klemm et al. 2020).

It is basic and is about 315 amino acid residue long polypeptide and has high cysteine residues in its peptide sequence. It is one of a target of SARS-CoV-2 as it plays role in the maturation of different viral proteins and also it induces the assembly of the replicase and the transcriptase enzyme and disrupts the host immune response (Osipiuk et al. 2021). Human coronavirus codes for two $PL^{pro}$ proteins, PLP1 and PLP2 by NL63, OCA4, HKU1, and 229E, whereas the SARS-CoV-2 genome is the same as that of MERS-CoV and SARS-CoV as the codes for a single $PL^{pro}$, but their enzymatic activity differs. It has a less efficient deubiquitinase enzyme but its deISGylase activity is more efficient (Freitas et al. 2020).

## 4.4 Herbal Formulation Effects On SARS-CoV-2 Infection

Herbal medicines are been used for curing disease for a long time. Both novel and old formulations effects are been studied for the symptoms in the infections caused by SARS-CoV-2. Hanshiyi Formula (HSYF) (Table 4.1; Fig. 4.2) a Chinese traditional medicinal formula was found to prevent the progression of infection from the mild or moderate to severe form of COVID-19 in the patients. The different plant components in this formulation are found to have antiviral activity and are also able to improve respiratory problems (Tian et al. 2020). Ayurvedic polyherbal formulation Kabasura kudineer (KK) (Fig. 4.2) is found to be effective against symptoms such as fever, cough, sore throat, and shortness of breath which are similar to the symptoms caused by SARS-CoV-2 infection (Table 4.1). This polyherbal formulation is composed of 15 herbs that are *Andrographis paniculata*, *Syzygium aromaticum*, *Zingiber officinale*, *Tragia involuerta*, *Hygrophila auriculata*, *Terminaila chebula*, *Ahatoda vasica*, *Coleus ambonicus*, *Saussurea lappa*, *Clerodendrum serratum*, *Cypreus rotundus*, *Tinospora cordifolia*, *Sida acuta*, *Piper longum*, and *Anacyclus pyrethrum*.

$3CL^{pro}$ or $M^{Pro}$ is one of the target proteins in SARS-CoV-2, 145 components of this formulation were chosen for molecular docking analysis against this targeted protein. The compounds that were found to be effective against the targeted proteins were acetoside, luteolin-7-rutinoside, rutin, chebulagic acid, syrigaresinol, acanthoside, violanthin, andrographidine C, myricetin, gingerenone-A, geraniol, nootkatone, asarianin, and sitosterol which were found to inhibit the $M^{pro}$ protein (Vincent et al. 2020). LCTE or Qing Fei Pai Du Tang a Chinese herbal formulation consisting of 20 herbal ingredients along with one mineral component are found to be effective in treating COVID-19 by alleviating inflammation (Table 4.1). Components present in this soup is found to have inhibiting potential against SARS-CoV-2 and the protein that are responsible for the prevalence of COVID-19 symptoms and also affect the key pathological process (Zhang et al. 2020).

Ayurvedic polyherbal formulations Pathyadi kwath (Table 4.1), Sanjeevani vati, Yastimadhu, Tribhuvan keeratiras, and Septillin are found to be more effective and can be potential treatments as anti-COVID-19 than formulations such as Samshamni vati, AYUSH-64 (Table 4.1; Fig. 4.2), and Trikatu (Table 4.1) in in silico studies

**Table 4.1** Composition of herbal formula

| Herbal formulation | Plants | Effects | Reference |
|---|---|---|---|
| Amalakyadiganakwatha bhavita churna | *Tinospora cardifolia, Zingiber officinale, Cyprus rotundus, Andrographis paniculata, Cissampelos pariera* | Antipyretic effect | Ping et al. (2022) |
| Ayush kwath | *Ocimum tenuiflorum, Cinnamomum verum, Zingiber officinale, Piper nigrum* | Antiviral | Sulaiman et al. (2020) |
| AYUSH-64 | *Alstonia scholaris, Picrorhiza kurroa, Swertia chirayita* | Antiviral, anti-asthmatic, immune boosting | Joshi et al. (2022) |
| Coronil | *Withania somnifera, Tinospora cordifolia, Ocimum sanctum* | Mitigate the cyto-kine response in SARS-CoV-2 infection | Balkrishna et al. (2021a, b) |
| Enterica decoction | *Spondias mombin, Persea americana, Psidium guajava, Trema orientalis, Cnestis ferruginea, Momordica charantia, Vernonia amygdalina, Latana carnara, Paullinia pinnata, Citrus aurantifolia, Morinda lucida, Bidens pilosa* | Typhoid fever | Kumadoh et al. (2015) |
| Gojihwadi Kwath | *Onosma bracteatum, Glycyrrhiza glabra, Anethum sowa, Vitis vinifera, Ficus hispida, Zizyphus jujube, Adhatoda vasica, Hyssopus officinalis, Cordia dichotoma, Sisymbrium irio, Cymbopogan martini, Viola odorata, Linum usitatissimum, Althaea officinalis, Solanum xanthcarpum, Piper nigrum* | Antiviral activity | Agrawal et al. (2022) |
| Hanshiyi formula (HSYF) | *Ephedrae herba, Gypsum fibrosum, Armeniacae semen, Rhizoma et Radix notopterygii, Cyrtomii rhizoma, Cynanchi paniculati, Pogostemonis herba, Eupatorii herba, Atractylodis rhizoma, Atractylodis macrocephalae, Crataegi fructus, Massa medi-cate, Fructus germinatus, Magnoliae officinalis, Arecae semen, Tsaoko fructus, Zingiberis rhizoma* | The antiviral effect improves asthma and respiratory symptoms | Tian et al. (2020) |
| Honitus | *Ocimum sanctum, Glycyrrhiza glabra, Viola odorata,* | Management of cough | Gupta et al. (2016) |

(continued)

**Table 4.1** (continued)

| Herbal formulation | Plants | Effects | Reference |
|---|---|---|---|
| | *Zingiber officinale, Adhatoda vasica, Piper longum* | | |
| Huangqi Jianzhong tang | *Radix paeoniae, Rhizoma zingiberis, Saccharum granorum, Fructus zizyphi, Radix glycyrrhixae, Cortex cinnamomi, Radix astragali* | Anti-fatigue | Bao et al. (2020) |
| JACOM | *Justicia adathoda, Carica papaya, Andrographis paniculata, Ocimum tenuiflorum, Melia azedarach* | Can be an inhibitor of COVID-19 infection | Kiran et al. (2020) |
| Jing guan fang (JGF) | *Forsythia suspensa, Scutellaria baicalensis, Bupleurum Chinese, Magnolia officinalis, Agastache rugose* | Reduces infection in COVID-19 and also reduces proliferation | Lau et al. (2005) |
| Kabasura kudineer (KK) | *Andrographis paniculata, Syzygium aromaticum, Zingiber officinale, Tragia involuerta, Hygrophila auriculata, Terminaila chebula, Ahatoda vasica, Coleus ambonicus, Saussurea lappa, Clerodendrum serratum, Cypreus rotundus, Tinospora cordifolia, Sida acuta, Piper longum,* and *Anacyclus pyrethrum* | Fever, cough, sore throat, shortness of breath | Vincent et al. (2020) |
| LCTE or Qing Fei Pai Du Tang | *Rhizoma atractylodis macrocephalae, Rhizoma pinelliae, Radix bupleuri, Pericarpium citri reticulatae, Farfarae flos poria, Radix glycyrrhizae, Cinnamomi ramulus, Radix scutellariae, Pogostemon cablin* *Herba ephedrae, Rhizoma dioscoreae* *Rhizoma belamcandae, Rhizoma zingiberis recens, Herba asari, Semen armeniacae amarum, Rhizoma alismatis, Fructus aurantii immaturus, Polyporus umbellatus, Radix asteria, Rudis gypsi miscueris* | Anti-inflammatory | Zhang et al. (2020) |
| LHQW | *Forsythiae fructus, Lonicerae japonicae flos, Ephedrae herba, Armeniacae semen amarum, Gypsum fibrosum,* | Antypiretic and anti-toxic effects | Zheng et al. (2020) |

(continued)

**Table 4.1** (continued)

| Herbal formulation | Plants | Effects | Reference |
|---|---|---|---|
| | *Isatidis radix, Dryopteridis crassirhizomatis rhizoma, Houttuyniae herba, Pogostemonis herba, Rhei radix et rhizoma, Rhodiolae crenulatae radix, menthol, Glycyrrhizae radix* | | |
| NRICM101 | *Scutellaria baicalensis, Houttuynia cordata, Morus alba, Saposhnikovia divaricata, Trichosanthes kirilowii, Isatis indigotica, Glycyrrhiza glabra, Magnolia officinalis, Mentha haplocalyx, Nepeta tenuifolia* | Antiviral and anti-inflammatory properties | Tsai et al. (2021) |
| Pathyadi kwath | *Terminalia chebula, Terminalia bellirica Phyllanthus emblica, Azadiracta indica Andrographis paniculata, Curcuma longa, Tinospora cordifolia* | Anti-COVID-19 | Joshi et al. (2022) |
| Qingkailing | *Calculus bovis, Cornu bubali, Radix scutellariae baicalensis, Flos lonicerae, Fructus gardeniae* | Reduced fever and also repaired amino acid metabolism | Gao et al. (2013) |
| Qinghua Zhixie | *Pteris multifida, Elsholtzia ciliata, Psoralea corylifolia, Coptis chinensis, Radix Aucklandiae, Baked ginger, Radix paeoniae alba, Radix Saposhnikoviae, Rhizoma atractylodis, Rhizoma Atractylodis Macrocephalae, Agrimonia pilosa* | Alleviate the diarrhea | Ji et al. (2022) |
| Sang Ju Yin | *Chrysanthemum morifolium, Forsythia suspensa, Glycyrrhiza uralensis, Semen Armeniacae amarum, Prunus armeniac, Platycodon grandiflorus, Phragmites australis, Mentha canadensis* | Common cold | Lau et al. (2005) |
| Sanjeevani vati | *Embelia ribes, Zingiber officinale, Piper longum, Terminalia chebula, Terminalia bellirica, Emblica officinalis, Acorus calamus Tinospora cordifolia,* | Anti-COVID-19 | Joshi et al. (2022) |

(continued)

**Table 4.1** (continued)

| Herbal formulation | Plants | Effects | Reference |
|---|---|---|---|
| | *Semecarpus anacardium*, *Aconitum ferox*, Cow's urine | | |
| Samshamni vati | *Tinospora cordifolia* | Anti-inflammatory and antipyretic properties | Joshi et al. (2022) |
| Shenling Baizhu San | *Panax ginseng*, *Poria cocos*, *Atractylodes macrocephala*, *Glycyrrhiza uralensis*, *Dolichos lablab*, *Dioscorea opposita*, *Nelumbo nucifera*, *Platycodon grandifloras*, *Amomum villosum*, *Coix lacryma-jobi* | Control chronic diarrhea | Wang et al. (2022) |
| Si Junzi | *Panax ginseng*, *Wolfiporia extensa*, *Atractylodes macrocephala*, *Glycyrrhiza uralensis* | Gastrointestinal problems | Yan et al. (2022) |
| Tao-Hong-Si-Wu-tang | *Rehmannia glutinosa*, *Paeonia lactiflora*, *Angelica sinensis*, *Ligusticum chuanxiong*, *Prunus persica*, *Carthamus tinctorius* | Anti-fatigue effect | Li et al. (2012) |
| Tongxie Yaofang (TXYF) | *Atractylodes macrocephala*, *Paeonia lactiflora*, *Citrus aurantium*, *Radix saposhnikoviae* | Irritable Bowel Syndrome-Disease | Liang et al. (2022) |
| Trikatu | *Piper nigrum*, *Piper longum*, *Zingiber officinale* | Antimicrobial and analgesic properties | Joshi et al. (2022) |
| Vipro™ | *Ocimum basilicum*, *Curcuma longa*, *Citrus lemon*, *Allium sativum*, *Plectranthus amboinicus*, *Momordica charantia*, *Cinnamomum verum*, *Zingiber officinale*, *Piper nigrum* | Alleviates clinical symptoms of COVID-19 | Arunkumar et al. (2021) |
| Vrihatsamaharker churna | *Syzygium aromaticum*, *Myristica fragrans*, *Piper longum*, *Zingiber officinale*, *Piper nigrum* | Treatment for treating nonproductive cough | Vats et al. (2022) |
| Yu Ping Feng San | *Astragalus mongholicus*, *Atractyloes macrocephala*, *Saposhnikovia divaricata* | Prevents influenza-like symptoms | Lau et al. (2005) |

(Joshi et al. 2022). Ayurvedic bhasma preparation is nanoparticles prepared from metals such as gold, silver, copper, zinc, and iron oxide, these compounds can modulate the levels of interleukins, interferons, and TNF-α levels in plasma, also known as swarna bhasma, rajata bhasma, tamra bhasma, and yashada bhasma could

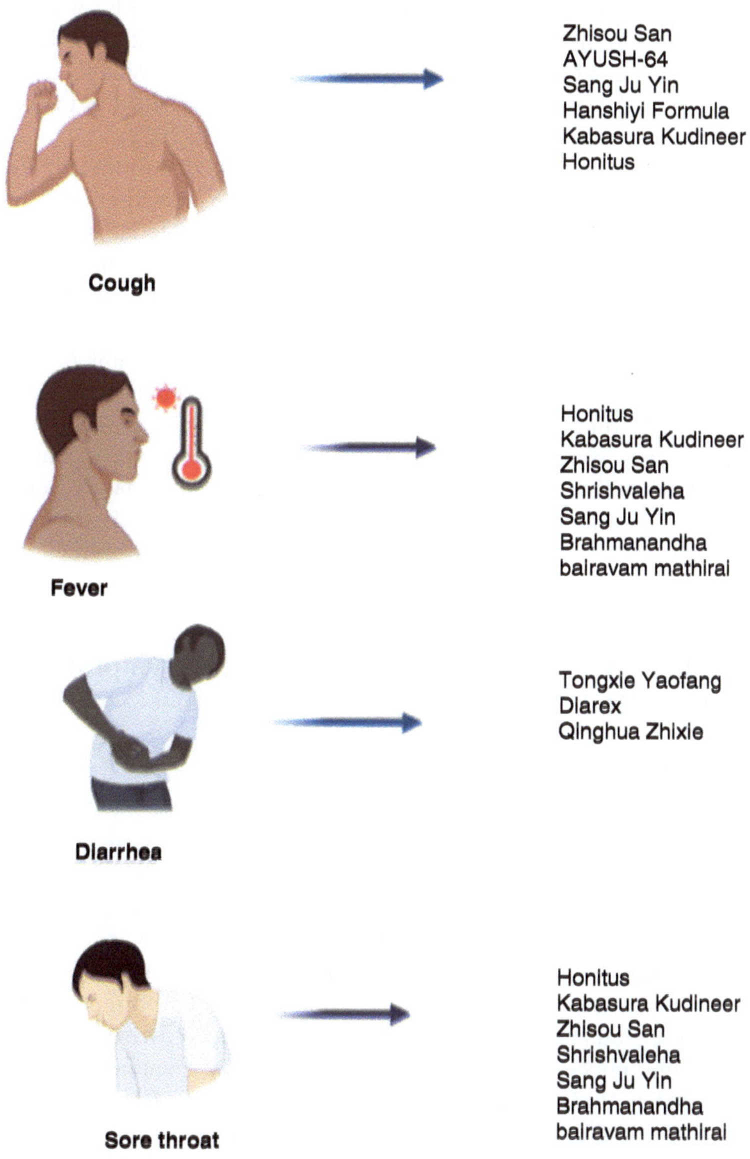


**Fig. 4.2** Different herbal formulation effects on COVID-19 symptoms cough, fever, diarrhea, and sore throat

be a possible treatment against infections caused by SARS-CoV-2 due to their activities that include immunomodulatory, anti-inflammatory, antiviral (Sarkar and Das Mukhopadhyay 2021). JGF (Table 4.1) also known as Jing Guan Fang a Chinese herbal formula is found to reduce the symptoms caused by COVID-19, it mechanistically induces the ACE2 degradation depending on the lysosomal process and suppresses both the mRNA and protein levels of TMPRSS2 in the human lung cell line of WI-38 and MRC-5, in in vivo study in mice this formula found to reduce the levels of both the proteins and also helped in the improvement of the symptoms caused in the disease on inhalation by the mice (Ping et al. 2022).

*Withania somnifera* a medicinal plant used in many herbal formulas, Withanone is one of the phytochemical compounds of this herb, by docking analysis it was found to bind with the complex of ACE2-RBD and also found to decrease the electrostatic component of binding free energy of the complex, which interrupts the interaction and weakens its infectivity, thus blocking its entry into the host cells (Balkrishna et al. 2021a, b). Persian herbal decoction and capsule treatment in 184 adults patients improved and decreased the clinical symptoms of COVID-19 like dry cough, dyspnea, muscle pain, headache, fatigue, anorexia, chills, runny nose sputum cough (Karimi et al. 2021). *Ayush kwath* (Table 4.1), an Ayurvedic herbal medicine prepared from specified parts of different medicinal plants contains many active components that show immunomodulatory, anti-inflammatory, and antioxidant properties that can be used to manage symptoms of COVID-19 (Sulaiman et al. 2020).

## 4.5 Herbal Formulation Effects on Different Symptoms of COVID-19

### 4.5.1 Fever

Fever is the most preserved evolutionary response over 600 years to infections in animals; it is the most common symptom in patients suffering from COVID-19 (Fig. 4.1). A high fever during COVID-19 is an indication of severe infection (Gul et al. 2021). Enterica decoction (Table 4.1) composed of plant materials belonging to 12 tropical plants was found to improve typhoid fever (Kumadoh et al. 2015). Chinese traditional medicine Qingkailing (Table 4.1) herbal formula composed of eight herbs has good antipyretic effects. Rat model-induced pyrexia by yeast treated with qingkailing injection reduced fever and also repaired amino acid metabolism (Gao et al. 2013). Brahmanandha bairavam mathirai (Table 4.1; Fig. 4.2) a Siddha formulation is found to be more effective against viral fever caused by infection of chikungunya virus than the polyherbal formulation Vishnu chakram by in vitro study (Jain et al. 2018). Amalakyadiganakwatha bhavita churna (Table 4.1) is a herbal formulation consisting of *Tinospora cardifolia*, *Zingiber officinale*, *Cyprus rotundus*, *Andrographis paniculata*, and *Cissampelos pariera* an antipyretic formulation can be therapeutically safe for the treatment of fever (Unnikrishnan and Nishteswar 2015). Two herbal formulas Sang Ju Yin (Table 4.1; Fig. 4.2) and Yu Ping Feng Sam (Table 4.1) consisting of traditional Chinese medicine were combined forming a polyherbal decoction, which was found to have good potential in preventing influenza-like symptoms and also found to effective in preventing the spread of disease caused by SARS (Lau et al. 2005). AYUSH-64 (Table 4.1; Fig. 4.2) a polyherbal formula was used for recovering from influenza-like symptoms, and is safe to use and can be used for different viral infections along with fever. It was practiced since 1980 for inflammation and in 1994 and 1996, this formulation was used in the malaria epidemic, thus it has some anti-malarial characteristics too (Gundeti et al. 2022).

### 4.5.2 Cough

Cough is a common symptom of SARS-CoV-2 infection (Fig. 4.1) and other respiratory diseases caused by microbes. Ethanolic extract of *Terminalia chebula* fruit, *Mentha piperita* leaves, *Adhatoda vasica* leaves, *Ocimum sanctum* leaves, *Zingiber officinale* rhizome, *Piper longum* fruit, *Glycyrrhiza glabra* roots, and *Withenia somnifera* roots formulated in a form of syrup are found to have an antitussive effect against cough, induced in the guinea pig by citric acid (Meher 2012). A polyherbomineral formulation prepared in the lab was found to be an effective antitussive effect against a sulfur dioxide-induced cough in albino mice, it was found to contain a major piperine active constituent which induces the cough suppressant effect of the components in the formulation (Reena et al. 2014). Traditional Chinese medicine prepared from nine commonly used medicinal plants was found to improve cough scores and also improve dysphonia (Wong et al. 2006). Shrishavaleha (Fig. 4.2) prepared from bark and heartwood of *Albizzia lebbeck* is evaluated for antitussive activity, this formulation was found to alleviate the sulfur dioxide-induced cough in the mice model (Singh et al. 2010). Honitus (Table 4.1; Fig. 4.2) a honey-based cough syrup comprised of herbs such as *Ocimim sanctum*, *Glycyrrhiza glabra*, *Viola odorata*, *Zingiber officinale*, *Adhatoda vasica*, and *Piper longum* are reported to reduce the acute nonproductive cough symptoms and throat irritation (Gupta et al. 2016). Extracts from the seeds of hedge mustard used for herbal cough syrup formulation are found to have an antimicrobial effect against bacterial strains such as *Staphylococcus aureus*, *Escherichia coli*, *Salmonella* spp., *Pseudomonas aeruginosa, and Bascillus subtillis* in studies (Sharma et al. 2020). A polyherbal formulation in a syrup base prepared by five namely plants is found to be effective against cough and prevent bronchial tract mucus secretion (Sharma et al. 2022). The herbal formula Zhisou San (Table 4.1; Fig. 4.2) is a potential drug for the treatment of cough variant asthma, the multiple components of this formulation targets inflammation associated with cough variant asthma and Th17/Treg immune balance. It was also found to affect lung function in in vivo studies in mice (Guo et al. 2022). Vrihatsamaharker churna (Table 4.1) is a polyherbal formulation prepared from plant parts like flower buds of *Syzygium aromaticum*, seeds of *Myristica fragrans*, the fruit of *Piper longum*, rhizome of *Zingiber officinale*, and fruit of *Piper nigrum* is used for the treatment for treating nonproductive cough (Vats et al. 2022).

### 4.5.3 Diarrhea

Diarrhea is one of the preliminary symptoms that develop in COVID-19 infection (Fig. 4.1) along with fever and cough (Fig. 4.1). Along with diarrhea other gastrointestinal symptoms that may appear in COVID-19 infection are nausea, poor appetite, pain in the abdomen, and vomiting. Chinese herbal formula Shenling Baizhu Sanis (Table 4.1) is used for treating chronic diarrhea, and the affected group treated with this formula has recovered (Wang et al. 2022). Diarrhea caused by irritable bowel movements reduces the quality of life of the patient suffering from it. It also affects the mental health of the sufferers and their daily activities are affected too. A Chinese herbal formula Tongxie Yaofang (TXYF) (Table 4.1; Fig. 4.2) consisting of plants *Atractylodes macrocephala*, *Paeonia lactiflora*, *Citrus aurantium*, and *Radix saposhnikoviae* administered to patients suffering from Irritable Bowel Syndrome-Disease, showed safe and effective therapeutic results in people who have another diagnosis like liver depression and spleen deficiency (Liang et al. 2022).

DIAREX a herbomineral Ayurvedic tablet is tested in children and resulted in a decrease in acute diarrhea symptoms (Verma et al. 2022). Chinese herbal decoction Qinghua Zhixie (Table 4.1) is been used for the treatment of diarrhea for a long time, it consists of some active components such as berberine, psoralenoside, quercitin, ehydrocostuslactone, and paeoniflorinv which reduced the 5-hydroxytryptamine and vasoactive intestinal polypeptide and thus alleviate diarrhea. It also balances intestine flora, regulates the brain-gut peptide and also improved intestinal digestion and absorption (Ji et al. 2022). Chinese decoction based on Pulsatilla and Si Junzi (Table 4.1) decoction improved the inflammation caused by diarrhea by inhibiting the signal pathway of PI3K/Akt/NF-κB in a mouse model-induced with diarrhea and the expression of the cytokines IL-1, IL-2, and IL-6 are decreased with increase in anti-inflammatory cytokine the IL-10 (Yan et al. 2022).

### 4.5.4 Sore Throat

One of the common manifestations of COVID-19 is sore throat (Fig. 4.1), SARS-CoV-2 mainly causes infection in the lower respiratory section but upper respiratory symptoms are also found in a patient infected by this virus (El-Anwar et al. 2020). Sore throats are been improved by plying herbal medicines. Ertong Qingyan Jiere Koufuye, Fufang Shuanghua Koufuye, Yanhouling mixture, and Qinganlan Liyan Hanpian are found to be effective against chronic and acute pharyngitis caused by inflammation due to sore throat (Shi et al. 2007). Lozenges a sweet hard candy containing an active ingredient used for the treatment of sore throat formulated with traditional herbs such as *Adhatoda vasica*, *Althaea officinalis*, *Glycyrrhiza glabra*, and *Trachyspermum ammi* that contain components that has the ability to treat throat infections (Kadirvel et al. 2022). Herbs such as *Adhatoda vasica*, *Adhatoda officinalis*, *Cordia latifolia*, *Origanum vulgare*, *Thymus vulgaris*, and *Ziziphus jujube* used in the traditional herbal formulation are found to exhibit antimicrobial activity due to their alkaloid and flavonoids content present in these plants (Mehreen et al. 2016). The extract of the roots of *Pelargonium sidoides* EPs 7630 is found to reduce the symptoms of sore throat and hoarseness (Kamin et al. 2022). The novel polyherbal formulation Vipro™ (Table 4.1) developed from traditional medicinal herbs, treatment of the patient COVID-19 suffering from symptoms such as sore throat, throat pain fever, and cough was able to alleviate the symptoms which are as similar to treatment with the standard drugs (Arunkumar et al. 2021).

### *4.5.5 Effect on Other Symptoms*

Many symptoms are found in patients suffering from infections caused by COVID-19. Fever, cough, diarrhea, and sore throat are the primary symptoms other than these, symptoms such as headache, fatigue, anosmia, myocardial injury (Fig. 4.1), and others were also found in certain cases. A Chinese herbal formulation Tao-Hong-Si-Wu-Tang (Table 4.1), administered to 32 male mice were found to have an anti-physical fatigue effect and also decrease the blood lactic acid and blood urea nitrogen (Li et al. 2012). Herbal tea formulations of *Astragalus mongholicus*, *Angelica gigas*, and *Ziziphus jijuba* reduce severe fatigue in adults (Baek et al. 2018). AG formulation prepared from two herbs *Antrodia camphorate* and *Panax ginseng* is a potent formulation with an anti-fatigue effect (Hsiao et al. 2018). Japanese herbal formula, Kampo can be one of the altered medications in treating headaches, but its efficiency of improvement from that of NSAID is not quite different (Katsuki et al. 2022). Mono herbal preparation of *Sambucus nigra* berry can reduce the symptoms in COVID-19 patients such as fever, headache, nasal congestion, and mucus discharge (Harnett et al. 2020). A Chinese herbal formula Huangqi Jianzhong Tang (Table 4.1) can improve the function of the cardiac system in a rat model by regulating the signaling pathway of apoptosis (Bao et al. 2020). Persian herbal formulation in the form of decoction and capsules on treatment in 174 adult patients improved dyspnea along with other symptoms of muscle pain, headache, anorexia, chills, runny nose sputum cough, and vertigo (Karimi et al. 2021).

## 4.6 Herbal Formulations Targeting SARS-CoV-2 Proteins

### *4.6.1 Effects on Spike Protein*

S protein is one of the important structural proteins of SARS-Cov-2, it binds with the host ACE2 transmembrane protein on infection to enter host cells. It is one of the main drug targets in COVID-19 infection. Coronil, a tri herbal formulation consisting of herbs *Withania somnifera*, *Tinospora cordifolia*, and *Ocimum sanctum*, is found to effectively inhibit the interaction between ACE2 and wild-type S protein in SARS-CoV-2 induced zebrafish and also inhibited the interaction between the S protein variant $S^{D614G}$ and $S^{W436R}$ which have a higher affinity toward ACE2. A549 cells induced by different variants of S protein, this formulation were able to reduce the levels of IL-6, IL-1β, and TNF-α (Balkrishna et al. 2021a, b). JACOM a novel herbal formulation and the Siddha formulation KK are chosen for in silico computation study against the S glycoprotein.

The phytocomponents chrysoeriol, and luteolin (Table 4.2) of KK (Fig. 4.2) and quercetin (Table 4.2) of JACOM are found to have a high binding affinity with S protein and are found to be free from cytotoxicity and by pharmacokinetic analysis, this compound is found to be absorbable in the gastrointestinal cavity and can be highly allocated to other tissues (Kiran et al. 2020). Atractylenoline III (Table 4.2) from *Atractylodes lancea* used in the traditional medicine of Thai Chinese and Japanese has a strong binding toward the ACE2 with the ASN at 149 positions, it has high inhibitory activity. This compound has anti-inflammatory properties and potential therapeutic properties against respiratory infection and inflammation induced by viral infection (Sun et al. 2020). On pharmacological assay of the Chinese herbal formula NRICM101 (Table 4.1) was found to inhibit the S protein, it was also found to inhibit the activity of $M^{pro}$, also inhibit the formation of viral plaques and interleukin-6 and TNF-α (Tsai et al. 2021). Epigallocatechin gallate, Curcumin, Apigenin, and Chrysophanol phytochemicals contained in different herbal formulations are found to have a binding affinity with S glycoprotein among them Epigallacatechin has the highest inhibitory effect against this glycoprotein (Subbaiyan et al. 2020).

### 4.6.2 Effects on the Main Protease

$M^{pro}$ is one of the drug targets in SARS-CoV-2. The components of the Triphala Ayurvedic formulation were investigated in in silico studies against the $M^{pro}$ protein. The components terflavin A, chebulinic acid, and corilagin (Table 4.2) from the formulation are found to inhibit this protein (Rudrapal et al. 2022). Aswagandha (*Withania somnifera*), Giloy (*Tinospora cordifolia*), and Tulsi (*Ocimim sanctum*) are used in the herbal formulation of traditional medicines. Six phytochemicals from these plants are found to be good inhibitors of $M^{pro}$ protein in molecular docking analysis these compounds are derived from Aswagandha, ithanosisde V, and Somniferine (Table 4.2), from Giloy Tinocordiside and Tulsi Vicenin, Isorientin 4′-*O*-glucoside 2″-*O*-p-hydroxybenzoate, and Ursolic acid (Table 4.2) (Shree et al. 2022). Glabridin, Catechin, and Fisetin (Table 4.2) are phytoconstituents and have a tendency to bind with the SARS-CoV-2 protein $M^{pro}$ by hydrophobic and hydrogen bond, this interaction leads to inhibition of this viral protease (Jamali et al. 2022). In autodoc analysis, it was found that Akuammicine N-oxide (Table 4.2), one of a components found in the AYUSH-64 formulation, has a higher affinity toward $M^{pro}$,

**Table 4.2** Phytoconstituents in the herbal formula effective against viral and host protein

| Phytoconstituents | SARS-CoV-2 proteins | Reference |
|---|---|---|
| Chrysoeriol | S protein | Kiran et al. (2020) |
| Luteolin | S protein | Kiran et al. (2020) |
| Quercetin | S protein | Kiran et al. (2020) |
| Atractylenoline III | S protein | Sun et al. (2020) |
| Epigallacatechin gallate | S protein | Subbaiyan et al. (2020) |
| Curcumin | S protein | Subbaiyan et al. (2020) |
| Apigenin | S protein | Subbaiyan et al. (2020) |
| Chrysophanol | S protein | Subbaiyan et al. (2020) |
| Terflavin A | $M^{pro}$ | Rudrapal et al. (2022) |
| Corilagin | $M^{pro}$ | Rudrapal et al. (2022) |
| Withanosisde V | $M^{pro}$ | Shree et al. (2022) |
| Somniferine | $M^{pro}$ | Shree et al. (2022) |
| Tinocordiside | $M^{pro}$ | Shree et al. (2022) |
| Vicenin | $M^{pro}$ | Shree et al. (2022) |
| Isorientin 4'-O-glucoside 2"-O-p-hydroxybenzoate | $M^{pro}$ | Shree et al. (2022) |
| Ursolic acid | $M^{pro}$ | Shree et al. (2022) |
| Glabridin | $M^{pro}$ | Jamali et al. (2022) |
| Catechin | $M^{pro}$ | Jamali et al. (2022) |
| Fisetin | $M^{pro}$ | Jamali et al. (2022) |
| Akuammicine N-oxide | $M^{pro}$ | Ram et al. (2022) |
| Amritoside | $M^{pro}$ | Murugesan et al. (2021) |
| Apigenin-6-C-glucosyl7-O-glucoside | $M^{pro}$ | Murugesan et al. (2021) |
| Pectolinarin | $M^{pro}$ | Murugesan et al. (2021) |
| Astragalin | $M^{pro}$ | Murugesan et al. (2021) |
| Glycyrrhizin | $M^{pro}$ | Choe et al. (2022) |
| Withaferin A | $M^{pro}$ | Choe et al. (2022) |
| Curcumin | $M^{pro}$ | Choe et al. (2022) |
| Nigellidine | $M^{pro}$ | Choe et al. (2022) |
| Chebulinic acid | $M^{pro}$ | Rudrapal et al. (2022) |
| Punicalin | $M^{pro}$ | Rakshit et al. (2022) |
| Cyanidine3-glucoside | $M^{pro}$ | Rakshit et al. (2022) |
| Quercetin3-O-rhamnoside | $M^{pro}$ | Rakshit et al. (2022) |
| Pelargonidin3-glucoside | $M^{pro}$ | Rakshit et al. (2022) |
| Cordifolioside A | $M^{pro}$ | Choe et al. (2022) |

(continued)

**Table 4.2** (continued)

| Phytoconstituents | SARS-CoV-2 proteins | Reference |
|---|---|---|
| Apigenin | Nucleocapsid | Husain et al. (2022) |
| Catechin | Nucleocapsid | Husain et al. (2022) |
| Apiin | Nucleocapsid | Husain et al. (2022) |
| Cinnamic acid | Nucleocapsid | Husain et al. (2022) |
| Fortunellin | Nucleocapsid | Agrawal et al. (2022) |
| Epicatechin | Membrane protein | Gupta et al. (2022) |
| Liquiritic acid | Membrane protein | Joshi et al. (2022) |
| Epicatechin | Papain-like protease | Gupta et al. (2022) |
| Hesperidin | ACE2 | Basu et al. (2020) |

it forms H-bonds with the Cysteine residue at 145 positions of the viral protein, Akuammicine N-oxide can be one of a possible candidate for management of the infections in COVID-19 (Ram et al. 2022).

Medicinal plants *Embilica officinalis*, *Phyllanthus niruri*, and *Tinospora cordifolia* are used for forming different herbal formulation, by docking and molecular dynamics study, the bioactive compounds Amritoside, Apigenin-6-C-glucosyl7-*O*-glucoside, Pectolinarin, and Astragalin (Table 4.2) of these plants are found to have better binding affinities with $M^{pro}$ protein and can be promising drug candidate as the inhibitor of this viral protein (Murugesan et al. 2021). Compounds such as Glycyrrhizin, Withaferin A, Curcumin, Nigellidine, and Cordifolioside A (Table 4.2) can inhibit the replication of the coronavirus and have a strong anti-inflammatory effect than the drugs that are been used for curing COVID-19 infections (Choe et al. 2022). Pomegranate peel has various medicinal properties like antimicrobial and anti-malarial and also controls infections in the respiratory system. Compounds contained in the peels such as Pelargonidin3-glucoside, Quercetin3-O-rhamnoside, Cyanidine3-glucoside, and Punicalin (Table 4.2) are found to have a greater affinity to $M^{pro}$ than the Curcumin (Rakshit et al. 2022).

### *4.6.3 Effects on Nucleocapsid (N), Membrane (M) and Papain-Like Protein ($PL^{pro}$)*

N and M proteins are the structural protein of SARS-CoV-2. These two proteins are found to have an important role in the replication of the nuclear material of the virus and also in its assembly and entry into the host cells. Ethanolic extracts prepared from *Cinnamomum zeylanicum*, *Cinnamomum tamala*, *Origanum vulgare*, and *Petroselinum crispum* were tested for their cytotoxic ability by in vitro study. The compounds found in the extracts such as Apigenin, Catechin, Apiin, and Cinnamic acid (Table 4.2) are found to have a stronger binding affinity with the N-terminal C-terminal domain of the N protein than the drug Remdesivir, hence these compounds can be one of the therapeutic options against COVID-19 infection (Husain et al. 2022). Kadha an Indian Ayurvedic preparation from herbs can control respiratory problems such as cough, flu, and cold. The phytochemicals present in it are found to have anti-inflammatory activities and are found to interact with the RNA binding domain of the N (Maurya and Sharma 2022). Gojihwadi Kwath an Ayurvedic preparation that is used for the treatment of fever, cough, and bronchitis, Fortunellin (Table 4.2) a phytochemical present in this formulation, is shown to have antiviral and immunomodulatory activity in patients of COVID-19. By computational approach, this compound is found to inhibit the virus replication, growth, invasion, and binding of the viral protein with the host cells (Agrawal et al. 2022).

Epicatechin (Table 4.2) is a component extracted from the anupana an Ayurvedic concoction of herbal ingredients. This herbal formula also contains Hesperidin (Table 4.2) and Mangiferin. Epicatechin (Table 4.2) on molecular interaction analysis is found to have better inhibition capability against the membrane protein of SARS-CoV-2 (Gupta et al. 2022). Liquiritic acid (Table 4.2) is one of a component found in herbal formulations, that have an inhibitory effect and by computational method, this compound is found to have a binding affinity of about −15.1 with this structural protein (Joshi et al. 2021). $PL^{pro}$ is an essential coronavirus enzyme required for the processing of the viral polyproteins for their function in replication and other processes Epicatechin (Table 4.2) a secondary metabolite of Anupana formulation is also able to inhibit the $PL^{pro}$. This component forms an H bond with the Aspartic acid and Asparagine residue at positions 108 and 267 (Gupta et al. 2022).

### *4.6.4 Effects on TMPRSS2 and ACE2*

TMPRSS2 is a cell surface protein in the host cell, VeroE6 and Calu-3 cells induced by SARS-CoV-2 infection resulted in enhancing its expression in infection which causes inefficient entry of the virus into the cells by endocytosis (Iwata-Yoshikawa et al. 2022). On long time treatment with JGF, a Chinese herbal formulation affected the expression of this protein in WI-38 cells and also reduced its mRNA within 24 hrs (Ping et al. 2022). Hesperidin (Table 4.2) is a phytochemical found in both Indian and Chinese traditional medicine. This phytochemical constituent was able to bind noncompetitively with ACE2 and bound the structure of ACE2 and S glycoprotein of SARS-CoV-2. The binding site of Hesperidin (Table 4.2) and S protein on ACE2 is in a different location. Binding of the this phytoconstituent to the ACE2 in a simultaneous bind with the S protein modulates their binding capacity and shows antiviral activity by destabilizing the S glycoprotein that interacts with ACE2 and preventing the interaction (Basu et al. 2020). LHQW a Chinese herbal formulation shows antiviral activities by regulating inflammation, and also repairs injury in the lungs and also improves the ACE2 expression which was cluttered in COVID-19 infection (Zheng et al. 2020).

## 4.7 Conclusions

Infections caused by SARS-CoV-2 develop serious symptoms in the people infected by the virus. Till now numerous medicines and vaccines are invented for alleviating the complications caused by these virus infections. Herbal medicines prepared from medicinal plants are been used for a long period to find the inhibiting capability of the compounds in these formulations to cure diseases. This chapter is based on herbal formulas and their effect on infections caused by SARS-CoV-2. Different herbal formulas were found to be effective against the infection caused by the SARS-CoV-2 and also found to be effective against the symptoms. In silico analysis was too carried out against the host protein S, $M^{pro}$, ACE2, etc. with phytochemicals consisting in these formulations. In the future, to conclude the usage of herbal formulas against COVID-19 infection, further clinical studies are needed to be done.

## References

Abubakar AR, Haque M (2020) Preparation of medicinal plants: basic extraction and fractionation procedures for experimental purposes. J Pharm Bioallied Sci 12(1):1

Agrawal S, Pathak E, Mishra R, Mishra V, Parveen A, Mishra SK, Byadgi PS, Dubey SK, Chaudhary AK, Singh V, Chaurasia RN (2022) Computational exploration of the dual role of the phytochemical fortunellin: antiviral activities against SARS-CoV-2 and immunomodulatory abilities against the host. Comput Biol Med 149:106049

Alharbi SN, Alrefaei AF (2021) Comparison of the SARS-CoV-2 (2019-nCoV) M protein with its counterparts of SARS-CoV and MERS-CoV species. J King Saud Univ Sci 33(2):101335

Alimohamadi Y, Tola HH, Abbasi-Ghahramanloo A, Janani M, Sepandi M (2021) Case fatality rate of COVID-19: a systematic review and meta-analysis. J Prev Med Hyg 62(2):E311

Anand AV, Balamuralikrishnan B, Kaviya M, Bharathi K, Parithathvi A, Arun M, Senthilkumar N, Velayuthaprabhu S, Saradhadevi M, Al-Dhabi NA, Arasu MV, Yatoo MI, Tiwari R, Dhama K (2021) Medicinal plants, phytochemicals, and herbs to combat viral pathogens including SARS-CoV-2. Molecules 26(6):1775

Arumugam VA, Sangeetha T, Zareena F, Pavithra R, Ann M, Sanjeev A, Sunantha B, Meyyazhagan A, Yatoo MI, Sharun K, Tiwari R, Pandey MK (2020) COVID-19 and the world with co-morbidities of heart disease, hypertension and diabetes. J Pure Appl Microbiol 14(3):1623–1638

Arunkumar R, Maignanakumar R, Duraivel M, Ahamedbasha A, Ruckmani A (2021) A single centre, prospective, randomized, open labelled clinical study to evaluate the effectiveness of Siddha Poly Herbal Formulation, Viprotm, towards the management of uncomplicated respiratory infection. Biomed Pharmacol J 14(2):793–802

Astuti I (2020) Severe Acute Respiratory Syndrome Coronavirus 2 (SARS-CoV-2): an overview of viral structure and host response. Diabetes Metab Syndr Clin Res Rev 14(4):407–412

Baek Y, Kim H, Mun S, Lee S (2018) Three-component herbal tea alleviates prolonged fatigue and improves sleep quality: a randomized controlled pilot study. Explore 14(6):420–423

Balkrishna A, Haldar S, Singh H, Roy P, Varshney A (2021a) Coronil, a tri-herbal formulation, attenuates spike-protein-mediated SARS-CoV-2 viral entry into human alveolar epithelial cells and pro-inflammatory cytokines production by inhibiting spike protein-ACE-2 interaction. J Inflamm Res 14:869

Balkrishna A, Pokhrel S, Singh J, Varshney A (2021b) Withanone from Withania somnifera may inhibit novel coronavirus (COVID-19) entry by disrupting interactions between viral S-protein receptor binding domain and host ACE2 receptor. Drug Des Devel Ther 15:1111–1133

Bao YR, Jiang WY, Yu JY, Chen JW, Zhang GX (2020) Traditional Chinese medicine formulation Huangqi Jianzhong tang improves cardiac function after myocardial infarction in rats. Evid Based Complement Alternat Med 2020:3106076

Basu A, Sarkar A, Maulik U (2020) Molecular docking study of potential phytochemicals and their effects on the complex of SARS-CoV-2 spike protein and human ACE2. Sci Rep 10(1):1–5

Bzówka M, Mitusińska K, Raczyńska A, Samol A, Tuszyński JA, Góra A (2020) Structural and evolutionary analysis indicate that the SARS-CoV-2 Mpro is a challenging target for small-molecule inhibitor design. Int J Mol Sci 21(9):3099

Cabrera-Garcia D, Bekdash R, Abbott GW, Yazawa M, Harrison NL (2021) The envelope protein of SARS-CoV-2 increases intra-Golgi pH and forms a cation channel that is regulated by pH. J Physiol 599(11):2851–2868

Cascarina SM, Ross ED (2020) A proposed role for the SARS-CoV-2 nucleocapsid protein in the formation and regulation of biomolecular condensates. FASEB J 34(8):9832–9842

Cascella M, Rajnik M, Aleem A, Dulebohn SC, Di Napoli R (2022) Features, evaluation, and treatment of coronavirus (COVID-19). Statpearls [internet]

Chandra Manivannan A, Malaisamy A, Eswaran M, Meyyazhagan A, Arumugam VA, Rengasamy KRR, Balasubramanian B, Liu WC (2022) Evaluation of clove phytochemicals as potential antiviral drug candidates targeting SARS-CoV-2 Main protease: computational docking, molecular dynamics simulation, and pharmacokinetic profiling. Front Mol Biosci 9:918101

Choe J, Har Yong P, Xiang NZ (2022) The efficacy of traditional medicinal plants in modulating the main protease of SARS-CoV-2 and cytokine storm. Chem Biodivers 19(11):e202200655

Dhama K, Khan S, Tiwari R, Sircar S, Bhat S, Malik YS (2020) Coronavirus disease 2019 COVID-19. Clin Microbiol Rev 33(4)

Dharmananda S (1997) The methods of preparation of herb formulas: decoctions, dried decoctions, powders, pills, tablets, and tinctures. Institute of Traditional Medicine, Portland, Oregon

Dobaño C, Santano R, Jiménez A, Vidal M, Chi J, Melero NR, Popovic M, López-Aladid R, Fernández-Barat L, Tortajada M, Carmona-Torre F (2021) Immunogenicity and crossreactivity of antibodies to the nucleocapsid protein of SARS-CoV-2: utility and limitations in seroprevalence and immunity studies. Transl Res 232:60–74

Du L, He Y, Zhou Y, Liu S, Zheng BJ, Jiang S (2009 Mar) The spike protein of SARS-CoV—a target for vaccine and therapeutic development. Nat Rev Microbiol 7(3):226–236

El-Anwar MW, Elzayat S, Fouad YA (2020) ENT manifestation in COVID-19 patients. Auris Nasus Larynx 47(4):559–564

Fougère BJ, Wynn SG (2007) Herb manufacture, pharmacy, and dosing. In: Veterinary herbal medicine. Mosby, pp 221–236

Freitas BT, Durie IA, Murray J, Longo JE, Miller HC, Crich D, Hogan RJ, Tripp RA, Pegan SD (2020) Characterization and noncovalent inhibition of the deubiquitinase and deISGylase activity of SARS-CoV-2 papain-like protease. ACS Infectious Diseases 6(8):2099–2109

Gao X, Guo M, Peng L, Zhao B, Su J, Liu H, Zhang L, Bai X, Qiao Y (2013) UPLC Q-TOF/MS-based metabolic profiling of urine reveals the novel antipyretic mechanisms of Qingkailing injection in a rat model of yeast-induced pyrexia. Evid Based Complement Alternat Med 2013: 864747

Govindharajan S, Thanapal P, Swaminathan P, Vijaya Anand A, Park S, Kim IH, Balamuralikrishnan B (2020) Influences of dietary inclusion of algae *Chaetomorpha aerea* enhanced growth performance, immunity, haematological response and disease resistance of Labeo rohita challenged with *Aeromonas hydrophila*. Aquacult Rep 2020(17):100353

Gul MH, Htun ZM, Inayat A (2021) Role of fever and ambient temperature in COVID-19. Expert Rev Respir Med 15(2):171–173

Gundeti MS, Bhurke LW, Mundada PS, Murudkar S, Surve A, Sharma R, Mata S, Rana R, Singhal R, Vyas N, Khanduri S, Sharma BS, Srikanth N, Dhiman KS (2022) AYUSH 64, a polyherbal

Ayurvedic formulation in Influenza-like illness - Results of a pilot study. J Ayurveda Integr Med 13:100325

Guo DH, Hao JP, Li XJ, Miao Q, Zhang Q (2022) Exploration in the mechanism of Zhisou san for the treatment of cough variant asthma based on network pharmacology. Evid Based Complement Alternat Med 2022:1698571

Gupta A, Gaikwad V, Kumar S, Srivastava R, Sastry JL (2016) Clinical validation of efficacy and safety of herbal cough formulation "Honitus syrup" for symptomatic relief of acute non-productive cough and throat irritation. Ayu 37(3–4):206

Gupta S, Singh V, Varadwaj PK, Chakravartty N, Katta AK, Lekkala SP, Thomas G, Narasimhan S, Reddy AR, Reddy Lachagari VB (2022) Secondary metabolites from spice and herbs as potential multitarget inhibitors of SARS-CoV-2 proteins. J Biomol Struct Dyn 40(5):2264–2283

Harnett J, Oakes K, Carè J, Leach M, Brown D, Cramer H, Pinder TA, Steel A, Anheyer D (2020) The effects of Sambucus nigra berry on acute respiratory viral infections: a rapid review of clinical studies. Adv Integrat Med 7(4):240–246

Hatmal MM, Alshaer W, Al-Hatamleh MA, Hatmal M, Smadi O, Taha MO, Oweida AJ, Boer JC, Mohamud R, Plebanski M (2020) Comprehensive structural and molecular comparison of spike proteins of SARS-CoV-2, SARS-CoV and MERS-CoV, and their interactions with ACE2. Cell 9(12):2638

Hoffmann M, Kleine-Weber H, Pöhlmann S (2020) A multibasic cleavage site in the spike protein of SARS-CoV-2 is essential for infection of human lung cells. Mol Cell 78(4):779–784

Hsiao CY, Hsu YJ, Tung YT, Lee MC, Huang CC, Hsieh CC (2018) Effects of Antrodia camphorata and Panax ginseng supplementation on anti-fatigue properties in mice. J Vet Med Sci 80(2):284–291

Husain I, Ahmad R, Siddiqui S, Chandra A, Misra A, Srivastava A, Ahamad T, Khan MF, Siddiqi Z, Trivedi A, Upadhyay S (2022) Structural interactions of phytoconstituent (s) from cinnamon, bay leaf, oregano, and parsley with SARS-CoV-2 nucleocapsid protein: a comparative assessment for development of potential antiviral nutraceuticals. J Food Biochem 46(10): e14262

Iwata-Yoshikawa N, Kakizaki M, Shiwa-Sudo N, Okura T, Tahara M, Fukushi S, Maeda K, Kawase M, Asanuma H, Tomita Y, Takayama I (2022) Essential role of TMPRSS2 in SARS-CoV-2 infection in murine airways. Nat Commun 13(1):1–1

Jain J, Pai S, Sunil S (2018) Standardization of *in vitro* assays to evaluate the activity of polyherbal siddha formulations against Chikungunya virus infection. Virus 29(1):32–39

Jamali N, Soureshjani EH, Mobini GR, Samare-Najaf M, Clark CC, Saffari-Chaleshtori J (2022) Medicinal plant compounds as promising inhibitors of coronavirus (COVID-19) main protease: an in silico study. J Biomol Struct Dyn 40(17):8073–8084

Ji L, Zhao X, Zhang Y, Zhao P, Gong R, Li F, Huang H (2022) Efficacy and safety of Qinghua Zhixie decoction against diarrhea-predominate irritable bowel syndrome: a protocol for a randomized controlled trial. Medicine 101(9):e28895

Joshi C, Chaudhari A, Joshi C, Joshi M, Bagatharia S (2022) Repurposing of the herbal formulations: molecular docking and molecular dynamics simulation studies to validate the efficacy of phytocompounds against SARS-CoV-2 proteins. J Biomol Struct Dyn 40(18):8405–8419

Kadirvel V, Vasuki MT, Narayana GP, Kulathooran R (2022) Formulation and evaluation of medicated lozenges using traditional herbs to treat sore throat infection. J Food Process Preserv 46(10):e16903

Kamin W, Lehmacher W, Zimmermann A, Brandes-Schramm J, Funk P, Seifert GJ, Kardos P (2022) Treatment of sore throat and hoarseness with Pelargonium sidoides extract EPs 7630: a meta-analysis. Pharmadvances 4(2):88–103

Karimi M, Zarei A, Soleymani S, Jamalimoghadamsiahkali S, Asadi A, Shati M, Jafari M, Rezadoost H, Kordafshar G, Naghizadeh A, Mardi R (2021) Efficacy of Persian medicine herbal formulations (capsules and decoction) compared to standard care in patients with COVID-19, a multicenter open-labeled, randomized, controlled clinical trial. Phytother Res 35(11):6295–6309

Katsuki M, Narita N, Matsumori Y, Ishida N, Watanabe O, Siqi C, Tominaga T (2022) Kampo (Japanese herbal) medicine for primary headache as an acute treatment–a retrospective investigation in Kesennuma City Hospital during 5 years

Kiran G, Karthik L, Devi MS, Sathiyarajeswaran P, Kanakavalli K, Kumar KM, Kumar DR (2020) In silico computational screening of Kabasura Kudineer-official Siddha formulation and JACOM against SARS-CoV-2 spike protein. J Ayurveda Integr Med 13:100324

Klemm T, Ebert G, Calleja DJ, Allison CC, Richardson LW, Bernardini JP, Lu BG, Kuchel NW, Grohmann C, Shibata Y, Gan ZY (2020) Mechanism and inhibition of the papain-like protease, PLpro, of SARS-CoV-2. EMBO J 39(18):e106275

Kuchi Bhotla H, Kaul T, Balasubramanian B, Easwaran M, Arumugam VA, Pappusamy M, Muthupandian S, Meyyazhagan A (2020) Platelets to surrogate lung inflammation in COVID-19 patients. Med Hypotheses 143:110098

Kuchi Bhotla H, Balasubramanian B, Arumugam VA, Pushparaj K, Murugesh E, Rathinasamy B (2021a) Insinuating cocktailed components in biocompatible-nanoparticles could act as an impressive neo-adjuvant strategy to combat COVID-19. Nat Resour Hum Health 1:3–7

Kuchi Bhotla H, Balasubramanian B, Meyyazhagan A, Pushparaj K, Easwaran M, Pappusamy M, Alwin Robert A, Arumugam VA, Tsibizova V, Msaad Alfalih A, Aljowaie RM, Saravanan M, Di Renzo GC (2021b) Opportunistic mycoses in COVID-19 patients/survivors: epidemic inside a pandemic. J Infect Public Health 14(11):1720–1726

Kumadoh D, Adotey J, Ofori-Kwakye K, Kipo SL, Prah T, Patterson S (2015) Development of oral capsules from Enterica herbal decoction-a traditional remedy for typhoid fever in Ghana. J Appl Pharmaceut Sci 5(4):083–088

Lau JT, Leung PC, Wong EL, Fong C, Cheng KF, Zhang SC, Lam CW, Wong V, Choy KM, Ko WM (2005) The use of an herbal formula by hospital care workers during the severe acute respiratory syndrome epidemic in Hong Kong to prevent severe acute respiratory syndrome transmission, relieve influenza-related symptoms, and improve quality of life: a prospective cohort study. J Alternat Complemen Med 11(1):49–55

Li SS, Chen ZC, Zhang CH (2012) Effects of Tao-Hong-Si-Wu-Tang, a traditional Chinese herbal medicine formula on glucose and lipid metabolism during. J Anim Vet Adv 11(15):2737–2740

Liang SB, Han M, Cheng HJ, Zhang QY, Zhang NW, Jia BY, Robinson N, Liu JP (2022) Chinese herbal formula Tongxie Yaofang for diarrhea-predominant irritable bowel syndrome: study protocol for a randomized, multiple-blind, placebo-controlled trial. Trials 23(1):1–9

Manikandan R, Anand AV (2016) Evaluation of antioxidant activity of Psidium guajava Linn. In streptozotocin–induced diabetic rats. Free Radic Res 6(1):72–76

Maurya DK, Sharma D (2022) Evaluation of traditional ayurvedic Kadha for prevention and management of the novel coronavirus (SARS-CoV-2) using in silico approach. J Biomol Struct Dyn 40(9):3949–3964

Meher A (2012) Antitussive evaluation of formulated polyherbal cough syrup. J Drug Deliv Ther 2(5)

Mehreen A, Waheed M, Liaqat I, Arshad N (2016) Phytochemical, antimicrobial, and toxicological evaluation of traditional herbs used to treat sore throat. Biomed Res Int 2016:8503426

Meyyazhagan A, Pushparaj K, Balasubramanian B, Kuchi Bhotla H, Pappusamy M, Arumugam VA, Easwaran M, Pottail L, Mani P, Tsibizova V, Di Renzo GC (2022) COVID-19 in pregnant women and children: insights on clinical manifestations, complexities, and pathogenesis. Int J Gynaecol Obstet 156(2):216–224

Murugesan S, Kottekad S, Crasta I, Sreevathsan S, Usharani D, Perumal MK, Mudliar SN (2021) Targeting COVID-19 (SARS-CoV-2) main protease through active phytocompounds of ayurvedic medicinal plants–Emblica officinalis (Amla), Phyllanthus niruri Linn.(Bhumi Amla) and Tinospora cordifolia (Giloy)–A molecular docking and simulation study. Comput Biol Med 136:104683

Nafiu MO, Hamid AA, Muritala HF, Adeyemi SB (2017) Preparation, standardization, and quality control of medicinal plants in Africa. In: Medicinal spices and vegetables from Africa, pp 171–204

Nagalingam A (2017) Drug delivery aspects of herbal medicines. Jpn Kampo Med Treat Common Dis Focus Inflammat 17:143

Nargis Begum T, Muhammad Ilyas MH, Burkanudeen A, Kalavathy S, Vijaya Anand A, Sampathkumar P, Jaswanth A (2009) Hypoglycemic and antihyperlipidemic activity of ethanolic leaf extract of *Azima tetracantha* Lam. on alloxan-induced diabetic rats. J Cell Tissue Res 9(1):1681–1685

Osipiuk J, Azizi SA, Dvorkin S, Endres M, Jedrzejczak R, Jones KA, Kang S, Kathayat RS, Kim Y, Lisnyak VG, Maki SL (2021) Structure of papain-like protease from SARS-CoV-2 and its complexes with non-covalent inhibitors. Nat Commun 12(1):1–9

Pal SK, Shukla Y (2003) Herbal medicine: current status and the future. Asian Pac J Cancer Prev 4(4):281–288

Parasuraman S, Thing GS, Dhanaraj SA (2014) Polyherbal formulation: concept of ayurveda. Pharmacogn Rev 8(16):73

Ping YH, Yeh H, Chu LW, Lin ZH, Hsu YC, Lin LC, Hsu CH, Fu SL, Lin TY (2022) The traditional Chinese medicine formula jing guan fang for preventing SARS-CoV-2 infection: from clinical observation to basic research. Front Pharmacol 13:744439

Pizzato M, Baraldi C, Boscato Sopetto G, Finozzi D, Gentile C, Gentile MD, Marconi R, Paladino D, Raoss A, Riedmiller I, Ur Rehman H (2022) SARS-CoV-2 and the host cell: a tale of interactions. Front Virol 1:815388

Pushparaj K, Kuchi Bhotla H, Arumugam VA, Pappusamy M, Easwaran M, Liu WC, Issara U, Rengasamy KRR, Meyyazhagan A, Balasubramanian B (2022) Mucormycosis (black fungus) ensuing COVID-19 and comorbidity meets—magnifying global pandemic grieve and catastrophe begins. Sci Total Environ 805:150355

Qiao J, Li YS, Zeng R, Liu FL, Luo RH, Huang C, Wang YF, Zhang J, Quan B, Shen C, Mao X (2021) SARS-CoV-2 Mpro inhibitors with antiviral activity in a transgenic mouse model. Science 371(6536):1374–1378

Rahman MS, Hoque MN, Islam MR, Islam I, Mishu ID, Rahaman MM, Sultana M, Hossain MA (2021) Mutational insights into the envelope protein of SARS-CoV-2. Gene reports 22:100997

Rakshit M, Muduli S, Srivastav PP, Mishra S (2022) Pomegranate peel polyphenols prophylaxis against SARS-CoV-2 main protease by in-silico docking and molecular dynamics study. J Biomol Struct Dyn 40(23):12917–12931

Ram TS, Munikumar M, Raju VN, Devaraj P, Boiroju NK, Hemalatha R, Prasad PV, Gundeti M, Sisodia BS, Pawar S, Prasad GP (2022) In silico evaluation of the compounds of the ayurvedic drug, AYUSH-64, for the action against the SARS-CoV-2 main protease. J Ayurveda Integrat Med 13(1):100413

Reena G, Kumar GM, Jitendra G, Khan PI (2014) Evaluation of antitussive activity of polyherbomineral formulation on cough reflex induced by different cough induced models in mice. Int J Drug Devel Res 6(4)

Rudrapal M, Celik I, Khan J, Ansari MA, Alomary MN, Yadav R, Sharma T, Tallei TE, Pasala PK, Sahoo RK, Khairnar SJ (2022) Identification of bioactive molecules from Triphala (Ayurvedic herbal formulation) as potential inhibitors of SARS-CoV-2 main protease (Mpro) through computational investigations. J King Saud Univ Sci 34(3):101826

Saravana Prabha P, Murugesh E, Balamuralikrishnan B, Mohan A, Arun M, Park S, Bhotla HK, Jeeva A, Vijaya Anand A, Sasikala K, Gopalakrishnan VK, Manikantan P, Tanushri K (2020) Employing bioactive compounds derived from *Ipomoea obscura* (L.) to evaluate potential inhibitor for SARS-CoV-2 main protease and ACE2 protein. Food Front 1:168–179

Sarkar PK, Das Mukhopadhyay C (2021) Ayurvedic metal nanoparticles could be novel antiviral agents against SARS-CoV-2. Int Nano Lett 11(3):197–203

Savastano A, Ibáñez de Opakua A, Rankovic M, Zweckstetter M (2020) Nucleocapsid protein of SARS-CoV-2 phase separates into RNA-rich polymerase-containing condensates. Nat Communications 11(1):6041

Shanmugam R, Thangavelu S, Fathah Z, Yatoo MI, Tiwari R, Pandey MK, Dhama J, Chandra R, Malik Y, Dhama K, Sha R, Chaicumpa W (2020) Shanmugam V Arumugam AV, SARS-CoV-2/COVID-19 pandemic-an update. J Exp Biol Agric Sci 8(1):S219–S245

Sharma V, Singh S, Dixit A, Saxena A (2020) Formulation and evaluation of herbal cough syrup from seeds extract of hedge mustard. Int J Res Pharm Chem 10(1):56–69

Sharma B, Bagchi A, Bhutia S, Sarkar BR, Pal P (2022) Formulation and evaluation of expectorant activity of poly herbal cough syrup from traditional medicinal plant extracts of north East India. Res J Pharm Technol 15(3):949–953

Sheik Amamuddy O, Verkhivker GM, Tastan Bishop O (2020) Impact of early pandemic stage mutations on molecular dynamics of SARS-CoV-2 Mpro. J Chem Inf Model 60(10):5080–5102

Shi Y, Gu R, Liu C, Ni J, Wu T, Yuan J (2007) Chinese medicinal herbs for sore throat. Cochrane Database Syst Rev 3:CD004877

Shin D, Mukherjee R, Grewe D, Bojkova D, Baek K, Bhattacharya A, Schulz L, Widera M, Mehdipour AR, Tascher G, Geurink PP (2020) Papain-like protease regulates SARS-CoV-2 viral spread and innate immunity. Nature 587(7835):657–662

Shree P, Mishra P, Selvaraj C, Singh SK, Chaube R, Garg N, Tripathi YB (2022) Targeting COVID-19 (SARS-CoV-2) main protease through active phytochemicals of ayurvedic medicinal plants–Withania somnifera (Ashwagandha), Tinospora cordifolia (Giloy) and Ocimum sanctum (Tulsi)–a molecular docking study. J Biomol Struct Dyn 40(1):190–203

Singh A, Mishra A (2021) Leucoefdin a potential inhibitor against SARS CoV-2 Mpro. J Biomol Struct Dyn 39(12):4427–4432

Singh YS, Prajapati PK, Ravishankar B, Ashok BK (2010) Evaluation of anti-tussive activity of Shirishavaleha-an ayurvedic herbal compound formulation in Sulphur dioxide induced cough in mice. Indian Drugs 47(9):38–41

Subbaiyan A, Ravichandran K, Singh SV, Sankar M, Thomas P, Dhama K, Malik YS, Singh RK, Chaudhuri P (2020) In silico molecular docking analysis targeting SARS-CoV-2 spike protein and selected herbal constituents. J Pure Appl Microbiol 14:989–998

Sui L, Zhao Y, Wang W, Wu P, Wang Z, Yu Y, Hou Z, Tan G, Liu Q (2021) SARS-CoV-2 membrane protein inhibits type I interferon production through ubiquitin-mediated degradation of TBK1. Front Immunol 12:662989

Sulaiman CT, Deepak M, Madhavankutty TS, Anandan EM, Balachandran I (2020) Chromatographic and Tandem mass spectroscopic characterization of Ayush Kwath Churnam-an Ayurvedic immunomodulatory drug during COVID-19

Sun Y, Yang AW, Hung A, Lenon GB (2020) Screening for a potential therapeutic agent from the herbal formula in the 4th edition of the Chinese national guidelines for the initial-stage management of COVID-19 via molecular docking. Evid Based Complement Alternat Med 24:2020

Sun S, Karki C, Aguilera J, Lopez Hernandez AE, Sun J, Li L (2021) Computational study on the function of palmitoylation on the envelope protein in SARS-CoV-2. J Chem Theory Comput 17(10):6483–6490

Teh BP, Ahmad N, Ibnu Rasid EN, Zolkifli NA, Sastu Zakaria UR, Mohamed Yusoff N, Zulkapli A, Japri N, Lee JC, Muhammad H (2021) Herbal-based formulation containing Eurycoma longifolia and Labisia pumila aqueous extracts: safe for consumption? Pharmaceuticals 14(2):142

Thomas S (2020) The structure of the membrane protein of SARS-CoV-2 resembles the sugar transporter semi SWEET. Pathogens Immun 5(1):342

Tian J, Yan S, Wang H, Zhang Y, Zheng Y, Wu H, Li X, Gao Z, Ai Y, Gou X, Zhang L (2020) Hanshiyi formula, a medicine for SARS-CoV-2 infection in China, reduced the proportion of mild and moderate COVID-19 patients turning to severe status: a cohort study. Pharmacol Res 161:105127

Tian F, Tong B, Sun L, Shi S, Zheng B, Wang Z, Dong X, Zheng P (2021) N501Y mutation of spike protein in SARS-CoV-2 strengthens its binding to receptor ACE2. elife 10:e69091

Toto A, Ma S, Malagrinò F, Visconti L, Pagano L, Stromgaard K, Gianni S (2020) Comparing the binding properties of peptides mimicking the envelope protein of SARS-CoV and SARS-CoV-2 to the PDZ domain of the tight junction-associated PALS1 protein. Protein Sci 29(10): 2038–2042

Tsai KC, Huang YC, Liaw CC, Tsai CI, Chiou CT, Lin CJ, Wei WC, Lin SJ, Tseng YH, Yeh KM, Lin YL (2021) A traditional Chinese medicine formula NRICM101 to target COVID-19 through multiple pathways: a bedside-to-bench study. Biomed Pharmacother 133:111037

Unnikrishnan V, Nishteswar K (2015) Antipyretic herbal formulation with special reference to Kwatha Kalpana of Sarngadhara Samhita. Glob J Res Med Plants Ind Med 4(7):147

Vats S, Agarwal K, Kumar A (2022) Preparation and standardization of Vrihatsamsharkar Churna: a polyherbal formulation. Research Journal of Pharmacy and Technology. 15(5):2244–2249

Verma DS, Dange SV, Nandagavli MH (2022) Comparative efficacy of antidiarrheal activity of DIAREX, an ayurvedic formulation vs. racecadotril in children with acute diarrhea. Europ J Mol Clin Med 9(6):1084–1089

Vijaya Anand A, Velayuthaprabhu S, Rengarajan RL, Sampathkumar P, Radhakrishnan R (2020) Bioactive Compounds of Guava (L.). In Bioactive compounds in underutilized fruits and nuts. Springer, Cham, pp 503–527

Vijayakumar K, Anand AV, Manikandan R (2015) *In vitro* antioxidant activity of ethanolic extract of *Psidium guajava* leaves. Int J Res Stud Biosci 3(5):145–149

Vijayakumar K, Rengarajan RL, Radhakrishnan R, Mathew S, Qadri I, Vijaya Anand A (2019) *Psidium guajava* leaf extracts and their quercetin protect HepG2 cell lines against $CCL_4$ induced cytotoxicity. Indian J Clin Biochem 34(3):324–329

Vincent S, Arokiyaraj S, Saravanan M, Dhanraj M (2020) Molecular docking studies on the anti-viral effects of compounds from kabasura kudineer on SARS-CoV-2 3CLpro. Front Mol Biosci 7:434

Wang MY, Zhao R, Gao LJ, Gao XF, Wang DP, Cao JM (2020) SARS-CoV-2: structure, biology, and structure-based therapeutics development. Front Cell Infect Microbiol 10:587269

Wang H, Hou YN, Yang M, Feng Y, Zhang YL, Smith CM, Hou W, Mao JJ, Deng G (2022) Herbal formula Shenling Baizhu san for chronic diarrhea in adults: a systematic review and meta-analysis. Integr Cancer Ther 21:15347354221081214

Wong WCW, Lee A, Lam AT, Li KT, Leung CYM, Leung PC, Wong ELY, Tang JL (2006) Effectiveness of a Chinese herbal medicine preparation in the treatment of cough in uncomplicated upper respiratory tract infection: a randomised double-blinded placebo-control trial. Cough 2(1):1–9

Wynn SG, Fougère BJ (2007) Clinical practice: getting started. Veterin Herb Med 2007:453457

Xia X (2021) Domains and functions of spike protein in Sars-Cov-2 in the context of vaccine design. Viruses 13(1):109

Yan Z, Zhang K, Zhang K, Wang G, Wang L, Zhang J, Qiu Z, Guo Z, Kang Y, Song X, Li J (2022) Huang Bai Jian PI decoction alleviates diarrhea and represses inflammatory injury via PI3K/Akt/NF-κB pathway: in vivo and in vitro studies. J Ethnopharmacol 292:115212

Yang Y (2010) Chinese herbal formulas: treatment principles and composition strategies E-book. Elsevier Health Sciences

Zeng W, Liu G, Ma H, Zhao D, Yang Y, Liu M, Mohammed A, Zhao C, Yang Y, Xie J, Ding C (2020) Biochemical characterization of SARS-CoV-2 nucleocapsid protein. Biochem Biophys Res Commun 527(3):618–623

Zhang DH, Zhang X, Peng B, Deng SQ, Wang YF, Yang L, Zhang KZ, Ling CQ, Wu KL (2020) Network pharmacology suggests biochemical rationale for treating COVID-19 symptoms with a Traditional Chinese Medicine. Commun Biol 3(1):1–9

Zhang Z, Nomura N, Muramoto Y, Ekimoto T, Uemura T, Liu K, Yui M, Kono N, Aoki J, Ikeguchi M, Noda T (2022) Structure of SARS-CoV-2 membrane protein essential for virus assembly. Nat Commun 13(1):1–2

Zheng S, Baak JP, Li S, Xiao W, Ren H, Yang H, Gan Y, Wen C (2020) Network pharmacology analysis of the therapeutic mechanisms of the traditional Chinese herbal formula Lian Hua Qing Wen in Corona virus disease 2019 (COVID-19), gives fundamental support to the clinical use of LHQW. Phytomedicine 79:153336

# Chapter 5
# Rejuvenation of Traditional Medicine in the Twenty-First Century against SARS-CoV-2

**Shristi Modanwal and Nidhi Mishra**

## 5.1 Introduction

SARS-CoV-2 is a positive-sense single-stranded RNA virus that ranges in size from 25 to 32 kb with 29,891 bases that are 96% identical to a bat coronavirus at the whole-genome level and that has 79.6% sequence identity with SARS-CoV (Wu et al. 2020). The structural spike (S), membrane (M), envelope (E), and nucleocapsid (N) proteins play a major role in the formation of the viral structure. The viral envelope, which is a lipid bilayer formed from the host cell membrane, contains the S, M, and E proteins, and the N protein interacts with the viral RNA inside the virion core (de Andrade Santos et al. 2020; Mithilesh et al. 2022). In December 2019, the Wuhan coronavirus, also known as the 2019 novel coronavirus, was detected in China. The virus was categorized as SARS-CoV-2 by the International Committee on Taxonomy of Viruses (ICTV) because it shared 79.5% genetic homology with SARS-CoV-1 and was the cause of the 2019 coronavirus disease (COVID-19) pandemic (Ahmad et al. 2021). COVID-19 spread from person to person when an infected individual coughs and sneezes releasing coronavirus droplets into the air. The coronavirus floats in the air or sticks to a surface that your hands touch before touching your eyes, nose, or mouth. The mucous membranes of the respiratory systems are typically infected with coronavirus in this way. The immune system of the body reacts in 2–14 days with symptoms like coughing, breathing issues, fever, chills, muscle discomfort, sore throat, and loss of taste or smell (Lee et al. 2021). These viruses can infect humans and have a wide range of other mammalian hosts, including bats and birds. The coronaviruses get their name

S. Modanwal · N. Mishra (✉)
Department of Applied Sciences, Indian Institute of Information Technology-Allahabad, Prayagraj, Uttar Pradesh, India
e-mail: rss2022001@iiita.ac.in; nidhimishra@iiita.ac.in



J.-T. Chen (ed.), *Ethnopharmacology and Drug Discovery for COVID-19: Anti-SARS-CoV-2 Agents from Herbal Medicines and Natural Products*,
https://doi.org/10.1007/978-981-99-3664-9_5

from the spikey surface proteins that are one of the structural proteins and resemble a crown (Latin corona) (Naserghandi et al. 2020; Nassar et al. 2021). Even though several industries have developed COVID-19 vaccines, community members and researchers are still looking for the most effective treatments, including traditional remedies (Paudyal et al. 2022).

Global demand for various traditional medicines as a potential COVID-19 treatment option is still quite significant due to the lack of appropriate medications. Severe acute respiratory infection (SARI), abnormal immunological response, and thrombosis are all associated with coronaviruses (CoV) in their most serious stages. The research is now concentrated on three approaches: first, testing broad-spectrum antiviral medicines; second, *in silico* screening; and third, drug repurposing (Singh et al. 2021). In addition, the use of herbal medicines and supplements, particularly those with anti-infective and immunomodulatory effects as well as those used as supportive therapies, has grown significantly as an adjunct in economically developed countries. However, this use is frequently not mentioned to medical professionals (Wang et al. 2021). Researchers can use natural remedies to create medical medicines that are secure and simple to use. For instance, plants used in traditional Chinese medicine (TCM), such as *Scutellaria baicalensis*, contain a variety of antiviral substances, such as phytochemicals that may be able to combat SARS-CoV-2 and viral replication inhibitors (Llivisaca-Contreras et al. 2021). A number of medicines have been offered by traditional Indian medical systems, including Siddha, to boost immunity and prevent the spread of COVID-19. Kabasura Kudineer Chooranam, one of the official Siddha formulations, is thought to be an effective remedy with antiviral and immunomodulatory effects (Parthasarathy et al. 2021). It has also been reported for the treatment of COVID-19 in children, TCM has a very high involvement rate (Duan et al. 2022).

Angiotensin converting enzyme 2 (ACE 2), a transmembrane protein in type-2 pneumocytes found in the alveoli, nasal goblet cells, and intestinal epithelial cells, is hypothesized to be the route by which the SARS-CoV-2 virus enters the cell (Gao et al. 2020). Although intestinal epithelial cells may also become infected, it is believed that the virus predominantly affects type-2 pneumocytes in the lungs (Capodice and Chubak 2021). For the treatment of several respiratory diseases, the use of nutritional supplements including herbal medicines is well known. COVID-19 has also attracted attention to the antiviral capabilities of these products. Various researchers have looked into the potential applications of natural products and compounds that are already approved for usage for another purpose (Mahaboob Ali et al. 2022). Combining herbal medicine with modern biomedicine has the potential to alleviate both hypoxemia and chronic obstructive pulmonary disease, as well as to have direct antiviral, immunomodulatory, and anti-inflammatory effects (Luo et al. 2020). Modern medicine has benefited from the introduction of single chemical entities rather than composite natural product sources due to substantial advancements in the field of ethnopharmacology (Yimer et al. 2021). The druggable targets for SARS-CoV-2 are RNA-dependent RNA polymerase, 3-chymotrypsin like protease (also known as $M^{pro}$) and papain-like protease ($PL^{pro}$), endoribonuclease Nsp15/NendoU, spike glycoprotein, transmembrane protease serine 2 (TMPRSS2), angiotensin-converting enzyme-2 (ACE-2), furin, cathepsin L (Faheem et al. 2020; Artese et al. 2020). These proteins/receptors can be targeted to treat COVID-19.

Since ancient times, people all over the world have used plants as medicines, particularly those from Asian countries like India, China, and Japan as well as some African nations. The widespread availability and very inexpensive cost of these plants among indigenous peoples are primarily responsible for their folkloric use. Even now, plants are still incredibly promising medicinal sources for the treatment of several diseases and infectious diseases as well as their effects, such as pain, oxidative stress, cancer, diarrhea, depression, fever, and thrombosis. This increases the possibility of developing medications with anti-COVID-19 properties derived from plant sources through a wide range of mechanisms of action (Alam et al. 2021). The World Health Organization supports the numerous benefits of traditional, complementary, and alternative medicine that is of excellent quality, safe, and potency (Umeta Chali et al. 2021). The patients' immunity is important in COVID-19. Standard drugs that have immunomodulatory effects could therefore be a possibility for COVID-19 patient prevention and treatment (Zhang and Liu 2020). Several studies have found that bioactive compounds derived from plants may have inhibitory effects on the SARS-CoV-2 coronavirus strain.

## 5.2 COVID-19 Treatment with Indian Traditional Remedies

The Government of India, Ministry of AYUSH (Ayurveda, Unani, Yoga Siddha, Naturopathy, and Homeopathy) organized an interdisciplinary AYUSH research and development special unit and guidance for AYUSH clinical studies in COVID-19 to develop an intervention to prevent or cure COVID-19 using concepts of AYUSH. This was done to promote research and the generation of evidence for various traditional medicines on COVID-19 (Charan et al. 2021). The science of Ayurveda has a system of knowledge in which Tattva (principles) translates to Sastra (theoretical constructs), and Sastra translates to Vyavahara (practical applications). The goal of the entire system is to use knowledge to create beneficial actions for society (Mathpati et al. 2020). Numerous ayurvedic plants have been explored for managing COVID-19 and its symptoms. A few of them are summarized below.

### *5.2.1* Allium sativum

There are two main categories of T cells: cytotoxic T cells ($CD8^+$ T cells) and helper T cells ($CD4^+$ T cells). Cytotoxic T cells kill tumors and virally infected cells, whereas helper T cells "help" other immune system cells. Both play a crucial role in

**Fig. 5.1** Structure of allicin from plant *Allium sativum*

the immune system. A type of immune cell called natural killer cell can kill tumor and virus-infected cells by cytolysis. These cells were shown to be suppressed in COVID-19-infected patients. They are even significantly lower in extreme cases. A decrease in IFN- $\gamma$ was also reported. Garlic increases $CD4^+$ and $CD8^+$ cells significantly and induces NK cells (Donma and Donma 2020). Using *in silico* non-covalent and covalent docking screening approaches, recent research suggests that allicin (Fig. 5.1) exhibits dual S-thioallylation of Cys-145 and solvent-exposed Cys-85/Cys-156 residue of SARS-CoV-2 $M^{pro}$, serving as a strong inhibitor of SARS-CoV-2 $M^{pro}$ (Shekh et al. 2021; Khubber et al. 2020).

### 5.2.2 *Cinnamon*

Target cells infected by SARS-CoV-2 express the angiotensin-converting enzyme (ACE2) and type 2 transmembrane serine protease (TMPRSS2) (Amawi et al. 2020). In sheep kidneys, lungs, and testis, *C. zeylanicum* methanolic extract was found to inhibit the angiotensin-converting enzyme. In the kidney, inhibiting ACE activity was as effective as the standard medicine, captopril. Caffeic acid, gallic acid, eugenol, and cinnamic acid (Fig. 5.2) are phenolic compounds, isolated from *Cinnamomum zeylanicum* and have an inhibitory effect on trypsin (a serine protease). The highest potential for inhibiting an enzyme was demonstrated by caffeine (IC50 = 84%) and cinnamic acid (IC50 = 53%) (Yakhchali et al. 2021). As a result of these findings, cinnamon may be recommended as an antiviral drug for SARS-CoV-2 treatment.

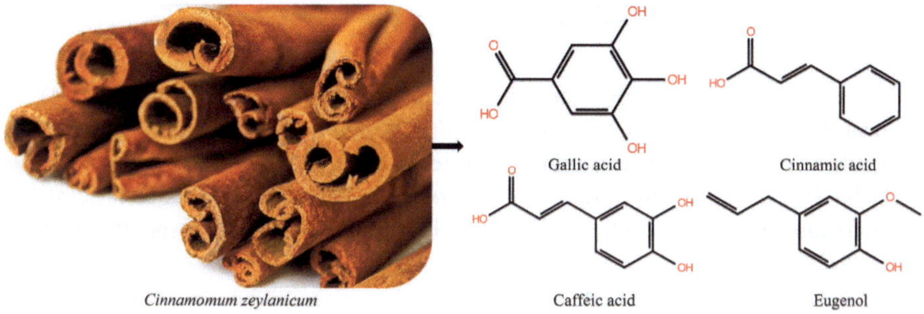


**Fig. 5.2** Structure of gallic acid, caffeic acid, eugenol, and cinnamic acid from plant *Cinnamomum zeylanicum*

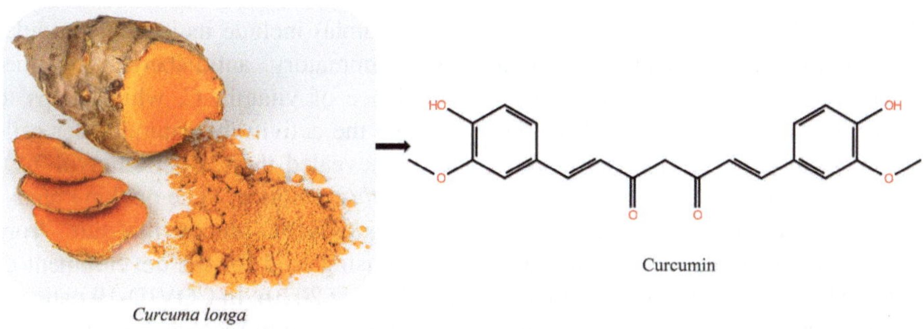


**Fig. 5.3** Structure of curcumin from plant *Curcuma longa*

## *5.2.3* Curcuma longa

The active component of *C. longa* (turmeric) rhizomes is curcumin, a hydrophobic polyphenol. The pharmacological effects of curcumin include anti-inflammatory, antioxidant, anticancer, anti-bacterial, antiviral, and antidiabetic properties (Babaei et al. 2020). Curcumin (Fig. 5.3) is an effective adjuvant treatment for COVID-19 due to the above and a number of additional benefits. Many in vitro and *in silico* studies suggest the effectiveness of curcumin against COVID-19 (Rattis et al. 2021; Rajagopal et al. 2020). Curcumin has been shown in *in silico*/computational studies to be able to alter SARS-CoV-2 cellular entrance events (spike protein, ACE2, TMPRSS2), their replication ($M^{pro}$), and the molecular cascade that results in pathologic COVID-19 effects (Rajagopal et al. 2020; Patel et al. 2022; Zahedipour et al. 2020). However, there is no direct experimental evidence to support the advantages of curcumin against SARS-CoV-2.

### 5.2.4 Linum usitatissimum

The seed of Linum is widely used for medicinal purposes. Herbacetin (Fig. 5.4) is a natural flavonoid from flax seed (*Linum usitatissimum*) (Solnier and Fladerer 2021) and plays a crucial role in inhibiting proteins of SARS-CoV-2. The infection of COVID-19 can be inhibited by *Linum usitatissimum* Secoisolariciresinol diglucoside and herbacetin by targeting ACE and SARS-3CL$^{pro}$, respectively (Jalali et al. 2021).

### 5.2.5 Phyllanthus emblica

The biological benefits of *Phyllanthus emblica* (amla) include its immunomodulatory, antioxidant, anticancer, antiviral, anti-inflammatory, antidiabetic, and other biological activities. Amla contains an abundance of vitamin C, which helps to lower oxidative stress and substantially enhance the activity of natural killer cells (Murugesan et al. 2021). The docking analysis revealed the (2S)-eriodictyol 7-O-(6′'-O-galloyl)-D-glucopyranoside from *Phyllanthus emblica*. (The docking analysis revealed the (2S)-eriodictyol 7-O-(6′'-O-galloyl)-D-glucopyranoside from Phyllanthus emblica. (Phyllanthaceae) is a promising lead for the development of anti-SARS-CoV-2 medicines (Orhan and Senol Deniz 2020). In COVID-19 patients, amla tea can reduce the length of time needed for symptom recovery and length of stay in hospitals (Varnasseri et al. 2022). Sixty-six phytoconstituents of *Phyllanthus emblica* were examined in an *in silico* study for potential antiviral effects against the SARS-CoV-2 proteins NSP15 endoribonuclease, main protease, and spike protein receptor binding domain. The promising ability of chlorogenic acid, quercitrin, and myricetin to inhibit the essential viral proteins was validated by docking studies and molecular dynamics simulation. The network pharmacology study showed that specific phytoconstituents were involved in regulating a number of signaling pathways that may be important for immunomodulation, regulating inflammation, and regulating cytokine (Chikhale et al. 2021). In another study, Murugesan S et al. docked 30 amlas bioactive with COVID-19 M$^{pro}$, of those, 7-ketositosterol (−

**Fig. 5.4** Structure of herbacetin from plant *Linum usitatissimum*

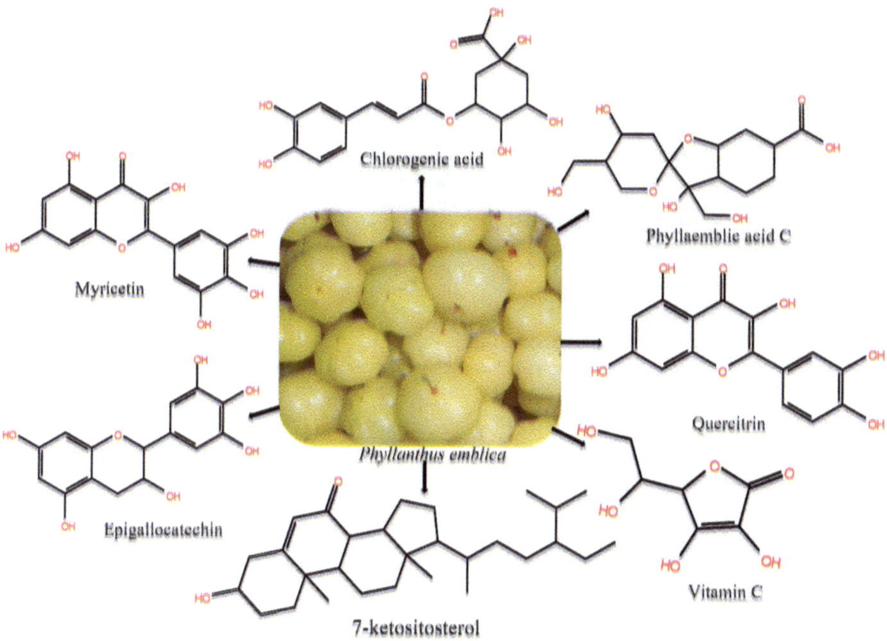


**Fig. 5.5** Structure of vitamin C, chlorogenic acid, quercitrin, myricetin, 7-ketositosterol, epigallocatechin, phyllaemblic acid C from plant *Phyllanthus emblica*

46.67 kcal/mol), epigallocatechin (32.69 kcal/mol), phyllaemblic acid C (11.98 kcal/mol), and quercetin (38.77 kcal/mol) were shown to fit the substrate binding cleft of $M^{pro}$ (Murugesan et al. 2021). The structure of some compounds from *Phyllanthus emblica* is shown in Fig. 5.5.

## 5.2.6 Nigella sativa

A sample of adult patients with mild COVID-19 symptoms who received oral *Nigella sativa* oil dosage supplementation at a dose of 500 mg twice daily for 10 days recovered more frequently than those who received only standard therapy. Low adverse effect profiles were also associated with a quicker return from COVID-19 symptoms and a lower hospitalization rate (Koshak et al. 2021).

## 5.2.7 Andrographis paniculata

*Andrographis paniculata*, commonly known as green chiretta, is a perennial herb that is widely distributed in India. According to Enmozhi et al. (2020),

**Fig. 5.6** Structure of andrographolide from plant *Andrographis paniculate*

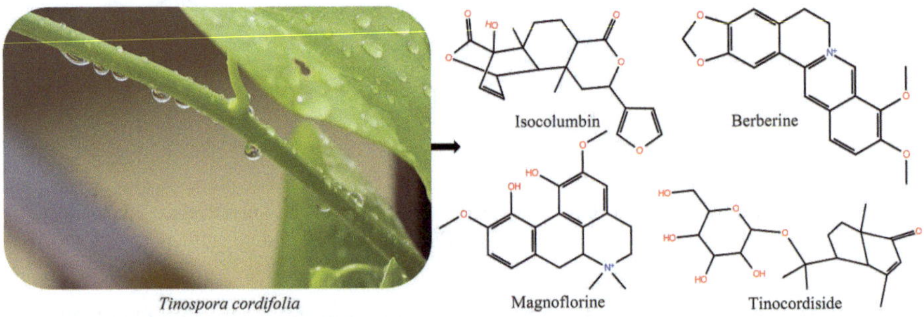


**Fig. 5.7** Structure of tinocordiside, berberine, isocolumbin, and magnoflorine from plant *Tinospora cordifolia*

andrographolide (an extract of *Andrographis paniculata*) is bind with the inhibitor site of the main protease ($M^{pro}$), one of the proteases essential to the SARS-CoV-2 virus. Andrographolide (Fig. 5.6) scored well in the docking study. Compared to synthetic compounds like disulfiram, oseltamivir, tideglusib, and a combination of lopinavir and ritonavir, the docking study was shown to be much higher. Considerable high druggability, permeability, solubility, and target specificity were seen in the andrographolide (Enmozhi et al. 2020).

## 5.2.8 Tinospora cordifolia

A shrub known as *Tinospora cordifolia*, sometimes known as heart-leaved moonseed, is found primarily at higher altitudes in India. Four compounds of *T. cordifolia*, namely berberine, isocolumbin, magnoflorine, and tinocordiside (Fig. 5.7) were found to have strong binding affinity for four crucial proteins of SARS-CoV-2 involved in the attachment (6VSb, 6M0J) and replication (6M71, 6Y84). The compounds' efficacy is either equal to or greater than that of some of the most widely used antiviral medications, including favipiravir, lopinavir, and remdesivir (Sagar and Kumar 2020). A study reported that the phytochemical saponarin, which is found in *T. cordifolia*, had a strong binding affinity with the main protease of about 8.75 kcal/mol (Mulpuru and Mishra 2021). In a study, only one molecule, tinocordiside (CID 177384), out of 28 active phytochemicals from *T. cordifolia* (Giloy), showed a maximum binding affinity for SARS-CoV-2 $M^{pro}$ compared to built-in ligand N3 as per YASARA score. Tinocordiside is reported to be a novel, restructured cadinane sesquiterpene glycoside from *T. cordifolia* (Shree et al. 2022). In another study, the components of *T. cordifolia*, such as β-sitosterol (C29H50O), coline (C5H14NO), berberine (C20H18NO4), tetrahydropalmatine (C21H25NO4), and octacosanol (C28H58O) were selected to examine the interactions between drugs and 3CLpro interaction. The *in silico* analysis showed that berberine, out of all the phytochemicals in *T. cordifolia* that were taken into consideration, can regulate the function of the $3CL^{pro}$ protein and can thus limit viral replication (Chowdhury 2021). The IgG antibody response, macrophage activation, induction of cell-regulated immunity, and humoral immunity can also be stimulated by giloy herbs.

### *5.2.9* Ocimum tenuiflorum

*Ocimum tenuiflorum* is extensively distributed in India, and Ayurveda in that country mentions its therapeutic uses. *In silico* research concluded that *Ocimum tenuiflorum* extracts may be able to inhibit the SARS-CoV-2 ACE-2 and TRMPSS2 proteins (Jindal and Rani 2022).

## 5.3 COVID-19 Treatment with Traditional Chinese medicine

Traditional Chinese medicine has historically been used to treat or cure a number of epidemic disorders. More than 3100 TCM medical professionals were sent to the province of Hubei during the COVID-19 treatment phase, and TCM procedures were incorporated into the COVID-19 diagnosis and therapy guidelines. Approximately 60,107 confirmed cases have been successfully treated by TCM (Ren et al. 2020). The World Health Organization recognizes the history of traditional medicine as an inherited legacy of knowledge, skills, and practices that are local to a community. The effectiveness of TCM against COVID-19 has been reported in a number of studies (Devpura et al. 2021). Chlorogenic acid, hyperoside, acaciin, p-hydroxyacetophenone, scopoletin, rutin, quercetin, (3R,4R,6S)- 3,6-dihydroxy-1-menthene, scoparone, luteolin, apigenin, quercetin, acacetin, aristolactam, and apigenin-7,4/-dimethyl ether are some of the phytocomponents from Chinese herbal extract and are employed in the treatment and prevention of SARS (Zhang and Chen 2008). The anti-SARS properties were also seen in the extracts of *Artemisia annua*, *Lindera aggregate*, *Isatis indigotica*, *Lycoris radiate*, *Torreya nucifera*, and *Houttuynia cordata*. Other TCM herbs that have been recommended for use in treating and preventing coronavirus include *Radix astragali* (Huangqi), *Radix saposhnikoviae* (Fangfeng), *Rhizoma Atractylodis Macrocephalae* (Baizhu), *Radix glycyrrhizae* (Ganacao), *Fructus forsythia* (Lianqiao) (Shahrajabian et al. 2020). It has been proved that TCM was extremely effective in treating COVID-19. There are currently several TCM medications available; however, only a few are covered in this chapter.

### 5.3.1 *Qingfei Touxie Fuzheng Recipe*

For the treatment of COVID-19, recent studies suggested using this mixture with Western therapy. The upregulation of antiviral components and the downregulation of pro-inflammatory mediators might be one strategic approach (Al-kuraishy et al. 2022). When taken by COVID-19 patients, the Qingfei Touxie Fuzheng recipe reduced IL-6 levels and improved lung inflammation or lesion absorption. Interestingly, in addition to their potent anti-SARS-CoV-2 activity, artemisinin and its derivatives controlled several immune cells, including macrophage, monocyte, dendritic cell, and T cell, to prevent the release of pro-inflammatory cytokines and the outbreak of cytokine storms, which protect tissues from damage (Lyu et al. 2021).

### 5.3.2 *Shufeng Jiedu*

Traditional Chinese medication called Shufeng Jiedu capsule (SFJDC) contains eight different medicinal herbs and is frequently used to treat influenza infection and other viral infections of the respiratory tract. It is now recommended for the management of COVID-19. When combined with arbidol, SFJDC shows promising results in the treatment of COVID-19 patients with no complications (Chen et al. 2020, 2021). In a study, Xia L et al. found that as compared to the model group, SFJDC significantly decreased the infection rate in the lung of HCoV-229E mice, reduced the levels of the inflammatory cytokines IL-6, IL-10, TNF-α, and IFN-γ, and enhanced the number of $CD4^+$ and $CD8^+$ cells in the blood. Network analysis showed that SFJDC inhibits NFκB activity across a number of signaling pathways. It has been clinically shown that adding SFJDC to conventional antiviral therapy (AVD) considerably sped up clinical recovery for COVID-19, as well as decreased fatigue and coughing when compared to AVD alone (Xia et al. 2021).

### 5.3.3 *Jinhua Qinggan*

Jinhua Qinggan was developed during the 2009 H1N1 pandemic. JH is composed of 12 herbs, including *Gypsum Fibrosum* (Shigao), *Ephedra sinica Stapf* (Mahuang), *Lonicera japonica Thunb.* (Jinyinhua), *Prunus armeniaca L.* (Kuxingren), *Forsythia suspensa* (Thunb.) Vahl (Lianqiao), *Fritillaria thunbergii Miq.* (Zhebeimu), *Scutellaria baicalensis Georgi* (Huangqin), *Anemarrhena asphodeloides Bunge* (Zhimu), *Artemisia annua L.* (Qinghao), *Mentha canadensis L.* (Bohe), *Glycyrrhiza inflata Batalin* (Gancao), and *Arctium lappa L.* (Niubangzi). In a molecular docking study, the constituents of JH, formononetin, stigmasterol, β-sitosterol, and anhydroicaritin were found to have a certain affinity for the SARS-CoV-2 3CL hydrolase and ACE2 enzymes (Chu et al. 2021). Jinhua Qinggan granules might stop SARS-CoV-2 by controlling cytokines through the metabolic pathway of arachidonic acid. Jinhua Qinggan granules played a role in apoptosis regulation via PI3K-Akt, MAPK, and Ras pathways (Huang et al. 2020).

### 5.3.4 *Huoxiang Zhengqi (HXZQ)*

The HXZQ components include *Agastache rugosus*, Perilla leaf, *Angelica dahurica*, and licorice, as well as six other TCM constituents. Due to its antiviral, anti-inflammatory, and immunomodulatory properties as well as its capacity to relieve gastrointestinal pain, HXZQ was suggested for the prevention of SARS-CoV-2. Additionally, clinical and pharmacological analyses have shown that HXZQ has therapeutic potential on fever, nausea or vomiting, stomach discomfort, and diarrhea linked to viral infection, indicating that HXZQ may potentially reduce COVID-19-related symptoms (Al-Romaima et al. 2020). For COVID-19 patients, using Huoxiang Zhengqi dropping tablets and Lianhua Qingwen granules in addition to western therapy may have therapeutic benefits. The inflammatory cytokine release induced by lipopolysaccharides was decreased by Huoxiang Zhengqi capsules. *Radix isatidis*, one of the key ingredients in Huoxiang Zhengqi, contains anti-inflammatory, antiviral, antipyretic, and immunity-boosting properties (Xiao et al. 2020).

### 5.3.5 *Lianhua Qingwen*

Traditional Chinese medicine known as Lianhua Qingwen (LHQW) is available in a number of dosage forms, such as capsules, granules, and decoction. LHQW is composed of 13 components. It has a wide range of antiviral properties, because of its immunomodulatory and as well as its inhibitory activity of pro-inflammatory cytokines and virus reproduction. The therapeutic properties of LHQW on

COVID-19 are predicated on its potent ability to bind to the SARS-CoV-2 therapeutic targets ACE2 and $M^{pro}$. As a result, it has proven to be beneficial for COVID-19 as a supplement and synergistic therapy strategy (Runfeng et al. 2020; Fan et al. 2022; Zhuang et al. 2021). The inhibitory effects and anti-inflammatory ability of Lianhua Qingwen against SARS-CoV-2 were analyzed by Runfeng et al. Lianhua Qingwen has traditionally been used to cure some ailments, including fever, cough, fatigue, influenza, bronchitis, pneumonia, and the early stages of measles. The Chinese National Health Commission recommended using this herbal preparation to treat or control COVID-19 (Runfeng et al. 2020; Ding et al. 2017; Yang et al. 2020). The herbal mixture had an IC50 of 411.2 g/ml and suppressed SARS-CoV-2 replication in a dose-dependent manner. A dose-dependent inhibition of the release of pro-inflammatory cytokines (TNFα, IL-6, CCL-2/MCP-1, and CXCL-10/IP-10) was also obtained by the mixture (Benarba and Pandiella 2020). Song et al. reported that the *Scutellaria baicalensis* extracts have baicalin, which is one of the main TCM herbal components, and hesperetin, an active compound found in tangerine peels. The COVID-19 symptoms have been treated with both bioactive compounds. The use of Xuebijing injection, a TCM medication has been widely accepted as a way to both reduce the time needed for patient ventilation in extreme situations and perhaps lower the risks of pneumonia spreading within the community (Song et al. 2019). According to a study, Lianhua Qingwen components such as 18β-glycyrrhetinic acid, indigo, β-sitosterol, naringenin, stigmasterol, luteolin, and quercetin may target ACE2 and protect the target organs of COVID-19 via the renin–angiotensin system (Huang et al. 2020).

### *5.3.6 Herbal Chinese Medicine Components Which Are Active Against SARS-Cov-2*

Chinese herbal medicine is comprised of a huge group of secondary metabolites (such as flavonoids) that show wide structural diversity and a plethora of compounds (such as flavones, flavonols, flavanonols, and isoflavones) that mediate a large number of useful bioactivities, including anti-browning, anti-tuberculosis, antimicrobial, anticancer, and antioxidant effects. An essential natural compound called scutellarein, a flavone monomer, could be extracted from the plant *Erigeron karvinskianus* and utilized as a traditional natural supplement. Scutellarein has IC50 values for SARS-CoV and HIV of 0.86 M and 2.50 M, respectively (Wang and Yang 2021). Scutellarein, dihydromyricetin, and quercetagetin (Fig. 5.8) are three natural flavonoids that have been reported by Liu et al. to efficiently suppress SARS-CoV-2 3CLpro activity in vitro and strongly reduce the replication of SARS-CoV-2 in vero cells, with IC50 values of 5.80 M, 1.27 M, and 1.20 M, respectively (Liu et al. 2020).

In addition to being immunomodulatory and adaptogenic substances, natural polysaccharides and terpenoids are also known for their antiviral,

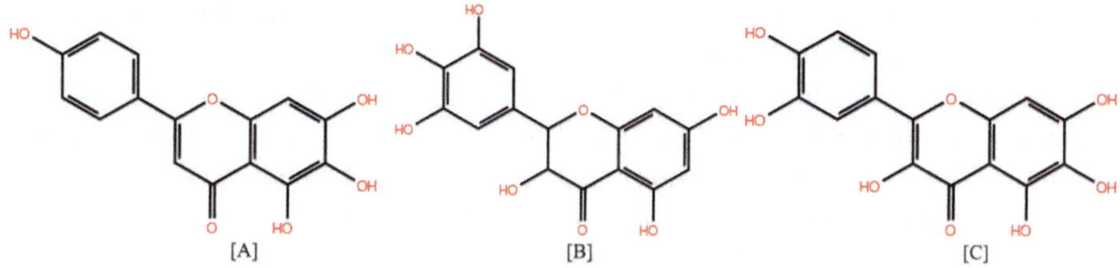


**Fig. 5.8** Structure of scutellarein (**a**), dihydromyricetin (**b**), and quercetagetin (**c**)

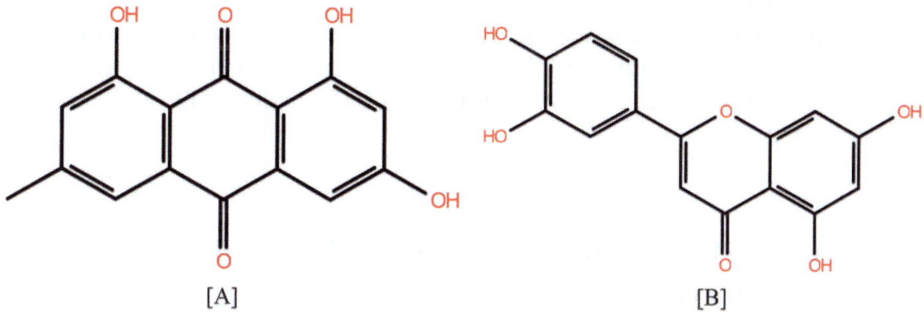


**Fig. 5.9** Structure of emodin (**a**) and luteolin (**b**)

immunomodulatory, anticancer, and anticoagulant bioactivities. Additionally, a number of plant triterpenes, including dammarenediol-II, hydroxyhopanone, ursonic acid, shoic acid, eichlerianic acid, and hydroxyoleanonic lactone, are crucial in the regulation of cellular metabolism (Llivisaca-Contreras et al. 2021). The *in silico* study showed that the 6-desacetyl nimbinene, isomeldenin, nimocinol, beta-sitosterol, quercetin, components of *Azadirachta indica* inhibit the Spike protein of SARS CoV-2 by generating proper pose inside the active site (Navabshan et al. 2021).

## 5.4 COVID-19 Treatment with Indonesian Medicinal Plants

The compounds emodin and luteolin (Fig. 5.9), which are present in a number of edible plants across Indonesia, may be able to protect against or lessen viral infection. The interactions between ACE-2 receptors and the S-protein in SARS-CoV can be blocked by these compounds. *Aloe vera* (L.) *Burm. F.*, *Rheum officinale Baill.*, Cassia plants such as *Cassia alata L.* and *Senna obtusifolia* (*L.*) are among the edible Indonesian plants that are rich sources of emodin and are utilized in folk medicine, mainly in tropical areas of the world. The plants, *Apium graveolens L.*, *Elephantopus scaber L.*, *Allium cepa L.*, and *Brassica oleracea L.* are high in luteolin. Many investigations have reported the presence of emodin or luteolin in various mangrove plants (Illian et al. 2021).

Potential antiviral Indonesian herbal candidates for SARS-CoV-2 include hesperidin, kaempferol-3,4′-di-O-methyl ether (Ermanin); myricetin-3-glucoside, peonidin 3-(4′-arabinosylglucoside); quercetin 3-(2G-rhamnosylrutinoside); and rhamnetin 3-mannosyl-(1–2)-alloside. The plants for COVID-19 prevention that include those constituents are *Moringa oleifera* and *Psidium guajava (*Erlina et al. 2022*)*. In a study, the Indonesian plants' compounds were screened and found 3-O-caffeoylquinic acid, justicidin D, 10-methoxycamptothecin, and inoxanthone as the potential SARS-CoV-2 RdRp inhibitors, emetine and carbine as potential SARS-CoV-2 spike protein inhibitors, and quercetin, quercetin 3-O-methyl ether, and dihydroquercetin as 3CLpro SARS-CoV-2 inhibitors. In vitro tests show that lycorine exhibits antiviral activity against SARS-CoV-2 with an EC50 of 0.31 μM (Alharthi et al. 2021).

## 5.5 COVID-19 Treatment with Ayurvedic Metal Nanoparticles

By preventing viral binding and entry into the cell, reducing viral replication, and directly killing viruses, nanoparticle (NP) based medications can fight against viral infections. For the treatment of COVID-19, various metal NPs, polylactic acid, etc. are frequently applied (Yasamineh et al. 2022). In modern science, green synthesized nanoparticles can be compared to the bhasmas mentioned in Ayurveda. Plant extracts are used to develop bhasma, which has high therapeutic value and is biocompatible, eco-friendly, stable, safe, and effective (Sreelakshmi et al. 2021). The significant antiviral potential is shown by various metal nanoparticles against a wide range of viral infections. Coronavirus can be successfully combatted with iron, gold, silver, copper, zinc, and other metal oxide nanoparticles. Disrupting the coronavirus's outer layers is one potential method of action for the metal nanoparticles to combat it. For COVID-19 treatment, Swarna Bhasma, Rajata Bhasma, Tamra Bhasma, and Yashada Bhasma are suggested due to their capacity to lower plasma interleukins, interferons, and TNFα (Sarkar and Das Mukhopadhyay 2021). Auranofin, a gold compound may be able to reduce SARS-CoV-2 infection and related lung damage because of its antiviral, anti-inflammatory, and anti-ROS activity. Auranofin may have a major impact on SARS-CoV-2 protein synthesis by inhibiting redox enzymes like thioredoxin reductase and induction of ER stress (Rothan et al. 2020). Thioredoxin reductase inhibition restricts NFκB DNA binding and NFκB-dependent gene expression. Additionally, auranofin inhibits the number of steps in the NFκB-IL-6-STAT3 signaling pathway, homodimerization of TLR4 (toll-like receptor), and viral replication and significantly lowers the expression of inflammatory response proteins like IL-6 (Sonzogni-Desautels and Ndao 2021).

## 5.6 Plants Studied for COVID-19 Treatments

Some researchers have turned to plant-based treatment modalities due to the absence of precise effective medicines against SARS-CoV-2. This is because many medications are made from plant materials or the bioactive components of herbs. Since plant-based treatments showed intriguing efficacy against a variety of viruses by boosting immunity, there is a significant interest in the discovery of potential anti-COVID-19 herbal medications (Al-kuraishy et al. 2022). The Chinese medicinal plants being studied include the *Radix astragali* (*Astragalus propinquus* Schischkin), *Radix glycyrrhizae* (*Glycyrrhiza uralensis* Fisch. ex DC.), *Radix saposhnikoviae* (*Saposhnikovia divaricata* (Turcz.) Schischk.), *Rhizoma atractylodis Macrocephala* (Chikowe et al. 2021). *Teucrium viscidum*, *Artemisia vulgaris*, *Verbena officinalis*, *Aster indicus*, *Pseudelephantopus spicatus*, *Hedyotis acutangula*, *Cyclea barbata*, *Alpinia malaccensis*, *Pothos scandens*, *Mimosa pigra*, *Plantago major*, *Artemisia verlotiorum*, *Tadehagi triquetrum*, *Blumea balsamifera*, *Zingiber ottensii*, *Ricinus communis*, *Tacca chantrieri*, *Zingiber officinale*, *Elephantopus scaber*, *Biancaea sappan*, *Scoparia dulcis*, *Cyclea barbata*, *Tinospora crispa*, *Celtis tetrandra*, and *Centella asiatica* are medicinal plants used for treatment of symptoms of COVID-19 among Hmong and Karen (Phumthum et al. 2021). Plants such as *Eucalyptus globulus Labill, Piper aduncum, Matricaria recutita*, and *Erythroxylum coca* were used to treat COVID-19-related respiratory problems (Villena-Tejada et al. 2021). *Thymus vulgaris*, *Achyranthes bidentata*, *Cinnamomum cassia*, *Cydonia oblonga*, *Embelin ribes*, *Justicia adhatoda*, *Momordica charantia*, *Withania somnifera*, *Althea officinalis*, and *ginseng* were used for a severe infection of COVID-19.

*Azadirachta indica* (Neem), the acetone–water extract of neem should be able to eliminate SARS-CoV-2 from vascular endothelium. If this procedure is successful, COVID-19 may not be circumvented by the formation of endothelial dysfunction or implied uncontrolled inflammation (Eze et al. 2022). *Gymnanthemum amygdalinum* increases the number of $CD4^+$ as well as white blood cells. Ethanolic extracts of *Nigella sativa* seeds, through the downregulation of the gene expression of various leukocyte transient receptor proteins (TRP) like TRPA1, TRPC4, TRPM6, TRPM7, TRPM8, and TRPV4 genes, in mice studies have confirmed anti-corona virus species of MHVA59 (mouse hepatitis virus-A59). *Eurycoma longifolia* (has NFκB inhibitory activity), *Mentha piperita Allium cepa*, *Malva sylvestris*, *Isatis indigotica*, *Psoralea corylifolia*, *Glycyrrhiza glabra* (Demeke et al. 2021) are medicinal plants which were studied or used against SARS CoV-2.

## 5.7 Phytoconstituents Being Studied for COVID-19 Treatment

The list of phytochemicals, targets, and references is given in Table 5.1.

**Table 5.1** Studied phytoconstituents against COVID-19

| Targets | Phytoconstituents | Study | References |
|---|---|---|---|
| Spike (S) proteins | Resveratrol | *In silico* | Islam et al. (2021), Wahedi et al. (2020) |
| | Saikosaponins U and V | *In silico* | Sinha et al. (2020) |
| | Luteolin-7-glucoside-3′-glucuronide | *In silico* | Prasanth et al. (2021) |
| | Melitric acid A | *In silico* | Prasanth et al. (2021) |
| Main protease (Mpro) | Withaferin A | *In silico* | Sudeep et al. (2020) |
| | Rutin | *In silico* | Johnson et al. (2021) |
| | Luteolin-7-glucoside-3′-glucuronide | *In silico* | Prasanth et al. (2021) |
| | Melitric acid A | In silico | Prasanth et al. (2021) |
| | Pycnanthuquinone C | *In silico* | Chtita et al. (2022) |
| | Pycnanthuquinone A | *In silico* | Chtita et al. (2022) |
| | Berberine | *In silico* | Chowdhury (2021) |
| RNA-dependent RNA polymerase | Theaflavin | *In silico* | Lung et al. (2020) |

## 5.8 Risks Associated with the Incorrect Use of Natural Products

Although several plant species have the ability to lessen or decrease COVID-19 symptoms, clinical investigations are still required to confirm their potential health benefits and uncover any negative effects. For example, despite the purported health benefits of ginkgo (*Ginkgo biloba*), at large doses can impact those who have peptic ulcers and clotting issues and increase cerebral blood flow (Llivisaca-Contreras et al. 2021). TCM for COVID-19 has the traits of many components and multiple targets, which were involved in various parts of the disease and had significant therapeutic value. TCM, however, also has certain drawbacks. For instance, it was challenging to detail the process behind TCM formulae because they included numerous components and targets yet were not targeted while exerting their therapeutic effect (Kang et al. 2022).

## 5.9 Conclusions

Research on the antiviral potential of herbal products has increased in the aftermath of the emergence of SARS-CoV-2 and the ongoing COVID-19 global pandemic. This is mainly due to the current shortage of efficient, secure, widely accessible antiviral medications, as well as the fact that herbal remedies are generally affordable and simple to transport without the need for special storage conditions. Furthermore, the widespread use of Ayurvedic remedies in India provided a chance to test these therapies in carefully managed and controlled clinical studies. Several possible traditional Indian medicinal plants and clinical studies of traditional Indian medicinal plants are now listed in several clinical trial databases for the development of pharmacological treatments against COVID-19. *Ocimum tenuiflorum* and *Tinospora cordifolia* are the two most promising traditional medicinal plants used in India for the treatment of COVID-19 because they showed strong immunomodulatory, thrombolytic, and binding affinity. Acute respiratory infections are known to be caused by a variety of viruses and several plants, including *Achyranthes bidentata*, *Cinnamomum cassia*, *Cydonia oblonga*, *Embelin ribes*, *Justicia adhatoda*, *Momordica charantia*, *Withania somnifera*, and *Zingiber officinale*, have strong antiviral properties. These herbs also have significant thrombolytic and immunomodulatory properties. Several *in silico* studies revealed that a traditional herb reported to bind with the crucial SARS-CoV-2 protease may show potential for the treatment of COVID-19. Although for the treatment of mild, severe, or critical COVID-19, some traditional medicinal plants cannot be recommended, more supportive studies are required for the recommendation. But these herbal remedies might be added with other treatments. Prospective cohort studies are urgently required to better understand the potential of herbal remedies. Since the population sample for COVID-19 is available, a series of prospective studies with well-designed research should soon begin yielding reliable evidence for this herbal treatment of COVID-19 or related respiratory infectious disorders. To ensure the safety and effectiveness of traditional medicine, in vitro and in vivo tests are required.

**Acknowledgments** The authors are thankful to the Indian Institute of Information Technology Allahabad, Prayagraj-India. Shristi Modanwal is grateful for the scholarship to pursue a doctorate from the Ministry of Education, Government of India.

## References

Ahmad I, Pawara R, Surana S, Patel H (2021) The repurposed ACE2 inhibitors: SARS-CoV-2 entry blockers of Covid-19. Top Curr Chem (Z) 379:40

Alam S, Sarker MMR, Afrin S, Richi FT, Zhao C, Zhou J-R, Mohamed IN (2021) Traditional herbal medicines, bioactive metabolites, and plant products against COVID-19: update on clinical trials and mechanism of actions. Front Pharmacol 12:671498

Alharthi N, Alsaeed M, Alsharif M, Almalki M, Alshehri W, Prabahar K (2021) Assessment of pharmacists' perception toward continuing education. J Adv Pharm Technol Res 12:368

Al-kuraishy HM, Al-Fakhrany OM, Elekhnawy E, Al-Gareeb AI, Alorabi M, De Waard M, Albogami SM, Batiha GE-S (2022) Traditional herbs against COVID-19: back to old weapons to combat the new pandemic. Eur J Med Res 27:186

Al-Romaima A, Liao Y, Feng J, Qin X, Qin G (2020) Advances in the treatment of novel coronavirus disease (COVID-19) with Western medicine and traditional Chinese medicine: a narrative review. J Thorac Dis 12:6054–6069

Amawi H, Abu Deiab GI, Aljabali AAA, Dua K, Tambuwala MM (2020) COVID-19 pandemic: an overview of epidemiology, pathogenesis, diagnostics and potential vaccines and therapeutics. Ther Deliv 11:245–268

Artese A, Svicher V, Costa G et al (2020) Current status of antivirals and druggable targets of SARS CoV-2 and other human pathogenic coronaviruses. Drug Resist Updat 53:100721

Babaei F, Nassiri-Asl M, Hosseinzadeh H (2020) Curcumin (a constituent of turmeric): new treatment option against COVID-19. Food Sci Nutr 8:5215–5227

Benarba B, Pandiella A (2020) Medicinal plants as sources of active molecules against COVID-19. Front Pharmacol 11:1189

Capodice JL, Chubak BM (2021) Traditional Chinese herbal medicine-potential therapeutic application for the treatment of COVID-19. Chin Med 16:24

Charan J, Bhardwaj P, Dutta S, Kaur R, Bist SK, Detha MD, Kanchan T, Yadav D, Mitra P, Sharma P (2021) Use of complementary and alternative medicine (CAM) and home remedies by COVID-19 patients: a telephonic survey. Ind J Clin Biochem 36:108–111

Chen X, Yin Y-H, Zhang M-Y, Liu J-Y, Li R, Qu Y-Q (2020) Investigating the mechanism of ShuFeng JieDu capsule for the treatment of novel coronavirus pneumonia (COVID-19) based on network pharmacology. Int J Med Sci 17:2511–2530

Chen J, Lin S, Niu C, Xiao Q (2021) Clinical evaluation of Shufeng Jiedu capsules combined with umifenovir (Arbidol) in the treatment of common-type COVID-19: a retrospective study. Expert Rev Respir Med 15:257–265

Chikhale RV, Sinha SK, Khanal P, Gurav NS, Ayyanar M, Prasad SK, Wanjari MM, Patil RB, Gurav SS (2021) Computational and network pharmacology studies of Phyllanthus emblica to tackle SARS-CoV-2. Phytomedicine Plus 1:100095

Chikowe I, Mtewa AG, Tembo D, Smith D, Ibrahim E, Mwamatope B, Nkhungulu J, Kumpalume P, Kumpalume P, Maroyi A (2021) Potential of Malawi's medicinal plants in Covid-19 disease management: a review. Mal Med J 33:85–107

Chowdhury P (2021) In silico investigation of phytoconstituents from Indian medicinal herb 'Tinospora cordifolia (giloy)' against SARS-CoV-2 (COVID-19) by molecular dynamics approach. J Biomol Struct Dyn 39:6792–6809

Chtita S, Fouedjou RT, Belaidi S et al (2022) In silico investigation of phytoconstituents from Cameroonian medicinal plants towards COVID-19 treatment. Struct Chem 33:1799–1813

Chu L, Huang F, Zhang M, Huang B, Wang Y (2021) Current status of traditional Chinese medicine for the treatment of COVID-19 in China. Chin Med 16:63

de Andrade Santos I, Grosche VR, Bergamini FRG, Sabino-Silva R, Jardim ACG (2020) Antivirals against coronaviruses: candidate drugs for SARS-CoV-2 treatment? Front Microbiol 11:1818

Demeke CA, Woldeyohanins AE, Kifle ZD (2021) Herbal medicine use for the management of COVID-19: a review article. Metabol Open 12:100141

Devpura G, Tomar BS, Nathiya D, Sharma A, Bhandari D, Haldar S, Balkrishna A, Varshney A (2021) Randomized placebo-controlled pilot clinical trial on the efficacy of ayurvedic treatment regime on COVID-19 positive patients. Phytomedicine 84:153494

Ding Y, Zeng L, Li R, Chen Q, Zhou B, Chen Q et al (2017) The Chinese prescription lianhuaqingwen capsule exerts anti-influenza activity through the inhibition of viral propagation and impacts immune function. BMC Complement Altern Med 17(1):130. https://doi.org/10.1186/s12906-017-1585-7

Donma MM, Donma O (2020) The effects of allium sativum on immunity within the scope of COVID-19 infection. Med Hypotheses 144:109934

Duan N, Liu B, Li X, Xiong Y, Li L, Ma Y, Lu C (2022) Traditional Chinese medicine in treating children with coronavirus disease 2019: a scoping review. Front Pediatr 10:935551

Enmozhi SK, Raja K, Sebastine I, Joseph J (2020) Andrographolide as a potential inhibitor of SARS-CoV-2 main protease: an in silico approach. J Biomol Struct Dyn:1–7

Erlina L, Paramita RI, Kusuma WA et al (2022) Virtual screening of Indonesian herbal compounds as COVID-19 supportive therapy: machine learning and pharmacophore modeling approaches. BMC Complement Med Ther 22:207

Eze MO, Ejike CECC, Ifeonu P, Udeinya IJ, Udenigwe CC, Uzoegwu PN (2022) Anti-COVID-19 potential of Azadirachta indica (Neem) leaf extract. Scient Afric 16:e01184

Faheem, Kumar BK, Sekhar KVGC, Kunjiappan S, Jamalis J, Balaña-Fouce R, Tekwani BL, Sankaranarayanan M (2020) Druggable targets of SARS-CoV-2 and treatment opportunities for COVID-19. Bioorg Chem 104:104269

Fan S-J, Liao J-K, Wei L, Wang B-Y, Kai L, Tan D-X (2022) Treatment efficacy of Lianhua Qingwen capsules for early-stage COVID-19. Am J Transl Res 14:1332–1338

Gao L, Xu J, Chen S (2020) In silico screening of potential Chinese herbal medicine against COVID-19 by targeting SARS-CoV-2 3CLpro and angiotensin converting enzyme II using molecular docking. Chin J Integr Med 26:527–532

Huang Y-F, Bai C, He F, Xie Y, Zhou H (2020) Review on the potential action mechanisms of Chinese medicines in treating Coronavirus Disease 2019 (COVID-19). Pharmacol Res 158:104939. https://doi.org/10.1016/j.phrs.2020.104939

Illian DN, Siregar ES, Sumaiyah S, Utomo AR, Nuryawan A, Basyuni M (2021) Potential compounds from several Indonesian plants to prevent SARS-CoV-2 infection: a mini-review of SARS-CoV-2 therapeutic targets. Heliyon 7:e06001

Islam F, Bibi S, Meem AFK et al (2021) Natural bioactive molecules: an alternative approach to the treatment and control of COVID-19. IJMS 22:12638

Jalali A, Dabaghian F, Akbrialiabad H, Foroughinia F, Zarshenas MM (2021) A pharmacology-based comprehensive review on medicinal plants and phytoactive constituents possibly effective in the management of COVID -19. Phytother Res 35:1925–1938

Jindal D, Rani V (2022) In silico studies of Phytoconstituents from Piper longum and Ocimum sanctum as ACE2 and TMRSS2 inhibitors: strategies to combat COVID-19. Appl Biochem Biotechnol 195:2618. https://doi.org/10.1007/s12010-022-03827-6

Johnson TO, Adegboyega AE, Adeyemi OE, Okonkwo FO, Ejembi SA, Oche J-RI, Oluwarotimi EA, Bashar F, Johnson GI, Adeyemi OS (2021) Computer-aided identification of bioactive compounds of Azadirachta indica (Neem) with potential activity against SARS-CoV-2 main protease. https://doi.org/10.21203/rs.3.rs-841180/v1

Kang X, Jin D, Jiang L et al (2022) Efficacy and mechanisms of traditional Chinese medicine for COVID-19: a systematic review. Chin Med 17:30

Khubber S, Hashemifesharaki R, Mohammadi M, Gharibzahedi SMT (2020) Garlic (Allium sativum L.): a potential unique therapeutic food rich in organosulfur and flavonoid compounds to fight with COVID-19. Nutr J 19:124

Koshak AE, Koshak EA, Mobeireek AF, Badawi MA, Wali SO, Malibary HM, Atwah AF, Alhamdan MM, Almalki RA, Madani TA (2021) Nigella sativa for the treatment of COVID-19: an open-label randomized controlled clinical trial. Complement Ther Med 61:102769

Lee DYW, Li QY, Liu J, Efferth T (2021) Traditional Chinese herbal medicine at the forefront battle against COVID-19: clinical experience and scientific basis. Phytomedicine 80:153337

Liu H, Ye F, Sun Q, Liang H, Li C, Lu R, Huang B, Tan W, Lai L (2020) Scutellaria baicalensis extract and baicalein inhibit replication of SARS-CoV-2 and its 3C-like protease in vitro. https://doi.org/10.1101/2020.04.10.035824

Llivisaca-Contreras SA, Naranjo-Morán J, Pino-Acosta A, Pieters L, Vanden Berghe W, Manzano P, Vargas-Pérez J, León-Tamariz F, Cevallos-Cevallos JM (2021) Plants and natural products with activity against various types of coronaviruses: a review with focus on SARS-CoV-2. Molecules 26:4099

Lung J, Lin Y, Yang Y, Chou Y, Shu L, Cheng Y, Liu HT, Wu C (2020) The potential chemical structure of anti-SARS-CoV-2 RNA-dependent RNA polymerase. J Med Virol 92:693–697

Luo L, Jiang J, Wang C, Fitzgerald M, Hu W, Zhou Y, Zhang H, Chen S (2020) Analysis on herbal medicines utilized for treatment of COVID-19. Acta Pharm Sin B 10:1192–1204

Lyu M, Fan G, Xiao G et al (2021) Traditional Chinese medicine in COVID-19. Acta Pharm Sin B 11:3337–3363

Mahaboob Ali AA, Bugarcic A, Naumovski N, Ghildyal R (2022) Ayurvedic formulations: potential COVID-19 therapeutics? Phytomedicine Plus 2:100286

Mathpati MM, Albert S, Porter JDH (2020) Ayurveda and medicalisation today: the loss of important knowledge and practice in health? J Ayurveda Integrat Med 11:89–94

Mithilesh S, Raghunandan D, Suresh PK (2022) In-silico identification of natural compounds from traditional medicine as potential drug leads against SARS-CoV-2 through virtual screening. Proc Natl Acad Sci, India, Sect B Biol Sci 92:81–87

Mulpuru V, Mishra N (2021) Computational identification of SARS-CoV-2 inhibitor in Tinospora cordifolia, Cinnamomum zeylanicum and Myristica fragrans. Virus Dis 32:511–517

Murugesan S, Kottekad S, Crasta I, Sreevathsan S, Usharani D, Perumal MK, Mudliar SN (2021) Targeting COVID-19 (SARS-CoV-2) main protease through active phytocompounds of ayurvedic medicinal plants – Emblica officinalis (Amla), Phyllanthus niruri Linn. (Bhumi Amla) and Tinospora cordifolia (Giloy) – a molecular docking and simulation study. Comput Biol Med 136:104683

Naserghandi A, Allameh SF, Saffarpour R (2020) All about COVID-19 in brief. New Microb New Infect 35:100678

Nassar A, Ibrahim IM, Amin FG et al (2021) A review of human coronaviruses' receptors: the host-cell targets for the crown bearing viruses. Molecules 26:6455

Navabshan I, Sakthivel B, Pandiyan R, Antoniraj MG, Dharmaraj S, Ashokkumar V, Khoo KS, Chew KW, Sugumaran A, Show PL (2021) Computational lock and key and dynamic trajectory analysis of natural Biophors against COVID-19 spike protein to identify effective Lead molecules. Mol Biotechnol 63:898–908

Orhan IE, Senol Deniz FS (2020) Natural products as potential leads against coronaviruses: could they be encouraging structural models against SARS-CoV-2? Nat Prod Bioprospect 10:171–186

Parthasarathy R, Mathew M, Koshy P, Babu M, Abraham G (2021) Traditional medicines prescribed for prevention of COVID -19: use with caution. Nephrology 26:961–964

Patel A, Rajendran M, Shah A, Patel H, Pakala SB, Karyala P (2022) Virtual screening of curcumin and its analogs against the spike surface glycoprotein of SARS-CoV-2 and SARS-CoV. J Biomol Struct Dyn 40:5138–5146

Paudyal V, Sun S, Hussain R, Abutaleb MH, Hedima EW (2022) Complementary and alternative medicines use in COVID-19: a global perspective on practice, policy and research. Res Soc Adm Pharm 18:2524–2528

Phumthum M, Nguanchoo V, Balslev H (2021) Medicinal plants used for treating mild Covid-19 symptoms among Thai Karen and Hmong. Front Pharmacol 12:699897

Prasanth DSNBK, Murahari M, Chandramohan V et al (2021) In-silico strategies of some selected phytoconstituents from Melissa officinalis as SARS CoV-2 main protease and spike protein (COVID-19) inhibitors. Mol Simul 47:457–470

Rajagopal K, Varakumar P, Baliwada A, Byran G (2020) Activity of phytochemical constituents of Curcuma longa (turmeric) and Andrographis paniculata against coronavirus (COVID-19): an in silico approach. Fut J Pharm Sci 6:104

Rattis BAC, Ramos SG, Celes MRN (2021) Curcumin as a potential treatment for COVID-19. Front Pharmacol 12:675287

Ren J, Zhang A-H, Wang X-J (2020) Traditional Chinese medicine for COVID-19 treatment. Pharmacol Res 155:104743

Rothan HA, Stone S, Natekar J, Kumari P, Arora K, Kumar M (2020) The FDA-approved gold drug auranofin inhibits novel coronavirus (SARS-COV-2) replication and attenuates inflammation in human cells. Virology 547:7–11

Runfeng L, Yunlong H, Jicheng H, Weiqi P, Qinhai M, Yongxia S et al (2020) Lianhuaqingwen exerts anti-viral and anti-inflammatory activity against novel coronavirus (SARS-CoV-2). Pharmacol Res 156:104761. https://doi.org/10.1016/j.phrs.2020.104761

Sagar V, Kumar AH (2020) Efficacy of natural compounds from Tinospora cordifolia against SARS-CoV-2 protease, surface glycoprotein and RNA polymerase. BEMS Rep 6:6–8

Sarkar PK, Das Mukhopadhyay C (2021) Ayurvedic metal nanoparticles could be novel antiviral agents against SARS-CoV-2. Int Nano Lett 11:197–203

Shahrajabian MH, Sun W, Shen H, Cheng Q (2020) Chinese herbal medicine for SARS and SARS-CoV-2 treatment and prevention, encouraging using herbal medicine for COVID-19 outbreak. Acta Agric Scand Sect B Soil Plant Sci 70:437–443

Shekh S, Reddy KKA, Gowd KH (2021) In silico allicin induced S-thioallylation of SARS-CoV-2 main protease. J Sulf Chem 42:109–120

Shree P, Mishra P, Selvaraj C, Singh SK, Chaube R, Garg N, Tripathi YB (2022) Targeting COVID-19 (SARS-CoV-2) main protease through active phytochemicals of ayurvedic medicinal plants – Withania somnifera (Ashwagandha), Tinospora cordifolia (Giloy) and Ocimum sanctum (Tulsi) – a molecular docking study. J Biomol Struct Dyn 40:190–203

Singh RS, Singh A, Kaur H et al (2021) Promising traditional Indian medicinal plants for the management of novel coronavirus disease: a systematic review. Phytother Res 35:4456–4484

Sinha SK, Shakya A, Prasad SK, Singh S, Gurav NS, Prasad RS, Gurav SS (2020) An in-silico evaluation of different Saikosaponins for their potency against SARS-CoV-2 using NSP15 and fusion spike glycoprotein as targets. J Biomol Struct Dyn:1–12

Solnier J, Fladerer J-P (2021) Flavonoids: a complementary approach to conventional therapy of COVID-19? Phytochem Rev 20:773–795

Song Z, Xu Y, Bao L, Zhang L, Yu P, Qu Y, Zhu H, Zhao W, Han Y, Qin C (2019) From SARS to MERS, thrusting coronaviruses into the spotlight. Viruses 11:59

Sonzogni-Desautels K, Ndao M (2021) Will Auranofin become a Golden new treatment against COVID-19? Front Immunol 12:683694

Sreelakshmi S, Vineeth PK, Mohanan A, Ramesh NV (2021) Ayurvedic bhasma and synthesized nanoparticles: a comparative review. Mater Today Proc 46:3079–3083

Sudeep HV, Gouthamchandra K, Shyamprasad K (2020) Molecular docking analysis of Withaferin a from Withania somnifera with the Glucose regulated protein 78 (GRP78) receptor and the SARS-CoV-2 main protease. Bioinformation 16:411–417

Sun X-H, Zhang S, Yang Z, Chen Z-L, Yue S-J, Zhang S, Tang Y-P (2022) Efficacy and safety of Lianhua Qingwen for patients with COVID-19: a systematic review and meta-analysis. Chin J Integr Med 28:650–660

Umeta Chali B, Melaku T, Berhanu N, Mengistu B, Milkessa G, Mamo G, Alemu S, Mulugeta T (2021) Traditional medicine practice in the context of COVID-19 pandemic: community claim in Jimma zone, Oromia, Ethiopia. IDR 14:3773–3783

Varnasseri M, Siahpoosh A, Hoseinynejad K, Amini F, Karamian M, Yad MJY, Cheraghian B, Khosravi AD (2022) The effects of add-on therapy of Phyllanthus Emblica (Amla) on laboratory confirmed COVID-19 cases: a randomized, double-blind, controlled trial. Complement Ther Med 65:102808

Villena-Tejada M, Vera-Ferchau I, Cardona-Rivero A, Zamalloa-Cornejo R, Quispe-Florez M, Frisancho-Triveño Z, Abarca-Meléndez RC, Alvarez-Sucari SG, Mejia CR, Yañez JA (2021) Use of medicinal plants for COVID-19 prevention and respiratory symptom treatment during the pandemic in Cusco, Peru: a cross-sectional survey. PLoS One 16:e0257165

Wahedi HM, Ahmad S, Abbasi SW (2020) Stilbene-based natural compounds as promising drug candidates against COVID-19. J Biomol Struct Dyn 39:1–10

Wang Z, Yang L (2021) Chinese herbal medicine: fighting SARS-CoV-2 infection on all fronts. J Ethnopharmacol 270:113869

Wang J, Andrade-Cetto A, Echeverria J, Wardle J, Yen H-R, Heinrich M (2021) Editorial: ethnopharmacological responses to the coronavirus disease 2019 pandemic. Front Pharmacol 12:798674

Wu R, Wang L, Kuo H-CD et al (2020) An update on current therapeutic drugs treating COVID-19. Curr Pharmacol Rep 6:56–70

Xia L, Shi Y, Su J et al (2021) Shufeng Jiedu, a promising herbal therapy for moderate COVID-19: antiviral and anti-inflammatory properties, pathways of bioactive compounds, and a clinical real-world pragmatic study. Phytomedicine 85:153390

Xiao M, Tian J, Zhou Y et al (2020) Efficacy of Huoxiang Zhengqi dropping pills and Lianhua Qingwen granules in treatment of COVID-19: a randomized controlled trial. Pharmacol Res 161:105126

Yakhchali M, Taghipour Z, Mirabzadeh Ardakani M, Alizadeh Vaghasloo M, Vazirian M, Sadrai S (2021) Cinnamon and its possible impact on COVID-19: the viewpoint of traditional and conventional medicine. Biomed Pharmacother 143:112221

Yang Y, Islam MS, Wang J, Li Y, Chen X (2020) Traditional Chinese medicine in the treatment of patients infected with 2019-new coronavirus (SARS-CoV-2CoV-2): a review and perspective. Int J Biol Sci 16(10):1708. https://doi.org/10.7150/ijbs.45538

Yasamineh S, Kalajahi HG, Yasamineh P et al (2022) An overview on nanoparticle-based strategies to fight viral infections with a focus on COVID-19. J Nanobiotechnol 20:440

Yimer G, Ekuadzi E, Fasinu P, Melo AC, Pillai G (Colin) (2021) Traditional medicines for COVID-19: perspectives from clinical pharmacologists. Br J Clin Pharmacol 87:3455–3458

Zahedipour F, Hosseini SA, Sathyapalan T, Majeed M, Jamialahmadi T, Al-Rasadi K, Banach M, Sahebkar A (2020) Potential effects of curcumin in the treatment of COVID -19 infection. Phytother Res 34:2911–2920

Zhang T, Chen D (2008) Anticomplementary principles of a Chinese multiherb remedy for the treatment and prevention of SARS. J Ethnopharmacol 117:351–361

Zhang L, Liu Y (2020) Potential interventions for novel coronavirus in China: a systematic review. J Med Virol 92:479–490

Zhuang J, Dai X, Wu Q, Cai H, Fu X, Zhang W, Chen B (2021) A meta-analysis for Lianhua Qingwen on the treatment of coronavirus disease 2019 (COVID-19). Complement Ther Med 60:102754

# Chapter 6
# Traditional Herbal Medicines and Their Active Constituents in Combating SARS-CoV-2 Infection

**Shivangani, N. Jaya Lakshmi, Lemnaro Jamir, and P. Hariprasad**

## 6.1 Introduction

Since its initial appearance in December 2019, a novel severe acute respiratory syndrome coronavirus 2 (SARS-CoV-2) causing coronavirus disease 2019 (COVID-19) has rapidly become a medical emergency and a global crisis (Dhama et al. 2020). Viruses have caused the majority of epidemic diseases over the last century. In 1918, an H1N1 (influenza) virus carrying avian genes caused a similar pandemic in humans. There were more than 500 million cases of viral infection, and 50 million deaths, ushering in the twentieth century. It is widely regarded as the deadliest pandemic in human history. New $H_2N_2$ influenza also known as "Asian Flu" emerged in 1957–1958 that caused a pandemic and claimed 1.1 million lives globally (Piret and Boivin 2021). Numerous cases of an unusual viral respiratory infection were recorded in Wuhan, China, in December 2019. A new coronavirus linked to betacoronaviruses was discovered in clinical specimens in January 2020 (Hu et al. 2022). The novel coronavirus and illness were known by the designations SARS-CoV-2 and COVID-19. Additionally, on March 11, 2020, the World Health Organization (WHO) classified COVID-19 as a pandemic (Cucinotta and Vanelli 2020).

According to WHO figures, more than 6.6 million deaths and over 649 million confirmed cases have been reported (as of 18 December 2022). Almost every country and territory on the planet has been affected by the disease. As of June 2022, the USA had reported the most deaths (1,002,946), followed by Brazil (668,693), India (524,873), the Russian Federation (380,517), and Mexico

Shivangani · N. J. Lakshmi · L. Jamir · P. Hariprasad (✉)
Environmental Biotechnology Lab, Centre for Rural Development and Technology, Indian Institute of Technology Delhi, New Delhi, India
e-mail: phari@iitd.ac.in



J.-T. Chen (ed.), *Ethnopharmacology and Drug Discovery for COVID-19: Anti-SARS-CoV-2 Agents from Herbal Medicines and Natural Products*,
https://doi.org/10.1007/978-981-99-3664-9_6

(325,271). Individuals, families, food security, healthcare systems, industries, and economies in a number of countries are all feeling the strain. The effect of COVID-19 has been catastrophic, turning the lives of many people on their heads and leading to tragedy for some (Saxena 2020). The destructive nature of SARS-CoV-2 stems from its ability to spread among hosts. The primary means of transmission is by respiratory droplets generated by sneezing and coughing. These droplets may enter the nose or mouths of those nearby, or they may be inhaled into the lungs. Additionally, the infection can be spread through fomites and their infectious secretions (Boone and Gerba 2007).

## 6.2 Pathogenicity and Life Cycle of SARS-CoV-2

SARS-CoV-2 is a spherical, single-stranded positive-sense RNA (+ssRNA) betacoronavirus with approximately 30 kb of genomic length (Klein et al. 2020; Lu et al. 2020). It consists of four structural proteins (envelope (E), nucleocapsid (N), membrane (M), and spike (S)) and sixteen non-structural ones. At the 5′ end, the RNA replication enzyme is encoded. The proofreading function of non-structural protein 14 (nsp14) keeps the mutation rate low. Viral multiplication and amplification occur in the host cell's endoplasmic reticulum and Golgi apparatus, as does virion assembly. Exocytosis is the final stage of virus release from the cell, allowing the infection to spread further (Cascella et al. 2022; Klein et al. 2020).

The life cycle of SARS-CoV-2 in the host cell is classified into various phases. To begin with, the viral Spike-Fc binds to the bronchial epithelial ACE2 receptor to initiate the infection cycle. This process also causes ACE2 expression to decrease, resulting in severe acute respiratory failure (Kuba et al. 2005). Small intestinal epithelial cells, like lung epithelial cells, have the ACE2 receptor protein on their surface and may act as a second site for SARS-CoV pathogenesis. Endocytosis is triggered when the S protein binds to the ACE2 receptor, which allows the virus to enter the host cell. Transmembrane serine protease 2 (TMPRSS2) then cleaves the S2′ site of the SARS-CoV-2 S protein, enabling the virus to fuse with the host membrane and infect the airways (Bestle et al. 2020; Fraser et al. 2022). When the virus's membrane merges with the membrane of the host cell, the virus can inject its genetic material into the host. The RNA of the virus must be translated by the host's machinery to produce replication and structural proteins. RNA-dependent RNA polymerase (RdRp) is one of the non-structural proteins that result from disassembling the replication enzyme. By sending signals to the host's ribosomal machinery, viral RNA is translated into a functional RNA polymerase protein. The plus ssRNA is replicated by the viral RNA polymerase into the dsRNA. The N, S, M, and E proteins are transcribed from RNA that is located outside of the genome. N is processed in cytoplasmic protein, while S, M, and E are modified in the rough endoplasmic reticulum after translation. Viruses assemble their nucleocapsid (N) and structural proteins E, M, and S inside their envelopes (McBride et al. 2014).

The Golgi vesicle is responsible for the final stages of virion development, where the lipid envelope is added to the assembled viral components. Exocytosis is the process by which viruses mature enough to infect other cells are released from their host (Shereen et al. 2020; Iqbal et al. 2020).

## 6.3 Covid Associated Conditions

It may take weeks to months to recover from coronavirus disease; however, infection and symptoms may persist in a few people even after the initial infection. Although years have passed since the first occurrence of COVID-19, people still have long-term difficulties. According to Kamal et al. (2021), only 10.8 % of all individuals have no manifestation following disease healing, despite the majority of subjects having many symptoms and illnesses. Possible risks include heart failure, lung illness, and neurological conditions like stroke. Brain fog and exhaustion may be brought on by cytokines that impact the brain and pass the blood–brain barrier (BBB). These symptoms may indicate a postviral syndrome caused by COVID-19 and should be taken seriously.

Furthermore, the most commonly reported symptoms are reading difficulties, dry skin, insomnia, nonspecific myalgia, and elevated anxiety (Moldofsky and Patcai 2011). Interferon-gamma and interleukin-7 are produced when lymphatic drainage is disrupted. Proinflammatory cytokines can cross the BBB in circumventricular structures such as the hypothalamus, causing autonomic dysfunction that begins as a fever and progresses to chronic fatigue syndrome/myalgic encephalomyelitis (CFS/ME) (Hives et al. 2017). CFS/ME is characterised by abnormal sleep-wake cycles, cognitive dysfunction, and extreme, ongoing exhaustion. The "post-COVID-19 syndrome," which is characterised by long-term chronic fatigue and post-exercise neuroimmune discomfort, was also observed by several COVID patients (Perrin et al. 2020). Persistent neurological symptoms, olfactory dysfunctions, and lung symptoms and dysfunctions are also common in COVID patients (Salamanna et al. 2021). A more severe and potentially fatal link exists between ENT and coronavirus: invasive fungal sinusitis caused by mucormycosis (DeShazo et al. 1997).

## 6.4 Variants of Concerns SARS-CoV-2 (VOCs)

Since the initial revelation of COVID-19 in December 2019, a number of SARS-CoV-2 variants have been discovered. However, due to their global impact on public health, WHO designated a small number of mutant strains as variants of concern (VOCs) (Table 6.1). The great majority of these alterations were expressed by the spike glycoprotein. This category also calls for increased sequence surveillance, laboratory characterisation, and epidemiological research to determine disease transmission rate, clinical manifestation, reinfectivity threat, and immunity to existing

**Table 6.1** Brief description of existing SARS-CoV-2 variants of concern declared by WHO

| VOC | Strain designation | Reported country | Mutations | Key features | Reference |
|---|---|---|---|---|---|
| Alpha | B.1.1.7 | United Kingdom Dec 2020 | Δ69-70 del, Δ144 del, N501Y, T716I, P681H, D1118H, A570D, and S982A | Amplified affinity to host hACE2 receptor due to spike mutations; 43% to 82% more transmissible and associated with high mortality | Galloway et al. (2021), Davies et al. (2021) |
| Beta | B.1.351 | South Africa Dec 2020 | D80A, L18F, D215G, K417N, R246I, E484K, D614G, N501Y, and A701V | The danger of spreading GOT increased and the efficacy of plasma therapy in neutralising it has reduced | Tegally et al. (2021), Wang et al. (2021) |
| Gamma | P.1 | Brazil Jan 2021 | L18F, T20N, P26S, D138Y, R190S, H655Y, T1027I V1176, K417T, E484K, and N501Y | Reduced neutralisation effects by monoclonal antibody treatment | Faria et al. (2021) |
| Delta | B.1.617.2 | India Dec 2020 | T19R,156del, 157del, L452R, R158G, P681R, T478K, D614G and D950N | Increased risk of mortality owing to 2nd wave of COVID-19 | Singanayagam et al. (2022), Twohig et al. (2022) |
| Omicron | B.1.1.529 | South Africa Nov 2021 | V70del, E31del, R32del, S33del, N211del/L212I, H69del, Y143del, Y144del, Y145del, G204R, D3G, A63T G142D, Q19E, T95I, P13L, R203K, A67V, Y505H, Q493R, N501Y, Q498R, G496S, E484A, N440K, T478K, S477N, G446S, K417N, S375F, S373P, S371L, G339D | Mutations in Spike protein led to vaccine breakthroughs; many subvariants such as BA.1-5 were also identified | Vaughan (2021) |

vaccines (Vasireddy et al. 2021; Meites et al. 2022). Given the recent global spread of heavily mutated Omicron VOC and the expected increase in viral diversity, the WHO has created a new category titled "omicron subvariants under monitoring" to alert global public health authorities to which VOC lineages may require additional attention and surveillance. These include the BF.7 and XBB omicron subvariants, which currently dominate the epidemiological scene of SARS-CoV-2 infection due to their increased transmissibility, high resistance to antibody neutralisation, and fractional resistance to COVID-19 vaccine-induced immunity (Torjesen 2021; Velavan et al. 2023). In the long run, mutational variations may result in vaccine-resistant viral strains. Such strains outperform, particularly in immunocompromised patients with long transmission cycles. These elements lengthen the time it takes for natural selection to operate on this novel variant and increase the frequency of new mutations (Cascella et al. 2022).

## 6.5 Druggable Molecular Targets of SARS-CoV-2

Agents used to treat SARS-CoV-2 infection can prevent the virus entry into host cells by interfering with the host recognition system, inhibition of structural proteins to prevent viral self-assembly, acting on the RNA synthesis through viral RNA, and blocking vital viral enzymes involved in the replication. This section briefly discusses the various molecular targets involved in SARS-CoV-2 pathogenesis.

**Spike Protein** The interaction of SARS-CoV-2's type I transmembrane glycoprotein (spike protein, approx. 200 kDa) facilitates host cell recognition and entry. Initially synthesised as a monomer, this protein undergoes homotrimerization after post-translational glycosylation in the endoplasmic reticulum (Tang et al. 2020; Ou et al. 2020). The SARS-CoV-2 spike proteins are approximately 1273 AA long and comprise an N-terminal S1 domain that binds to host cell receptors and a C-terminal S2 domain that is involved in host cell membrane fusion (Wrapp et al. 2020). The S1 subunit has a receptor-binding domain (RBD) and an N-terminal domain (14-305 amino acids) (Liu et al. 2020; Xia et al. 2020). The RBD is made up of subdomains that generate the trimer and interact with the receptor (Lu et al. 2013; Li et al. 2005). The S2 subdomain is made up of two heptad repeat domains: a fusion peptide domain and a transmembrane domain. After the S1 domain of spike protein interacts with the host cell's receptor, the S2 domain goes through conformational changes. The heptad domains are important targets for fusion inhibitors because they work together to form a helical bundle core, which improves the tight contact between the cellular and viral membranes during fusion. Figure 6.1a depicts all of the structural components of spike glycoprotein. Small compounds that block Spike-ACE2 interaction via RBD (shown in Fig. 6.1b) are being investigated to prevent and inhibit viral attachment in host cells.

**Non-Structural Protein (NSP)** SARS-CoV-2 has 16 highly conserved NSPs that share structural similarities with other coronaviruses. While any NSP in principle

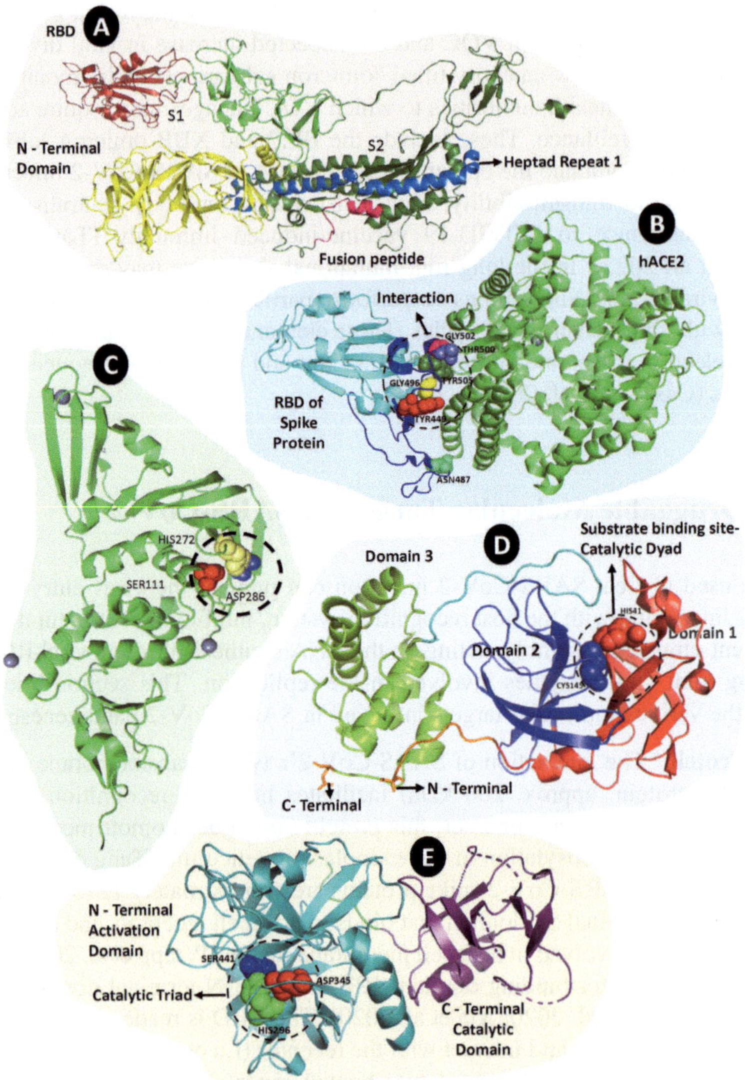


**Fig. 6.1** Cartoon structure of different SARS-CoV-2 and human proteins identified as potential drug targets along with their key/catalytic amino acid residues. (**a**) Spike glycoprotein (PDB ID: 7TGW). (**b**) Interactions between receptor-binding domain of spike-ACE2 (PDB ID: 6M0J). (**c**) $PL^{pro}$ (PDB ID:7CJM). (**d**) $M^{pro}$ (PDB ID:6LU7). (**e**) Interactions between receptor-binding domain of spike-transmembrane protease serine 2 (PDB ID: 7MEQ)

could be used as a drug target, the reported ligand, 3D crystal structure, and function in viral infection are particularly promising indicators of future success. Only the three chymotrypsin proteases ($3CL^{pro}$, nsp5), papain-like protease ($PL^{pro}$, nsp3),

RdRp (nsp12), and the methyltransferase-stimulating factor complex (NSP10 and NSP16) have crystal structures. RdRp, helicase, and two intracellular proteases, $M^{pro}$ and $PL^{pro}$, have all been extensively studied as potential therapeutic targets based on this logic.

**Papain-Like Protease ($PL^{pro}$)** PLpro is typically found in two copies, PL1pro and PL2pro, and has a total of 1945 amino acid residues (212 kDa) as part of NSP3, a large multidomain protein that is important to the RTC (Lei et al. 2018; Woo et al. 2010). PLpro's three-dimensional structure is shown in Figure 6.1c. The enzyme activity is found in NSP3 between a nucleic acid-binding domain and a SARS-specific domain (SUD/HVR). By cleaving peptide links, this cysteine protease releases NSP1, NSP2, and NSP3. Proteases Pp1a and Pp1b, whose substrate locations are located near the LXGG motif, are essential for PLpro proteolytic activity (Harcourt et al. 2004). NSP1 is a protein with 180 amino acid residues that constrains host translation by interacting with the 80S ribosome (Schubert et al. 2020; Thoms et al. 2020). NSP2 is a protein with 638 amino acid residues that may have an effect on host cell survival (Zhao et al. 2020). It is hypothesised that SARS-COV-2 PLpro has a deubiquitinating function because PLpro is highly conserved and present in all coronaviruses (Ratia et al. 2014). For its essential function in RTC assembly, viral polypeptide chain cleavage and maturation, and host immunity suppression, PLpro is an attractive therapeutic target (Osipiuk et al. 2021).

**Main Protease ($M^{pro}$)** The $M^{pro}$ responsible for the multiplication of SARS-CoV-2 is a cysteine protease with 306 residues and three domains. It is found in a dimer form, with both the N- and C-terminal catalytic sites present in each monomer (Lee et al. 2005). It is clear from the 3D structures of other CoVs that the 15-residue loop connecting domains II and III contains a Cys-His catalytic dyad and a Mpro substrate binding region located in the pocket between domains I and II (Fig. 6.1d) (Jin et al. 2020). The $M^{pro}$ or $3CL^{pro}$ first cleaves the polypeptide chain to generate mature enzymes, forming an RTC. In addition, it splits NSP4 and NSP16 off of the downstream end of peptide chain 1ab at 11 NSP sites. The maturation of NSPs is dependent on 3CLpro and is vital for the viral life cycle (Yang et al. 2005). Several molecules, peptides, and peptidomimetics with $3CL^{pro}$ inhibitory activity have been reported, with small molecules of plant origin discussed later in the manuscript.

**RNA-Dependent RNA Polymerase (RdRp)** RdRp, or nuclear envelope protein 12 (NSP12), is a component of the viral replication and transcription complex. Its a must-have for the transcription of viral RNA replication. While NSP12 has negligible catalytic activity on its own, when combined with cofactors NSP7 and NSP8, it becomes a powerful polymerase (Subissi et al. 2014; Kirchdoerfer and Ward 2019). Therefore, the NSP12-NSP7-NSP8 complex has been postulated to play a crucial function in viral propagation. The RdRp complex of SARS-CoV-2 includes a dimeric NSP8 subunit, a heterodimer complex, and a core catalytic NSP12 subunit. In addition, SARS-CoV-2 RdRp is structurally identical to SARS-CoV RdRp. A recently discovered RdRp mutation in Europe, 14408C $>$ T, that occurred via an unknown route has been linked to higher mutation rates than viral genomes from

Asia (Pachetti et al. 2020). Ribavirin, remdesivir, favipiravir, and hydroxychloroquine are four antiviral drugs that target RdRp and have been approved by the Food and Drug Administration (FDA) for use against HIV and other viruses; they can be used as rescue treatments for COVID-19. Repurposing current RdRp-based antiviral medicines for this purpose is likewise the subject of intensive study (Ko et al. 2020; Gangadharan et al. 2022; Xie et al. 2021).

**Human Angiotensin-Converting Enzyme-2 (hACE-2)** Spike proteins of SARS-CoV-2 are associated with hACE-2 for host cell recognition and entry, making hACE-2 a potential target in managing COVID-19 (Hoffmann et al. 2020; Zhou et al. 2020). Homeostasis of vascular processes is controlled by hACE2, a membrane protein, through its catalytic activity in the renin–angiotensin system (RAS) (Tikellis and Thomas 2012). Angiotensin II is the primary vasoactive protein produced by the RAS system. It produces angiotensin 1e7, an inhibitor of angiotensin II activity, and acts as a potent vasoconstrictor through its angiotensin II type I receptor (AT1R) (Clarke and Turner 2012). Increased ACE-2 cellular expression has been observed during long-term treatment with an angiotensin receptor blocker. Following virus attachment and membrane fusion, researchers found a decrease in ACE-2 levels. Since AT1R causes increased angiotensin expression due to increased ACE-1 action and inhibited ACE-2 action, it leads to significant pathophysiological changes in the lungs (Gurwitz 2020). Despite the mystery, it appears that treating COVID-19 with ARBs might be beneficial rather than harmful to lung function.

**Transmembrane Protease Serine 2 (TMPRSS2)** Proteases are enzymes that specifically and rapidly degrade proteins and peptides, thereby activating and regulating a wide range of fundamental physiological processes. As can be seen in Fig. 6.1e and 6.4, TMPRSS2 is primarily localised in the epithelial cell membrane of the lungs, where it plays a role in priming spike proteins. This mechanism induces the union of viral membrane with cellular membranes and their interactions. According to previous studies, SARS-CoV spike proteins need to be primed by TMPRSS2 before they can enter host cells (Matsuyama et al. 2010; Glowacka et al. 2011). One study indicates that camostat is an approved therapeutic TMRSS2 inhibitor that blocks TMPRSS2's priming activity (Hoffmann et al. 2020). Because of this, TMPRSS2-based small molecule therapies could be effective against COVID-19.

**Furin** The trans-Golgi network is home to the proprotein convertase furin (Feliciangeli et al. 2006). Furin is a proteinase that, when activated by an acidic pH, breaks specific precursor proteins with defined motifs, resulting in biologically active mature proteins that have been linked to a wide range of cardiovascular diseases, cancers, and COVID-19 infections. The viral spike glycoprotein must be proteolyzed at the S1/S2 and S2 cleavage regions before the virus can enter host cells (Wu et al. 2020). The proprotein convertase furin eases membrane fusion and viral replication by cleaving the S1/S2 site (PRRAR). Inhibition of furin reduced viral replication in Calu-3 cells, suggesting it may be a target (Bestle et al. 2020).

## 6.6 Conventional Management and Treatment Measures for COVID-19

Several nations have taken precautions against the virus by including social distancing and self-isolation, which is contact between people. The best ways to prevent the spread of disease are still the old standbys: washing one's hands often and wearing protective gear like a surgical mask (Chu et al. 2020; Sardar et al. 2020). National closures of public places were enacted in many countries as a means by which governments hoped to limit the spread of the disease by reducing people's opportunities for interpersonal contact (Balachandar et al. 2020). However, long-term physical distancing is difficult where some countries reported infection waves after relaxing the lockdown rules. Prolonged lockdowns also cause income loss and poverty, especially in developing nations (UNDP 2020; Bonaccorsi et al. 2020).

Because of the striking resemblance between their genomic sequences and that of SARSCoV-2, antiviral drugs originally developed to treat SARS and MERS are now routinely prescribed to treat SARS-CoV-2 infection. Extensive scientific studies on COVID-19 patients are currently being conducted with antiviral drugs like remdesivir, lopinavir, hydroxychloroquine, and dexamethasone (Adhikari et al. 2021; Lamontagne et al. 2020; Balkhair 2020). Due to their promising results, the WHO has recommended these medications (Adhikari et al. 2021; Lamontagne et al. 2020; Balkhair 2020). While hydroxychloroquine is frequently prescribed for asymptomatic patients, especially those caring for suspected or confirmed COVID-19 cases, dexamethasone is usually reserved for patients who are critically ill (Lamontagne et al. 2020). The majority of antiviral medications that are currently on the market also cause virus resistance, side effects, recurrence, and delayed response. This scenario dissatisfies numerous public health institutions, researchers, and healthcare practitioners who continue to seek a cure for COVID-19. Many research studies are focussed on medications and vaccines evaluated at the experimental and field levels around the world to combat the SARS-CoV-2 pandemic (Lamontagne et al. 2020; Alam et al. 2021). Antiviral medications used for treating SARS and MERS are today routinely prescribed to treat COVID-19 due to the significant similarity between their genomic sequence and SARSCoV-2. Dexamethasone is typically prescribed for severely ill patients, whereas hydroxychloroquine is commonly prescribed for asymptomatic conditions, particularly for those involved in the care of suspected or confirmed SARS-CoV-2 infection (Lamontagne et al. 2020). The majority of antiviral medications that are currently available also cause virus resistance, side effects, recurrence, and delayed response. Numerous public health organisations, academics, and medical professionals who are still working to find a cure for COVID-19 are unhappy with this situation. Numerous studies focusing on drugs and vaccines tested at the experimental and field levels worldwide are being carried out during the SARS-CoV-2 pandemic (Lamontagne et al. 2020; Alam et al. 2021).

## 6.7 Vaccines for COVID-19

Since the beginning of the current pandemic, there has been an increase in the international effort to develop vaccines to prevent SARS-CoV-2 infection. People of all ages receive vaccinations to build up their humoral and cell-mediated immunity to prevent illness. Currently, 13 vaccines have been developed and authorised/ approved, while 58 vaccines are through various rounds of testing. The covid vaccines most widely administered are discussed below,

**mRNA-1273 (Moderna TX, Inc) (Moderna TX, Inc)** A synthetic liposome-encapsulated mRNA that encodes for the full-length stabilised spike glycoprotein of the SARS-CoV-2 virus makes up this vaccine. It is capable of inducing a focused antiviral response against a spike protein. Since it lacks living or inactive pathogen subunits, it is also regarded as being relatively safe (Tu et al. 2020). In the USA, this mRNA vaccine is now widely used after receiving expedited FDA approval for phase II studies (Kaur and Gupta 2020).

**BNT162b1 (BioNTech-Fosun Pharma-Pfizer)** A crucial target of the virus-neutralising antibody, the trimerized SARS-CoV-2 RBD is encoded for by the codon-optimised mRNA vaccine known as BNT162b1. The RBD antigen is more immunogenic than other vaccines because it includes a trimerization domain. The mRNA is effectively transported because it is encased in ionizable cationic lipid nanoparticles with a diameter of 80 nm. According to phase 1/2 clinical trials, the concentration of IgG antibodies specific to RBD ranges from 8 to 46.3 times that of convalescent serum. This vaccine was assessed to have a geometric mean titre of SARS-CoV-2 neutralising antibodies between 1.8 and 2.8 times higher than that of convalescent serum (Mulligan et al. 2020).

**ChAdOx1 (University of Oxford) AstraZeneca** The recombinant adenovirus vaccine ChAdOx1 was created using a codon-improved S glycoprotein that has a leader sequence at the 5′ end that codes for tissue plasminogen activator. The tPA leader and the SARS-CoV-2 amino acid coding sequence (2–1273) were propagated in the shuttle plasmid. Between the Gateway recombination cloning site and the human cytomegalovirus promoter region, tetracycline-acting sites, and a polyadenylation signal from bovine growth hormone are all encoded on this transport plasmid (Falsey et al. 2021).

## 6.8 Role of Medicinal Plants and Their Secondary Metabolites in Managing COVID-19

The Earth is home to more than 500,000 species of plants, among which 10–15% are used as sources of pharmaceuticals and 10% as food (Borris 1996). A significant portion of the global population relies on plant-based medications and phytochemicals for primary healthcare (Farnsworth 2007). Throughout human history, separating and characterising biologically active compounds and molecules from nature have resulted in the development of new remedies, propelling the health care and pharmaceutical sectors forward (Pye et al. 2017). In China, India, Japan, Africa, and several other countries, ethnomedicine has been the primary therapeutic approach since antiquity (Chang and But 2014; Dev 1999; Schultes and Raffauf 1990).

In traditional medicine and folk remedies, secondary plant metabolites play a crucial part in treating a variety of ailments. Usually, primary metabolites are converted into secondary metabolites in small quantities during a firm growth phase and for a definite function. They give plants resistance to biotic and abiotic stress. Thus, defence mechanisms used by plants vary depending on their unique requirements and are influenced by climatic variations, physiological traits, and environmental factors (Ballhorn et al. 2009). They provided the active pharmaceutical ingredients for drugs used to treat conditions as diverse as migraine, malaria, nervous disorders, and cancer. Several of these chemicals have been utilised as whole plants or plant extracts for human consumption or medicinal purposes. The scientific rationale for using plant substances in traditional medicine from diverse ancient civilizations can be traced back to the finding that secondary metabolites have various biological effects. They can prevent plant diseases and have antibacterial, antifungal, anti-inflammatory, anti-allergic, and antiviral activities. The chemical structures of secondary plant metabolites are used to categorise them into distinct classes. The major classes of plant secondary metabolites consist of alkaloids, terpenes, phenolics, and saponins. Potent pharmacological activities of this class of secondary metabolites are frequently reported (Hussein and El-Anssary 2019).

Plants' ability to create a wide variety of secondary metabolites stifled protein and enzyme function and reduced viral propagation in several ways. For a long time, people have turned to medicinal plants for relief from and protection against many sorts of ailments, including respiratory virus infections (Kiran et al. 2020; Park et al. 2016; ul Qamar et al. 2020). It is advantageous to use these herbs to treat viral infections because they stop viral replication and, through their effects on the immune system and inflammation regulation, indirectly enhance the health of the host. While there have been numerous studies on the efficacy of medicinal plants against SARS-CoV and MERS-CoV, the same cannot be said for COVID-19 illness, for which the available research is mostly limited to in silico experiments with phytometabolites (Mohammadi and Shaghaghi 2020). Figure 6.2 shows the various classes of plant secondary metabolites that target the host and viral elements involved in SARS-CoV-2 pathogenesis.

In this chapter, a review of the relevant literature was done using several sources, including PubMed, Science Direct, Web of Science, Google Scholar, and Scopus, to determine the role of medicinal floras and secondary metabolites in combating SARS-CoV-2. Different combinations of keywords related to "antiviral plants", "plants used against COVID", "plants used against SARS-CoV-2", and "plant molecule inhibitors targeting SARS-CoV-2" were used to retrieve the relevant

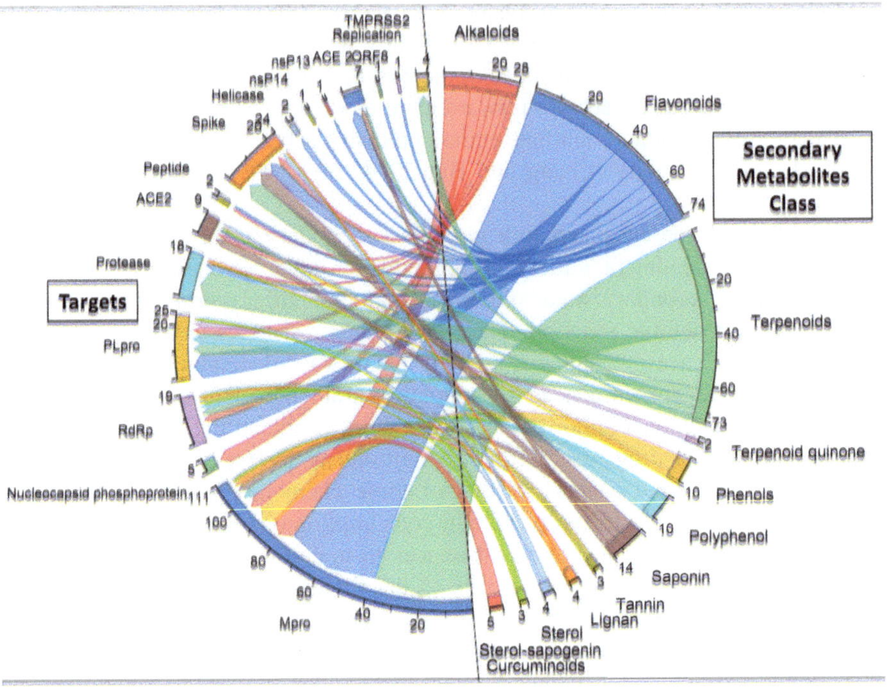


**Fig. 6.2** The chord graph indicates the class of secondary metabolites and their SARS-CoV-2 targets

publications. A total of 64 papers were referred, further divided into two parts "pre-covid" before 2019 and "post-covid" from 2019 to 2022 (depicted schematically in Fig. 6.3). Based on this survey, the highest reports of antiviral plants against COVID-19 were identified from India owing to its rich diversity in medicinal plants, followed by Korea and China, respectively. Plants from the Fabaceae family are found to be the key inhibitors for SARS-CoV-2, following the family Lamiaceae, Asteraceae, and Zingiberaceae, respectively.

### *6.8.1 Use of Phyto-Crudes in Treating COVID-19 and Their Mode of Action*

Many medicinal plants' crude extracts have shown antiviral efficacy against SARS-CoV, MERS-CoV, and SARS-CoV-2, which have the characters to advance drug development and is useful in treating COVID-19 (depicted in Fig. 6.4).

Wen et al. (2011) tested nearly 200 crude preparations of Chinese medicinal herbs in an in vitro study using Vero E6 cells to study the SARS-induced cytopathogenic effect (CPE). At dosages ranging from 25 to 200 g/ml, six herbal extracts from the

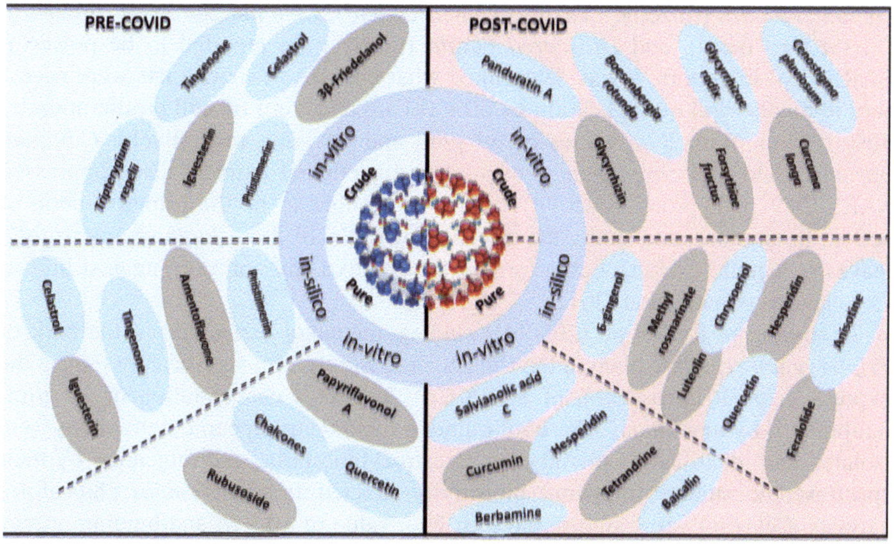


**Fig. 6.3** Pre- and post-COVID studies involving plants, extracts, and purified metabolites targeting many structural and non-structural proteins of the SARS/MERS virus

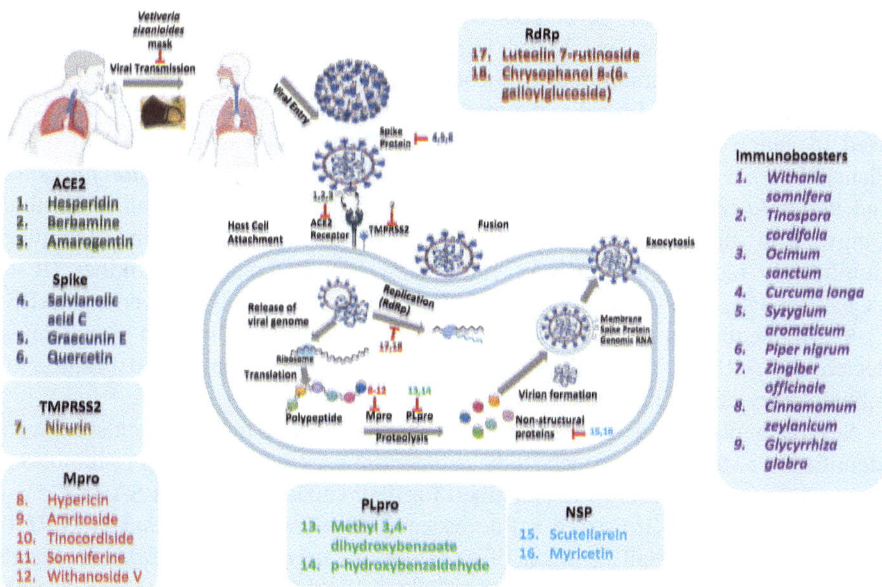


**Fig. 6.4** Different well-established drug targets of SARS-CoV-2 and potential plant secondary metabolites were reported against those targets

*Gentiana scabra* (rhizome), *Taxillus chinensis* (leaf), *Cibotium barometz* (rhizome), *Cassia tora* (seed), and *Dioscorea batata* (tuber) were reported to be powerful SARS-CoV inhibitors. It was determined what dosages of six extracts were necessary to hinder viral multiplication by 50% ($EC_{50}$) and Vero E6 cell proliferation by 50% ($CC_{50}$). With $IC_{50}$ values of 39 g/ml and 44 g/ml, respectively, *Cibotium barometz* and *Dioscorea batata* ethanol and methanol extracts also suppressed SARS-CoV 3CL protease activity, which is thought to be the primary mode of action. In an associated study, the terpene 3-friedelanol, from *Euphorbia neriifolia* leaves (ethanolic extract), outperformed the positive antiviral drug against human Coronavirus (Chang et al. 2012).

According to Ryu et al. (2010b)'s studies, methanol extracts from the bark of *Tripterygium regelii*, a plant in the Celastraceae family, can significantly reduce the expression of 3CLpro action of SARS-CoV-2, with a concentration of 30 g/mL inhibiting it by more than 70%. Four quinone-methide triterpenoid derivatives were isolated and identified as pristimerin, celastrol, iguesterin, and tingenone by their spectroscopic studies following bioactivity-directed fractionation of chloroform extracts. Celastrol was found to have an $IC_{50}$ value of 10.3 M and binding energy of $-9.58$ kcal/mol in additional in vitro and in silico studies with 3CLpro. Similar to this, pristimerin has a binding energy of $-9.87$ kcal/mol and an $IC_{50}$ value of 5.5 M, whereas tingenone and iguesterin had $IC_{50}$ values of 9.9 and 2.6 M and binding energies of $-9.75$ and $-9.18$ kcal/mol, respectively.

Several phyto-crudes were reported for their antiviral activity following the SARS-CoV-2 outbreak in 2020. The anti-SARS-CoV-2 activity of Lianhua Qingwen, a patented traditional Chinese medicine, was demonstrated by Runfeng et al. (2020). On Vero E6 cells, cytopathic effect and plaque reduction assays were used to gauge Lianhua Qingwen's antiviral effectiveness against SARS-CoV-2. Plants like *Forsythia fructus*, *Lonicera japonica*, *Ephedra sinica*, *Gypsum fibrosum*, and *Rhodiola crenulate*, among others, are included in the composition of Lianhua Qingwen. In support of this, Hu et al. (2022) studies found that Lianhua Qingwen significantly reduced the mRNA synthesis of proinflammatory cytokines (IL-6, CCL-2/MCP-1, TNF-α, and CXCL-10/IP-10) and successfully prevented SARS-CoV-2 replication in Vero E6 cells. Furthermore, Lianhua Qingwen significantly decreases SARS-COV-2 multiplication, modifies virus shape, and exhibits anti-inflammatory effects in vitro.

Seventeen plant products used in both contemporary and traditional cuisines were identified in a strategic approach study as potential SARS-CoV-2 3CLPro inhibitors. Plant methanolic extracts were examined for their capacity to inhibit SARS-CoV-2 3CLpro activity in vitro using a fluorescence resonance energy transfer (FRET) evaluation. At 500 μg/mL, extracts of rhizomes of *Curcuma longa* (turmeric), seeds of *Brassica nigra* (mustard), and *Diplotaxis erucoides* (wall rocket) significantly inhibited $3CL^{pro}$ activity, with residual enzyme activities of 0%, 9.4%, and 14.9%, respectively. Among all the plant extracts, turmeric exhibited a 15.74 μg/mL of $IC_{50}$ value, indicating the potential role of turmeric metabolites in reducing $3CL^{pro}$ activity (Guijarro-Real et al. 2021). Kanjanasirirat et al. (2020) examined 122 Thai natural plant products using fluorescence-based SARS-CoV-2 nucleoprotein detection in Vero E6 cells in conjunction with a plaque reduction assay. With $IC_{50}$ values of 3.62 g/mL and 0.08 M, respectively, treatment of Vero E6 cells with extracts of *Boesenbergia rotunda* and its metabolite panduratin A significantly reduced SARS-CoV-2 infectivity after viral infection. The pre-entry phase of SARS-CoV-2 spread was also prevented by panduratin A treatment with an $IC_{50}$ of 5.30 M. This study showed that panduratin A can treat and prevent SARS-CoV-2 infection. Furthermore, panduratin A's antiviral activity was confirmed by the discovery that it prevented viral replication in human airway epithelial cells.

The bioactive saponin glycyrrhizin from *Glycyrrhiza glabra* was found to have potent antiviral activity against SARS-CoV-2. Glycyrrhizin prevents viral replication by blocking viral Mpro. At concentrations as low as 0.024 mg/mL (30 M), glycyrrhizin completely inhibited Mpro activity. At 1 mg/mL (post-entry) or 0.5 mg/mL (combined pre- and post-entry), glycyrrhizin completely inhibited SARS-CoV-2 proliferation (van de Sand et al. 2021). Pattaro-Júnior et al. (2022) used computational studies with Mpro as the target to evaluate the antiviral efficacy of a crude extract of Cenostigma pluviosum var. peltophoroides stem bark against SARS-CoV-2. Further, a recombinant Mpro was used to evaluate the inhibitory potential of proteolytic activity, and antiviral efficacy was analysed with infected Vero cells as experimental proof. In proteolytic kinetics experiments, pentagalloylglucose generated the best results, reducing the activity of recombinant $M^{pro}$ by around 60%.

Nilavembu Kudineer (NVK) is a polyherbal Siddha decoction used as an effective antiviral therapy against chikungunya and dengue in India (Jain et al. 2020). It has been verified that this traditional treatment is safe, effective, and widely accepted. It is composed of *Andrographis paniculata*, *Trichosanthes cucumerina*, *Plectranthus vettiveroides*, *Hedyotis corymbose*, *Cyperus rotundus*, *Piper nigrum*, *Zingiber officinale*, *Santalum album*, and *Vetiveria zizanioides*. In their study, Srivastava et al. (2021) examined the relative effectiveness of Nilavembu Kudineer and placebo in treating mild to moderate SARS-CoV-2 illness. Patients treated with Nilavembu Kudineer demonstrated a significant decrease in SARS-CoV-2 load, length of hospitalisation time, and time needed to become asymptomatic as compared to placebo-treated patients.

### *6.8.2 Use of Purified Plant Secondary Metabolites in Treating COVID-19 and Their Mode of Action*

As shown in Fig. 6.4, several purified PSMs are evaluated for their potential to inhibit the replication of SARS-CoV, MERS-CoV, and SARS-CoV-2 by inhibiting the viral proteins such as main protease (Mpro), papain-like protease (PLpro), RNA-dependent RNA polymerase (RdRp), spike proteins, and angiotensin-converting enzyme (ACE2).

### 6.8.3 *Main Protease ($M^{pro}$)*

Biflavonoids are broadly diversified metabolites in bryophytes, gymnosperms, and angiosperms. Dimers of flavones, flavonones, and flavone-flavonones make up biflavonoids. The scientific community has started paying more attention to biflavonoids in recent years because of the pharmacological activities they exhibit, such as their ability to fight cancer, bacteria, viruses, inflammation, etc. (He et al. 2021). *Torreya nucifera* is a traditional Asian medicine that has been used for centuries. Eight diterpenoids and four biflavonoids were extracted from the plant and tested for their ability to inhibit SARS-CoV 3CLpro using FRET. Inhibitory activity against 3CLpro was most prominent for the biflavone amentoflavone ($IC_{50}$ = 8.3 M). Three other purified flavones (luteolin, quercetin, and apigenin) were studied to determine the underlying structure–activity relationship of biflavones. The $IC_{50}$ values for inhibiting 3CLpro activity were 280.8, 20.2, and 23.8 M for apigenin, luteolin, and quercetin, respectively. Results from enzymatic experiments were confirmed by binding energy values predicted by a molecular docking analysis (Ryu et al. 2010a).

Alkaloids are naturally occurring phytochemicals with established biological properties, a number of which have been competently investigated for their wide-ranging antiviral actions against diverse RNA and DNA viruses (Abookleesh et al. 2022). Abdallah et al. (2022) investigated the potential to inhibit SARS-CoV-2 Mpro under in vitro conditions by testing 33 plants from 17 different Saudi Arabian plant groups that were rich in alkaloids and phenols. To determine how effective the bioactive extracts were at inhibiting this enzyme, their key components were isolated. Additional computational analyses were performed, and their antiviral activity against the Egyptian SARS-CoV-2 variant was assessed. SARS-CoV-2 Mpro was inhibited by a high percentage in tests with extracts from *Nigella sativa*, *Aframomum melegueta*, and *Psiadia punctulata*. High percentages of inhibition (73.80%, 65.2%, 71.8%, and 63.21%) were found for gardenins A and B (*Psiadia punctulata*), 6-gingerol and 6-paradol (*Aframomum melegueta*), and thymoquinone (*Nigella sativa*), respectively. Anisotine, a naturally occurring alkaloid isolated from *Justicia adhatoda* leaves, has been shown to inhibit the Mpro of SARS-CoV-2 in a computational study. As shown by a binding energy of −7.9 kcal/mol, anisotine has a strong interaction with the catalytic residues of Mpro (His41 and Cys145) (Ghosh et al. 2021). In addition, *Justicia adhatoda*, also known as Vasaka, synthesises several alkaloids (vasicoline, anisotine, vasicinone, adhatodine, vasicine, and vasicolinone). These alkaloids are widely recognised for antiviral action against herpes simplex viruses and influenza (Chavan and Chowdhary 2014).

Flavonoids represent one of the most diverse classes of secondary metabolites, with over 9000 structurally distinct forms. Vegetables, tea, seeds, nuts, wine, and fruits are common sources of flavonoids. They are small, 15-carbon-backbone molecules with a variety of health benefits. Anti-inflammatory, antioxidant, antibacterial, antiviral, anticancer, and neuroprotective effects are just some of the many biological effects of flavonoids that have been discovered since their discovery in 1938 (Cazarolli et al. 2008; Kumar and Pandey 2013). Kandeil et al. (2021)

reported that the antioxidant compounds curcumin, quercetin, and hesperidin exhibited anti-SARS-CoV-2 activity. These phytochemicals inhibited virus replication in Vero E6 cells significantly. Their efficacy was also confirmed by molecular docking methods. The compounds are said to inhibit Mpro and to act directly on the virion to trigger inactivation. SARS-CoV-2 was tested against a Chinese herbal remedy for respiratory infections called "Shuanghuanglian" (Su et al. 2020). Shuanghuanglian's bioactive components inhibited 3CLpro and SARS-CoV-2 replication in a Vero E6 cell model in a dose-dependent manner. Bioactive components of Shuanghuanglian identified as baicalin and baicalein have been found to have potential anti-SARS-CoV-2 activity.

Andrographolide is a flavonoid discovered in the Acanthaceae plant, Andrographis paniculata. The antiviral properties of andrographolide have been reported in studies on several different viruses, including dengue, herpes simplex virus type 1, and Chikungunya (Gupta et al. 2017). In a study that relied on computational methods like molecular docking, target analysis, toxicity prediction, and ADME prediction, andrographolide's potential to inhibit SARS-major COV-2's protease (Mpro) was evaluated. When docked against the catalytic site of Mpro using computational docking tools, Andrographolide scored a much lower −3.09 Kcal/mol than synthetic compounds like disulfiram (−46.16 Kcal/mol), tideglusib (−61.79 Kcal/mol), and shikonin (−17.35 Kcal/mol) (Enmozhi et al. 2021; Jin et al. 2020).

In another study, alkaloids, flavonoids, and terpenoids from ayurvedic medicinal herbs Amla (*Emblica officinalis*), Bhumi Amla (*Phyllanthus niruri*), and Giloy (*Tinospora cordifolia*) were docked, and a simulation study was performed targeting $M^{pro}$ of SARS-CoV-2. A total of 96 bioactive molecules, docked with Mpro of SARS-COV-2 and selected molecules were recognised and further confirmation was made through a molecular dynamics investigation. The least binding energy of −10.32 kcal/mol was recorded by withanoside V, followed by somniferine (−9.62 kcal/mol), tinocordiside (−9.10 kcal/mol), vicenin (−8.97 kcal/mol), isorientin 4′-*O*-glucoside 2″-*O*-p-hydroxybenzoate (−8.55 kcal/mol), and ursolic acid (−8.52 kcal/mol), respectively. The most promising drug candidates were analysed by the DruLiTo, pkCSM, and Molinspiration servers for their drug-likeness, ADEMT, and bioactivity score prediction (Murugesan et al. 2021). The Mpro inhibitory potential of Moroccan medicinal plants against SARS-CoV-2 was evaluated by computational investigations conducted by Aanouz et al. (2021). A total of 67 compounds of plant origin were docked, among which Crocin, Digitoxigenin, and β-Eudesmol recorded the least binding energy which is equal to −8.2 kcal/mol, −7.2 kcal/mol, and −7.1 kcal/mol, respectively, as compared to the chloroquine.

### 6.8.4 Papain-Like Protease ($PL^{pro}$)

*Alnus japonica*, a member of the family Betulaceae, is an endemic Alnus species in Korea, North China, and Japan that have well-reported antiviral activity (Tung et al. 2010). In a scientific investigation made by Park et al. (2012a), secondary metabolites isolated from *Alnus japonica* were screened for their antiviral activity against the $PL^{pro}$, which regulates the replication of SARS coronavirus. Among the isolated hirsutenone, diarylheptanoids recorded the most promising suppression of PLpro activity which was evident by lower $IC_{50}$ (4.1 M). When the structure–activity link was examined, it became clear that the compound's catechol and unsaturated carbonyl moiety were crucial for inhibiting the SARS-CoV protease. Another perennial herb with antiviral properties against the zika virus is *Angelica keiskei* (Umbelliferae), which has traditional uses as a mild cathartic, diuretic, and antiviral (Kil et al. 2017; Mottin et al. 2022). Four coumarins and nine alkylated chalcones from *Angelica keiskei* (leaf) were found to inhibit $3CL^{pro}$ and $PL^{pro}$ of SARS-CoV. Cell-free trans-cleavage inhibition assays and deubiquitination activity assays were used to determine the degree of inhibition. The perhydroxyl group on chalcone 6 was the most effective inhibitor of 3CLpro and PLpro, with $IC_{50}$ values of 11.4 M and 1.2 M, respectively (Park et al. 2016).

Flavonoids isolated from *Psoralea corylifolia* seeds were investigated to inhibit the SARS-CoV's $PL^{pro}$. The crude ethanol extract showed promising results, having 15 μg/ml as an $IC_{50}$ value. Six aromatic molecules named bavachinin, isobavachalcone, neobavaisoflavone, 4′-*O*-methylbavachalcone, psoralidin, and corylifol A were identified by bioactivity-guided fractionation of ethanolic extracts. With $IC_{50}$ concentrations ranging between 4.2 and 38.4 μM, all flavonoids recorded dose-dependent $PL^{pro}$ inhibition. Isobavachalcone ($IC_{50,}$7.3 μM) and psoralidin ($IC_{50,}$ 4.2 μM) were identified as the two most promising inhibitors of $PL^{pro}$ (Kim et al. 2014).

Natural phenolic acids and their derivatives inhibited the replication of several human viruses. These viruses included hepatitis C virus, hepatitis B virus, influenza virus, herpes simplex virus, human immunodeficiency virus, and respiratory syncytial virus (Wu et al. 2017). In a scientific investigation conducted by Srinivasan et al. (2022), several phenolic secondary metabolites were analysed for their antiviral property against $PL^{pro}$ of SARS-CoV-2. They identified three phenolic chemicals, viz. 4-(2-hydroxyethyl) phenol (YRL), 4-hydroxybenzaldehyde (HBA), and methyl 3, 4-dihydroxybenzoate (HE9), that bind to $PL^{pro}$ catalytic site, thereby inhibiting the enzyme activity. The molecules that were found through X-ray screening and then functionalized were able to inhibit PLpro in a deISGylation activity experiment. Further, the antiviral potential of these compounds is confirmed with Vero cell line experiments.

### *6.8.5 RNA-Dependent RNA Polymerase (RdRp)*

Our review of the available literature suggests that there are few in vitro and in vivo studies looking for RdRp inhibitory plant secondary metabolites. The drugs from Chinese traditional medicinal with documented anti-SARS-CoV properties and structurally comparable chemicals were screened by computational studies targeting SARS-CoV-2's RdRp. Similarly, the most common phenolic compound, theaflavin in *Camellia sinensis* (family, Theaceae) was proposed as a potential inhibitor of SARS-CoV-2's RdRp. In comparison to other flavonoids, theaflavin recorded the least idock score of −9.11 kcal/mol when interacting with RdRp's catalytic pocket of SARS-CoV-2. However, theaflavin recorded −8.03 kcal/mol and −8.26 kcal/mol of idock score with RdRp's of SARS-CoV and MERS-CoV, respectively. Further investigation showed that hydrophobic interactions were the primary contributors to binding, and the interaction was stabilised by multiple hydrogen bonds between theaflavin and AAs of RdRp's catalytic site (Lung et al. 2020). In a similar attempt, Hesperidin, a flavonoid isolated from fruits of *Citrus sinensis*, a member of the family Rutaceae, was reported as a potential anti-SARS-CoV-2 via inhibiting RdRp (Goyal et al. 2020). Koulgi et al. (2021) performed computational studies to find potential RdRp inhibitory plant secondary metabolites. They found four potential metabolites swertiapuniside, cordifolide A, sitoindoside IX, and amarogentin, commonly found *Swertia chirayita*, *Tinospora cordifolia,* and *Withania somnifera* as inhibitors of RdRp through MD simulation studies. The glycosidic residue of these phytochemicals was found to form favourable interactions with RdRp's seven conserved motifs containing polar and charged amino acids.

In a different attempt, Ahmad et al. (2020) screened drugs approved by FDA following computational methods against RdRp, aiming to repurpose these drugs to manage COVID-19. Here, Ornipressin, Lypressin, Examorelin, Polymyxin B1, Nacortocin, Cistinexine, and Cisatracurium were reported as potential inhibitors of RdRp.

### *6.8.6 Spike Protein*

Risener et al. (2023) screened 1867 extracts primarily derived from 632 plants, 27 fungi, and one chromista and 18 pure compounds to inhibit the attachment/entry of a pseudotyped lentivirus model expressing SARS-COV-2 spike into human HEK-293T-hACE2 cells. After initial screening and eliminating the extracts with cytotoxicity and containing cardiac glycosidase, two extracts from the *Solidago altissima* flower and *Pteridium aquilinum* rhizome were found to have significant potential to inhibit the entry of four pseudotyped variants and infectious SARS-CoV-2 with $EC_{50} < 5$ μg/mL. In a similar attempt, Yang et al. (2020) studied the efficacy of a flavonoid compound Salvianolic acid C in a cell line assay using a pseudovirus model system containing the S protein of SARS-CoV-2. When the experiment was carried out utilising HEK293T cells expressing hACE2, Salvianolic acid C was found to inhibit the viral entry into the host cell with $IC_{50}$ of 3.85 μM. Further, the researchers validated the results with SARS-CoV-2 and Vero-E6 cells, where they found similar results with $EC_{50}$ of 3.41 μM.

Extracts from seven medicinal plants grown in high-altitude from Nepal were tested for their capabilities to block interactions following enzyme-linked immunosorbent assay (Basnet et al. 2022). *Tinospora cordifolia* recorded a 50% reduction in SARS-CoV-2 S protein–hACE2 interactions at 1.25 mg/mL, followed by extracts of

*Heracleum nepalense* and *Pogostemon benghalensis*. Meanwhile, positive control molnupiravir required 50 μM concentration to inhibit the interaction by 25%. Further computational studies with secondary metabolites of *Tinospora cardifolia* highlighted cordifolioside-A as a potential metabolite which interferes with host recognition by SARS-COV-2 by blocking critical amino acids of S1-RBD.

For colds, coughs, and fever, Kabasura Kudineer Chooranam (KKC), a Siddha remedy, is extensively prescribed in Southern India. 32 plant secondary metabolites of herbs found in KKC and 5 from JACOM herbal formulation were molecular docked against the S protein of SARS-CoV-2. Chrysoeriol, Luteolin, and Scutellarein belong to the flavonoids class of secondary metabolites, and these are some of the main phytochemicals present in Chooranam and exhibit a good binding affinity for the S protein of SARS-CoV-2. It has been discovered that all these compounds exhibit hydrogen bond interactions with at least four amino acid residues in S protein, indicating their stronger affinity for S protein (Kiran et al. 2020).

Another study tested the antiviral properties of seselin extracted from *Aegle marmelos* leaf extracts. In silico evaluation of the efficiency of seselin against multiple SARS-COV-2 targets, including S protein's S2 and $M^{pro}$were performed. Having binding energies of −6.3 kcal/mol, −6.9 kcal/mol, and −6.7 kcal/mol, respectively, seselin demonstrated the highest interaction with receptors, spike protein S2, and SARS-CoV-2 $M^{pro}$. Docking studies involving three unique receptors showed that all the protein–ligand complexes with the minimum energy were stabilised by hydrogen bonds and stacking interactions. According to the findings of the pharmacokinetic study, this secondary metabolite exhibited desirable druggability characteristics. These observations demonstrated the potential of seselin to inhibit many targets of SARS-COV-2 and a potential candidate to be a feasible therapeutic treatment for COVID-19 (Nivetha et al. 2021).

### 6.8.7 Angiotensin-Converting Enzyme (ACE2)

Organosulfur are signature compounds present in Allium plants, including onions (*Allium cepa*) and garlic (*Allium sativum*). Organosulfur compounds from *Allium sativum*, allyl disulfide and trisulfide, were studied in silico and demonstrated considerable binding potential with human ACE2 (−9.07 kcal/mol). Results indicate that the 17 major organosulfur compounds from the essential oil of garlic recorded an energetically favourable interaction with ACE2 protein, thereby reducing the capabilities of SARS-CoV-2 to recognise the host cell and subsequent entry into host cells (Thuy et al. 2020). A similar computational study has reported that hesperidin, a flavonoid present in *Citrus sinensis* inhibits ACE2, RdRp, and open reading frame 8 (ORF8) of SARS-CoV-2 (Goyal et al. 2020). Hesperidin bounds tightly at the active position of ACE2 with the binding energy of −9.5 kcal/mol and also bounds tightly in the active position of RdRp with a docking score of −10.01. Similarly, it is bound tightly to an active site of ORF8 with a binding energy of −14.45 kcal/mol. Huang et al. (2021) demonstrated the SARS-CoV-2 suppressive nature of berbamine alkaloid by interfering with the endolysosomal trafficking of ACE2. They evaluated the anti-SARS-CoV-2 action of berbamine using Vero-E6 cells and discovered that berbamine considerably suppressed viral load having $EC_{50}$ 2.4 μM. Some reported plant extracts and their secondary metabolites targeting SARS-CoV-2, SARS-CoV, and MERS-CoV are listed in Table 6.2.

## 6.9 Herbal Formulations Used for COVID-19 Management

Medicinal plants have been widely utilised to improve health by enhancing the host's resistance to certain diseases. At the start of the epidemic, the nature of the coronavirus was poorly understood, and the type of medications to be consumed was inaccessible due to the lockdown. Much of the world resorted to home remedies that could alleviate the symptoms (cold, cough, headache, and fatigue) of COVID 19 which was regarded as a pneumonia-like disease. This section discusses community-based medicine, which includes various herbal formulations used worldwide to manage COVID-19. Ethnobotanical studies reveal that different medicinal plants were used for their antioxidant, anti-inflammatory, and immunity-boosting characters based on geographical location.

A literature survey of medicinal plants consumed in Morocco fez city reported 49 plant species representing 28 families were suggested to explore further for COVID-19 management. Among them, the most representative families were Lamiaceae (12.5 %), containing many aromatic plants such as *Mentha pulegium, Origanum majorana, Rosmarinus officinalis, Salvia officinalis,* and *Thymus vulgaris* used as essential oils prepared from aerial parts. Followed by Asteraceae (11%), a family of plants *Artemisia vulgaris, Artemisia absinthium, Chamaemelum nobile,* and *Saussurea costus* were orally consumed (Benkhaira et al. 2021). Another research from Morocco found that oral ingestion of infusion or decoction of aerial components from Lamiaceae and Asteraceae species, seeds from Apiaceae species, rhizomes from *Zingiber officinale* and *Alpinia officinarum*, and dried powder from young Cupressaceae leaves. The fruits of *Phoenix dactylifera*, the seeds of *Lepidium sativum* and *Nigella sativa*, the rhizome powder of *Curcuma xanthorrhiza*, the oil of *Olea europaea*, and the bulbs of the Liliaceae family are also consumed directly. For inhalation, a hot infusion of Eucalyptus globulus is utilised (Alamri et al. 2020). The Ethiopian medicinal plants such as *Azadirachta indica, Moringa borziana, Acacia abyssinica, Acanthus polystachyus, Acacia etbaica, Acacia nigra,* and *Osyris quadripartite* used to treat viral infections were reported to be rich in flavonoids, terpenoids, and polyphenols (Tegen et al. 2021).

It is no exaggeration to say that the traditional medicine system of Indian is one of the world's ancient medical disciplines and has been instrumental in the development of healthcare worldwide ever since it was first practised thousands of years ago. Indigenously recognised traditional medicinal systems of India are Ayurveda, Yoga, Unani, Siddha, and Homoeopathy (AYUSH), which takes a holistic approach to treat a variety of diseases. The pharmacological modalities of AYUSH rely on

**Table 6.2** Compilation of medicinal plants, class of compounds, and secondary metabolites reported to be anti-SARS-CoV and anti-SARS-CoV-2

| S. No | Plant | Family | Plant part | Nature of study | Relevant SM | SM class | Target/mode of action | Reference |
|---|---|---|---|---|---|---|---|---|
| 1 | *Torreya nucifera* | Taxaceae | Leaves | A | Ferruginol, amentoflavone | Terpenoids, flavonoids | SARS-CoV Mpro | Ryu et al. (2010a) |
| 2 | *Tripterygium regelii* | Celastraceae | Bark | B | Celastrol | Terpenoids | SARS-CoV Mpro | Ryu et al. (2010b) |
| | | | | | Pristimerin | | | |
| | | | | | Tingenone | | | |
| | | | | | Iguesterin | | | |
| 3 | *Alnus japonica* | Betulaceae | Bark | A | Hirsutenone | Diarylheptanoids | SARS-CoV PLpro | Park et al. (2012b) |
| 4 | *Scutellaria baicalensis* | Lamiaceae | Whole plant | A | Scutellarein | Flavone | SARS-CoV helicase, nsP13 | Yu et al. (2012) |
| | *Aglaia perviridis* | Meliaceae | Whole plant | C | Myricetin | Flavonoids | SARS-CoV helicase, nsP14 | |
| 5 | *Psoralea corylifolia* | Fabaceae | Seeds | A | Isobavachalcone | Flavonoids | SARS-CoV PLpro | Kim et al. (2014) |
| | | | | | Psoralidin | | | |
| | | | | | 4′-*O*-methylbavachalcone | | | |
| 6 | *Paulownia tomentosa* | Paulowniaceae | Fruits | A | Tomentin E | Flavonoids | SARS-CoV PLpro | Cho et al. (2013) |
| | | | | | Tomentin A | | | |
| | | | | | Tomentin B | | | |
| | | | | | 3′-*O*-methyldiplacol | | | |
| | | | | | 4′-*O*-methyldiplacol | | | |
| 7 | *Tribulus terrestris* | Zygophyllaceae | Fruits | A | Terrestriamide | Phenylpropanoids | SARS-CoV PLpro | Song et al. (2014) |
| | | | | | N-trans-feruloyloctopamine | | | |

| | | | | | | | | |
|---|---|---|---|---|---|---|---|---|
| 8 | *Cibotium barometz* | Dicksoniaceae | Rhizome | D | NA | NA | SARS-CoV Mpro | Wen et al. (2011) |
| | *Dioscorea batatas* | | Tuber | | | | | |
| 9 | *Salvia miltiorrhiza* | Lamiaceae | Roots | A | Cryptotanshinone | Terpenoid quinone | SARS-CoV—PLpro | Park et al. (2012b) |
| | | | | | Tanshinone IIA | | | |
| 10 | *Euphorbia neriifolia* | Euphorbiaceae | Leaves | D | 3β-Friedelanol | Terpenoids | Antiviral activity against SARS-CoV | Chang et al. (2012) |
| 11 | *Angelica keiskei* | Apiaceae | Leaf | A | Chalcones | Flavonoids | SARS-CoV Mpro and PL pro | Park et al. (2016) |
| 12 | *Broussonetia papyrifera* | Moraceae | Roots | A | Papyriflavonol A | Flavonoids | SARS-CoV—PLpro | Park et al. (2017) |
| 13 | *Echinacea purpurea* | Asteraceae | Roots | A | NA | NA | MERS-CoV | Engler et al. (2017) |
| 14 | NA | NA | NA | C | Rubusoside | Terpenoids | SARS-CoV Mpro | Nguyen et al. (2015) |
| | | | | | Quercetin | Flavonoids | | |
| | | | | | Rubusoside+Quercetin | | | |
| 15 | *Zingiber officinale* | Zingiberaceae | Rhizome | E | 6-gingerol | Phenolic | SARS-CoV 5R7Y protease | Rathinavel et al. (2020) |
| 16 | *Plectranthus amboinicus* | Lamiaceae | NA | E | Chrysoeriol | Flavonoids | Spike (S) protein of SARS-CoV-2 | Kiran et al. (2020) |
| | *Costus speciosus* | Costaceae | | | Luteolin | | | |
| | *Carica Papaya* | Caricaceae | | | Quercetin | | | |
| 17 | *Citrus sinensis* | Rutaceae | Fruit | E | Hesperidin | Flavonoids | SARS-CoV-2 ACE2 | Goyal et al. (2020) |
| | | | | | | | SARS-CoV-2 RdRp | |
| | | | | | | | SARS-CoV-2 ORF8 | |

(continued)

**Table 6.2** (continued)

| S. No | Plant | Family | Plant part | Nature of study | Relevant SM | SM class | Target/mode of action | Reference |
|---|---|---|---|---|---|---|---|---|
| 18 | *Justicia adhatoda* | Acanthaceae | Leaf | E | Anisotine | Alkaloids | Mpro of SARS-CoV-2 | Ghosh et al. (2021) |
| 19 | *Aloe vera* | Asphodelaceae | Leaf | E | Feralolide | curcuminoids | SARS-CoV-2 Mpro | Mpiana et al. (2020) |
| 20 | *Psorothamnus arborescens* | Fabaceae | NA | E | 5, 7, 30, 40-Tetrahydroxy-2′-(3,3-dimethylallyl) isoflavon | Flavonoids | SARS-CoV-2 Mpro | ul Qamar et al. (2020) |
| | *Myrica cerifera* | Myricaceae | NA | NA | Myricitrin | | | |
| | *Hyptis atrorubens* | Lamiaceae | NA | NA | Methyl rosmarinate | Phenolic | | |
| 21 | *Forsythiae Fructus* | Oleaceae | NA | D | NA | NA | SARS-CoV-2 replication | Runfeng et al. (2020) |
| 22 | NA | NA | NA | E | Herbacetin | Flavonoids | SARS-CoV-2 Mpro | Jo et al. (2020) |
| | | | | | Pectolinarin | | | |
| | | | | | Rhoifolin | | | |
| 23 | NA | NA | NA | E | Naringenin | Flavonoids | SARS-CoV-2 Mpro | Khaerunnisa et al. (2020) |
| | | | | | Kaempferol | Polyphenols | | |
| | | | | | Quercetin | Flavonoids | | |
| | | | | | Allicin | Thiosulfonates | | |
| | | | | | Catechin | Flavonoids | | |
| | | | | | Gingerol | Phenolic | | |
| | | | | | Zingerol | Phenolic | | |
| | | | | | Epicatechin-gallate | Flavonoids | | |
| | | | | | Curcumin | Curcuminoids | | |
| | | | | | Oleuropein | Phenolic | | |
| | | | | | Apigenin-7-glucoside | Flavonoids | | |
| | | | | | Demethoxycurcumin | Curcuminoids | | |
| | | | | | Luteolin-7-glucoside | Flavonoids | | |

| | | | | | | | | |
|---|---|---|---|---|---|---|---|---|
| 24 | *Andrographis paniculata* | Acanthaceae | NA | E | Andrographolide | Flavonoids | SARS-CoV-2 Mpro | Enmozhi et al. (2021) |
| 25 | NA | NA | NA | E | Gallic acid | Polyphenol | SARS-CoV-2 Mpro | Alrasheid et al. (2021) |
| | | | | | Quercetin | Flavonoids | | |
| | | | | | Naringin | Flavonoids | | |
| | | | | | Capsaicin | Alkaloids | | |
| | | | | | Psychotrine | | | |
| 26 | *Crocus Sativus* | Iridaceae | NA | E | Crocin | NA | SARS-CoV-2 Mpro | Aanouz et al. (2021) |
| | *Nerium Oleander* | Apocynaceae | | | Digitoxigenin | Cardenolide glycoside | | |
| | *Laurus Nobilis* | Lauraceae | | | β-Eudesmol | Sesquiterpenoid | SARS-CoV-2 Mpro | |
| 27 | *Camellia sinensis* | Theaceae | NA | E | Theaflavin | Phenolic | SARS-CoV-2 RdRp | Lung et al. (2020) |
| 28 | *Aegle marmelos* | Rutaceae | NA | E | Seselin | Curcuminoids | SARS-CoV-2 Mpro, spike protein S2 | Nivetha et al. (2021) |
| 29 | *Curcuma longa* | Zingiberaceae | Rhizomes | D | NA | NA | SARS-CoV-2 Mpro | Guijarro-Real et al. (2021) |
| 31 | NA | NA | NA | NA | Quercetin | Flavonoids | SARS-CoV-2 Mpro | Bahun et al. (2022) |
| | | | | | Ellagic acid | Polyphenol | | |
| | | | | | Curcumin | Curcuminoids | | |
| | | | | | Epigallocatechin gallate | Polyphenol | | |
| 32 | *Echinacea purpurea* | Asteraceae | Herb and roots | C | Echinaforce® | | Antiviral activity against MERS-CoV | Signer et al. (2020) |
| 33 | *Withania somnifera* | Solanaceae | NA | E | Withanoside V | Alkaloids | SARS-CoV-2 Mpro | Shree et al. (2022) |
| | | | | | Somniferine | | | |
| | *Tinospora cordifolia* | Menispermaceae | | | Tinocordiside | Sesquiterpenoid | | |
| | *Ocimum sanctum* | Lamiaceae | | | Vicenin | Flavonoids | | |
| | | | | | Isorientin 4′-*O*-glucoside 2″-*O*-p-hydroxybenzoate | | | |
| | | | | | Ursolic acid | Terpenoids | | |

(continued)

**Table 6.2** (continued)

| S. No | Plant | Family | Plant part | Nature of study | Relevant SM | SM class | Target/mode of action | Reference |
|---|---|---|---|---|---|---|---|---|
| 34 | NA | NA | NA | NA | Methyl 3,4-dihydroxybenzoate | Polyphenol | SARS-CoV-2 PLpro | Srinivasan et al. (2022) |
| | | | | | p-hydroxybenzaldehyde | | | |
| | | | | | 4-(2-hydroxyethyl) phenol | | | |
| 35 | *Picrorhiza kurroa* | Plantaginaceae | NA | E | Acetovanillone | Phenylketones | SARS-CoV-2 PLpro, RdRp | Khuntia et al. (2021) |
| | *Azadirachta indica* | Meliaceae | | | Myrtenol | Terpenoids | | |
| | *Cyperus rotundus* | Cyperaceae | | | Nimbochalcin | NA | | |
| 36 | NA | NA | NA | E | 10-Hydroxyusambarensine | Alkaloids | SARS-CoV-2 Mpro | Gyebi et al. (2021) |
| | | | | | Cryptoquindoline | | | |
| | | | | | 6-Oxoisoiguesterin | Terpenoids | | |
| | | | | | 22-Hydroxyhopan-3-one | | | |
| 37 | *Tinospora cordifolia* | Menispermaceae | NA | E | Amritoside | Tannin | SARS-CoV-2 Mpro | Murugesan et al. (2021) |
| | | | | | Apigenin-6-*C*-glucosyl7-*O*-glucosid | Flavonoids | | |
| | *Emblica officinalis* | Phyllanthaceae | NA | | Pectolinarin | | | |
| | | | | | Astragalin | | | |

| | | | | | | | | |
|---|---|---|---|---|---|---|---|---|
| 38 | *Calotropis gigantean* | Apocynaceae | NA | E | Juniper camphor | Terpenoids | SARS-CoV-2 Mpro | Dutta et al. (2021) |
| | | | | | Ethyl 4-fluoro-1-methyl-1H-imidazole-5-carboxylate | | | |
| | | | | | Bicyclo[4.3.0]nonane, 1 isopropenyl-4,5-dimethyl-5-phenylsulfonylmethyl | | | |
| | | | | | Stearic acid | | | |
| 39 | *Rheum emodi* | Polygonaceae | NA | E | Emodin | Alkaloids | N-terminal domain (NTD) of nucleo-capsid phosphoprotein | Rolta et al. (2021) |
| | | | | | Aloe-emodin | | | |
| | | | | | Anthrarufin | | | |
| | | | | | Alizarine | | | |
| | | | | | Dantron | | | |
| 40 | *Marrubium vulgare* | Lamiaceae | NA | E | Luteolin 7-rutinoside | Flavonoids | SARS-CoV-2 RdRp | El Alami et al. (2020) |
| | *Rumex dentatus* | Polygonaceae | | | Chrysophanol 8-(6-galloylglucoside) | Alkaloids | | |
| | *Acacia* | Fabaceae | | | Kaempferol 7-(600-galloylglucoside) | Flavonoids | | |
| | *Rumex dentatus* | Polygonaceae | | | Chrysophanol 8-(6-galloylglucoside) | Alkaloids | | |
| | *Guiera senegalensis* | Combretaceae | | | 3,4,5-tri-*O*-galloylquinic acid | Tannin | | |
| | *Chenopodium ambrosioides* | Amaranthaceae | | | Mulberrofuran G | Flavonoids | | |
| | *Datura innoxia* | Solanaceae | | | Withanolide A | Alkaloids | | |
| | *Capparis decidua* | Capparaceae | | | Isocodonocarpine | Alkaloids | | |
| | *Senna obtusifolia* | Fabaceae | | | Calonysterone | Sterol | | |

(continued)

**Table 6.2** (continued)

| S. No | Plant | Family | Plant part | Nature of study | Relevant SM | SM class | Target/mode of action | Reference |
|---|---|---|---|---|---|---|---|---|
| 41 | *Nyctanthes Arbor-tristis* | Oleaceae | NA | E | Astragalin | Flavonoids | SARS-CoV-2 protease | Srivastava et al. (2021) |
| | | | | | Lupeol | Terpenoids | | |
| | *Aloe barbadensis* | Asphodelaceae | | | Aloenin | Glycoside | | |
| | | | | | Aloesin | | | |
| | *Tinospora cordifolia* | Menispermaceae | | | Berberine | Alkaloids | | |
| | | | | | Sitosterol | Sterol | | |
| | *Curcuma longa* | Zingiberaceae | | | Curcumin | Curcuminoids | | |
| | *Azadirachta indica* | Meliaceae | | | Nimbin | Terpenoids | | |
| | *Withania somnifera* | Solanaceae | | | Withanolide | Alkaloids | | |
| | | | | | Withaferin A | Saponin | | |
| | *Zingiber officinale* | Zingiberaceae | | | Gingerol | Phenolic | | |
| | | | | | Shogaol | Phenolic | | |
| | *Allium cepa* | Amaryllidaceae | | | Quercetin | Flavonoids | | |
| | *Ocimum tenuiflorum* | Lamiaceae | | | Ursolic acid | Terpenoids | | |
| | | | | | Apigenin | Flavonoids | | |
| | *Cannabis sativa* | Cannabaceae | | | Cannabidiol | Cannabinoids | | |
| | *Piper nigrum* | Piperaceae | | | Piperine | Alkaloids | | |
| 42 | *Boesenbergia rotunda* | Zingiberaceae | Rhizome | F | Panduratin A | Flavonoids | SARS-CoV-2 infectivity in Vero E6 cells | Kanjanasirirat et al. (2020) |

| | | | | | | | | |
|---|---|---|---|---|---|---|---|---|
| 43 | *Trigonella foenum-graecum* | Fabaceae | NA | E | Graecunin E | Saponins | SARS-CoV-2 spike protein | Puttaswamy et al. (2020) |
| | | | | | Trigofoenoside E1 | Saponins | | |
| | | | | | Uttronin B | Saponins | | |
| | | | | | Stigmasteryl glucoside | Steroid | | |
| | | | | | Yuccagenin | Terpenoids | | |
| | *Veratrum album* | Liliaceae | | | Pseudojervine | Saponins | | |
| | *Mallotus philippensis* | Euphorbiaceae | | | Kamalachalcone C | Flavonoids | | |
| | *Phyllanthus niruri* | Phyllanthaceae | | | Nirurin | Flavonoids | SARS-CoV-2 TMPRSS2 | |
| | *Foeniculum vulgare* | Apiaceae | | | Cis-Miyabenol C | Flavonoids | | |
| | *Plumbago zeylanica* | Plumbaginaceae | | | Plumbagin | Alkaloids | | |
| | *Punica granatum* | Punicaceae | | | Granatin B | Tannins | | |
| | *Hypericum perforatum* | Hypericaceae | | | Hypericin | Alkaloids | SARS-CoV-2 Mpro | |
| | *Ginkgo biloba* | Ginkgoaceae | | | Amentoflavone | Flavonoids | | |
| | | | | | Ginkgetin | Flavonoids | | |
| | | | | | Isoginkgetin | Flavonoids | | |
| | | | | | Agathisflavone | Flavonoids | | |
| | *Terminalia chebula* | Combretaceae | | | Terflavin B | Flavonoids | | |
| | *Paeonia delavayi* | Paeoniaceae | | | Mudanpioside J | Terpenoids | | |
| | *Justicia purpurea* | Acanthaceae | | | Justalakonin | Lignans | | |

(continued)

**Table 6.2** (continued)

| S. No | Plant | Family | Plant part | Nature of study | Relevant SM | SM class | Target/mode of action | Reference |
|---|---|---|---|---|---|---|---|---|
| | *Chrysobalanus icaco* | Chrysobalanaceae | | | Myricetin 3-rutinoside | Flavonoids | SARS-CoV-2 RdRp | |
| | *Citrus limon* | Rutaceae | | | Eriodictyol-7-*O*-rutinoside | Flavonoids | | |
| | *Bupleurum rotundifolium* | Apiaceae | | | Rotundioside B | Terpenoids | | |
| | *Emblica officinalis* | Phyllanthaceae | | | Phyllanemblinin C | Tannins | | |
| | | | | | Emblicanin A | Tannins | | |
| | *Hippomane mancinella* | Euphorbiaceae | | | Hippomanin A | Flavonoids glycoside | | |
| 44 | NA | NA | NA | C | Salvianolic acid C | Flavonoids | SARS-CoV-2 spike protein | Yang et al. (2020) |
| 45 | *Scutellaria Baicalensis* | Lamiaceae | NA | C | Baicalin | Flavonoids | SARS-CoV-2 Mpro | Huang et al. (2020) |
| 46 | NA | NA | NA | C | Berbamine | Alkaloids | SARS-CoV-2 ACE2 | Huang et al. (2021) |
| 47 | *Andrographis paniculata* | Acanthaceae | NA | E | Neoandrographolide | Terpenoids | Mpro, PLpro, RdRp, Spike-ACE2 | Murugan et al. (2021) |
| 48 | *Ocimum sanctum* | Lamiaceae | NA | E | Chlorogenic acid | Phenolic | SARS-CoV-2 Mpro | Mohapatra et al. (2020) |
| | | | | | Luteolin-7-*O*-glucuronide | Flavonoids | | |
| 49 | *Allium sativum* | Amaryllidaceae | NA | E | | | SARS-CoV-2 ACE2 | Thuy et al. (2020) |
| 50 | *Moringa oleifera* | Moringaceae | NA | E | Kaempferol | Flavonoids | SARS-CoV-2 peptide | Hamza et al. (2021) |
| | | | | | Anthraquinone | Alkaloids | | |

| | | | | | | | | |
|---|---|---|---|---|---|---|---|---|
| 51 | NA | NA | NA | E | Luteolin | Flavonoids | SARS-CoV-2 Mpro | Ansari et al. (2020) |
| 52 | *Citri reticulatae* | Rutaceae | NA | E | Hesperidin | Flavonoids | Mpro, ACE 2 receptor | Yang et al. (2020) |
| | *Bupleuri radix* | Apiaceae | | | Saikosaponin A | Saponin | | |
| | *Mori folium* | Asteraceae | | | Rutin | Flavonoids | | |
| | *Eriobotryae folium* | Rosaceae | | | Corosolic acid | Terpenoids | | |
| | *Rehmanniae radix* | Orobanchaceae | | | Verbascoside | Polyphenol | | |
| | *Scutellariae radix* | Lamiaceae | | | Baicalin | Flavonoids | | |
| | *Glycyrrhizae radix* | Fabaceae | | | Glycyrrhizin | Saponin | | |
| | *Mori ramulus* | Moraceae | | | Mulberroside A | Stilbenoid | | |
| | *Lonicerae flos* | Caprifoliaceae | | | Cynaroside | Flavonoids | | |
| 53 | *Citrus Sp.* | Rutaceae | NA | E | Hesperidin | Flavonoids | SARS-CoV-2 Mpro | Adem et al. (2020) |
| | | | | | Rutin | | | |
| | | | | | Diosmin | | | |
| | | | | | Apiin | | | |
| | | | | | Diacetylcurcumin | Curcuminoids | | |
| 54 | NA | NA | NA | E | Glabridin | Flavonoids | SARS-CoV-2 Mpro | Jamali et al. (2021) |
| | | | | | Catechin | | | |
| | | | | | Fisetin | | | |
| 55 | *Acacia seyal* | Fabaceae | NA | E | Rutin | Flavonoids | Mpro, SARS-CoV-2 receptor binding domain | Ahmed et al. (2020) |
| | *Cymbopogon commutatus* | Poaceae | | | Catechin | | | |
| | *Indigofera caerulea* | Fabaceae | | | Kaempferol | | | |

(continued)

**Table 6.2** (continued)

| S. No | Plant | Family | Plant part | Nature of study | Relevant SM | SM class | Target/mode of action | Reference |
|---|---|---|---|---|---|---|---|---|
| 56 | *Swertia chirata* | Gentianaceae | NA | E | Amarogentin | Phenolic | Spike glycoprotein, ACE2 receptor | Maurya et al. (2022) |
| | *Artemisia vulgaris* | Asteraceae | | | Eufoliatorin | Saponin | | |
| | *Alstonia scholaris* | Apocynaceae | | | α-amyrin | Terpenoids | | |
| | *Caesalpinia crista* | Fabaceae | | | Caesalpinins | Saponin | | |
| | *Picrorhiza kurroa* | Plantaginaceae | | | Kutkin | Saponin | | |
| | *Alstonia scholaris* | Apocynaceae | | | β-sitosterol | Sterol | | |
| | *Atropa belladonna* | Solanaceae | | | Belladonnine | Alkaloids | | |
| 57 | *Justicia procumbens* | Acanthaceae | | E | Justicidin D | Lignan | RdRp, Mpro, and spike | Gani et al. (2021) |
| | *Camptotheca acuminata* | Nyssaceae | | | 10-methoxycamptothecin | Alkaloids | | |
| | *Calophyllum inophyllum* | Clusiaceae | | | Inoxanthone | Flavonoids | | |
| | *Hymenocallis arenicola* | Amaryllidaceae | | | Caribine | Alkaloids | | |
| 58 | *Eurycoma harmandiana* | Simaroubaceae | NA | E | Canthin-6-one 9-*O*-beta-glucopyranoside | Alkaloids | Mpro, Mpro, Plpro | Verma et al. (2021) |

| | | | | | | | | |
|---|---|---|---|---|---|---|---|---|
| 59 | *Glycyrrhiza glabra* | Fabaceae | NA | E | Rocymosin B | Polyphenol | Mpro, Mpro, Plpro | Shawky et al. (2020) |
| | *Cichorium intybus* | Asteraceae | | | Verbascoside | | | |
| | *Anastatica hierochuntica* | Brassicaceae | | | Rutin | Flavonoids | | |
| | *Trigonella foenum-graecum* | Fabaceae | | | Fenugreekine | Sterol-sapogenin | | |
| 60 | *Nigella sativa* | Ranunculaceae | NA | A | Thymoquinone | Alkaloids | SARS-CoV-2 Mpro | Abdallah et al. (2022) |
| | *Aframomum melegueta* | Zingiberaceae | | | 6-gingerol | Phenolic | | |
| | | | | | 6-paradol | Alkaloids | | |
| | *Psiadia punctulata* | Asteraceae | | | Gardenin A | Flavonoids | | |
| 61 | *Glycyrrhiza glabra* | Fabaceae | NA | D | Glycyrrhizin | Saponin | SARS-CoV-2 Mpro | van de Sand et al. (2021) |
| 62 | *Cenostigma pluviosum* | Fabaceae | Stem bark | D | | | SARS-CoV-2 Mpro | Pattaro-Júnior et al. (2022) |
| 63 | NA | NA | NA | C | Curcumin | Curcuminoids | SARS-CoV-2 Mpro | Kandeil et al. (2021) |
| | | | | | Hesperidin | Flavonoids | | |
| 64 | *Apium graveolens* | NA | NA | E | Apigenin | NA | SARS-CoV-2 Mpro | Sampangi-Ramaiah et al. (2020) |
| | *Coriandrum sativum* | | | | Coriandrin | | | |
| | *Salvia officinalis* | | | | Safficinolide | | | |

natural compounds of plant and animal or mineral base (Adhikari and Paul 2018). Post trials, Ayush has approved Zingi-Vir and Virulina as antiviral agents, Astha 15 as an immunity booster and Kabasura Kudineer, Arogya Kashayam, Tulsi drops, and Ayudh as health supplements (Ahmad et al. 2021). Amongst Chinese formulations such as Qing-Fei-Pai-Du-Tang (QFPDT), Lianhua Qingwen capsule (LHC), Taiwan Chingguan Yihau (NRICM101), and Jing Si herbal drink (JSHD) are approved for COVID-19 (Su et al. 2022). NLC-001, a 3CL protease inhibitory product of plant origin was approved by US FDA. Currently, clinical trials as oral dietary supplements of NLC-001 are being conducted in Israel, targeting SARS-CoV-2 infection. Meanwhile, Todos Medical Ltd. evaluates potential global commercialization opportunities (Golodetz et al. 2020).

In Table 6.3 a list of herbal remedies is compiled that can be used for the treatment and management of COVID-19.

## 6.10 Employing Computational Tools to Discover PSM Aiming COVID-19 Management

COVID-19 vaccine and drug candidates were developed through well-planned scientific investigations (Liu et al. 2021; Dong et al. 2020). Nevertheless, full immunisation treatment and therapeutic efficiency assessment under actual conditions (in vivo and clinical trials) remain challenges, necessitating the continuing research and development of the best management strategies for SARS-CoV-2 infection.

The exponential growth of bioinformatics in recent years suggests that this type of progress may be hastened by the discipline's ability to tackle problems on previously insurmountable scales. The relatively new field of computational and structural biology can "decode" pathogens and hosts based on genomic sequences, allowing researchers to predict and accelerate the mechanism of pathogen insights and investigate ways to stop the spread of pathogens. The rapid development of computational technology, especially in the fields of artificial intelligence (AI) and machine learning (ML), has increased the scope of this technology's use in the fields of biology, medicine, and public health and prompted a rethinking of the best ways to treat various diseases. The use of this assay has been widespread during the current COVID-19 pandemic for the purposes of drug screening, vaccine/drug design, and disease prediction (Liu et al. 2022). Computational drug discovery methods such as molecular docking, dynamic simulations, and various analyses such as ADMET are extensively used to find the potential of synthetic or natural compounds that can target structural and function proteins of SARS-CoV-2.

Through computation approaches, terpenes (NPACT01552, NPACT01557, and NPACT00631) bind specifically to SARS-CoV-2 S RBD, Mpro inhibitors such as tinosponone, montelukast, ChEMBL275592, ChEMBL288347, biflavone amentoflavone, quercetin-3-*O*-rhamnoside, and RdRp inhibitory compound

**Table 6.3** Compilation of herbal remedies available as a product in the global market to manage COVID-19 disease

| Ref. | Categories | Product name | Ingredients | Mode of action | Target |
|---|---|---|---|---|---|
| 1 | A | Herbal Strategi | *Cymbopogon citratus*; *Azadirachta indica*; *Citrus limon*; *Ocimum tenuiflorum* | Defensive mechanism against COVID-19 | Removes the virus from hands |
| 2 | A | ARAR Neem care sanitizer | 0.20% (v/v) *Azadirachta indica*, 0.20% (v/v) *Citrus medica* mixed with 80% v/v ethanol, 0.125% (v/v) of $H_2O_2$ and 1.45% (v/v) of glycerol. | Defensive mechanism against COVID-19 | NA |
| 3 | A | Himalaya pure hands | *Aloe barbadensis* | Defensive mechanism against COVID-19 | NA |
| 4 | A | Mangalam CamPure Camphor Hand Sanitizer | *Cinnamomum camphora* | Defensive mechanism against COVID-19 | NA |
| 5 | B | Jing-Si-Herbal-Tea | *Anisomeles indica*; *Artemisia argyi folium*; *Chrysanthemi Flos*; *Glycyrrhizae radix*; *Perillae folium*; *Houttuyniae herba cum radice*; *Ophiopogonis radix*; *Platycodonis Radix*; microcrystalline cellulose and maltodextrin | Disease management for COVID-19 | NA |
| 6 | B | Ayush Kwath -Dabur | *Zingiber officinale*; *Ocimum sanctum*; *Piper nigrum*; *Cinnamomum zeylanicum* | Immunity against COVID-19 | NA |
| 7 | B | Coronil tablet | *Withania somnifera*; *Ocimum sanctum*; *Tinospora cordifolia* | Disruption of S Protein-ACE-2 Interaction | Recognition and entry of the virus into the host cell |
| 8 | B | Guduchi Ghanavati (Supplement)-Himalaya | *Tinospora cordifolia* | Defensive mechanism against COVID-19 | |

(continued)

**Table 6.3** (continued)

| Ref. | Categories | Product name | Ingredients | Mode of action | Target |
|---|---|---|---|---|---|
| 9 | B | Golden Milk Turmeric Latte—CARE | Haldi (turmeric; *Curcuma longa*) powder in milk | Defensive mechanism against COVID-19 | NA |
| 10 | C | Azadvir | *Ocimum sanctum*; *Cedrus deodara*; *Boswellia serrate*; *Azadirachta indica*; *Mentha piperita*; *Thymus vulgaris*; *Cinnamomum camphora* | Management of COVID-19 | NA |
| 11 | C | Ayush Kadha -YOGAFY | *Ocimum sanctum*; *Curcuma longa*; *Syzygium aromaticum*; *Elettaria cardamomum*; *Citrus limon*; *Withania somnifera*; *Tinospora cordifolia*; *Piper nigrum*; *Zingiber officinale* | Prevention and management of the novel Coronavirus (SARS-CoV-2) | NA |
| 12 | C | Taiwan Chingguan Yihau (NRICM101) decoction | *Scutellaria baicalensis*; *Houttuynia cordata*; *Morus alba*; *Saposhnikovia divaricate*; *Trichosanthes kirilowii*; *Isatis indigotica*; *Glycyrrhiza glabra*; *Magnolia officinalis*; *Mentha haplocalyx*; *Nepeta tenuifolia* | Management of COVID-19 | Target S protein/ACE2 interaction, $3CL^{pro}$ activity, viral plaque formation, and cytokines interleukin (IL)-6 and tumour necrosis factor (TNF)-α |
| 13 | C | Ayush Kwath | *Ocimum sanctum*; *Cinnamomum zeylanicum*; *Piper nigrum* | Defensive mechanism against COVID-19 | NA |
| 14 | C | Chyawanprash | *Phyllanthus emblica*; *Piper longum*; *Tribulus terrestris*; *Pistacia integerrima*; *Phyllanthus Amarus*; *Vitis vinifera*; *Inula racemose*; *Terminalia Chebula*; *Tinospora Cordifolia* | Defensive mechanism against COVID-19 | NA |

| | | | | | |
|---|---|---|---|---|---|
| 15 | C | Huo-Xiang-Zheng-Qi | *Atractylodis Rhizoma*; *Citri Reticulatae Pericarpium*; *Magnoliae Officinalis*; *Angelicae dahuricae radix*; *Arecae Pericarpium*; *Pinelliae rhizoma*; *Glycyrrhizae radix*; *Pogostemonis herba*; *Perillae Folium* | Management of COVID-19 | PTGS2, HSP90AB1, mPGES-1, LTA4H, NOS2, PTGS2 |
| 16 | D | ShuFeng JieDu capsule | *Polygonum cuspidatum*; *Forsythia suspense*; *Isatis indigotica*; *Bupleurum chinense*; *Patrinia scabiosifolia*; *Verbena officinalis*; *Phragmites communis*; *Glycyrrhiza uralensis* | SFJDC treats COVID-19 by multi-compounds and multi-pathways | RELA, MAPK1, MAPK14, CASP3, CASP8, and IL6 were the key target genes |
| 17 | D | DivyaSwasari-Vati (DSV) (Patanjali) | *Glycyrrhiza glabra*; *Pistacia integerrima*; *Cressa cretica*; *Zingiber officinale*; *Piper longum*; *Piper nigrum*; *Cinnamomum zylancium*; *Anacyclus pyrethrum*; *Syzygium aromaticum*; Salt crease, Gall plant | Disease management for COVID-19 Suppressed the proinflammatory cytokines, IL-6 and TNF-α | Target spike protein of SARS-CoV-2 |
| 18 | D | Qingfei Paidu Tang (QFPDT) | *Ephedra sinica*; *Bupleurum chinense*; *Pogostemon cablin*; *Cinnamomum cassia*; *Scutellaria baicalensis* | Treatment of coronavirus (COVID-19) | Cytokine storm prevention and angiotensin-converting enzyme 2 (ACE2) receptor-binding regulation |
| 19 | D | Nilavembu Kudineer Chooranam | *Andrographis paniculate*; *Vetiveria zizanioides*; *Plectranthus vettiveroides*; *Santalum album*; *Trichosanthes dioica*; *Zingeber officinale*; *Piper nigrum*; *Mollugo cerviana* | Disease management for COVID-19 | spike glycoprotein, protease, and angiotensin-converting enzyme 2 (ACE2) receptor, |
| 20 | D | Divya Swasari Ras (Patanjali) | *Glycyrrhiza glabra*; *Syzygium aromaticum*; *Cinnamomum zeylanicum*; *Pistacia integerrima*; *Cressa cretica*; *Zingiber officinale*; *Piper nigrum*; *Piper longum*; *Anacyclus pyrethrum* | Treatment against COVID-19 | SARS-CoV-2 S protein and human ACE 2 |

(continued)

**Table 6.3** (continued)

| Ref. | Categories | Product name | Ingredients | Mode of action | Target |
|---|---|---|---|---|---|
| | | | Herbally processed calcined mica ash, Herbally processed calcined shell of Pearl oyster ash, Herbally processed calcium rich gypsum ash, Herbally processed calcined cowry shell ash of *Cypraea moneta* | | |
| 21 | D | Kaba Sura Kudineer | *Zingiber officinale*; *Piper longum*; *Syzygium aromaticum*; *Tragia involucrate*; *Anacyclus pyrethrum*; *Andrographis paniculata*; *Hygrophilla auriculata*; *Terminalia chebula*; *Justicia adhatoda*; *Plectranthus amboinicus*; *Costus speciosus*; *Tinospora cordifolia*; *Clerodendrum serratum*; *Sida acuta*; *Cypreus rotundus* | Disease management for COVID-20 | SARS-CoV-2 S protein |
| 22 | D | Lianhuaqingwen capstule | *Forsythia suspense*; *Ephedra sinica*; *Lonicera japonica*; *Mentha haplocalyx*; *Dryopteris crassirhizoma*; *Rhodiola rosea*; *Gypsum fibrosum*; *Pogostemon cablin*; *Rheum palmatum*; *Houttuynia cordata*; *Glycyrrhiza uralensis*; *Armeniaca sibirica* | Treatment against COVID-19 | Target pro-inflammatory cytokines (TNF-α, IL-6, CCL-2/MCP-1 and CXCL-10/IP-10) |
| 23 | D | Marshmallow root's syrup | *Althaea officinalis* | Treatment against COVID-19 | NA |
| 24 | D | Divya Anu Taila-Patanjali | *Aegle marmelos*; *Nymphaea stellate*; *Solanum indicum*; *Solanum virginianum*; *Pluchea lanceolate*; *Desmodium gangeticum*; *Uraria* | Treatment against COVID-19 | Ameliorate proinflammatory cytokines such as IL-1β,TNF-α, IFN-γ, IP-10, MCP-1, IL-4, and IL-8 |

| | | | | | |
|---|---|---|---|---|---|
| | | | *picta*; *Embelia ribes*; *Cinnamomum tamala*; *Amomum aromaticum*; *Vitex negundo*; *Mallotus philippensis* Goat milk and Sesame oil | | |
| 25 | D | Saliravira | *Echinacea purpurea*; *Glycyrrhiza glabra*; *Rheum palmatum*; *Hyssopus officinalis*; *Rosmarinus officinalis*; *Panax ginseng* | Antiviral, anti-inflammatory, and immune-enhancing agent | NA |
| 26 | D | Koflet syrup-Himalaya | *Glycyrrhiza glabra* | Immunomodulatory, anti-inflammatory | Viral fusion protein |
| 27 | D | Fah talai jone capstule | *Andrographis paniculata* | Treatment against COVID-19 | NA |
| 28 | D | Phela | *Clerodendrum glabrum*; *Gladiolus dalenii*; *Rotheca myricoides*; *Senna occidentalis* | Treatment against COVID-19 | Inhibits >90% of SARS-CoV-2 and SARS-CoV infection |
| 29 | D | Su He Xiang Pill (SHXP) | *Liquidambar orientalis*; *Styrax tonkinensis*; *Cyperus rotundus*; *Aucklandia lappa*; *Cyperus rotundus*; *Eugenia caryophyllata*; *Aquilaria sinensis*; *Santalum album*; *Boswellia carterii* | Treatment against COVID-20 | NA |
| 30 | D | Tanreqing Capsule | *Scutellariae radix*; *Lonicerae japonicae*; *Forsythiae fructus* bear bile powder and goat horn | Treatment against COVID-20 | COX-2, sEH, IL6, CCL2, CASP3, IL4, MAPK1, RELA, FOS, NOS2, IL1B, CXCL10, MAPK14, EGFR |

A: COVID-19 Preventive products, B: Immunity booster, C: Prevention and treatment of COVID-19, D: Treatment against COVID-20, NA: Information not available

Reference: (1) https://www.amazon.in/, (2–4) Malabadi et al. (2021), (5) Liu et al. (2021), (6) https://www.amazon.in/DABUR-Kaadha-Ayurvedic-Immunity-Boosting/dp/B08C5T9FSL, (7) Balkrishna et al. (2021), (8) Thakar et al. (2022), (9) https://www.amazon.in/Golden-Turmeric-Latte-Saffron-Herbs/dp/B08B7WNTMD?th=1, (10) https://www.ijbcp.com/index.php/ijbcp/article/view/4574/3151, (11) Maurya and Sharma (2022), (12) Tsai et al. (2021), (13) Gautam et al. (2022), (14) Godatwar et al. (2021), (15) Li et al. (2022), (16) Chen et al. (2020), (17) Balkrishna et al. (2020), (18) Zhong et al. (2020), (19) Srivastava et al. (2021), (20) Balkrishna et al. (2022); Balkrishna et al. (2021), (21) Srivastava et al. (2021), (22) Srivastava et al. (2021), (23-24) Balkrishna et al. (2021), (25) Khorshiddoust et al. (2022), (26) Gomaa and Abdel-Wadood (2021), (27) Yearsley (2021), (28) Matsabisa et al. (2022), (29) Li et al. (2021), (30) Zhang et al. (2021)

galidesivir, have been identified from various studies (Muhseen et al. 2020; Krupanidhi et al. 2021; Abu-Saleh et al. 2020; Liu et al. 2022; Ryu et al. 2010a; Aftab et al. 2020).

According to the results of an investigation conducted by Puttaswamy et al. (2020), several different classes of PSM, anthocyanidins, flavonoids, glycosides, biflavonoids, ellagitannins, and triterpenes, have the potential to inhibit TMPRSS2, S protein of SARS-CoV-2, Mpro, and RdRp. The triterpenoid saponin glycyrrhizic acid, also known as glycyrrhizin, has demonstrated antiviral activity against both the SARS-CoV and SARS-CoV-2 viruses. Clinical trials are currently being conducted (Murck 2020). Molecular docking and machine learning tools in a study have been used to screen over 2000 natural compounds for 3CLpro inhibitor, where the compound rutin exhibited the highest AUC score and binding affinity, suggesting rutin as a potential drug candidate against COVID-19 (Xu et al. 2020).

Computational analysis can also design viral protein-targeting small molecules or peptides. In vitro inhibition assays, molecular docking, and X-ray crystallography were utilised by Zhang et al. (2020) to enhance ketoamide class $M^{pro}$ inhibitors with numerous functional groups. This was done to identify the most promising potential candidates. Specific inhibitors of Mpro's backbone were developed as part of the research conducted by Dai et al. (2020). Several peptidyl inhibitors targeting viral proteins have also been studied, in addition to small molecules. A common strategy was using the structure of the hACE2 and SARS-CoV-2 S RBD complex as a basis for designing peptide inhibitors that encompass essential ACE2 residues and bind to the RBD to avoid the RBD from interacting with host cells. These peptide inhibitors are intended to stop the RBD from interacting with the host cells (Han and Král 2020; Pomplun et al. 2020).

## 6.11 Conclusion and Future Scope

In the past two decades, SARS-CoV-2 casual agent of COVID-19 is the 3$^{rd}$ coronavirus to cause a worldwide epidemic in consort with MERS and SARS. Almost every nation and region on the planet has been impacted severely by the disease. Physical isolation, hygiene, medication, and vaccination were all used to reduce the spread of disease. Vaccination is the most successful preventive measure for this disease, but as new COVID variants evolved, the protection provided by the original vaccines did not. Scientists want to develop broadly protective vaccines that can target future variants and even related coronaviruses before they arise. Moreover, SARS-CoV-2 and its emerging variants are the outcome of evolving nature of the virus, posing a challenge to researchers. Therefore, in this chapter, herbal drugs, remedies, and practices for treatment and management of SARS-CoV-2 have been focused upon. Plants of the family Fabaceae and Lamiaceae are testified to have potent antiviral properties against SARS-CoV-2, along with plants of the family Asteraceae, Zingiberaceae, and Rutaceae. Plant secondary metabolites classes like alkaloids, flavonoids, and terpenoids majorly affect several targets of SARS-CoV-2, which includes $M^{pro}$, $PL^{pro}$, RdRp, spike, helicase, nucleocapsid, peptide, ACE 2, and TMPRSS. Looking into the plant diversity and huge amount of antiviral molecules they possess, there is a lack of in vivo studies and clinical trials. Computational studies are more in number as compared to in vitro studies. So, in future, more focus should be shifted to in vitro and in vivo studies for proper utilisation of antiviral plant molecules.

## References


Aanouz I, Belhassan A, El-Khatabi K, Lakhlifi T, El-Ldrissi M, Bouachrine M (2021) Moroccan Medicinal plants as inhibitors against SARS-CoV-2 main protease: computational investigations. J Biomol Struct Dyn 39(8):2971–2979

Abdallah HM, El-Halawany AM, Darwish KM, Algandaby MM, Mohamed GA, Ibrahim SR, Koshak AE, Elhady SS, Fadil SA, Alqarni AA, Abdel-Naim AB (2022) Bio-guided isolation of SARS-CoV-2 main protease inhibitors from medicinal plants: *in-vitro* assay and molecular dynamics. Plants 11(15):1914

Abookleesh FL, Al-Anzi BS, Ullah A (2022) Potential antiviral action of alkaloids. Molecules 27(3):903

Abu-Saleh AAAA, Awad IE, Yadav A, Poirier RA (2020) Discovery of potent inhibitors for SARS-CoV-2's main protease by ligand-based/structure-based virtual screening, MD simulations, and binding energy calculations. Phys Chem Chem Phys 22(40):23099–23106

Adem S, Eyupoglu V, Sarfraz I, Rasul A Ali M (2020) Identification of potent COVID-19 main protease (Mpro) inhibitors from natural polyphenols: an *in-silico* strategy unveils a hope against CORONA

Adhikari PP, Paul SB (2018) History of Indian traditional medicine: a medical inheritance. History 11(1)

Adhikari B, Marasini BP, Rayamajhee B, Bhattarai BR, Lamichhane G, Khadayat K, Adhikari A, Khanal S, Parajuli N (2021) Potential roles of medicinal plants for the treatment of viral diseases focusing on COVID-19: a review. Phytother Res *35*(3):1298–1312

Aftab SO, Ghouri MZ, Masood MU, Haider Z, Khan Z, Ahmad A, Munawar N (2020) Analysis of SARS-CoV-2 RNA-dependent RNA polymerase as a potential therapeutic drug target using a computational approach. J Transl Med 18(1):1–15

Ahmad J, Ikram S, Ahmad F, Rehman IU, Mushtaq M (2020) SARS-CoV-2 RNA dependent RNA polymerase (RdRp)–a drug repurposing study. Heliyon *6*(7):e04502

Ahmad S, Zahiruddin S, Parveen B, Basist P, Parveen A, Parveen R, Ahmad M (2021) Indian medicinal plants and formulations and their potential against COVID-19–preclinical and clinical research. Front Pharmacol:2470

Ahmed M (2020) Natural compounds from Djiboutian medicinal plants as inhibitors of COVID-19 by *in-silico* investigations

Alam S, Sarker MMR, Afrin S, Richi FT, Zhao C, Zhou JR (2021) Traditional herbal medicines, bioactive metabolites, and plant products against COVID-19: update on clinical trials and mechanism of actions. Front Pharmacol 2021(12):671498

Alamri MA, Altharawi A, Alabbas AB, Alossaimi MA, Alqahtani SM (2020) Structure-based virtual screening and molecular dynamics of phytochemicals derived from Saudi medicinal plants to identify potential COVID-19 therapeutics. Arab J Chem 13(9):7224–7234

Alrasheid AA, Babiker MY, Awad TA (2021) Evaluation of certain medicinal plants compounds as new potential inhibitors of novel corona virus (COVID-19) using molecular docking analysis. In-Silico Pharmacol 9(1):1–7

Ansari WA, Ahamad T, Khan MA, Khan ZA, Khan MF (2020) Luteolin: a dietary molecule as potential anti-COVID-19 agent

Bahun M, Jukić M, Oblak D, Kranjc L, Bajc G, Butala M, Bozovičar K, Bratkovič T, Podlipnik Č, Ulrih NP (2022) Inhibition of the SARS-CoV-2 3CLpro main protease by plant polyphenols. Food Chem 373:131594

Balachandar V, Mahalaxmi I, Kaavya J, Vivekanandhan G, Ajithkumar S, Arul N, Singaravelu G, Kumar NS, Devi SM (2020) COVID-19: emerging protective measures. Eur Rev Med Pharmacol Sci 24(6):3422–3425

Balkhair AA (2020) COVID-19 pandemic: a new chapter in the history of infectious diseases. Oman Med J 35(2):e123

Balkrishna A, Verma S, Solleti SK, Khandrika L, Varshney A (2020) Calcio-herbal medicine Divya-Swasari-Vati ameliorates SARS-CoV-2 spike protein-induced pathological features and inflammation in humanized zebrafish model by moderating IL-6 and TNF-α cytokines. J Inflamm Res:1219–1243

Balkrishna A, Haldar S, Singh H, Roy P, Varshney A (2021) Coronil, a tri-herbal formulation, attenuates spike-protein-mediated SARS-CoV-2 viral entry into human alveolar epithelial cells and pro-inflammatory cytokines production by inhibiting spike protein-ACE-2 interaction. J Inflamm Res 14:869

Balkrishna A, Goswami S, Singh H, Gohel V, Dev R, Haldar S, Varshney A (2022) Herbo-mineral formulation, Divya-Swasari-Vati averts SARS-CoV-2 pseudovirus entry into human alveolar epithelial cells by interfering with spike protein-ACE 2 interaction and IL-6/TNF-α/NF-κB signaling. Front Pharmacol 13:1024830

Ballhorn DJ, Kautz S, Heil M, Hegeman AD (2009) Analyzing plant defenses in nature. Plant Signal Behav 4(8):743–745

Basnet S, Marahatha R, Shrestha A, Bhattarai S, Katuwal S, Sharma KR, Marasini BP, Dahal SR, Basnyat RC, Patching SG, Parajuli N (2022) In-vitro and in-silico studies for the identification of potent metabolites of some high-altitude medicinal plants from Nepal inhibiting SARS-CoV-2 spike protein. Molecules 27(24):8957

Benkhaira N, Koraichi SI, Fikri-Benbrahim K (2021) Ethnobotanical survey on plants used by traditional healers to fight against COVID-19 in Fez city, Northern Morocco. Ethnobot Res Appl 21:1–18

Bestle D, Heindl MR, Limburg H, Pilgram O, Moulton H, Stein DA, Hardes K, Eickmann M, Dolnik O, Rohde C, Klenk HD (2020) TMPRSS2 and furin are both essential for proteolytic activation of SARS-CoV-2 in human airway cells. Life Sci Allian 3(9)

Bonaccorsi G, Pierri F, Cinelli M, Flori A, Galeazzi A, Porcelli F, Schmidt AL, Valensise CM, Scala A, Quattrociocchi W, Pammolli F (2020) Economic and social consequences of human mobility restrictions under COVID-19. Proc Natl Acad Sci 117(27):15530–15535

Boone SA, Gerba CP (2007) Significance of fomites in the spread of respiratory and enteric viral disease. Appl Environ Microbiol 73(6):1687–1696

Borris RP (1996) Natural products research: perspectives from a major pharmaceutical company. J Ethnopharmacol 51(1-3):29–38

Cascella M, Rajnik M, Aleem A, Dulebohn SC, Di Napoli R (2022) Features, evaluation, and treatment of coronavirus (COVID-19). StatPearls [internet]

Cazarolli LH, Zanatta L, Alberton EH, Bonorino Figueiredo MSR, Folador P, Damazio RG, Pizzolatti MG, Barreto Silva FRM (2008) Flavonoids: prospective drug candidates. Mini-Rev Med Chem 8(13):1429–1440

Chang HM, But PPH (2014) Pharmacology and applications of chinese materia medica, vol I

Chang FR, Yen CT, Ei-Shazly M, Lin WH, Yen MH, Lin KH, Wu YC (2012) Anti-human coronavirus (anti-HCoV) triterpenoids from the leaves of Euphorbia neriifolia. Nat Prod Commun 7(11):1415–1417

Chavan R, Chowdhary A (2014) *In-vitro* inhibitory activity of Justicia adhatoda extracts against influenza virus infection and hemagglutination. Int J Pharm Sci Rev Res 25(2):231–236

Chen X, Yin YH, Zhang MY, Liu JY, Li R, Qu YQ (2020) Investigating the mechanism of ShuFeng JieDu capsule for the treatment of novel Coronavirus pneumonia (COVID-19) based on network pharmacology. Int J Med Sci 17(16):2511

Cho JK, Curtis-Long MJ, Lee KH, Kim DW, Ryu HW, Yuk HJ, Park KH (2013) Geranylated flavonoids displaying SARS-CoV papain-like protease inhibition from the fruits of Paulownia tomentosa. Bioorg Med Chem 21(11):3051–3057

Chu DK, Akl EA, Duda S, Solo K, Yaacoub S, Schünemann HJ, El-Harakeh A, Bognanni A, Lotfi T, Loeb M, Hajizadeh A (2020) Physical distancing, face masks, and eye protection to prevent person-to-person transmission of SARS-CoV-2 and COVID-19: a systematic review and meta-analysis. Lancet 395(10242):1973–1987

Clarke NE, Turner AJ (2012) Angiotensin-converting enzyme 2: the first decade. Int J Hypertens 2012:307315

Cucinotta D, Vanelli M (2020) WHO declares COVID-19 a pandemic. Acta Bio Medica Atenei Parmensis 91(1):157

Dai W, Zhang B, Jiang XM, Su H, Li J, Zhao Y, Xie X, Jin Z, Peng J, Liu F, Li C (2020) Structure-based design of antiviral drug candidates targeting the SARS-CoV-2 main protease. Science 368(6497):1331–1335

Davies NG, Abbott S, Barnard RC, Jarvis CI, Kucharski AJ, Munday JD, Pearson CA, Russell TW, Tully DC, Washburne AD, Wenseleers T (2021) Estimated transmissibility and impact of SARS-CoV-2 lineage B. 1.1. 7 in England. Science 372(6538):3055

DeShazo RD, Chapin K, Swain RE (1997) Fungal sinusitis. N Engl J Med 337(4):254–259

Dev S (1999) Ancient-modern concordance in Ayurvedic plants: some examples. Environ Health Perspect 107(10):783–789

Dhama K, Khan S, Tiwari R, Sircar S, Bhat S, Malik YS, Singh KP, Chaicumpa W, Bonilla-Aldana DK, Rodriguez-Morales AJ (2020) Coronavirus disease 2019–COVID-19. Clin Microbiol Rev *33*(4):e00028–e00020

Dong Y, Dai T, Wei Y, Zhang L, Zheng M, Zhou F (2020) A systematic review of SARS-CoV-2 vaccine candidates. Sig Transduct Target Therap *5*(1):1–14

Dutta M, Nezam M, Chowdhury S, Rakib A, Paul A, Sami SA, Uddin MZ, Rana MS, Hossain S, Effendi Y, Idroes R (2021) Appraisals of the Bangladeshi medicinal plant Calotropis gigantea used by folk medicine practitioners in the management of COVID-19: a biochemical and computational approach. Front Mol Biosci 8:625391

El Alami A, Fattah A, Chait A (2020) Medicinal plants used for the prevention purposes during the covid-19 pandemic in Morocco. J Analy Sci Appl Biotechnol 2(1):2

Engler O, Strasser M, Signer J, Schoop R (2017) Neutralizing activity of Echinacea purpurea on coronaviruses including highly pathogenic Middle-East-Respiratory Syndrome Virus (MERS-CoV). Planta Med Int Open 4(Suppl 1):S1–S202

Enmozhi SK, Raja K, Sebastine I, Joseph J (2021) Andrographolide as a potential inhibitor of SARS-CoV-2 main protease: an *in-silico* approach. J Biomol Struct Dyn 39(9):3092–3098

Falsey AR, Sobieszczyk ME, Hirsch I, Sproule S, Robb ML, Corey L, Neuzil KM, Hahn W, Hunt J, Mulligan MJ, McEvoy C (2021) Phase 3 safety and efficacy of AZD1222 (ChAdOx1 nCoV-19) Covid-19 vaccine. N Engl J Med 385(25):2348–2360

Faria NR, Mellan TA, Whittaker C, Claro IM, Candido DDS, Mishra S, Crispim MA, Sales FC, Hawryluk I, McCrone JT, Hulswit RJ (2021) Genomics and epidemiology of the P. 1 SARS-CoV-2 lineage in Manaus, Brazil. Science 372(6544):815–821

Farnsworth NR (2007) The role of ethnopharmacology in drug development. In: Ciba foundation symposium 154-bioactive compounds from plants: bioactive compounds from plants: ciba foundation symposium 154. Wiley, Chichester, UK, pp 2–21

Feliciangeli SF, Thomas L, Scott GK, Subbian E, Hung CH, Molloy SS, Jean F, Shinde U, Thomas G (2006) Identification of a pH sensor in the furin propeptide that regulates enzyme activation. J Biol Chem 281(23):16108–16116

Fraser BJ, Beldar S, Seitova A, Hutchinson A, Mannar D, Li Y, Kwon D, Tan R, Wilson RP, Leopold K, Subramaniam S (2022) Structure and activity of human TMPRSS2 protease implicated in SARS-CoV-2 activation. Nat Chem Biol 18(9):963–971

Galloway SE, Paul P, MacCannell DR, Johansson MA, Brooks JT, MacNeil A, Slayton RB, Tong S, Silk BJ, Armstrong GL, Biggerstaff M (2021) Emergence of SARS-CoV-2 b. 1.1. 7 lineage—United States, December 29, 2020–January 12, 2021. Morb Mortal Wkly Rep 70(3): 95

Gangadharan S, Ambrose JM, Rajajagadeesan A, Kullappan M, Patil S, Gandhamaneni SH, Veeraraghavan VP, Nakkella AK, Agarwal A, Jayaraman S, Surapaneni KM (2022) Repurposing of potential antiviral drugs against RNA-dependent RNA polymerase of SARS-CoV-2 by computational approach. J Infect Public Health 15(11):1180–1191

Gani MA, Nurhan AD, Maulana S, Siswodihardjo S, Shinta DW, Khotib J (2021) Structure-based virtual screening of bioactive compounds from Indonesian medical plants against severe acute respiratory syndrome coronavirus-2. J Adv Pharmaceut Technol Res 12(2):120

Gautam S, Gautam A, Chhetri S, Bhattarai U (2022) Immunity against COVID-19: potential role of Ayush Kwath. J Ayurveda Integr Med 13(1):100350

Ghosh R, Chakraborty A, Biswas A, Chowdhuri S (2021) Identification of alkaloids from Justicia adhatoda as potent SARS CoV-2 main protease inhibitors: an *in-silico* perspective. J Mol Struct 1229:129489

Glowacka I, Bertram S, Müller MA, Allen P, Soilleux E, Pfefferle S, Steffen I, Tsegaye TS, He Y, Gnirss K, Niemeyer D (2011) Evidence that TMPRSS2 activates the severe acute respiratory syndrome coronavirus spike protein for membrane fusion and reduces viral control by the humoral immune response. J Virol 85(9):4122–4134

Godatwar PK, Deshpande S, JoshiDeshmukh PS, Deshpande VS, Ghungralekar R, Tamoli S, Gupta A, Vedula S, Rugvedi P, Rai RK (2021) Clinical evaluation of Chyawanprash as a preventive measure during the COVID-19 pandemic: an open-label, multicentric, randomized, comparative, prospective, and interventional community-based clinical study on healthy individuals. J Indian Syst Med 9(2):104

Golodetz KS (2020) 3CL protease inhibitor NLC-001 added to COVID-19-focused joint venture between Todos medical and NLC pharma. Available at: https://www.globenewswire.com/news-release/2020/09/17/2095292/0/en/3CL-Protease-Inhibitor-NLC-001-Added-to-COVID-19-focused-Joint-Venture-Between-Todos-Medical-and-NLC-Pharma.html. Accessed 17 Sept 2020

Gomaa AA, Abdel-Wadood YA (2021) The potential of glycyrrhizin and licorice extract in combating COVID-19 and associated conditions. Phytomed Plus 1(3):100043

Goyal RK, Majeed J, Tonk R, Dhobi M, Patel B, Sharma K, Apparsundaram S (2020) Current targets and drug candidates for prevention and treatment of SARS-CoV-2 (COVID-19) infection. Rev Cardiovasc Med 21(3):365–384

Guijarro-Real C, Plazas M, Rodríguez-Burruezo A, Prohens J, Fita A (2021) Potential *in-vitro* inhibition of selected plant extracts against SARS-CoV-2 chymotrypsin-like protease (3CLPro) activity. Foods 10(7):1503

Gupta S, Mishra KP, Ganju L (2017) Broad-spectrum antiviral properties of andrographolide. Arch Virol 162(3):611–623

Gurwitz D (2020) Angiotensin receptor blockers as tentative SARS-CoV-2 therapeutics. Drug Dev Res 81(5):537–540

Gyebi GA, Ogunro OB, Adegunloye AP, Ogunyemi OM, Afolabi SO (2021) Potential inhibitors of coronavirus 3-chymotrypsin-like protease (3CLpro): an *in-silico* screening of alkaloids and terpenoids from African medicinal plants. J Biomol Struct Dyn 39(9):3396–3408

Hamza M, Ali A, Khan S, Ahmed S, Attique Z, Ur Rehman S, Khan A, Ali H, Rizwan M, Munir A, Khan AM (2021) nCOV-19 peptides mass fingerprinting identification, binding, and blocking of inhibitors flavonoids and anthraquinone of Moringa oleifera and hydroxychloroquine. J Biomol Struct Dyn 39(11):4089–4099

Han Y, Král P (2020) Computational design of ACE2-based peptide inhibitors of SARS-CoV-2. ACS Nano *14*(4):5143–5147

Harcourt BH, Jukneliene D, Kanjanahaluethai A, Bechill J, Severson KM, Smith CM, Rota PA, Baker SC (2004) Identification of severe acute respiratory syndrome coronavirus replicase products and characterization of papain-like protease activity. J Virol 78(24):13600–13612

He X, Yang F, Huang XA (2021) Proceedings of chemistry, pharmacology, pharmacokinetics and synthesis of biflavonoids. Molecules 26(19):6088

Hives L, Bradley A, Richards J, Sutton C, Selfe J, Basu B, Maguire K, Sumner G, Gaber T, Mukherjee A, Perrin RN (2017) Can physical assessment techniques aid diagnosis in people with chronic fatigue syndrome/myalgic encephalomyelitis? A diagnostic accuracy study. BMJ Open 7(11):e017521

Hoffmann M, Kleine-Weber H, Schroeder S, Krüger N, Herrler T, Erichsen S, Schiergens TS, Herrler G, Wu NH, Nitsche A, Müller MA (2020) SARS-CoV-2 cell entry depends on ACE2 and TMPRSS2 and is blocked by a clinically proven protease inhibitor. Cell 181(2):271–280

https://www.amazon.in/

https://www.amazon.in/DABUR-Kaadha-Ayurvedic-Immunity-Boosting/dp/B08C5T9FSL

https://www.amazon.in/Golden-Turmeric-Latte-Saffron-Herbs/dp/B08B7WNTMD?th=1

https://www.ijbcp.com/index.php/ijbcp/article/view/4574/3151

Hu C, He B, Gong F, Liang M, Zhao D, Zhang G (2022) The adverse reactions of Lianhua Qingwen capsule/granule compared with conventional drug in clinical application: a meta-analysis. Front Pharmacol 13:764774

Huang S, Liu YE, Zhang Y, Zhang R, Zhu C, Fan L, Pei G, Zhang B, Shi Y (2020) Baicalein inhibits SARS-CoV-2/VSV replication with interfering mitochondrial oxidative phosphorylation in a mPTP dependent manner. Sig Transduct Target Therap 5(1):1–3

Huang L, Yuen TTT, Ye Z, Liu S, Zhang G, Chu H, Yue J (2021) Berbamine inhibits SARS-CoV-2 infection by compromising TRPMLs-mediated endolysosomal trafficking of ACE2. Sig Transduct Target Therap 6(1):1–3

Hussein RA, El-Anssary AA (2019) Plants secondary metabolites: the key drivers of the pharmacological actions of medicinal plants. Herb Med 1(3)

Iqbal HM, Romero-Castillo KD, Bilal M, Parra-Saldivar R (2020) The emergence of novel-coronavirus and its replication cycle-an overview. J Pure Appl Microbiol 14(1):13–16

Jain J, Kumar A, Narayanan V, Ramaswamy RS, Sathiyarajeswaran P, Devi MS, Kannan M, Sunil S (2020) Antiviral activity of ethanolic extract of Nilavembu Kudineer against dengue and chikungunya virus through *in-vitro* evaluation. J Ayurveda Integrat Med 11(3):329–335

Jamali N, Soureshjani EH, Mobini GR, Samare-Najaf M, Clark CC, Saffari-Chaleshtori J (2021) Medicinal plant compounds as promising inhibitors of coronavirus (COVID-19) main protease: an *in-silico* study. J Biomol Struct Dyn:1–12

Jin Z, Du X, Xu Y, Deng Y, Liu M, Zhao Y, Zhang B, Li X, Zhang L, Peng C, Duan Y (2020) Structure of Mpro from SARS-CoV-2 and discovery of its inhibitors. Nature 582(7811): 289–293

Jo S, Kim S, Shin DH, Kim MS (2020) Inhibition of SARS-CoV 3CL protease by flavonoids. J Enzyme Inhibit Med Chem 35(1):145–151

Kamal M, Abo Omirah M, Hussein A, Saeed H (2021) Assessment and characterisation of post-COVID-19 manifestations. Int J Clin Pract 75(3):e13746

Kandeil A, Mostafa A, Kutkat O, Moatasim Y, Al-Karmalawy AA, Rashad AA, Kayed AE, Kayed AE, El-Shesheny R, Kayali G, Ali MA (2021) Bioactive polyphenolic compounds showing strong antiviral activities against severe acute respiratory syndrome coronavirus 2. Pathogens 10(6):758

Kanjanasirirat P, Suksatu A, Manopwisedjaroen S, Munyoo B, Tuchinda P, Jearawuttanakul K, Seemakhan S, Charoensutthivarakul S, Wongtrakoongate P, Rangkasenee N, Pitiporn S (2020) High-content screening of Thai medicinal plants reveals Boesenbergia rotunda extract and its component Panduratin A as anti-SARS-CoV-2 agents. Sci Rep 10(1):1–12

Kaur SP, Gupta V (2020) COVID-19 Vaccine: a comprehensive status report. Virus Res 288: 198114

Khaerunnisa S, Kurniawan H, Awaluddin R, Suhartati S, Soetjipto S (2020) Potential inhibitor of COVID-19 main protease (Mpro) from several medicinal plant compounds by molecular docking study. Preprints 2020:2020030226

Khorshiddoust RR, Khorshiddoust SR, Hosseinabadi T, Mottaghitalab F, Mokhtari F, Azadinia F, Mozdarani H, Shabani M, Emadi-Kouchak H, Taheri B, Khani-Juyabad F (2022) Efficacy of a multiple-indication antiviral herbal drug (Saliravira®) for COVID-19 outpatients: a pre-clinical and randomized clinical trial study. Biomed Pharmacother 149:112729

Khuntia BK, Sharma V, Qazi S, Das S, Sharma S, Raza K, Sharma G (2021) Ayurvedic medicinal plants against COVID-19: an *in-silico* analysis. Nat Prod Commun 16(11):9

Kil YS, Pham ST, Seo EK, Jafari M (2017) Angelica keiskei, an emerging medicinal herb with various bioactive constituents and biological activities. Arch Pharm Res 40:655–675

Kim DW, Seo KH, Curtis-Long MJ, Oh KY, Oh JW, Cho JK, Lee KH, Park KH (2014) Phenolic phytochemical displaying SARS-CoV papain-like protease inhibition from the seeds of Psoralea corylifolia. J Enzyme Inhibit Med Chem 29(1):59–63

Kiran G, Karthik L, Devi MS, Sathiyarajeswaran P, Kanakavalli K, Kumar KM, Kumar DR (2020) *In-silico* computational screening of Kabasura Kudineer-official Siddha formulation and JACOM against SARS-CoV-2 spike protein. J Ayurveda Integr Med 13(1):100324

Kirchdoerfer RN, Ward AB (2019) Structure of the SARS-CoV nsp12 polymerase bound to nsp7 and nsp8 co-factors. Nat Commun 10(1):2342

Klein S, Cortese M, Winter SL, Wachsmuth-Melm M, Neufeldt CJ, Cerikan B, Stanifer ML, Boulant S, Bartenschlager R, Chlanda P (2020) SARS-CoV-2 structure and replication characterized by in situ cryo-electron tomography. Nat Commun 11(1):1–10

Ko WC, Rolain JM, Lee NY, Chen PL, Huang CT, Lee PI, Hsueh PR (2020) Arguments in favour of remdesivir for treating SARS-CoV-2 infections. Int J Antimicrob Agents 55(4):105933

Koulgi S, Jani V, Uppuladinne VN, Sonavane U, Joshi R (2021) Natural plant products as potential inhibitors of RNA dependent RNA polymerase of Severe Acute Respiratory Syndrome Coronavirus-2. PLoS One 16(5):e0251801

Krupanidhi S, Abraham Peele K, Venkateswarulu TC, Ayyagari VS, Nazneen Bobby M, John Babu D, Venkata Narayana A, Aishwarya G (2021) Screening of phytochemical compounds of Tinospora cordifolia for their inhibitory activity on SARS-CoV-2: an *in-silico* study. J Biomol Struct Dyn 39(15):5799–5803

Kuba K, Imai Y, Rao S, Gao H, Guo F, Guan B, Huan Y, Yang P, Zhang Y, Deng W, Bao L (2005) A crucial role of angiotensin converting enzyme 2 (ACE2) in SARS coronavirus–induced lung injury. Nat Med 11(8):875–879

Kumar S, Pandey AK (2013) Chemistry and biological activities of flavonoids: an overview. Sci World J 2013:162750

Lamontagne F, Agoritsas T, Macdonald H, Leo YS, Diaz J, Agarwal A, Appiah JA, Arabi Y, Blumberg L, Calfee CS, Cao B (2020) A living WHO guideline on drugs for covid-19. BMJ 370:m3379

Lee TW, Cherney MM, Huitema C, Liu J, James KE, Powers JC, Eltis LD, James MN (2005) Crystal structures of the main peptidase from the SARS coronavirus inhibited by a substrate-like aza-peptide epoxide. J Mol Biol 353(5):1137–1151

Lei J, Kusov Y, Hilgenfeld R (2018) Nsp3 of coronaviruses: structures and functions of a large multi-domain protein. Antivir Res 149:58–74

Li F, Li W, Farzan M, Harrison SC (2005) Structure of SARS coronavirus spike receptor-binding domain complexed with receptor. Science 309(5742):1864–1868

Li J, Huang Z, Lu S, Luo H, Tan Y, Ye P, Liu X, Wu Z, Wu C, Stalin A, Wang H (2021) Exploring potential mechanisms of Suhexiang Pill against COVID-19 based on network pharmacology and molecular docking. Medicine 100(51):e27112

Li DH, Wu L, Chen FQ, Liu XK, Mei JQ (2022) Mechanism of Huo-Xiang-Zheng-Qi in preventing and treating COVID-19: a study based on network pharmacology and molecular docking techniques. Nat Prod Commun 17(7):1934578X221102028

Liu Z, Xiao X, Wei X, Li J, Yang J, Tan H, Zhu J, Zhang Q, Wu J, Liu L (2020) Composition and divergence of coronavirus spike proteins and host ACE2 receptors predict potential intermediate hosts of SARS-CoV-2. J Med Virol 92(6):595–601

Liu CH, Lu CH, Wong SH, Lin LT (2021) Update on antiviral strategies against COVID-19: unmet needs and prospects. Front Immunol 11:616595

Liu CH, Lu CH, Lin LT (2022) Pandemic strategies with computational and structural biology against COVID-19: a retrospective. Comput Struct Biotechnol J 20:187–192

Lu G, Hu Y, Wang Q, Qi J, Gao F, Li Y, Zhang Y, Zhang W, Yuan Y, Bao J, Zhang B (2013) Molecular basis of binding between novel human coronavirus MERS-CoV and its receptor CD26. Nature 500(7461):227–231

Lu R, Zhao X, Li J, Niu P, Yang B, Wu H, Wang W, Song H, Huang B, Zhu N, Bi Y (2020) Genomic characterisation and epidemiology of 2019 novel coronavirus: implications for virus origins and receptor binding. Lancet 395(10224):565–574

Lung J, Lin YS, Yang YH, Chou YL, Shu LH, Cheng YC, Liu HT, Wu CY (2020) The potential chemical structure of anti-SARS-CoV-2 RNA-dependent RNA polymerase. J Med Virol 92(6): 693–697

Malabadi RB, Kolkar KP, Meti NT, Chalannavar RK (2021) Role of plant based hand sanitizers during the recent outbreak of coronavirus (SARS-CoV-2) disease (Covid-19). Sign Bioengineer Biosci 5(1):458–468

Matsabisa MG, Alexandre K, Ibeji CU, Tripathy S, Erukainure OL, Malatji K, Chauke S, Okole B, Chabalala HP (2022) *In-vitro* study on efficacy of PHELA, an African traditional drug against SARS-CoV-2. Sci Rep 12(1):1–12

Matsuyama S, Nagata N, Shirato K, Kawase M, Takeda M, Taguchi F (2010) Efficient activation of the severe acute respiratory syndrome coronavirus spike protein by the transmembrane protease TMPRSS2. J Virol 84(24):12658–12664

Maurya DK, Sharma D (2022) Evaluation of traditional ayurvedic Kadha for prevention and management of the novel Coronavirus (SARS-CoV-2) using in silico approach. J Biomol Struct Dyn 40(9):3949–3964. https://doi.org/10.1080/07391102.2020.1852119. Epub 2020 Nov 30. PMID: 33251972; PMCID: PMC7754934

Maurya VK, Kumar S, Bhatt ML, Saxena SK (2022) Antiviral activity of traditional medicinal plants from Ayurveda against SARS-CoV-2 infection. J Biomol Struct Dyn 40(4):1719–1735

McBride R, Van Zyl M, Fielding BC (2014) The coronavirus nucleocapsid is a multifunctional protein. Viruses *6*(8):2991–3018

Meites, E., Knuth, M., Hall, K., Dawson, P., Wang, T.W., Wright, M., Yu, W., Senesie, S., Stephenson, E., Imachukwu, C. and Sayi, T., 2022. COVID-19 scientific publications from the centers for disease control and prevention, January 2020–January 2022. Public Health Reports, 138(2):241–247.

Mohammadi N, Shaghaghi N (2020) Inhibitory effect of eight secondary metabolites from conventional medicinal plants on COVID_19 virus protease by molecular docking analysis

Mohapatra PK, Chopdar KS, Dash GC, Raval MK (2020) *In-silico* screening of phytochemicals of Ocimum sanctum against main protease of SARS-CoV-2

Moldofsky H, Patcai J (2011) Chronic widespread musculoskeletal pain, fatigue, depression and disordered sleep in chronic post-SARS syndrome; a case-controlled study. BMC Neurol 11(1): 1–7

Mottin M, Caesar LK, Brodsky D, Mesquita NC, de Oliveira KZ, Noske GD, Sousa BK, Ramos PR, Jarmer H, Loh B, Zorn KM (2022) Chalcones from Angelica keiskei (ashitaba) inhibit key Zika virus replication proteins. Bioorg Chem 120:105649

Mpiana PT, Tshibangu DS, Kilembe JT, Gbolo BZ, Mwanangombo DT, Inkoto CL, Lengbiye EM, Mbadiko CM, Matondo A, Bongo GN, Tshilanda DD (2020) Identification of potential inhibitors of SARS-CoV-2 main protease from Aloe vera compounds: a molecular docking study. Chem Phys Lett 754:137751

Muhseen ZT, Hameed AR, Al-Hasani HM, ul Qamar MT, Li G (2020) Promising terpenes as SARS-CoV-2 spike receptor-binding domain (RBD) attachment inhibitors to the human ACE2 receptor: integrated computational approach. J Mol Liq 320:114493

Mulligan MJ, Lyke KE, Kitchin N (2020) Phase 1/2 study to describe the safety and immunogenicity of a COVID-19 RNA vaccine candidate (BNT162b1) in adults 18 to 55 years of age: interim report. preprint

Murck H (2020) Symptomatic protective action of glycyrrhizin (licorice) in COVID-19 infection? Front Immunol 11:1239

Murugan NA, Pandian CJ, Jeyakanthan J (2021) Computational investigation on Andrographis paniculata phytochemicals to evaluate their potency against SARS-CoV-2 in comparison to known antiviral compounds in drug trials. J Biomol Struct Dyn 39(12):4415–4426

Murugesan S, Kottekad S, Crasta I, Sreevathsan S, Usharani D, Perumal MK, Mudliar SN (2021) Targeting COVID-19 (SARS-CoV-2) main protease through active phytocompounds of ayurvedic medicinal plants–Emblica officinalis (Amla), Phyllanthus niruri Linn. (Bhumi Amla) and Tinospora cordifolia (Giloy)–A molecular docking and simulation study. Comput Biol Med 136:104683

Nguyen TTH, Yu SH, Kim J, An E, Hwang K, Park JS, Kim D (2015) Enhancement of quercetin water solubility with steviol glucosides and the studies of biological properties. Funct Foods Health Dis 5(12):437–449

Nivetha R, Bhuvaragavan S, Muthu Kumar T, Ramanathan K, Janarthanan S (2021) Inhibition of multiple SARS-CoV-2 proteins by an antiviral biomolecule, seselin from Aegle marmelos deciphered using molecular docking analysis. J Biomol Struct Dyn:1–12

Osipiuk J, Azizi SA, Dvorkin S, Endres M, Jedrzejczak R, Jones KA, Kang S, Kathayat RS, Kim Y, Lisnyak VG, Maki SL (2021) Structure of papain-like protease from SARS-CoV-2 and its complexes with non-covalent inhibitors. Nat Commun 12(1):1–9

Ou X, Liu Y, Lei X, Li P, Mi D, Ren L, Guo L, Guo R, Chen T, Hu J, Xiang Z (2020) Characterization of spike glycoprotein of SARS-CoV-2 on virus entry and its immune cross-reactivity with SARS-CoV. Nat Commun 11(1):1–12

Pachetti M, Marini B, Benedetti F, Giudici F, Mauro E, Storici P, Masciovecchio C, Angeletti S, Ciccozzi M, Gallo RC, Zella D (2020) Emerging SARS-CoV-2 mutation hot spots include a novel RNA-dependent-RNA polymerase variant. J Transl Med 18:1–9

Park JY, Kim JH, Kim YM, Jeong HJ, Kim DW, Park KH, Kwon HJ, Park SJ, Lee WS, Ryu YB (2012a) Tanshinones as selective and slow-binding inhibitors for SARS-CoV cysteine proteases. Bioorg Med Chem 20(19):5928–5935

Park J-Y, Jeong HJ, Kim JH, Kim YM, Park S-J, Kim D, Park KH, Lee WS, Ryu YB (2012b) Diarylheptanoids from Alnus japonica inhibit papain-like protease of severe acute respiratory syndrome coronavirus. Biol Pharm Bull 35(11):2036–2042

Park JY, Ko JA, Kim DW, Kim YM, Kwon HJ, Jeong HJ, Kim CY, Park KH, Lee WS, Ryu YB (2016) Chalcones isolated from Angelica keiskei inhibit cysteine proteases of SARS-CoV. J Enzyme Inhibit Med Chem 31(1):23–30

Park JY, Yuk HJ, Ryu HW, Lim SH, Kim KS, Park KH, Ryu YB, Lee WS (2017) Evaluation of polyphenols from Broussonetia papyrifera as coronavirus protease inhibitors. J Enzyme Inhibit Med Chem 32(1):504–512

Pattaro-Júnior JR, Araújo IG, Moraes CB, Barbosa CG, Philippsen GS, Freitas-Junior LH, Guidi AC, de Mello JCP, Peralta RM, Fernandez MA, Teixeira RR (2022) Antiviral activity of Cenostigma pluviosum var. peltophoroides extract and fractions against SARS-CoV-2. J Biomol Struct Dyn:1–12

Perrin R, Riste L, Hann M, Walther A, Mukherjee A, Heald A (2020) Into the looking glass: post-viral syndrome post COVID-19. Med Hypotheses 144:110055

Piret J, Boivin G (2021) Pandemics throughout history. Front Microbiol 11:631736

Pomplun S, Jbara M, Quartararo AJ, Zhang G, Brown JS, Lee YC, Ye X, Hanna S, Pentelute BL (2020) De novo discovery of high-affinity peptide binders for the SARS-CoV-2 spike protein. ACS Cent Sci 7(1):156–163

Puttaswamy H, Gowtham HG, Ojha MD, Yadav A, Choudhir G, Raguraman V, Kongkham B, Selvaraju K, Shareef S, Gehlot P, Ahamed F (2020) *In-silico* studies evidenced the role of structurally diverse plant secondary metabolites in reducing SARS-CoV-2 pathogenesis. Sci Rep 10(1):1–24

Pye CR, Bertin MJ, Lokey RS, Gerwick WH, Linington RG (2017) Retrospective analysis of natural products provides insights for future discovery trends. Proc Natl Acad Sci 114(22): 5601–5606

Rathinavel T, Palanisamy M, Palanisamy S, Subramanian A, Thangaswamy S (2020) Phytochemical 6-gingerol–a promising drug of choice for COVID-19. Int J Adv Sci Eng 6(4):1482–1489

Ratia K, Kilianski A, Baez-Santos YM, Baker SC, Mesecar A (2014) Structural basis for the ubiquitin-linkage specificity and deISGylating activity of SARS-CoV papain-like protease. PLoS Pathog 10(5):e1004113

Risener CJ, Woo S, Samarakoon T, Caputo M, Edwards E, Klepzig K, Applequist W, Zandi K, Goh SL, Downs-Bowen JA, Schinazi RF (2023) Botanical inhibitors of SARS-CoV-2 viral entry: a phylogenetic perspective. Sci Rep 13(1):1244

Rolta R, Yadav R, Salaria D, Trivedi S, Imran M, Sourirajan A, Baumler DJ, Dev K (2021) *In-silico* screening of hundred phytocompounds of ten medicinal plants as potential inhibitors of nucleocapsid phosphoprotein of COVID-19: an approach to prevent virus assembly. J Biomol Struct Dyn 39(18):7017–7034

Runfeng L, Yunlong H, Jicheng H, Weiqi P, Qinhai M, Yongxia S, Chufang L, Jin Z, Zhenhua J, Haiming J, Kui Z (2020) Lianhuaqingwen exerts anti-viral and anti-inflammatory activity against novel coronavirus (SARS-CoV-2). Pharmacol Res 156:104761

Ryu YB, Jeong HJ, Kim JH, Kim YM, Park JY, Kim D, Naguyen TTH, Park SJ, Chang JS, Park KH, Rho MC (2010a) Biflavonoids from Torreya nucifera displaying SARS-CoV 3CLpro inhibition. Bioorg Med Chem 18(22):7940–7947

Ryu YB, Park SJ, Kim YM, Lee JY, Seo WD, Chang JS, Park KH, Rho MC, Lee WS (2010b) SARS-CoV 3CLpro inhibitory effects of quinone-methide triterpenes from Tripterygium regelii. Bioorg Med Chem Lett 20(6):1873–1876

Salamanna F, Veronesi F, Martini L, Landini MP, Fini M (2021) Post-COVID-19 syndrome: the persistent symptoms at the post-viral stage of the disease. A systematic review of the current data. Front Med 8:653516

Sampangi-Ramaiah MH, Vishwakarma R, Shaanker RU (2020) Molecular docking analysis of selected natural products from plants for inhibition of SARS-CoV-2 main protease. Curr Sci 118(7):1087–1092

Sardar T, Nadim SS, Rana S, Chattopadhyay J (2020) Assessment of lockdown effect in some states and overall India: a predictive mathematical study on COVID-19 outbreak. Chaos, Solitons Fractals *139*:110078

Saxena SK (ed) (2020) Coronavirus disease 2019 (COVID-19): epidemiology, pathogenesis, diagnosis, and therapeutics. Springer

Schubert K, Karousis ED, Jomaa A, Scaiola A, Echeverria B, Gurzeler LA, Leibundgut M, Thiel V, Mühlemann O, Ban N (2020) SARS-CoV-2 Nsp1 binds the ribosomal mRNA channel to inhibit translation. Nat Struct Mol Biol 27(10):959–966

Schultes RE, Raffauf RF (1990) The healing forest: medicinal and toxic plants of the Northwest Amazonia. Dioscorides Press

Shawky E, Nada AA, Ibrahim RS (2020) Potential role of medicinal plants and their constituents in the mitigation of SARS-CoV-2: identifying related therapeutic targets using network pharmacology and molecular docking analyses. RSC Adv 10(47):27961–27983

Shereen MA, Khan S, Kazmi A, Bashir N, Siddique R (2020) COVID-19 infection: emergence, transmission, and characteristics of human coronaviruses. J Adv Res 24:91–98

Shree P, Mishra P, Selvaraj C, Singh SK, Chaube R, Garg N, Tripathi YB (2022) Targeting COVID-19 (SARS-CoV-2) main protease through active phytochemicals of ayurvedic medicinal plants–Withania somnifera (Ashwagandha), Tinospora cordifolia (Giloy) and Ocimum sanctum (Tulsi)–a molecular docking study. J Biomol Struct Dyn 40(1):190–203

Signer J, Jonsdottir HR, Albrich WC, Strasser M, Züst R, Ryter S, Ackermann-Gäumann R, Lenz N, Siegrist D, Suter A, Schoop R (2020) *In-vitro* antiviral activity of Echinaforce®, an Echinacea purpurea preparation, against common cold coronavirus 229E and highly pathogenic MERS-CoV and SARS-CoV. Virol J 17(1):136

Singanayagam A, Hakki S, Dunning J, Madon KJ, Crone MA, Koycheva A, Derqui-Fernandez N, Barnett JL, Whitfield MG, Varro R, Charlett A (2022) Community transmission and viral load kinetics of the SARS-CoV-2 delta (B. 1.617. 2) variant in vaccinated and unvaccinated individuals in the UK: a prospective, longitudinal, cohort study. Lancet Infect Dis 22(2): 183–195

Song YH, Kim DW, Curtis-Long MJ, Yuk HJ, Wang Y, Zhuang N, Lee KH, Jeon KS, Park KH (2014) Papain-like protease (PLpro) inhibitory effects of cinnamic amides from Tribulus terrestris fruits. Biol Pharm Bull 37(6):1021–1028

Srinivasan V, Brognaro H, Prabhu PR, de Souza EE, Günther S, Reinke PY, Lane TJ, Ginn H, Han H, Ewert W, Sprenger J (2022) Antiviral activity of natural phenolic compounds in complex at an allosteric site of SARS-CoV-2 papain-like protease. Commun Biol 5(1):805

Srivastava A, Rengaraju M, Srivastava S, Narayanan V, Gupta V, Upadhayay R, Kumar J, Parameswaran S (2021) Efficacy of two siddha polyherbal decoctions, Nilavembu Kudineer and Kaba Sura Kudineer, along with standard allopathy treatment in the management of mild to moderate symptomatic COVID-19 patients—a double-blind, placebo-controlled, clinical trial. Trials 22(1):1–11

Su HX, Yao S, Zhao WF, Li MJ, Liu J, Shang WJ, Xie H, Ke CQ, Hu HC, Gao MN, Yu KQ (2020) Anti-SARS-CoV-2 activities in-vitro of Shuanghuanglian preparations and bioactive ingredients. Acta Pharmacol Sin 41(9):1167–1177

Su YC, Huang GJ, Lin JG (2022) Chinese herbal prescriptions for COVID-19 management: special reference to Taiwan Chingguan Yihau (NRICM101). Front Pharmacol 13:928106

Subissi L, Posthuma CC, Collet A, Zevenhoven-Dobbe JC, Gorbalenya AE, Decroly E, Snijder EJ, Canard B, Imbert I (2014) One severe acute respiratory syndrome coronavirus protein complex integrates processive RNA polymerase and exonuclease activities. Proc Natl Acad Sci 111(37): E3900–E3909

Tang T, Bidon M, Jaimes JA, Whittaker GR, Daniel S (2020) Coronavirus membrane fusion mechanism offers a potential target for antiviral development. Antivir Res 178:104792

Tegally H, Wilkinson E, Giovanetti M, Iranzadeh A, Fonseca V, Giandhari J, Doolabh D, Pillay S, San EJ, Msomi N, Mlisana K (2021) Detection of a SARS-CoV-2 variant of concern in South Africa. Nature 592(7854):438–443

Tegen D, Dessie K, Damtie D (2021) Candidate anti-COVID-19 medicinal plants from Ethiopia: a review of plants traditionally used to treat viral diseases. Evid Based Complement Alternat Med 2021(2021):6622410

Thakar A, Panara K, Shah H, Kalsariya B, Ruparel S, Jain N, Bhatt P, Jani D, Dodia R, Patel F, Salve R (2022) Guduchi Ghanavati (Ayurveda medication) improves the perceived immunity in individuals at risk of SARS-CoV-2: a multicentred, controlled, before-and-after study. Eur J Integrat Med 53:102131

Thoms M, Buschauer R, Ameismeier M, Koepke L, Denk T, Hirschenberger M, Kratzat H, Hayn M, Mackens-Kiani T, Cheng J, Straub JH (2020) Structural basis for translational shutdown and immune evasion by the Nsp1 protein of SARS-CoV-2. Science 369(6508): 1249–1255

Thuy BTP, My TTA, Hai NTT, Hieu LT, Hoa TT, Loan TP, Triet NT, Anh TT, Quy PT, Tat PV, Hue NV (2020) Investigation into SARS-CoV-2 resistance of compounds in garlic essential oil. ACS Omega 5(14):8312–8320

Tikellis C, Thomas MC (2012) Angiotensin-converting enzyme 2 (ACE2) is a key modulator of the renin angiotensin system in health and disease. Int J Pept 2012:256294

Torjesen I (2021) Covid-19: Omicron may be more transmissible than other variants and partly resistant to existing vaccines, scientists fear

Tsai KC, Huang YC, Liaw CC, Tsai CI, Chiou CT, Lin CJ, Wei WC, Lin SJS, Tseng YH, Yeh KM, Lin YL (2021) A traditional Chinese medicine formula NRICM101 to target COVID-19 through multiple pathways: a bedside-to-bench study. Biomed Pharmacother 133:111037

Tu YF, Chien CS, Yarmishyn AA, Lin YY, Luo YH, Lin YT, Lai WY, Yang DM, Chou SJ, Yang YP, Wang ML (2020) A review of SARS-CoV-2 and the ongoing clinical trials. Int J Mol Sci 21(7):2657

Tung NH, Kwon HJ, Kim JH, Ra JC, Kim JA, Kim YH (2010) An anti-influenza component of the bark of Alnus japonica. Arch Pharm Res 33:363–367

Twohig KA, Nyberg T, Zaidi A, Thelwall S, Sinnathamby MA, Aliabadi S, Seaman SR, Harris RJ, Hope R, Lopez-Bernal J, Gallagher E (2022) Hospital admission and emergency care attendance risk for SARS-CoV-2 delta (B. 1.617. 2) compared with alpha (B. 1.1. 7) variants of concern: a cohort study. Lancet Infect Dis 22(1):35–42

ul Qamar MT, Alqahtani SM, Alamri MA, Chen LL (2020) Structural basis of SARS-CoV-2 3CLpro and anti-COVID-19 drug discovery from medicinal plants. J Pharmaceut Analy 10(4): 313–319

United Nations Development Programme (2020) Addressing the COVID-19 economic crisis in Asia through social protection. United States of America

van de Sand L, Bormann M, Alt M, Schipper L, Heilingloh CS, Steinmann E, Todt D, Dittmer U, Elsner C, Witzke O, Krawczyk A (2021) Glycyrrhizin effectively inhibits SARS-CoV-2 replication by inhibiting the viral main protease. Viruses 13(4):609

Vasireddy D, Vanaparthy R, Mohan G, Malayala SV, Atluri P (2021) Review of COVID-19 variants and COVID-19 vaccine efficacy: what the clinician should know? J Clin Med Res 13(6):317

Vaughan A (2021) Omicron emerges. New Sci 252(3363):7

Velavan TP, Ntoumi F, Kremsner PG, Lee SS, Meyer CG (2023) Emergence and geographic dominance of Omicron subvariants XBB/XBB. 1.5 and BF. 7–the public health challenges. Int J Infect Dis 128:307–309

Verma D, Mitra D, Paul M, Chaudhary P, Kamboj A, Thatoi H, Janmeda P, Jain D, Panneerselvam P, Shrivastav R, Pant K (2021) Potential inhibitors of SARS-CoV-2 (COVID 19) proteases PLpro and Mpro/3CLpro: molecular docking and simulation studies of three pertinent medicinal plant natural components. Curr Res Pharmacol Drug Discov 2:100038

Wang P, Casner RG, Nair MS, Wang M, Yu J, Cerutti G, Liu L, Kwong PD, Huang Y, Shapiro L, Ho DD (2021) Increased resistance of SARS-CoV-2 variant P. 1 to antibody neutralization. Cell Host Microbe 29(5):747–751

Wen CC, Shyur LF, Jan JT, Liang PH, Kuo CJ, Arulselvan P, Wu JB, Kuo SC, Yang NS (2011) Traditional Chinese medicine herbal extracts of Cibotium barometz, Gentiana scabra, Dioscorea batatas, Cassia tora, and Taxillus chinensis inhibit SARS-CoV replication. J Tradit Complement Med 1(1):41–50

WHO (n.d.) WHO warns Omicron variant poses 'very high' global risk. World Health Organization News. Al Jazeera. https://www.aljazeera.com/news/2021/11/29/whocovid-omicron-variant-very-high-global-risk. Accessed 13 Dec 2021

Woo PC, Huang Y, Lau SK, Yuen KY (2010) Coronavirus genomics and bioinformatics analysis. Viruses 2(8):1804–1820

Wrapp D, Wang N, Corbett KS, Goldsmith JA, Hsieh CL, Abiona O, Graham BS, McLellan JS (2020) Cryo-EM structure of the 2019-nCoV spike in the prefusion conformation. Science 367(6483):1260–1263

Wu YH, Zhang BY, Qiu LP, Guan RF, Ye ZH, Yu XP (2017) Structure properties and mechanisms of action of naturally originated phenolic acids and their derivatives against human viral infections. Curr Med Chem 24(38):4279–4302

Wu C, Zheng M, Yang Y, Gu X, Yang K, Li M, Liu Y, Zhang Q, Zhang P, Wang Y, Wang Q (2020) Furin: a potential therapeutic target for COVID-19. iScience 23(10):101642

Xia S, Zhu Y, Liu M, Lan Q, Xu W, Wu Y, Ying T, Liu S, Shi Z, Jiang S, Lu L (2020) Fusion mechanism of 2019-nCoV and fusion inhibitors targeting HR1 domain in spike protein. Cell Mol Immunol 17(7):765–767

Xie Y, Yin W, Zhang Y, Shang W, Wang Z, Luan X, Tian G, Aisa HA, Xu Y, Xiao G, Li J (2021) Design and development of an oral remdesivir derivative VV116 against SARS-CoV-2. Cell Res 31(11):1212–1214

Xu Z, Yang L, Zhang X, Zhang Q, Yang Z, Liu Y, Wei S, Liu W (2020) Discovery of potential flavonoid inhibitors against COVID-19 3CL proteinase based on virtual screening strategy. Front Mol Biosci 7:556481

Yang H, Xie W, Xue X, Yang K, Ma J, Liang W, Zhao Q, Zhou Z, Pei D, Ziebuhr J, Hilgenfeld R (2005) Design of wide-spectrum inhibitors targeting coronavirus main proteases. PLoS Biol 3(10):e324

Yang C, Pan X, Xu X, Cheng C, Huang Y, Li L, Jiang S, Xu W, Xiao G, Liu S (2020) Salvianolic acid C potently inhibits SARS-CoV-2 infection by blocking the formation of six-helix bundle core of spike protein. Signal Transduct Target Ther 5(1):1–3

Yao X, Ye F, Zhang M, Cui C, Huang B, Niu P, Liu X, Zhao L, Dong E, Song C, Zhan S (2020) *In-vitro* antiviral activity and projection of optimized dosing design of hydroxychloroquine for the treatment of severe acute respiratory syndrome coronavirus 2 (SARS-CoV-2). Clin Infect Dis 71(15):732–739

Yearsley C (2021) Thailand approves Asian herb Andrographis to treat COVID-19. HerbalGram:35–37

Yu MS, Lee J, Lee JM, Kim Y, Chin YW, Jee JG, Keum YS, Jeong YJ (2012) Identification of myricetin and scutellarein as novel chemical inhibitors of the SARS coronavirus helicase, nsP13. Bioorg Med Chem Lett 22(12):4049–4054

Zhang L, Lin D, Sun X, Curth U, Drosten C, Sauerhering L, Becker S, Rox K, Hilgenfeld R (2020) Crystal structure of SARS-CoV-2 main protease provides a basis for design of improved α-ketoamide inhibitors. Science 368(6489):409–412

Zhang X, Xue Y, Chen X, Wu JM, Su ZJ, Sun M, Liu LJ, Zhang YB, Xu GH, Shi MY, Song XM (2021) Effects of Tanreqing Capsule on the negative conversion time of nucleic acid in patients with COVID-19: a retrospective cohort study. J Integrat Med 19(1):36–41

Zhao J, Sun J, He WT, Ji X, Gao Q, Zhai X, Suchard MA, Hong SL, Baele G, Su S, Zhou J (2020) Snapshot of the evolution and mutation patterns of SARS-CoV-2. BioRxiv

Zhong LLD, Lam WC, Yang W, Chan KW, Sze SCW, Miao J, Yung KKL, Bian Z, Wong VT (2020) Potential targets for treatment of coronavirus disease 2019 (COVID-19): a review of Qing-Fei-Pai-Du-Tang and its major herbs. Am J Chin Med 48(05):1051–1071

Zhou Z, Ren L, Zhang L, Zhong J, Xiao Y, Jia Z, Guo L, Yang J, Wang C, Jiang S, Yang D (2020) Heightened innate immune responses in the respiratory tract of COVID-19 patients. Cell Host Microbe 27(6):883–890

# Chapter 7
# Antiviral Phytocompounds Against Animal-to-Human Transmittable SARS-CoV-2

**Partha Biswas, Anwar Parvez, Asif Abdullah, Tanjim Ishraq Rahaman, Dipta Dey, Shakil Ahmmed, Md. Abdur Rashid Mia, Ranjit Chandra Das, Sharifa Sultana, and Shabana Bibi**

P. Biswas
Laboratory of Pharmaceutical Biotechnology and Bioinformatics, Department of Genetic Engineering and Biotechnology, Jashore University of Science and Technology, Jashore, Bangladesh

ABEx Bio-Research Center, Dhaka, Bangladesh

A. Parvez · S. Sultana
Department of Pharmacy, Faculty of Allied Health Sciences, Daffodil International University, Dhaka, Bangladesh

A. Abdullah
Department of Biomedical Engineering, Jashore University of Science and Technology, Jashore, Bangladesh

T. I. Rahaman
Department of Biotechnology and Genetic Engineering, Faculty of Life Science, Bangabandhu Sheikh Mujibur Rahman Science and Technology University, Gopalganj, Bangladesh

D. Dey · S. Ahmmed
Biochemistry and Molecular Biology Department, Life Science Faculty, Bangabandhu Sheikh Mujibur Rahman Science and Technology University, Gopalgonj, Bangladesh

Md. A. R. Mia
Department of Pharmaceutical Technology, Faculty of Pharmacy, International Islamic University Malaysia, Kuantan, Pahang, Malaysia

R. C. Das
Materials Science and Engineering, Florida State University, Tallahassee, FL, USA

S. Bibi (✉)
Department of Bioscience, Shifa Tameer-e-Millat University, Islamabad, Pakistan

Yunnan Herbal Laboratory, College of Ecology and Environmental Sciences, Yunnan University, Kunming, China

 
J.-T. Chen (ed.), *Ethnopharmacology and Drug Discovery for COVID-19: Anti-SARS-CoV-2 Agents from Herbal Medicines and Natural Products*,
https://doi.org/10.1007/978-981-99-3664-9_7

## 7.1 Background

COVID-19 which causes disease in the human respiratory system is brought on by coronavirus-2, which culminates in SARS-CoV-2. Since COVID-19's debut in Wuhan, China, in December 2019, almost 230 nations have joined (Chauhan 2020). Because of its genetic resemblance to SARS-CoV-1, the disastrous near-pandemic virus that hit in 2002–2003, SARS-CoV-2, the COVID-19 agent, was given this moniker (Ksiazek et al. 2003). The genetic sequences of SARS-CoV-2 and viruses from either humans or animals have never been detected before 2019. Both the SARS and MERS coronavirus (MERS-CoV; 70% similarity) are beta coronaviruses (Cov) with the same genetic makeup and viral structure (SARS-CoV; 349 fatalities in China between 2002 and 2003) (Biswas et al. 2021a; Zilani et al. 2021). Several healthcare facilities in Wuhan, China, reported a large number of patients with pneumonia whose origin was unknown in the latter part of December 2019. These people had signs of viral pneumonia with those who had SARS and MERS, such as coughing, fever, and also feeling mild to acute pain in the chest, as well as more severe symptoms such as dyspnea and bilateral lung infiltration (Zhu et al. 2020; Gralinski and Menachery 2020). Recent research studies reported that the SARS-CoV-2 spike (S) mutant protein D614G exhibited lesser infection to the upper respiratory tract compared to the B.1.1.7 (now known as the alpha variation, the first epidemic in the UK) (Plante et al. 2021).

The lack of licensed medications for the treatment of virulent diseases has come to light consequently for COVID-19. Numerous natural phytochemicals (NPs) possess bioactive characteristics that have antiviral (AV) and anti-inflammatory effects. Several vascular NPs, comprising alkaloids, polyphenols, tannins, and flavonoids, have been demonstrated to be effective as AVs, particularly for the treatment purpose of COVID-19 (Españo et al. 2021). Inflammation is one of the primary COVID-19 indications and symptoms. The anti-inflammatory properties of several NPs offer a workable solution for accelerating the recovery from anosmia and ageusia brought on by COVID-19. Even without considering the essential oils' chemical composition, scent training has already been shown to speed anosmia recovery (Koyama et al. 2021). NPs utilized in olfactory training are among those found in essential oils (lemon, rose, eucalyptus, and clove) (Koyama and Heinbockel 2020; Paul et al. 2022).

Recent research has placed a strong emphasis on the AV effects of diverse NPs generated from different resources, such as vegetables, fruits, plants, and cereals, as possible sources of various bioactive components. In addition, a lot of natural substances have polypharmacology or many mechanisms of action. Due to their polypharmacology, natural products may be more successful than traditional single-target pharmaceuticals (Tomas et al. 2022). The AV activity of those components may also be impacted by the additive or synergistic actions of many components; for instance, griffithsin and carrageenan together were successful in preventing COVID infections, along with the most recent SARS-CoV-2 mutations (Alsaidi et al. 2021; Khan et al. 2021a; Ahmed et al. 2022).

Before considering using the medicine, one should be aware of any possible adverse effects, such as hyperglycemia, ulceration, hepatic liver failure, angioedema, stomach irritation, hemolytic anemia, and immune system problems (Shakoor et al. 2021). Natural medications with high absorption and low cytotoxicity are the best options. During the initial phase of COVID-19 infection, these NPs can effectively halt the connection of the virus and its penetration in the cell by blocking enzymes ACE2, papain-like protease (PLpro), and 3CLpro (Dey et al. 2022a; Munshi et al. 2022; Sohel et al. 2022). The synergistic action of several NPs is opening the door for the design and recognition of novel SARS-CoV-2 AV treatment drugs. Combining azithromycin and hydroxychloroquine is reported to be somewhat more efficient in diminishing the viral efficiency in cases of COVID-19 (Gautret et al. 2020). The plants *Scutettaria baicalensis* and *Asplenium belangeri* contain the flavonoids myricetin and scutellarein respectively, which halt the action of ATPase of the nsP13 which is the helicase of SARS-CoV-2 (Yu et al. 2012). The Government of India has suggested a composition comprised of 15 different plants, including *Hygrophilla auriculata*, Zingiber *officinale*, *Tragia involucrata*, *Plectranthus amboinicus*, *Saussurea costus*, *Adhatoda vasica*, *Syzygium aromaticum*, *Terminalia chebula*, *Anacyclus pyrethrum*, *Piper longum*, and *Tinospora cordifolia* (Arefin et al. 2021; Paul et al. 2021; Rahman et al. 2021; Hossain et al. 2022).

## 7.2 Structural Composition of SARS-CoV-2

Bioinformatic research about the structure composition of SARS-CoV-2 has revealed that it has characteristics in common with the family of coronavirus. It is a member of the 2B lineage of beta-CoVs (Lai et al. 2020). Five SARS-CoV-2 infected individuals had their full genomes sequenced at the beginning of the Wuhan pneumococcal outbreak. The study found 79.5 percent of the information in these genome sequences and the sequencing of SARS-CoV is identical though these are different viruses. It is believed that the virus is a novel human betacoronavirus (Zhou et al. 2020). According to a study, the strain of bat coronavirus BatCov RaTG13 and SARS-CoV-2 is 96% identical. These findings imply that SARS-CoV-2 arose spontaneously, a major source of infections was bats Cov RaTG13 and may have come from a bat (Zhou et al. 2020; Zhang et al. 2020; Wang et al. 2020a).

There is a widespread consensus that CoVs, a diverse collection of single-stranded RNA viruses that can infect both people and animals, can cause respiratory, gastrointestinal, hepatic, and neurological problems (Weiss and Leibowitz 2011). CoVs are classified into four genera: alpha, beta, gamma, and delta CoVs. The six human coronaviruses (HCoVs) that have been identified so far are the HCoVs-OC43, beta-CoVs, SARS-CoV, HCoVs-HKU1 (Hossain et al. 2022; Lai et al. 2020), and MERS-CoV (Yang and Leibowitz 2015). Alpha-CoVs include HCoVs-229E, HCoVs-NL63, and HCoVs-OC (Drosten et al. 2003).

### 7.2.1 Spike (S) Proteins

The transmembrane (TM) region embedded in the viral envelope, an extracellular N-terminus, and a brief intracellular C-terminal stretch comprise the structural S protein (180 to 200 k daltons) (Liu et al. 2014). When the virus encounters the host cell, the S protein undergoes substantial structural changes, allowing the virus to cling to the host cell's membrane. The S protein often has a metastable, prefusion structure. Polysaccharide molecules are used to stop the immune system from identifying the spikes as soon as they enter (Bosch et al. 2003; Watanabe et al. 2020).

The SARS-CoV-2 S is composed of the S1 and S2 subunits, which contain a total of 1273 amino acids. However, unlike the S1 subunit, which has 14–685 amino acid residues, the S2 subunit contains 686–1273 amino acid residues (Fig. 7.1). The other two do this task by binding to receptors and fusing membranes. In contrast to the S1 subunit, which contains the receptor-binding domain (RBD, 319–541 amino acid residues), the S2 subunit contains the fusion peptide (FP), the heptapeptide repeat

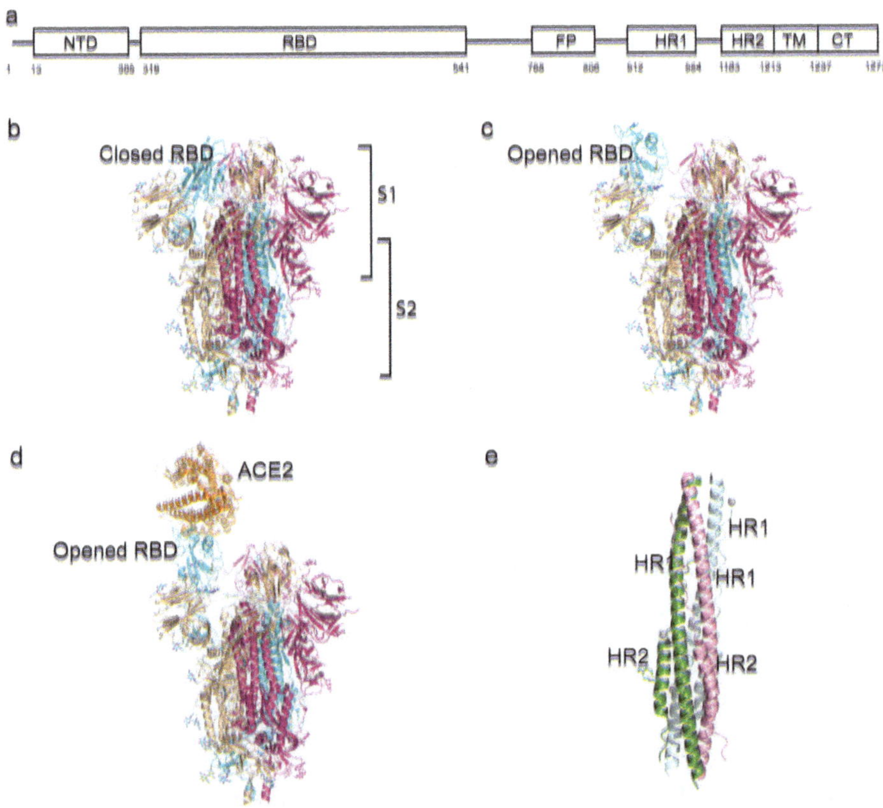


**Fig. 7.1** (**a**) Schematic representation of the SARS-CoV-2 spike. (**b–c**) open and close—S protein's RBD, (**d**) the opened RBD [S1 subunit] of S protein allows it to connect to ACE2. (**e**). The S2 subunits HR1 and HR2 work together to generate a six-helix structure

sequence 1 (HR1), HR2 (1163–1213 amino acid residues), the TM domain (1213–1237 amino acid residues), and the cytoplasm domain (12 amino acid residues) (Xia et al. 2020; Biswas et al. 2022a; Dey et al. 2022b). Around the virus particle, the S protein trimers create an interesting halo that resembles a bulbous crown. Using the structure of coronavirus S protein monomers as a guide, we can see that the S1 and S2 subunits create the virus' characteristic bulbous head and stalk region (Tang et al. 2020; Huang et al. 2020a). When it first makes an appearance, the CoV S protein is an inactive progenitor. To activate the membrane fusion domain, the S protein must be cleaved into the S1 and S2 subunits by proteases in the target cells (Bertram et al. 2013). As cellular proteases cleave the S protein into the S1 and S2 subunits, SARS-CoV-2 employs the serine protease TMPRSS2 as a protein primer. Although the location of the SARS-cleavage CoV is unknown, the location of the SARS-CoV-2 S is known to exist (Hoffmann et al. 2020; Du et al. 2007).

### 7.2.2 *Membrane Protein (M)*

The M protein is 223 amino acids long and is divided into three primary domains: the TMH1, TMH2, and TMH3 transmembrane helices, the long C-terminal endodomain found on the cytoplasmic face of virions (Zhang et al. 2022). There isn't much research on M protein homodimers and the ones that do only partly explain how M-M interactions lead to homodimer formation. The juxtamembrane hinge region, N-terminal three transmembrane helices, and the inward-facing C-terminal sheet sandwich domain (BD), which is made up of an outer sheet (residues 1, 2, and 6), an inner sheet (residues 1, 2, and 6), make up the three structural segments that make up the M protein (residues 3, 4, and 5). When the long form is transformed into the shortened form, the two bundles spin in opposite directions: toward one another on the extraversion side, and against one another on the introversion side. As a result, the two bundles shift their interface, drawing closer together in the extraversion lobe and further apart in the introversion lobe. Because of this, the axis of dimerization is strongly twisted in the opposite direction of the linker that connects TM1 and TM2 (Zhang et al. 2022; Dey et al. 2022c; Hasan et al. 2022a).

Because they regulate how the helical bundles and BDs are arrayed in relation to one another, the hinge region and the TM1-TM2 loop are expected to be significant in the structural change between the two types. The hollow that forms at the top of the bundle is deeper and inserted into the hinge area. Every other structural protein is capable of interacting with the M protein. By ensuring that the N protein-RNA complex remains contained within the internal virion, the binding of M protein contributes to the stabilization of N proteins and encourages the completion of viral assembly. Because the M protein cooperates with the S protein to establish a connection with and enter the host cell, mutations in either protein could have an impact on viral attachment and entry. Glycosylation of the virus's S protein may aid in its ability to evade the immune system (Thomas 2020; Hasan et al. 2022b; Morshed et al. 2022; Rahman et al. 2022). However, the mechanism of glycosylation of the structural protein (S) and the physiological function of the membrane protein (M) protein is still unknown.

### 7.2.3 Nonstructural Protein

In contrast to the genes for structural proteins, the 5′ region of the viral RNA genome also includes the genes for various other types of proteins (Yadav et al. 2021). The SCoV nsp1 protein, which is the gene 1 polyprotein's most N-terminal product and is present in both infected and expressing cells, works to silence the expression of host genes. Nsp1 obstructs the hosts' AV signaling pathways, preventing infected cells from producing type I interferon. The nsp1 protein of the nearly similar mouse hepatitis virus functions as a virulence factor, and the type I interferon system is also disrupted. Additionally, host gene expression is also suppressed. When Nsp1 attaches to 40S ribosomes, host protein synthesis is inhibited. Capped mRNA is further modified by ribosome-bound nsp1, which prevents translation (Huang et al. 2011). The interaction of the NSp2 protein with internal host communication during SARS-CoV infections is demonstrated by the link between the host proteins PHB2 and prohibitin 1 (PHB1) (Cornillez-Ty et al. 2009).

Nsp3 is the largest predicted protein from the coronavirus genome, with a molecular weight of around 200 kD. The replication/transcription complex must contain Nsp3. It is composed of multiple domains, the organization of which varies between CoV genera due to domain duplication or absence. There are eight different Nsp3 domains, and all known CoVs contain the same four: Ubl1, PL2pro, Nsp3 ectodomain (3Ecto), and Glu-rich acidic domain (also known as the "hypervariable region"). Furthermore, all CoVs have the TM1 and TM2 transmembrane domains (Lei et al. 2018; Dey et al. 2021; Bibi et al. 2022; Biswas et al. 2022b).

Numerous Nsp6, Nsp4, and Nsp3 trans-membranes exist. Nsp3, Nsp4, and Nsp6 are only found in the ER of certain cells following transfection with expression plasmids, despite being expressed in the RTC during infection. Studies involving immunoprecipitation have shown that there is a connection between nsp4 and the C-terminal third of nsp3, which is responsible for MHV (nsp3C). After being co-transfected with expression plasmids, these proteins showed a similar pattern of colocalization in the perinuclear foci as they did in infected cells. Researchers found that when Nsp3 and Nsp4 are expressed together, a membrane remodeling process takes place that results in the formation of cytoplasmic and dynein-myosin vesicles (CMs and DMVs), respectively. The replication complex, the creation of DMVs, and the virus-caused membrane rearrangement all depend on the glycosylated MHV nsp4 protein. Studies using electron and confocal microscopy show that SARS-CoV nsp3, nsp4, and nsp6 expression promotes the development of DMVs in the transfected cells. However, it is uncertain how the SARS-nsp3 CoVs and nsp4 interact as well as how they might prevent viral replication by changing membrane structure (Sakai et al. 2017; Baral et al. 2022; Singh et al. 2023).

The smallest RNA-polymerase complex consists of nsp7 and nsp8 cofactor proteins, the C-terminal region of nsp12 encoding the RNA-dependent RNA-polymerase (RdRp) catalytic domain, and other proteins. Cryo-electron microscopy (cryo-EM), which was used to obtain the structures, showed that the stoichiometry of the nsp12:nsp7:nsp8 complex is 1:1:2. Studies done in vitro have demonstrated that at least nsp7 and nsp8 must be bound for nsp12 to function as a polymerase. Studies on the nsp7 and nsp8 complexes without nsp12 have received minimal attention in terms of their structure and biology (Courouble et al. 2021). In contrast to SARS-CoV nsp12's conventional RdRp, which efficiently replicates a very large RNA genome in vivo, RNA processing is subpar in vitro. SARS-CoV nsp7 and nsp8 initiate the RNA-synthesizing activity of nsp12 and increase the severity of SARS-CoV pathogenesis. Biochemical and reverse genetic analyses were used to determine the importance of conserved nsp7 and nsp8 residues. When two nsp8 residues (P183 and R190) required for interaction with nsp12 and a third (K58) required for the polymerase complex's interface with RNA were replaced, the virus perished, whereas numerous nsp7 mutations had little effect on virus replication. The nsp7/nsp8/nsp12 complex can interact with nsp14, a dual-function enzyme that has 3′-5′ exoribonuclease activity as well as RNA cap N7-guanine methyltransferase activity. This enzyme is associated with replication fidelity as well as 5′-RNA capping. This connection can be made without compromising efficiency. The discovery of this tripartite polymerase complex, which associates with the nsp14 proofreading enzyme, reveals new information regarding how coronaviruses construct RNA-synthesizing machinery to replicate the largest known RNA genomes. This protein complex provides remarkable examples of the integration of RNA polymerase, capping, and proofreading functions (Subissi et al. 2014; Al Azad et al. 2022; Sarker et al. 2022).

## 7.3 Limitations of Present Antiviral Solution Against SARS-CoV-2

There are several COVID-19 vaccinations available, but the effectiveness of COVID-19 immunizations across a range of populations and their potential for persistent protection is still in doubt. As a result, developing SARS-CoV-2 AV drugs is crucial. However, the effectiveness of AVs has not yet been shown beyond a reasonable doubt. The AV may function by blocking viral enzyme activity, preventing viral entrance, concentrating on a specific host need to restrict the creation of viral particles, or by viral replication. Some AV drugs work by obstructing the entry, suppressing the protease, and through many additional methods. AV drugs are employed to treat diseases caused by viral infections. To ensure that the medications used to treat COVID-19 or other viruses function properly and according to label instructions, more research and testing are required.

Future effective treatment of similar viruses may be possible with these drugs (Biswas et al. 2021a, 2023; Rahman et al. 2020a).

Favipiravir, an AV drug, is used to treat people in Japan who have contracted the flu. Favipiravir is a purine nucleoside derivative that inhibits viral RNA-dependent RNA polymerase, thereby preventing viral genome replication and transcription (Furuta et al. 2017). However, due to the lack of a control group and its open-label design, it is unable to demonstrate the statistical effectiveness of favipiravir. The study only covered hospitalized patients due to a lack of funding and conflicting methodologies, and a sizable portion of follow-up SARS-CoV-2 (RT-PCR) assays was not acquired. The trial's early completion could have made it more difficult to share further information about clinical results. Additionally, patients in the SoC group underwent extra AV therapy (Bosaeed et al. 2021; Rahman et al. 2020b; Islam et al. 2022).

Remdesivir, a nucleoside monophosphoramidate prodrug for coronavirus infection, was approved by the FDA in 2019. There are concerns about the treatment effectiveness of remdesivir for COVID-19 since no trials have discovered substantial differences in mortality or the time it takes for a clinical improvement between control groups and redeliver-treated groups. Remdesivir was unable to outperform other experimental medications in terms of therapeutic benefit in COVID patients, in contrast to the significant, therapeutic preclinical data reached in rhesus macaque models. The dramatic contrast between the promising preclinical results for COVID-19 and the poor clinical effectiveness of remdesivir has baffled many (Yan and Muller 2021; Huang et al. 2021).

Due to the limited quality and confidence of the available data as well as the high heterogeneity, it is now challenging to make a solid determination regarding the advantages of umifenovir for SARS-CoV-2. Hospital length of stay (LOS), nucleus acid negative conversion time, or the chance of developing a disease could not be significantly reduced. In conclusion, there is no proof that umifenovir can improve patients' vital outcomes in COVID-19 patients (Huang et al. 2021). A retrospective analysis of real data from hospitalized patients with mild to severe COVID infection found that the clinical and virological outcomes were the same for lopinavir/ritonavir, hydroxychloroquine, lopinavir/ritonavir + hydroxychloroquine, and standard of care (Gagliardini et al. 2021). Thus, for COVID-19, finding new, powerful AV medications is crucial.

## 7.4 Plant-Based Bioactive Compounds and Synergistic Effects Thereof

Traditional plant-based medications were used effectively to treat patients afflicted with SARS-CoV in Guangdong Province, China, from the latter half of 2002 until the middle of 2003 (Brovelli et al. 2004; Peng et al. 2004). Secondary metabolites received a lot of attention because they have bioactive properties, even though there isn't much known about the nature and composition of these plants or how they work (Gourinat et al. 2015). For instance, the use of the alkaloid quinine, extracted from the Cinchona officinalis tree bark, in the treatment of malaria dates back to the 1960s (Achan et al. 2011). Similar in form to quinine is the drugs chloroquine (Cq) and hydroxychloroquine (Huq). The viral load of SARS-CoV-2 is reduced more effectively when Huq is combined with azithromycin (Prasad et al. 2020). Equally promising against SARS-CoV is glycyrrhizin, a saponin derived from the roots of the Glycyrrhiza glabra plant (Gavrilov 1975). Because SARS-CoV and SARS-CoV-2 have the same structures and ways of spreading, glycyrrhizin could also be used to treat the current pandemic. The 3CLpro and RdRp proteins that are present in SARS-CoV are inhibited from functioning by the water extract of *Houttuynia cordata*. Flavonoids myricetin (from *Myrica ruba*) and scutellarein (from *Scutettaria baicalensis* and *Asplenium belangeri*) are known to inhibit ATPase production by the SARS-CoV helicase nsP13, making the water extract active against the virus (Huang et al. 2020b). It has also been demonstrated that flavones from *Torreya nucifera* inhibit 3CLpro function. These flavones include amentoflavone, quercetin, luteolin, and apigenin (Ryu et al. 2010a). Lycorine is an alkaloid that is derived from *Lycoris radiata*. It has antiviral (AV) activity against the poliovirus and herpes simplex virus, and it has shown efficacy against SARS-CoV (Li et al. 1912). Emodin, sinigrin, and hesperetin, which were found in *Isatis indigotica*, also showed that they stopped 3CLpro from working (Lin et al. 2005a). In addition, the lectins that are found in plants can impede the activity of viruses. In a study, the antiviral (AV) properties of 33 lectins derived from different plant species were evaluated for their potential efficacy against SARS-CoV and feline coronavirus (FCoV). By targeting mannose, they discovered that mannose-binding lectin possessed potent anti-coronaviral activity by inhibiting both the entry and release of virus particles (Keyaerts et al. 2007; Ferdausi et al. 2022; Al Saber et al. 2022). Synthetic drug nelfinavir and agglutinin extracted from *Galanthus nivalis* both proved effective against FCov when administered in combination (Browning et al. 2010). This shows how important it is to study how NPs and man-made molecules work together to get around the viral load in the host system. But not much was done in this direction to study how biomolecules and drugs work together to fight viruses. Recently, a natural stilbene derivative called resveratrol (trans-3, 5, 4′-trihydroxystilbene) that is found in high amounts in *Vitis vinifera*, *Polygonum cuspidatum*, and *Vaccinium macrocarpon* was shown to stop MERS-Cov infection (Lin et al. 2017). Ayurvedic, Siddha, and Unani (nonnative) medicines are the three branches of traditional Indian medicine, and each of these schools prescribes NPs to their patients. Ayurveda is the oldest school of traditional Indian medicine (Thileepan and Prasad 2018). The Indian Government's Ministry of AYUSH (Ayurvedic medicine, Yoga and Homeopathy, Unani, Siddha, and Naturopathy) has recommended a formulation containing 15 plants to treat SARS-CoV-2. These plants are *Zingiber officinale*, *Piper longum*, *Syzygium aromaticum*, *Tragia involucrata*, *Anacyclus pyrethrum*, *Hygrophilla auriculata*, and *Terminalia* (Parada et al. 2020). Although Sivaraman, Pradeep, and Vellingiri have emphasized the benefits of this plant-based cocktail that reduces infection levels, no exhaustive research has been conducted to determine the chemical composition or mode of action of these plants (Sivaraman and Pradeep 2020; Vellingiri et al. 2020). Artemisinin derivatives isolated from *Artemisia annua* are being studied for their effect on SARS-CoV-2 at the Max Planck Institute for Colloids and Interfaces in Germany, in collaboration with ArtemiLife Inc. in the United States (Prasad et al. 2020).

## 7.5 Insight into the Molecular Mechanisms of Natural Bioactive Compounds Targeting SARS-CoV-2

### *7.5.1 In Vitro Studies*

#### 7.5.1.1 Curcuma Longa

Antiviral medications that target numerous viral subtypes, Curcumin's strength lies in initiating cellular signaling pathways associated with lung disease, metabolic disease, liver disease, and inflammation (Marx et al. 2020). Curcumin interacts directly with over thirty proteins, including DNA polymerase, thioredoxin reductase, focal adhesion kinase, protein kinase (PK), tubulin, and lipoxygenase (LOX), according to previous research. Other examples of these interactions include LOX. In addition, curcumin influences crucial intercellular signaling cascades, such as the inhibition of NF-B and PI3K/Akt signaling, which are both essential for efficient viral replication. Curcumin also has anti-inflammatory properties. As a result of influencing viral replication-related processes as well as cellular post-transcriptional and post-translational modifications, viral multiplication is stifled. This includes the replication of the viral genome and the attachment of the virus (Pourhajibagher et al. 2021). It has also been demonstrated that curcumin therapy modifies the shape of the S protein in viruses, preventing their entry into cells. Docking studies revealed that curcumin binds to RBD-S, PD-ACE2, and SARS-CoV-2 protease (Grosshans and Cerf 2004; Soni et al. 2021; Hosseini et al. 2021).

#### 7.5.1.2 *Saponin* Derivatives

Human fetal lung fibroblasts were tested for their susceptibility to SARS-CoV-2 replication inhibition using saikosaponins A, B2, C, and D. At concentrations up to 25 mol/L, saikosaponins A, B2, C, and D showed no cytotoxicity. At this concentration, however, all of the NPs showed significant SARS-CoV-2 inhibition. One of the most efficient blockers was found to be saikosaponin B2, which produced a result of 100 0.2% inhibition at a concentration of 25 mol/L. The half-maximal effective concentrations (EC50) of saikosaponins A, B2, C, and D were calculated to be 8.6, 0.3 (SI = 26.6), 1.7, 19.9, and 13.2. Further research was performed, and the results showed that saikosaponin B2 was the most effective virus-host adhesion and penetration blocker (Bray et al. 2006).

### 7.5.2 Propolis

Honeybees produce propolis, a resinous NP used in herbal medicine that is typically employed as a medical aid and defensive enhancer. Propolis has been demonstrated to have significant anti-inflammatory and pain-relieving effects by modulating a variety of metabolic pathways. Several recent studies demonstrate that propolis extract and some of its constituents act against several important SARS-CoV-2 targets, including reducing TMPRSS2 expression and reducing ACE2 anchorage, as well as immuno-modulating monocytes/macrophages (reducing production and removing IL-1 beta and IL-6), minimizing the signaling molecules NF-KB and JAK2/STAT3, and blocking PAK1, which determine inflammatory actions and fibrosis caused by SARS-CoV-2 (Diurno et al. 2020; Mehari et al. 2012; Piñero et al. 2020).

### 7.5.3 Algae-Derived Bioactive Metabolites

Carrageenan, nostoflan, microvirin, galactans, and cyanovirin are some of the bioactive metabolites derived from algae that have been proposed as potential treatments for SARS-CoV-2. They prevent viral infection by attaching themselves to the SARS-CoV-2 S protein and preventing the S protein from attaching itself to the heparin sulfate co-receptor in host tissues. This is how they function as diversions (Pradhan et al. 2022). Nagle et al. demonstrated that exopolysaccharides composed of carrageenan and sulfated polysaccharides derived from Porphyridium inhibit virus attachment to the host cell. Modified chitosan is highly effective against four different types of human coronavirus: HCoV-OC43, HCoV-NL63, HCoV-229E, and HCoV-HKU1 (Pereira and Critchley 2020). In an in vitro cell-based study, rhamnan sulfate from the green seaweed *Monostroma nitidum* was more effective than heparin at stopping S protein RBD or pseudoviral particles from attaching to immobilized heparin. It also showed strong AV activity against both the wild-type and delta form of SARS-CoV-2 (Song et al. 2021). Furthermore, fucoidan and rhamnan sulfate inhibit the expression of genes involved in the epidermal growth factor receptor pathway, which aids in the fight against coronavirus (Cai et al. 2020). Griffinsin, which was isolated from red algae, was very effective against SARS-CoV-2 by preventing viral entry, integrating viral DNA into RNA, inhibiting viral protease activity, and inhibiting viral reverse transcriptase activity (Ahan et al. 2022). Folic acid, phycoerythrobilins, and phycocyanobilins, which were found in Arthrospira, showed that they might be effective against SARS-CoV-2 (Petit et al. 2021). Ulvansare, a polysaccharide isolated from a green alga, is also being studied as a possible therapeutic agent against SARS-CoV-2 (Pereira and Critchley 2020). It is anticipated that the high-mannose-binding lectin griffithsin, which originates from red algae, will prevent SARS-CoV from entering the cell by binding to S glycoprotein. What's more, griffithsin did not harm the host cell even when present in extremely high concentrations (Zumla et al. 2016; Biswas et al. 2021b).

Algae of the red variety *Porphyridium* sp.-derived sulfated polysaccharides are promising AV NPs that can be used to coat sanitary products for SARS-CoV-2 prophylaxis (Alam et al. 2021). A phaeophyte called *Ecklonia cava* produces a product called dieckol, which has been shown to prevent the trans−/cis-cleavage of the SARS-CoV-2 3CLpro (chymotrypsin-like cysteine protease) (Zaporozhets and Besednova 2020). Moreover, the *Ecklonia kurome*-derived NPs dieckol targets 3CLpro and prevents SARS-CoV-2 from replicating. Several in silico investigations on the AV effects of metabolites generated from *Arthrospira* and other marine algae including *Grateloupia filicina, Gracilaria corticate, and Laurencia papillosa* on several target proteins of SARS-CoV-2 were carried out by various researchers (Kalasariya et al. 2022). Reactive oxygen species (ROS) are released because of SARS-CoV-2 infection, making neighboring cells much more susceptible to virus infection. Strong antioxidant capabilities have been documented for some bioactive substances produced from marine algae. The carotenoid fucoxanthin, which was obtained from *Sargassum siliquastrum*, prevented $H_2O_2$-induced DNA damage by boosting GSH synthesis and SOD gene expression (Pangestuti et al. 2013; Singh et al. 2021). In RAW264.7 cells stimulated with tert-butyl hydroperoxide (t-BHP), fucosterol extracted from *Ecklonia cava* has been shown to effectively reduce ROS generation (Jung et al. 2013). The powders of *Sargassum* spp., of which *C. vulgaris* and *A. maxima* as well as the fucoidan Alquimar were purified, are powerful inhibitors of SARS-CoV-2 in vitro. The findings of this research suggest that these algae might be useful in the treatment of SARS-CoV-2 infections; however, additional comprehensive preclinical and clinical research is still necessary to verify this finding (Garcia-Ruiz et al. 2022).

Recent studies have shown that SARS-CoV-2 and SARS-CoV share a staggering 83% similarity in the sequence of their PLpro gene in addition to a staggering 96% similarity in the sequence of their 3CLpro and RdRp genes (Morse et al. 2020; Li and De Clercq 2020). Therefore, the current SARS-CoV inhibitors may be effective against SARS-CoV-2 (Thiel et al. 2003, 2001). Table 7.1 and Fig. 7.2 represent the potential natural bioactive NPs for SARS-CoV-2.

## *7.5.4 In Vivo Studies*

### 7.5.4.1 Artemisia Annua

It has been demonstrated that in people with SARS-CoV-2 infection, interleukin-1 (IL-I) induction after virus binding to TLR promotes cellular damage. Fever and fibrosis are both mediated by IL-I, making it a highly inflammatory pleiotropic cytokine. In pulmonary fibrosis, the growth factor transforming growth factor beta

**Table 7.1** Natural bioactive compounds targeting SARS-CoV viral proteins

| Compound | Viral strain | $IC_{50}/EC_{50}$ | Target | Reference |
|---|---|---|---|---|
| Hesperetin | SARS-CoV | 60.00 μM | 3CLpro | Lin et al. (2005b) |
| Quercetin-3-β-galactoside | SARS-CoV | 42.79 μM | 3CLpro | Chen et al. (2006) |
| 18-Hydroxyferruginol | SARS-CoV | 45.80 μM | 3CLpro | Ryu et al. (2010b) |
| Kayadiol | SARS-CoV | 75.20 μM | 3CLpro | Ryu et al. (2010b) |
| Amentoflavone | SARS-CoV | 8.30 μM | 3CLpro | Ryu et al. (2010b)) |
| Scutellarein | COVID-19 | 3.02 μM | 3CLpro | Su et al. 2020) |
| Quercetin | SARS-CoV | 73.00 μM | 3CLpro | Nguyen et al. (2012)) |
| Quercetin | SARS-CoV | 23.80 μM | 3CLpro | Ryu et al. (2010b) |
| Quercetin | COVID-19 | $K_i$ = 7.00 μM | 3CLpro | Abian et al. (2020) |
| Isoforsythiaside | COVID-19 | 5.85 μM | 3CLpro | Su et al. (2020) |
| Curcumin | SARS-CoV | 23.50 μM | 3CLpro | Ryu et al. (2010c) |
| Tingenone | SARS-CoV | 9.90 μM | 3CLpro | Ryu et al. (2010c) |
| Pectolinarin | SARS-CoV | 37.78 μM | 3CLpro | Jo et al. (2020) |
| Andrographolide | SARS-CoV | 15.05 μM | 3CLpro | Shi et al. (2020) |
| Kaempferol | COVID-19 | 34.46 μM (antiviral activity) | 3CLpro | Khan et al. (2021b) |
| Broussochalcone A | SARS-CoV | 9.20 μM | PLpro | Park et al. (2017) |
| Anacardic acid | COVID-19 | 17.08 μM | PLpro | Chen et al. (2021) |
| Ginkgolic acid | COVID-19 | 16.30 μM | PLpro | Chen et al. (2021) |
| Kobophenol A | COVID-19 | 1.81 μM | S protein | Gangadevi et al. (2021) |
| Rhamnan sulfate | COVID-19 | 1.6 ng/mL | S protein | Song et al. (2021) |
| Tannic acid | COVID-19 | 13.4 μM | Mpro | Wang et al. (2020b) |
| Daidzein | COVID-19 | 56 μM | Mpro | Ghosh et al. (2022) |
| Puerarin | COVID-19 | 42 ± 2 μM | Mpro | Hu et al. (2020) |
| Ampelopsin | COVID-19 | 128 ± 5 μM | Mpro | Nguyen et al. (2021) |
| Silymarin | COVID-19 | 46.88 μM | Mpro | Saravanan et al. (2022) |
| Ursolic acid | COVID-19 | 12.6 μM | Mpro | Tripathi et al. (2021) |
| Chlorogenic acid | COVID-19 | 39.48 ± 5.51 μM | Mpro | Su et al. (2020) |
| Ellagic acid | COVID-19 | 11.8 ± 5.7 μM | Mpro | Bahun et al. (2022) |
| Curcumin | COVID-19 | 11.9 μM | Mpro | Antonopoulou et al. (2022) |
| Herbacetin | COVID-19 | 33.1 μM | Mpro | Tallei et al. (2020) |
| Rhoifolin | COVID-19 | 27.4 μM | Mpro | Choudhry et al. (2020) |

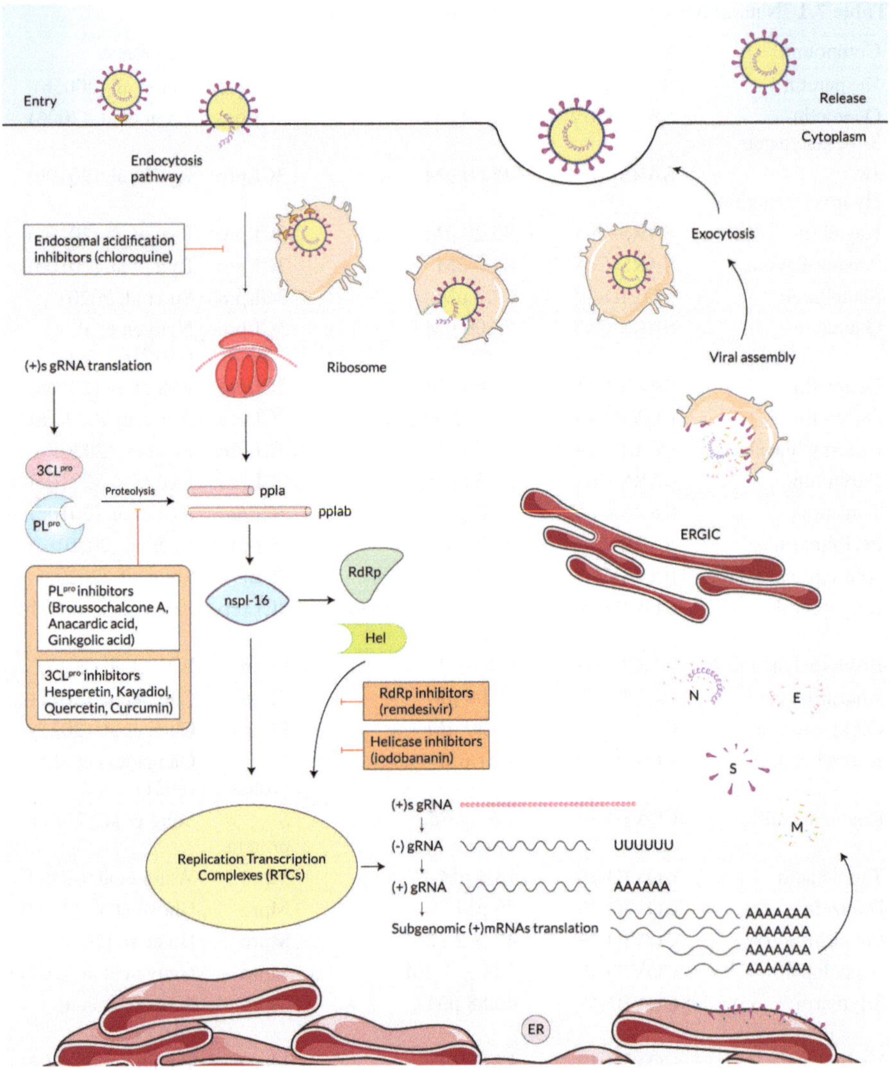


**Fig. 7.2** Molecular mechanisms of natural bioactive compounds targeting SARS-CoV-2

1 (TGF-1) and its downstream components Smad3 and SMA are over-expressed. Additionally, the heat shock protein HSP47 is associated with an increase in collagen deposition. *Artemisia annua*, a possible SARS-CoV-2 treatment, has been shown in rats to inhibit the production of pulmonary fibrosis by blocking the aforementioned molecules and their relevant pathways (Fig. 7.3) (Peeri et al. 2020).

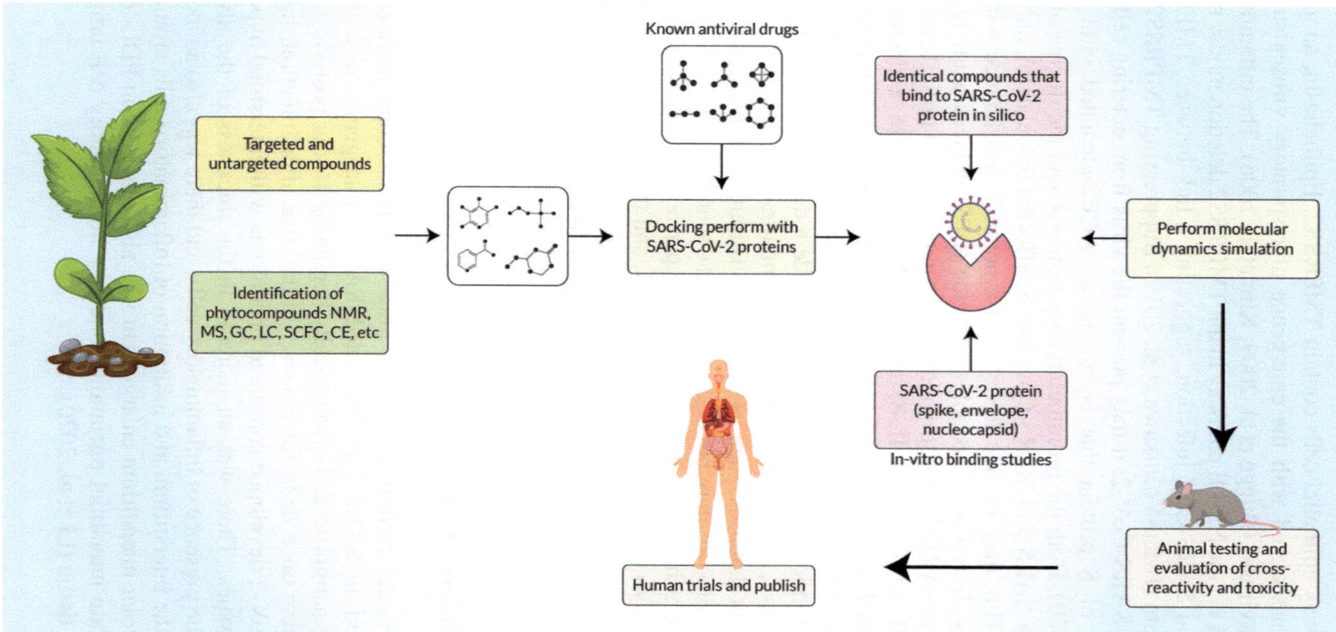


**Fig. 7.3** Isolation of phytocompounds from different plant routes using NMR, GC, LC, MS, etc. Molecular Docking studies can be done on the targeted compound to see how it interacts with a specific part of the SARS-CoV-2 proteins. If it is given the potential binding affinity and strong interaction, then will perform MD simulations to check the binding behavior and drug-protein mechanism. These can be used in animal models to test the efficacy of anti-SARS-CoV-2 action, with results confirmed by clinical trials. *NMR* nuclear magnetic resonance, *MS* mass spectrometry, *GC* gas chromatography, *LC* liquid chromatography, *MD* molecular dynamics

#### 7.5.4.2 Tripterygium Wilfordii

Lung and upper airway epithelial cells contain TMPRSS2. Epitheliasin, an airway protease, has been associated with the emergence of respiratory viruses such as influenza and coronaviruses (Bugge et al. 2009; Kim et al. 2006). The inflammatory response is activated via the NF-kB pathway in the TMPRSS2 deletion mouse model with MERS-Cov and SARS-CoV infection, suggesting a link between TMPRSS2 and the NF-kB system (Iwata-Yoshikawa et al. 2019). Suppressing TMPRSS2 may have two effects on SARS-CoV-2: it may prevent the virus from entering cells and reduce the amount of S protein that is cut during ACE2-mediated viral entry (Hoffmann et al. 2020). Both the proinflammatory response and the degree of lung pathology caused by SARS and MERS-Cov may be affected by celastrol, although in different ways. This may be due to celastrol's possible interference with the NF-kappaB pathway. In rats with lipopolysaccharide (LPS)-induced acute respiratory distress syndrome (ARDS), celastrol reduced the expression of proinflammatory cytokines (TNF-, IL-1, IL-6, and IL-8) and NF-B, resulting in a decrease in inflammation-mediated damage (Sahebnasagh et al. 2020). In addition, celastrol's anti-inflammatory effects have shown promise in reducing chronic obstructive pulmonary disease in mice. These mechanisms include a decrease in inflammatory cytokines, such as IL-8, TNF-, and monocyte chemoattractant protein-1, and an increase in antioxidant defenses, such as superoxide dismutase and catalase. These results indicate that celastrol may be an effective treatment for COPD (Zeng et al. 2018). Patients with asthma or improved airway hyper-responsiveness showed a positive outcome in vivo after receiving oral celastrol, and this effect was accompanied by a reduction in disease severity and a dampening of Th17 proinflammatory cells. However, celastrol has been shown to alleviate the severity of ventilator-induced lung injury in mice (Wei et al. 2018; Ren et al. 2018).

#### 7.5.4.3 Pudilan Xiaoyan Oral Liquid

Compared to the model control group, the lung viral RNA copy number was significantly reduced in SARS-CoV-2-infected hACE2 transgenic mice treated with PDL (4 ml/kg) administered intragastrically beginning 1-hour post-virus inoculation and continuing once daily for 5 days. On day five after infection, hACE2 mice with SARS-CoV-2 developed controlled pneumonia with interstitial hyperplasia in their lung tissues. There was an increase in the thickness of the alveolar interstitium due to the presence of inflammatory cells, and there was also an increase in the thickness of the interstitium and infiltration of inflammatory cells around the blood vessels. Alveolar interstitium enlargement was also observed in PDL-treated animals, but moderate interstitial pneumonia and few inflammatory cell infiltrates were found in lung tissue (Li et al. 2020).

## 7.6 Bio-farming: To Develop Plant-Based Vaccines and Active Metabolites

Recombinant molecule technology and its expression in plants using bio-farming are well established, and numerous studies have already demonstrated the efficacy of plant-based vaccinations (Prasad et al. 2020). Cloning and protein expression in plants are now commonplace tasks in labs all over the world due to the rapid advancement of plant biotechnology. This facilitates the bio-farming process used to generate recombinant biomolecules such as antibodies, vaccines, hormones, and enzymes in plant systems (Rosales-Mendoza 2020). To exploit plants as bioreactors for practical vaccine production, it is necessary to (a) achieve a high level of expression of recombinant genes, (b) can quickly and efficiently design and generate new antigens in response to novel pathogen subtypes, (c) detect the genes to be transfected, and (d) make sure the safety of produced proteins to be used in people or animals (Takeyama et al. 2015). In the past 30 years, techniques for genetically altering plants have been created to enable the production of heterologous proteins in plant cells. Initially, nuclear transformation mediated by Agrobacterium and later chloroplast via the biolistic approach were used to genetically modify plants. These core techniques laid the groundwork for a variety of transformation techniques that are currently used to manufacture various vaccine designs in plants (whole organisms, particular tissues, and cell culture), as well as microalgae (Gelvin 2003; Monreal-Escalante et al. 2022). Employing the plant system solves the post-translational modification issue when using bacteria as a host (Takeyama et al. 2015). By using plants as a host instead of bacteria, the issue of post-translational changes is also resolved (Shoseyov et al. 2014). There is a lot of variation among plant species or within a single plant for the generation of commercial recombinant proteins. Recombinant proteins have been created by a variety of plants, including the hepatitis B vaccine made in lettuce, as well as others like potato, maize, rice, and soybean (Kapusta et al. 1999). To generate a rabies vaccine, tomatoes have undergone genetic modification, making them a unique crop for the production of vaccines (Twyman et al. 2003). Clinical trials are being conducted on a number of plant-produced vaccines against viruses like the influenza viruses (H1N1, H5N1, and H7N9), hepatitis B virus, norovirus, and rabies virus (Takeyama et al. 2015). Additionally, vaccinations based on plant biotechnology are being developed to prevent the propagation of tuberculosis. Seven oral plant-bioengineering-based TB vaccines have so far undergone comprehensive testing in experimental, preclinical, and phase I clinical studies (Zenteno-Cuevas 2017).

Plant biotechnology enables the expression of foreign proteins in plants, suggesting a rapid development of a possible SARS-CoV-2 vaccine candidate. To what extent this antigen is expressed depends on the type of antigen being attacked. For the next section, we will discuss the possibility of developing the SARS-CoV-2 vaccine by using the plant biotechnology-based platform shown in Fig. 7.4.

Even though there are already a number of COVID-19 vaccines available, various companies are working on plant-based vaccines, some of which are currently

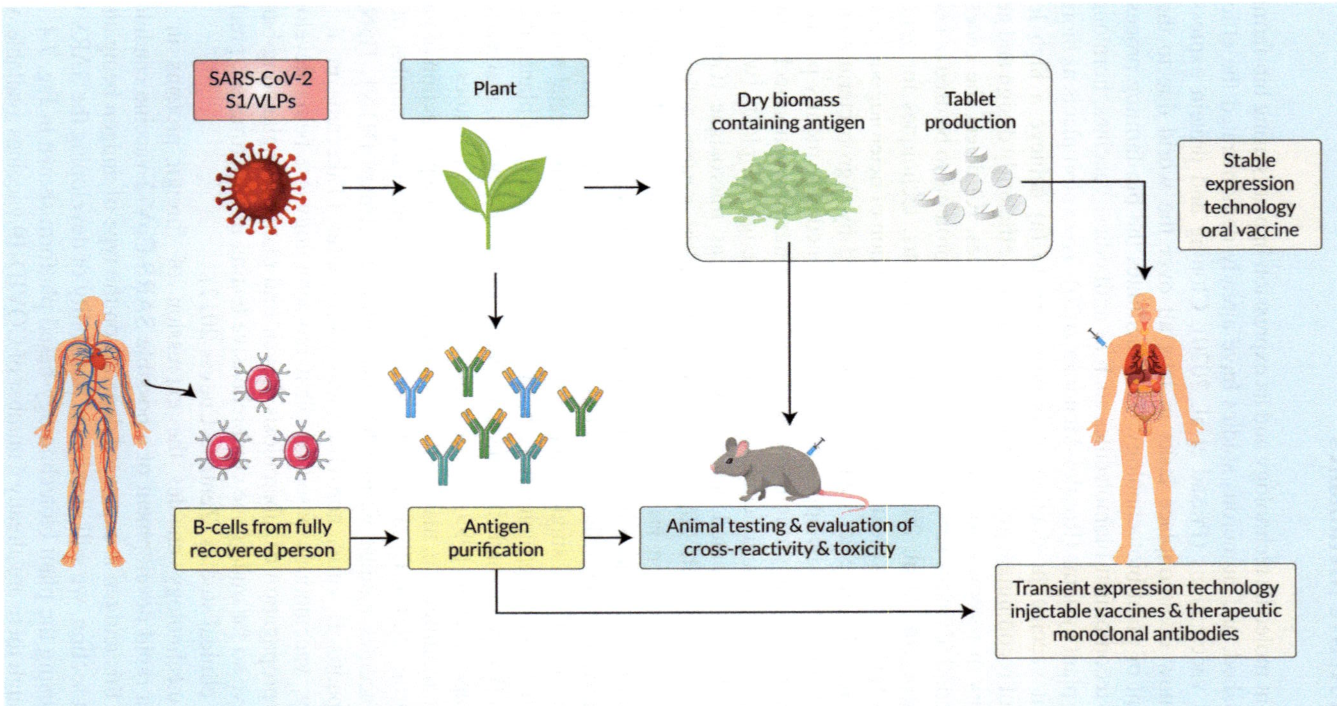


**Fig. 7.4** Applications of plant-based biotechnology toward the development of SARS-CoV-2 vaccine candidates and diagnostic tools

undergoing clinical trials (Monreal-Escalante et al. 2022; Maharjan and Choe 2021). Researchers have succeeded to get mice to produce virus-specific Immunoglobin-A (IgA) by transforming *Solanum lycopersicum* fruit with S glycoprotein of SARS-CoV (Pogrebnyak et al. 2005). Another study examined the SARS-CoV N recombinant protein's immunogenicity when it was transiently generated in *Nicotiana benthamiana*. A significant humoral immune response was induced after the third parental injection with tobacco-produced recombinant N protein (Shanmugaraj et al. 2021). Recently, Medicago (Quebec, Canada), a biopharmaceutical company, announced that they had developed a plant-derived CoVLP (Coronavirus Like Particle) vaccine targeting COVID-19 that was approved by the Food and Drug Administration when administered alone or in combination with AS03 or CpG1018 adjuvants from the GSK Company (Brentford, Middlesex, UK). In this pilot study, we aimed to measure the immunogenicity of CoVLP formulations using a neutralizing antibody (NAb) and cellular responses, as well as the short-term safety of these products (Ward et al. 2021). Antigenic differentiation between circulating VOI and VOC has contributed to a decline in vaccine efficacy, at least in part because immune responses mounted against the ancestral S protein present in all administered vaccines are less likely to cross-react with the superior SARS-CoV-2 variants (Cai et al. 2021; Fiolet et al. 2022). Understanding CoVLP+AS03's place in the changing viral landscape requires research into its ability to produce NAbs to VOI/VOC. In the efficacy part of the Phase 3 trial, efficacies of 75.3% (Delta), 88.6% (Gamma), and 100% (Alpha) were seen, even though the number of participants was small. This fits well with the strong and long-lasting cross-reactivity to the Alpha, Delta, and Gamma variants that were seen (Charland et al. 2022). The most advanced clinical trials have been conducted by Medicago and Kentucky BioProcessing Inc. (Owensboro, KY, USA) (Uthaya Kumar et al. 2021). Since the expression of hemagglutinin protein in plants has resulted in virus-like particles (VLPs), scientists believe that S protein expression may be necessary for the production of SARS-CoV-2 VLPs. In this situation, if the nucleus makes a protein that targets the trans-Golgi secretion route, it may make a protein that can be used to make VLPs for SARS-CoV-2 through the processes of secretion and glycosylation (Satija and Lal 2007). Geminiviruses, like the Bean yellow dwarf virus, have been selectively bred to rapidly increase plant production of medicinal proteins. In addition, a novel synthetic COVID-19 antibody derived from a phage display library is currently being evaluated using geminivirus vector technology (Rattanapisit et al. 2020; Mahmood et al. 2020).

Some health risks, such as the possibility of mutation and recombination (in the case of attenuated viruses), have been linked to the use of viral vectors in human vaccines (in the case of viral vectors). Due to potential side effects, the production of monoclonal antibodies against SARS-CoV-2 may not be a viable long-term treatment. Thus, a vaccine based on SARS-CoV-2 VLPs expressed in plants has a chance of succeeding in the future. In conclusion, metabolite engineering has great potential for the creation of antiviral drugs against SARS-CoV-2 and other infections in plants and other heterologous species.

## 7.7 Antiviral Bioactive-Based Nanocarrier Approaches

The plant provides us with multiple NPs, which fulfill synergistic effects for human beings (Goyal et al. 2022). Most of the drug products that are inhaled are delivered with a size <5 μm. It can be effective by applying nanoparticle formulation with a size <1 μm and used as a carrier to enhance the bioavailability, and delayed metabolism as well as improve therapeutic activity (Goyal et al. 2022; Varahachalam et al. 2021). For instance, catechin contains chitosan nanoparticles resulting in an increased level of intestinal absorption (Dube et al. 2010). Natural flavonoids such as myricetin load a nanoparticle carrier, which noticeably raises its solubility line (Sindhu et al. 2022). The incorporation of flavonoids into erythrocytes can boost bioavailability and antiviral activity. The flavonoids have shown positive results to protect the red blood cell membrane from oxidative damage (Table 7.2) (Sims Jr et al. 2020). In the bloodstream, chitosan flavonoids keep antioxidant action and combat free radicals (Fiorani et al. 2003). Polylactic acid-4 is used as a nanoparticle to encapsulate quercetin, which allows the slow release of quercetin (Zhang et al. 2016). Some studies have shown that methanolic extract from ginger (*Zingiber officinale*) is used for silver nanoparticle synthesis to observe the inhibitory strength of SARS-CoV-2 (Kurniawan and Ikhsanudin 2020).

One of the problematic issues is flavonoids, which have low water solubility and play a function in less bioavailability. Nanotechnology has shown a revolution to solve this problem. Fisetin, a typical plant flavonoid from polyphenol present in numerous plant-like fruits and vegetables (in strawberries, pineapples, onions, and persimmons), possesses less aqueous solubility as well as resultant little bioavailability. Despite this condition, it is effective to improve health and show potent therapeutic results against SARS-CoV-2 (Antika and Dewi 2021; Pandey et al. 2021). Therefore it is significant to improve its bioavailability. Some nanotechnology approaches have proposed that based on fisetin's lipophilic nature, it can be utilized in multifarious techniques including liposomes (a short artificial vesicle in a

**Table 7.2** List of antiviral phytoconstituents and their formulation

| Active phytocompounds | Formulation | Applications | References |
|---|---|---|---|
| Oxymatrine | Phytosome | Increases the bioavailability | Jiang et al. (2014) |
| *Artemisia arborescens* | Liposomal | Improve the ability to suppress the viral infection | Yue et al. (2010) |
| Hypocrellins | Nanoparticles | Improve the level of hydrophilicity | Sinico et al. (2005) |
| Mayrine | Emulsion | Improves the sustainability | Wang et al. (2010) |
| Quercetin | Microsphere | Enhances the bioavailability | Sun and Yeh (2005) |
| Honokiol | Cyclodextrin inclusion complexes | Enhances the solubility and bioavailability | Casettari et al. (2012) |

spherical shape), nanoemulsion (nano-form emulsions) for enhancing the release of active pharmaceutical ingredients (API) (Mignet et al. 2012; Ragelle et al. 2012; Feng et al. 2019).

Gold nanoparticles (AuNP) have antiviral activity and are widely used for COVID-19 treatment and stabilize with a particular biocompatible polymer (Kerry et al. 2019). Carbon-based nanoparticles are potentially used in these fields to deliver various nanomedicines against different types of diseases for their diverse structure, high drug-loading efficiency, low immunogenicity, enhanced bioavailability, and optical and physiological properties (Debnath and Srivastava 2021). Carbon nanomaterials (CBNs) are effective against almost 13 single-stranded RNA viruses, like SARS-CoV-2 (Serrano-Aroca et al. 2021). Carbon is a necessary element for our human body, covering the maximum surface, and could assist to associate with various types of polymers to improve efficacy. When SARS-CoV-2 attacks a specific organ in our body, it exhibits a potent physiological disbalance of the most biochemical cell cycles and metabolism in our body. CBNs help the generation of angiogenic features that induce signaling pathways by the propagation of endothelial and mesenchymal stem cells. SARS-CoV-2 cannot infect those cells (Mallakpour et al. 2021).

Albumin is a vital protein in the blood and reflects multiple physical properties. Albumin nanoparticles are nontoxic and nonimmunogenic. The residue of albumin protein has several pharmacophore entities significant in the binding of drugs. When Covid-19 infects our body, the cytokine level is increased, and diminishes the albumin level. Drugs loaded with albumin nanoparticles can be effective for this purpose. Polyethylene glycol (PGE) coating the surface of albumin nanoparticle bound was developed to enhance the bioactivity of natural steroid ginsenoside saponins found in plants (Park et al. 2021; Debnath and Srivastava 2022). The different categories of nanomaterials (NMs) and nanoparticles (NPs) have been used for vaccine development against SARS-CoV-2 as in Fig. 7.5.

## 7.8 Challenges and Limitations

SARS-CoV-2 infections can be a dangerous threat, and diagnoses have been challenging when faced with a lack of proper management systems and selective therapies. Hence, the identification of disease is essential for the management of COVID-19 (Yang et al. 2020). Although some report ribavirin, promazine, and inosine-5-monophosphate (IMP) dehydrogenase were used to inhibit viral replication, these were later proven invalid (Barnard et al. 2006a, b; Morrey et al. 2008). Real-time polymerase chain reaction (RT-PCR) and other potentially high-accuracy techniques can be used to detect SARS-CoV-2 viral genetic materials, but this is a costly method that requires a time-consuming protocol and diagnostic kits that are not available in all countries. In contrast, less expensive tests can be used for detection, but their accuracy level is limited (Mattioli et al. 2020). Many countries, especially with poorer settings, have failed to implement COVID-19 testing and

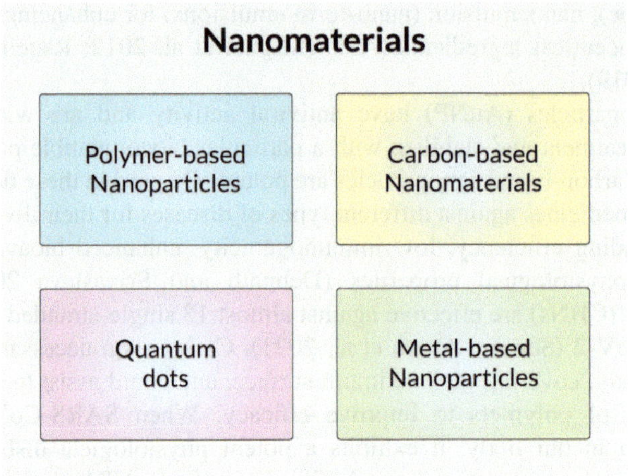


**Fig. 7.5** The different types of nanoparticles used for the treatment of SARS-CoV-2. Carbon-based nanomaterials have excellent characteristics for the detection of a pathogen like SARS-CoV-2. For example, graphene and graphene oxide and carbon nanotubes. They demonstrate various potential applications such as biosensor diagnosis, airborne virus filtration, antiviral coating, and delivery of different drugs (Ferrari et al. 2015). Quantum dots such as carbon quantum dots and zirconium quantum dots can inhibit viral RNA replication (Mullard 2020). Metal-based nanoparticles, including gold nanoparticles (AuNPs), copper nanoparticles, silver nanoparticles, and zinc nanoparticles are widely used against SARS-CoV-2 (Weng and Neethirajan 2018; Warnes et al. 2015). Polymer-based nanoparticles include synthetics polymer, nitrocellulose, chitosan nanoparticles, and lipid nanoparticles used for empowering antiviral vaccine design (Sivasankarapillai et al. 2020)

highlighted the demand for extra efforts as well as to diagnose the disease technical development is needed. First-generation vaccines have high activity against SARS-CoV-2, but the long-term effect of vaccines is changeable (Grady 2020). Several research studies have reported that the flavonol glycosides have been used as a drug but had a few adverse effects. The sugar moieties of flavonoids aglycone raise bioavailability. Flavonols with glucose are fast absorbed which is the limitation of bioavailability (Xiao 2017). Various research institutes and multinational companies have been performing the synthesis of clinical trial drugs to rapidly combat SARS-CoV-2 (Rudra et al. 2017). At present, antiviral drug development is the first challenge that successfully can shut down viral replication when a virus admits into the human body. The main target for drug development is receptor protein where the virus binds and helps uncoating, as well as loosening the salvation of its contents (Schaechter 2009). Angiotensin-converting enzyme 2 (ACE2) interacts with SARS-CoV-2, which is linked to the highest rates of mortality worldwide. ACE2 is not only expressed in the lung, kidneys, and heart but also present in the nasal mucosa, thymus, liver, and brain. The renin-angiotensin system (RAS) controls the mechanism of ACE2 (Patel et al. 2016). It is a critical challenge to regulate the harmful effect of ACE2. The antiviral drug can prevent the mortality rate if the virus has not replicated and infected the number of cells in the human body (Goyal et al. 2020).

## 7.9 Conclusion and Future Prospective

The SARS-CoV-2 infection has been challenging for all countries. The pathogenesis and transmission rates are more rapid than SARS-CoV-1, causing high mortality on the planet. Some precautions are needed to control virus expansion, for example, changing lifestyle, maintaining social distance, and movement limitations. SARS-CoV-2 can cause diverse mutations over time and show high infection volume. After a long time, at the end of 2020, many scientific laboratories and passionate researchers have succeeded in publishing the first approved vaccine against COVID-19. Many populous countries like the USA, China, and India have contributed to the discovery of the vaccine and that has already been approved. Best therapeutics must be promoted to regulate the present manifestation. Considering this occurrence, a huge number of natural bioactive NPs are developed against SARS-CoV-2, which are capable of suppressing the virus replication process and metastasis as well as evaluating their activity and side effects. This on-going pandemic situation shows adverse effects on the economy, so we require a minimum-cost vaccine that can show large-scale immunity. NPs are suitable for discovering AV vaccines for this purpose and recently, some are engaged in clinical trials. Multiple natural products such as essential oils and NPs are extracted from the plant, observing the activity of these NPs against viral infection. Cooperation between scientific laboratories and the government is necessary for the synthesis of plant-based AV vaccines. The synergistic effects should be tested for determination of effectiveness based on the future. It is possible to develop potential therapeutics against the SARS-CoV-2 infection through NPs targeted delivery of the therapeutics after evaluating its effectiveness in vivo clinical trial. In addition, phytopharmaceuticals also will play a vital role in contributing to the production of safe and effective drugs.

## References

Abian O, Ortega-Alarcon D, Jimenez-Alesanco A, Ceballos-Laita L, Vega S, Reyburn HT, Rizzuti B, Velazquez-Campoy A (2020) Structural stability of SARS-CoV-2 3CLpro and identification of quercetin as an inhibitor by experimental screening. Int J Biol Macromol 164:1693–1703. https://doi.org/10.1016/j.ijbiomac.2020.07.235

Achan J, Talisuna AO, Erhart A, Yeka A, Tibenderana JK, Baliraine FN, Rosenthal PJ, D'Alessandro U (2011) Quinine, an old anti-malarial drug in a modern world: role in the treatment of malaria. Malar J 10:144. https://doi.org/10.1186/1475-2875-10-144

Ahan RE, Hanifehnezhad A, Kehribar E, Oguzoglu TC, Földes K, Özçelik CE, Filazi N, Öztop S, Palaz F, Önder S et al (2022) A highly potent SARS-CoV-2 blocking lectin protein. ACS Infect Dis 8:1253–1264. https://doi.org/10.1021/acsinfecdis.2c00006

Ahmed H, Mahmud AR, Siddiquee MFR, Shahriar A, Biswas P, Shimul MEK, Ahmed SZ, Ema TI, Rahman N, Khan MA et al (2022) Role of T cells in cancer immunotherapy: opportunities and challenges. Cancer Pathogen Ther. https://doi.org/10.1016/j.cpt.2022.12.002

Al Azad S, Ahmed S, Biswas P, Mia MAR, Farjana M, Arshe FA, Mily SJ, Ankhi AB, Shaikat MM, Sultana S (2022) Quantitative analysis of the factors influencing IDA and TSH downregulation in correlation to the fluctuation of activated vitamin D3 in women. JABET 5(2):320–333

Al Saber M, Biswas P, Dey D, Kaium MA, Islam MA, Tripty MIA, Rahman MH, Rahaman TI, Biswas MY, Paul P et al (2022) A comprehensive review of recent advancements in cancer immunotherapy and generation of CAR T cell by CRISPR-Cas9. Processes 10:16

Alam MA, Parra-Saldivar R, Bilal M, Afroze CA, Ahmed MN, Iqbal HMN, Xu J (2021) Algae-derived bioactive molecules for the potential treatment of SARS-CoV-2. Molecules (Basel, Switzerland) 26:2134. https://doi.org/10.3390/molecules26082134

Alsaidi S, Cornejal N, Mahoney O, Melo C, Verma N, Bonnaire T, Chang T, O'Keefe BR, Sailer J, Zydowsky TM et al (2021) Griffithsin and carrageenan combination results in antiviral synergy against SARS-CoV-1 and 2 in a pseudoviral model. Mar Drugs 19. https://doi.org/10.3390/md19080418

Antika LD, Dewi RM (2021) Pharmacological aspects of fisetin. Asian Pac J Trop Biomed 11:1

Antonopoulou I, Sapountzaki E, Rova U, Christakopoulos P (2022) Inhibition of the main protease of SARS-CoV-2 (M(pro)) by repurposing/designing drug-like substances and utilizing nature's toolbox of bioactive compounds. Comput Struct Biotechnol J 20:1306–1344. https://doi.org/10.1016/j.csbj.2022.03.009

Arefin A, Ismail Ema T, Islam T, Hossen S, Islam T, Al Azad S, Uddin Badal N, Islam A, Biswas P, Alam NU et al (2021) Target specificity of selective bioactive compounds in blocking α-dystroglycan receptor to suppress Lassa virus infection: an in silico approach. J Biomed Res 35:459–473. https://doi.org/10.7555/jbr.35.20210111

Bahun M, Jukić M, Oblak D, Kranjc L, Bajc G, Butala M, Bozovičar K, Bratkovič T, Podlipnik Č, Poklar Ulrih N (2022) Inhibition of the SARS-CoV-2 3CL(pro) main protease by plant polyphenols. Food Chem 373:131594. https://doi.org/10.1016/j.foodchem.2021.131594

Baral SK, Biswas P, Kaium MA, Islam MA, Dey D, Saber MA et al (2022) A comprehensive discussion in vaginal cancer based on mechanisms, treatments, risk factors and prevention. Front Oncol 12:883805. https://doi.org/10.3389/fonc.2022.883805

Barnard DL, Day CW, Bailey K, Heiner M, Montgomery R, Lauridsen L, Chan PK, Sidwell RW (2006a) Evaluation of immunomodulators, interferons and known in vitro SARS-CoV inhibitors for inhibition of SARS-CoV replication in BALB/c mice. Antivir Chem Chemother 17: 275–284

Barnard DL, Day CW, Bailey K, Heiner M, Montgomery R, Lauridsen L, Winslow S, Hoopes J, Li JK-K, Lee J (2006b) Enhancement of the infectivity of SARS-CoV in BALB/c mice by IMP dehydrogenase inhibitors, including ribavirin. Antivir Res 71:53–63

Bertram S, Dijkman R, Habjan M, Heurich A, Gierer S, Glowacka I, Welsch K, Winkler M, Schneider H, Hofmann-Winkler H et al (2013) TMPRSS2 activates the human coronavirus 229E for cathepsin-independent host cell entry and is expressed in viral target cells in the respiratory epithelium. J Virol 87:6150–6160. https://doi.org/10.1128/jvi.03372-12

Bibi S, Hasan MM, Biswas P, Shkodina A, Shah MA, Shah GM, Khan A, Al-Harrasi A (2022) Chapter 7 - Phytonutrients in the management of lipids metabolism. In: Khan H, Akkol EK, Daglia M (eds) The role of phytonutrients in metabolic disorders. Academic Press, pp 195–236

Biswas P, Hasan MM, Dey D, Dos Santos Costa AC, Polash SA, Bibi S, Ferdous N, Kaium MA, Rahman MDH, Jeet FK et al (2021a) Candidate antiviral drugs for COVID-19 and their environmental implications: a comprehensive analysis. Environ Sci Pollut Res Int 28:59570–59593. https://doi.org/10.1007/s11356-021-16096-3

Biswas P, Dey D, Rahman A, Islam MA, Susmi TF, Kaium MA, Hasan MN, Rahman MDH, Mahmud S, Saleh MA et al (2021b) Analysis of SYK gene as a prognostic biomarker and suggested potential bioactive phytochemicals as an alternative therapeutic option for colorectal cancer: an in-silico pharmaco-informatics investigation. J Personal Med 11. https://doi.org/10.3390/jpm11090888

Biswas P, Hany Rumi O, Ahmed Khan D, Ahmed MN, Nahar N, Jahan R, Hasan Zilani MN, Paul TK, Hasan A, Bondhon TA et al (2022a) Evaluation of Melongosides as potential inhibitors of NS2B-NS3 activator-protease of dengue virus (serotype 2) by using molecular docking and dynamics simulation approach. J Trop Med 2022:7111786. https://doi.org/10.1155/2022/7111786

Biswas P, Dey D, Biswas PK, Rahaman TI, Saha S, Parvez A, Khan DA, Lily NJ, Saha K, Sohel M et al (2022b) A comprehensive analysis and anti-cancer activities of quercetin in ROS-mediated cancer and cancer stem cells. Int J Mol Sci 23(19):11746. https://doi.org/10.3390/ijms231911746

Biswas P, Polash SA, Dey D, Kaium MA, Mahmud AR, Yasmin F, Baral SK, Islam MA, Rahaman TI, Abdullah A et al (2023) Advanced implications of nanotechnology in disease control and environmental perspectives. Biomed Pharmacother 158:114172. https://doi.org/10.1016/j.biopha.2022.114172

Bosaeed M, Mahmoud E, Alharbi A, Altayib H, Albayat H, Alharbi F, Ghalilah K, Al Arfaj A, AlJishi J, Alarfaj A et al (2021) Favipiravir and hydroxychloroquine combination therapy in patients with moderate to severe COVID-19 (FACCT trial): an open-label, multicenter, randomized, controlled trial. Infect Dis Therap 10:2291–2307. https://doi.org/10.1007/s40121-021-00496-6

Bosch BJ, van der Zee R, de Haan CA, Rottier PJ (2003) The coronavirus spike protein is a class I virus fusion protein: structural and functional characterization of the fusion core complex. J Virol 77:8801–8811. https://doi.org/10.1128/jvi.77.16.8801-8811.2003

Bray M, French H, Shur YJP, Processes P (2006) Further cryostratigraphic observations in the CRREL permafrost tunnel, fox, Alaska. Permafr Periglac Proc 17:233–243

Brovelli A, Ding M, Ledberg A, Chen Y, Nakamura R, Bressler SL (2004) Beta oscillations in a large-scale sensorimotor cortical network: directional influences revealed by granger causality. Proc Natl Acad Sci U S A 101:9849–9854. https://doi.org/10.1073/pnas.0308538101

Browning LM, Hsieh SD, Ashwell M (2010) A systematic review of waist-to-height ratio as a screening tool for the prediction of cardiovascular disease and diabetes: 0·5 could be a suitable global boundary value. Nutr Res Rev 23:247–269. https://doi.org/10.1017/s0954422410000144

Bugge TH, Antalis TM, Wu Q (2009) Type II transmembrane serine proteases. J Biol Chem 284:23177–23181. https://doi.org/10.1074/jbc.R109.021006

Cai Y, Xu W, Gu C, Cai X, Qu D, Lu L, Xie Y, Jiang S (2020) Griffithsin with A broad-spectrum antiviral activity by binding glycans in viral glycoprotein exhibits strong synergistic effect in combination with A pan-coronavirus fusion inhibitor targeting SARS-CoV-2 spike S2 subunit. Virol Sin 35:857–860. https://doi.org/10.1007/s12250-020-00305-3

Cai C, Liu Y, Zeng S, Shen H, Han Y (2021) The efficacy of COVID-19 vaccines against the B.1.617.2 (delta) variant. Mol Ther 29:2890–2892. https://doi.org/10.1016/j.ymthe.2021.09.024

Casettari L, Gennari L, Angelino D, Ninfali P, Castagnino E (2012) ORAC of chitosan and its derivatives. Food Hydrocoll 28:243–247

Charland N, Gobeil P, Pillet S, Boulay I, Séguin A, Makarkov A, Heizer G, Bhutada K, Mahmood A, Trépanier S et al (2022) Safety and immunogenicity of an AS03-adjuvanted plant-based SARS-CoV-2 vaccine in adults with and without comorbidities. NPJ Vaccin 7:142. https://doi.org/10.1038/s41541-022-00561-2

Chauhan S (2020) Comprehensive review of coronavirus disease 2019 (COVID-19). Biom J 43:334–340. https://doi.org/10.1016/j.bj.2020.05.023

Chen L, Li J, Luo C, Liu H, Xu W, Chen G, Liew OW, Zhu W, Puah CM, Shen X et al (2006) Binding interaction of quercetin-3-beta-galactoside and its synthetic derivatives with SARS-

CoV 3CL(pro): structure-activity relationship studies reveal salient pharmacophore features. Bioorg Med Chem 14:8295–8306. https://doi.org/10.1016/j.bmc.2006.09.014

Chen Z, Cui Q, Cooper L, Zhang P, Lee H, Chen Z, Wang Y, Liu X, Rong L, Du R (2021) Ginkgolic acid and anacardic acid are specific covalent inhibitors of SARS-CoV-2 cysteine proteases. Cell Biosci 11:45. https://doi.org/10.1186/s13578-021-00564-x

Choudhry N, Zhao X, Xu D, Zanin M, Chen W, Yang Z, Chen J (2020) Chinese therapeutic strategy for fighting COVID-19 and potential small-molecule inhibitors against severe acute respiratory syndrome coronavirus 2 (SARS-CoV-2). J Med Chem 63:13205–13227. https://doi.org/10.1021/acs.jmedchem.0c00626

Cornillez-Ty CT, Liao L, Yates JR 3rd, Kuhn P, Buchmeier MJ (2009) Severe acute respiratory syndrome coronavirus nonstructural protein 2 interacts with a host protein complex involved in mitochondrial biogenesis and intracellular signaling. J Virol 83:10314–10318. https://doi.org/10.1128/jvi.00842-09

Courouble VV, Dey SK, Yadav R, Timm J, Harrison J, Ruiz FX, Arnold E, Griffin PR (2021) Revealing the structural plasticity of SARS-CoV-2 nsp7 and nsp8 using structural proteomics. J Am Soc Mass Spectrom 32:1618–1630. https://doi.org/10.1021/jasms.1c00086

Debnath SK, Srivastava R (2021) Drug delivery with carbon-based nanomaterials as versatile nanocarriers: progress and prospects. Front Nanotechnol 3:644564

Debnath SK, Srivastava R (2022) Potential application of bionanoparticles to treat severe acute respiratory syndrome coronavirus-2 infection. Frontiers in Nanotechnology 3:813847

Dey D, Ema TI, Biswas P, Aktar S, Islam S, Rinik UR et al (2021) Antiviral effects of bacteriocin against animal-to-human transmittable mutated SARS-CoV-2: a systematic review. Front Agr Sci Eng 8:603–622

Dey D, Hossain R, Biswas P, Paul P, Islam MA, Ema TI, Gain BK, Hasan MM, Bibi S, Islam MT et al (2022a) Amentoflavone derivatives significantly act towards the main protease (3CL (PRO)/M(PRO)) of SARS-CoV-2: in silico admet profiling, molecular docking, molecular dynamics simulation, network pharmacology. Mol Divers:1–15. https://doi.org/10.1007/s11030-022-10459-9

Dey D, Hasan MM, Biswas P, Papadakos SP, Rayan RA, Tasnim S, Bilal M, Islam MJ, Arshe FA, Arshad EM et al (2022b) Investigating the anticancer potential of Salvicine as a modulator of topoisomerase II and ROS signaling cascade. Front Oncol 12:899009. https://doi.org/10.3389/fonc.2022.899009

Dey D, Biswas P, Paul P, Mahmud S, Ema TI, Khan AA, Ahmed SZ, Hasan MM, Saikat ASM, Fatema B et al (2022c) Natural flavonoids effectively block the CD81 receptor of hepatocytes and inhibit HCV infection: a computational drug development approach. Mol Divers. https://doi.org/10.1007/s11030-022-10491-9

Diurno F, Numis FG, Porta G, Cirillo F, Maddaluno S, Ragozzino A, De Negri P, Di Gennaro C, Pagano A, Allegorico E et al (2020) Eculizumab treatment in patients with COVID-19: preliminary results from real life ASL Napoli 2 Nord experience. Eur Rev Med Pharmacol Sci 24:4040–4047. https://doi.org/10.26355/eurrev_202004_20875

Drosten C, Günther S, Preiser W, van der Werf S, Brodt HR, Becker S, Rabenau H, Panning M, Kolesnikova L, Fouchier RA et al (2003) Identification of a novel coronavirus in patients with severe acute respiratory syndrome. N Engl J Med 348:1967–1976. https://doi.org/10.1056/NEJMoa030747

Du L, Kao RY, Zhou Y, He Y, Zhao G, Wong C, Jiang S, Yuen KY, Jin DY, Zheng BJ (2007) Cleavage of spike protein of SARS coronavirus by protease factor Xa is associated with viral infectivity. Biochem Biophys Res Commun 359:174–179. https://doi.org/10.1016/j.bbrc.2007.05.092

Dube A, Nicolazzo JA, Larson I (2010) Chitosan nanoparticles enhance the intestinal absorption of the green tea catechins (+)-catechin and (−)-epigallocatechin gallate. Eur J Pharm Sci 41:219–225

Españo E, Kim J, Lee K, Kim JK (2021) Phytochemicals for the treatment of COVID-19. J Microbiol (Seoul, Korea) 59:959–977. https://doi.org/10.1007/s12275-021-1467-z

Feng C, Yuan X, Chu K, Zhang H, Ji W, Rui M (2019) Preparation and optimization of poly (lactic acid) nanoparticles loaded with fisetin to improve anti-cancer therapy. Int J Biol Macromol 125: 700–710

Ferdausi N, Islam S, Rimti FH, Quayum ST, Arshad EM, Ibnat A, Islam T, Arefin A, Ema TI, Biswas P et al (2022) Point-specific interactions of isovitexin with the neighboring amino acid residues of the hACE2 receptor as a targeted therapeutic agent in suppressing the SARS-CoV-2 influx mechanism. J Adv Veter Anim Res 9:230–240. https://doi.org/10.5455/javar.2022.i588

Ferrari AC, Bonaccorso F, Fal'Ko V, Novoselov KS, Roche S, Bøggild P, Borini S, Koppens FH, Palermo V, Pugno N (2015) Science and technology roadmap for graphene, related two-dimensional crystals, and hybrid systems. Nanoscale 7:4598–4810

Fiolet T, Kherabi Y, MacDonald CJ, Ghosn J, Peiffer-Smadja N (2022) Comparing COVID-19 vaccines for their characteristics, efficacy and effectiveness against SARS-CoV-2 and variants of concern: a narrative review. Clin Microbiol Infect 28:202–221. https://doi.org/10.1016/j.cmi.2021.10.005

Fiorani M, Accorsi A, Cantoni O (2003) Human red blood cells as a natural flavonoid reservoir. Free Radic Res 37:1331–1338

Furuta Y, Komeno T, Nakamura T (2017) Favipiravir (T-705), a broad spectrum inhibitor of viral RNA polymerase. Proc Japan Acad B Phys Biol Sci 93:449–463. https://doi.org/10.2183/pjab.93.027

Gagliardini R, Cozzi-Lepri A, Mariano A, Taglietti F, Vergori A, Abdeddaim A, Di Gennaro F, Mazzotta V, Amendola A, D'Offizi G et al (2021) No efficacy of the combination of lopinavir/ritonavir plus hydroxychloroquine versus standard of care in patients hospitalized with COVID-19: a non-randomized comparison. Front Pharmacol 12:621676. https://doi.org/10.3389/fphar.2021.621676

Gangadevi S, Badavath VN, Thakur A, Yin N, De Jonghe S, Acevedo O, Jochmans D, Leyssen P, Wang K, Neyts J et al (2021) Kobophenol A inhibits binding of host ACE2 receptor with spike RBD domain of SARS-CoV-2, a Lead compound for blocking COVID-19. J Phys Chem Lett 12:1793–1802. https://doi.org/10.1021/acs.jpclett.0c03119

Garcia-Ruiz D, Villalobos-Sánchez E, Alam-Escamilla D, Elizondo-Quiroga D (2022) In vitro inhibition of SARS-CoV-2 infection by dry algae powders. Sci Rep 12:17101. https://doi.org/10.1038/s41598-022-22148-6

Gautret P, Lagier JC, Parola P, Hoang VT, Meddeb L, Mailhe M, Doudier B, Courjon J, Giordanengo V, Vieira VE et al (2020) Hydroxychloroquine and azithromycin as a treatment of COVID-19: results of an open-label non-randomized clinical trial. Int J Antimicrob Agents 56:105949. https://doi.org/10.1016/j.ijantimicag.2020.105949

Gavrilov VI (1975) Acta Virol 19:510

Gelvin SB (2003) Improving plant genetic engineering by manipulating the host. Trends Biotechnol 21:95–98. https://doi.org/10.1016/s0167-7799(03)00005-2

Ghosh R, Chakraborty A, Biswas A, Chowdhuri S (2022) Depicting the inhibitory potential of polyphenols from Isatis indigotica root against the main protease of SARS CoV-2 using computational approaches. J Biomol Struct Dyn 40:4110–4121. https://doi.org/10.1080/07391102.2020.1858164

Gourinat AC, O'Connor O, Calvez E, Goarant C, Dupont-Rouzeyrol M (2015) Detection of zika virus in urine. Emerg Infect Dis 21:84–86. https://doi.org/10.3201/eid2101.140894

Goyal RK, Majeed J, Tonk R, Dhobi M, Patel B, Sharma K, Apparsundaram S (2020) Current targets and drug candidates for prevention and treatment of SARS-CoV-2 (COVID-19) infection. Rev Cardiovasc Med 21:365–384

Goyal R, Bala R, Sindhu RK, Zehravi M, Madaan R, Ramproshad S, Mondal B, Dey A, Rahman M, Cavalu S (2022) Bioactive based nanocarriers for the treatment of viral infections and SARS-CoV-2. Nano 12:1530

Grady D (2020) Early data show Moderna's coronavirus vaccine is 94.5% effective. New York Times:11

Gralinski LE, Menachery VD (2020) Return of the coronavirus: 2019-nCoV. Viruses 12. https://doi.org/10.3390/v12020135

Grosshans F, Cerf NJ (2004) Continuous-variable quantum cryptography is secure against non-gaussian attacks. Phys Rev Lett 92:047905

Hasan MM, Zilani MNH, Akter S, Nasrin P, Shajib GMA, Islam MA, Biswas P, Mahmud S, Saleh MA, Hasan MN et al (2022a) UHPLC-Q/orbitrap/MS based chemical fingerprinting and hepatoprotective potential of a medicinal plant, Morinda angustifolia Roxb. S Afr J Bot 148: 561–572. https://doi.org/10.1016/j.sajb.2022.05.037

Hasan A, Biswas P, Bondhon TA, Jannat K, Paul TK, Paul AK, Jahan R, Nissapatorn V, Mahboob T, Wilairatana P et al (2022b) Can artemisia herba-alba be useful for managing COVID-19 and comorbidities? Molecules:27. https://doi.org/10.3390/molecules27020492

Hoffmann M, Kleine-Weber H, Schroeder S, Krüger N, Herrler T, Erichsen S, Schiergens TS, Herrler G, Wu NH, Nitsche A et al (2020) SARS-CoV-2 cell entry depends on ACE2 and TMPRSS2 and is blocked by a clinically proven protease inhibitor. Cell 181:271–280.e278. https://doi.org/10.1016/j.cell.2020.02.052

Hossain R, Dey D, Biswas P, Paul P, Ahmed SZ, Khan AA, Ema TI, Islam MT (2022) Chlorophytum borivilianum (Musli) and Cimicifuga racemosa (black cohosh). In: Herbs, shrubs, and trees of potential medicinal benefits. CRC Press, pp 45–82

Hosseini SA, Zahedipour F, Mirzaei H, Kazemi Oskuee R (2021) Potential SARS-CoV-2 vaccines: concept, progress, and challenges. Int Immunopharmacol 97:107622. https://doi.org/10.1016/j.intimp.2021.107622

Hu X, Cai X, Song X, Li C, Zhao J, Luo W, Zhang Q, Ekumi IO, He ZJFV (2020) Possible SARS-coronavirus 2 inhibitor revealed by simulated molecular docking to viral main protease and host toll-like receptor. Future Virol 15:359–368

Huang C, Lokugamage KG, Rozovics JM, Narayanan K, Semler BL, Makino S (2011) SARS coronavirus nsp1 protein induces template-dependent endonucleolytic cleavage of mRNAs: viral mRNAs are resistant to nsp1-induced RNA cleavage. PLoS Pathog 7:e1002433. https://doi.org/10.1371/journal.ppat.1002433

Huang Y, Yang C, Xu XF, Xu W, Liu SW (2020a) Structural and functional properties of SARS-CoV-2 spike protein: potential antivirus drug development for COVID-19. Acta Pharmacol Sin 41:1141–1149. https://doi.org/10.1038/s41401-020-0485-4

Huang C, Wang Y, Li X, Ren L, Zhao J, Hu Y, Zhang L, Fan G, Xu J, Gu X et al (2020b) Clinical features of patients infected with 2019 novel coronavirus in Wuhan, China. Lancet (London, England) 395:497–506. https://doi.org/10.1016/s0140-6736(20)30183-5

Huang D, Yu H, Wang T, Yang H, Yao R, Liang Z (2021) Efficacy and safety of umifenovir for coronavirus disease 2019 (COVID-19): a systematic review and meta-analysis. J Med Virol 93: 481–490. https://doi.org/10.1002/jmv.26256

Islam MA, Zilani MNH, Biswas P, Khan DA, Rahman MH, Nahid R, Nahar N, Samad A, Ahammad F, Hasan MN (2022) Evaluation of in vitro and in silico anti-inflammatory potential of some selected medicinal plants of Bangladesh against cyclooxygenase-II enzyme. J Ethnopharmacol 285:114900. https://doi.org/10.1016/j.jep.2021.114900

Iwata-Yoshikawa N, Okamura T, Shimizu Y, Hasegawa H, Takeda M, Nagata N (2019) TMPRSS2 contributes to virus spread and immunopathology in the airways of murine models after coronavirus infection. J Virol 93. https://doi.org/10.1128/jvi.01815-18

Jiang Y, Wang F, Xu H, Liu H, Meng Q, Liu W (2014) Development of andrographolide loaded PLGA microspheres: optimization, characterization and in vitro–in vivo correlation. Int J Pharm 475:475–484

Jo S, Kim S, Shin DH, Kim MS (2020) Inhibition of SARS-CoV 3CL protease by flavonoids. J Enzyme Inhib Med Chem 35:145–151. https://doi.org/10.1080/14756366.2019.1690480

Jung HA, Jin SE, Ahn BR, Lee CM, Choi JS (2013) Anti-inflammatory activity of edible brown alga Eisenia bicyclis and its constituents fucosterol and phlorotannins in LPS-stimulated RAW264.7 macrophages. Food Chem Toxicol 59:199–206. https://doi.org/10.1016/j.fct.2013.05.061

Kalasariya HS, Patel NB, Gacem A, Alsufyani T, Reece LM, Yadav VK, Awwad NS, Ibrahium HA, Ahn Y, Yadav KK et al (2022) Marine alga Ulva fasciata-derived molecules for the potential treatment of SARS-CoV-2: an in silico approach. Mar Drugs 20:586. https://doi.org/10.3390/md20090586

Kapusta J, Modelska A, Figlerowicz M, Pniewski T, Letellier M, Lisowa O, Yusibov V, Koprowski H, Plucienniczak A, Legocki AB (1999) A plant-derived edible vaccine against hepatitis B virus. FASEB J 13:1796–1799. https://doi.org/10.1096/fasebj.13.13.1796

Kerry RG, Malik S, Redda YT, Sahoo S, Patra JK, Majhi S (2019) Nano-based approach to combat emerging viral (NIPAH virus) infection. Nanomedicine 18:196–220

Keyaerts E, Vijgen L, Pannecouque C, Van Damme E, Peumans W, Egberink H, Balzarini J, Van Ranst M (2007) Plant lectins are potent inhibitors of coronaviruses by interfering with two targets in the viral replication cycle. Antivir Res 75:179–187. https://doi.org/10.1016/j.antiviral.2007.03.003

Khan RA, Hossain R, Siyadatpanah A, Al-Khafaji K, Khalipha ABR, Dey D, Asha UH, Biswas P, Saikat ASM, Chenari HA et al (2021a) Diterpenes/diterpenoids and their derivatives as potential bioactive leads against dengue virus: a computational and network pharmacology study. Molecules 26:6821. https://doi.org/10.3390/molecules26226821

Khan A, Heng W, Wang Y, Qiu J, Wei X, Peng S, Saleem S, Khan M, Ali SS, Wei DQ (2021b) In silico and in vitro evaluation of kaempferol as a potential inhibitor of the SARS-CoV-2 main protease (3CLpro). Phytoth Res 35:2841–2845. https://doi.org/10.1002/ptr.6998

Kim TS, Heinlein C, Hackman RC, Nelson PS (2006) Phenotypic analysis of mice lacking the Tmprss2-encoded protease. Mol Cell Biol 26:965–975. https://doi.org/10.1128/mcb.26.3.965-975.2006

Koyama S, Heinbockel T (2020) The effects of essential oils and terpenes in relation to their routes of intake and application. Int J Mol Sci 21. https://doi.org/10.3390/ijms21051558

Koyama S, Kondo K, Ueha R, Kashiwadani H, Heinbockel T (2021) Possible use of phytochemicals for recovery from COVID-19-induced anosmia and ageusia. Int J Mol Sci 22. https://doi.org/10.3390/ijms22168912

Ksiazek TG, Erdman D, Goldsmith CS, Zaki SR, Peret T, Emery S, Tong S, Urbani C, Comer JA, Lim W et al (2003) A novel coronavirus associated with severe acute respiratory syndrome. N Engl J Med 348:1953–1966. https://doi.org/10.1056/NEJMoa030781

Kurniawan DW, Ikhsanudin A (2020) Potential of jamu in nanotechnology perspective as an alternative treatment for Covid-19. Pharmaceut Sci Res 7:1

Lai CC, Shih TP, Ko WC, Tang HJ, Hsueh PR (2020) Severe acute respiratory syndrome coronavirus 2 (SARS-CoV-2) and coronavirus disease-2019 (COVID-19): the epidemic and the challenges. Int J Antimicrob Agents 55:105924. https://doi.org/10.1016/j.ijantimicag.2020.105924

Lei J, Kusov Y, Hilgenfeld R (2018) Nsp3 of coronaviruses: structures and functions of a large multi-domain protein. Antivir Res 149:58–74. https://doi.org/10.1016/j.antiviral.2017.11.001

Li G, De Clercq E (2020) Therapeutic options for the 2019 novel coronavirus (2019-nCoV). Nat Rev Drug Discov 19:149–150. https://doi.org/10.1038/d41573-020-00016-0

Li P, Li J, Wu C, Wu Q, Li JJN (1912) Synergistic antibacterial effects of β-lactam antibiotic combined with silver nanoparticles. Nanotechnology 2005:16

Li Q, Guan X, Wu P, Wang X, Zhou L, Tong Y, Ren R, Leung KSM, Lau EHY, Wong JY et al (2020) Early transmission dynamics in Wuhan, China, of novel coronavirus-infected pneumonia. N Engl J Med 382:1199–1207. https://doi.org/10.1056/NEJMoa2001316

Lin J, Handschin C, Spiegelman BM (2005a) Metabolic control through the PGC-1 family of transcription coactivators. Cell Metab 1:361–370. https://doi.org/10.1016/j.cmet.2005.05.004

Lin CW, Tsai FJ, Tsai CH, Lai CC, Wan L, Ho TY, Hsieh CC, Chao PD (2005b) Anti-SARS coronavirus 3C-like protease effects of Isatis indigotica root and plant-derived phenolic compounds. Antivir Res 68:36–42. https://doi.org/10.1016/j.antiviral.2005.07.002

Lin K-YA, Chen Y-C, Lin Y-FJCES (2017) LaMO3 perovskites (M= Co, Cu, Fe and Ni) as heterogeneous catalysts for activating peroxymonosulfate in water. Chem Eng Sci 160:96–105

Liu DX, Fung TS, Chong KK, Shukla A, Hilgenfeld R (2014) Accessory proteins of SARS-CoV and other coronaviruses. Antivir Res 109:97–109. https://doi.org/10.1016/j.antiviral.2014.06.013

Maharjan PM, Choe S (2021) Plant-based COVID-19 vaccines: current status, design, and development strategies of candidate vaccines. Vaccines 9(9):992. https://doi.org/10.3390/vaccines9090992

Mahmood N, Nasir SB, Hefferon K (2020) Plant-based drugs and vaccines for COVID-19. Vaccine 9. https://doi.org/10.3390/vaccines9010015

Mallakpour S, Azadi E, Hussain CM (2021) Fight against COVID-19 pandemic with the help of carbon-based nanomaterials. New J Chem 45:8832–8846

Marx U, Akabane T, Andersson TB, Baker E, Beilmann M, Beken S, Brendler-Schwaab S, Cirit M, David R, Dehne EM et al (2020) Biology-inspired microphysiological systems to advance patient benefit and animal welfare in drug development. ALTEX 37:365–394. https://doi.org/10.14573/altex.2001241

Mattioli IA, Hassan A, Oliveira ON Jr, Crespilho FN (2020) On the challenges for the diagnosis of SARS-CoV-2 based on a review of current methodologies. ACS Sens 5:3655–3677

Mehari A, Gladwin MT, Tian X, Machado RF, Kato GJ (2012) Mortality in adults with sickle cell disease and pulmonary hypertension. JAMA 307:1254–1256. https://doi.org/10.1001/jama.2012.358

Mignet N, Seguin J, Romano MR, Brullé L, Touil YS, Scherman D, Bessodes M, Chabot GG (2012) Development of a liposomal formulation of the natural flavonoid fisetin. Int J Pharm 423:69–76

Monreal-Escalante E, Ramos-Vega A, Angulo C, Bañuelos-Hernández B (2022) Plant-based vaccines: antigen design, diversity, and strategies for high level production. Vaccine 10. https://doi.org/10.3390/vaccines10010100

Morrey JD, Motter NE, Taro B, Lay M, Fairman J (2008) Efficacy of cationic lipid–DNA complexes (CLDC) on hepatitis B virus in transgenic mice. Antivir Res 79:71–79

Morse JS, Lalonde T, Xu S, Liu WR (2020) Learning from the past: possible urgent prevention and treatment options for severe acute respiratory infections caused by 2019-nCoV. Chembiochem 21:730–738. https://doi.org/10.1002/cbic.202000047

Morshed A, Al Azad S, Mia MAR, Uddin MF, Ema TI, Yeasin RB, Srishti SA, Sarker P, Aurthi RY, Jamil F et al (2022) Oncoinformatic screening of the gene clusters involved in the HER2-positive breast cancer formation along with the in silico pharmacodynamic profiling of selective long-chain omega-3 fatty acids as the metastatic antagonists. Mol Divers. https://doi.org/10.1007/s11030-022-10573-8

Mullard A (2020) Flooded by the torrent: the COVID-19 drug pipeline. Lancet 395:1245–1246

Munshi M, Zilani MNH, Islam MA, Biswas P, Das A, Afroz F, Hasan MN (2022) Novel compounds from endophytic fungi of Ceriops decandra inhibit breast cancer cell growth through estrogen receptor alpha in in-silico study. Inform Med Unlock 32:101046. https://doi.org/10.1016/j.imu.2022.101046

Nguyen TT, Woo HJ, Kang HK, Nguyen VD, Kim YM, Kim DW, Ahn SA, Xia Y, Kim D (2012) Flavonoid-mediated inhibition of SARS coronavirus 3C-like protease expressed in Pichia pastoris. Biotechnol Lett 34:831–838. https://doi.org/10.1007/s10529-011-0845-8

Nguyen TTH, Jung JH, Kim MK, Lim S, Choi JM, Chung B, Kim DW, Kim D (2021) The inhibitory effects of plant derivate polyphenols on the msain protease of SARS coronavirus 2 and their structure-activity relationship. Molecules (Basel, Switzerland) 26:1924. https://doi.org/10.3390/molecules26071924

Pandey P, Rane JS, Chatterjee A, Kumar A, Khan R, Prakash A, Ray S (2021) Targeting SARS-CoV-2 spike protein of COVID-19 with naturally occurring phytochemicals: an in silico study for drug development. J Biomol Struct Dyn 39:6306–6316

Pangestuti R, Vo TS, Ngo DH, Kim SK (2013) Fucoxanthin ameliorates inflammation and oxidative responses in microglia. J Agric Food Chem 61:3876–3883. https://doi.org/10.1021/jf400015k

Parada CM, Yang B, Campbell CG, Storey R (2020) Synthesis, characterization, and photopolymerization of (meth) acrylate-functional polyisobutylene macromers produced by cleavage/alkylation of butyl rubber. J Polym Sci 58:2807–2822

Park JY, Yuk HJ, Ryu HW, Lim SH, Kim KS, Park KH, Ryu YB, Lee WS (2017) Evaluation of polyphenols from Broussonetia papyrifera as coronavirus protease inhibitors. J Enzyme Inhib Med Chem 32:504–515. https://doi.org/10.1080/14756366.2016.1265519

Park HH, Kim H, Lee HS, Seo EU, Kim J-E, Lee J-H, Mun Y-H, Yoo S-Y, An J, Yun M-Y (2021) PEGylated nanoparticle albumin-bound steroidal ginsenoside derivatives ameliorate SARS-CoV-2-mediated hyper-inflammatory responses. Biomaterials 273:120827

Patel VB, Zhong J-C, Grant MB, Oudit GY (2016) Role of the ACE2/angiotensin 1–7 axis of the renin–angiotensin system in heart failure. Circ Res 118:1313–1326

Paul PK, Al Azad S, Rahman MH, Farjana M, Uddin MR, Dey D, Mahmud S, Ema TI, Biswas P, Anjum M et al (2022) Catabolic profiling of selective enzymes in the saccharification of non-food lignocellulose parts of biomass into functional edible sugars and bioenergy: an in silico bioprospecting. J Adv Veter Anim Res 9:19–32. https://doi.org/10.5455/javar.2022.i565

Paul P, Biswas P, Dey D, Saikat ASM, Islam MA, Sohel M, Hossain R, Mamun AA, Rahman MA, Hasan MN et al (2021) Exhaustive plant profile of "Dimocarpus longan lour" with significant phytomedicinal properties: a literature based-review. Processes 9:1803

Peeri NC, Shrestha N, Rahman MS, Zaki R, Tan Z, Bibi S, Baghbanzadeh M, Aghamohammadi N, Zhang W, Haque U (2020) The SARS, MERS and novel coronavirus (COVID-19) epidemics, the newest and biggest global health threats: what lessons have we learned? Int J Epidemiol 49: 717–726. https://doi.org/10.1093/ije/dyaa033

Peng S, Huang J, Sheehy JE, Laza RC, Visperas RM, Zhong X, Centeno GS, Khush GS, Cassman KG (2004) Rice yields decline with higher night temperature from global warming. Proc Natl Acad Sci U S A 101:9971–9975. https://doi.org/10.1073/pnas.0403720101

Pereira L, Critchley AT (2020) The COVID-19 novel coronavirus pandemic 2020: seaweeds to the rescue? Why does substantial, supporting research about the antiviral properties of seaweed polysaccharides seem to go unrecognized by the pharmaceutical community in these desperate times? J Appl Phycol 32:1875–1877. https://doi.org/10.1007/s10811-020-02143-y

Petit L, Vernès L, Cadoret JP (2021) Docking and in silico toxicity assessment of Arthrospira compounds as potential antiviral agents against SARS-CoV-2. J Appl Phycol 33:1579–1602. https://doi.org/10.1007/s10811-021-02372-9

Piñero J, Ramírez-Anguita JM, Saüch-Pitarch J, Ronzano F, Centeno E, Sanz F, Furlong LI (2020) The DisGeNET knowledge platform for disease genomics: 2019 update. Nucleic Acids Res 48: D845–d855. https://doi.org/10.1093/nar/gkz1021

Plante JA, Liu Y, Liu J, Xia H, Johnson BA, Lokugamage KG, Zhang X, Muruato AE, Zou J, Fontes-Garfias CR et al (2021) Spike mutation D614G alters SARS-CoV-2 fitness. Nature 592: 116–121. https://doi.org/10.1038/s41586-020-2895-3

Pogrebnyak N, Golovkin M, Andrianov V, Spitsin S, Smirnov Y, Egolf R, Koprowski H (2005) Severe acute respiratory syndrome (SARS) S protein production in plants: development of recombinant vaccine. Proc Natl Acad Sci U S A 102:9062–9067. https://doi.org/10.1073/pnas.0503760102

Pourhajibagher M, Rahimi-Esboei B, Ahmadi H, Bahador A (2021) The anti-biofilm capability of nano-emodin-mediated sonodynamic therapy on multi-species biofilms produced by burn wound bacterial strains. Photodiagn Photodyn Ther 34:102288. https://doi.org/10.1016/j.pdpdt.2021.102288

Pradhan B, Nayak R, Patra S, Bhuyan PP, Dash SR, Ki JS, Adhikary SP, Ragusa A, Jena M (2022) Cyanobacteria and algae-derived bioactive metabolites as antiviral agents: evidence, mode of action, and scope for further expansion; a comprehensive review in light of the SARS-CoV-2 outbreak. Antioxidants (Basel, Switzerland) 11:354. https://doi.org/10.3390/antiox11020354

Prasad A, Muthamilarasan M, Prasad M (2020) Synergistic antiviral effects against SARS-CoV-2 by plant-based molecules. Plant Cell Rep 39:1109–1114. https://doi.org/10.1007/s00299-020-02560-w

Ragelle H, Crauste-Manciet S, Seguin J, Brossard D, Scherman D, Arnaud P, Chabot GG (2012) Nanoemulsion formulation of fisetin improves bioavailability and antitumour activity in mice. Int J Pharm 427:452–459

Rahman MA, Rahman MDH, Hossain MS, Biswas P, Islam R, Uddin MJ, Rahman MH, Rhim H (2020a) Molecular insights into the multifunctional role of natural compounds: autophagy modulation and cancer prevention. Biomedicine 8:517. https://doi.org/10.3390/biomedicines8110517

Rahman MA, Rahman MH, Biswas P, Hossain MS, Islam R, Hannan MA, Uddin MJ, Rhim H (2020b) Potential therapeutic role of phytochemicals to mitigate mitochondrial dysfunctions in Alzheimer's disease. Antioxidants (Basel, Switzerland) 10:23. https://doi.org/10.3390/antiox10010023

Rahman MS, Zilani MNH, Islam MA, Hasan MM, Islam MM, Yasmin F, Biswas P, Hirashima A, Rahman MA, Hasan MN et al (2021) In vivo neuropharmacological potential of Gomphandra tetrandra (wall.) Sleumer and in-silico study against β-amyloid precursor protein. PRO 9:1449

Rahman MDH, Biswas P, Dey D, Hannan MA, Sahabuddin M, Araf Y, Kwon Y, Emran TB, Ali MS, Uddin MJ (2022) An in-silico identification of potential flavonoids against kidney fibrosis targeting TGFβR-1. Life (Basel, Switzerland) 12:1764. https://doi.org/10.3390/life12111764

Rattanapisit K, Shanmugaraj B, Manopwisedjaroen S, Purwono PB, Siriwattananon K, Khorattanakulchai N, Hanittinan O, Boonyayothin W, Thitithanyanont A, Smith DR et al (2020) Rapid production of SARS-CoV-2 receptor binding domain (RBD) and spike specific monoclonal antibody CR3022 in Nicotiana benthamiana. Sci Rep 10:17698. https://doi.org/10.1038/s41598-020-74904-1

Ren B, Yan F, Kuang Y, Li N, Zhang D, Zhou X, Lin H, Zhou H (2018) Improved base editor for efficiently inducing genetic variations in rice with CRISPR/Cas9-guided hyperactive hAID mutant. Mol Plant 11:623–626. https://doi.org/10.1016/j.molp.2018.01.005

Rosales-Mendoza S (2020) Will plant-made biopharmaceuticals play a role in the fight against COVID-19? Expert Opin Biol Ther 20:545–548. https://doi.org/10.1080/14712598.2020.1752177

Rudra S, Kalra A, Kumar A, Joe W (2017) Utilization of alternative systems of medicine as health care services in India: evidence on AYUSH care from NSS 2014. PLoS One 12:e0176916

Ryu Y, Nilson T, Kobayashi H, Sonnentag O, Law BE, Baldocchi DDJA, Meteorology F (2010a) On the correct estimation of effective leaf area index: does it reveal information on clumping effects? Agric For Meteorol 150:463–472

Ryu YB, Jeong HJ, Kim JH, Kim YM, Park JY, Kim D, Nguyen TT, Park SJ, Chang JS, Park KH et al (2010b) Biflavonoids from Torreya nucifera displaying SARS-CoV 3CL(pro) inhibition. Bioorg Med Chem 18:7940–7947. https://doi.org/10.1016/j.bmc.2010.09.035

Ryu YB, Park SJ, Kim YM, Lee JY, Seo WD, Chang JS, Park KH, Rho MC, Lee WS (2010c) SARS-CoV 3CLpro inhibitory effects of quinone-methide triterpenes from Tripterygium regelii. Bioorg Med Chem Lett 20:1873–1876. https://doi.org/10.1016/j.bmcl.2010.01.152

Sahebnasagh A, Saghafi F, Avan R, Khoshi A, Khataminia M, Safdari M, Habtemariam S, Ghaleno HR, Nabavi SM (2020) The prophylaxis and treatment potential of supplements for COVID-19. Eur J Pharmacol 887:173530. https://doi.org/10.1016/j.ejphar.2020.173530

Sakai Y, Kawachi K, Terada Y, Omori H, Matsuura Y, Kamitani W (2017) Two-amino acids change in the nsp4 of SARS coronavirus abolishes viral replication. Virology 510:165–174. https://doi.org/10.1016/j.virol.2017.07.019

Saravanan KM, Zhang H, Senthil R, Vijayakumar KK, Sounderrajan V, Wei Y, Shakila H (2022) Structural basis for the inhibition of SARS-CoV2 main protease by Indian medicinal plant-derived antiviral compounds. J Biomol Struct Dyn 40:1970–1978. https://doi.org/10.1080/07391102.2020.1834457

Sarker MT, Saha S, Biswas P, Islam MT, Sheikh MA, Hasan MN, Islam N, Rabbe MMI, Rafi MO (2022) Identification of blood-based inflammatory biomarkers for the early-stage detection of acute myocardial infarction. Netw Model Anal Health Inform Bioinform 11:28. https://doi.org/10.1007/s13721-022-00371-5

Satija N, Lal SK (2007) The molecular biology of SARS coronavirus. Ann N Y Acad Sci 1102:26–38. https://doi.org/10.1196/annals.1408.002

Schaechter M (2009) Encyclopedia of microbiology. Academic Press

Serrano-Aroca Á, Takayama K, Tuñón-Molina A, Seyran M, Hassan SS, Pal Choudhury P, Uversky VN, Lundstrom K, Adadi P, Palù G (2021) Carbon-based nanomaterials: promising antiviral agents to combat COVID-19 in the microbial-resistant era. ACS Nano 15:8069–8086

Shakoor H, Feehan J, Mikkelsen K, Al Dhaheri AS, Ali HI, Platat C, Ismail LC, Stojanovska L, Apostolopoulos V (2021) Be well: a potential role for vitamin B in COVID-19. Maturitas 144: 108–111. https://doi.org/10.1016/j.maturitas.2020.08.007

Shanmugaraj B, Siriwattananon K, Malla A, Phoolcharoen W (2021) Potential for developing plant-derived candidate vaccines and biologics against emerging coronavirus infections. Pathogens (Basel, Switzerland) 10:1051. https://doi.org/10.3390/pathogens10081051

Shi TH, Huang YL, Chen CC, Pi WC, Hsu YL, Lo LC, Chen WY, Fu SL, Lin CH (2020) Andrographolide and its fluorescent derivative inhibit the main proteases of 2019-nCoV and SARS-CoV through covalent linkage. Biochem Biophys Res Commun 533:467–473. https://doi.org/10.1016/j.bbrc.2020.08.086

Shoseyov O, Posen Y, Grynspan F (2014) Human collagen produced in plants: more than just another molecule. Bioengineered 5:49–52. https://doi.org/10.4161/bioe.26002

Sims KR Jr, He B, Koo H, Benoit DS (2020) Electrostatic interactions enable nanoparticle delivery of the flavonoid myricetin. ACS Omega 5:12649–12659

Sindhu RK, Gupta R, Wadhera G, Kumar P (2022) Modern herbal nanogels: formulation, delivery methods, and applications. Gels 8:97

Singh R, Chauhan N, Kuddus M (2021) Exploring the therapeutic potential of marine-derived bioactive compounds against COVID-19. Environ Sci Pollut Res Int 28:52798–52809. https://doi.org/10.1007/s11356-021-16104-6

Singh R, Khalid M, Batra N, Biswas P, Singh L, Bhatti R (2023) Exploring the anticonvulsant activity of aqueous extracts of *Ficus benjamina* L. figs in experimentally induced convulsions. J Chem 2023:6298366. https://doi.org/10.1155/2023/6298366

Sinico C, De Logu A, Lai F, Valenti D, Manconi M, Loy G, Bonsignore L, Fadda AM (2005) Liposomal incorporation of Artemisia arborescens L. essential oil and in vitro antiviral activity. Eur J Pharm Biopharm 59:161–168

Sivaraman D, Pradeep PS (2020) Revealing anti-viral potential of bio-active therapeutics targeting SARS-CoV2-polymerase (RdRp) in combating COVID-19: Molecular investigation on Indian traditional medicines. Acad Ther

Sivasankarapillai VS, Pillai AM, Rahdar A, Sobha AP, Das SS, Mitropoulos AC, Mokarrar MH, Kyzas GZ (2020) On facing the SARS-CoV-2 (COVID-19) with combination of nanomaterials and medicine: possible strategies and first challenges. Nano 10:852

Sohel M, Biswas P, Al Amin M, Hossain MA, Sultana H, Dey D, Aktar S, Setu A, Khan MS, Paul P et al (2022) Genistein, a potential phytochemical against breast cancer treatment-insight into the molecular mechanisms. Processes 10:415

Song Y, He P, Rodrigues AL, Datta P, Tandon R, Bates JT, Bierdeman MA, Chen C, Dordick J, Zhang F et al (2021) Anti-SARS-CoV-2 activity of Rhamnan sulfate from Monostroma nitidum. Mar Drugs 19. https://doi.org/10.3390/md19120685

Soni V, Raizada P, Singh P, Cuong HN, Rangabhashiyam S, Saini A, Saini RV, Le QV, Nadda AK, Le TT et al (2021) Sustainable and green trends in using plant extracts for the synthesis of biogenic metal nanoparticles toward environmental and pharmaceutical advances: a review. Environ Res 202:111622. https://doi.org/10.1016/j.envres.2021.111622

Su HX, Yao S, Zhao WF, Li MJ, Liu J, Shang WJ, Xie H, Ke CQ, Hu HC, Gao MN et al (2020) Anti-SARS-CoV-2 activities in vitro of Shuanghuanglian preparations and bioactive ingredients. Acta Pharmacol Sin 41:1167–1177. https://doi.org/10.1038/s41401-020-0483-6

Subissi L, Posthuma CC, Collet A, Zevenhoven-Dobbe JC, Gorbalenya AE, Decroly E, Snijder EJ, Canard B, Imbert I (2014) One severe acute respiratory syndrome coronavirus protein complex integrates processive RNA polymerase and exonuclease activities. Proc Natl Acad Sci U S A 111:E3900–E3909. https://doi.org/10.1073/pnas.1323705111

Sun S-W, Yeh P-C (2005) Analysis of rhubarb anthraquinones and bianthrones by microemulsion electrokinetic chromatography. J Pharm Biomed Anal 36:995–1001

Takeyama N, Kiyono H, Yuki Y (2015) Plant-based vaccines for animals and humans: recent advances in technology and clinical trials. Therap Adv Vaccin 3:139–154. https://doi.org/10.1177/2051013615613272

Tallei TE, Tumilaar SG, Niode NJ, Fatimawali, Kepel BJ, Idroes R, Effendi Y, Sakib SA, Emran TB (2020) Potential of plant bioactive compounds as SARS-CoV-2 main protease (M(pro)) and spike (S) glycoprotein inhibitors: a molecular docking study. Scientifica 2020:6307457. https://doi.org/10.1155/2020/6307457

Tang T, Bidon M, Jaimes JA, Whittaker GR, Daniel S (2020) Coronavirus membrane fusion mechanism offers a potential target for antiviral development. Antivir Res 178:104792. https://doi.org/10.1016/j.antiviral.2020.104792

Thiel V, Herold J, Schelle B, Siddell SG (2001) Viral replicase gene products suffice for coronavirus discontinuous transcription. J Virol 75:6676–6681. https://doi.org/10.1128/jvi.75.14.6676-6681.2001

Thiel V, Ivanov KA, Putics Á, Hertzig T, Schelle B, Bayer S, Weißbrich B, Snijder EJ, Rabenau H, Doerr HW et al (2003) Mechanisms and enzymes involved in SARS coronavirus genome expression. J Gen Virol 84:2305–2315. https://doi.org/10.1099/vir.0.19424-0

Thileepan T, Prasad VMJJRTM (2018) Literature review of suram (fever) in siddha medicine. J Res Tradit Med 4:21–25

Thomas S (2020) The structure of the membrane protein of SARS-CoV-2 resembles the sugar transporter SemiSWEET. Pathog Immun 5:342–363. https://doi.org/10.20411/pai.v5i1.377

Tomas M, Capanoglu E, Bahrami A, Hosseini H, Akbari-Alavijeh S, Shaddel R, Rehman A, Rezaei A, Rashidinejad A, Garavand F et al (2022) The direct and indirect effects of bioactive compounds against coronavirus. Food Front 3:96–123. https://doi.org/10.1002/fft2.119

Tripathi MK, Singh P, Sharma S, Singh TP, Ethayathulla AS, Kaur P (2021) Identification of bioactive molecule from Withania somnifera (ashwagandha) as SARS-CoV-2 main protease inhibitor. J Biomol Struct Dyn 39:5668–5681. https://doi.org/10.1080/07391102.2020.1790425

Twyman RM, Stoger E, Schillberg S, Christou P, Fischer R (2003) Molecular farming in plants: host systems and expression technology. Trends Biotechnol 21:570–578. https://doi.org/10.1016/j.tibtech.2003.10.002

Uthaya Kumar A, Kadiresen K, Gan WC, Ling APK (2021) Current updates and research on plant-based vaccines for coronavirus disease 2019. Clin Exp Vac Res 10:13–23. https://doi.org/10.7774/cevr.2021.10.1.13

Varahachalam SP, Lahooti B, Chamaneh M, Bagchi S, Chhibber T, Morris K, Bolanos JF, Kim N-Y, Kaushik A (2021) Nanomedicine for the SARS-CoV-2: state-of-the-art and future prospects. Int J Nanomedicine 16:539

Vellingiri B, Jayaramayya K, Iyer M, Narayanasamy A, Govindasamy V, Giridharan B, Ganesan S, Venugopal A, Venkatesan D, Ganesan H et al (2020) COVID-19: a promising cure for the global panic. Sci Total Environ 725:138277. https://doi.org/10.1016/j.scitotenv.2020.138277

Wang F, Zhou L, Zhou J, Gu X, Feng Y (2010) Characterization of anticancer hypocrellin A encapsulated with silica nanoparticles: thermal analysis. J Therm Anal Calorim 102:69–74

Wang MY, Zhao R, Gao LJ, Gao XF, Wang DP, Cao JM (2020a) SARS-CoV-2: structure, biology, and structure-based therapeutics development. Front Cell Infect Microbiol 10:587269. https://doi.org/10.3389/fcimb.2020.587269

Wang S-C, Chen Y, Wang Y-C, Wang W-J, Yang C-S, Tsai C-L, Hou M-H, Chen H-F, Shen Y-C, Hung, M.-C.J.A.J.o.C.R. (2020b) Tannic acid suppresses SARS-CoV-2 as a dual inhibitor of the viral main protease and the cellular TMPRSS2 protease. Am J Cancer Res 10:4538

Ward BJ, Gobeil P, Séguin A, Atkins J, Boulay I, Charbonneau PY, Couture M, D'Aoust MA, Dhaliwall J, Finkle C et al (2021) Phase 1 randomized trial of a plant-derived virus-like particle vaccine for COVID-19. Nat Med 27:1071–1078. https://doi.org/10.1038/s41591-021-01370-1

Warnes SL, Little ZR, Keevil CW (2015) Human coronavirus 229E remains infectious on common touch surface materials. MBio 6:e01697-15

Watanabe Y, Allen JD, Wrapp D, McLellan JS, Crispin M (2020) Site-specific glycan analysis of the SARS-CoV-2 spike. Science (New York, N.Y.) 369:330–333. https://doi.org/10.1126/science.abb9983

Wei X, Zhao X, Chen C, Lu J, Cheng W, Li B, Li H, Lin W, Tian C, Zhao J, et al (2018) Klebsiella pneumonia, one of potential chief culprits of non-alcoholic fatty liver disease: through generation of endogenous ethanol. bioRxiv, 446765

Weiss SR, Leibowitz JL (2011) Coronavirus pathogenesis. Adv Virus Res 81:85–164. https://doi.org/10.1016/b978-0-12-385885-6.00009-2

Weng X, Neethirajan S (2018) Immunosensor based on antibody-functionalized MoS 2 for rapid detection of avian coronavirus on cotton thread. IEEE Sensors J 18:4358–4363

Xia S, Zhu Y, Liu M, Lan Q, Xu W, Wu Y, Ying T, Liu S, Shi Z, Jiang S et al (2020) Fusion mechanism of 2019-nCoV and fusion inhibitors targeting HR1 domain in spike protein. Cell Mol Immunol 17:765–767. https://doi.org/10.1038/s41423-020-0374-2

Xiao J (2017) Dietary flavonoid aglycones and their glycosides: which show better biological significance? Crit Rev Food Sci Nutr 57:1874–1905

Yadav R, Chaudhary JK, Jain N, Chaudhary PK, Khanra S, Dhamija P, Sharma A, Kumar A, Handu S (2021) Role of structural and non-structural proteins and therapeutic targets of SARS-CoV-2 for COVID-19. Cell 10:821. https://doi.org/10.3390/cells10040821

Yan VC, Muller FL (2021) Why remdesivir failed: preclinical assumptions overestimate the clinical efficacy of remdesivir for COVID-19 and Ebola. Antimicrob Agents Chemother 65:e0111721. https://doi.org/10.1128/aac.01117-21

Yang D, Leibowitz JL (2015) The structure and functions of coronavirus genomic 3′ and 5′ ends. Virus Res 206:120–133. https://doi.org/10.1016/j.virusres.2015.02.025

Yang X, Yu Y, Xu J, Shu H, Liu H, Wu Y, Zhang L, Yu Z, Fang M, Yu T (2020) Clinical course and outcomes of critically ill patients with SARS-CoV-2 pneumonia in Wuhan, China: a single-centered, retrospective, observational study. Lancet Respir Med 8:475–481

Yu MS, Lee J, Lee JM, Kim Y, Chin YW, Jee JG, Keum YS, Jeong YJ (2012) Identification of myricetin and scutellarein as novel chemical inhibitors of the SARS coronavirus helicase, nsP13. Bioorg Med Chem Lett 22:4049–4054. https://doi.org/10.1016/j.bmcl.2012.04.081

Yue P-F, Yuan H-L, Li X-Y, Yang M, Zhu W-F (2010) Process optimization, characterization and evaluation in vivo of oxymatrine–phospholipid complex. Int J Pharm 387:139–146

Zaporozhets TS, Besednova NN (2020) Biologically active compounds from marine organisms in the strategies for combating coronaviruses. AIMS Microbiol 6:470–494. https://doi.org/10.3934/microbiol.2020028

Zeng X, Tian G, Shi J, Deng G, Li C, Chen H (2018) Vaccination of poultry successfully eliminated human infection with H7N9 virus in China. Sci China Life Sci 61:1465–1473. https://doi.org/10.1007/s11427-018-9420-1

Zenteno-Cuevas R (2017) Successes and failures in human tuberculosis vaccine development. Expert Opin Biol Ther 17:1481–1491. https://doi.org/10.1080/14712598.2017.1378641

Zhang H, Yang X, Zhao L, Jiao Y, Liu J, Zhai G (2016) In vitro and in vivo study of baicalin-loaded mixed micelles for oral delivery. Drug Deliv 23:1933–1939

Zhang C, Zheng W, Huang X, Bell EW, Zhou X, Zhang Y (2020) Protein structure and sequence reanalysis of 2019-nCoV genome refutes snakes as its intermediate host and the unique similarity between its spike protein insertions and HIV-1. J Proteome Res 19:1351–1360. https://doi.org/10.1021/acs.jproteome.0c00129

Zhang Z, Nomura N, Muramoto Y, Ekimoto T, Uemura T, Liu K, Yui M, Kono N, Aoki J, Ikeguchi M et al (2022) Structure of SARS-CoV-2 membrane protein essential for virus assembly. Nat Commun 13:4399. https://doi.org/10.1038/s41467-022-32019-3

Zhou P, Yang XL, Wang XG, Hu B, Zhang L, Zhang W, Si HR, Zhu Y, Li B, Huang CL et al (2020) A pneumonia outbreak associated with a new coronavirus of probable bat origin. Nature 579:270–273. https://doi.org/10.1038/s41586-020-2012-7

Zhu N, Zhang D, Wang W, Li X, Yang B, Song J, Zhao X, Huang B, Shi W, Lu R et al (2020) A novel coronavirus from patients with pneumonia in China, 2019. N Engl J Med 382:727–733. https://doi.org/10.1056/NEJMoa2001017

Zilani MNH, Islam MA, Biswas P, Anisuzzman M, Hossain H, Shilpi JA, Hasan MN, Hossain MG (2021) Metabolite profiling, anti-inflammatory, analgesic potentials of edible herb Colocasia gigantea and molecular docking study against COX-II enzyme. J Ethnopharmacol 281:114577. https://doi.org/10.1016/j.jep.2021.114577

Zumla A, Chan JF, Azhar EI, Hui DS, Yuen KY (2016) Coronaviruses - drug discovery and therapeutic options. Nat Rev Drug Discov 15:327–347. https://doi.org/10.1038/nrd.2015.37

# Chapter 8
# Plant-Derived Bioactive Compounds as Potential ACE-2 Inhibitors Against SARS-CoV-2 Infection

**Muhammad Adil, Pragya Tiwari, Jen-Tsung Chen, and Shamsa Kanwal**

## 8.1 Introduction

SARS-COV-2 is the causative agent of the COVID-19 pandemic that resulted in extensive morbidity, mortality and economic losses throughout the world (Kanmodi and Kanmodi 2020). Moreover, the remarkable evolutional capacity of etiological agents caused the emergence of different variant strains (such as omicron and delta) and sequential waves of disease across the diverse parts of the world (Dhawan et al. 2022). This single-stranded RNA virus possesses a nucleocapsid with a characteristic helical configuration (Khan et al. 2020). Apart from its several non-structural proteins, the SARS-COV-2 also exhibits four distinct structural proteins that are referred to as envelope (E), nucleocapsid (N), membrane (M), and spike (S) proteins (Rakib et al. 2020). Spike protein comprising of $S_1$ and $S_2$ subunits is used by SARS-COV-2 for binding with the ACE-2 receptors of human cells. The $S_1$ subunit binds with the ACE-2 receptor of host cells, whereas, the viral cell membrane is fused with the host cell through the $S_2$ subunit (de Matos et al. 2022).

M. Adil (✉)
Pharmacology and Toxicology Section, University of Veterinary and Animal Sciences, Lahore, Jhang Campus, Jhang, Pakistan
e-mail: muhammad.adil@uvas.edu.pk

P. Tiwari (✉)
Department of Biotechnology, Yeungnam University, Gyeongsan, Gyeongbuk, Republic of Korea
e-mail: pragyatiwari@ynu.ac.kr

J.-T. Chen
Department of Life Sciences, National University of Kaohsiung, Kaohsiung, Taiwan
e-mail: jentsung@nuk.edu.tw

S. Kanwal
Institute of Microbiology, University of Veterinary and Animal Sciences, Lahore, Pakistan



J.-T. Chen (ed.), *Ethnopharmacology and Drug Discovery for COVID-19: Anti-SARS-CoV-2 Agents from Herbal Medicines and Natural Products*,
https://doi.org/10.1007/978-981-99-3664-9_8

Despite the development of effective vaccines, the global COVID-19 control status highly varies on account of uneven distribution attributed to technological, socioeconomic, political and geographical factors (Ma et al. 2021). Therapeutic management of COVID-19 infection was primarily anticipated using either traditional medicines or different antiviral and anti-inflammatory agents leading to variable success rates. Besides, the technique of drug repurposing was also employed to find the prospective therapeutic utility of typical drugs against COVID-19 infection. However, despite the extensive research and considerable advancement in the development of therapeutic modalities, highly effective and FDA-approved therapies against COVID-19 infection are still lacking (Raman et al. 2022). Therefore, the discovery and development of more effective and safe therapeutic agents are also inevitable along with strengthening the vaccine-based preventive strategies.

The relatively lower burden of SARS-COV-2 infection in Asian and African regions has been correlated with the widespread use of ethnomedicinal natural products (Abubakar et al. 2021). Plant-based natural compounds constitute the oldest and most significant source of pharmaceutical products and contribute to almost 60% of drugs presently available in the market (Beutler 2009). Currently, plant-derived bioactive compounds have gained substantial attention for the therapeutic management of SARS-COV-2 infection on account of their relatively lower side effects and equivalent efficacy in comparison with typical drugs (Malviya et al. 2022). Various plant-derived bioactive products including alkaloids, phenolic compounds and terpenoids have revealed promising efficiency against SARS-COV-2 infection through their anti-inflammatory, antioxidant and antimicrobial activities (Raman et al. 2022). Keeping in view the rapidly changing genomic pattern of SARS-COV-2, uneven availability of effective vaccines and lack of disease-specific, FDA-approved drugs, plant-derived bioactive products with promising ACE-2 inhibitory activity may provide new lead compounds for the development of more efficient and safe therapeutic agents against the devastating SARS-COV-2 infection.

## 8.2 Therapeutic Rationale for ACE-2 Blockage in the Context of SARS-CoV-2 Infection

The pivotal role of renin angiotensin aldosterone system (RAAS) in the maintenance of fluid homeostasis and regulation of blood pressure is well-established for decades. Apart from the classical pathway of RAAS comprising of angiotensin-II and aldosterone, the angiotensin 1–7-associated new cascade has also been described in cardiovascular, renal and central nervous systems (Simões e Silva and Teixeira 2016). ACE-2 converts angiotensin-II into angiotensin 1–7 which exerts anti-inflammatory, antifibrotic and vasodilatory effects upon receptor occupancy (de Matos et al. 2022). Additionally, it also inhibits the vasodilation and enhancement of vascular permeability by altering the pulmonary metabolism of bradykinin

(Tikellis and Thomas 2012; Sanchis-Gomar et al. 2020). Previously, the antagonism of RAAS for the treatment of cardiovascular diseases was primarily achieved through the blockage of aldosterone, renin or angiotensin type 1 ($AT_1$) receptors (Bernardi et al. 2016; Ferrario and Mullick 2017). However, the occurrence of COVID-19 outbreak shifted the future research on RAAS toward ACE-2 as a potential target for disrupting the viral interaction with human cells (Junior et al. 2021).

Organs characterized by higher ACE-2 expression are more frequently affected by SARS-COV-2 infection (Letko et al. 2020). Besides principally targeting the respiratory tract, SARS-COV-2 also invades the cardiovascular, renal, digestive and central nervous systems (To et al. 2004). It predominantly targets the type II alveolar cells due to higher ACE-2 expression (Verdecchia et al. 2020). The presence of ACE-2 in the pancreas and small intestine, and its implication in the manifestation of digestive signs (such as vomiting and diarrhea) during the early stage of COVID-19 infection has been described (Gheblawi et al. 2020; Yan et al. 2020). ACE-2 regulates the homeostasis of antimicrobial peptides, intestinal microflora and amino acid concentrations in the gastrointestinal tract (Abubakar et al. 2021). The critical role of ACE-2 in the intracellular movement of SARS-COV-2 was corroborated in 2003 (Kuhn et al. 2004). Owing to its strong affinity for ACE-2, the viral S protein undergoes activation by the transmembrane protease serine 2 and the viral membrane is fused with the host cell membrane for subsequent release of viral RNA into the host cells via a receptor-based endocytosis process (Hoffmann et al. 2020). Strong intermolecular interaction between the lysine 31 fraction of host ACE-2 enzyme and 394-glutamine fragment associated with the receptor binding motif of SARS-COV-2 S protein occurs through significant Van der Waal forces (Zhou et al. 2020).

ACE-2 consists of two distinct isoforms termed as sACE-2 and mACE-2. Its soluble type, called sACE-2 is found in trace quantities in the bloodstream and is devoid of any membrane anchor (Wysocki et al. 2010). Conversely, the mACE-2 exhibits distinguishable extracellular and transmembrane domains and lies adjacent to the cell membrane (Ma et al. 2021). The extracellular domain of mACE-2 is used by SARS-COV-2 as a binding site for its intracellular transport (Lu et al. 2020). Whereas, sACE-2 is known to antagonize the binding of SARS-COV-2 with mACE-2 (Batlle et al. 2020). Moreover, the interaction of sACE-2 with viral S protein and resultant endocytosis via $AT_1$ receptors also reflect its significant role in the pathophysiological pathway of COVID-19 infection (Yeung et al. 2021). The vasoconstrictor, angiotensin-II is typically hydrolyzed by ACE-2 into angiotensin-(1–7), a vasodilator that exerts a hypotensive effect (Verdecchia et al. 2020). Contrarily, the ACE-2 dysfunction occurring during the SARS-COV-2 infection results in hypertension and inflammatory changes in the affected organs (Letko et al. 2020). Therefore, ACE-2 has been established as the principal target of therapeutic intervention for blocking the intracellular transport of SARS-COV-2.

## 8.3 Screening of Natural Products for Potential ACE2 Inhibition against SARS-CoV-2

A wide range of assays is presently available for the screening of natural products in terms of ACE-2 inhibitory activity. Moreover, continuous advancement in the fields of computational drug designing, medicinal chemistry and analytical techniques is expected to contribute appreciably toward the addition of more superior, economical and convenient testing methods.

### *8.3.1 Biochromatography*

Biochromatography has been developed in 1980s and widely applied for determining the bioactive constituents and mechanisms of action of potential drugs (Zhao and Jiang 2010). In this method, bioactive materials (living cells, active macromolecules) attached to chromatographic carriers are employed to induce interference between biological macromolecules and the respective drug, and further screened for in vivo drug expression during pathological and physiological conditions (Chen et al. 2013). The activity of drug is influenced by its retention in a bio-chromatographic column and bioactivity is demonstrated when the drug binds to cells or biological macromolecules (Beigi and Lundahl 1999; Lagerquist et al. 2001). This technique can be classified into biomembrane chromatography, molecular-biology chromatography and cell membrane chromatography.

#### 8.3.1.1 Cell Membrane Chromatography

Cell membrane chromatography includes the selective interaction between the ligands and membrane receptors. Specific receptors are present on the cell membrane and the powerful adhesion of silicon hydroxyl on $SiO_2$ and self-fusion of phospholipids comprise the stationary phase (He et al. 2001). Moreover, the binding parameters between ligands and membrane receptors are effective in determining the desired constituents of complex samples (Chen et al. 2013). Current advances in bioengineering and molecular biology have improved the screening of bioactive components in complex samples, further facilitating the biomimetic studies on in vitro drug interaction (Hou et al. 2009). A recent study discussed the collection, of ACE-$2^h$ cells, cell membrane separation, mixing with $SiO_2$ and loading into an ACE-2/cell membrane chromatography column (Lv et al. 2021). The bioactive components, methylephedrine, ephedrine and pseudoephedrine were screened through ACE-2/cell membrane chromatography, bioaffinity chromatography and HPLC-IT-TOF-MS system (Gao et al. 2021).

#### 8.3.1.2 Molecular-Biology Chromatography

This technology is widely applied to screen and identify bioactive constituents in a complex sample. In this method, the interaction of drugs in the body with biological molecules, namely enzymes, receptors and antibodies is taken into account. It employs receptors, transporters, enzymes and other biomolecules (with key functions) as stationary phases for the separation and purification of bioactive constituents and determination of the biochemical parameters, respectively (Wang et al. 2000; Yoo et al. 2009). When coupled with nuclear magnetic resonance, mass spectroscopy and other methods, the molecular-biology chromatography forms an integrated system that can provide better identification and structural assessment of bioactive components.

### *8.3.2 Computational Screening Methods*

#### 8.3.2.1 Virtual Screening

Virtual screening represents a frequently used method to identify the active compounds having high binding affinity for target proteins. The databases including multiple molecules are screened for determining the best lead molecules in drug development (Haga and Ichikawa 2016). It is classified into structure-based virtual screening and ligand-based virtual screening. The ligand-based virtual screening is mainly used when structural information and properties of bioactive compounds are available, forming a basis for the selection of new potential compounds (Wang et al. 2020), while, structure-based virtual screening is conducted to predict the stable molecular complexes (Lionta et al. 2014).

#### 8.3.2.2 Molecular Docking Strategies

Molecular docking comprises a virtual drug-designing approach, based upon the interaction between the drug molecule and it's receptor. This method defines a feasible strategy for screening natural products as ACE-2 inhibitors against various diseases (Hirayama 2017), including SARS-CoV-2 infection. In this technique, the scoring functions are utilized to ascertain the ligand binding site and conformational state between the ligand and the target, and subsequently predict the binding energies (Li et al. 2013). The binding affinities of the bioactive ligands and the receptor sites are properly identified and further subjected to clinical validation. Moreover, the potential compounds which show a certain bioavailability and are absorbed, metabolized and excreted by the body, show key prospects in drug development. In addition, the absorption, distribution, metabolism and excretion (ADME) attributes of the compounds are virtually determined using a mathematical algorithm (Gimeno et al. 2019). A useful study was conducted to screen the potential

ACE-2 inhibitors as therapeutic targets of SARS-CoV-2, applying molecular docking, drug-likeness, and ADME predictions of each compound (Vardhan and Sahoo 2020). The study showed that ursolic acid, obacunone, glycyrrhizic acid and maslinic acid showed good binding affinities at the ACE-2 catalytic site.

#### 8.3.2.3 Molecular Dynamics Simulations

The molecular docking method has limitations in dealing with flexible molecules like proteins, that restrict it's wider applications (Okimoto et al. 2009). Several methods including local optimization of the side chains and flexible butting were explored but these methods do not allow rearrangement and extension of protein structures and conformational changes (Friesner et al. 2004; Huang and Zou 2010; De Vivo et al. 2016). Although a key challenge accounts for receptors' flexibility, molecular dynamics work based on interatomic interactions demonstrate variations with time. Moreover, the molecular dynamics method provides information on structural dynamics and energy information about protein–ligand interactions (Ma et al. 2021).

### *8.3.3 Surface Plasmon Resonance*

Surface plasmon resonance is gaining momentum to study various biomolecule interactions and is based on biosensor technology. It is widely employed in screening drugs and is a highly sensitive and label-free method (Patching 2014). This method involves the linking of protein to the chip (as a receptor) surface and the flowing of the analyte solution through it. The binding of compounds (regarded as ligands) with receptors leads to an increase in sensor mass and change in refractive index. Moreover, this is converted into resonance and signal graphs over time. By the assessment of signal graphs, the kinetic properties and the binding affinity of protein and ligand molecules can be known and compounds having good affinities can be identified (Prabowo et al. 2018). Moreover, the detection of compounds is based on the change in refractive index and labeling of small molecules is not required.

### *8.3.4 Gene Chip Technology*

The technique, also referred to as DNA microarray, is a method in which DNA fragments are linked to glass, silicon or other phases in a particular arrangement. In this method, fluorescent or isotope is used to label DNA fragments and a number of gene expressions are detected via base hybridization (Zhu 2003; Huo et al. 2007). This assay has gained widespread recognition and is used to understand the pathogenesis of diseases, gene function and high-throughput drug screening, among others (Ma et al. 2021). In this technology, a comparison between normal human cells and diseased cells can be made by analyzing differential gene expression (Wang et al. 2008). Furthermore, the method can be directly employed to study the changes in gene expression in different tissues before and after taking medication and identify new lead compounds for drug development (Scherf et al. 2000).

## 8.4 Major Bioactive Natural Products as ACE-2 Inhibitors against SARS-CoV-2

Natural products characterized by low molecular weight that partially or completely inhibit the enzyme functions, are gaining recognition as ACE-2 inhibitors/modulators. In the present time, natural products with ACE-2 modulatory activity are being discovered and studied for their prospects in drug discovery. Bioactive natural products screened through in vitro, in silico, and in vivo assays and associated with promising ACE-2 inhibitory potential have been summarized in Tables 8.1, 8.2, and 8.3, respectively.

### *8.4.1 Alkaloids*

Alkaloids are classified as nitrogen-containing organic compounds with different structures and constitute a prospective source of antiviral drugs (Ma et al. 2021). Alkaloids obtained from African plants were evaluated through molecular dynamics and docking techniques, for the inhibition of ACE-2 and the association between SARS-CoV-2 and target human cells (Gyebi et al. 2022). The results of this study indicated a high affinity of cryptoquindoline and cryptospirolepine for ACE-2, TMPRSS2, and SARS-CoV-2 S protein. Several other alkaloids, namely uncarine F, bicuculline, speciophylline and anisodamine bind to the S protein and ACE-2. In addition, bis-benzylisoquinoline alkaloids effectively blocked the S protein-ACE-2-associated cell membrane fusion through their effect on the host cell calcium channels. These compounds hindered the calcium channels, Ca2þ--mediated fusion and entry of viruses into the cell (Ma et al. 2021).

### *8.4.2 Flavonoids*

Flavonoids are commonly found in a number of fruits and vegetables, including the citrus varieties (Tutunchi et al. 2020). This class of compounds represents the largest group of ACE-2 inhibitors showing potent activity. The flavonoids, namely

**Table 8.1** List of bioactive natural products screened through in vitro assays and associated with promising ACE-2 inhibitory potential

| Bioactive natural product(s) | Reference |
|---|---|
| Anthocyanins | Ojeda et al. (2010) |
| Berberine | Kang et al. (2002) |
| Chlorogenic acid | Chiou et al. (2017) |
| Echinatin | Ke et al. (2017) |
| Epigallocatechin-3-gallate | |
| Licochalcone A | |
| Apigenin | Al Shukor et al. (2013) |
| Benzoic acid | |
| Catechol | |
| Chlorogenic acid | |
| Ellagic acid | |
| Epicatechin | |
| Gallic acid | |
| Kaempferol | |
| p-coumaric acid | |
| Phloretin | |
| p-hydroxybenzoic acid | |
| Protocatechuic acid | |
| Pyrogallol | |
| Resveratrol | |
| Rutin | |
| Syringic acid | |
| Tannic acid | |
| Resveratrol | Moran et al. (2017), Kim et al. (2018), Tiao et al. (2018) |
| Oenothein B | Kiss et al. (2004) |
| Acteoside | Geng et al. (2010) |
| Isoacteoside | |
| Plantainoside D | |
| Plantamajoside | |
| Oleanolic acid | Gutiérrez-Román et al. (2021) |
| Ursolic acid | |
| Quercetin | Guerrero et al. (2012) |

hesperidin, naringin, quercetin, kaempferol, and naringenin showed ACE-2 binding affinities for the S protein of SARS-CoV-2 (Chen and Du 2020). Besides, hypertensive rats treated with apigenin showed significant upregulation of ACE-2 gene expression (Sui et al. 2010). Likewise, a natural flavone, baicalin, modulated the ACE-2 expression and diminished angiotensin-II-induced endothelial dysfunction (Wei et al. 2015). Thus the anti-inflammatory phytochemical constituent, baicalin also disrupts the entry of SARS-COV-2 into the host cells (Goyal and Goyal 2020). Another flavanone found in grapefruit, called naringenin downregulated the ACE-2

**Table 8.2** List of bioactive natural products screened through in silico assays and associated with promising ACE-2 inhibitory potential

| Bioactive natural product(s) | Reference(s) |
|---|---|
| Baicalin | Islam et al. (2021) |
| Caffeic acid | Elfiky (2021) |
| Chlorogenic acid | |
| p-coumaric acid | |
| Curcumin | Utomo et al. (2020) |
| Hesperidin | Wu et al. (2020) |
| Andrographolide | Maurya et al. (2020) |
| Berberine | |
| Gallic acid | |
| Luteolin | |
| Mangiferin | |
| Nimbin | |
| Piperine | |
| Thebaine | |
| Withaferin A | |
| Zingiberene | |
| Oleanolic acid | Kumar et al. (2021) |
| Catechol | Al Shukor et al. (2013) |
| Epicatechin | |
| Protocatechuic acid | |
| Pyrogallol | |
| Resveratrol | |
| Rutin | |
| Syringic acid | |
| Quercetin | Muhammad and Fatima (2015), Smith and Smith (2020) |
| Ursolic acid | Kumar et al. (2021) |

expression and reduced nephron-vascular impairment in an animal model of hypertension (Wang et al. 2019). Myricitrin produced by *Myrica rubra* blocked the entry of SARS-COV-2 into the host cells through the induction of conformational modification in the ACE-2 enzyme (Ngwa et al. 2020). Moreover, several reports emphasize the potential of flavonoids in modulating the ACE-2 functions to counteract the SARS-CoV-2 infection. Rutin, isolated from viola, tobacco, hydrangea and buckwheat provokes the endocytosis of SARS-COV-2 following the proteolytic breakdown of ACE-2 at the junction of $S_1$ and $S_2$ subunits (Balmeh et al. 2020).

**Table 8.3** List of bioactive natural products screened through in vivo assays and associated with promising ACE-2 inhibitory potential

| Bioactive natural product(s) | Reference |
|---|---|
| Baicalin | Deng et al. (2012) |
| Caffeic acid | Chiou et al. (2017), Agunloye et al. (2019) |
| Curcumin | Abd Allah and Gomaa (2015) |
| Esculetin | Kadakol et al. (2015) |
| Gallic acid | Lee et al. (2015) |
| H-Ginsenoside Rb1 | Zheng et al. (2017) |
| Hesperidin | Wunpathe et al. (2018) |
| Magnolol | Chang et al. (2018) |
| Naringenin | Gao et al. (2018) |
| Nicotine | Oakes et al. (2018) |
| Oxymatrine | Huang and Chen (2013) |
| Quercetin | Luo et al. (2017), Adefegha et al. (2018) |
| Resveratrol | Kim et al. (2018) |
| Rosmarinic acid | Liu et al. (2016) |
| Tanshinone IIA | Wu et al. (2014) |
| Trigonelline | Hamden et al. (2013) |

### 8.4.3 *Phenols*

Phenols constitute a category of metabolites, widely found in nature and used in the medicine and food sector, owing to their antitumor, antioxidant, and other functions (Dai and Mumper 2010). The phenolic carboxylic acids, salvianolic acid C, salvianolic acid B and salvianolic acid A were studied for their toxicity in ACE-2 cells (found to be low) and their binding affinities with ACE-2 and RBD were evaluated by molecular docking studies (Hu et al. 2021). Another study employed a 2D ACE-2 column/C18 column/TOFMS system and screened active binding components to ACE-2, further validated by surface plasmon resonance. Moreover, the in vitro inhibitory capacity of the constituents and binding affinities of the active components were determined by the ACE-2 inhibitor screening kit and molecular docking studies, respectively (Chen et al. 2021).

### 8.4.4 *Steroids and Steroid Glycosides*

Ginsenosides, the triterpene saponins, are commonly present in the rhizome of ginseng. Ginsenoside Rg3 was shown to suppress the Ang II-induced renal damage through the induction of ACE-2 level (Liu et al. 2019). Moreover, azukisaponin I, ginsenoside Rg6 and arundoin also modulated the ACE-2 functions (Zi et al. 2020). These bioactive constituents showed low inhibition when analysed through ACE-2 kinase inhibition assay. Glycyrrhizin, isolated from *Glycyrrhiza glabra*, prevented the intracellular viral entry after binding to ACE-2 (Chen and Du 2020).

### *8.4.5 Terpenoids*

Terpenoids are volatile substances and include diterpenes, monoterpenes, triterpenes and sesquiterpenes. This class of metabolites shows multiple pharmacological functions including antiviral, antitumor, anti-inflammatory and antibacterial effects, among others (Jaeger and Cuny 2016). Other studies on triterpenoids, namely maslinic acid, ursolic acid and astragaloside IV showed significance as S protein-ACE-2 inhibitors. Moreover, other diterpenoids like limonene and citronellol, present in lemon and geranium, downregulated ACE-2 expression in epithelial cells and showed cytotoxicity on Ht-29 cells, suggesting antiviral properties. The anti-SARS-COV-2 activity of terpenoids such as glycyrrhizic acid and its analogs is gaining considerable attention. This is supported by virtual screening methods, namely docking, molecular dynamics and network pharmacology that showed glycyrrhizic acid as an inhibitor of S protein-ACE-2 (Vardhan and Sahoo 2020).

## 8.5 Conclusions and Future Perspectives

Regardless of the striking structural resemblance between ACE-I and ACE-2, the conventional ACE inhibitors are ineffective for ACE-2 blockage (Tikellis and Thomas 2012). Unlike the cardioselective ACE-2 typical blockers, natural products with potential ACE-2 inhibitory activity are known for their unique multitarget pharmacological effects. The pleiotropic pharmacological actions and resultant anti-inflammatory, antimicrobial and bronchodilatory effects of bioactive compounds can be translated into clinically useful drugs against SARS-COV-2 infection (Junior et al. 2021). Consequently, plant-derived scaffolds can be efficiently employed for designing novel ACE-2 inhibitors against COVID-19 infection (Muchtaridi et al. 2020). Additionally, the association of human ACE-2 with the S protein of other genetically identical coronaviruses has also been postulated (Kai and Kai 2020). Therefore, the ACE-2-inhibitory natural products may also exhibit therapeutic utility against other coronaviruses (Ma et al. 2021).

Currently, natural products are primarily evaluated for ACE-2 inhibitory potential using virtual screening, molecular dynamics simulation and molecular docking techniques. Regardless of its improved and rapid assessment capacity, virtual screening can sometimes lead to false positive results on account of variability in receptor conformation and flexibility of the binding sites located on protein molecules (Ma et al. 2021). Accordingly, the combination of virtual screening with bioassays is recommended for enhancing its efficacy and avoidance of false positive results. A combination strategy involving the co-application of two different techniques such as molecular dynamics simulation with molecular docking, and virtual screening with surface plasmon resonance can help to overcome the constraints in effectiveness and accuracy caused by a single method (Ma et al. 2021). In addition to the aforementioned techniques, fluorescence polarization assay and affinity mass spectrometry can also be employed to screen the natural products for potential ACE-2 blocking capacity.

So far, the affinity of several compounds including alkaloids, phenolics and terpenoids with ACE-2 has been demonstrated by means of in silico studies. Curcumin, quercetin, limonin, dithymoquinone, isothymol, and anabsinthin represent the plant-derived secondary metabolites that have been extensively screened for their ability to inhibit ACE-2 in the context of COVID-19 treatment. Despite the currently available numerous in vitro and in silico reports on ACE-2 inhibition attributed to natural products, confirmatory evidence from clinical studies is still insufficient (Junior et al. 2021). Accordingly, such promising substances should be further subjected to preclinical and clinical investigations for subsequent development as future drugs.

## References

Abd Allah ES, Gomaa AM (2015) Effects of curcumin and captopril on the functions of kidney and nerve in streptozotocin-induced diabetic rats: role of angiotensin converting enzyme 1. Appl Physiol Nutr Metab 40(10):1061–1067

Abubakar MB, Usman D, El-Saber Batiha G, Cruz-Martins N, Malami I, Ibrahim KG, Abubakar B, Bello MB, Muhammad A, Gan SH (2021) Natural products modulating angiotensin converting enzyme 2 (ACE2) as potential COVID-19 therapies. Front Pharmacol 12:629935

Adefegha SA, Oyeleye SI, Dada FA, Olasehinde TA, Oboh G (2018) Modulatory effect of quercetin and its glycosylated form on key enzymes and antioxidant status in rats penile tissue of paroxetine-induced erectile dysfunction. Biomed Pharmacother 107:1473–1479

Agunloye OM, Oboh G, Ademiluyi AO, Ademosun AO, Akindahunsi AA, Oyagbemi AA, Omobowale TO, Ajibade TO, Adedapo AA (2019) Cardio-protective and antioxidant properties of caffeic acid and chlorogenic acid: mechanistic role of angiotensin converting enzyme, cholinesterase and arginase activities in cyclosporine induced hypertensive rats. Biomed Pharmacother 109:450–458

Al Shukor N, Van Camp J, Gonzales GB, Staljanssens D, Struijs K, Zotti MJ, Raes K, Smagghe G (2013) Angiotensin-converting enzyme inhibitory effects by plant phenolic compounds: a study of structure activity relationships. J Agric Food Chem 61(48):11832–11839

Balmeh N, Mahmoudi S, Mohammadi N, Karabedianhajiabadi A (2020) Predicted therapeutic targets for COVID-19 disease by inhibiting SARS-CoV-2 and its related receptors. Inform Med Unlock 20:100407

Batlle D, Wysocki J, Satchell K (2020) Soluble angiotensin-converting enzyme 2: a potential approach for coronavirus infection therapy? Clin Sci 134(5):543–545

Beigi F, Lundahl P (1999) Immobilized biomembrane chromatography of highly lipophilic drugs. J Chromatogr A 852(1):313–317

Bernardi S, Michelli A, Zuolo G, Candido R, Fabris B (2016) Update on RAAS modulation for the treatment of diabetic cardiovascular disease. J Diabetes Res 2016:8917578

Beutler JA (2009) Natural products as a foundation for drug discovery. Curr Protoc Pharmacol 46(1):9.11.11–19.11.21

Chang H, Chang C-Y, Lee H-J, Chou C-Y, Chou T-C (2018) Magnolol ameliorates pneumonectomy and monocrotaline-induced pulmonary arterial hypertension in rats through inhibition of angiotensin II and endothelin-1 expression. Phytomedicine 51:205–213

Chen H, Du Q (2020) Potential natural compounds for preventing 2019-nCoV infection. Europe PMC

Chen C, Yang F-Q, Zuo H-L, Song Y-L, Xia Z-N, Xiao W (2013) Applications of biochromatography in the screening of bioactive natural products. J Chromatogr Sci 51(8): 780–790

Chen X, Wu Y, Chen C, Gu Y, Zhu C, Wang S, Chen J, Zhang L, Lv L, Zhang G (2021) Identifying potential anti-COVID-19 pharmacological components of traditional Chinese medicine Lianhuaqingwen capsule based on human exposure and ACE2 biochromatography screening. Acta Pharm Sin B 11(1):222–236

Chiou S-Y, Sung J-M, Huang P-W, Lin S-D (2017) Antioxidant, antidiabetic, and antihypertensive properties of Echinacea purpurea flower extract and caffeic acid derivatives using in vitro models. J Med Food 20(2):171–179

Dai J, Mumper RJ (2010) Plant phenolics: extraction, analysis and their antioxidant and anticancer properties. Molecules 15(10):7313–7352

de Matos PH, da Silva TP, Mansano AB, Gancedo NC, Tonin FS, Pelloso FC, Petruco MV, de Melo EB, Fernandez-Llimos F, Sanches ACC, de Mello JCP, Chierrito D, de Medeiros Araújo DC (2022) Bioactive compounds as potential angiotensin-converting enzyme II inhibitors against COVID-19: a scoping review. Inflamm Res 71(12):1489–1500

De Vivo M, Masetti M, Bottegoni G, Cavalli A (2016) Role of molecular dynamics and related methods in drug discovery. J Med Chem 59(9):4035–4061

Deng Y, Aluko R, Jin Q, Zhang Y, Yuan L (2012) Inhibitory activities of baicalin against renin and angiotensin-converting enzyme. Pharm Biol 50(4):401–406

Dhawan M, Priyanka, Choudhary OP (2022) Omicron SARS-CoV-2 variant: reasons of emergence and lessons learnt. Int J Surg 97:106198

Elfiky AA (2021) Natural products may interfere with SARS-CoV-2 attachment to the host cell. J Biomol Struct Dyn 39(9):3194–3203

Ferrario CM, Mullick AE (2017) Renin angiotensin aldosterone inhibition in the treatment of cardiovascular disease. Pharmacol Res 125:57–71

Friesner RA, Banks JL, Murphy RB, Halgren TA, Klicic JJ, Mainz DT, Repasky MP, Knoll EH, Shelley M, Perry JK (2004) Glide: a new approach for rapid, accurate docking and scoring. 1. Method and assessment of docking accuracy. J Med Chem 47(7):1739–1749

Gao Y, Wang Z, Zhang Y, Liu Y, Wang S, Sun W, Guo J, Yu C, Wang Y, Kong W (2018) Naringenin inhibits NG-nitro-L-arginine methyl ester-induced hypertensive left ventricular hypertrophy by decreasing angiotensin-converting enzyme 1 expression. Exp Ther Med 16(2):867–873

Gao J, Ding Y, Wang Y, Liang P, Zhang L, Liu R (2021) Oroxylin A is a severe acute respiratory syndrome coronavirus 2-spiked pseudotyped virus blocker obtained from Radix Scutellariae using angiotensin-converting enzyme II/cell membrane chromatography. Phytother Res 35(6): 3194–3204

Geng F, Yang L, Chou G, Wang Z (2010) Bioguided isolation of angiotensin-converting enzyme inhibitors from the seeds of Plantago asiatica L. Phytother Res 24(7):1088–1094

Gheblawi M, Wang K, Viveiros A, Nguyen Q, Zhong J-C, Turner AJ, Raizada MK, Grant MB, Oudit GY (2020) Angiotensin-converting enzyme 2: SARS-CoV-2 receptor and regulator of the renin-angiotensin system: celebrating the 20th anniversary of the discovery of ACE2. Circ Res 126(10):1456–1474

Gimeno A, Ojeda-Montes MJ, Tomás-Hernández S, Cereto-Massagué A, Beltrán-Debón R, Mulero M, Pujadas G, Garcia-Vallvé S (2019) The light and dark sides of virtual screening: what is there to know? Int J Mol Sci 20(6):1375

Goyal B, Goyal D (2020) Targeting the dimerization of the main protease of coronaviruses: a potential broad-spectrum therapeutic strategy. ACS Comb Sci 22(6):297–305

Guerrero L, Castillo J, Quinones M, Garcia-Vallve S, Arola L, Pujadas G, Muguerza B (2012) Inhibition of angiotensin-converting enzyme activity by flavonoids: structure-activity relationship studies. PLoS One 7(11):e49493

Gutiérrez-Román AS, Gonzalez-Cortazar M, Trejo-Tapia G, Herrera-Ruiz M, Zamilpa A, Sanchéz-Mendoza E, De la Cruz-Sanchez NG, Jiménez-Ferrer E (2021) Angiotensin-converting enzyme inhibitors from Salvia elegans Vahl. Nat Prod Res 35(23):5344–5349

Gyebi GA, Adegunloye AP, Ibrahim IM, Ogunyemi OM, Afolabi SO, Ogunro OB (2022) Prevention of SARS-CoV-2 cell entry: insight from in silico interaction of drug-like alkaloids with spike glycoprotein, human ACE2, and TMPRSS2. J Biomol Struct Dyn 40(5):2121–2145

Haga J, Ichikawa K (2016) Virtual screening techniques and current computational infrastructures. Curr Pharm Des 22(23):3576–3584

Hamden K, Bengara A, Amri Z, Elfeki A (2013) Experimental diabetes treated with trigonelline: effect on key enzymes related to diabetes and hypertension, β-cell and liver function. Mol Cell Biochem 381(1):85–94

He L, Wang S, Geng X (2001) Coating and fusing cell membranes onto a silica surface and their chromatographic characteristics. Chromatographia 54(1):71–76

Hirayama N (2017) Docking simulations between drugs and HLA molecules associated with idiosyncratic drug toxicity. Drug Metab Pharmacokinet 32(1):31–39

Hoffmann M, Kleine-Weber H, Schroeder S, Krüger N, Herrler T, Erichsen S, Schiergens TS, Herrler G, Wu N-H, Nitsche A, Müller MA, Drosten C, Pöhlmann S (2020) SARS-CoV-2 cell entry depends on ACE2 and TMPRSS2 and is blocked by a clinically proven protease inhibitor. Cell 181(2):271–280.e278

Hou X, Zhou M, Jiang Q, Wang S, He L (2009) A vascular smooth muscle/cell membrane chromatography–offline-gas chromatography/mass spectrometry method for recognition, separation and identification of active components from traditional Chinese medicines. J Chromatogr A 1216(42):7081–7087

Hu S, Wang J, Zhang Y, Bai H, Wang C, Wang N, He L (2021) Three salvianolic acids inhibit 2019-nCoV spike pseudovirus viropexis by binding to both its RBD and receptor ACE2. J Med Virol 93(5):3143–3151

Huang XY, Chen CX (2013) Effect of oxymatrine, the active component from radix Sophorae flavescentis (Kushen), on ventricular remodeling in spontaneously hypertensive rats. Phytomedicine 20(3–4):202–212

Huang S-Y, Zou X (2010) Advances and challenges in protein-ligand docking. Int J Mol Sci 11(8): 3016–3034

Huo J, Miao Y, Zeng Y (2007) Gene chip technology and its application. Let Biotech 2:329–332

Islam R, Parves MR, Paul AS, Uddin N, Rahman MS, Mamun AA, Hossain MN, Ali MA, Halim MA (2021) A molecular modeling approach to identify effective antiviral phytochemicals against the main protease of SARS-CoV-2. J Biomol Struct Dyn 39(9):3213–3224

Jaeger R, Cuny E (2016) Terpenoids with special pharmacological significance: a review. Nat Prod Commun 11(9):1934578X1601100946

Junior AG, Tolouei SEL, Dos Reis Lívero FA, Gasparotto F, Boeing T, de Souza P (2021) Natural agents modulating ACE-2: a review of compounds with potential against SARS-CoV-2 infections. Curr Pharm Des 27(13):1588–1596

Kadakol A, Malek V, Goru SK, Pandey A, Bagal S, Gaikwad AB (2015) Esculetin attenuates alterations in Ang II and acetylcholine mediated vascular reactivity associated with hyperinsulinemia and hyperglycemia. Biochem Biophys Res Commun 461(2):342–347

Kai H, Kai M (2020) Interactions of coronaviruses with ACE2, angiotensin II, and RAS inhibitors—lessons from available evidence and insights into COVID-19. Hypertens Res 43(7):648–654

Kang DG, Sohn EJ, Kwon EK, Han JH, Oh H, Lee HS (2002) Effects of berberine on angiotensin-converting enzyme and NO/cGMP system in vessels. Vasc Pharmacol 39(6):281–286

Kanmodi KK, Kanmodi PA (2020) COVID-19 stigmatization: a devil and a deep blue sea. Populat Med 2:10. https://doi.org/10.18332/popmed/121098

Ke Z, Su Z, Zhang X, Cao Z, Ding Y, Cao L, Ding G, Wang Z, Liu H, Xiao W (2017) Discovery of a potent angiotensin converting enzyme inhibitor via virtual screening. Bioorg Med Chem Lett 27(16):3688–3692

Khan S, Siddique R, Shereen MA, Ali A, Liu J, Bai Q, Bashir N, Xue M (2020) Emergence of a novel coronavirus, severe acute respiratory syndrome coronavirus 2: biology and therapeutic options. J Clin Microbiol 58(5):e00187–e00120

Kim EN, Kim MY, Lim JH, Kim Y, Shin SJ, Park CW, Kim Y-S, Chang YS, Yoon HE, Choi BS (2018) The protective effect of resveratrol on vascular aging by modulation of the renin–angiotensin system. Atherosclerosis 270:123–131

Kiss A, Kowalski J, Melzig MF (2004) Compounds from Epilobium angustifolium inhibit the specific metallopeptidases ACE, NEP and APN. Planta Med 70(10):919–923

Kuhn JH, Li W, Choe H, Farzan M (2004) Angiotensin-converting enzyme 2: a functional receptor for SARS coronavirus. Cell Mol Life Sci 61(21):2738–2743

Kumar A, Choudhir G, Shukla SK, Sharma M, Tyagi P, Bhushan A, Rathore M (2021) Identification of phytochemical inhibitors against main protease of COVID-19 using molecular modeling approaches. J Biomol Struct Dyn 39(10):3760–3770

Lagerquist C, Beigi F, Karlén A, Lennernäs H, Lundahl P (2001) Effects of cholesterol and model transmembrane proteins on drug partitioning into lipid bilayers as analysed by immobilized-liposome chromatography. J Pharm Pharmacol 53(11):1477–1487

Lee J-W, Baek N-I, Lee D-Y (2015) Inhibitory effects of seco-triterpenoids from Acanthopanax sessiliflorus fruits on HUVEC invasion and ACE activity. Nat Prod Commun 10(9): 1934578X1501000907

Letko M, Marzi A, Munster V (2020) Functional assessment of cell entry and receptor usage for SARS-CoV-2 and other lineage B betacoronaviruses. Nat Microbiol 5(4):562–569

Li G-B, Yang L-L, Wang W-J, Li L-L, Yang S-Y (2013) ID-score: a new empirical scoring function based on a comprehensive set of descriptors related to protein–ligand interactions. J Chem Inf Model 53(3):592–600

Lionta E, Spyrou G, Vassilatis DK, Cournia Z (2014) Structure-based virtual screening for drug discovery: principles, applications and recent advances. Curr Top Med Chem 14(16): 1923–1938

Liu Q, Tian J, Xu Y, Li C, Meng X, Fu F (2016) Protective effect of RA on myocardial infarction-induced cardiac fibrosis via AT1R/p38 MAPK pathway signaling and modulation of the ACE2/ ACE ratio. J Agric Food Chem 64(35):6716–6722

Liu H, Jiang Y, Li M, Yu X, Sui D, Fu L (2019) Ginsenoside Rg3 attenuates angiotensin II-mediated renal injury in rats and mice by upregulating angiotensin-converting enzyme 2 in the renal tissue. Evid Based Complement Alternat Med 2019:1–11

Lu R, Zhao X, Li J, Niu P, Yang B, Wu H, Wang W, Song H, Huang B, Zhu N (2020) Genomic characterisation and epidemiology of 2019 novel coronavirus: implications for virus origins and receptor binding. Lancet 395(10224):565–574

Luo J, Zhang C, Liu Q, Ou S, Zhang L, Peng X (2017) Combinative effect of sardine peptides and quercetin alleviates hypertension through inhibition of angiotensin I converting enzyme activity and inflammation. Food Res Int 100:579–585

Lv Y, Wang S, Liang P, Wang Y, Zhang X, Jia Q, Fu J, Han S, He L (2021) Screening and evaluation of anti-SARS-CoV-2 components from Ephedra sinica by ACE2/CMC-HPLC-IT-TOF-MS approach. Anal Bioanal Chem 413(11):2995–3004

Ma L, Liu H, Liu X, Yuan X, Xu C, Wang F, Lin J, Xu R, Zhang D (2021) Screening S protein-ACE2 blockers from natural products: strategies and advances in the discovery of potential inhibitors of COVID-19. Eur J Med Chem 226:113857–113857

Malviya S, Parihar A, Parihar DS, Khan R (2022) Natural products as a therapy to combat against SARS-CoV-2 virus infection. In: Parihar A, Khan R, Kumar A, Kaushik AK, Gohel H (eds)

Computational approaches for novel therapeutic and diagnostic designing to mitigate SARS-CoV-2 infection. Academic Press, pp 115–145
Maurya VK, Kumar S, Prasad AK, Bhatt ML, Saxena SK (2020) Structure-based drug designing for potential antiviral activity of selected natural products from Ayurveda against SARS-CoV-2 spike glycoprotein and its cellular receptor. Virus 31(2):179–193
Moran CS, Biros E, Krishna SM, Wang Y, Tikellis C, Morton SK, Moxon JV, Cooper ME, Norman PE, Burrell LM (2017) Resveratrol inhibits growth of experimental abdominal aortic aneurysm associated with upregulation of angiotensin-converting enzyme 2. Arterioscler Thromb Vasc Biol 37(11):2195–2203
Muchtaridi M, Fauzi M, Khairul Ikram NK, Mohd Gazzali A, Wahab HA (2020) Natural flavonoids as potential angiotensin-converting enzyme 2 inhibitors for anti-SARS-CoV-2. Molecules 25(17):3980
Muhammad SA, Fatima N (2015) In silico analysis and molecular docking studies of potential angiotensin-converting enzyme inhibitor using quercetin glycosides. Pharmacogn Mag 11(42): 123
Ngwa W, Kumar R, Thompson D, Lyerly W, Moore R, Reid T-E, Lowe H, Toyang N (2020) Potential of flavonoid-inspired Phytomedicines against COVID-19. Molecules 25(11):2707
Oakes JM, Fuchs RM, Gardner JD, Lazartigues E, Yue X (2018) Nicotine and the renin-angiotensin system. Am J Phys Regul Integr Comp Phys 315(5):R895–R906
Ojeda D, Jiménez-Ferrer E, Zamilpa A, Herrera-Arellano A, Tortoriello J, Alvarez L (2010) Inhibition of angiotensin converting enzyme (ACE) activity by the anthocyanins delphinidin- and cyanidin-3-O-sambubiosides from Hibiscus sabdariffa. J Ethnopharmacol 127(1):7–10
Okimoto N, Futatsugi N, Fuji H, Suenaga A, Morimoto G, Yanai R, Ohno Y, Narumi T, Taiji M (2009) High-performance drug discovery: computational screening by combining docking and molecular dynamics simulations. PLoS Comput Biol 5(10):e1000528
Patching SG (2014) Surface plasmon resonance spectroscopy for characterisation of membrane protein–ligand interactions and its potential for drug discovery. Biochim BiophysActa (BBA) Biomemb 1838(1):43–55
Prabowo BA, Purwidyantri A, Liu K-C (2018) Surface plasmon resonance optical sensor: a review on light source technology. Biosensors 8(3):80
Rakib A, Paul A, Chy MNU, Sami SA, Baral SK, Majumder M, Tareq AM, Amin MN, Shahriar A, Uddin MZ, Dutta M, Tallei TE, Emran TB, Simal-Gandara J (2020) Biochemical and computational approach of selected phytocompounds from Tinospora crispa in the management of COVID-19. Molecules 25(17):3936
Raman K, Rajagopal K, Islam F, Dhawan M, Mitra S, Aparna B, Varakumar P, Byran G, Choudhary OP, Emran TB (2022) Role of natural products towards the SARS-CoV-2: a critical review. Ann Med Surg (Lond) 80:104062
Sanchis-Gomar F, Lavie CJ, Perez-Quilis C, Henry BM, Lippi G (2020) Angiotensin-converting enzyme 2 and antihypertensives (angiotensin receptor blockers and angiotensin-converting enzyme inhibitors) in coronavirus disease 2019. Mayo Clin Proc 95(6):1222–1230
Scherf U, Ross DT, Waltham M, Smith LH, Lee JK, Tanabe L, Kohn KW, Reinhold WC, Myers TG, Andrews DT (2000) A gene expression database for the molecular pharmacology of cancer. Nat Genet 24(3):236–244
Simões e Silva AC, Teixeira MM (2016) ACE inhibition, ACE2 and angiotensin-(1-7) axis in kidney and cardiac inflammation and fibrosis. Pharmacol Res 107:154–162
Smith M, Smith JC (2020) Repurposing therapeutics for COVID-19: supercomputer-based docking to the SARS-CoV-2 viral spike protein and viral spike protein-human ACE2 interface
Sui H, Yu Q, Zhi Y, Geng G, Liu H, Xu H (2010) Effects of apigenin on the expression of angiotensin-converting enzyme 2 in kidney in spontaneously hypertensive rats. J Hygiene Res 39(6):693–696, 700
Tiao M-M, Lin Y-J, Yu H-R, Sheen J-M, Lin I, Lai Y-J, Tain Y-L, Huang L-T, Tsai C-C (2018) Resveratrol ameliorates maternal and post-weaning high-fat diet-induced nonalcoholic fatty liver disease via renin-angiotensin system. Lipids Health Dis 17(1):1–11

Tikellis C, Thomas MC (2012) Angiotensin-converting enzyme 2 (ACE2) is a key modulator of the renin angiotensin system in health and disease. Int J Pept 2012:256294

To K, Tong JH, Chan PK, Au FW, Chim SS, Allen Chan K, Cheung JL, Liu EY, Tse GM, Lo AW (2004) Tissue and cellular tropism of the coronavirus associated with severe acute respiratory syndrome: an in-situ hybridization study of fatal cases. J Pathol 202(2):157–163

Tutunchi H, Naeini F, Ostadrahimi A, Hosseinzadeh-Attar MJ (2020) Naringenin, a flavanone with antiviral and anti-inflammatory effects: a promising treatment strategy against COVID-19. Phytother Res 34(12):3137–3147

Utomo RY, Ikawati M, Meiyanto E (2020) Revealing the potency of citrus and galangal constituents to halt SARS-CoV-2 infection

Vardhan S, Sahoo SK (2020) In silico ADMET and molecular docking study on searching potential inhibitors from limonoids and triterpenoids for COVID-19. Comput Biol Med 124:103936

Verdecchia P, Cavallini C, Spanevello A, Angeli F (2020) The pivotal link between ACE2 deficiency and SARS-CoV-2 infection. Eur J Intern Med 76:14–20

Wang H, Zou H, Kong L, Ni J (2000) Analysis of bioactive components in traditional Chinese medicines by molecular biochromatography with α1-acid glycoprotein stationary phase. J Basic Clin Physiol Pharmacol 11(2):155–172

Wang XW, Gao HJ, Fang DC (2008) Advances in gene chip technique in Barrett's metaplasia and adenocarcinoma. J Dig Dis 9(2):68–71

Wang Z, Wang S, Zhao J, Yu C, Hu Y, Tu Y, Yang Z, Zheng J, Wang Y, Gao Y (2019) Naringenin ameliorates renovascular hypertensive renal damage by normalizing the balance of renin-angiotensin system components in rats. Int J Med Sci 16(5):644

Wang Z, Sun H, Shen C, Hu X, Gao J, Li D, Cao D, Hou T (2020) Combined strategies in structure-based virtual screening. Phys Chem Chem Phys 22(6):3149–3159

Wei X, Zhu X, Hu N, Zhang X, Sun T, Xu J, Bian X (2015) Baicalin attenuates angiotensin II-induced endothelial dysfunction. Biochem Biophys Res Commun 465(1):101–107

Wu H, Li Y, Wang Y, Xu D, Li C, Liu M, Sun X, Li Z (2014) Tanshinone IIA attenuates bleomycin-induced pulmonary fibrosis via modulating angiotensin-converting enzyme 2/angiotensin-(1-7) axis in rats. Int J Med Sci 11(6):578

Wu C, Liu Y, Yang Y, Zhang P, Zhong W, Wang Y, Wang Q, Xu Y, Li M, Li X (2020) Analysis of therapeutic targets for SARS-CoV-2 and discovery of potential drugs by computational methods. Acta Pharm Sin B 10(5):766–788

Wunpathe C, Potue P, Maneesai P, Bunbupha S, Prachaney P, Kukongviriyapan U, Kukongviriyapan V, Pakdeechote P (2018) Hesperidin suppresses renin-angiotensin system mediated NOX2 over-expression and sympathoexcitation in 2K-1C hypertensive rats. Am J Chin Med 46(04):751–767

Wysocki J, Ye M, Rodriguez E, González-Pacheco FR, Barrios C, Evora K, Schuster M, Loibner H, Brosnihan KB, Ferrario CM (2010) Targeting the degradation of angiotensin II with recombinant angiotensin-converting enzyme 2: prevention of angiotensin II–dependent hypertension. Hypertension 55(1):90–98

Yan R, Zhang Y, Li Y, Xia L, Guo Y, Zhou Q (2020) Structural basis for the recognition of SARS-CoV-2 by full-length human ACE2. Science 367(6485):1444–1448

Yeung ML, Teng JLL, Jia L, Zhang C, Huang C, Cai J-P, Zhou R, Chan K-H, Zhao H, Zhu L (2021) Soluble ACE2-mediated cell entry of SARS-CoV-2 via interaction with proteins related to the renin-angiotensin system. Cell 184(8):2212–2228. e2212

Yoo MJ, Smith QR, Hage DS (2009) Studies of imipramine binding to human serum albumin by high-performance affinity chromatography. J Chromatogr B 877(11–12):1149–1154

Zhao H-Y, Jiang J-G (2010) Application of chromatography technology in the separation of active components from nature derived drugs. Mini Rev Med Chem 10(13):1223–1234

Zheng X, Wang S, Zou X, Jing Y, Yang R, Li S, Wang F (2017) Ginsenoside Rb1 improves cardiac function and remodeling in heart failure. Exp Anim 66(3):217–228

Zhou P, Yang X-L, Wang X-G, Hu B, Zhang L, Zhang W, Si H-R, Zhu Y, Li B, Huang C-L, Chen H-D, Chen J, Luo Y, Guo H, Jiang R-D, Liu M-Q, Chen Y, Shen X-R, Wang X, Zheng X-S,

Zhao K, Chen Q-J, Deng F, Liu L-L, Yan B, Zhan F-X, Wang Y-Y, Xiao G-F, Shi Z-L (2020) A pneumonia outbreak associated with a new coronavirus of probable bat origin. Nature 579(7798):270–273

Zhu T (2003) Global analysis of gene expression using GeneChip microarrays. Curr Opin Plant Biol 6(5):418–425

Zi C, Zhang N, Yang L, Wang L, Wu Y, Su Y, Wang X (2020) Discovery of a potent angiotensin converting enzyme 2 inhibitor from Chinese Medicinal and Edible Plant via Docking-based Virtual Screening. Bioorg Med Chem Lett

# Chapter 9
# Insights into In Silico Methods to Explore Plant Bioactive Substances in Combating SARS-CoV-2

**Sourish Pramanik , Dibyendu Seth , and Sandip Debnath**

## 9.1 Introduction

Coronavirus disease 2019 (COVID-19) is caused by the severe acute respiratory syndrome coronavirus 2 (SARS-CoV-2) and can range from acute respiratory failure to a mildly attacking state with cough and fever, or it can be completely symptomless. Early on, the disease manifests as interstitial pneumonia and acute respiratory distress syndrome (ARDS) in approximately 10–20% of all patients, generally with co-morbidities (Huang et al. 2020; Paules et al. 2020). On March 11, 2020, the World Health Organization (WHO) declared the coronavirus outbreak a pandemic; a series of deaths and illnesses have been diagnosed. After 2 years, the virus seemed to have subsided but ended incorrectly. Since then, a few vaccines have been developed, but none can eradicate the virus; thus, a particular motive for drug development has been needed during this crisis (Mukerjee et al. 2022). Nonetheless, Chang et al. 2020 have suggested that FDA-approved antiviral medications against the same types of viruses may produce positive outcomes (Chang et al. 2020). Other researchers have observed that hydroxychloroquine is effective against COVID-19, and in silico approaches revealed Remdesivir and Ivermectin to be potential drugs too.

Especially in India, where Ayurveda has been practiced for so long, it is very likely to search for plant-derived biomolecules as drugs against the virus. Also, even at higher concentrations, plant-derived biomolecules do not cause cellular toxicity, making them a good candidate for an antiviral drug (Singh et al. 2021; Maitra et al. 2022). The virus genome is a 29.7 kB long positive-sense single-stranded RNA that

S. Pramanik · D. Seth · S. Debnath (✉)
Department of Genetics and Plant Breeding, Palli-Siksha Bhavana (Institute of Agriculture), Visva-Bharati University, Birbhum, West Bengal, India
e-mail: sandip.debnath@visva-bharati.ac.in



J.-T. Chen (ed.), *Ethnopharmacology and Drug Discovery for COVID-19: Anti-SARS-CoV-2 Agents from Herbal Medicines and Natural Products*,
https://doi.org/10.1007/978-981-99-3664-9_9

consists of around 12 open reading frames (ORFs), which produce two polyproteins that are further processed by protease enzymes 3CLpro and PLpro to yield 16 non-structural proteins. Apart from the non-structural proteins, four structural proteins are also produced for the virion, namely the S, E, M, and N proteins, which are discussed in detail in this chapter (Boopathi et al. 2021). Thus, after extensive research and testing, several biochemical compounds were discovered to be suitable as antivirals, and several points in the virus were identified to be targeted. Focusing on drug targets has aided in understanding and improving potential antiviral drug design methods. Synthetic and semi-synthetic drug development, as well as drug repurposing of potential molecules, is a very time-consuming process, which is shifting the focus to computational methods such as virtual molecular docking apart from molecular dynamic simulation, which can be used more effectively (Singh and Florez 2020; Debnath et al. 2022). Thus, with a solid knowledge of the molecular-structural characteristics of the drug and the potential drug targets, the drug design can be assessed in no time. The following chapter discusses the different molecular characteristics and potential drug targets of the coronavirus and the plant bioactive compounds to be used as potential drugs with the help of in silico drug design methods.

## 9.2 Coronavirus: Introduction and Molecular Basics

The COVID virus that has recently caused problems is SARS-CoV-2, which belongs to the order Nidovirales, family Coronaviridae, and genus coronavirus. Coronaviruses (CoVs) consist of a positive-sense single-stranded RNA (ssRNA) ranging between 26.2 and 31.7 kb as genetic material, which is further enclosed inside a protein envelope (Yang et al. 2006). Coronavirus is known to infect various hosts, including birds—avian coronavirus; pigs—porcine epidemic diarrhea virus; mice—mouse hepatitis virus and most importantly, human—human coronavirus, which includes severe acute respiratory syndrome coronavirus (SARS-CoV), SARSCoV-2, Middle East Respiratory Syndrome-CoV (MERS-CoV), HCoV-OC43, HCoV-NL63, and HCoV-229E.

The coronavirus genome is considered one of the most significant viruses with RNA as its genome (Belouzard et al. 2012). The genome of SARS-CoV-2 has approximately 13–15 ORFs, which are limited on two sides by 5′ and 3′ untranslated regions (UTRs) (Lu et al. 2020). The genome translates into two types of proteins, structural and non-structural proteins (NSPs), of which there are 27 distinct ones for SARS-CoV-2 (Liu et al. 2020). The first ORF, ORF 1 a/b, consumes 2/3 of the genome and translates into 7096 residues of the long polypeptide (PP). PP1a and PP1b are encoded by ORF1a and ORF1b, respectively, resulting in 16 NSPs after further processing (Wu et al. 2020). These NSPs are involved in a range of viral processes, including the construction of the replicase transcriptase complex, which is one of those processes. The portion of the virus' genome that has not yet been replicated encodes the messenger RNA (mRNA) that develops the structural

**Table 9.1.** Different functionalities of NSP of SARS-CoV-2

| Sl. No. | Protein | Function | Reference |
|---|---|---|---|
| 1. | NSP1 | Specific function not known, but is assumed to be involved in mRNA degradation and characteristic immune response antagonism | Almeida et al. (2007), Lei et al. (2020) |
| 2. | NSP2 | Replicase products which involved in host cell variation in signaling pathway as a result of interaction with PHB and PHB2 | Harcourt et al. (2004) |
| 3. | NSP3 | Plays part in the processing of polyproteins by attaching to nucleocapsids, viral genome, and other proteins | Lei et al. (2018) |
| 4. | NSP4 | Viral membrane rearrangement and replication in association with NSP3 | Lei et al. (2018) |
| 5. | NSP5 | Named as main protease or 3C-like protease which cleaves viral protein | Shamsi et al. (2021) |
| 6. | NSP6 | Has part in preliminary initiation of autophagosome from host endoplasmic reticulum as a potential transmembrane scaffold protein | Shamsi et al. (2021) |
| 7. | NSP7 | RNA-dependent RNA polymerase (RdRp) binds to ssRNA forming a hexadecameric compound in attachment with NSP8 acting as a clamp for polymerase and NSP12 | Yin et al. (2020) |
| 8. | NSP8 | Forms a hexadecameric complex in association with NSP7 and cofactor of NSP12. NSP12-NSP7-NSP8 complex performs RNA polymerization activity on the poly U tail with the addition of ATP | Yin et al. (2020) |
| 9. | NSP9 | Single-stranded binding protein to facilitate virulence and replication | Zhang et al. (2020a, b) |
| 10. | NSP10 | Acts as a cofactor of NSP16 and NSP14 to form a heterodimer to perform MTase activity and has a role in mRNA cap methylation, ExoN for proofreading of viral genome and 2-*O*-methyltransferase | Joseph et al. (2006) |
| 11. | NSP11 | Not known | Gupta et al. (2020) |
| 12. | NSP12 | RdRp and nucleotidyltransferase | Kirchdoerfer and Ward (2019) |
| 13. | NSP13 | RNA helicase and 5′ triphosphatase activity | Gupta et al. (2020) |
| 14. | NSP14 | 3′ to 5′ exonuclease activity, ExoN activity, N7 methyltransferase (addition of 5′ caps to the viral genome) | von Grotthuss et al. (2003) |
| 15. | NSP15 | $Mn^{2+}$ assisted endoribonuclease activity | Guarino et al. (2005) |
| 16. | NSP16 | Protein 5 recognition and 2′-*O*-ribose methyltransferase activity by defending viral genome from melanoma differentiation | Chen et al. (2011) |

proteins. These structural proteins include the spike, envelope, membrane, and nucleocapsid. Other auxiliary proteins are also generated (McBride et al. 2014). The details of 16 non-structural proteins are given in Table 9.1..

Pp1a and Pp1b take over the host ribosome and form the replication–translation complex for the virus's good (Te Velthuis et al. 2012; Stobart et al. 2013). 16 NSPs are produced after proteases break down PP. NSP1 and NSP2 decrease the expression of host genes, NSP3 forms a multidomain complex, NSP5 is an M protease that aids in replication, NSP4 and NSP6 are transmembrane (TM) proteins, NSP7 and NSP8 operate as primases, and other NSPs have unique roles (Te Velthuis et al. 2012). NSP9 is an RNA-binding protein that is essential for viral infection in its dimeric form. CoV infection can be avoided by disrupting NSP9 dimerization (Egloff et al. 2004; Hu et al. 2017). NSP10 functions as a cofactor for the replicative enzyme's activation (Bouvet et al. 2014). The NSPs NSP12, NSP13, NSP14, NSP15, and NSP16 all exhibit RNA-dependent RNA polymerase activity, helicase activity, exoribonuclease activity, endoribonuclease activity, and methyltransferase activity. All NSPs play a significant role in transcription and replication (Wang et al. 2016). Spike surface glycoprotein (S), a small envelope protein (E), membrane (M), and nucleocapsid (N) proteins are the four primary structural proteins that are also synthesized. CoV genetic material is frequently subject to recombination, which can result in the emergence of new strains with altered virulence (Hilgenfeld 2014).

Angiotensin-converting enzyme 2 (ACE2), a monocarboxypeptidase found on the cell membranes of various organs such as the heart, lungs, arteries, intestine, kidney, gastrointestinal tract, esophagus, colon, and small intestine, but absent from the stomach and testes, eventually functions as a cell receptor for SARS-CoV (Oudit et al. 2009, Rehman et al. This enzyme allows the SARS-CoV to enter; thereafter, its genome is translated into PPs, which in turn encode for NSPs. The major protease (Mpro) and papain-like protease (PLpro) mediate the processing of PP, chopping it into smaller pieces to facilitate replication and aid in the formation of new virions (Zhang et al. 2020a, b). ORF1a and ORF1b in SARS-CoV-2 are of 486 kDa and 790 kDa, respectively, and are processed by proteases at three sites, ending up with NSP1, NSP2, and NSP3. The other NSPs are processed by 3-chymotrypsin-like cysteine protease (3CLpro). The nucleocapsid packages of the genome are made up of the soluble structural protein N. The M protein, which is the most common protein in the virion envelope (Masters 2006; Hogue and Machamer 2007), interacts with the tiny E protein, which is just 79–109 amino acids in length. The CoV enters the host cell by endocytosis when the S protein binds to the ACE2 enzyme. It is a three-region type I-TM protein with a clove shape (Belouzard et al. 2012). The ectodomain (ED) region, the TM region, and the intracellular domain, which make up the intracellular short tail section, are collectively formed by the receptor-binding S1 domain (tri-headed) and membrane fusion component S2 (trimeric stalk) on the C-terminal. Even before the genetic code is translated into NSPs such as RNA-dependent RNA polymerase (RdRp), 3CLpro, and PLpro, followed by other structural proteins (N, M, E, and S) of the virus, the S1 domain binds to the host receptor, and the S2 segment aids in the merging of host and viral membranes for the transport of the RNA genome. The spike protein, which gives the virus its moniker of "crown virus," seems to be in the trimeric form on the exterior of the virion.

## 9.3 Plant Bioactive Compounds as Therapeutics

Natural substances have been a significant source of possible pharmacological hits and lead throughout the last few decades (Petrovska 2012; Ivanova et al. 2018; Gantait et al. 2014). Potential drugs for various epidemics, endemics, and sporadic diseases around the world are highly concentrated, specifically targeted, and have a number of side effects. This is where plant-derived bioactive natural compounds come into play. A specific amount within the host body is not reactive to the established immune system and thus does not attempt to demolish the host's already achieved health. Plant-derived biochemicals have been in use since ancient times, far before chemically synthesized drugs were available in markets, and they have been successful in keeping the human lineage alive from the Stone Age till now. India, specifically, being so rich in its biodiversity, is believed to hold numerous medicinal plants that ensure the cure of many diseases, the practice of which in ancient India has been called "Ayurveda."

According to the World Health Organization (WHO), traditional plants are used by 65–80% of the population in underdeveloped countries to treat medical conditions (Robinson and Zhang 2011; Ghosh et al. 2022; Maniruzzaman et al. 2022). According to Mukhtar et al. (2008), approximately 25% of commonly used medications, including those used to treat malaria, illnesses such as human immunodeficiency virus (HIV), tuberculosis, cancer, cardiovascular disease, diabetes, inflammatory diseases, and so on, have all been developed from naturally occurring substances.

In nature or produced at the time of processing food or medicinal plants, bioactive compounds are necessary or nonessential substances that have the ability to modulate a variety of biological processes for the betterment of human health (Biesalski et al. 2009). While nonessential chemicals are neither necessary for any biological activity nor is their absence harmful to the host body, essential compounds meet essential biological functions in the host body, and their absence results in sickness (Biesalski et al. 2013). Vitamins, such as vitamins A, B, C, D, and E, have emerged as being among the most researched and important components. According to Chambial et al. (2013), they have shown promise in the treatment of conditions including allergies, bone development, wound repair, conservation of connective tissue, and many others (Chambial et al. 2013; Das et al. 2022). The vast majority of bioactive substances that are nonessential are phytochemicals. According to their chemical composition, they are classified as polyphenols, terpenoids, phytosterols, alkaloids, and other substances (Somani et al. 2015). There are more categories for phytochemicals that take into consideration factors like the number of rings and carbons, among other things (Tsao 2010). Vegetables, fruits, algae, cereal grains, and medicinal plants are rich sources of this group of compounds since they are generated in the secondary metabolism of plant cells (Zhang et al. 2015). The vegetables with the highest levels of phytochemicals are broccoli, spinach, broccoli, red peppers, and other foods with high levels of polyphenols and antioxidant capacity (Liu 2013). Purple rice, red rice, and black rice are three grains that are regarded as having a high phytochemical content. Many different plants and flowers contain phytochemicals that may have positive impacts on health, including medicinal plants like *Camellia sinensis*, *Ephedra sinica,* and *Dioscorea bulbifera* (Zhang et al. 2015).

The effectiveness of certain plant-derived compounds used in both ayurveda and the west against a number of viral infections, including dengue, chikungunya, hepatitis, Ebola, and others, has been documented (Bhuiyan et al. 2020; Boukhatem and Setzer 2020). Four Ayurvedic herbs were used in COVID-19 clinical studies by the Ministry of Ayush (India): ashwagandha (*Withania somnifera* L.) Dunal; guduchi (*Tinospora cordifolia* (Willd.) Miers); yashtimadhu (*Glycyrrhiza glabra* L.); and pipli (*Piper longum* L.) (Kotecha 2021). Glycyrrhetinic acid (Huang et al. 2016), guggulsterone (Shah et al. 2012), boswellic acid (Darshan and Doreswamy 2004), withaferin A (Bale et al. 2018), and diosgenin (Haratake et al. 2017) are examples of such plant-derived anti-inflammatory steroids (Gyebi et al. 2021). Table 9.2. shows many plant extracts that have been examined and shown to be effective against SARS-CoV.

## 9.4 In Silico Drug Designing: Structure-Based Molecular Docking

Drug repurposing of prospective compounds is a time-consuming process in creating synthetic and semi-synthetic drugs. As a result, computing virtual molecular docking and molecular dynamic simulation are examples of in *silico* technologies that could be put to good use. Then, to start the medication development process, promising pharmaceutical compounds can be later verified by an in vitro or in vivo study. In that regard, molecular docking has grown significantly as a method for developing new pharmaceutical drugs.

Numerous potentially useful therapeutic targets for various disorders have been discovered since the human genome project was completed. Furthermore, advances in nuclear magnetic resonance spectroscopy, high-throughput protein purification, and crystallography have contributed to the definition and improvement of protein and protein–ligand interaction and structural elements. Because of these changes, in silico techniques can continue and progress in drug discovery.

In addition to being more effective, direct, and rational in finding approaches with lower costs than the conventional high-throughput screening (HTS) methods, virtual screening (VS) techniques have been used for hit identification and lead optimization (Moitessier et al. 2008). Ligand-based and structure-based strategies are available for VS. Quantitative structure–activity relationship (QSAR) or pharmacophore modeling are two ligand-based methods that may be utilized when there is a collection of active ligand molecules accessible despite limited or no structural knowledge about the targets. Since the 1980s, molecular docking has been the most popular and

**Table 9.2.** A few phytochemicals with antiviral properties against CoV

| Sl. No. | Phytochemicals | Mode/target of inhibition | The scientific name of the source plants | Reference |
|---|---|---|---|---|
| 1. | Abietane | 3CL$^{pro}$ | *Lycopus europaeus* L. | Wen et al. (2007) |
| 2. | Cepharanthine | 3CLpro | *Catharanthus roseus* (L.) G. Don | Zhang et al. (2005) |
| 3. | Tomentin B | 3CLpro | *Paulownia tomentosa* Steud. | Chen et al. (2005) |
| 4. | Apigenin | 3CLpro | *Torreya nucifera* (L.) Siebold & Zucc. | Ryu et al. (2010) |
| 5. | Tomentin C | 3CLpro | *Paulownia tomentosa* Steud. | Chen et al. (2005) |
| 6. | β-sitosterol | 3CLpro | *Isatis tinctoria* L. (Brassicaceae) | Lin et al. (2005) |
| 7. | Betulonic acid | 3CLpro | *Juniperus formosana* Hayata | Wen et al. (2007) |
| 8. | Rhein | Restrict viral entry | *Houttuynia cordata* Thunb | Ho et al. (2007) |
| 9. | Broussochalcone B | 3CLpro and PLpro | *Broussonetia papyrifera* (L.) L'Her. ExVent. | Park et al. (2017) |
| 10. | Tomentin D | 3CLpro | *Paulownia tomentosa* Steud. | Chen et al. (2005) |
| 11. | Catechin gallate | Viral replication | *Camellia sinensis* (L.) Kuntze | Roh (2012) |
| 12. | Aloe emodin | 3CLpro | *Isatis tinctoria* L. | Lin et al. (2005) |
| 13. | Cinnamtannin B | Pseudovirus infection | *Cinnamomum verum* J. Presl | Zhuang et al. (2009) |
| 14. | Chrysin | Restrict viral entry | *Scutellaria baicalensis* Georgi | Ho et al. (2007) |
| 15. | Corylifol A | PLpro | *Cullen corylifolium* (L.) Medik. | Kim et al. (2014) |
| 16. | Tomentin E | 3CLpro | *Paulownia tomentosa* Steud. | Chen et al. (2005) |
| 17. | Diplacone | PLpro | *Paulownia tomentosa* Steud. | Cho et al. (2013) |
| 18. | Emodin | Restrict viral entry | *Rheum officinale* Baill. | Ho et al. (2007) |
| 19. | Ferruginol | CoV-replication | *Chamaecyparis obtuse* var. *formosana* Hayata | Wen et al. (2007) |
| 20. | Gallocatechin gallate | Viral replication | *Camellia sinensis* (L.) Kuntze | Roh (2012) |
| 21. | Hesperetin | 3CLpro | *Isatis tinctoria* L. | Lin et al. (2005) |
| 22. | Hinokinin | 3CLpro | *Phyllanthus amarus* Schumach. &Thonn. | Wen et al. (2007) |

(continued)

**Table 9.2.** (continued)

| Sl. No. | Phytochemicals | Mode/target of inhibition | The scientific name of the source plants | Reference |
|---|---|---|---|---|
| 23. | 4-Hydroxy-isolonchocarpin | PLpro | *Broussonetia papyrifera* (L.) L'Her. ExVent. | Park et al. (2017) |
| 24. | Indigo | 3CLpro | *Isatis tinctoria* L. | Lin et al. (2005) |
| 25. | Isobavachalcone | PLpro | *Cullen corylifolium* (L.) Medik. | Kim et al. (2014) |
| 26. | Juglanin | Blocks the 3a channel | *Taxus caespitosa* Nakai | Schwarz et al. (2014) |
| 27. | Psoralidin | PLpro | *Cullen corylifolium* (L.) Medik. | Kim et al. (2014) |
| 28. | Neobavaisoflavone | PLpro | *Cullen corylifolium* (L.) Medik. | Kim et al. (2014) |
| 29. | Myricetin | 3CLpro | *Isatis tinctoria* L. | Yu et al. (2012) |
| 30. | Rosmariquinone | CoV-replication | *Salvia miltiorrhiza* Bunge | Park et al. (2012) |
| 31. | Savinin | 3CLpro | *Chamaecyparis obtusa* var. formosana | Wen et al. (2007) |
| 32. | Broussochalcone A | 3CLpro and PLpro | *Broussonetia papyrifera* (L.) L'Her. Ex vent. | Park et al. (2017) |
| 33. | Scutellarein | 3CLpro | *Scutellaria baicalensis* Georgi | Yu et al. (2012) |
| 34. | Silvestrol | Cap-mRNA Translation | *Aglaia foveolata* Pannell | Müller et al. (2018) |
| 35. | Tanshinone I | PLpro | *Salvia miltiorrhiza* Bunge | Park et al. (2012) |
| 36. | Tetra-*O*-galloyl-b-D-glucose | Restrict viral entry | *Phyllanthus emblica* L. | Yi et al. (2004) |
| 37. | Sinigrin | 3CLpro | *Isatis tinctoria* L. | Lin et al. (2005) |
| 38. | Tylophorine | 3CLpro | *Tylophora indica* (Burm. f.) Merr. | Yang et al. (2010) |
| 39. | 3′-*O*-methyl diplacone | PLpro | *Paulownia tomentosa* Steud. | Cho et al. (2013) |
| 40. | 4′-*O*-methyl diplacone | PLpro | *Paulownia tomentosa* Steud. | Cho et al. (2013) |
| 41. | 3′-*O*-methyl diplacol | PLpro | *Paulownia tomentosa* Steud. | Cho et al. (2013) |
| 42. | 4′-*O*-methyl diplacol | PLpro | *Paulownia tomentosa* Steud. | Cho et al. (2013) |
| 43. | 7-Methoxy-cryptopleurine | 3CLpro | *Tylophora indica* (Burm. f.) Merr. | Yang et al. (2010) |
| 44. | 6-geranyl-40,5,7-trihy-droxy- | PLpro | *Paulownia tomentosa* Steud. | Cho et al. (2013) |

(continued)

**Table 9.2.** (continued)

| Sl. No. | Phytochemicals | Mode/target of inhibition | The scientific name of the source plants | Reference |
|---|---|---|---|---|
| | 30,50-di-methoxy flavanone | | | |
| 45. | 8b-hydroxy abieta-9 (11),13-dien-12-one | CoV-replication | *Chamaecyparis obtusa* var. *formosana* | Chang et al. (2012) |
| 46. | Tomentin A | 3CLpro | *Paulownia tomentosa* Steud. | Swain et al. (2021) |
| 47. | Amentoflavone | 3CLpro | *Torreya nucifera* (L.) Siebold &Zucc. | Ryu et al. (2010) |
| 48. | Bavachinin | PLpro | *Cullen corylifolium* (L.) Medik. | Kim et al. (2014) |
| 49. | Broussoflavan A | 3CLpro and PLpro | *Broussonetia papyrifera* (L.) L'Her. ExVent. | Park et al. (2017) |
| 50. | Curcumin | 3CLpro | *Curcuma longa* L. | Wen et al. (2007) |

widely acknowledged approach compared to methods that use structures (Kuntz et al. 1982).

Docking approaches are used to model the molecular-level interaction between molecules and proteins in order to understand the behavior of molecule and protein binding at the binding site as well as the underlying biochemical process. In general, the docking procedure is divided into two steps: first, determining the pose, which includes the ligand's conformation, location, and orientation; and second, determining the binding affinity, which is connected to the sampling technique and scoring schemes, respectively. If the location of the binding site is known, docking efficiency can be increased. This can be done by comparing proteins from related families, those with similar activities, or those that have crystallized along with other ligands. GRID (Goodford 1985; Kastenholz et al. 2000), POCKET (Levitt and Banaszak 1992), SurfNet (Laskowski 1995; Glaser et al. 2006), PASS (Brady and Stouten 2000), and MMC (Mezei 2003), among others, can be used to discover active sites within proteins when no information about binding sites is available. Blind docking is the term used when docking is performed without knowledge of the binding location.

Fischer's lock-and-key hypothesis (Fischer 1894), according to which ligands fit into receptors like a lock and key, was the basis for docking in the past. Then, when Koshland's induced fit theory was developed, it replaced the lock-and-key hypothesis by arguing that the protein's active sites are constantly changing due to interactions with ligands. In the end, this hypothesis gave a more accurate description of what happens during docking when ligands and receptors are seen as flexible instead of rigid.

Utilizing computational methods, molecular docking is used to predict the complex ligand–protein structure, and it has two parts: sampling algorithms and scoring functions. The first step in sampling is matching the protein and ligand conformations, after which the conformations are rated using the scoring function. It should be ranked among all the created conformations by the testing functions, and the sampling methods should ideally be the same as the experimental binding conditions.

Concerning the six degrees of translational, rotational, and conformational freedom of the proteins and ligands, it is possible to come up with countless numbers of modes of binding among the two, which, if done, can be extremely expensive to generate computationally. Thus, various sampling algorithms are produced to determine the most suitable. The MA process relies on the shape features and chemical information of the ligand molecule's shape map in the active site of a protein. Protein and ligand are accepted as pharmacophores, and each distance between the two is calculated for a match, which results in new ligand conformations with the help of the distance matrix between pharmacophores and matching ligands. Few characteristics are taken into account while samplings, such as the chemical properties of hydrogen donors or acceptors. The matching algorithms, if documented, save a lot of time and labor, thus enriching the active compound libraries, i.e. database enrichment (Brint and Willett 1987; Fischer et al. 1993; Norel et al. 1994). Incremental construction, or IC (Des Jarlais et al. 1986; Leach and Kuntz 1992; Rarey et al. 1996), is a fragmental approach that matches the ligand in a fragmented and incremental manner. The ligands are broken into several fragments using the rotatable bonds, which are used to dock at the active site. One fragment, being the largest, is interacted with first based on its significant functional role, and other fragments are added on an incremental basis. In terms of ligand flexibility, different ligand orientations are taken into account. MCSS or multiple copy sequential search (Miranker and Karplus 1991; Eisen et al. 1994) and LUDI (Böhm 1992a) are fragment-based methods for de novo design well and for modification of known ligands to improve binding to the active site of the protein. In the case of MCSS, the ligand is fragmented into 1000–5000 copies of the functional group. It is placed randomly on the binding site, for which instantaneous minimization of energy or suppressed molecular dynamics are targeted. In that sense, a series of energetically suitable binding sites and orientations are identified based on the energy of interactions, and new molecules are designed to link up those functional groups. In LUDI, small molecules are put into the active sites of protein molecules in a way that lets hydrogen bonds form, and hydrophobic groups are put into the pockets that are left over. First, the interaction points that can form H-bonds or be filled by groups that do not like water are counted. So, the interaction sites are filled with parts that fit well, and then the parts that fit well are joined by bridge fragments to make a single molecule (Böhm 1992b). Monte Carlo (MC) and genetic algorithm (GA) are types of random methods that look for conformational space by changing the orientation or number of ligands in a population. MC creates different conformations of ligands by rigid body translation or bond rotations, and poses generated by this are assessed by energy-based selection. Conformations are not completed unless a predetermined level of quality is obtained. MC enables the ligand to cross the energy barriers on the potential energy surface by generating a large amount of change. Conversely, GA is known to have derived the concept of evolution from Darwin's ideas. The degree of

**Table 9.3** Various software using different sampling algorithms

| Sl. No. | Algorithm | Software | Reference |
|---|---|---|---|
| 1. | Matching algorithms (MA) | DOCK | Kuntz et al. (1982) |
| | | FLOG | Miller et al. (1994) |
| | | LibDock | Diller and Merz Jr (2001) |
| | | SANDOCK | Burkhard et al. (1998) |
| 2. | Incremental construction (IC) | DOCK 4.0 | Ewing et al. (2001) |
| | | FlexX | Rarey et al. (1996) |
| | | Hammerhead | Welch et al. (1996) |
| | | SLIDE | Schnecke and Kuhn (2000) |
| | | eHiTS | Zsoldos et al. (2006) |
| 3. | LUDI | – | Böhm (1992a) |
| 4. | Monte Carlo (MC) | AutoDock | Goodsell and Olson (1990) |
| | | ICM | Abagyan et al. (1994) |
| | | QXP | Mcmartin and Bohacek (1997) |
| | | Affinity | – |
| 5. | Genetic algorithms (GA) | AutoDock | Morris et al. (1998) |
| | | GOLD | Verdonk et al. (2003) |
| | | DIVALI | Clark (1995) |
| | | DARWIN | Taylor and Burnett (2000) |

freedom of the ligand is represented by genes, which collectively form the ligand's conformation, represented by a chromosome. Mutation occurs when random changes are made to the genes, and crossing over is the exchange of genes between chromosomes, both of which are genetic operations on the genetic algorithm, which in turn operates on the structure of the ligand. Scoring functions evaluate brand-new structures, and the ones that exceed the threshold are used for the next generation. Molecular dynamics (MD) are used to move each atom independently among the other atoms, which indicates the flexibility of both the ligand and protein. Because MD is associated with difficulties in overcoming high energy and conformational barriers, which lead to short sampling, it is operated for local optimization. Thus, MD is performed to identify subtle orientations after the random identification conformation of ligands. Table 9.3 enlists different software using different sampling methods.

Scoring is accomplished by guessing rather than calculating, employing assumptions and simplifications of protein and ligand binding selectivity to distinguish correct from incorrect conformations and binders from incorrect locations to an active site in a reasonable amount of time. There are three methods for scoring: force-field, knowledge-based, and empirical. The traditional scoring method, force-field-based scoring, evaluates the binding energy through non-bonded interactions like electrostatics and van der Waals interactions and contributions from hydrogen bonds, solvation, and entropy. While empirical scoring is straightforward to assess, its applicability for ligand–protein complexes outside of the training set is only partially evident. In the knowledge-based scoring of ligand and protein complexes, statistical analysis is carried out to determine the contact frequencies between atoms or the distance between the ligand and the protein. It assumes that the more favorable a contact is, the more likely it will occur. For each atom in the ligand and protein, the score is computed by counting the number of favorable connections and penalizing the number of repulsive interactions below a threshold. The distribution of frequencies is converted into paired atom-type potentials. Due to their computational simplicity, knowledge-based functions have the benefit of being able to scan enormous compound datasets. By integrating multiple distinct scores to evaluate the docking confirmation, consensus scoring has been established to reduce enrichment in virtual screening drastically.

Aside from the complexity of receptor flexibility, particularly the flexibility of the backbone and the movement of several essential secondary elements of the receptor that involve ligand binding and the catalyst, real-world applications show how computational methods can be used to find hits in an extensive database and create new molecules. For computational studies and figuring out how the structure of a protein interacts with ligands, the structure of the protein can be obtained from experiments or databases like the Protein Data Bank (PDB).

## 9.5 Potential Drug Targets of SARS-CoV-2

SARS-CoV-2 seemed to have a spherical form and certain pleomorphic traits when studied under an electron microscope. When the virus was seen in the epithelial regions of the human airway, it manifested as inclusion bodies and membrane-bound vesicles in the cytoplasm. As determined earlier by genome sequencing (Hall Jr and Ji 2020), SARS-CoV-2 shares more than 85% of its genetic makeup with the bat SARS-like CoV (bat-SL-CoVZC45, MG772933.1) genome. Scientists have found that SARS-CoV-2 attacks the lower airways by binding to the ACE2 enzyme and causing the pulmonary system to make more inflammatory cytokines and chemokines, which can cause flu-like symptoms within 2 weeks of infection (Jiang et al. 2020).

The first area to prevent CoV infection is where it attaches to the host cells, which is the best spot to stop the infection. The viral nucleoprotein is released in the host cells once the virions are bound to cell membranes, and the protein covering has been removed. As the primary contributor to the systemic occurrence of the events, the structural protein S is one of the crucial targets of inhibition by creating possible drugs while retaining the S protein as a prospective pharmacological target. Its many components' structures and molecular uniqueness have already been addressed. The N-terminal domain (NTD) and the C-terminal domain (CTD) are the critical components of the S1 domain, which functions as the principal antigen on the surface of viruses with the receptor-binding domain (RBD) (Du et al. 2009). The R453 residue of RBD and K341 of ACE2 both play a crucial role in the ACE2 enzyme's interaction with the S protein's RBD, which has 18 residues and 14 amino acids (Prabakaran et al. 2004). Two heptad repeat sections (HR 1 and 2) and a hydrophobic fusion peptide (Du et al. 2009) are present in the S2 subunit. Ultimately, the RBD has been the primary therapeutic target for S proteins. In a study involving the sequence similarity of a peptide to the RBD of the S protein, SARS-CoV entry into Vero cells was successfully inhibited (IC50 of about 40 mM) (Du et al. 2009). Han et al. (2006) discovered that chloroquine inhibits SERS-CoV by raising endosomal pH, which changes the terminal glycosylation of ACE-2 and eventually prevents the virus from attaching to its receptor (Vincent et al. 2005).

The structural protein E is the minor transmembrane structural protein of CoV and consists of two domains: a hydrophobic domain and a charged cytoplasmic tail (8.4–12 KDa). It appears to have a significant role in viral morphogenesis during the entry and exiting of the host cell. Additionally, protein E inhibition has been demonstrated in lower viral tiers and immature and ineffective progeny (Kuo et al. 2007; Schoeman and Fielding 2019). Since protein E forms a pentamer that functions as an ion channel, giving it the name E channel or viroporin, it can be a potential therapeutic target (Verdiá-Báguena et al. 2012; Nieto-Torres et al. 2014). This function is critical for the interaction between viruses and their hosts and is advantageous to the pathogenicity of the virus. Additionally, it has been discovered that protein E participates in the regulation and trafficking of intercellular proteins, which increases its potential as a significant therapeutic target.

By interacting with other CoV proteins, integrating the Golgi complex into new virions, and stabilizing the nucleocapsid protein, the membrane protein, also known as the M protein, preserves the form of the viral envelope (Schoeman and Fielding 2019; Arndt et al. 2010). It has three transmembrane domains with long C- and short N-termini on the inside and outside, respectively (Hogue and Machamer 2007). This protein also plays a significant role in cellular homeostasis, which is controlled by the interactions of other proteins. M and S protein contact is necessary for the S protein to interact with the ERGIC complex. M and N protein interactions control viral uptake and release and dictate the virus's form (Schoeman and Fielding 2019). M protein aids in the humoral response to certain viruses and can produce potent neutralizing antibodies in SARS patients (Pang et al. 2004).

The structural protein N is composed of three intrinsically disordered regions (IDRs): the N-terminal RNA-binding domain (NTD), a central linker region rich in Ser/Arg (SR), and a C-terminal dimerization domain (CTD). But these proteins aid in packaging the viral genome into the viral nucleocapsid through their roles in RNA-binding, dimerization, N-M protein interactions, and nucleocapsid protein oligomerization (Chang et al. 2014). The ribonucleoprotein complex that governs viral genome replication and transcription, cytoskeleton remodeling, and host cell death is maintained and formed by the N protein, which increases the protein's suitability as a therapeutic target. These N proteins facilitate viral genome release into the endoplasmic reticulum-Golgi intermediate compartment network (ERGIC). The N protein uses a region of its structure that contains around 140 amino acids to bind to the viral RNA. To treat COVID-19, it may be feasible to target the RNA-binding protein.

In addition to producing structural proteins, the virus also generates 16 different types of non-structural proteins. These 16 NSPs are composed of two polyproteins encoded by two ORFs: ORF 1a and ORF 1b, which together create the PP1a and PP1b polyproteins. The 16 NSPs are produced when two cysteine proteases break down the polyproteins. NSP1-NSP3 are cleaved by a papain-like protease, or PLpro, whereas NSP4-NSP16 are cleaved by chymotrypsin-like cysteine protease, 3CLpro, or Main protease (Mpro) (Lindner et al. 2005; Zhang et al. 2020a, b). Since 3CLpro could create 12 major non-structural proteins (NSP 4 to NSP 16), including the viral RdRp (NSP 12) and helicase (NSP 13), it has been the subject of far more research than PLpro. The homodimer form of 3CLpro has a Cys-His dyad on the active site, which displays protease. The protease is crucial for RNA replication, transcription, and subsequent viral proliferation, making 3CLpro one of the most crucial targets for therapeutic research. The 306 amino acid residues are divided into domains (I and II) in the N-terminal region: domain I has amino acid residues 8–101, and domain II has 102–184. Domain III is in the C-terminal part and has residues 201–303. The active dimer structure of the protease is provided by a 15-residue long loop (residues 185–200), which connects domains II and III and is composed of a globular cluster of five helices. Studies have revealed that nucleoside analogs, terpenes, peptides, flavonoids, and quinolines have considerable anti-SARS-CoV-2 action. More surprisingly, because humans lack a homolog of 3CLpro, it has been continually investigated as a crucial target for therapeutic research (Zhang et al. 2020a, b). Like other CoVs, SARS-CoV-2 has a substrate-binding site for 3CLpro in the gap between domains I and II, which breaks down polyproteins to produce NSPs and creates a complex called a replicase transcriptase (RTC). Less than 11 Leu-Gl (Ser, Ala, and Gly) cleavage sites of polyprotein replicase 1ab are recognized by 3CLpro. Targeting conserved enzyme residues in 3CLpro can create safe and effective SARS-CoV-2 treatments.

Several polymerase inhibitors, including Remdesivir, have been developed to treat various viral illnesses because RNA-dependent RNA polymerase (RdRp) has been recognized as one of the most crucial enzymes for transcription and replication. Its central part is the multi-subunit replication and transcription complex of RdRp's NSP12. For NSP12 to operate correctly, NSP7 and NSP8 are needed (Kirchdoerfer and Ward 2019).

The endosomal cysteine proteases cathepsin B and L (CatB/L) and the serine protease TMPRSS2 (Simmons et al. 2005; Glowacka et al. 2011), which are known to play a role in the priming of S protein, have been reported to be involved in the binding of the virions in addition to the viral proteases 3CLpro and PLpro. Since it was found to be necessary for the virus to cause disease and spread, TMPRSS2 has become a possible therapeutic target.

A class of polypeptide-signaling molecules known as cytokines interacting with cell surface receptors controls a wide range of biological activities (Ratia et al. 2006). A condition known as "cytokine storm" is characterized by elevated immune system activation and is connected to several viral infections (Yuen and Wong 2005). It has been clear that one of the processes causing acute respiratory distress syndrome and multiple organ failure in COVID-19 is due to excessive cytokine production in response to viral infection (Tay et al. 2020). According to studies, individuals with COVID-19 have higher levels of inflammatory cytokines than healthy people do, including interleukin (IL)-1, IL-2, IL-6, IL-7, IL-8, IL-9, IL-10, IL-18, tumor necrosis factor (TNF), granulocyte-macrophage colony-stimulating factor (G-CSF), and macrophages (Huang et al. 2020). In any case, the interaction between glucocorticoid receptors (GR) and their ligands, glucocorticoids (GCs), influences the production of cytokines in both acute and chronic inflammatory diseases (Chow and Simons Jr 2018). The activated GR complex has the potential to bind to promotor-responsive elements, which would result in the inactivation of important pro-inflammatory transcription factors (such as AP-1 and NF-kappa B). Alternatively, the activated GR complex could either upregulate the expression of cytokine inhibitory proteins, such as I kappa B, which inactivates the transcription factor NF-kappa B, thereby suppressing the secondary expression of many cytokines (Brattsand and Linden 1996). Unfortunately, adverse side effects include osteoporosis, glucose and lipid metabolism abnormalities, and hypertension, which limit the usage of GCs (Sannarangappa and Jalleh 2014). As a result, developing medicines targeting the active areas of human glucocorticoid receptors (hGRs) has significant potential.

## 9.6 Conclusion

From December 2019 onward, instances of quickly spreading viral pneumonia were identified in Wuhan, China, which were eventually identified as a novel strain of coronavirus dubbed SARS-CoV-2 because of its 82% resemblance to the previously known SARS coronavirus (SARS-CoV). The virus resulted in a terrible epidemic that affected the whole world and appears to have faded but not ceased entirely. There are currently no effective antiviral drugs or vaccines, which opens the door to alternate solutions. Medications derived from plants have been employed by ancient civilizations to control early epidemics for a considerable amount of time and are even promising in our contemporary emergency. Several biomolecules produced by plants are efficient against the coronavirus in vivo and in vitro investigations; thus, the creation of specialized antiviral drugs is essential. In the current context of the urgent need for COVID virus medications, the computational in silico technique is both time- and resource-efficient. This chapter briefly describes the molecular docking technique as a possible weapon in a drug design system employing bioactive chemicals originating from plants. This study is anticipated to be completed rapidly with the use of in silico approaches such as molecular docking and plant bioactive compounds that can be introduced into the human body in large quantities without adverse side effects.

# References

Abagyan R, Totrov M, Kuznetsov D (1994) ICM—A new method for protein modeling and design: applications to docking and structure prediction from the distorted native conformation. J Comput Chem 15(5):488–506

Almeida MS, Johnson MA, Herrmann T, Geralt M, Wüthrich K (2007) Novel β-barrel fold in the nuclear magnetic resonance structure of the replicase nonstructural protein 1 from the severe acute respiratory syndrome coronavirus. J Virol 81(7):3151–3161

Arndt AL, Larson BJ, Hogue BG (2010) A conserved domain in the coronavirus membrane protein tail is important for virus assembly. J Virol 84(21):11418–11428

Bale S, Pulivendala G, Godugu C (2018) Withaferin A attenuates bleomycin-induced scleroderma by targeting FoxO3a and NF-κβsignaling: connecting fibrosis and inflammation. Biofactors 44(6):507–517

Belouzard S, Millet JK, Licitra BN, Whittaker GR (2012) Mechanisms of coronavirus cell entry mediated by the viral spike protein. Viruses 4(6):1011–1033

Bhuiyan FR, Howlader S, Raihan T, Hasan M (2020) Plants metabolites: possibility of natural therapeutics against the COVID-19 pandemic. Front Med 7:444

Biesalski HK, Dragsted LO, Elmadfa I, Grossklaus R, Müller M, Schrenk D, Walter P, Weber P (2009) Bioactive compounds: definition and assessment of activity. Nutrition 25(11–12): 1202–1205

Biesalski HK, Erdman JW, Hathcock J, Ellwood K, Beatty S, Johnson E, Marchioli R, Lauritzen L, Rice HB, Shao A, Griffiths JC (2013) Nutrient reference values for bioactives: new approaches needed? A conference report. Eur J Nutr 52(1):1–9

Böhm HJ (1992a) LUDI: rule-based automatic design of new substituents for enzyme inhibitor leads. J Comput Aided Mol Des 6(6):593–606

Böhm HJ (1992b) The computer program LUDI: a new method for the de novo design of enzyme inhibitors. J Comput Aided Mol Des 6(1):61–78

Boopathi S, Poma AB, Kolandaivel P (2021) Novel 2019 coronavirus structure, mechanism of action, antiviral drug promises and rule out against its treatment. J Biomol Struct Dyn 39(9): 3409–3418

Boukhatem MN, Setzer WN (2020) Aromatic herbs, medicinal plant-derived essential oils, and phytochemical extracts as potential therapies for coronaviruses: future perspectives. Plan Theory 9(6):800

Bouvet M, Lugari A, Posthuma CC, Zevenhoven JC, Bernard S, Betzi S, Imbert I, Canard B, Guillemot JC, Lécine P, Pfefferle S (2014) Coronavirus Nsp10, a critical co-factor for activation of multiple replicative enzymes. J Biol Chem 289(37):25783–25796

Brady GP, Stouten PF (2000) Fast prediction and visualization of protein binding pockets with PASS. J Comput Aided Mol Des 14(4):383–401

Brattsand R, Linden M (1996) Cytokine modulation by glucocorticoids: mechanisms and actions in cellular studies. Aliment Pharmacol Ther 10(Sup2):81–90

Brint AT, Willett P (1987) Algorithms for the identification of three-dimensional maximal common substructures. J Chem Inf Comput Sci 27(4):152–158

Burkhard P, Taylor P, Walkinshaw MD (1998) An example of a protein ligand found by database mining: description of the docking method and its verification by a 2.3 Å X-ray structure of a thrombin-ligand complex. J Mol Biol 277(2):449–466

Chambial S, Dwivedi S, Shukla KK, John PJ, Sharma P (2013) Vitamin C in disease prevention and cure: an overview. Indian J Clin Biochem 28(4):314–328

Chang FR, Yen CT, Ei-Shazly M, Lin WH, Yen MH, Lin KH, Wu YC (2012) Anti-human coronavirus (anti-HCoV) triterpenoids from the leaves of Euphorbia neriifolia. Nat Prod Commun 7(11):1415–1417

Chang CK, Hou MH, Chang CF, Hsiao CD, Huang TH (2014) The SARS coronavirus nucleocapsid protein–forms and functions. Antivir Res 103:39–50

Chang YC, Tung YA, Lee KH, Chen TF, Hsiao YC, Chang HC et al (2020) Potential therapeutic agents for COVID-19 based on the analysis of protease and RNA polymerase docking (preprint)
Chen CN, Lin CP, Huang KK, Chen WC, Hsieh HP, Liang PH, Hsu JTA (2005) Inhibition of SARS-CoV 3C-like protease activity by theaflavin-3, 3′-digallate (TF3). Evid Based Complement Alternat Med 2(2):209–215
Chen Y, Su C, Ke M, Jin X, Xu L, Zhang Z, Wu A, Sun Y, Yang Z, Tien P, Ahola T (2011) Biochemical and structural insights into the mechanisms of SARS coronavirus RNA ribose 2′-O-methylation by nsp16/nsp10 protein complex. PLoS Pathog 7(10):e1002294
Cho JK, Curtis-Long MJ, Lee KH, Kim DW, Ryu HW, Yuk HJ, Park KH (2013) Geranylated flavonoids displaying SARS-CoV papain-like protease inhibition from the fruits of Paulownia tomentosa. Bioorg Med Chem 21(11):3051–3057
Chow CC, Simons SS Jr (2018) An approach to greater specificity for glucocorticoids. Front Endocrinol 9:76
Clark KP (1995) Flexible ligand docking without parameter adjustment across four ligand–receptor complexes. J Comput Chem 16(10):1210–1226
Darshan S, Doreswamy R (2004) Patented antiinflammatory plant drug development from traditional medicine. Phytother Res 18(5):343–357
Das T, Mukerjee N, Ghosh A, Lorenzo JM, Dhama K, Dey A (2022) Growing risk of aristolochic acid nephropathy in the era of COVID-19–correspondence. Int J Surg (Lond Engl) 108:106992
Debnath S, Seth D, Pramanik S, Adhikari S, Mondal P, Sherpa D, Sen D, Mukherjee D, Mukerjee N (2022) A comprehensive review and meta-analysis of recent advances in biotechnology for plant virus research and significant accomplishments in human health and the pharmaceutical industry. Biotechnol Genet Eng Rev:1–33
Des Jarlais RL, Sheridan RP, Dixon JS, Kuntz ID, Venkataraghavan R (1986) Docking flexible ligands to macromolecular receptors by molecular shape. J Med Chem 29(11):2149–2153
Diller DJ, Merz KM Jr (2001) High throughput docking for library design and library prioritization. Proteins 43(2):113–124
Du L, He Y, Zhou Y, Liu S, Zheng BJ, Jiang S (2009) The spike protein of SARS-CoV—a target for vaccine and therapeutic development. Nat Rev Microbiol 7(3):226–236
Egloff MP, Ferron F, Campanacci V, Longhi S, Rancurel C, Dutartre H, Snijder EJ, Gorbalenya AE, Cambillau C, Canard B (2004) The severe acute respiratory syndrome-coronavirus replicative protein nsp9 is a single-stranded RNA-binding subunit unique in the RNA virus world. Proc Natl Acad Sci 101(11):3792–3796
Eisen MB, Wiley DC, Karplus M, Hubbard RE (1994) HOOK: a program for finding novel molecular architectures that satisfy the chemical and steric requirements of a macromolecule binding site. Proteins 19(3):199–221
Ewing TJ, Makino S, Skillman AG, Kuntz ID (2001) DOCK 4.0: search strategies for automated molecular docking of flexible molecule databases. J Comput Aided Mol Des 15(5):411–428
Fischer E (1894) Einfluss der Configuration auf die Wirkung der Enzyme. Berichte der deutschenchemischen Gesellschaft 27(3):2985–2993
Fischer D, Norel R, Wolfson H, Nussinov R (1993) Surface motifs by a computer vision technique: searches, detection, and implications for protein–ligand recognition. Proteins 16(3):278–292
Gantait S, Debnath S, Ali N (2014) Genomic profile of the plants with pharmaceutical value. 3. Biotech 4(6):563–578
Ghosh A, Mukerjee N, Sharma B, Pant A, Kishore Mohanta Y, Jawarkar RD, Bakal RL, Terefe EM, Batiha GES, Mostafa-Hedeab G, ArefAlbezrah NK (2022) Target specific inhibition of protein tyrosine kinase in conjunction with cancer and SARS-COV-2 by olive nutraceuticals. Front Pharmacol 12:812565
Glaser F, Morris RJ, Najmanovich RJ, Laskowski RA, Thornton JM (2006) A method for localizing ligand binding pockets in protein structures. Proteins 62(2):479–488
Glowacka I, Bertram S, Müller MA, Allen P, Soilleux E, Pfefferle S, Steffen I, Tsegaye TS, He Y, Gnirss K, Niemeyer D (2011) Evidence that TMPRSS2 activates the severe acute respiratory

syndrome coronavirus spike protein for membrane fusion and reduces viral control by the humoral immune response. J Virol 85(9):4122–4134
Goodford PJ (1985) A computational procedure for determining energetically favorable binding sites on biologically important macromolecules. J Med Chem 28(7):849–857
Goodsell DS, Olson AJ (1990) Automated docking of substrates to proteins by simulated annealing. Proteins 8(3):195–202
Guarino LA, Bhardwaj K, Dong W, Sun J, Holzenburg A, Kao C (2005) Mutational analysis of the SARS virus Nsp15 endoribonuclease: identification of residues affecting hexamer formation. J Mol Biol 353(5):1106–1117
Gupta R, Paswan RR, Saikia R, Borar BK (2020) Insights into the severe acute respiratory syndrome coronavirus-2: transmission, genome composition, replication, diagnostics and therapeutics. Curr J Appl Sci Technol 39(21):71–91
Gyebi GA, Ogunyemi OM, Ibrahim IM, Afolabi SO, Adebayo JO (2021) Dual targeting of cytokine storm and viral replication in COVID-19 by plant-derived steroidal pregnanes: an in silico perspective. Comput Biol Med 134:104406
Hall DC Jr, Ji HF (2020) A search for medications to treat COVID-19 via in silico molecular docking models of the SARS-CoV-2 spike glycoprotein and 3CL protease. Travel Med Infect Dis 35:101646
Han DP, Penn-Nicholson A, Cho MW (2006) Identification of critical determinants on ACE2 for SARS-CoV entry and development of a potent entry inhibitor. Virology 350(1):15–25
Haratake A, Watase D, Setoguchi S, Nagata-Akaho N, Matsunaga K, Takata J (2017) Effect of orally ingested diosgenin into diet on skin collagen content in a low collagen skin mouse model and its mechanism of action. Life Sci 174:77–82
Harcourt BH, Jukneliene D, Kanjanahaluethai A, Bechill J, Severson KM, Smith CM, Rota PA, Baker SC (2004) Identification of severe acute respiratory syndrome coronavirus replicase products and characterization of papain-like protease activity. J Virol 78(24):13600–13612
Hilgenfeld R (2014) From SARS to MERS: crystallographic studies on coronaviral proteases enable antiviral drug design. FEBS J 281(18):4085–4096
Ho TY, Wu SL, Chen JC, Li CC, Hsiang CY (2007) Emodin blocks the SARS coronavirus spike protein and angiotensin-converting enzyme 2 interaction. Antivir Res 74(2):92–101
Hogue BG, Machamer CE (2007) Coronavirus structural proteins and virus assembly. Nidoviruses:179–200
Hu T, Chen C, Li H, Dou Y, Zhou M, Lu D, Zong Q, Li Y, Yang C, Zhong Z, Singh N (2017) Structural basis for dimerization and RNA binding of avian infectious bronchitis virus nsp9. Protein Sci 26(5):1037–1048
Huang QC, Wang MJ, Chen XM, Yu WL, Chu YL, He XH, Huang RY (2016) Can active components of licorice, glycyrrhizin and glycyrrhetinic acid, lick rheumatoid arthritis? Oncotarget 7(2):1193
Huang C, Wang Y, Li X, Ren L, Zhao J, Hu Y, Zhang L, Fan G, Xu J, Gu X, Cheng Z (2020) Clinical features of patients infected with 2019 novel coronavirus in Wuhan, China. Lancet 395(10223):497–506
Ivanova L, Karelson M, Dobchev DA (2018) Identification of natural compounds against neurodegenerative diseases using in silico techniques. Molecules 23(8):1847
Jiang F, Deng L, Zhang L, Cai Y, Cheung CW, Xia Z (2020) Review of the clinical characteristics of coronavirus disease 2019 (COVID-19). J Gen Intern Med 35(5):1545–1549
Joseph JS, Saikatendu KS, Subramanian V, Neuman BW, Brooun A, Griffith M, Moy K, Yadav MK, Velasquez J, Buchmeier MJ, Stevens RC (2006) Crystal structure of nonstructural protein 10 from the severe acute respiratory syndrome coronavirus reveals a novel fold with two zinc-binding motifs. J Virol 80(16):7894–7901
Kastenholz MA, Pastor M, Cruciani G, Haaksma EE, Fox T (2000) GRID/CPCA: a new computational tool to design selective ligands. J Med Chem 43(16):3033–3044

Kim DW, Seo KH, Curtis-Long MJ, Oh KY, Oh JW, Cho JK, Lee KH, Park KH (2014) Phenolic phytochemical displaying SARS-CoV papain-like protease inhibition from the seeds of Psoralea corylifolia. J Enzyme Inhib Med Chem 29(1):59–63
Kirchdoerfer RN, Ward AB (2019) Structure of the SARS-CoV nsp12 polymerase bound to nsp7 and nsp8 co-factors. Nat Commun 10(1):1–9
Kotecha R (2021) The journey with COVID-19: initiatives by ministry of AYUSH. J Ayurved Integrat Med 12(1):1
Kuntz ID, Blaney JM, Oatley SJ, Langridge R, Ferrin TE (1982) A geometric approach to macromolecule-ligand interactions. J Mol Biol 161(2):269–288
Kuo L, Hurst KR, Masters PS (2007) Exceptional flexibility in the sequence requirements for coronavirus small envelope protein function. J Virol 81(5):2249–2262
Laskowski RA (1995) SURFNET: a program for visualizing molecular surfaces, cavities, and intermolecular interactions. J Mol Graph 13(5):323–330
Leach AR, Kuntz ID (1992) Conformational analysis of flexible ligands in macromolecular receptor sites. J Comput Chem 13(6):730–748
Lei J, Kusov Y, Hilgenfeld R (2018) Nsp3 of coronaviruses: structures and functions of a large multi-domain protein. Antivir Res 149:58–74
Lei X, Dong X, Ma R, Wang W, Xiao X, Tian Z, Wang C, Wang Y, Li L, Ren L, Guo F (2020) Activation and evasion of type I interferon responses by SARS-CoV-2. Nat Commun 11(1): 1–12
Levitt DG, Banaszak LJ (1992) POCKET: a computer graphics method for identifying and displaying protein cavities and their surrounding amino acids. J Mol Graph 10(4):229–234
Lin CW, Tsai FJ, Tsai CH, Lai CC, Wan L, Ho TY, Hsieh CC, Chao PDL (2005) Anti-SARS coronavirus 3C-like protease effects of Isatis indigotica root and plant-derived phenolic compounds. Antivir Res 68(1):36–42
Lindner HA, Fotouhi-Ardakani N, Lytvyn V, Lachance P, Sulea T, Ménard R (2005) The papain-like protease from the severe acute respiratory syndrome coronavirus is a deubiquitinating enzyme. J Virol 79(24):15199–15208
Liu RH (2013) Dietary bioactive compounds and their health implications. J Food Sci 78(s1):A18–A25
Liu Z, Xiao X, Wei X, Li J, Yang J, Tan H, Zhu J, Zhang Q, Wu J, Liu L (2020) Composition and divergence of coronavirus spike proteins and host ACE2 receptors predict potential intermediate hosts of SARS-CoV-2. J Med Virol 92(6):595–601
Lu R, Zhao X, Li J, Niu P, Yang B, Wu H, Wang W, Song H, Huang B, Zhu N, Bi Y (2020) Genomic characterisation and epidemiology of 2019 novel coronavirus: implications for virus origins and receptor binding. Lancet 395(10224):565–574
Maitra S, Mukerjee N, Dey A, Ghosh A, Alexiou A (2022) Drug development strategies and immunological aspects of SARS-CoV-2. Open Public Health J 15(1)
Maniruzzaman M, Islam M, Ali M, Mukerjee N, Maitra S, Kamal MA, Ghosh A, Castrosanto MA, Alexiou A, Ashraf GM, Tagde P (2022) COVID-19 diagnostic methods in developing countries. Environ Sci Pollut Res 29:1–14
Masters PS (2006) The molecular biology of coronaviruses. Adv Virus Res 66:193–292
McBride R, Van Zyl M, Fielding BC (2014) The coronavirus nucleocapsid is a multifunctional protein. Viruses 6(8):2991–3018
Mcmartin C, Bohacek RS (1997) QXP: powerful, rapid computer algorithms for structure-based drug design. J Comput Aided Mol Des 11(4):333–344
Mezei M (2003) A new method for mapping macromolecular topography. J Mol Graph Model 21(5):463–472
Miller MD, Kearsley SK, Underwood DJ, Sheridan RP (1994) FLOG: a system to select 'quasi-flexible' ligands complementary to a receptor of known three-dimensional structure. J Comput Aided Mol Des 8(2):153–174
Miranker A, Karplus M (1991) Functionality maps of binding sites: a multiple copy simultaneous search method. Proteins 11(1):29–34

Moitessier N, Englebienne P, Lee D, Lawandi J, Corbeil AC (2008) Towards the development of universal, fast and highly accurate docking/scoring methods: a long way to go. Br J Pharmacol 153(S1):S7–S26

Morris GM, Goodsell DS, Halliday RS, Huey R, Hart WE, Belew RK, Olson AJ (1998) Automated docking using a Lamarckian genetic algorithm and an empirical binding free energy function. J Comput Chem 19(14):1639–1662

Mukerjee N, Maitra S, Das P, Malik S, Alexiou A, Ghosh A (2022) Omicron with threatened antagonistic consequences and are convoyed by new-fangled risks–correspondence. Int J Surg (Lond, Engl) 106:106947

Mukhtar M, Arshad M, Ahmad M, Pomerantz RJ, Wigdahl B, Parveen Z (2008) Antiviral potentials of medicinal plants. Virus Res 131(2):111–120

Müller C, Schulte FW, Lange-Grünweller K, Obermann W, Madhugiri R, Pleschka S, Ziebuhr J, Hartmann RK, Grünweller A (2018) Broad-spectrum antiviral activity of the eIF4A inhibitor silvestrol against corona-and picornaviruses. Antivir Res 150:123–129

Nieto-Torres JL, DeDiego ML, Verdiá-Báguena C, Jimenez-Guardeño JM, Regla-Nava JA, Fernandez-Delgado R, Castaño-Rodriguez C, Alcaraz A, Torres J, Aguilella VM, Enjuanes L (2014) Severe acute respiratory syndrome coronavirus envelope protein ion channel activity promotes virus fitness and pathogenesis. PLoS Pathog 10(5):e1004077

Norel R, Fischer D, Wolfson HJ, Nussinov R (1994) Molecular surface recognition by a computer vision-based technique. Protein Eng Des Sel 7(1):39–46

Oudit GY, Kassiri Z, Jiang C, Liu PP, Poutanen SM, Penninger JM, Butany J (2009) SARS-coronavirus modulation of myocardial ACE2 expression and inflammation in patients with SARS. Eur J Clin Investig 39(7):618–625

Pang H, Liu Y, Han X, Xu Y, Jiang F, Wu D, Kong X, Bartlam M, Rao Z (2004) Protective humoral responses to severe acute respiratory syndrome-associated coronavirus: implications for the design of an effective protein-based vaccine. J Gen Virol 85(10):3109–3113

Park JY, Kim JH, Kim YM, Jeong HJ, Kim DW, Park KH, Kwon HJ, Park SJ, Lee WS, Ryu YB (2012) Tanshinones as selective and slow-binding inhibitors for SARS-CoV cysteine proteases. Bioorg Med Chem 20(19):5928–5935

Park JY, Yuk HJ, Ryu HW, Lim SH, Kim KS, Park KH, Ryu YB, Lee WS (2017) Evaluation of polyphenols from Broussonetia papyrifera as coronavirus protease inhibitors. J Enzyme Inhib Med Chem 32(1):504–512

Paules CI, Marston HD, Fauci AS (2020) Coronavirus infections—more than just the common cold. JAMA 323(8):707–708

Petrovska BB (2012) Historical review of medicinal plants' usage. Pharmacogn Rev 6(11):1

Prabakaran P, Xiao X, Dimitrov DS (2004) A model of the ACE2 structure and function as a SARS-CoV receptor. Biochem Biophys Res Commun 314(1):235–241

Rarey M, Kramer B, Lengauer T, Klebe G (1996) A fast flexible docking method using an incremental construction algorithm. J Mol Biol 261(3):470–489

Ratia K, Saikatendu KS, Santarsiero BD, Barretto N, Baker SC, Stevens RC, Mesecar AD (2006) Severe acute respiratory syndrome coronavirus papain-like protease: structure of a viral deubiquitinating enzyme. Proc Natl Acad Sci 103(15):5717–5722

Robinson MM, Zhang X (2011) The world medicines situation 2011, traditional medicines: global situation, issues and challenges. World Health Organization, Geneva, pp 1–2

Roh C (2012) A facile inhibitor screening of SARS coronavirus N protein using nanoparticle-based RNA oligonucleotide. Int J Nanomedicine 7:2173

Ryu YB, Jeong HJ, Kim JH, Kim YM, Park JY, Kim D, Naguyen TTH, Park SJ, Chang JS, Park KH, Rho MC (2010) Biflavonoids from Torreya nucifera displaying SARS-CoV 3CLpro inhibition. Bioorg Med Chem 18(22):7940–7947

Sannarangappa V, Jalleh R (2014) Inhaled corticosteroids and secondary adrenal insufficiency. Open Respir Med J 8:93

Schnecke V, Kuhn LA (2000) Virtual screening with solvation and ligand-induced complementarity. In: Virtual screening: an alternative or complement to high throughput screening? Springer, Dordrecht, pp 171–190

Schoeman D, Fielding BC (2019) Coronavirus envelope protein: current knowledge. Virol J 16(1): 1–22

Schwarz S, Sauter D, Wang K, Zhang R, Sun B, Karioti A, Bilia AR, Efferth T, Schwarz W (2014) Kaempferol derivatives as antiviral drugs against the 3a channel protein of coronavirus. Planta Med 80(02/03):177–182

Shah R, Gulati V, Palombo EA (2012) Pharmacological properties of guggulsterones, the major active components of gum guggul. Phytother Res 26(11):1594–1605

Shamsi A, Mohammad T, Anwar S, Amani S, Khan MS, Husain FM, Rehman MT, Islam A, Hassan MI (2021) Potential drug targets of SARS-CoV-2: from genomics to therapeutics. Int J Biol Macromol 177:1–9

Simmons G, Gosalia DN, Rennekamp AJ, Reeves JD, Diamond SL, Bates P (2005) Inhibitors of cathepsin L prevent severe acute respiratory syndrome coronavirus entry. Proc Natl Acad Sci 102(33):11876–11881

Singh S, Florez H (2020) Coronavirus disease 2019 drug discovery through molecular docking. F1000Res 9:502

Singh S, Sk MF, Sonawane A, Kar P, Sadhukhan S (2021) Plant-derived natural polyphenols as potential antiviral drugs against SARS-CoV-2 via RNA-dependent RNA polymerase (RdRp) inhibition: an in-silico analysis. J Biomol Struct Dyn 39(16):6249–6264

Somani SJ, Modi KP, Majumdar AS, Sadarani BN (2015) Phytochemicals and their potential usefulness in inflammatory bowel disease. Phytother Res 29(3):339–350

Stobart CC, Sexton NR, Munjal H, Lu X, Molland KL, Tomar S, Mesecar AD, Denison MR (2013) Chimeric exchange of coronavirus nsp5 proteases (3CLpro) identifies common and divergent regulatory determinants of protease activity. J Virol 87(23):12611–12618

Swain SS, Panda SK, Luyten W (2021) Phytochemicals against SARS-CoV as potential drug leads. Biom J 44(1):74–85

Tay MZ, Poh CM, Rénia L, MacAry PA, Ng LF (2020) The trinity of COVID-19: immunity, inflammation and intervention. Nat Rev Immunol 20(6):363–374

Taylor JS, Burnett RM (2000) DARWIN: a program for docking flexible molecules. Proteins 41(2): 173–191

Te Velthuis AJ, Van Den Worm SH, Snijder EJ (2012) The SARS-coronavirus nsp7+ nsp8 complex is a unique multimeric RNA polymerase capable of both de novo initiation and primer extension. Nucleic Acids Res 40(4):1737–1747

Tsao R (2010) Chemistry and biochemistry of dietary polyphenols. Nutrients 2(12):1231–1246

Verdiá-Báguena C, Nieto-Torres JL, Alcaraz A, DeDiego ML, Torres J, Aguilella VM, Enjuanes L (2012) Coronavirus E protein forms ion channels with functionally and structurally-involved membrane lipids. Virology 432(2):485–494

Verdonk ML, Cole JC, Hartshorn MJ, Murray CW, Taylor RD (2003) Improved protein–ligand docking using GOLD. Proteins 52(4):609–623

Vincent MJ, Bergeron E, Benjannet S, Erickson BR, Rollin PE, Ksiazek TG, Seidah NG, Nichol ST (2005) Chloroquine is a potent inhibitor of SARS coronavirus infection and spread. Virol J 2(1): 1–10

von Grotthuss M, Wyrwicz LS, Rychlewski L (2003) mRNA cap-1 methyltransferase in the SARS genome. Cell 113(6):701–702

Wang H, Xue S, Yang H, Chen C (2016) Recent progress in the discovery of inhibitors targeting coronavirus proteases. Virol Sin 31(1):24–30

Welch W, Ruppert J, Jain AN (1996) Hammerhead: fast, fully automated docking of flexible ligands to protein binding sites. Chem Biol 3(6):449–462

Wen CC, Kuo YH, Jan JT, Liang PH, Wang SY, Liu HG, Lee CK, Chang ST, Kuo CJ, Lee SS, Hou CC (2007) Specific plant terpenoids and lignoids possess potent antiviral activities against severe acute respiratory syndrome coronavirus. J Med Chem 50(17):4087–4095

Wu F, Zhao S, Yu B, Chen YM, Wang W, Song ZG, Hu Y, Tao ZW, Tian JH, Pei YY, Yuan ML (2020) A new coronavirus associated with human respiratory disease in China. Nature 579(7798):265–269

Yang H, Bartlam M, Rao Z (2006) Drug design targeting the main protease, the Achilles' heel of coronaviruses. Curr Pharm Des 12(35):4573–4590

Yang CW, Lee YZ, Kang IJ, Barnard DL, Jan JT, Lin D, Huang CW, Yeh TK, Chao YS, Lee SJ (2010) Identification of phenanthroindolizidines and phenanthroquinolizidines as novel potent anti-coronaviral agents for porcine enteropathogenic coronavirus transmissible gastroenteritis virus and human severe acute respiratory syndrome coronavirus. Antivir Res 88(2):160–168

Yi L, Li Z, Yuan K, Qu X, Chen J, Wang G, Zhang H, Luo H, Zhu L, Jiang P, Chen L (2004) Small molecules blocking the entry of severe acute respiratory syndrome coronavirus into host cells. J Virol 78(20):11334–11339

Yin W, Mao C, Luan X, Shen DD, Shen Q, Su H, Wang X, Zhou F, Zhao W, Gao M, Chang S (2020) Structural basis for inhibition of the RNA-dependent RNA polymerase from SARS-CoV-2 by remdesivir. Science 368(6498):1499–1504

Yu MS, Lee J, Lee JM, Kim Y, Chin YW, Jee JG, Keum YS, Jeong YJ (2012) Identification of myricetin and scutellarein as novel chemical inhibitors of the SARS coronavirus helicase, nsP13. Bioorg Med Chem Lett 22(12):4049–4054

Yuen KY, Wong SSY (2005) Human infection by avian influenza A H5N1. Hong Kong Med J 11: 189

Zhang CH, Wang YF, Liu XJ, Lu JH, Qian CW, Wan ZY, Yan XG, Zheng HY, Zhang MY, Xiong S, Li JX (2005) Antiviral activity of cepharanthine against severe acute respiratory syndrome coronavirus in vitro. Chin Med J 118(06):493–496

Zhang YJ, Gan RY, Li S, Zhou Y, Li AN, Xu DP, Li HB (2015) Antioxidant phytochemicals for the prevention and treatment of chronic diseases. Molecules 20(12):21138–21156

Zhang C, Chen Y, Li L, Yang Y, He J, Chen C, Su D (2020a) Structural basis for the multimerization of nonstructural protein nsp9 from SARS-CoV-2. Mol Biomed 1(1):1–9

Zhang L, Lin D, Sun X, Curth U, Drosten C, Sauerhering L, Becker S, Rox K, Hilgenfeld R (2020b) Crystal structure of SARS-CoV-2 main protease provides a basis for design of improved α-ketoamide inhibitors. Science 368(6489):409–412

Zhuang M, Jiang H, Suzuki Y, Li X, Xiao P, Tanaka T, Ling H, Yang B, Saitoh H, Zhang L, Qin C (2009) Procyanidins and butanol extract of Cinnamomi cortex inhibit SARS-CoV infection. Antivir Res 82(1):73–81

Zsoldos Z, Reid D, Simon A, Sadjad BS, Peter Johnson A (2006) eHiTS: an innovative approach to the docking and scoring function problems. Curr Prot Peptide Sci 7(5):421–435

# Chapter 10
# Dietary Plants, Spices, and Fruits in Curbing SARS-CoV-2 Virulence

**Mohandass Kaviya, Irudhayaraj Peatrise Geofferina, Prasad Poornima, Anand Prem Rajan, Balasubramanian Balamuralikrishnan, Meyyazhagan Arun, Naif Abdullah Al-Dhabi, Mariadhas Valan Arasu, Pushparaj Karthika, Subramaniam Kallidass, Shanmugam Ramya, and Arumugam Vijaya Anand**

## Abbreviations

| | |
|---|---|
| $3CL^{Pro}$ | 3C like protease |
| ACE-2 | Angiotensin-converting enzyme 2 |
| AgNPs | Silver nanoparticles |
| AIDS | Acquired immune deficiency syndrome |

M. Kaviya · I. P. Geofferina · P. Poornima · S. Ramya (✉) · A. V. Anand (✉)
Department of Human Genetics and Molecular Biology, Bharathiar University, Coimbatore, Tamil Nadu, India
e-mail: avahgmb@buc.edu.in

A. P. Rajan
Department of Bio-Medical Sciences, School of Biosciences and Technology, Vellore Institute of Technology, Vellore, Tamil Nadu, India

B. Balamuralikrishnan
Department of Food Science and Biotechnology, Sejong University, Seoul, South Korea

M. Arun
Department of Life Sciences, CHRIST Deemed to be University, Bengaluru, India

N. A. Al-Dhabi · M. V. Arasu
Department of Botany and Microbiology, College of Science, King Saud University, Riyadh, Saudi Arabia

P. Karthika
Department of Zoology, Avinashilingam Institute for Home Science and Higher Education for Women, Coimbatore, Tamil Nadu, India

S. Kallidass
Department of Animal Science, Manonmaniam Sundaranar University, Tirunelveli, Tamil Nadu, India



J.-T. Chen (ed.), *Ethnopharmacology and Drug Discovery for COVID-19: Anti-SARS-CoV-2 Agents from Herbal Medicines and Natural Products*,
https://doi.org/10.1007/978-981-99-3664-9_10

| | |
|---|---|
| APG | Acylphloroglucinol |
| C-domain | C-terminal domain |
| CHIKV | Chikungunya virus |
| CLR | C-type lectin receptors |
| CoVs | Corona viruses |
| CVB1 | Coxsackievirus B1 virus |
| DENV-2 | Dengue 2 virus |
| DPP4 | Dipeptidyl peptidase 4 |
| EB | Ebselen |
| EBV | Epstein-Barr virus |
| ER | Endoplasmic reticulum |
| FDA | Food and Drug Administration |
| FP | Fluorescence polarization |
| HA | Hemagglutination |
| HAT | Human airway trypsin-like protease |
| HAV | Hepatitis A virus |
| HBV | Hepatitis B virus |
| HCV | Hepatitis C virus |
| HIV | Human immunodeficiency virus |
| HR1 | Heptad repeat 1 |
| HSV-1 | Herpes simplex type I virus |
| mAbs | Monoclonal antibodies |
| MF | Meliae Fructus |
| $M^{Pro}$ | Main proteases |
| NAs | Nucleotide analogs |
| NSPs | Non-structural proteins |
| NTD | N-terminal domain |
| PAV | Pavetannin C1 |
| PIV | Parainfluenza virus |
| $PL^{Pro}$ | Papain-like proteases |
| PRRSV | Porcine reproductive and respiratory syndrome virus |
| PRV | Pseudorabies virus |
| PV | Poliovirus |
| RBD | Receptor binding domain |
| RLV | Rauscher murine leukemia virus |
| RSV | Respiratory syncytial virus |
| RT | Reverse transcriptase |
| RV | Rhinoviruses |
| SFN | Sulforaphane |
| TEN | Tenuifolin |
| TMPRSS-2 | Transmembrane serine protease 2 |
| TRV | Tobacco rattle virus |

## 10.1 Introduction

The recently emerged SARS-CoV-2 virus reported during the year 2019 in December has created a global pandemic and has become a global concern in this present era. It globally affected human health and economic activities (Shanmugam et al. 2020; Kuchi Bhotla et al. 2020; Kuchi Bhotla et al. 2021a, b; Pushparaj et al. 2022; Meyyazhagan et al. 2022). To date, there are no known therapeutic agents to treat this virus except a few Food and Drug Administration (FDA) recommended medications and vaccination (Cannalire et al. 2020). Many reports suggest that good nutritive healthier food will help overcome the infection of SARS-CoV-2 and its prolonged pathological conditions. The phytotherapy (herbalism) is generally defined as the incorporation of plants or herbs in the daily diet that is expected to enhance or modulate the individual's immune system for preventing disease, this phytomedicine therapeutic approach is found to be an affordable and safer alternative which may also minimize the occurrence of drug resistance and will typically improve the effort of the world to fight against deadly SARS-CoV-2 which is illustrated in Fig. 10.1 (Goyal and Goyal 2020). It has been well confirmed in peer research that natural phytochemicals have the potentials to combat COVID-19 and its pathogenicity without any side effects. During this period of global fretfulness, it is pertinent to implement enduring measures to battle these awful viruses. A better understanding of the basic structural targets on the virus surface facilitates the invention of the drug, which is a novel safe and effective targeted therapeutic

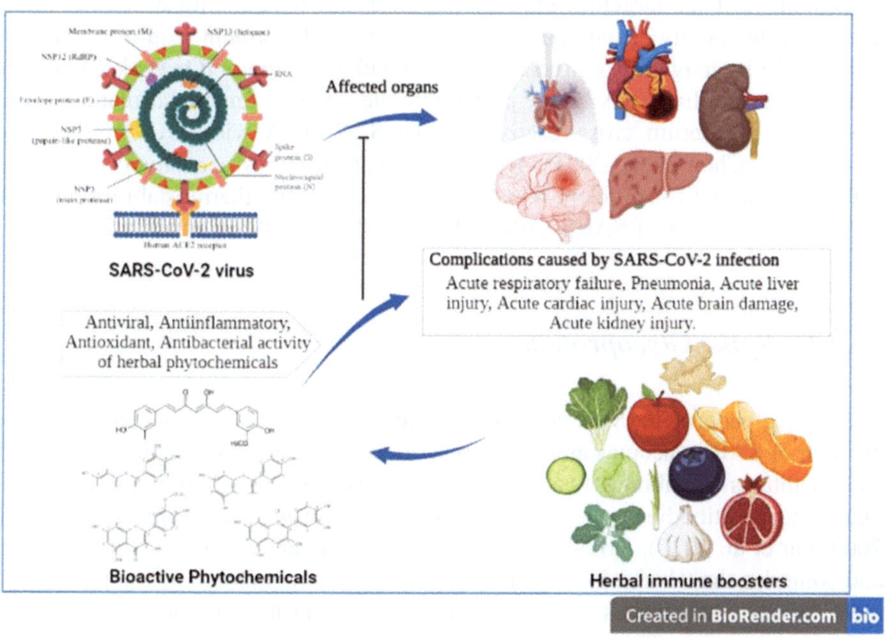


**Fig. 10.1** The dietary plants, spices, and fruits in curbing SARS-CoV-2 virulence

approach. Hence, this current chapter aims to summarize the herbal immune boosters, fruits, culinary spices, and their nutritive compounds facilitating the immune defensive mechanism for counteracting infection of SARS-CoV-2 and this chapter will elaborate on the basic structural targets of SARS-CoV-2 virus (Ou et al. 2020).

## 10.2 Major Structural Targets of the SARS-CoV-2 Virus

The viral protease enzymes are the vital targets effectively render over SARS-CoV-2 focus on the main protease enzyme ($M^{Pro}$) enzyme, RNA-dependent RNA polymerase (RdRp), spike glycoprotein, 3 chymotrypsin-like proteases ($3CL^{Pro}$), and papain-like protease ($PL^{Pro}$) which comes up with the high degree of conservation across corona viruses emerging for the identification of the broad spectrum of inhibitors (Cannalire et al. 2020). The two crucial proteases, $M^{Pro}$ and $PL^{Pro}$ are notable for viral replication (Goyal and Goyal 2020). $M^{Pro}$ enzyme is recognized as a $3CL^{Pro}$ and contains a high level of structural and sequential resemblance to SARS-CoV $M^{Pro}$ (Zhang et al. 2020). $3CL^{Pro}$ plays a significant role in the enzymatic action of replicase polyproteins to inhibit SARS-CoV (Peele et al. 2020). $PL^{Pro}$ a viral protease enzyme that are essential for viral polypeptide (Thiel et al. 2003) processing and conduct the antiviral reactions ubiquitin enzyme in the host defense mechanisms (Keng et al. 2011). The spike (S) glycoprotein vital role mediates the entry of the virus to the host cell target for immune responses and neutralizing antibodies of the Coronaviridae family (Tang et al. 2020). The viral spike protein requires a protease cleavage for its activation to proceed to fusion with the receptor (Ou et al. 2016). The $3CL^{Pro}$ cleavage functions as the proteolytic cleavage of functional maturation of pro-activity antiserum cross-reactive against SARS-CoV virus $3CL^{pro}$ (Froggatt et al. 2020; Chen et al. 2021). RdRp represents itself as a catalytic subunit to cofactors required for efficient boosting which lowers the thermostability with the polymerase complex of SARS-CoV-2 (Peng et al. 2020; Wang et al. 2020).

### *10.2.1 Spike Glycoprotein*

The spike glycoprotein (S) of corona viruses is one of the most crucial targets for the creation of SARS vaccines and treatments because of efficient receptor recognition and facilitates membrane fusion and viral entry and also a primary target for neutralizing antibodies and also a key factor in coronavirus tropism (Du et al. 2009; Ou et al. 2020). Trimeric S protein monomers are roughly 180 kDa in size and comprise the S1 and S2 subunits, which mediate membrane fusion and attachment, respectively. Both C and N terminal sections in the S1 fold are two distinguished structures. The receptor-binding domain of a virus can either be the N-terminal domain (NTD) or the C-domain (receptor binding domain). In case of activation of the spike protein for SARS-CoV-2 and Middle East respiratory syndrome coronavirus (MERS-CoV) the protease cleavage is involved in a sequential manner where there exists initial cleavage between S1 and S2 and next is the cleavage that activates the S2' site.

The cleavage process takes place utilizing the host protein-cleaving enzymes like trypsin, furin, transmembrane serine protease 2 (TMPRSS-2), cathepsins, TMPRSS-4, or human airway trypsin-like (HAT) protease, based on viral strains and cell types (Ou et al. 2020). The ability of the SARS-CoV-2 to migrate the species barrier is also significantly influenced by spike protein since the adaptive evolution of S protein can end up transmitting the SARS-CoV from animals to humans. The S protein, similar to other proteins seen in various other viral strains, is produced and placed in an inactive state in viral particles. During viral infection, a host protease cleaves the S1 and S2 subunits, stimulating them and permitting membrane attachment including a viral entry in the formed precursor that is not yet activated and present inside the envelope of the virus. A cellular protease cleaves the S protein at the time of infection of the cell, releasing it. The S1 subunit is an essential factor in host cell identification and adhesion and the most plausible target for neutralizing antibodies and creating medicines due to its host the receptor binding domain (RBD) which is essential for angiotensin-converting enzyme 2 (ACE-2) binding. The S2 subunit is involved in viral entrance and membrane fusion.

It has three important regions to establish this function: fusion peptide (FP, 788–806 residues), heptapeptide repeated sequence 1 (HR1) (912–984 residues), and heptapeptide repeated sequence 2 (HR2) (1163–1213) (Souza et al. 2022). SARS-CoV-2 S protein attaches the virus to the host cellular receptor, ACE-2. This contact is directed by a designated RBD on S. The crystal structure of the RBD associated with the human ACE-2 is said to be a protease domain at 2.9-angstrom resolution revealing that the RBD has a mild concave-like surface that carries the peptidase's N-terminal lobe. The atomic features at the two proteins interface highlight the significance of residue modifications that allow for effective cross-species associated infection and same species transmission (Li et al. 2005a, b). Modification of the spike can modify cell and tissue tropism and, in rare situations, can change virus pathogenicity when combined with other viral and host variables (Belouzard et al. 2012). The S protein is the key determinant of CoV tropism. The trimeric structure of the S glycoproteins on the viral surface allows for the attachment to a cell surface receptor and membrane fusion (Hulswit et al. 2016; Du et al. 2017). The viruses belong to the category of intracellular parasites, where they interact with the host receptors to establish their infection, the SARS-CoV-2 also initiates its entry by employing interaction with the ACE-2 receptor of host ACE-2.

ACE-2 has also been discovered as a SARS-CoV and human coronavirus NL63 functional receptor. The RBD of the S1 subunit attaches to the ACE-2 and this is a reason for S1 to be primed at the S1/S2 junction through host proteases which include furin or endosomal proteases, revealing a secondary site (S2′) is cut by the TMPRSS2. The cleavage irreversibly changes the shape of the removed protein S, allowing the subunit of S2 to enter into the host membrane and direct the fusing of the virus and host membranes (Bakhiet and Taurin 2021). The S protein is notable for its extensive N-linked glycosylation, which covers a high part of the protein's surface area and exhibits conformation-dependent dynamic alterations. Aside from shielding, N-glycans, particularly at the N165 and N234 locations, are implicated in the modification of RBD confirmation, which might be utilized as a therapeutic target. Furthermore, it has a disulfide-bonded external N-terminus, a transmembrane domain, and a palmitoylated short intracellular C-terminus. Since it is a key determinant of the host immune response and is involved in the pathogenesis process through the virus through the activation of the endoplasmic reticulum (ER) stress response, any alterations due to mutations may result in altered pathogenesis (Yadav et al. 2021a, b). The activation of cathepsin L inside the endocytic vesicle results in the priming of the spike proteins, which results in the release of the viral RNA genome into the cytoplasm (endocytic pathway).

The TMPRSS2however, is crucial in priming the spike proteins at the cell surface, allowing the viral DNA to enter (non-endocytic pathway). Once inside the cytoplasm, the viral RNA ORF1ab portion is converted to the replicase polyprotein PP1ab, which is reacted at by enzymes secreted by the virus such as $PL^{pro}$ and $3CL^{Pro}$ ($M^{Pro}$) to create nsps 1–16, including RdRp and helicase, which are essential for virus replication. Protein S binds to ACE-2 and assaults target cells including bronchial cells, nasal epithelial cells, and pneumocytes. The RBM-ACE-2 link activates the S protein, revealing the fusion peptide and causing it to assume a more favorable shape with epistatic areas, permitting the entry of the virus to the host cell. The TMPRSS2 protein is a protease enzyme that works similarly to the ACE-2 receptor in binding to the spike protein of SARS-CoV-2. The furin site is present in the usual coronavirus which contains a unique sequence of amino acids between S1 and S2 in SAR-CoV-2, consisting of Pro-Arg-Arg-Ala-Arg (Pro: proline; Arg: Arginine; Ala: Alanine). This sequence separation is critical for the effective absorption of lung cells. Without the furin site, the virus appears less infectious (Candido et al. 2022).

The interaction of the RBD in the S protein of the SARS-CoV-2 ACE-2 host receptor, as well as subsequent replication in the coronavirus S protein, leads the virus to invade the host cell. Locus E (LY6E) of the human lymphocyte antigen 6 complex restricts CoV entrance inside the host membrane receptor by interaction with the gene of humans and promoting S protein-directed membrane attachment. It has also been demonstrated that some natural formulations block the S protein from attaching to the human receptor present in the cell (Nandi et al. 2021). SARS-CoV-2 S protein may bind to various innate immunological receptors, including C-type lectin receptors (CLRs). Through a $Ca^{2+}$-dependent interaction, CLRs bind to particular glycans. Several receptors, including L-SIGN/CD209L/CLEC4M, DC-SIGN/CD209, mannose receptor/MR/MRC1/CD206, Dectin-2/CLEC6A, and MGL/CLEC10A/CD301 are found to have a maximum expression in the including monocytes, dendritic cells, immune system, and macrophages of the humans and act as the first line of defense against invading pathogens. CLRs such as DC-SIGN can influence toll-like receptor-induced activation and so direct glycan-specific reactions of host immune systems against infections (Shajahan et al. 2021).

### 10.2.2 Main Protease Enzyme ($M^{Pro}$) Enzymes

SARS-CoV-2 $M^{Pro}$ is an important protease of CoV-2 which has been essential for mediating the replication of the virus and its transcription. It has also served as a potent target for the designing and development of antiviral agents against SARS-CoV-2 (Sabbah et al. 2021). The inhibition of the main protease helps prevent the multiplications of viruses and so can be used in the discovery of targeted drugs (Rudrapal et al. 2022). Pfizer's newly approved nirmatrelvir serves as a potent inhibitor of viral main protease functioning in the therapeutic front as well as the context of prophylaxis against infection (Mótyán et al. 2022). The mimetic of the peptide and inhibitor of the covalent coronavirus protease has the ability the inhibition of active sites. The drug invention for inhibiting the potential protease possesses various strategies including the observation of minimum molecular weight compounds for the inhibition of protease, good fragments from peptidomimetic compounds, including the optimal fragments for the lead monitorization and final selection and improvisation of hybrid molecules for the required pharmacokinetic and pharmacodynamics properties. The good inhibitors of the main protease of SARS-CoV-2 are derived with the help of in silico fragment-based drug design (Teli et al. 2022). It is strongly inhibited by nine ebselen derivatives (EBs) with $IC_{50}$ 0.07–0.38 μM.

Among this EB2–7 a greater inhibition of the replication of virus SARS-CoV-2 of HPAepiC cells with an $IC_{50}$ value of 4.08 μM. Thus, it is concluded that EBs may end up in the discovery of drugs for COVID-19 and SARS-CoV-2 infections (Qiao et al. 2021). Three phytochemical components, namely with $M^{Pro}$ binding site licoleafol, epi-catechin gallate, and silibinin are said to be effective against CoVID-19 resulting in higher binding affinity. Thus, they can be used as an effective anti-CoVID-19 drug (Mukherjee et al. 2022). His41, Cys145, and Glu166 are the three compounds identified and subjected to in-depth augmentation deep reinforcement learning, docking for covalency, and imitations of molecular dynamics through which they are tested for their potential ability of inhibition against SARS-CoV-2 by forming high-frequency contacts with key amino acids (Zhang et al. 2022). Four cyclic peptides were scrutinized utilizing virtual screening, of which MN-2 exhibited a greater affinity to $M^{Pro}$. Thus, a cyclic peptide that is not covalent MN-2 serves as a specific factor to inhibit omicron (Yin et al. 2022).

### 10.2.3 Papain-Like Protease ($PL^{Pro}$)

$PL^{pro}$ is the coronavirus proteins that are considered specific agents against the virus as they are needed for the replication of the virus. Thus, targeting of $PL^{pro}$ with various antiviral drugs prevents the replication of viruses including the inhibition of their regulation of the signaling cascades of the cells which may lead to lethality as a result of infection (Baez-Santos et al. 2015). It also targets the immune response including posttranslational ubiquitin and ISG15 modifications. The replication cycle takes place in the virus and the responses of the host immune in association with $PL^{pro}$ lie as a specific chance in the drug delivery (Ullrich and Nitsche 2022). It also plays a significant activity in the cleaving and the maturation of viral proteins, the arrangement of the replicase-transcriptase complex along with the destruction of host responses (Osipiuk et al. 2021). Furthermore, it provides a virulence trait among members of the same species of SARS-CoV (Niemeyer et al. 2018). The proteolytic processing of SARS-CoV-2 is carried out by the $PL^{pro}$ and is composed of one of the main targets examined for the pharmacological intervention of in silico methodologies (Rieder et al. 2022). It can be reversibly inactivated by various covalent inhibitors such as VIR251 ligand through the development of covalent bonds with remnants of catalytic site (Cys111) employing Michael addition reaction.

The mechanism of inhibition undergoes four stages including entry of the ligand inside the pocket for protease pocket, Cys111 removal of protons from the thiol group through Brønsted-Lowry base, the addition of Cys111-S-at the ligand and exchange of proton from the protonated base to the covalently bound ligand (Hognon et al. 2022). Three phenolic compounds bound with $PL^{pro}$ were found an increase of $PL^{pro}$ mutation resulting in the evolution of recent variants of SARS-CoV-2 preventing the preliminary molecular communications to ISG15 by evaluating a natural compound library which was scrutinized by X-ray screening and complexed to $PL^{pro}$ revealed best inhibitor $PL^{pro}$ in an observation by deISGylation activity assay. Increased mutation of $PL^{pro}$ results in generation of novel variants of SARS-CoV-2 in spite of some natural compounds found to restore the immune response against the virus of the host cell in COVID-19 infections (Srinivasan et al. 2022).

A zinc finger domain is present in $PL^{pro}$ which is essential for the attachment of the substrate and maintaining the stability of the structure. Zinc had a greater curbing potential on $PL^{pro}$, followed by calcium, manganese, iron, and magnesium that had minimal or null effects on the activity of $PL^{pro}$ which was confirmed by the attachment of zinc and manganese with $PL^{pro}$ through tryptophan intrinsic fluorescence analysis (Shetler et al. 2022). The invention of inhibitors of $PL^{pro}$ is carried out by vigorous and simple sandwich-like fluorescence polarization (FP) screening assay and anacardic acid was found to be an effective inhibitor against $PL^{pro}$ in vitro (Yan et al. 2022). The molecular rules the $PL^{pro}$ substrate specificity and provides an outline for the development of inhibitors with an effective therapeutic value of drug designing (Rut et al. 2020).

### *10.2.4 RNA-Dependent RNA Polymerase*

RdRp role in transcription and replication of the genetic material of the virus is identified as an appealing target for developing innovative antiviral methods. Though the genetic resemblance of SARS-CoV-2 to SARS-CoV (79%) and MERS-CoV (50%) is less, the RdRps of these three variants is largely conserved, indicating this protein is a promising broad-spectrum antiviral target for corona viruses (Wang et al. 2021a, 2021b). RdRp, also known as nsp12, catalyzes the synthesis of viral RNA and hence plays a critical role in both replication and transcription of SARS-CoV-2 genetic material, perhaps with the help of cofactors nsp7 and nsp8. As a result, nsp12 is regarded as a prime target for nucleotide analog antiviral agents (Gao et al. 2020). The smallest core component for viral RNA replication is specified as nsp12-nsp7-nsp8. The SARS-CoV-2 RdRp complex is made up of core catalytic unit nsp12, a heterodimer of nsp7-nsp8 (nsp8–1), and an extra subunit nsp8 (nsp8–2). A b-hairpin (residues V31 to K50) and a domain of nidovirus-specific extension are found at the N-terminus of nsp12 (NiRAN, residues S115 to A250). RdRp core, palm subdomain, and NiRAN sandwich the b-hairpin.

The C-terminal catalytic domain (residues L366 to F920) of nsp12 communicates with the interface subdomain of NiRAN (residues L251 to R365). The RdRp active site in SARS-CoV-2 is constituted by seven conserved A to G catalytic motifs. The RdRp, palm subdomain has five of these A-E motifs, whereas the remaining two F and G motifs are contained in the finger subdomain (Jiang et al. 2021). Most of the viral polymerases, notably Hepatitis C virus (HCV) NS5B (residue D220) and poliovirus (PV) polymerase include the typical residue of divalent-cation-binding D618, which is found to be conserved in motif A (residue D233). From 759 to 761 catalytic residues of RdRp protein is necessary for viral transcription and are mostly conserved in motif C of viral RdRps, which binds to 3 prime ends of RNA (from 317 to 319 in HCV and from 327 to 329 in PV) (Vicenti et al. 2021). When the virus infects the host cell, it contributes to the creation of the replication machinery of the viral genome by interacting with different components.

RdRP controls the elongation of the RNA strand, which involves the insertion of nucleotides of hundreds to thousands. Once incorporated into the freshly produced RNA strand, nucleotide analogs (NAs) can block the RNA elongation mediated by RdRP. Transcription of RNA by RdRP from the 3′ ends of RNA as a template can be primer-dependent or primer-independent. Primer-independent RNA synthesis by RdRP may start synthesizing with nucleotides without primer. Primer-dependent, RNA synthesis by RdRP however, makes use of primers that are short oligonucleotide or a protein that covalently attaches to nucleotides as a primer; the primer anneals with the template RNA to give a starting point for RNA synthesis (Zhu et al. 2020). RdRp inhibitors are classified into two types based on their structures and mechanisms of action nucleoside analog inhibitors that act at the substrate site and non-nucleoside inhibitors that engage with allosteric binding sites (Xu et al. 2022). RdRp incorporates NAs into nascent viral RNA, which subsequently either terminates or slows down RNA production, exerting antiviral effects.

Many inhibitors of NA are investigated against SARS-CoV-2 causing COVID-19 from the beginning of the outbreak. After entering the host cell, antiviral nucleoside prodrugs targeting the viral protein RdRp must convert into 5′-triphosphates, which compete with nucleotide triphosphates that are endogenous in origin as viral RdRp substrates (Zhao et al. 2021). Between its fingers and thumb subdomains, the first turn of the RNA is bound to the nsp12 subunit. The active site of the palm subdomain of RdRp is made up of five conserved motif components of nsp12,

A-E. The 3′ ends of the RNA are attached to motif C which contains the RNA synthesis residues D760 and D76110, 14. The extra motifs F and G found in the fingers subdomain of nsp12 are responsible for RNA template positioning. The interactions of nsp12 with newly transcribed RNA strands suggest that short RNA is retained throughout the early stages of RNA production (Hillen et al. 2020).

### 10.2.5 3-Chymotrypsin-like Protease ($3CL^{Pro}$)

$3CL^{Pro}$ also called $M^{Pro}$ is a structurally conserved protease among CoVs responsible for the SARS-CoV-2 genome replication (Anand et al. 2003; Mody et al. 2021; Nascimento Junior et al. 2020). Two replicative polyproteins (PP1a and PP 1 ab) are cleaved by $3CL^{Pro}$, this protein is encoded by the viral replicase gene along with $PL^{pro}$ to generate non-structural proteins (NSPs) that are pivotal for viral replication (Anand et al. 2002; Mody et al. 2021). NSP4 to NSP16 are generated via the activity of $3CL^{Pro}$ (Deng et al. 2014). The functional replication complex is produced by the activity of the proteinases (Anand et al. 2002). It is an attractive target for therapeutic approaches against the SARS coronavirus (Anand et al. 2003). The use of $3CL^{Pro}$ proteinase inhibitors is the most common treatment and several covalent and non-covalent inhibitors have been found (Ghosh et al. 2008; Nascimento Junior et al. 2020). They can successfully prevent viral replication and many such protease inhibitors have been created and found to be effective in CoV cell cultures (Deng et al. 2014).

Peptoids and non-peptidomimetics are the two types of $3CL^{Pro}$ inhibitors (Liu et al. 2020). The peptidomimetic $3CL^{Pro}$ inhibitors bind covalently to Cys-145 residue in the active site of $3CL^{Pro}$ by utilizing electrophilic warhead groups (Ghosh et al. 2008; Liu et al. 2020). Both aldehyde compounds 11 and 12 and a-ketoamide compounds 25–27 are some peptidomimetic inhibitors with anti-SARS-CoV-2 effects, while compounds 29a-d (decahydroisoquinoline and octahydroisochromene derivatives), compounds 39, 40a, 40b (inhibitor with 3-pyridyl or triazole moiety) are some of the nonpeptidic $3CL^{Pro}$ inhibitors (Liu et al. 2020). Antiviral activity against SARS-CoV is exhibited by a 3-chloropyridyl ester-based $3CL^{Pro}$ inhibitor developed by Ghosh et al. in cell culture assays (Ghosh et al. 2008). PF-07321332 also called nirmatrelvir is a derivative of PF-00835231 (a hydroxymethyl ketone derivative) with the activity of anti-COVID-19 in 89% of patients (Reina and Iglesias 2022). Some of the drugs that target the $3CL^{Pro}$ enzyme with high specificity include boceprevir, ombitasvir, paritaprevir, tipranavir, ivermectin, and micafungin which can be used either alone or in combination with other drugs (Mody et al. 2021). But corona viruses that encode double-mutant $3CL^{Pros}$ are found to be resistant to these $3CL^{Pro}$ inhibitors as they exhibit delayed processing of the replicative polyproteins along with reduced sensitivity to the viral inhibitors and their mechanism of pathogenicity has been demonstrated in the study by Deng and their colleagues (Deng et al. 2014).

## 10.3 The Plausible Role of Plants and Herbs in Counteracting the Risk and Improving the Treatment of COVID-19 by Targeting SARS-CoV-2

It has been a timeworn practice of utilizing herbal compounds in developing drugs. Medicinal plants that biologically yield active compounds have great interest among scientists, as they are being significant in preventing the human epidemic as illustrated in Table 10.1 (Umesh et al. 2021). In the entire world, the medicinal plants, herbs, and spices that belong to South Asian countries especially India have greater therapeutic usefulness and hence are contemplated as a generous pivot of ingredients (Halder et al. 2022; Natesh et al. 2021). The bioactive compound from plants has exhibited antiviral, anti-inflammatory, and antibacterial activities against various viruses, including the Zika virus, chikungunya virus, hepatitis, dengue virus, influenza viruses, and SARS-CoV. In vitro studies have reported that those compounds obstruct the replication of SARS-CoV. Recently, the data generated using computational approaches gave very encouraging results that the potential bioactive compounds from the herbal sources function as opposed to the targets of SARS-CoV-2 (Natesh et al. 2021; Halder et al. 2022). The discovery of a new drug is time-intensive and normally takes a definite number of years for clinical approval. In this deadly illness, such medicinal plants are particularly attractive in such cases where there is no known possible substitute medications or vaccination (Nargis Begum et al. 2009; Vijayakumar et al. 2015, 2019; Manikandan and Anand 2016; Govindharajan et al. 2020; Vijaya Anand et al. 2020; Halder et al. 2022). The current study tries to explore, whether Indian spices can be used as novel agents for controlling this pandemic (Umesh et al. 2021). It has been proclaimed that target protein inhibition may cause attenuation of viral replication or minimize the virus infectivity. Further it requires confirmation on the effect of bioactive spices on minimizing the virus infectivity from in vivo and in vitro studies (Arumugam et al. 2020; Saravana Prabha et al. 2020; Anand et al. 2021; Kuchi Bhotla et al. 2021a, b; Chandra Manivannan et al. 2022a, b; Halder et al. 2022).

### *10.3.1* Glycyrrhiza glabra

*Glycyrrhiza glabra* is a blooming plant that belongs to the bean family called Fabaceae, which is denoted by the common names of Liquorice, Licorice, or Black Sugar. In India, *Glycyrrhiza glabra* is known as Athimaduram. The native liquorice plant is in the Western part of Asia, North part of Africa, and the Southern European parts. The root of liquorice contains glycyrrhizin, a substance that is 50–170 times sweeter than sucrose. *Glycyrrhiza glabra* root is used in the form of powder to treat SARS-COVID (Lim 2015). It is reported that the *Glycyrrhiza glabra* has an in vitro anti-SARS effect against SARS-COVID and the natural compounds

**Table 10.1** Herbal immune boosters, active compounds, and their mechanism of action against viruses

| Botanical name | Common name | Major chemical constituent | Edible parts | Virus type | Mechanism of action | Reference |
|---|---|---|---|---|---|---|
| *Glycyrrhiza glabra* | Liquorice, mulethi | Glycyrrhizin | Leaves, roots | SARS-CoV | Inhibits viral adsorption, penetration, and replication | Patel et al. (2021), Cinatl et al. (2003), Malabadi et al. (2021) |
| *Areca catechu* | Areca nut | Arecoline, arecaidine, guvacine, guvacoline | Dried ripe seed | SARS-CoV-2 | Inhibits viral RNA levels and infectious titers | Khan et al. (2021) |
| *Armeniaca sibirica* | Apricot | Sibiriaester A, sibiriaester B | Leaves, seed | SARS-CoV-2 | PGE2 gets suppressed and produces NO | Wu et al. (2018), Khan et al. (2021), Mukherjee et al. (2022) |
| *Rauvolfia serpentine* | Indian snakeroot | Reserpine | Dried root | SARS-CoV-2 | Inhibits the SARS-CoV-2 replication | Khan et al. (2021), Malabadi et al. (2021) |
| *Curcuma* spp. | Turmeric | Curcumin, eugenol | Rhizome | SARS-CoV-2 | Inhibits $3CL^{Pro}$ of SARS-CoV-2; has good binding affinity with $M^{Pro}$ and S protein | Malabadi et al. (2021) |
| *Allium cepa* | Onion | Flavonols: Quercetin, quercetin glycosides, kaempferol | Bulb | SARS-CoV-2 | Interfere with entry and replication cycles such as $PL^{pro}$, $3CL^{Pro}$, and NTPase/helicase | Bondhon et al. (2021) |
| *Brassica oleracea* | Broccoli | Glucosinolate type sinigrin, sulforaphane | Flower | SARS-CoV-2 | Blocks the cleavage process of $3CL^{Pro}$ | Gasparello et al. (2021) |
| *Psidium guajava* | Guava | Eugenol | Leaves, fruits | SARS-CoV-2 | Binds with $M^{Pro}$ and S protein to interfere with virus | Malabadi et al. (2021) |
| *Cinnamomum verum* | Cinnamon | Butanol, procyanidins | Dried bark | SARS-CoV-2 | Blocks the entry of virus cells through endocytosis | Malabadi et al. (2021) |
| *Salvia officinalis* | Sage | b-ocimene, 1,8-cineole, a-pinene, b-pinene | Essential oils | SARS-CoV-2 | Obstructs the interaction of virus and host receptor and also attachment of cellular membrane and the viral envelope | Babaeekhou et al. (2021) |

| | | | | | | |
|---|---|---|---|---|---|---|
| *Artemisia annua* | Annual wormwood | Artemisinin, deoxyartemisinin, artemisinic acid, arteannuin-B, stigmasterol, friedelin, friedelan-3-β-ol, artemetin, quercetagetin, 6,7,3′,4′--tetramethyl ether | Whole plant | SARS-CoV-2 | Inhibits the $3CL^{Pro}$ tease of the virus and blocks the viral RdRp activity | Lin et al. (2014), Khan et al. (2021) Malabadi et al. (2021) |
| *Lycoris radiate* | Red spider lily | Lycorine, glycyrrhizin | Stem | SARS-CoV-2 | Against viral strains BJ-001 and BJ-006 activity | Li et al. (2005a, 2005b), Khan et al. (2021), Malabadi et al. (2021) |
| *Glycyrrhiza uralensis* Fisch. | Licorice | Echinatin | Root | SARS-CoV-2 Human respiratory syncytial virus (HRS) | Disrupts the viral uptake into the host cells and the interaction between RBD of SARS-CoV2 and ACE-2 | Khan et al. (2021) |
| *Rosmarinus officinalis* | Rosemary | Oleanolic acid | Leaves | SARS-CoV-2, herpes viruses, HIV, influenza | Inhibit the proliferation of SARS-CoV-2 | Nolkemper et al. (2006), Khan et al. (2021) |
| *Zingiber officinale* | Adrak, ginger rhizome | Arcurcumene, β-sesquiphellandrene, α-zingiberene and β-bisabolene, flavan, and 4, 6-dichloroflavan [6]-gingerol | Rhizome | SARS-CoV-2 *Rhinovirus* IB | Blocks TMPRSS2 receptor | Khan et al. (2021), Malabadi et al. (2021) |
| *Origanum vulgare* | Oregano | Carvacrol | Essential oil | SARS-CoV-2, HRS virus | Inhibits SARS-CoV-2 spike RBD crystal structure | Gilling et al. (2014), Khan et al. (2021), Fallah et al. (2021) |
| *Hyoscyamus niger* | Nightshade; ajwain kharasani | Acyclovir | Flower | SARS-CoV-2 Influenza A | Acts as a bronchodilator and has inhibitory effects on the $Ca^{2+}$ channel. | Khan et al. (2021), Kosari et al. (2021) |

(continued)

**Table 10.1** (continued)

| Botanical name | Common name | Major chemical constituent | Edible parts | Virus type | Mechanism of action | Reference |
|---|---|---|---|---|---|---|
| *Allium sativum* | Garlic | Allicin, alliin, diallyl sulfide, diallyl disulfide | Root, leaves | SARS-CoV-2, HIV, HSV-1, influenza A and B, | Formation of hydrogen bonds between viruses and bioactive compound | Bayan et al. (2014), Khan et al. (2021), Khubber et al. (2020) |
| *Solanum nigrum* | Black nightshade | Solanine, solamargine | Leaves | HCV | Inhibits NS3 protease | Patel et al. (2021) |
| *Moringa oleifera* | Drumstick tree | Quercetin, linolenic acid | Leaves, seeds, tubers, flowers, pods | HIV type 1 | Inhibits viral replication | Patel et al. (2021), Malabadi et al. (2021) |
| *Justicia adhatoda* | Bansa, malabar nut, adhatoda | Phenols, tannins, alkaloids, anthraquinone, saponins, flavonoids, | Leaves | Influenza virus, herpes simplex virus-2 (HSV 2) | Inhibit the cleavage activity of the virus | Chavan and Chowdhary (2014), Khan et al. (2021), Malabadi et al. (2021) |

present in *Glycyrrhiza glabra* will hinder the host fusion steps of the virus through the viral envelope. Glycyrrhizin acts on the cellular signaling pathways like protein kinase C; casein kinase II; and with the activator protein I and nuclear factor κB (Cinatl et al. 2003). Antiviral compounds of *Glycyrrhiza glabra* can inhibit or block the viral replication cycle during the virus attachment to the cells, also interfere with the viral enzymes, or suspend the replication of the viral genome (Malabadi et al. 2021).

### *10.3.2* Areca catechu

*Areca catechu* is a palm species that belongs to the family called Arecaceae. It grows in the region of the Pacific, Asia, and east Africa. The common names of areca are areca palm, betel nut palm, supari, Indian nut, Pinang palm, and catechu. This palm, in English, is known as the betel tree as its fruit, the areca nut is frequently used with the betel leaf. The seeds of *Areca catechu* (areca nut) have been in use for clinical practice in China and India. Raw areca and charred-type areca nuts are very commonly used (Peng et al. 2015). It is a widespread traditional medicine that is chewed to separate the fluid collected in the alimentary canal and to kill the worms (Ansari et al. 2021). It inhibits the levels of viral RNA and infectious titers of SARS-CoV-2 (Ngwe Tun et al. 2022). Kaempferol, a compound of areca, showed the best binding affinity to the 7BQY and 6Y2FB enzymes (Emon et al. 2021).

### *10.3.3* Armeniaca sibirica

*Armeniaca sibirica* belongs to the Rosaceae family. It is being cultivated as a medicinal edible plant, which is well known as "apricot." It is domestic to the climates of temperate, continental, and mountainous which includes the regions of Northeast China, Southeast Mongolia, and the East part of Siberia. The dried leaves are used for medicinal purposes (Wu et al. 2018). Wild apricot *Armeniaca sibirica* is the kind of bitter apricot. Leaves of *Armeniaca sibirica* were taken up as a natural tea, as it is rich in antioxidants, also it is rich in dietary fiber, which has health-promoting properties (Zhang et al. 2018). *Armeniaca sibirica* is said to be a SARS-CoV-2 protease inhibitor, an in silico approach called AutoDock Vina, had revealed that *Armeniaca sibirica*, has the highest binding affinity with SARS-CoV-2 spike (S) protein and may help to prevent COVID-19. It has antiviral properties and can prevent SARS-CoV-2 from infiltrating host cells (Yunus 2021). It inhibits the replication of SARS-CoV-2 and induces apoptosis in infected cells (Qin et al. 2019).

### *10.3.4* Rauvolfia serpentine

*Rauvolfia serpentine* belongs to the family Apocynaceae. This family contains 50 species that are found worldwide in the Himalayas, Indian peninsula, Burma, Indonesia, and Sri Lanka and are indigenous to India, Bangladesh, and Asia. This plant is generally known as Sarpagandha, Chandrabagha, Snakeroot plant, Chotachand, Chandrika, and Harkaya (Kumari et al. 2013). *Rauvolfia serpentine*, its dried root can be used as green tea and is a common plant that can be used by COVID-19 infected patients and not serious patients during the isolation period. Reserpine, an alkaloid isolated from the roots of *Rauvolfia serpentina* (Indian snakeroot) inhibits the activity of the $3CL^{Pro}$ enzyme, inside the viral cells (Khan et al. 2021). The molecules of *Rauvolfia serpentina* showed anti-CoV activity via multiple mechanisms such as suppression of $PL^{pro}$, the relation between ACE-2 and SARS-CoV (S) protein, viral polymerase, virus replication/cell division, viral attachment, and penetration stages. Besides, the compounds were found to be active in blocking the virus binding to the host cell, 3a channel, and the viral entry to host cells (Semwal et al. 2020).

### *10.3.5* Curcuma *spp.*

*Curcuma* belongs to the family called Zingiberaceae, which comprises perennial and rhizomatous species, and is naturally found in tropical and subtropical regions. Turmeric is an evergreen plant that grows mostly in India, China, Thailand, Malaysia, Indonesia, and Northern Australia (Ayati et al. 2019). Curcumin, a compound found in *Curcuma* is shown to have an antiviral activity; it is an effective inhibitor of Epstein-Barr virus (EBV) and human immunodeficiency virus (HIV). It has anti-SARS-CoV 2 components and the strong binding affinity of those active components to the COVID-19 6 LU7 and 6Y2E proteases was expressed (Manzano-Santana et al. 2021). The mechanism of actions as anti-SARS-CoV-2 of the molecules of *Curcuma* causes the inhibition action in viral transcription and replication by binding to the N protein; prevents the invasion of the virus into the host cell by inhibiting the binding of S protein to ACE-2 receptors; inhibits the viral transcription and maturation by inhibiting the function of $3CL^{Pro}$; and binds directly to ACE-2 receptor (Hendrayana et al. 2021). The docking studies states that the compounds of *Curcuma* (curcumin, quinine, and demothoxycurcumin) exhibit strong binding affinity toward RdRp with the binding energy of −7.80, −7.80, and −7.64 kcal/mol, respectively, *Curcuma* has potential to develop to be the future anti-SARS-CoV-2 (El-Aziz et al. 2022).

### *10.3.6* Allium cepa

*Allium cepa* comes from the family of Amaryllidaceae. This plant is generally referred to as the onion bulb or common onion. It is exclusively cultivated in Central Asia and Iran. The phytochemical compounds of this plant exhibit viral inhibition of non-enveloped viruses by prohibiting the process of absorption or penetration. The primary mechanism is the blockage of cellular receptor sites, thus inhibiting the virus invasion. The secondary mechanism is the exhibition of neutralizing effects caused by the inhibitors that interact with the viruses (Harazem et al. 2019). Quercetin and kaempferol show crucial antiviral activity against herpes simplex type I virus (HSV-1), rabies virus, PV, mengo virus, pseudorabies virus (PRV), sindbis virus, parainfluenza type 3virus. The in vitro study data reveals that flavonol quercetin obstructs pulmonary viral replication by minimizing the viral titer (Pareek et al. 2017). Amarogentin represents the greater affinity toward the spike glycoprotein of coronavirus with almost 9 hydrogen bonds together with amino acid residues, namely Thr739, Thr768, Gln314, Val736, Cys738, Ser735 and Arg765 also it exhibited the greater MolDock score of 149.76 kcal/mole (Hendrayana et al. 2021).

### *10.3.7* Brassica oleracea

*Brassica oleracea* belongs to the Brassicaceae family, which is previously called the Cruciferae. It is native to the Mediterranean region and Southwestern Europe, Northward to Southern England, growing on Seaside cliffs (Ferreres et al. 2005). It is a biennial life cycle type of plant (Rodríguez et al. 2015) Astragalin and coumaroylquinic acids and the compounds from *Brassica oleracea* consist of prospective inhibitory ability averse to (S) protein. The compounds of *Brassica oleracea* exhibit a remarkable affinity toward the S2 domain of spike protein (Jose et al. 2022; Illian et al. 2021). Sulforaphane (SFN), a naturally occurring nutritional supplement found in abundant quantities in cruciferous vegetables was said to have antiviral activity. The in vitro studies reveal that SFN administration is a propitious therapeutic approach as it obstructs the viral replication of four strains of SARS-CoV-2 and the seasonal coronavirus HCoV-OC43. On earlier intranasal infection of the SARS-CoV-2 virus, the prophylactic SFN supplementation to mice remarkably reduced the viral titer in the pulmonary tract and also minimized the pulmonary pathology and injury (Ordonez et al. 2021).

### *10.3.8* Psidium guajava

*Psidium guajava* belongs to the family called Myrtaceae. The common names of guava are yellow guava, lemon guava, or apple guava, which is a small evergreen shrub or tree indigenous to the Caribbean, Central America, and South America. The fruits and leaves are part of the plant that has antiviral properties. The vital compound of this plant was Asiatic acid which has a binding energy score of – 8.31 kcal/mol (Fitriani et al. 2020). It has been established that the major bioactive compounds from this plant may inhibit viral replication which includes pheophytin a, pheophytin b, β-caryophyllene, and Jejuguajavone A which were examined using animal models. The combined therapeutic approach with the combination of pheophytins with β-caryophyllene or Jejuguajavone A expressed essential interdependent antiviral effects which results in the conservation of affected animals even from variants of SARS-CoV-2 including omicron (Kim et al. 2022). The potent inhibitor luteolin obstructs the furin protein which is induced by the cleavage of the S protein (Wardani 2022).

### *10.3.9* Cinnamomum verum

*Cinnamomum verum* belongs to a family called Lauraceae. The inner side bark of *Cinnamomum verum* has been considered the true Cinnamon. It is indigenous to Sri Lanka and Southern India and is further extensively grown in Southeast Asia, China, Burma, Indonesia, Madagascar, the Caribbean, Australia, and Africa. The extract of this plant remarkably inhibits the pepsin enzyme which leads to significant HIV protease inhibition which exhibited a remarkable $IC_{50}$ value that revealed the anti-HIV potential of this plant (Singh et al. 2021a, 2021b). The computative enumeration has established that kaempferol 3-O-rutinoside, baicalin, and hesperidin may debilitate cellular viral absorption and prevent SARS-CoV-2 infection. At the level of RNA, the bioactive compounds such as procyanidins and butanol from this plant extract exhibited antiviral activity. The tea prepared using the bark of Cinnamomum improves immunity in COVID-19 patients. Furthermore, the phenolic compounds of this plant displayed potent trypsin enzyme inhibition ability which includes majorly caffeic acid and cinnamic acid additionally includes gallic acid and eugenol these findings have led to recommending Cinnamon as a herbal medicine to act against SARS-CoV-2 infection and also to other complementary viral infections (Yakhchali et al. 2021).

### *10.3.10* Justicia adhatoda

*Justicia adhatoda* belongs to the family called Acanthaceae. It is a perennial, evergreen, and highly branched shrub, widely distributed throughout the tropical regions of Southeast Asia and in India, Sri Lanka, Malaysia, and Burma (Sobia et al. 2018). It is extensively referred to as Vasaka or Malabar nut. It is a beneficial Ayurvedic plant employed to treat colds, asthma, cough, and tuberculosis. The importance of the Vasaka plant for treating pulmonary ailments is followed since the ancient Indian saying, "As long as the Vasaka plant exists there will be no human suffering from phthisis" (Pa and Mathew 2012). It indicates that about 51.5% of herbal bioactive compounds from this plant are effective to treat COVID-19. It has been reported that phytocompounds such as quercetin, xanthoxylol, chinensinaphthol methyl ether, podophyllotoxin, and apigenin may function against multiple target proteins of SARS-CoV-2 (Suryanarayana et al. 2021). The crude extracts of *Justicia adhatoda* reported the antiviral effect against the influenza virus by hemagglutination (HA) reduction. The studies suggest that *Justicia adhatoda* has a strong anti-influenza virus activity (Jamwal et al. 2022).

*Salvia officinalis* generally named sage belongs to the family called Lamiaceae. It is an evergreen perennial subshrub, with hard stems, grayish leaves, and blue to purplish flowers, it is indigenous to the Mediterranean region (Miraj and Kiani 2016). The phytocompounds from *Salvia officinalis* extract exhibit greater binding energy for kaempferol 3-O-rutinoside and neodiosmin with CoV-2-SP (Babaeekhou et al. 2021). The notable flavonoid Rutin creates indestructible hydrogen bond interactions with crucial $M^{Pro}$ residues. Molecular docking studies have revealed that flavonoids of this plant extract possess anti-coronavirus ability (Moezzi 2022). It was reported that the methanol extract and ethanolic extract and essential oils of *Salvia* sp. consist of antiviral potential against various deadly viruses. It can minimize the period and critical symptoms in patients with pulmonary ailments (Abou Baker et al. 2021).

### *10.3.11* Artemisia annua

*Artemisia annua* belongs to the family of Asteraceae, which is considered to be an annual herb with tremendous woody unistem consisting of aromatic leaves, native to Asia, but naturalized in many countries including North America. Otherwise referred to as annual mugwort, annual and sweet wormwood, sweet Annie, and sweet sagewort (Ferreira and Janick 1996). The ethanolic extract of *Artemisia annua*, in vivo test, states that it contains antiviral properties counter to SARS-CoV-2 (Li et al. 2005a, b). As the extract exhibits high phenolic content they are associated with pulmonary diseases and effective over lung fibrosis with the highest formulates toward the antiviral activity of respiratory problems. It is the first report that shows the boiled or hot water *Artemisia annua* extracts exhibited antiviral activity toward SARS-CoV-2 and its variants B1.1.7 and B1.351 which were performed in humans. The upcoming fortunate clinical trials may hold the hope that *Artemisia annua* can be prospectively used as a prudent therapeutic approach that can be practiced globally at an accessible cost which may be a substitute for vaccines (Nair et al. 2021).

## 10.3.12 Lycoris radiate

*Lycoris radiate* belongs to the family called Amaryllidaceae. *Lycoris radiate* Herbert (Amaryllidaceae, L'Her) is extensively found in China, Japan, and Korea (Liu et al. 2012). The lycorine was purified from *Lycoris* spp. and was identified as an auspicious bioactive compound opposing the antiviral activity of SARS-CoV with an $EC_{50}$ value of 34.5 ± 2.6 μg/mL. Herbal medicines of *Lycoris radiate* have been second-hand in coronavirus outbreaks like SARS-CoV and MERS-CoV. *Lycoris radiate extract* showed anti-SARS effects with a 50% effective concentration (Lawan et al. 2020).

## 10.3.13 Rosmarinus officinalis

*Rosmarinus officinalis* is an aromatic plant allied to the family called Lamiaceae in which the leaves are needle-shaped. It is cultivated throughout the world. Southeast Spain is one of the major producers and importers of rosemary. Rosemary is a unique spice and an economically available antioxidant that has antiviral properties and is popularly utilized as a therapeutic dosage, in the USA and Europe (Nieto et al. 2018). Some flavonoids like hesperidin, genkwanin, isoscutellarein 7-*O*-glucoside, hispidulin 7-*O*-glucoside, eriocitrin, luteolin 3'-*O*- *β*-D-glucuronide, and diosmin have been identified in *Rosmarinus officinalis*. In molecular docking studies, bioactive compounds such as ursolic acid, rosmarinic acid, and oleanolic acid are determined to possess antiviral activity countering SARS-CoV-2. The binding affinity of active molecules toward the COVID-19 6LU7 and 6Y2E proteases was found to be greater (Manzano-Santana et al. 2021). The phenolic compounds such as carnosol and carnosic acid along with two abietane-type diterpenoids have been analyzed to found as the greater amount in Rosemary extracts, which exhibit antiviral activity by inhibiting the viral reverse transcriptase. It exhibits the functions such as inhibition of viral RNA replication and synthesis either before the binding of the virus to the host cell or after the viral entry into the host cell (Zrig 2022).

## 10.3.14 Zingiber officinale

*Zingiber officinale* comes under the family called Zingiberaceae. The tea prepared using the *Zingiber officinale* improves immunity in COVID-19 patients. It has been greatly recognized for its medicinal value since more than a thousand years cultivated as a spice. It is used extensively in Traditional Chinese Medicine to treat headaches, nausea, and colds (Jiang et al. 2006). SARS-CoV entry inhibitors should be encoded with the molecules that attach with the surface target receptors such as ACE-2 and TMPRSS2 that obstruct the membrane fusion and interaction with the cell receptors. The molecule called gingerol from *Zingiber officinale* prevents the progression of the coronavirus by obstructing the cell's TMPRSS2 receptor. The phytochemical cyanin exhibits extensive inhibitory activity against the $M^{Pro}$ of corona viruses and their variants (Manzano-Santana et al. 2021).

### *10.3.15* Origanum vulgare

*Origanum vulgare* belongs to the family called Lamiaceae and it is indigenous to the Mediterranean region and western Eurasia (Pezzani et al. 2017). *Origanum vulgare* was established as the prospective substitute for constraining of SARS-CoV-2 delta variant (Torres Neto et al. 2022). Some studies exhibited that phenolic compounds of *Origanum vulgare* have the inhibitory activity to counter the replication of the respiratory syncytial virus (RSV). The compounds such as carvacrol, a monoterpenoid phenol found in *Origanum vulgare*, can consort with SARS-CoV-2 $M^{Pro}$ and endeavor its resistance process (Belmouhoub and Aberkane 2021). The gamma-terpineol was referred to as a lead compound of *Origanum vulgare* and was said to have the best binding affinity and pharmacological properties. The findings suggest that the identified compound may serve the inhibiting property. Hence, *Origanum vulgare* was recommended as a potential inhibitor of $3CL^{Pro}$ of COVID-19 (Saif 2022).

### *10.3.16* Hyoscyamus niger

*Hyoscyamus niger* also referred to as a traditional hallucinogen belongs to the family called the family Solanaceae (Al-Snafi 2018). *Hyoscyamus niger* was indigenous to Europe, North Asia, and North and west Africa. It was later cultivated in East Asia, Australia, and North America (Hu et al. 2022). The utilization of herbs in Indian traditional therapeutics has been suggested to cure COVID-19. It showed antiviral properties against bronchitis and was used as a novel therapeutic agent against SARS-CoV-2 (Alburae 2020). Herbal extracts of *Hyoscyamus niger* were demonstrated to minimize the inflammatory cascade induced by SARS-CoV, in respiratory distress. *Hyoscyamus niger* extract was adopted to target the $Ca^{2+}$ channels which are inferred to induce downstream pathways associated with viral infections. The studies suggest that the methanolic extract of *Hyoscyamus niger* may represent an affordable naturally obtained therapeutic approach (Kosari et al. 2021).

## *10.3.17* Allium sativum

*Allium sativum* comes under the family called Amaryllidaceae. The plant was cultivated as a common vegetable and spice to develop an immune system. It functions opposing by inhibiting the penetration or absorption of non-enveloped viruses which acts on the cell surface and obstructs the receptor site which is necessary for numerous viral invasions whereas various other inhibitors produce a neutralizing effect on reacting with the virus (Harazem et al. 2019). The tea made of leaves and roots of *Allium sativum* are used to improve the immunity in COVID-19 patients. The bioactive components of garlic such as S-cysteine, L-cysteine, S-propyl, S-ally-mercapto-cysteine, and alliin exhibit the inhibitory effect against $M^{Pro}$ of SARS-CoV-2 by hindering the hydrogen bonds. Molecular docking studies have revealed that the alliin bioactive component itself or in interaction with a major therapeutic drug could exhibit a stronger antiviral potential to eradicate SARS-CoV spread with minimum toxicity and side effects (Khubber et al. 2020). Garlic supplementation leads to the prominent provoke of CD4+ and CD8+ cells and also induces the production of NK cells (Donma and Donma 2020).

## *10.3.18* Solanum nigrum

A therapeutic plant *Solanum nigrum* belongs to the family of Solanaceae. *Solanum nigrum* often referred to as Makoi or black nightshade, generally breeds as a weed in wetland habitats. The favorable soil conditions of this plant include stony, deep soils, dry or shallow, and also can be propagated in tropical and subtropical regions (Jain et al. 2011). Spirostan-3-ol, Solasodine, Diosgenin, N-methylsolasodine, and Solanocapsine are the phytochemicals found in this weed that was effective in inhibiting the surface targets of SARS-CoV-2 such as NSP16-NSP10, NSP9, and $M^{Pro}$ (Manzano-Santana et al. 2021). At the non-toxic concentration, the methanol and chloroform extracts of this weed provided approximately 37% and greater than 50% inhibition of HCV, respectively. The peer research results revealed that the chloroform extract of seeds of this weed exhibits the antiviral effect to counter the expression and activity of HCV NS3 protease by mediating the transfer of HCV NS3 protease plasmid into the liver cells (Javed et al. 2011). Moreover, the extracts of this weed are expected to exhibit an inhibitory effect on significant enzymes which includes protease and α-glucosidase and reverse transcriptase (RT) and obstruct HIV-1 viral replication (Yu 2004).

### *10.3.19* Moringa oleifera

*Moringa oleifera* comes under the family called Moringaceae. It is generally referred to as Drumstick. Its plant is rarely found in marginal land and is widely cultivated for its leaves and fruits and used as a superfood due to its antioxidant and antiviral properties. Earlier reports have established that several bioactive components are present in *Moringa oleifera* which exhibits effective binding with the receptors of SARS-CoV-2 which includes $M^{Pro}$ and Spike protein (Sen et al. 2021). Since the bioactive compounds of this plant associate with $M^{Pro}$ and inhibit viral multiplication, the inhibition of $M^{Pro}$ is anticipated to sarcastically minimize the viral concentration in the patients infected with coronavirus (Kharisma et al. 2021; Nair and James 2020).

### *10.3.20* Justicia adhatoda

*Justicia adhatoda* belongs to the family called Acanthaceae. It is a perennial, evergreen, and highly branched shrub, widely distributed throughout the tropical regions of Southeast Asia and in India, Sri Lanka, Malaysia, and Burma (Sobia et al. 2018). It is extensively referred to as Vasaka or Malabar nut. It is a beneficial Ayurvedic plant employed to treat colds, asthma, cough, and tuberculosis. The importance of the Vasaka plant for treating pulmonary ailments is followed since the ancient Indian saying, “As long as the Vasaka plant exists there will be no human suffering from phthisis” (Pa and Mathew 2012). It indicates that about 51.5% of herbal bioactive compounds from this plant are effective to treat COVID-19. It has been reported that phytocompounds such as quercetin, xanthoxylol, chinensinaphthol methyl ether, podophyllotoxin, and apigenin may function against multiple target proteins of SARS-CoV-2 (Suryanarayana et al. 2021). The crude extracts of *Justicia adhatoda* reported the antiviral effect against the influenza virus by HA reduction. The studies suggest that *Justicia adhatoda* has a strong anti-influenza virus activity (Jamwal et al. 2022).

## 10.4 Fruits Against the Vulnerability of SARS-CoV-2 Infection

As the world returns to immune-boosting traditional treatments, there is renewed interest in the Indian medical system, which is endowed with a plethora of herbal drugs and therapies. Citrus fruits and liquids with potential health benefits are among the most abundant sources of chemicals that were illustrated in Table 10.2. Flavanones (such as hesperetin, isosakuranetin, naringenin, eriodictyol, and their related glycosides), which occur in levels ranging from 180 to 740 mg/L (depending on

**Table 10.2** Fruits, their chemical compounds, and their potential health benefits

| Scientific name | Common name | Major chemical constituent | Virus type | Mechanistic property | Reference |
|---|---|---|---|---|---|
| *Phyllanthus emblica* | Amla | Amritoside, pectolinarin, astragalin, quercetin | SARS-CoV-2 | Immunity booster, natural cure for respiratory disorders | Zannella et al. (2021), Ben-Othman et al. (2021) |
| *Vitis vinifera* | Grapes | Luteolin, kaempferol, apigenin, isorhamnetin, myricetin, chrysoeriol, biochanin, isookanin, scutellarein, resveratrol | HSV-1, SARS-CoV-2 | Inhibits viral replication in the early stages of infection by directly blocking the proteins enriched on the viral surface, anti-inflammatory and anticarcinogenic properties | Zannella et al. (2021), Orallo (2006) |
| *Malus domestica* | Apple | Phloretin, procyanidin, quercetin | Influenza virus, SARS-CoV-2 | Antioxidant properties, antiviral properties | Shimamura et al. (2017), Papadakis (2022), Ben-Othman et al. (2021) |
| *Citrus sinensis* | Oranges | Hesperidin | SARS-CoV-2, HIV-1 | The inhibitory effect of methylglyoxal is higher in the late stage of HIV-1 infection | Elghani et al. (2022), Behbahani (2014) |
| *Capsicum annuum* | Red bell pepper | Dichloromethane (DCM), butanol (BUT) | Tobacco rattle virus (TRV) | Anti-inflammatory in vivo effects | Allemand et al. (2016), Tian et al. (2014) |
| *Citrus limon* | Lemon | Limonene, limonin, naringenin | SARS-CoV-2, DEN virus | Secondary metabolites exhibit a wide range of biological properties, including anticancer, antibacterial, antifungal, anti-malarial, and antiviral activities | Tundis et al. (2014), Frabasile et al. (2017) |
| *Citrus limetta* | Mosambi | Nano-Ag and Fe | CHIKV | Antiviral property | Choudhary et al. (2020) |
| *Carica papaya* | Papaya | Quercetin, protocatechuic acid, p-coumaric acid, caffeic acid, chlorogenic acid, kaempferol, 5,7-dimethoxycoumarin, nelfinavir, amprenavir, lopinavir-ritonavir | DENV 2, SARS-CoV-2, HIV, HCV | Inhibitory activity on NS2B-NS3 protease is very essential for viral assembly | Senthilvel et al. (2013), Hariyono et al. (2021) |
| *Fragaria ananassa* | Strawberries | Acylphloroglucinol (APG) glucosides, anthocyanin, phytomelatonin | SARS-CoV-2 | Antiviral and antioxidant activity | Song et al. (2016), Okeke et al. (2022) |

citrus species and cultivar) play a variety of biological roles. These compounds strengthen the body's defenses against oxidative stress and help to prevent cardiovascular disease, atherosclerosis, and cancer. They also have anti-inflammatory, antiviral, and antibacterial properties (Barreca et al. 2017). Apigenin is a flavonoid present in many plants, including the majority of vegetables and fruits. Apigenin has been proven in studies to have anti-inflammatory, antioxidative, antibacterial, antiviral, antitumor, and cardiovascular protective activities (Xu et al. 2021).

Betulinic acid, which is found in a variety of medicinal plants and fruits, has a variety of biological effects, including its exceptional anti-HIV activity, which has piqued the interest of many pharmacists. Some betulinic acid derivatives displayed inhibitory effects at nanomolar concentrations and have started phase II clinical trials (Huang et al. 2018). *Meliae fructus* is the dried ripe fruit of the Meliaceae family. *Meliae fructus* has antibacterial, antioxidant, anticancer, anti-inflammatory, and analgesic effects, and it is commonly used in traditional medicine to treat inflammation and helminthic infection (Basu et al. 2018). Quercetin may be found in a variety of fruits and vegetables, including apples and citrus, as well as onion, broccoli, kale, and tomato. Papaya leaves, green and black tea leaves, buckwheat, seeds, and grains all contain quercetin. It was discovered that quercetin inhibited viral reverse transcriptases in the Rauscher murine leukemia virus (RLV) and HIV.

Quercetin also suppresses the replication of negative and positive-strand RNA as well as the translation of capsid proteins in Rhinoviruses that cause colds (RV). SARS-CoV-3CLp can be inhibited by the same quercetin that inhibits SARS-CoV-3CL$^{Pro}$ (Saakre et al. 2021). The lentil lectin from *Lens culinaris*, which binds to oligomannose-type glycans and GlcNAc at the non-reducing end terminus, was found to have the most potent and broad antiviral activity against a panel of mutant strains and variants, which include artificial mutants at the N-/O-linked glycosylation site, naturally occurring amino acid mutants, as well as epidemic variants B.1.1.7, B.1.351, and P. Lentil lectin has also been demonstrated to have antiviral properties against SARS-CoV and MERS-CoV (Wang et al. 2021a, b). Researchers have recently focused on hesperidin, a flavonoid that interacts with the major proteins of the SARS-CoV-2.

Several computational methodologies utilized by different investigations independently demonstrated that hesperidin has a low binding energy with the coronavirus "spike" protein as well as the primary protease that transforms the virus's early proteins (pp1a and pp1b) into the complex responsible for viral replication. The binding energy of hesperidin to these important components is lower than that of lopinavir, ritonavir, and indinavir, indicating that it may have an antiviral effect. Furthermore, both hesperidin and ascorbic acid protect cells against the oxidative stress generated by a viral infection and inflammation (Bellavite and Donzelli 2020).

### *10.4.1* Phyllanthus emblica

*Phyllanthus emblica* is a tree native to Southeast Asia's tropical climatic regions (Zhao et al. 2015). It is popularly known as "Aonla," "Amla," or "Indian gooseberry" and is a popular fruit tree that belongs to the order Geraniales and family Euphorbiaceae. According to ancient Indian mythology, it was the very first tree to originate on Earth. Scientific data supports the antiviral effectiveness of amla in the treatment of hepatitis B virus (HBV), HIV, Coxsackie VB3, and HSV. The fruits of *Emblica officinalis* are used extensively in Rasayana to cure a variety of infectious and non-infectious disorders. Docking studies and MDS verified quercitrin, chlorogenic acid, and Myricetin's promising ability to inhibit the n-CoV-2 major viral proteins (Chikhale et al. 2021). Some Phyllanthus species are inhibitory to HBV, HCV, HIV, and HSV, while *Phyllanthus emblica* is inhibitory to respiratory syndrome virus (PRRSV). Pedunculagin, azadirachtin, chebulagic acid, and nimbolide may be ACE-2 receptor and $M^{Pro}$ inhibitors for SARS-CoV-2. Seselin inhibited multiple SARS-CoV-2 targets, including $M^{Pro}$ and spike protein. However, there is no evidence that these medications can be used to treat COVID-19. As a consequence, the efficacy of these drugs to inhibit SARS-CoV-2 RdRp was tested (Pandey et al. 2021).

### *10.4.2* Vitis vinifera

*Vitis vinifera* (grape) is a fleshy non-climacteric fruit (Alem et al. 2021). The grapevine is native to Western Asia and Southern Europe. Grape seed and skin include flavonoids, proanthocyanidins, polyphenols, anthocyanins, procyanidins, and the stilbene derivative resveratrol. Grape seed extract, in particular, has been demonstrated to offer several pharmacological and therapeutic effects, including antioxidative, anti-inflammatory, and antibacterial capabilities, as well as cardioprotective, hepatoprotective, and neuroprotective qualities (Nassiri-Asl and Hosseinzadeh 2016). The fractioned component produced from chloroform extract inhibited the Parainfluenza virus more effectively (PIV) and HSV-1. Leaf extracts at pH 7.00 and 13.00 suppressed the early stages of HSV-1 infection as well as the pandemic and currently ubiquitous SARS-CoV-2.

The leaf extract significantly outperformed the Greco extract applied as a positive control against HSV-1 and SARS-CoV-2 (Zannella et al. 2021). The chemicals 3-galloylcatechin, proanthocyanidin B1, and luteolin 7-galactoside discovered in African flora grapes have been identified as hit compounds against various SARS-CoV-2. These chemicals might be potential leads and nutraceuticals used for the prevention or treatment of COVID-19. These compounds' scaffolds can be optimized to overcome minor flaws in their metabolism and toxicity (Iheagwam and Rotimi 2020).

### *10.4.3* Malus domestica

In temperate climates, the apple (*Malus domestica*) is one of the most extensively produced and commercially significant fruits (Chen et al. 2021). The apple, like other important temperate fruit tree species, is a member of the Rosaceae family (Cornille et al. 2014). Chinese quince had the most phenolics, which were mostly polymeric procyanidins. Quince included significant levels of hydroxycinnamic derivatives, including 5-caffeoylquinic acid, polymeric procyanidins, and 3-caffeoylquinic acid. The apple had the fewest phenolics, primarily monomeric and oligomeric procyanidins, and 5-caffeoylquinic acid (Hamauzu et al. 2005). According to a recent review paper, screening of conserved plant miRNAs against the HBV and HCV revealed six HCV gene sequences that are most likely to be targeted by miR-156, miR-157, miR-166, miR-172, and miR-390. While five conserved plant miRNAs have been demonstrated to be especially likely to align and interact with the six HBV gene sequences: miR-166, miR-169, miR-172, miR-390, and miR-399 (Mangukia et al. 2022).

### *10.4.4* Citrus sinensis

*Citrus sinensis* is one of the most significant citrus cultivar groups, accounting for more than 70% of total annual Citrus species output. It is native to Asia, but it has spread to the Pacific and other tropical places throughout the world. It is an evergreen flowering tree (Favela-Hernández et al. 2016*)*. *Citrus sinensis* is a member of the Rutaceae family, and its juice and pulp portions have been demonstrated to have hypolipidemic effects in rats with diet-induced hypercholesterolemia (Mallick and Khan 2016). Researchers have recently focused on hesperidin, a flavonoid that interacts with the major proteins of the SARS-CoV-2. Several computational methodologies utilized by different investigations independently demonstrated that hesperidin has a low binding energy with the coronavirus "spike" protein as well as the primary protease that transforms the virus's early proteins (pp1a and pp1b) into the complex responsible for viral replication. The binding energy of hesperidin to these important components is lower than that of lopinavir, ritonavir, and indinavir, indicating that it may have an antiviral effect. Furthermore, both hesperidin and ascorbic acid protect cells against the oxidative stress generated by a viral infection and inflammation (Bellavite and Donzelli 2020). *Citrus sinensis* monofloral honey exhibited substantial anti-HIV-1 activity, with $EC_{50}$ values of 70 g/ml (Behbahani 2014).

### 10.4.5 Capsicum annum

*Capsicum annuum* (Solanaceae) was domesticated in Mexico, where wild and cultivated (chile pepper populations (>60 landraces)) coexist, and wild-like individuals develop spontaneously in human situations (Pérez-Martínez et al. 2022). *Capsicum annuum*, a tropical and subtropical fruit plant, has a variety of vital nutrients and bioactive chemicals that have been shown to have a variety of bioactivities such as free radical scavenging (antioxidant), anti-inflammatory, antibacterial, antiviral, and anticancer properties (Khan et al. 2014). Several edible plants found across Indonesia contain emodin and luteolin compounds, which can prevent or reduce viral infection. In SARS-CoV, luteolin inhibits connections between S protein and ACE-2 receptors (Illian et al. 2021).

### 10.4.6 Citrus limon

*Citrus limon* is a Rutaceae tree with evergreen foliage and yellow edible fruits. The diverse chemical makeup of *Citrus limon* determines its medicinal potential. Flavonoids are the most abundant secondary metabolites in the fruit, although there are also coumarins, phenolic acids, carboxylic acids, amino acids, and vitamins. Monoterpenoids, particularly D-limonene, are the primary constituents of essential oils (Klimek-Szczykutowicz et al. 2020). When compared to the typical antiviral medication, diosmetin had the best docking values against the $M^{Pro}$ of SARS-CoV-2. The order associated with biochemical reactivity in DFT calculations is as follows: eriodictyol > quercetin > spinacetin > apigenin > diosmetin > luteolin, whereas the regions of oxygen and hydrogen atoms from the selected isolated compounds are suitable for electrophilic and nucleophilic attacks, respectively.

Furthermore, the HOMO-LUMO and global descriptor values of these compounds indicated a good outcome (Khan et al. 2021). Because the ACE-2 receptor, a host cell receptor, has been found to play a vital role in virus cell entry, ACE-2 blockers may be a viable antiviral intervention target. Geranium and lemon oils strongly reduced ACE-2 in epithelial cells. Furthermore, immunoblotting and qPCR analysis revealed that geranium and lemon oils had potent ACE-2 inhibitory properties. Geranium and lemon essential oils, as well as their derivative compounds, are key natural antiviral medicines that may aid in the prevention of COVID-19 in humans (Senthil Kumar et al. 2020). Citrus essential oils (Family Rutaceae) are extremely effective against the dengue mosquito *Aedes albopictus*. Under experimental circumstances, *Citrus limon* oils showed high action against *Aedes albopictus*. These oils include secondary metabolites that are extremely beneficial to plants in terms of insect pest deterrence. Many researchers have discovered limonoids in citrus oils to be particularly active against mosquitoes (Khan et al. 2014).

### *10.4.7* Citrus limetta

*Citrus limetta* (Rutaceae) is a commercial citrus fruit crop that is utilized in the juice processing industry. *Citrus limetta* peels are a perishable waste item, posing a significant difficulty in the juice processing industry. The ethanol extract (ClPs) exhibits promising anti-inflammatory action and is high in hesperidin concentration, according to the preliminary pharmaco-chemical profile of peel extracts (Babu et al. 2021). In a study, several nanometals (Ag, Zn, and Fe) were infused into raw *Citrus limetta* peels. The plaque assay, real-time PCR, and immunofluorescence assays all demonstrated that the three nano-biomaterials reduced the Chikungunya virus (CHIKV) viral titer and viral RNA levels significantly. The findings indicated that silver and iron nano-biomaterials have more antiviral activity against CHIKV than zinc (Choudhary et al. 2020).

### *10.4.8* Carica papaya

Papaya (*Carica papaya*) is commonly grown in tropical and subtropical regions. While ripe fruit is widely consumed across the world, unripe fruit is only consumed in a few Asian nations (Hiraga et al. 2021). *Carica papaya* belongs to the Caricaceae family. According to one study, quercetin is a good candidate for the development of effective anti-dengue medicines. *Carica papaya* flavonoid quercetin may have antiviral activity by interfering with the viral assembly process of the Dengue virus (DENV2) virus (Senthilvel et al. 2013). The presence of several secondary metabolites, including flavonoids, alkaloids, sterols, triterpenoids, isothiocyanates, tannins, and other phenolic chemicals, explains the broad range of activity. Recently, it was shown that papaya fruit pulp possesses antiviral properties against Zika and DEN virus. Betulinic acid, a pentacyclic triterpene, has antiviral action against HIV and HSV. The n-hexane fraction not only had the highest selectivity index, but it also had the strongest anti-SARS-CoV-2 activity (Adel et al. 2022).

### *10.4.9* Fragaria ananassa

Wild Fragaria (Rosaceae) species are found across the northern hemisphere and in southern South America. *Fragaria ananassa*, the contemporary cultivated strawberry, arose in the eighteenth century in Europe as a result of hybridization between two species brought from North and South America (Liston et al. 2014). Strawberry methanolic extracts were used to make silver nanoparticles (AgNPs) to test their SARS-CoV-2 inhibitory activity. The strawberry methanolic extract was the most effective against SARS-CoV-2. The extraction of secondary metabolites from strawberry crude methanolic extracts resulted in the identification of several compounds, including phenolic, flavonoids, fatty acids, sesquiterpenes, triterpenes, sterols, and others. The docking study suggested different patterns of interaction between strawberry compounds and seven SARS-CoV-2 protein targets, including five viral proteins ($M^{Pro}$, ADP ribose phosphatase, NSP14, NSP16, and $PL^{pro}$) and two human proteins (AAK1, Cathepsin L). The study of molecular docking and dynamics modeling demonstrated that neo hesperidin can bind to both human AAK1 and SARS-CoV-2 NSP16 proteins (Al-Sanea et al. 2021).

### *10.4.10* Citrus paradisi

Citrus (Rutaceae) is an old, widely traded and widely consumed crop. Grapefruit essential oil is known as the "dieter's buddy" because of its anti-obesity properties. At 80 ppm, grapefruit EO is also a strong larvicide against *Anopheles stephensi* (Dosoky and Setzer 2018). Citrus essential oils include 85–99% volatile components such as sesquiterpenes, monoterpene (limonene), and hydrocarbons; aldehydes (citral), alcohols (linalool), ketones, acids, and esters. The most effective essential oil for decreasing viral titer on berries was rosemary cineole, followed by grapefruit EO. One of the most common chemicals found in citrus trees is limonene (LMN). LMN has been shown to affect many signaling pathways as well as suppress inflammatory mediators such as cytokines, adhesion molecules, prostanoids, chemokines, and eicosanoids. Given the etiology of COVID-19, which includes infection, inflammation, and immunity, we hypothesized that LMN, with its anti-inflammatory, immunomodulatory, and antiviral properties, might reduce the severity and progression of the condition. The use of LMN in COVID-19 treatment looks speculative, but considering its good physiochemical and druggable properties, it cannot be disregarded (Adel et al. 2022).

## 10.5 Spices Involved in Traditional Indian Cooking to Step Ahead the Battle of COVID-19 Infection

The Indian style of cooking involves the usage of ingredients that adds not only taste but also medicinal properties. The spices pose a strong source of therapeutic in traditional ailments which are also involved in regular diets by incorporating them in cooking. Turmeric and pepper are two major ingredients used in Indian cuisine and they possess medicinal values like anti-inflammatory, anticancer, and antiviral properties (Rattis et al. 2021; Priya and Kumari 2017) that are the needs of the current social scenario. This section deals with a brief elaboration of the medicinal values of different spices used in Indian cuisine and their activity against SARS-CoV-2 depicted in Table 10.3. Spices (the seed, leaf, flower, fruit, bud, root, and rhizome of several plants used to give taste and flavor to food) are particularly

**Table 10.3** Spices used in Indian cuisine and their activity against SARS-CoV-2

| Spices | Scientific name | Major chemical constituent | Virus type | Medicinal properties | Reference |
|---|---|---|---|---|---|
| Turmeric | *Curcuma longa* | Curcumin | Influenza A virus, DEN virus, SARS-CoV-2 | Antibacterial, antiviral, antitumor, and hepatoprotective activity | Rattis et al. (2021) |
| Ginger | *Zingiber officinale* | Zingiberol, paradol, shogoals, and gingerol | HSV, HRSV, SARS-CoV-2 | Colds, nausea, arthritis, migraines, and hypertension | Bode and Dong (2011) |
| Coriander seeds | *Coriandrum sativum* | Linalool, rutin, lopinavir, quercetin | SARS-CoV-2 | Anti-diabetic, anticancer, antifungal, antiulcer, hepatoprotective, antimutagenic sedative-hypnotic, anticonvulsant, and antifungal properties | Nair et al. (2012) |
| Cumin seeds | *Cuminum cyminum* | Cumin aldehyde, γ-terpinene, p-cymene, β-pinene, and estragole | HSV, SARS-CoV-2 | Fertility, respiratory and gastrointestinal disorders, toothache, diarrhea, and epilepsy | Saab et al. (2021) |
| Black pepper | *Piper nigrum* | Piperidine | SARS-CoV-2 | Acute respiratory infection, asthma, chronic indigestion, and flu | Priya and Kumari (2017) |
| Cloves | *Syzygium aromaticum* | Eugenol, eugenol acetate, and gallic acid | SARS-CoV-2 | Anti-inflammatory, antiviral, insect repellent, antimicrobial and antiparasitic | Xu et al. (2016) |
| Mustard | *Brassica juncea* | Curcumenol, N-desmethylselegiline, phentermine, and sphingolipid | Peritonitis, neuralgia, and SARS-CoV-2 | Anti-inflammatory, antibacterial, and antiviral | Tian and Deng (2020) |
| Fenugreek | *Trigonella foenum-graecum* | Trigoneoside | Malaria, SARS-CoV-2 | In the treatment of diabetes, cancer, hypercholesterolemia, myocardial infarction, skin irritation, indigestion and flatulence, inflammation, anemia, immune deficiency, and kidney diseases | Priya and Kumari (2017) |

(continued)

**Table 10.3** (continued)

| Spices | Scientific name | Major chemical constituent | Virus type | Medicinal properties | Reference |
|---|---|---|---|---|---|
| Cinnamon | *Cinnamomum verum* | Cinnamaldehyde, eugenol, caryophyllene, cinnamyl acetate, and cinnamic acid | H1N1, HSV, SARS-CoV-2 | Upregulation of inhibitory receptors | Connell et al. (2016) |
| Fennel | *Foeniculum vulgare* | Trans-anethole, LMN, pinene, trans-ocimene, fenchyl acetate, and fenchone | HAV and HCV, HSV | Antiviral activity | Suleiman and Helal (2022) |

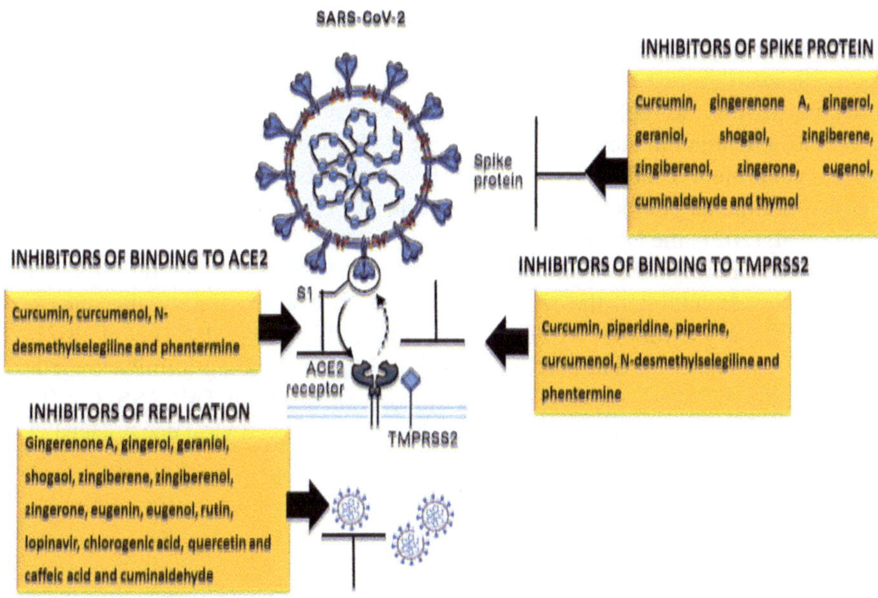


**Fig. 10.2** Compounds identified in dietary spices and their role in SARS-CoV-2 Inhibition

medicinal. A profusion of trials involving clinical and preclinical research have documented the efficacy of numerous spices with a variety of diseases. Spices are a recent interest of phytoresearch including immune-boosting home treatments in the period of uncertain times due to their possible immune-boosting capabilities and great safety profiles illustrated in Fig. 10.2 (Devan et al. 2022; Natesh et al. 2021).

### *10.5.1 Turmeric*

Turmeric (*Curcuma longa*) is from the Zingiberaceae family and is one of the native plants of India. Turmeric is one of the main ingredients used often in all types of Indian cuisine. Even though it is one of the primary ingredients, it is used only in small amounts. Curcumin is a polyphenol mainly found in turmeric and it is recommended that about 12 g/day of curcumin can be safe without any side effects (Gupta et al. 2013). The main bioactive constituents identified in the rhizomes of turmeric are sesquiterpenes, steroids, curcuminoids, and polyphenols (Omosa et al. 2017). Turmeric is identified to have various therapeutic properties like antioxidant, anti-inflammatory, antibacterial, antiviral, antitumor, and hepatoprotective activity for which it is being used in traditional medicine (Rattis et al. 2021).

Curcumin is enhanced with antiviral properties and it has been explored against a wide range of viruses like influenza A virus (Chen et al. 2010; Ou et al. 2013), DEN virus, Zika (Balasubramanian et al. 2019), CHIKV (Mounce et al. 2017), HCV

(Chen et al. 2012), RSV (Obata et al. 2013; Yang et al. 2016), Japanese encephalitis virus (Dutta et al. 2009), EBV (Hergenhahn et al. 2002), human cytomegalovirus (Lv et al. 2014a,b), and HIV (Gupta et al. 2011). Curcumin in Vero-E6 cells was observed to stop the replication of SARS CoV at a 3–10 μM concentration (Wen et al. 2007). Curcumin stops the infection of SARS-CoV-2 in cardiac myocytes by acting with protein S and ACE 2 receptors because of its higher affinity and ability to control endothelial cell activation (Rattis et al. 2021). In the work done by Marín-Palma et al. (2021) the D614G and Delta variation of SARS-CoV-2 was used for the evaluation of curcumin as therapeutic, it was found that the curcumin at lesser to cytotoxic levels inhibited the replicative cycle and also minimized the release of pro-inflammatory cytokines like IL-8, IL-6, and IL-1β (Marín-Palma et al. 2021). It has also been identified that curcumin displayed higher binding affinity toward the SARS-CoV-2 S protein and also can control the expression of TMPRSS, Cat B and L which are some of the key molecules in the viral entry and viral decay (Jena et al. 2021; Hoffmann et al. 2020; Padmanabhan et al. 2020).

### *10.5.2 Ginger*

Ginger is an ingredient used in Indian cuisine to enhance the flavor of food. Apart from the culinary uses, it is also employed in medicines practiced traditionally for the ailment of various diseases like colds, nausea, arthritis, migraines, and hypertension (Bode and Dong 2011). The ginger (*Zingiber officinale)* belongs to the family Zingiberaceae similar to turmeric. Ginger is enriched with the phytochemicals like phenols, steroids, and alkaloids. The rhizome possesses the aromatic compound zingiberol which has analogs such as zingerone, paradol, and shogoals along with the sub-bioactive compounds like 6-gingerol, 4-gingerol, 10-gingerols, 8-gingerol, 6-shogaols, and 14-shogaols (Singh et al. 2021a, b; Ali et al. 2008). Ginger is found to have activity against viruses like HRSV, HSV, CHIKV, and influenza virus (Dorra et al. 2019). The ginger is also observed to possess an inhibitory effect for SARS-CoV-2 in a docking study, where its compounds like geraniol, gingerenone A, shogaol, zingiberene, gingerol, zingerone, and zingiberenol had interaction with both $M^{Pro}$ and Spike protein (Ahkam et al. 2020). The inhibition of $M^{Pro}$ is crucial as it is involved in the viral replication process by mediating polyproteins pp1a and pp1ab (Hilgenfeld 2014).

### *10.5.3 Coriander Seeds*

Coriander (*Coriandrum sativum*) leaves and dried seeds are used in Indian cooking (Baliga et al. 2015). The coriander belongs to the family Apiaceae. The coriander seeds have been used to treat disorders of the digestive, urinary, and respiratory systems (Momin et al. 2012). The properties that enhance the seeds are diuretic, carminative, stimulant, and diaphoretic effects along with antioxidant, anti-diabetic, anticancer, antifungal, antiulcer, hepatoprotective, antimutagenic sedative-hypnotic, anticonvulsant, and antifungal properties (Nair et al. 2012). Linalool is one of the volatile compounds as well as a plant-specific compound found in different parts of coriander including its seeds, and it is observed to stimulate the immune system through inhalation or by the use of essential oil obtained from coriander (Dissanayake et al. 2020). Coriandrum was found to inhibit the nucleocapsid phosphoprotein which is a conserved structure of coronavirus by the interaction of phytoligands such as (+)−germacrene A, alpha-thujene, and geranyl acetate (Muthumanickam et al. 2021). The compounds like rutin, lopinavir, chlorogenic acid, quercetin, and caffeic acid, identified in the coriander seeds were observed to possess maximum binding affinity toward the $M^{Pro}$ enzyme of SARS-CoV-2 (Kumar et al. 2021).

### *10.5.4 Cumin Seeds*

Cumin seeds (*Cuminum cyminum*) belong to the family Apiaceae and are one of the most common ingredients in Indian and Pakistani dishes. Cumin seeds are used in traditional medicine as an ailment for fertility, respiratory, and gastrointestinal disorders and also to treat toothache, diarrhea, and epilepsy (Saab et al. 2021). Plant-specific compounds like cumin aldehyde, γ-terpinene, p-cymene, β-pinene, and estragole are identified in the cumin seeds, which are found to have antiviral activity (Bonesi et al. 2020). The cumin seeds' activity inhibited the infections of viruses like HSV1 and HSV2 as they inhibited the plaque developed by viral infection (El-Serehy et al. 2016). The cumin aldehyde present in the essential oils of cumin seeds was identified to be one of the potent compounds in inhibiting SARS-CoV-2 infection (Saab et al. 2021). The steaming of cumin seeds with other plants like eucalyptus, etc., could help in the inhibition of infectivity of SARS-CoV-2, as the volatile compounds like cumin aldehyde, thymol, and terpenes in association with various other compounds can interact with the amine group of the surface protein of SARS-CoV-2 and heal the lungs and throat (Marwah and Marwah 2020). The cumin aldehyde had also shown the minimum binding energy against the $M^{Pro}$ of SARS-CoV-2 in the in silico analyses (Rout et al. 2022).

### *10.5.5 Black Pepper*

The black pepper (*Piper nigrum*) is a spice that is primarily used in most Indian dishes and it comes under the Piperaceae family. Pepper is used in the ailment of digestive and respiratory diseases due to the infection of viruses that cause pathogenesis such as acute respiratory infection, chronic indigestion, asthma, and flu (Priya and Kumari 2017). The piperidine present in the pepper helps in rising the curcumin and catechin levels in the plasma and curbs the entry of 2019 n-CoV into the cell (Joshi et al. 2018). Pepper in combination with other dietary spices like turmeric and ginger can prevent the SARS-CoV-2 pathogenesis and this action is directed by the inhibition of TMPRSS2 of the host cell which is the underlying reason for SARS-CoV-2 entry (Yadav et al. 2022). The compounds like piperidine and piperine observed in the black pepper have been found to have the potential to inhibit SARS-CoV-2 by in silico analysis (Rajagopal et al. 2020).

### 10.5.6 Cloves

The cloves (*Syzygium aromaticum)* are the flowers that are obtained from the tree belonging to the Myrtaceae family (Chandra Manivannan et al. 2022a, b). Cloves are one of the ingredients used in Indian cooking. The cloves are enriched with bio-activities like anti-inflammatory, antioxidant antiviral, antimicrobial, insect repellent, and antiparasitic properties (Xu et al. 2016). The cloves are abundant in phenolic content with the presence of phenolic contents like eugenol, eugenol acetate, hydroxybenzoic acids, hydroxycinnamic acids, hydroxyphenylpropene, and gallic acid. The aspects of cloves like strong, pungent, and spicy order make it to be used in commercial applications like perfumes, cleaning agents, and also to anesthetize certain fishes (Taylor and Roberts 1999). Eugenol is one of the abundant phenol present in cloves and its excess amount can lead to cytotoxicity. The eugenol in a computational study against SARS-CoV-2 was found to inhibit the spike protein (Kiran et al. 2020). Eugeniin and eugenol found in the cloves were observed to exhibit a better binding effect on $M^{Pro}$ of SARS-CoV-2 (Chandra Manivannan et al. 2022a, b).

### 10.5.7 Mustard

The mustard (*Brassica juncea*) belongs to the family Brassicaceae which is an often used spice in Indian cuisine. The mustard seeds are enriched with various minerals like iron, magnesium, calcium, zinc, and phosphorous which are essential for healthy living (Divakaran and Babu 2016). The phytochemicals like vitamins, dietary fibers, chlorophylls, and glucosinolates along with their degradation products make mustard seeds good in antioxidant, anti-inflammatory, antibacterial, and antiviral activity (Tian and Deng 2020). In traditional folk, medicine mustard has been used as a stimulant, diuretic as well as purgative to manage diseases like peritonitis and neuralgia, etc. (Chan et al. 2014). The compounds identified in mustard like curcumenol, N-desmethylselegiline, phentermine, and sphingolipid derivatives were effective in the ACE-2 and TMPRSS2 receptor modifications for the prevention of the SARS-CoV-2 infection in computational analysis (Dave et al. 2020).

### 10.5.8 Fenugreek

The fenugreek (*Trigonella foenum-graecum*) has also got its application in the Indian style of cooking. It belongs to the family Fabaceae (Flammang et al. 2004). The fiber, phospholipids, glycolipids, linolenic acid, oleic acid, choline, nicotinic acid, vitamin A, B1, B2, C, and niacin add to the potential aspects of fenugreek (Chatterjee et al. 2010; Leela and Shafeekh 2008). The pharmacological properties of fenugreek seeds help in the treatment of cancer, diabetes, myocardial infarction, hypercholesterolemia, skin irritation, indigestion and flatulence, anemia, inflammation, immune deficiency, kidney diseases, aging, etc. Fenugreek seeds have been used to treat malaria (Priya et al. 2011). The compounds identified in the fenugreek seeds are observed to have inhibitory properties against $M^{Pro}$ of SARS-CoV-2 (Sen et al. 2022). The trigoneoside found in fenugreek was identified to be an efficient candidate against the active sites of SARS-CoV-2 in an in silico analysis (Pandey et al. 2022).

### 10.5.9 Fennel

Common *Foeniculum vulgare* sometimes known as fennel is an ancient medicine utilized to treat various illnesses affecting distinguished organs. It is a medicinal plant of the Umbelliferae (Apiaceae) family that has been recognized and utilized by humans since antiquity because of its taste. It was grown in practically every nation (Badgujar et al. 2014). The research found 67 volatile components containing phenylpropenes and monoterpenes as main ingredients, including trans-anethole, limonene, pinene, trans-ocimene, fenchyl acetate, and fenchone. In fennel hydro-distilled oils, phenylpropenes predominated, while monoterpenes predominated in the majority of the headspace scent. The unsupervised multivariate data analysis techniques PCA and HCA were used to analyze infraspecific variability. The plaque reduction test, HAV 3C proteinase, and HCV NS5B polymerase inhibitory assays were used to study the antiviral activity of samples against HAV and HCV, with percentage inhibition ranging from 66% to 85% and $IC_{50}$ values ranging from 1.8 to 26.7 g mL-1. In silico molecular docking scores in later enzyme binding pockets indicated critical allosteric interactions with trans-ocimene and fenchyl acetate, which had the lowest Gibb's free energy (Ibrahim and Moussa 2021). *Foeniculum vulgare* was shown to have a high percentage of inhibition value at the maximal noncytotoxic dose, with 83 percent inhibition against the influenza virus. The essential oil from the fruit of this spice in combination with 12 distinct Turkish therapeutic plants exhibited antiviral activity against HSV-1 and CoxB4 viruses at 21.95% and 13.14%, respectively (Suleiman and Helal 2022).

### *10.5.10 Cinnamon*

*Cinnamomum verum* has been used as a spice for both medicinal and culinary purposes since ancient times. It is indigenous to Sri Lanka and southern India, although it is also found in several Asian, Caribbean, Australian, and African nations. It is also used to treat asthma, bronchitis, diarrhea, headaches, inflammation, and heart problems. The main chemicals identified in its essential oil are cinnamaldehyde, eugenol, caryophyllene, cinnamyl acetate, and cinnamic acid (Singh et al. 2021a). Various researches have demonstrated *Cinnamomum verum's* anti-HIV efficacy in the treatment of acquired immune deficiency syndrome (AIDS) (Semenya et al. 2012). In a study using a customized cinnamon combination, effectiveness against H1N1 and HSV was demonstrated. It has anti-HSV1 action by directly inactivating free virus particles and may interfere with virion envelope structures essential for host cell entrance (Brochot et al. 2017). Cinnamon bark extract and its nanoparticles were evaluated in Vero cells against the H7N3 influenza A virus, and cell viability was estimated using the tetrazolium dye (MTT) assay. The silver nanoparticle obtained from the extract of this plant exhibited potent antiviral activity (Fatima et al. 2016). Tenuifolin (TEN) and Pavetannin C1 (PAV) have been found as hit drugs against the COVID-19 primary protease enzyme (Prasanth et al. 2021; Connell et al. 2016). IND02, a cinnamon-derived type A procyanidin polyphenol with trimeric and pentameric forms, has anti-HIV-1 action averse to CXCR4 and CCR5 viruses, with a 1–7 M ED50 for the trimer (Connell et al. 2016).

## 10.6 Conclusions

The SARS-CoV-2 infection has caused a global decline in population and economic status. The management of SARS-CoV-2 infection had been a challenge during the time of the pandemic and different modes of medication were practiced at different places. The allopathic treatment was used as a curative, but the preventive measure was provided by the practice of traditional medicine. The traditional medicine in India and other countries is incorporated with their traditional cooking style. The plants, fruits, and spices of medicinal importance used in various traditional dishes are briefly elaborated on in this chapter along with their phytoconstituents and their mode of action. The plants, fruits, and spices used on a culinary basis are identified to possess great potential as a preventive and curative measure for SARS-CoV-2 infection and virulence. The plants, fruits, and spices elucidated in this chapter further need investigation in proper clinical trials.

## References


Abou Baker DH, Amarowicz R, Kandeil A, Ali MA, Ibrahim EA (2021) Antiviral activity of Lavandula angustifolia L. and Salvia officinalis L. essential oils against avian influenza H5N1 virus. J Agric Food Res 4:100135

Adel A, Elnaggar MS, Albohy A, Elrashedy AA, Mostafa A, Kutkat O, Abdelmohsen UR, Al-Sayed E, Rabeh MA (2022) Evaluation of antiviral activity of Carica papaya leaves against SARS-CoV-2 assisted by metabolomic profiling. RSC Adv 12(51):32844–32852

Ahkam AH, Hermanto FE, Alamsyah A, Aliyyah IH, Fatchiyah F (2020) Virtual prediction of antiviral potential of ginger (Zingiber officinale) bioactive compounds against spike and MPro of SARS-CoV2 protein. Berkala Penelitian Hayati J Biol Res 25(2):52–57

Alburae NA (2020) Bioscience research

Alem H, Ojeda H, Rigou P, Schneider R, Torregrosa L (2021) The reduction of plant sink/source does not systematically improve the metabolic composition of Vitis vinifera white fruit. Food Chem 345:128825

Ali BH, Blunden G, Tanira MO, Nemmar A (2008) Some phytochemical, pharmacological and toxicological properties of ginger (Zingiber officinale Roscoe): a review of recent research. Food Chem Toxicol 46(2):409–420. https://doi.org/10.1016/j.fct.2007.09.085

Allemand A, Leonardi BF, Zimmer AR, Moreno S, Romao PR, Gosmann G (2016) Red pepper (Capsicum baccatum) extracts present anti-inflammatory effects in vivo and inhibit the production of TNF-α and NO in vitro. J Med Food 19(8):759–767

Al-Sanea MM, Abelyan N, Abdelgawad MA, Musa A, Ghoneim MM, Al-Warhi T, Aljaeed N, Alotaibi OJ, Alnusaire TS, Abdelwahab SF, Helmy A (2021) Strawberry and ginger silver nanoparticles as potential inhibitors for SARS-CoV-2 assisted by in silico modeling and metabolic profiling. Antibiotics 10(7):824

Al-Snafi AE (2018) Therapeutic importance of Hyoscyamus species grown in Iraq (Hyoscyamus albus, Hyoscyamus Niger and Hyoscyamus reticulates)-a review. IOSR J Pharm 8(6):18–32

Anand K, Palm GJ, Mesters JR, Siddell SG, Ziebuhr J, Hilgenfeld R (2002) Structure of coronavirus main proteinase reveals combination of a chymotrypsin fold with an extra α-helical domain. EMBO J 21(13):3213–3224

Anand K, Ziebuhr J, Fau-Wadhwani P, Wadhwani P, Fau-Mesters JR, Mesters Jr Fau-Hilgenfeld R, Hilgenfeld R (2003) Coronavirus main proteinase (3CLpro) structure: basis for design of anti-SARS drugs. Science 300(5626):1763–1767

Anand AV, Balamuralikrishnan B, Kaviya M, Bharathi K, Parithathvi A, Arun M, Senthilkumar N, Velayuthaprabhu S, Saradhadevi M, Al-Dhabi NA, Arasu MV, Yatoo MI, Tiwari R, Dhama K (2021) Medicinal plants, phytochemicals, and herbs to combat viral pathogens including SARS-CoV-2. Molecules 26(6):1775

Ansari A, Mahmood T, Bagga P, Ahsan F, Shamim A, Ahmad S, Shariq M, Parveen S (2021) Areca catechu: a phytopharmacological legwork. Food Front 2(2):163–183

Arumugam VA, Sangeetha T, Zareena F, Pavithra R, Ann M, Sanjeev A, Sunantha B, Meyyazhagan A, Yatoo MI, Sharun K, Tiwari R, Pandey MK (2020) COVID-19 and the world with co-morbidities of heart disease, hypertension and diabetes. J Pure Appl Microbiol 14(3):1623–1638

Ayati Z, Ramezani M, Amiri MS, Moghadam AT, Rahimi H, Abdollahzade A, Sahebkar A, Emami SA (2019) Ethnobotany, phytochemistry and traditional uses of Curcuma spp. and pharmacological profile of two important species (C. longa and C. zedoaria): a review. Curr Pharm Des 25(8):871–935

Babaeekhou L, Ghane M, Abbas-Mohammadi M (2021) In silico targeting SARS-CoV-2 spike protein and main protease by biochemical compounds. Biologia 76(11):3547–3565

Babu V, Binwal M, Kumari R, Sen S, Kumar A, Mugale MN, Shanker K, Kumar N, Bawankule DU (2021) Hesperidin-rich ethanol extract from waste peels of Citrus limetta mitigates rheumatoid arthritis and related complications. Phytother Res 35(6):3325–3336

Badgujar SB, Patel VV, Bandivdekar AH (2014) Foeniculum vulgare Mill: a review of its botany, phytochemistry, pharmacology, contemporary application, and toxicology. Biomed Res Int 2014:842674

Baez-Santos YM, St John SE, Mesecar AD (2015) The SARS-coronavirus papain-like protease: structure, function and inhibition by designed antiviral compounds. Antivir Res 115:21–38. https://doi.org/10.1016/j.antiviral.2014.12.015

Bakhiet M, Taurin S (2021) SARS-CoV-2: targeted managements and vaccine development. Cytok Growth Factor Rev 58:16–29

Balasubramanian A, Pilankatta R, Teramoto T, Sajith AM, Nwulia E, Kulkarni A, Padmanabhan R (2019) Inhibition of dengue virus by curcuminoids. Antivir Res 162:71–78

Baliga MS, Mane PP, Nallemgera JT, Thilakchand KR, Kalekhan F (2015) Dietary spices in the prevention of rheumatoid arthritis: past, present, and future. In: Foods and dietary supplements in the prevention and treatment of disease in older adults. Academic Press, pp 41–49

Barreca D, Gattuso G, Bellocco E, Calderaro A, Trombetta D, Smeriglio A, Laganà G, Daglia M, Meneghini S, Nabavi SM (2017) Flavanones: citrus phytochemical with health-promoting properties. Biofactors 43(4):495–506

Basu A, Schell J, Scofield RH (2018) Dietary fruits and arthritis. Food Funct 9(1):70–77

Bayan L, Koulivand PH, Gorji A (2014) Garlic: a review of potential therapeutic effects. Avicen J Phytomed 4(1):1

Behbahani M (2014) Anti-HIV-1 activity of eight monofloral Iranian honey types. PLoS One 9(10): e108195. https://doi.org/10.1371/journal.pone.0108195

Bellavite P, Donzelli A (2020) Hesperidin and SARS-CoV-2: new light on the healthy function of citrus fruits. Antioxidants 9(8):742

Belmouhoub M, Aberkane B (2021) Ethnopharmacological survey on medicinal plants used by Algerian population to prevent SARS-CoV-2 infection. Ethnobot Res Appl 14(22):1–3

Belouzard S, Millet JK, Licitra BN, Whittaker GR (2012) Mechanisms of coronavirus cell entry mediated by the viral spike protein. Viruses 4(6):1011–1033

Ben-Othman S, Kaldmäe H, Rätsep R, Bleive U, Aluvee A, Rinken T (2021) Optimization of ultrasound-assisted extraction of phloretin and other phenolic compounds from apple tree leaves (Malus domestica Borkh.) and comparison of different cultivars from Estonia. Antioxidants 10(2):189

Bode AM, Dong Z (2011) The amazing and mighty ginger. Herbal Medicine: biomolecular and clinical aspects, 2nd edn

Bondhon TA, Fatima A, Jannat K, Hasan A, Jahan R, Nissapatorn V, Wiart C, Pereira ML, Rahmatullah M (2021) RESEARCH ARTICLE in silico screening of Allium cepa phytochemicals for their binding abilities to SARS and SARS-CoV-2 3C-like protease and COVID-19 human receptor ACE-2. Trop Biomed 38(2):214–221

Bonesi M, Saab AM, Tenuta MC, Leporini M, Saab MJ, Loizzo MR, Tundis R (2020) Screening of traditional Lebanese medicinal plants as antioxidants and inhibitors of key enzymes linked to type 2 diabetes. Plant Biosyst 154(5):656–662

Brochot A, Guilbot A, Haddioui L, Roques C (2017 Aug) Antibacterial, antifungal, and antiviral effects of three essential oil blends. Microbiology 6(4):e00459

Candido KL, Eich CR, de Fariña LO, Kadowaki MK, da Conceição Silva JL, Maller A, Simão RD (2022) Spike protein of SARS-CoV-2 variants: a brief review and practical implications. Braz J Microbiol 53(3):1133–1157

Cannalire R, Cerchia C, Beccari AR, Di Leva FS, Summa V (2020) Targeting SARS-CoV-2 proteases and polymerase for COVID-19 treatment: state of the art and future opportunities. J Med Chem 65(4):2716–2746

Chan K, Zhang HW, Lin ZX (2014) Treatments used in complementary and alternative medicine. In: InSide effects of drugs annual, vol 35. Elsevier, pp 889–898

Chandra Manivannan A, Malaisamy A, Eswaran M, Meyyazhagan A, Arumugam VA, Rengasamy KRR, Balasubramanian B, Liu WC (2022a) Evaluation of clove phytochemicals as potential antiviral drug candidates targeting SARS-CoV-2 Main protease: computational docking, molecular dynamics simulation, and pharmacokinetic profiling. Front Mol Biosci 9:918101

Chandra Manivannan A, Malaisamy AK, Eswaran M, Meyyazhagan A, Arumugam VA, Rengasamy KR, Balasubramanian B, Liu WC (2022b) Evaluation of clove phytochemicals as potential antiviral drug candidate targeting SARS-CoV-2 Main protease: an computational docking, molecular dynamics simulation and pharmacokinetic profiling. Front Mol Biosci 28: 604

Chatterjee S, Variyar PS, Sharma A (2010) Bioactive lipid constituents of fenugreek. Food Chem 119(1):349–353

Chavan R, Chowdhary A (2014) In vitro inhibitory activity of Justicia adhatoda extracts against influenza virus infection and hemagglutination. Int J Pharm Sci Rev Res 25(2):231–236

Chen DY, Shien JH, Tiley L, Chiou SS, Wang SY, Chang TJ, Lee YJ, Chan KW, Hsu WL (2010) Curcumin inhibits influenza virus infection and haemagglutination activity. Food Chem 119(4): 1346–1351

Chen MH, Lee MY, Chuang JJ, Li YZ, Ning ST, Chen JC, Liu YW (2012) Curcumin inhibits HCV replication by induction of heme oxygenase-1 and suppression of AKT. Int J Mol Med 30(5): 1021–1028

Chen Z, Yu L, Liu W, Zhang J, Wang N, Chen X (2021 May) Research progress of fruit color development in apple (Malus domestica Borkh.). Plant Physiol Biochem 1(162):267–279

Chikhale RV, Sinha SK, Khanal P, Gurav NS, Ayyanar M, Prasad SK, Wanjari MM, Patil RB, Gurav SS (2021) Computational and network pharmacology studies of Phyllanthus emblica to tackle SARS-CoV-2. Phytomed Plus 1(3):100095

Choudhary S, Kumar R, Dalal U, Tomar S, Reddy SN (2020) Green synthesis of nanometal impregnated biomass–antiviral potential. Mater Sci Eng C 112:110934

Cinatl J, Morgenstern B, Bauer G, Chandra P, Rabenau H, Doerr H (2003) Glycyrrhizin, an active component of liquorice roots, and replication of SARS-associated coronavirus. Lancet 361(9374):2045–2046

Connell BJ, Chang SY, Prakash E, Yousfi R, Mohan V, Posch W, Wilflingseder D, Moog C, Kodama EN, Clayette P, Lortat-Jacob H (2016) A cinnamon-derived procyanidin compound displays anti-hiv-1 activity by blocking heparan sulfate-and co-receptor-binding sites on gp120 and reverses T cell exhaustion via impeding tim-3 and PD-1 upregulation. PLoS One 11(10): e0165386

Cornille A, Giraud T, Smulders MJ, Roldán-Ruiz I, Gladieux P (2014) The domestication and evolutionary ecology of apples. Trends Genet 30(2):57–65

Dave GS, Rakholiya KD, Kaneria MJ, Galvadiya BP, Vyas SR, Kanbi VH, Patel MP (2020) High affinity interaction of Solanum tuberosum and Brassica juncea residue smoke water compounds with proteins involved in coronavirus infection. Phytother Res 34(12):3400–3410

Deng X, StJohn SE, Osswald HL, O'Brien A, Banach BS, Sleeman K, Ghosh AK, Mesecar AD, Baker SC (2014) Coronaviruses resistant to a 3C-like protease inhibitor are attenuated for replication and pathogenesis, revealing a low genetic barrier but high fitness cost of resistance. J Virol 88(20):11886–11898

Devan AR, Nair B, Kumar AR, Gorantla JN, Aishwarya TS, Nath LR (2022) Unravelling the immune modulatory effect of Indian spices to impede the transmission of COVID-19: a promising approach. Curr Pharm Biotechnol 23(2):201–220

Dissanayake KC, Fernando WS, Perera WP (2020) Investigation of the phytochemistry of Coriandrum sativum to combat against viral infections. Int J Innov Pharm Sci Res 8(6):1

Divakaran M, Babu KN (2016) Mustard. In: Caballero B, Finglas PM, Toldrá F (eds) Encyclopedia of food and health. Academic Press, pp 9–1

Donma MM, Donma O (2020) The effects of allium sativum on immunity within the scope of COVID-19 infection. Med Hypotheses 144:109934

Dorra N, El-Berrawy M, Sallam S, Mahmoud R (2019) Evaluation of antiviral and antioxidant activity of selected herbal extracts. J High Inst Public Health 49(1):36–40

Dosoky NS, Setzer WN (2018) Biological activities and safety of Citrus spp. essential oils. Int J Mol Sci 19(7):1966

Du L, He Y, Zhou Y, Liu S, Zheng BJ, Jiang S (2009) The spike protein of SARS-CoV—a target for vaccine and therapeutic development. Nat Rev Microbiol 7(3):226–236

Du L, Yang Y, Zhou Y, Lu L, Li F, Jiang S (2017) MERS-CoV spike protein: a key target for antivirals. Expert Opin Therapeut Target 21(2):131–143

Dutta K, Ghosh D, Basu A (2009) Curcumin protects neuronal cells from Japanese encephalitis virus-mediated cell death and also inhibits infective viral particle formation by dysregulation of ubiquitin–proteasome system. J NeuroImmune Pharmacol 4(3):328–337

El-Aziz A et al (2022) Docking analysis of some bioactive compounds from traditional plants against SARS-CoV-2 target proteins. Molecules 27(9):2662

Elghani EM, Omar FA, Emam MM, Al-Mahallawi AM, Tadros SH, Soliman FM, ElSayed AM (2022) Hesperidin hexosomal loaded nanodispersion: insights of its antimycobacterial, cytotoxic and anti-HCoV effects. Nat Prod Res 26:1–6

El-Serehy HA, Al-Misned FA, Irshad R, Ismail MM (2016) In vitro antioxidant and anti-herpes activities of Cuminum cyminum seeds extract. Biomed Res 27:1255–1260

Emon NU, Alam M, Akter I, Akhter S, Sneha AA, Irtiza M, Afroj M, Munni A, Chowdhury MH, Hossain S (2021) Virtual screenings of the bioactive constituents of tea, prickly chaff, catechu, lemon, black pepper, and synthetic compounds with the main protease (Mpro) and human angiotensin-converting enzyme 2 (ACE 2) of SARS-CoV-2. Future Journal of Pharmaceutical Sciences 7(1):1–2

Fallah MS, Bayati M, Najafi A, Behmard E, Davarpanah SJ (2021) Molecular docking investigation of antiviral herbal compounds as potential inhibitors of SARS-CoV-2 spike receptor. Biointerf Res Appl Chem 11(5):12916–12924

Fatima M, Sadaf Zaidi NU, Amraiz D, Afzal F (2016) In vitro antiviral activity of Cinnamomum cassia and its nanoparticles against H7N3 influenza a virus. J Microbiol Biotechnol 26(1): 151–159

Favela-Hernández JM, González-Santiago O, Ramírez-Cabrera MA, Esquivel-Ferriño PC, Camacho-Corona MD (2016) Chemistry and Pharmacology of Citrus sinensis. Molecules 21(2):247

Ferreira JF, Janick J (1996) Distribution of artemisinin in Artemisia annua. Prog New Crops:579–584

Ferreres F, Valentão P, Llorach R, Pinheiro C, Cardoso L, Pereira JA, Sousa C, Seabra RM, Andrade PB (2005) Phenolic compounds in external leaves of tronchuda cabbage (Brassica oleracea L. var. costata DC). J Agric Food Chem 53(8):2901–2907

Fitriani IN et al (2020) In silico approach of potential phytochemical inhibitor from Moringa oleifera, Cocos nucifera, Allium cepa, Psidium guajava, and Eucalyptus globulus for the treatment of COVID-19 by molecular docking

Flammang AM, Cifone MA, Erexson GL, Stankowski LF Jr (2004) Genotoxicity testing of a fenugreek extract. Food Chem Toxicol 42(11):1769–1775

Frabasile S, Koishi AC, Kuczera D, Silveira GF, Verri WA, Duarte dos Santos CN, Bordignon J (2017) The citrus flavanone naringenin impairs dengue virus replication in human cells. Sci Rep 7(1):1

Froggatt HM, Heaton BE, Heaton NS (2020) Development of a fluorescence-based, high-throughput SARS-CoV-2 3CLpro reporter assay. J Virol 94(22):e01265-20

Gao Y, Yan L, Huang Y, Liu F, Zhao Y, Cao L, Wang T, Sun Q, Ming Z, Zhang L, Ge J (2020) Structure of the RNA-dependent RNA polymerase from COVID-19 virus. Science 368(6492): 779–782

Gasparello J, D'Aversa E, Papi C, Gambari L, Grigolo B, Borgatti M, Finotti A, Gambari R (2021) Sulforaphane inhibits the expression of interleukin-6 and interleukin-8 induced in bronchial epithelial IB3–1 cells by exposure to the SARS-CoV-2 Spike protein. Phytomedicine 87: 153583

Ghosh AK, Gong G, Grum-Tokars V, Mulhearn DC, Baker SC, Coughlin M, Prabhakar BS, Sleeman K, Johnson ME, Mesecar AD (2008) Design, synthesis and antiviral efficacy of a series of potent chloropyridyl ester-derived SARS-CoV 3CLpro inhibitors. Bioorg Med Chem Lett 18(20):5684–5688

Gilling DH, Kitajima M, Torrey JR, Bright KR (2014) Antiviral efficacy and mechanisms of action of oregano essential oil and its primary component carvacrol against murine norovirus. J Appl Microbiol 116(5):1149–1163

Govindharajan S, Thanapal P, Swaminathan P, Vijaya Anand A, Park S, Kim IH, Balamuralikrishnan B (2020) Influences of dietary inclusion of algae *Chaetomorpha aerea* enhanced growth performance, immunity, haematological response and disease resistance of Labeo rohita challenged with *Aeromonas hydrophila*. Aquacult Rep 17:100353

Goyal B, Goyal D (2020) Targeting the dimerization of the main protease of coronaviruses: a potential broadspectrum therapeutic strategy. ACS Comb Sci 22:297–305

Gupta P, Garg P, Roy N (2011) Comparative docking and CoMFA analysis of curcumine derivatives as HIV-1 integrase inhibitors. Mol Divers 15(3):733–750

Gupta SC, Patchva S, Aggarwal BB (2013) Therapeutic roles of curcumin: lessons learned from clinical trials. AAPS J 15(1):195–218

Halder P, Pal U, Paladhi P, Dutta S, Paul P, Pal S, Das D, Ganguly A, Dutta I, Mandal S, Ray A (2022) Evaluation of potency of the selected bioactive molecules from Indian medicinal plants with MPro of SARS-CoV-2 through in silico analysis. J Ayurved Integrat Med 13(2):100449

Hamauzu Y, Yasui H, Inno T, Kume C, Omanyuda M (2005) Phenolic profile, antioxidant property, and anti-influenza viral activity of Chinese quince (Pseudocydonia sinensis Schneid.), quince (Cydonia oblonga Mill.), and apple (Malus domestica Mill.) fruits. J Agric Food Chem 53(4): 928–934

Harazem R, El Rahman SA, El-Kenawy A (2019) Evaluation of antiviral activity of allium cepa and allium sativum extracts against newcastle disease virus. Alexandria J Vet Sci 61(1)

Hariyono P, Patramurti C, Candrasari DS, Hariono M (2021) An integrated virtual screening of compounds from Carica papaya leaves against multiple protein targets of SARS-Coronavirus-2. Result Chem 3:100113

Hendrayana MA, Jawi IM, Sukrama DM, Suprapta DN (2021) The potential of polyphenols from natural ingredients against SARS-CoV-2 infection: a review. Indonesian J Pharmacol Therap 2(3):152–161

Hergenhahn M, Soto U, Weninger A, Polack A, Hsu CH, Cheng AL, Rösl F (2002) The chemopreventive compound curcumin is an efficient inhibitor of Epstein-Barr virus BZLF1 transcription in Raji DR-LUC cells. Mol Carcinog 33(3):137–145

Hilgenfeld R (2014) From SARS to MERS: crystallographic studies on coronaviral proteases enable antiviral drug design. FEBS J 281(18):4085–4096

Hillen HS, Kokic G, Farnung L, Dienemann C, Tegunov D, Cramer P (2020) Structure of replicating SARS-CoV-2 polymerase. Nature 584(7819):154–156

Hiraga Y, Ara T, Sato N, Akimoto N, Sugiyama K, Suzuki H, Kera K (2021) Metabolic analysis of unripe papaya (Carica papaya L.) to promote its utilization as a functional food. Biosci Biotechnol Biochem 85(5):1194–1204

Hoffmann M, Kleine-Weber H, Schroeder S, Krüger N, Herrler T, Erichsen S, Schiergens TS, Herrler G, Wu NH, Nitsche A, Müller MA (2020) SARS-CoV-2 cell entry depends on ACE2 and TMPRSS2 and is blocked by a clinically proven protease inhibitor. Cell 181(2):271–280

Hognon C, Marazzi M, García-Iriepa C (2022) Atomistic-level description of the covalent inhibition of SARS-CoV-2 papain-like protease. Int J Mol Sci 23(10):5855

Hu Z, Bo X, Sun G, Zhao L, Shi C, Huang L, Tian X (2022) Identification of vanillic acid and its new amide derivative from Hyoscyamus niger and their modes of action in controlling tobacco mosaic virus. Ind Crop Prod 189:115853

Huang QX, Chen HF, Luo XR, Zhang YX, Yao X, Zheng X (2018) Structure and anti-HIV activity of betulinic acid analogues. Curr Med Sci 38(3):387–397

Hulswit RJ, de Haan CA, Bosch BJ (2016) Advances in virus research. Coronav Spike Prot Trop Chang 96:29–57

Ibrahim N, Moussa AY (2021) A comparative volatilomic characterization of Florence fennel from different locations: antiviral prospects. Food Funct 12(4):1498–1515

Iheagwam FN, Rotimi SO (2020) Computer-aided analysis of multiple SARS-CoV-2 therapeutic targets: identification of potent molecules from African medicinal plants. Scientifica 12:2020

Illian DN, Siregar ES, Sumaiyah S, Utomo AR, Nuryawan A, Basyuni M (2021) Potential compounds from several Indonesian plants to prevent SARS-CoV-2 infection: a mini-review of SARS-CoV-2 therapeutic targets. Heliyon 7(1):e06001

Jain R, Sharma A, Gupta S, Sarethy IP, Gabrani R (2011) Solanum nigrum: current perspectives on therapeutic properties. Altern Med Rev 16(1):78–85

Jamwal M, Sharma S, Puri S (2022) A review on therapeutic potential of Justicia adhatoda L. Int J Recent Adv Multidiscipl Top 3(6):26–37

Javed T, Ashfaq UA, Riaz S, Rehman S, Riazuddin S (2011) In-vitro antiviral activity of Solanum nigrum against Hepatitis C Virus. Virol J 8(1):1–7

Jena AB, Kanungo N, Nayak V, Chainy GB, Dandapat J (2021) Catechin and curcumin interact with S protein of SARS-CoV2 and ACE2 of human cell membrane: insights from computational studies. Sci Rep 11(1):1–4

Jiang H, Xie Z, Koo HJ, McLaughlin SP, Timmermann BN, Gang DR (2006) Metabolic profiling and phylogenetic analysis of medicinal Zingiber species: tools for authentication of ginger (Zingiber officinale Rosc.). Phytochemistry 67(15):1673–1685

Jiang Y, Yin W, Xu HE (2021) RNA-dependent RNA polymerase: structure, mechanism, and drug discovery for COVID-19. Biochem Biophys Res Commun 538:47–53

Jose S, Gupta M, Sharma U, Quintero-Saumeth J, Dwivedi M (2022) Potential of phytocompounds from Brassica oleracea targeting S2-domain of SARS-CoV-2 spike glycoproteins: structural and molecular insights. J Mol Struct 1254:132369

Joshi DR, Shrestha AC, Adhikari N (2018) A review on diversified use of the king of spices: Piper nigrum (black pepper). Korea

Keng CT, Åkerström S, Leung CS, Poon LL, Peiris JM, Mirazimi A, Tan YJ (2011) SARS coronavirus 8b reduces viral replication by down-regulating E via an ubiquitin-independent proteasome pathway. Microbes Infect 13(2):179–188

Khan FA, Mahmood T, Ali M, Saeed A, Maalik A (2014) Pharmacological importance of an ethnobotanical plant: Capsicum annuum L. Nat Prod Res 28(16):1267–1274

Khan T, Khan MA, Ullah N, Nadhman A (2021) Therapeutic potential of medicinal plants against COVID-19: the role of antiviral medicinal metabolites. Biocatalysis Agric Biotechnol 31: 101890

Kharisma VD, Agatha A, Ansori AN, Widyananda MH, Rizky WC, Dings TG, Derkho M, Lykasova I, Antonius Y, Rosadi I, Zainul R (2021) Herbal combination from Moringa oleifera Lam. and Curcuma longa L. as SARS-CoV-2 antiviral via dual inhibitor pathway: a viroinformatics approach. J Pharm Pharmacogn Res 10(1):138–146

Khubber S, Hashemifesharaki R, Mohammadi M, Gharibzahedi SM (2020) Garlic (Allium sativum L.): a potential unique therapeutic food rich in organosulfur and flavonoid compounds to fight with COVID-19. Nutr J 19(1):1–3

Kim EH, Lee BW, Ryu B, Cho HM, Kim SM, Jang SG, Casel MA, Rollon R, Yoo JS, Poo H, Oh WK (2022) Inhibition of a broad range of SARS-CoV-2 variants by antiviral phytochemicals in hACE2 mice. Antivir Res 204:105371

Kiran G, Karthik L, Devi MS, Sathiyarajeswaran P, Kanakavalli K, Kumar KM, Kumar DR (2020) In silico computational screening of Kabasura Kudineer-official Siddha formulation and JACOM against SARS-CoV-2 spike protein. J Ayurveda Integrat Med 13:100324

Klimek-Szczykutowicz M, Szopa A, Ekiert H (2020) Citrus limon (Lemon) phenomenon—a review of the chemistry, pharmacological properties, applications in the modern pharmaceutical, food, and cosmetics industries, and biotechnological studies. Plants 9(1):119

Kosari M, Noureddini M, Khamechi SP, Najafi A, Ghaderi A, Sehat M, Banafshe HR (2021) The effect of propolis plus Hyoscyamus niger L. methanolic extract on clinical symptoms in patients with acute respiratory syndrome suspected to COVID-19: a clinical trial. Phytother Res 35(7): 4000–4006

Kuchi Bhotla H, Kaul T, Balasubramanian B, Easwaran M, Arumugam VA, Pappusamy M, Muthupandian S, Meyyazhagan A (2020) Platelets to surrogate lung inflammation in COVID-19 patients. Med Hypotheses 143:110098

Kuchi Bhotla H, Balasubramanian B, Arumugam VA, Pushparaj K, Murugesh E, Rathinasamy B (2021a) Insinuating cocktailed components in biocompatible-nanoparticles could act as an impressive neo-adjuvant strategy to combat COVID-19. Nat Resour Hum Health:3–7

Kuchi Bhotla H, Balasubramanian B, Meyyazhagan A, Pushparaj K, Easwaran M, Pappusamy M, Alwin Robert A, Arumugam VA, Tsibizova V, Msaad Alfalih A, Aljowaie RM, Saravanan M, Di Renzo GC (2021b) Opportunistic mycoses in COVID-19 patients/survivors: epidemic inside a pandemic. J Infect Public Health 14(11):1720–1726

Kumar GS, Manivannan R, Nivetha B (2021) In Silico identification of flavonoids from coriandrum sativum seeds against coronavirus Covid-19 main protease. J Drug Deliv Therap 11(2):145–152

Kumari R, Rathi B, Rani A, Bhatnagar S (2013) Rauvolfia serpentina L. Benth. ex Kurz.: phytochemical, pharmacological and therapeutic aspects. Int J Pharm Sci Rev Res 23(2): 348–355

Lawan KM, Bharti J, Kargo MA, Bello UR (2020) Impact of medicinal plants on treatment of SARS-CoV, SARS-CoV-2 and influenza virus in India. Asian J Pharm Pharmacol 6(5):306–311

Leela N, Shafeekh K (2008) Fenugreek, chemistry of spices. CAB International, Pondicherry

Li F, Li W, Farzan M, Harrison SC (2005a) Structure of SARS coronavirus spike receptor-binding domain complexed with receptor. Science 309(5742):1864–1868

Li SY, Chen C, Zhang HQ, Guo HY, Wang H, Wang L, Zhang X, Hua SN, Yu J, Xiao PG, Li RS (2005b) Identification of natural compounds with antiviral activities against SARS-associated coronavirus. Antivir Res 67(1):18–23

Lim TK (2015) Glycyrrhiza glabra. In: Edible medicinal and non-medicinal plants, pp 354–457. Published 2015 Oct 22. https://doi.org/10.1007/978-94-017-7276-1_18

Lin LT, Hsu WC, Lin CC (2014) Antiviral natural products and herbal medicines. J Tradit Complement Med 4(1):24–35

Liston A, Cronn R, Ashman TL (2014) Fragaria: a genus with deep historical roots and ripe for evolutionary and ecological insights. Am J Bot 101(10):1686–1699

Liu S, Ding Z, Zhang L, Gu Z, Wang X, Sun X, Shi G (2012) Ethanol production from Lycoris radiata Herbert (Amaryllidaceae) residues as a new resource. Biomass Bioenergy 37:237–242

Liu Y, Liang C, Xin L, Ren X, Tian L, Ju X, Li H, Wang Y, Zhao Q, Liu H, Cao W (2020) The development of coronavirus 3C-Like protease (3CLpro) inhibitors from 2010 to 2020. Eur J Med Chem 206:112711

Lv Y, An Z, Chen H, Wang Z, Liu L (2014a) Mechanism of curcumin resistance to human cytomegalovirus in HELF cells. BMC Complement Altern Med 14(1):1–7

Lv Y, Lei N, Wang D, An Z, Li G, Han F, Liu H, Liu L (2014b) Protective effect of curcumin against cytomegalovirus infection in Balb/c mice. Environ Toxicol Pharmacol 37(3):1140–1147

Malabadi RB, Kolkar KP, Meti NT, Chalannavar RK (2021) Traditional herbal folk medicine used for controlling corona virus (sars-cov-2) disease (covid-19). Int J Innov Sci Res Rev 3(7): 1507–1517

Mallick N, Khan RA (2016) Antihyperlipidemic effects of Citrus sinensis, Citrus paradisi, and their combinations. J Pharm Bioall Sci 8(2):112

Mangukia N, Rao P, Patel K, Pandya H, Rawal RM (2022) Unveiling the nature's fruit basket to computationally identify Citrus sinensis csi-mir169–3p as a probable plant miRNA against Reference and Omicron SARS-CoV-2 genome. Comput Biol Med 146:105502

Manikandan R, Anand AV (2016) Evaluation of antioxidant activity of Psidium guajava Linn. in streptozotocin–induced diabetic rats. Free Radical Antioxid 6(1):72–76

Manzano-Santana PI, Tivillin JP, Chóez-Guaranda IA, Lucas AD, Orellana-Manzano AK, Rastrelli L (2021) Potential bioactive compounds of medicinal plants against new Coronavirus (SARS-CoV-2): a review. Rev Bionatura 6(1):1653–1658

Marín-Palma D, Tabares-Guevara JH, Zapata-Cardona MI, Flórez-Álvarez L, Yepes LM, Rugeles MT, Zapata-Builes W, Hernandez JC, Taborda NA (2021) Curcumin inhibits in vitro SARS-CoV-2 infection in vero E6 cells through multiple antiviral mechanisms. Molecules 26(22):6900

Marwah A, Marwah P (2020) Coronavirus (COVID-19): a protocol for prevention, treatment and control. J Appl Nat Sci 12(2):119–123

Meyyazhagan A, Pushparaj K, Balasubramanian B, Kuchi Bhotla H, Pappusamy M, Arumugam VA, Easwaran M, Pottail L, Mani P, Tsibizova V, Di Renzo GC (2022) COVID-19 in pregnant women and children: insights on clinical manifestations, complexities, and pathogenesis. Int J Gynaecol Obstet 156(2):216–224

Miraj S, Kiani S (2016) A review study of therapeutic effects of Salvia officinalis L. Pharm Lett 8(6):299–303

Mody V, Ho J, Wills S, Mawri A, Lawson L, Ebert MC, Fortin GM, Rayalam S, Taval S (2021) Identification of 3-chymotrypsin like protease (3CLPro) inhibitors as potential anti-SARS-CoV-2 agents. Commun Biol 4(1):1–10

Moezzi MS (2022) Comprehensive in silico screening of flavonoids against SARS-COV-2 main protease. J Biomol Struct Dyn:1–4

Momin AH, Acharya SS, Gajjar AV (2012) Coriandrum sativum-review of advances in phytopharmacology. Int J Pharm Sci Res 3(5):1233

Mótyán JA, Mahdi M, Hoffka G, Tőzsér J (2022) Potential resistance of SARS-CoV-2 main protease (Mpro) against protease inhibitors: lessons learned from HIV-1 protease. Int J Mol Sci 23(7):3507

Mounce BC, Cesaro T, Carrau L, Vallet T, Vignuzzi M (2017) Curcumin inhibits Zika and chikungunya virus infection by inhibiting cell binding. Antivir Res 142:148–157

Mukherjee PK, Efferth T, Das B, Kar A, Ghosh S, Singha S, Debnath P, Sharma N, Bhardwaj P, Haldar PK (2022) Role of medicinal plants in inhibiting SARS-CoV-2 and in the management of post-COVID-19 complications. Phytomedicine 98:153930

Muthumanickam S, Kamaladevi A, Boomi P, Gowrishankar S, Pandian SK (2021) Indian ethnomedicinal phytochemicals as promising inhibitors of RNA-binding domain of SARS-CoV-2 nucleocapsid phosphoprotein: an in silico study. Front Mol Biosci 8:114

Nair A, James TJ (2020) Computational screening of phytocompounds from Moringa oleifera leaf as potential inhibitors of SARS-CoV-2 Mpro

Nair V, Singh S, Gupta YK (2012) Evaluation of disease modifying activity of Coriandrum sativum in experimental models. Indian J Med Res 135(2):240

Nair MS, Huang Y, Fidock DA, Polyak SJ, Wagoner J, Towler MJ, Weathers PJ (2021) Artemisia annua L. extracts inhibit the in vitro replication of SARS-CoV-2 and two of its variants. J Ethnopharmacol 274:114016

Nandi S, Roy H, Gummadi A, Saxena AK (2021) Exploring spike protein as potential target of novel coronavirus and to inhibit the viability utilizing natural agents. Curr Drug Targets 22(17): 2006–2020

Nargis Begum T, Muhammad Ilyas MH, Burkanudeen A, Kalavathy S, Vijaya Anand A, Sampathkumar P, Jaswanth A (2009) Hypoglycemic and antihyperlipidemic activity of ethanolic leaf extract of *Azima tetracantha* Lam. on alloxan-induced diabetic rats. J Cell Tissue Res 9(1):1681–1685

Nascimento Junior JA, Santos AM, Quintans-Junior LJ, Walker CI, Borges LP, Serafini MR (2020) SARS, MERS and SARS-CoV-2 (COVID-19) treatment: a patent review. Expert Opin Therap Patent 30(8):567–579

Nassiri-Asl M, Hosseinzadeh H (2016) Review of the pharmacological effects of Vitis vinifera (Grape) and its bioactive constituents: an update. Phytother Res 30(9):1392–1403

Natesh J, Mondal P, Penta D, Salam AA, Meeran SM (2021) Culinary spice bioactives as potential therapeutics against SARS-CoV-2: computational investigation. Comput Biol Med 128:104102

Ngwe Tun MM, Toume K, Luvai E, Nwe KM, Mizukami S, Hirayama K, Komatsu K, Morita K (2022) The discovery of herbal drugs and natural compounds as inhibitors of SARS-CoV-2 infection in vitro. J Nat Med 76(2):402–409

Niemeyer D, Mösbauer K, Klein EM, Sieberg A, Mettelman RC, Mielech AM, Dijkman R, Baker SC, Drosten C, Müller MA (2018) The papain-like protease determines a virulence trait that varies among members of the SARS-coronavirus species. PLoS Pathog 14(9):e1007296

Nieto G, Ros G, Castillo J (2018) Antioxidant and antimicrobial properties of rosemary (Rosmarinus officinalis, L.): a review. Fortschr Med 5(3):98

Nolkemper S, Reichling J, Stintzing FC, Carle R, Schnitzler P (2006) Antiviral effect of aqueous extracts from species of the Lamiaceae family against Herpes simplex virus type 1 and type 2 in vitro. Planta Med 72(15):1378–1382

Obata K, Kojima T, Masaki T, Okabayashi T, Yokota S, Hirakawa S, Nomura K, Takasawa A, Murata M, Tanaka S, Fuchimoto J (2013) Curcumin prevents replication of respiratory syncytial virus and the epithelial responses to it in human nasal epithelial cells. PLoS One 8(9):e70225

Okeke ES, Ogugofor MO, Nkwoemeka NE, Nweze EJ, Okoye CO (2022) Phytomelatonin: a potential phytotherapeutic intervention on COVID-19-exposed individuals. Microbes Infect 24(1):104886

Omosa et al (2017) Medicinal spices and vegetables from Africa. Therapeutic potential against metabolic, inflammatory, infectious and systemic diseases, 1st edn. Victor Kuete. ISBN: 9780128092866, eBook ISBN: 9780128094419

Orallo F (2006) Comparative studies of the antioxidant effects of cis-and trans-resveratrol. Curr Med Chem 13(1):87–98

Ordonez AA et al (2021) Sulforaphane exhibits in vitro and in vivo antiviral activity against pandemic SARS-CoV-2 and seasonal HCoV-OC43 coronaviruses. BioRxiv

Osipiuk J, Azizi SA, Dvorkin S, Endres M, Jedrzejczak R, Jones KA, Kang S, Kathayat RS, Kim Y, Lisnyak VG, Maki SL (2021) Structure of papain-like protease from SARS-CoV-2 and its complexes with non-covalent inhibitors. Nat Commun 12(1):1–9

Ou JL, Mizushina Y, Wang SY, Chuang DY, Nadar M, Hsu WL (2013) Structure–activity relationship analysis of curcumin analogues on anti-influenza virus activity. FEBS J 280(22): 5829–5840

Ou X, Zheng W, Shan Y, Mu Z, Dominguez SR, Holmes KV, Qian Z (2016) Identification of the fusion peptide-containing region in betacoronavirus spike glycoproteins. J Virol 90(12): 5586–5600

Ou X, Liu Y, Lei X, Li P, Mi D, Ren L, Guo L, Guo R, Chen T, Hu J, Xiang Z (2020) Characterization of spike glycoprotein of SARS-CoV-2 on virus entry and its immune cross-reactivity with SARS-CoV. Nat Commun 11(1):1–2

Pa R, Mathew L (2012) Antimicrobial activity of leaf extracts of Justicia adhatoda L. in comparison with vasicine. Asian Pac J Trop Biomed 2(3):S1556–S1560

Padmanabhan P, Desikan R, Dixit NM (2020) Targeting TMPRSS2 and Cathepsin B/L together may be synergistic against SARS-CoV-2 infection. PLoS Comput Biol 16(12):e1008461

Pandey K, Lokhande KB, Swamy KV, Nagar S, Dake M (2021) In silico exploration of phytoconstituents from Phyllanthus emblica and Aegle marmelos as potential therapeutics against SARS-CoV-2 RdRp. Bioinform Biol Insights 15:11779322211027403

Pandey S, Kushwaha R, Singh M, Varma JK, Singh S (2022) phytochemical composition and medicinal uses of Trigonella foenum-graecum (fenugreek) with special assessment against sars-cov-2 infection. Europ J Biomed 9(10):375–379

Papadakis M (2022) Why comparing coronavirus disease 2019 (COVID-19) and seasonal influenza fatality rates is like comparing apples to pears. Infect Control Hosp Epidemiol 43(7):958–959

Pareek S, Sagar NA, Sharma S, Kumar V (2017) Onion (Allium cepa L.). Fruit and vegetable phytochemicals: chemistry and human. Health 2:1145–1162

Patel B, Sharma S, Nair N et al (2021) Therapeutic opportunities of edible antiviral plants for COVID-19. Mol Cell Biochem 476(6):2345–2364. https://doi.org/10.1007/s11010-021-04084-7. PMID: 33587232; PMCID: PMC7882868

Peele KA, Durthi CP, Srihansa T, Krupanidhi S, Ayyagari VS, Babu DJ, Indira M, Reddy AR, Venkateswarulu TC (2020) Molecular docking and dynamic simulations for antiviral compounds against SARS-CoV-2: a computational study. Informat Med Unlock 19:100345

Peng W, Liu YJ, Wu N, Sun T, He XY, Gao YX, Wu CJ (2015) Areca catechu L.(Arecaceae): a review of its traditional uses, botany, phytochemistry, pharmacology and toxicology. J Ethnopharmacol 164:340–356

Peng Q, Peng R, Yuan B, Zhao J, Wang M, Wang X, Wang Q, Sun Y, Fan Z, Qi J, Gao GF (2020) Structural and biochemical characterization of the nsp12-nsp7-nsp8 core polymerase complex from SARS-CoV-2. Cell Rep 31(11):107774

Pérez-Martínez AL, Eguiarte LE, Mercer KL, Martínez-Ainsworth NE, McHale L, van der Knaap E, Jardón-Barbolla L (2022) Genetic diversity, gene flow, and differentiation among wild, semiwild, and landrace chile pepper (Capsicum annuum) populations in Oaxaca. Mexico Am J Bot 109(7):1157–1176

Pezzani R, Vitalini S, Iriti M (2017) Bioactivities of Origanum vulgare L.: an update. Phytochem Rev 16(6):1253–1268

Prasanth DS, Murahari M, Chandramohan V, Panda SP, Atmakuri LR, Guntupalli C (2021) In silico identification of potential inhibitors from Cinnamon against main protease and spike glycoprotein of SARS CoV-2. J Biomol Struct Dyn 39(13):4618–4632

Priya NC, Kumari PS (2017) Antiviral activities and cytotoxicity assay of seed extracts of Piper longum and Piper nigrum on human cell lines. Int J Pharm Sci Rev Res 44(1):197–202

Priya V, Jananie RK, Vijayalakshmi K (2011) GC/MS determination of bioactive components of Trigonella foenum-graecum. J Chem Pharm Res 3(5):35–40

Pushparaj K, Kuchi Bhotla H, Arumugam VA, Pappusamy M, Easwaran M, Liu WC, Issara U, Rengasamy KRR, Meyyazhagan A, Balasubramanian B (2022) Mucormycosis (black fungus) ensuing COVID-19 and comorbidity meets - Magnifying global pandemic grieve and catastrophe begins. Sci Total Environ 805:150355

Qiao Z, Wei N, Jin L, Zhang H, Luo J, Zhang Y, Wang K (2021) The Mpro structure-based modifications of ebselen derivatives for improved antiviral activity against SARS-CoV-2 virus. Bioorg Chem 117:105455

Qin F, Yao L, Lu C, Li C, Zhou Y, Su C, Chen B, Shen Y (2019) Phenolic composition, antioxidant and antibacterial properties, and in vitro anti-HepG2 cell activities of wild apricot (Armeniaca Sibirica L. Lam) kernel skins. Food Chem Toxicol 129:354–364

Rajagopal K, Byran G, Jupudi S, Vadivelan R (2020) Activity of phytochemical constituents of black pepper, ginger, and garlic against coronavirus (COVID-19): an in silico approach. Int J Health Allied Sci 9(5):43–50

Rattis BA, Ramos SG, Celes M (2021) Curcumin as a Potential Treatment for COVID-19. Front Pharmacol 12:1068

Reina J, Iglesias C (2022) Nirmatrelvir más ritonavir (Paxlovid) una potente combinación inhibidora de la proteasa 3CLpro del SARS-CoV-2. Revista Española de Quimioterapia 35(3):236

Rieder AS, Deniz BF, Netto CA, Wyse AT (2022) A review of in silico research, SARS-CoV-2, and neurodegeneration: focus on papain-like protease. Neurotox Res:1–7

Rodríguez VM, Soengas P, Alonso-Villaverde V, Sotelo T, Cartea ME, Velasco P (2015) Effect of temperature stress on the early vegetative development of Brassica oleracea L. BMC Plant Biol 15(1):1–9

Rout J, Swain BC, Tripathy U (2022) In silico investigation of spice molecules as potent inhibitor of SARS-CoV-2. J Biomol Struct Dyn 40(2):860–874

Rudrapal M, Celik I, Chinnam S, Ansari MA, Khan J, Alghamdi S, Almehmadi M, Zothantluanga JH, Khairnar SJ (2022) Phytocompounds as potential inhibitors of SARS-CoV-2 Mpro and PLpro through computational studies. Saudi J Biol Sci 29(5):3456–3465

Rut W, Lv Z, Zmudzinski M, Patchett S, Nayak D, Snipas SJ, El Oualid F, Huang TT, Bekes M, Drag M, Olsen SK (2020) Activity profiling and crystal structures of inhibitor-bound SARS-

CoV-2 papain-like protease: a framework for anti–COVID-19 drug design. Sci Adv 6(42): eabd4596

Saab AM, Tacchini M, Sacchetti G, Contini C, Schulz H, Lampronti I, Gambari R, Makhlouf H, Tannoury M, Venditti A, Bianco A (2021) Phytochemical analysis and potential natural compounds against SARS-CoV-2/COVID-19 in essential oils derived from medicinal plants originating from Lebanon. An information note. Plant Biosyst 156(5):1–17

Saakre M, Mathew D, Ravisankar V (2021) Perspectives on plant flavonoid quercetin-based drugs for novel SARS-CoV-2. Beni-Suef Univ J Basic Appl Sci 10(1):1–3

Sabbah DA, Hajjo R, Bardaweel SK, Zhong HA (2021) An updated review on SARS-CoV-2 main Proteinase (MPro): protein structure and small-molecule inhibitors. Curr Top Med Chem 21(6): 442–460

Saif A (2022) Study of essential oil of different spices as inhibitor of 3CL protease of SARS-CoV-2. Doctoral dissertation, Capital University

Saravana Prabha P, Murugesh E, Balamuralikrishnan B, Mohan A, Arun M, Park S, Bhotla HK, Jeeva A, Vijaya Anand A, Sasikala K, Gopalakrishnan VK, Manikantan P, Tanushri K (2020) Employing bioactive compounds derived from *Ipomoea obscura* (L.) to evaluate potential inhibitor for SARS-CoV-2 main protease and ACE2 protein. Food Front 1:168–179

Semenya S, Potgieter M, Erasmus L (2012) Ethnobotanical survey of medicinal plants used by Bapedi healers to treat diabetes mellitus in the Limpopo Province, South Africa. J Ethnopharmacol 141(1):440–445

Semwal DK, Chauhan A, Semwal RB, Sircar D, Roy P, Lehmann J (2020) Natural molecules having anti-SARS-CoV activity-cannot they be effective against SARS-CoV-2? Curr Sci 119(5)

Sen D, Bhaumik S, Debnath P, Debnath S (2021) Potentiality of Moringa oleifera against SARS-CoV-2: identified by a rational computer aided drug design method. J Biomol Struct Dyn:1–8

Sen D, Debnath P, Debnath B, Bhaumik S, Debnath S (2022) Identification of potential inhibitors of SARS-CoV-2 main protease and spike receptor from 10 important spices through structure-based virtual screening and molecular dynamic study. J Biomol Struct Dyn 40(2):941–962

Senthil Kumar KJ, Gokila Vani M, Wang CS, Chen CC, Chen YC, Lu LP, Huang CH, Lai CS, Wang SY (2020) Geranium and lemon essential oils and their active compounds downregulate angiotensin-converting enzyme 2 (ACE2), a SARS-CoV-2 spike receptor-binding domain, in epithelial cells. Plants. 9(6):770

Senthilvel P, Lavanya P, Kumar KM, Swetha R, Anitha P, Bag S, Sarveswari S, Vijayakumar V, Ramaiah S, Anbarasu A (2013) Flavonoid from Carica papaya inhibits NS2B-NS3 protease and prevents Dengue 2 viral assembly. Bioinformation 9(18):889

Shajahan A, Pepi LE, Rouhani DS, Heiss C, Azadi P (2021) Glycosylation of SARS-CoV-2: structural and functional insights. Anal Bioanal Chem 413(29):7179–7193

Shanmugam R, Thangavelu S, Fathah Z, Yatoo MI, Tiwari R, Pandey MK, Dhama J, Chandra R, Malik Y, Dhama K, Sha R, Chaicumpa W, Shanmugam V, Arumugam AV (2020) SARS-CoV-2/COVID-19 pandemic-an update. J Exp Biol Agric Sci 8(1):S219–S245

Shetler CL, Ferreira JC, Cardoso TH, Silva EM, Saksena NK, Rabeh WM (2022) Therapeutic potential of metal ions for COVID-19: insights from the papain-like protease of SARS-CoV-2. Biochem J 479(20):2175–2193

Shimamura Y, Hirai C, Sugiyama Y, Utsumi M, Yanagida A, Murata M, Ohashi N, Masuda S (2017) Interaction between various apple procyanidin and staphylococcal enterotoxin A and their inhibitory effects on toxin activity. Toxins 9(8):243

Singh N, Rao AS, Nandal A, Kumar S, Yadav SS, Ganaie SA, Narasimhan B (2021a) Phytochemical and pharmacological review of Cinnamomum verum J. Presl-a versatile spice used in food and nutrition. Food Chem 338:127773. https://doi.org/10.1016/j.foodchem.2020.127773

Singh NA, Kumar P, Kumar N (2021b) Spices and herbs: potential antiviral preventives and immunity boosters during COVID-19. Phytother Res 35(5):2745–2757

Sobia HN, Khan S, Nadeem F (2018) Use of malabar nut (Justicia adhatoda L.) from traditional medicine to current pharmacopeia–a review study. Int J Chem Biochem Sci 13:46–51

Song C, Zhao S, Hong X, Liu J, Schulenburg K, Schwab W (2016) A UDP-glucosyltransferase functions in both acylphloroglucinol glucoside and anthocyanin biosynthesis in strawberry (Fragaria× ananassa). Plant J 85(6):730–742

Souza PF, Mesquita FP, Amaral JL, Landim PG, Lima KR, Costa MB, Farias IR, Belém MO, Pinto YO, Moreira HH, Magalhaes IC (2022) The spike glycoproteins of SARS-CoV-2: a review of how mutations of spike glycoproteins have driven the emergence of variants with high transmissibility and immune escape. Int J Biol Macromol 208:105–125

Srinivasan V, Brognaro H, Prabhu PR, de Souza EE, Günther S, Reinke PY, Lane TJ, Ginn H, Han H, Ewert W, Sprenger J (2022) Antiviral activity of natural phenolic compounds in complex at an allosteric site of SARS-CoV-2 papain-like protease. Commun Biol 5(1):1–2

Suleiman WB, Helal EE (2022) Chemical constituents and potential pleiotropic activities of Foeniculum vulgare (Fennel) ethanolic extract; in vitro approach. Egypt J Chem 65(7):617–626

Suryanarayana AA, Shankar S, Manoharan JP, Subramanian V (2021) In-Silico analysis and molecular docking studies of phytoconstituents of Justicia adhatoda as potential inhibitors of SARS-CoV2 target proteins. In: Proceedings of the first international conference on combinatorial and optimization, ICCAP 2021, December 7–8 2021, Chennai, India

Tang T, Bidon M, Jaimes JA, Whittaker GR, Daniel S (2020) Coronavirus membrane fusion mechanism offers a potential target for antiviral development. Antivir Res 178:104792

Taylor PW, Roberts SD (1999) Clove oil: an alternative anaesthetic for aquaculture. N Am J Aquac 61(2):150–155

Teli DM, Patel B, Chhabria MT (2022) Fragment-based design of SARS-CoV-2 Mpro inhibitors. Struct Chem 33(6):2155–2168

Thiel V, Ivanov KA, Putics A, Hertzig T, Schelle B, Bayer S, Weißbrich B, Snijder EJ, Rabenau H, Doerr HW, Gorbalenya AE (2003) Mechanisms and enzymes involved in SARS coronavirus genome expression. J Gen Virol 84(9):2305–2315

Tian Y, Deng F (2020) Phytochemistry and biological activity of mustard (Brassica juncea): a review. CyTA-J Food 18(1):704–718

Tian SL, Li L, Chai WG, Shah SN, Gong ZH (2014) Effects of silencing key genes in the capsanthin biosynthetic pathway on fruit color of detached pepper fruits. BMC Plant Biol 14(1):1–2

Torres Neto L, Monteiro ML, Fernández-Romero J, Teleshova N, Sailer J, Conte Junior CA (2022) Essential oils block cellular entry of SARS-CoV-2 delta variant. Sci Rep 12(1):1–9

Tundis R, Loizzo MR, Menichini F (2014) An overview on chemical aspects and potential health benefits of limonoids and their derivatives. Crit Rev Food Sci Nutr 54(2):225–250

Ullrich S, Nitsche C (2022) SARS-CoV-2 papain-like protease: structure, function and inhibition. Chembiochem 23(19):e202200327

Umesh, Kundu D, Selvaraj C, Singh SK, Dubey VK (2021) Identification of new anti-nCoV drug chemical compounds from Indian spices exploiting SARS-CoV-2 main protease as target. J Biomol Struct Dyn 39(9):3428–3434

Vicenti I, Zazzi M, Saladini F (2021) SARS-CoV-2 RNA-dependent RNA polymerase as a therapeutic target for COVID-19. Expert Opin Therapeut Patent 31(4):325–337

Vijaya Anand A, Velayuthaprabhu S, Rengarajan RL, Sampathkumar P, Radhakrishnan R (2020) Bioactive compounds of Guava (L.). In: Bioactive compounds in underutilized fruits and nuts. Springer, Cham, pp 503–527

Vijayakumar K, Anand AV, Manikandan R (2015) In vitro antioxidant activity of ethanolic extract of *Psidium guajava* leaves. Int J Res Stud Biosci 3(5):145–149

Vijayakumar K, Rengarajan RL, Radhakrishnan R, Mathew S, Qadri I, Vijaya Anand A (2019) *Psidium guajava* leaf extracts and their quercetin protect HepG2 cell lines against $CCL_4$ induced cytotoxicity. Indian J Clin Biochem 34(3):324–329

Wang Q, Wu J, Wang H, Gao Y, Liu Q, Mu A, Ji W, Yan L, Zhu Y, Zhu C, Fang X (2020) Structural basis for RNA replication by the SARS-CoV-2 polymerase. Cell 182(2):417–428

Wang W, Li Q, Wu J, Hu Y, Wu G, Yu C, Xu K, Liu X, Wang Q, Huang W, Wang L (2021a) Lentil lectin derived from Lens culinaris exhibit broad antiviral activities against SARS-CoV-2 variants. Emerg Microb Infect 1(1):1519–1529

Wang Y, Anirudhan V, Du R, Cui Q, Rong L (2021b) RNA-dependent RNA polymerase of SARS-CoV-2 as a therapeutic target. J Med Virol 93(1):300–310

Wardani AK (2022) Some natural flavonoids from guava (psidium guajava l.) against SARS-CoV-2, in silico prediction and drugs evaluation. Jurnal Farmasi Sains dan Praktis:170–179

Wen CC, Kuo YH, Jan JT, Liang PH, Wang SY, Liu HG, Lee CK, Chang ST, Kuo CJ, Lee SS, Hou CC (2007) Specific plant terpenoids and lignoids possess potent antiviral activities against severe acute respiratory syndrome coronavirus. J Med Chem 50(17):4087–4095

Wu YN, Wang W, Yao GD, Jiang XF, Zhang Y, Song SJ (2018) Antioxidant phenolic acids from the leaves of Armeniaca sibirica. J Asian Nat Prod Res 20(10):969–976

Xu JG, Liu T, Hu QP, Cao XM (2016) Chemical composition, antibacterial properties and mechanism of action of essential oil from clove buds against Staphylococcus aureus. Molecules 21(9):1194

Xu L, Zaky MY, Yousuf W, Ullah A, Abdelbaset GR, Zhang Y, Ahmed OM, Liu S, Liu H (2021) The anticancer potential of apigenin via immunoregulation. Curr Pharm Des 27(4):479–489

Xu X, Chen Y, Lu X, Zhang W, Fang W, Yuan L, Wang X (2022) An update on inhibitors targeting RNA-dependent RNA polymerase for COVID-19 treatment: promises and challenges. Biochem Pharmacol 205:115279

Yadav PK, Jaiswal A, Singh RK (2021a) In silico study on spice-derived antiviral phytochemicals against SARS-CoV-2 TMPRSS2 target. J Biomol Struct Dyn 40:1

Yadav R, Chaudhary JK, Jain N, Chaudhary PK, Khanra S, Dhamija P, Sharma A, Kumar A, Handu S (2021b) Role of structural and non-structural proteins and therapeutic targets of SARS-CoV-2 for COVID-19. Cell 10(4):821

Yadav PK, Jaiswal A, Singh RK (2022) In silico study on spice-derived antiviral phytochemicals against SARS-CoV-2 TMPRSS2 target. J Biomol Struct Dyn 40(22):11874–11884

Yakhchali M, Taghipour Z, Ardakani MM, Vaghasloo MA, Vazirian M, Sadrai S (2021) Cinnamon and its possible impact on COVID-19: the viewpoint of traditional and conventional medicine. Biomed Pharmacother 143:112221

Yan H, Liu Z, Yan G, Liu X, Liu X, Wang Y, Chen Y (2022) A robust high-throughput fluorescence polarization assay for rapid screening of SARS-CoV-2 papain-like protease inhibitors. Virology 574:18–24

Yang M, Lee G, Si J, Lee SJ, You HJ, Ko G (2016) Curcumin shows antiviral properties against norovirus. Molecules 21(10):1401

Yin S, Mei S, Li Z, Xu Z, Wu Y, Chen X, Liu D, Niu MM, Li J (2022) Non-covalent cyclic peptides simultaneously targeting Mpro and NRP1 are highly effective against Omicron BA. 2.75. Front Pharmacol 13:1037993

Yu YB (2004) The extracts of Solanum nigrum L. for inhibitory effects on HIV-1 and its essential enzymes. Korean J Orient Med 10(1):119–126

Yunus G (2021) Herbal compounds from Syzygium aromaticum and Cassia acutifolia as a shield against SARS-CoV-2 Mpro: a molecular docking approach. Biointerf Res Appl Chem 11: 14853–14865

Zannella C, Giugliano R, Chianese A, Buonocore C, Vitale GA, Sanna G, Sarno F, Manzin A, Nebbioso A, Termolino P, Altucci L (2021) Antiviral activity of Vitis vinifera leaf extract against SARS-CoV-2 and HSV-1. Viruses 13(7):1263

Zhang Y, Wu YN, Wang W, Yao GD, Jiang XF, Song SJ (2018) Flavones from a natural tea (the leaves of Armeniaca sibirica L.) prevent oxidative stress-induced neuronal death. J Food Biochem 42(6):e12657

Zhang L, Lin D, Sun X, Curth U, Drosten C, Sauerhering L, Becker S, Rox K, Hilgenfeld R (2020) Crystal structure of SARS-CoV-2 main protease provides a basis for design of improved α-ketoamide inhibitors. Science 368(6489):409–412

Zhang LC, Zhao HL, Liu J, He L, Yu RL, Kang CM (2022) Design of SARS-CoV-2 Mpro, PLpro dual-target inhibitors based on deep reinforcement learning and virtual screening. Future Med Chem 14(6):393–405

Zhao T, Sun Q, Marques M, Witcher M (2015) Anticancer properties of Phyllanthus emblica. Oxidative Med Cell Longev 2015:950890

Zhao J, Guo S, Yi D, Li Q, Ma L, Zhang Y, Wang J, Li X, Guo F, Lin R, Liang C (2021) A cell-based assay to discover inhibitors of SARS-CoV-2 RNA dependent RNA polymerase. Antivir Res 190:105078

Zhu W, Chen CZ, Gorshkov K, Xu M, Lo DC, Zheng W (2020) RNA-dependent RNA polymerase as a target for COVID-19 drug discovery. SLAS Discov Adv Sci Drug Discov 25(10): 1141–1151

Zrig A (2022) The effect of phytocompounds of medicinal plants on coronavirus (2019-NCOV) infection. Pharm Chem J 55(10):1080–1084

# Chapter 11
# Therapeutic Potential of Selected Medicinal Plants for Neurological Disorders after the Infection of COVID-19

Ravindra Verma, Prakash S. Bisen, and Mònica Bulló

## 11.1 Introduction

World Health Organization statistics indicate that 14% of the global disease burden is caused by neurological disorders, and 85% of global deaths are caused by cerebrovascular diseases. Lower-middle-income countries have a higher death rate (16.8%) than high-income countries (13.2%) (https://www.who.int/mental_health/neurology/neurological_disorders_report_web.pdf). WHO also stated that COVID-19-related mental health system upheavals are mostly affecting preexisting mental health issues (Lancet 2020). According to a United Nations study, COVID-19 affects more than one in five people mentally and is associated with a number of neurological disorders (Brief 2020). The mechanism behind the evolution of SARS-CoV-2, which causes the severe acute respiratory syndrome, is unclear. Several of the claims are based on research conducted at Wuhan University, and others are based on ecological evidence (Boni et al. 2020; Xiao et al. 2020; Lam et al. 2020; Latinne et al. 2020). Several neurological disorders have been reported in patients infected with the SARS-CoV-2 virus, including heart, liver, and brain infections (Karuppan et al. 2021). Blood clotting and cytokine storms in the peripheral and

R. Verma (✉) · P. S. Bisen
School of Studies in Biotechnology, Jiwaji University, Gwalior, India

M. Bulló
Department of Biochemistry and Biotechnology, University Rovira i Virgili (URV), Tarragona, Spain

Nutrition and Metabolic Health Research Group, Institute of Health Pere Virgili (IISPV), Tarrangona, Spain

CIBER Physiology of Obesity and Nutrition (CIBEROBN), Carlos III Health Institute, Madrid, Spain
e-mail: monica.bullo@urv.cat

 
J.-T. Chen (ed.), *Ethnopharmacology and Drug Discovery for COVID-19: Anti-SARS-CoV-2 Agents from Herbal Medicines and Natural Products*,
https://doi.org/10.1007/978-981-99-3664-9_11

central nervous systems are responsible for a number of these neurological disorders (Fotuhi et al. 2020). SARS-CoV-2 may cause neuroinvasive respiratory disease in COVID-19 patients who exhibit respiratory failure (Sellner et al. 2020). In the nasal mucosa, lamina cribrosa, and olfactory bulb, as well as retrogressive axonal movement, ACE-2 receptors enable SARS-CoV-2 entry into the central nervous system (Hartung and Aktas 2020). The high levels of angiotensin II observed in COVID-19-infected patients are associated with several disorders, including vasoconstriction, kidney failure, heart disease, apoptosis, and oxidative stress, all of which accelerate aging and lead to cognitive decline (Vidrio et al. 2022; Elrayess et al. 2022; Amezcua-Guerra et al. 2022). There are a number of biomarkers associated with COVID-19, including interleukin-2 (IL-2), interleukin-6 (IL-6), interleukin-7 (IL-7), granulocyte colony-stimulating factor (GCSF), and tumor necrosis factor-alpha (TNF-alpha) (Azizi Jalilian et al. 2022; Chaumont et al. 2020). The exact mechanisms responsible for coronavirus entry into the nervous system are unknown (Susilawathi et al. 2021).

Several patients have reported blood clots in their brains and multiple organs as a result of COVID-19 infection. This protocol does not specifically treat neurological disorders associated with COVID-19 infection (Tang et al. 2021). In this study, herbs were examined for their potential to aid the early recovery of COVID-19 patients from neurological disorders. The discovery of new drugs for infectious diseases, such as cancer, kidney and heart disease, and neurological disorders, can be attributed to nature and its derivatives. Plant derivatives are thought to contribute to early recovery. It is critical to identify potential drugs from traditional medicines without harming nature's database. Scientists must rethink and rearrange nature's database compassionately. More than 50 million people are suffering from dementia, a neurological disease associated with aging (https://www.alz.co.uk/research/WorldAlzheimerReport2019.pdf). In addition to ensuring that the elderly and corona warriors maintain social distance, it is also critical to provide mental health and psychosocial support (Wang et al. 2020). According to COVID-19 data in Wuhan, fatigue (63%), sleep disorder (26%), and anxiety/depression (13%) were the most commonly reported disorders (Nalbandian et al. 2021).

## 11.2 Methodology

In this study, we reviewed the available literature related to the clinical and neuropathological findings of COVID-19 patients. This study examined specific herbs associated with neurological complications observed in patients with SARS-CoV-2 infection and afterward. PubMed, MEDLINE, Scopus, EMBASE, Google Scholar, EBSCO, Web of Science, Cochrane Library, WHO database, and ClinicalTrials.gov were searched for the neuroprotective potential of herbs for post-COVID-19 patients. For Boolean search, the terms are herbs and neuroprotective potential, post-COVID-19 symptoms, etc. Assessing the pathobiological mechanisms involved in neurological complications associated with COVID-19 using biomarkers that indicate an immune response to SARS-CoV-2.

## 11.3 Discussion

Studies have been done on COVID-19 and post-COVID-infected patients (Leach et al. 2021). Unfortunately, there is no scientific evidence proving the effectiveness of herbs in treating neurological disorders in COVID-19 patients. However, there are widely recognized literature about the use of herbal drugs in many neurodegenerative disorders and cerebrovascular accidents (Ding and He 1986; Akhondzadeh 2007; Wu et al. 2010; Makkar et al. 2020). The most common drug for treating Parkinson's disease is Leva-dopa, which is found in the seeds of the *Mucuna pruriens* plant (Katzenschlager et al. 2004). It is expected that the findings of this study will have a significant impact on the management of COVID-19 patients and post-COVID patients. Some neurological disorders like migraine headaches, loss of smell, etc., might be cured by the use of herbal remedies (Sen 2020).

### *11.3.1 Neurological Disturbances in COVID-19 Patients*

The majority of people with COVID-19 infections experience neurological symptoms such as muscle pain, headaches, confusion, dizziness, and loss of taste and smell (Koralnik and Tyler 2020; Acharya et al. 2020; Liguori et al. 2020). In a study of 236,379 patients with COVID-19 disease, 14 neurological and psychiatric disorders were observed in six months, including intracranial hemorrhage (0.56%), ischemic stroke (2.10%), parkinsonism (0.11%), Guillain-Barré syndrome (0.08%), nerve, nerve root, and plexus disorders (2.85%), myoneural junction and muscle disease (0.45%), encephalitis (0.10%), dementia (0.67%), mood disorder (13.66%), anxiety disorder (17.39%), psychotic disorders (1.40%), substance use disorder (6.58%), and insomnia (5.42%) (Taquet et al. 2021). The neurological symptoms of patients who have a higher risk of neurological disorders due to COVID-19 are described in Table 11.1 and Fig. 11.1.

## 11.4 Immune Response and Brain

An inherent immune response exists in the central nervous system (CNS). A CNS response to infection or disease involves an innate immune reaction. Several types of injuries to the CNS system are observed, including acute injuries (mechanical trauma, strokes, etc.), chronic neurodegenerative diseases (multiple sclerosis, Alzheimer's, etc.), brain tumors (gliomas), and infections (HIV, E. coli, etc.)

**Table 11.1** Neurological patients at higher risk due to COVID-19

| S. No. | Name of disease | COVID-19 effects | Reference |
|---|---|---|---|
| 1 | Alzheimer's disease and related dementias (ADRD) | Mental hygiene, trauma, and stigma | Brown et al. (2020) |
| 2 | Parkinson's disease (PD) | Mitochondrial dysfunction, proteostasis, lipid metabolism, and stress responses | Gonzalez-Latapi et al. (2021) |
| 3 | Motor neuron disease (pwMND) | Proximal muscle weakness of acute or subacute onset and high creatine kinase levels | Barp et al. (2021) |
| 4 | CNS disorder with reduced mobility or immobility | Movement disorders, frontal syndrome, brainstem impairment | Lambrecq et al. (2021) |
| 5 | Neuromuscular disorder with reduced mobility and compromised respiratory function | Exacerbation, ventilatory muscle weakness, or cardiomyopathy | Tseng and Chen (2021) |
| 6 | Autoimmune condition<br>(a) Multiple sclerosis<br>(b) Neuromyelitis optica spectrum disorder<br>(c) Myasthenia gravis<br>(d) Guillain-Barre syndrome<br>(e) Chronic dysimmune neuropathies | (a) Comorbidities, obesity, and black/African ancestry<br>(b) Vomiting and hiccoughs (area postrema syndrome)<br>(c) Diaphragmatic task failure<br>(d) Presence of *Campylobacter jejuni*<br>(e) – | Reder et al. (2021), Ghosh et al. (2021), Galassi and Marchioni (2021), Vogrig et al. (2021), Younger (2021) |

(Lampron et al. 2013). During the course of SARS-CoV-2 infection, the endothelium is able to detect circulating IFN type 1 signaling and thus limit the entry of SARS-CoV-2 into the central nervous system (Iadecola et al. 2020). In terms of stress and neurological disorders, nuclear factor erythroid 2-related factor 2 (Nrf2) is a dominant biomarker (Johnson and Johnson 2015). Study results suggest that deregulation of the NF-κB pathway may modulate T cell activation, which may lead to autoimmune and inflammatory responses. It is the activation of $CD4^+$ T-helper cells that is the main feature of inflammation (Guisado-Vasco et al. 2020). In pathogenic diseases, immune response promotes the development of macrophages, dendritic cells, and proinflammatory cytokines (IL-1, IL-6, IL-12, and tumor necrosis factor-α) (Zhang et al. 2017a). Parkinson's disease (PD) is related to immune responses in the central nervous system (Su and Federoff 2014). The presence of $CD4^+$ and $CD8^+$ T cells is a strong immune response biomarker associated with PD. Researchers have observed that regulatory T cells (Tregs) in PD patients are less effective in suppressing effectors T cells (Saunders et al. 2012).

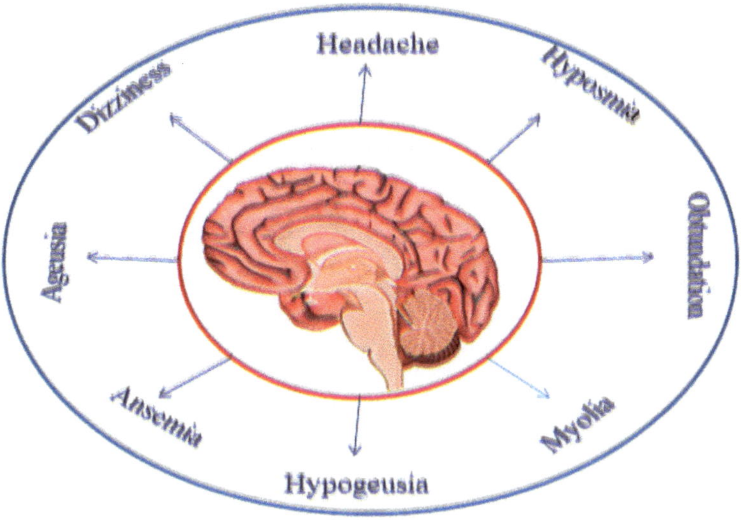


**Fig. 11.1** Neurological symptoms in COVID-19 patients

Amyloid-β directly activates microglia in Alzheimer's disease (AD) through the NALP3 inflammasome pathway and production of IL-1β. In Alzheimer's disease, amyloid-β directly activates microglia through the NALP3 inflammasome pathway and production of IL-1β. This may intensify neurodegeneration (Fuhrmann et al. 2010; Masters and O'Neill 2011). Pathologic inflammation in the CNS is integral to diseases like multiple sclerosis (MS), AD, PD, stroke, and traumatic brain injury (Ransohoff and Brown 2012).

### *11.4.1 Evidence of Clinical Trials for Neurological Disorders in COVID-19 Patients*

The clinical trial of patients with neurological symptoms concluded that altered mental status (42%), headache (42%), central (21%), and peripheral weakness (32%). For mild to critical COVID-19 patients, increases in NfL protein, total tau, and GFAp levels were observed in 63%, 37%, and 16% of patients, respectively (Virhammar et al. 2021). A study that examined the neurological manifestations of 901 patients found encephalopathy, Guillain-Barré syndrome, and stroke to be the most prevalent disorders among COVID-19 patients (Ellul et al. 2020). Figures 11.2 and 11.3 show the effects of cytokine strokes on the brain and neurological disorders associated with COVID-19.

A study found some COVID-19 neurological patients exhibit SARS-CoV-2 in the central nervous system, while others do not (Mitra et al. 2022). The eight drugs (hydroxychloroquine, baricitinib, ruxolitinib, remdesivir, tocilizumab, lopinavir/

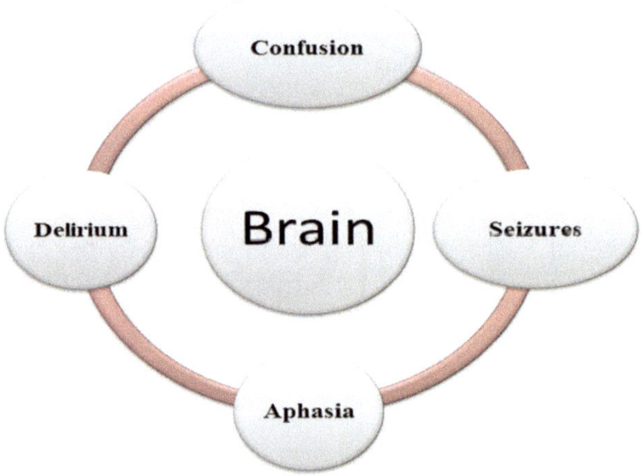


**Fig. 11.2** Cytokine storm effects on the brain

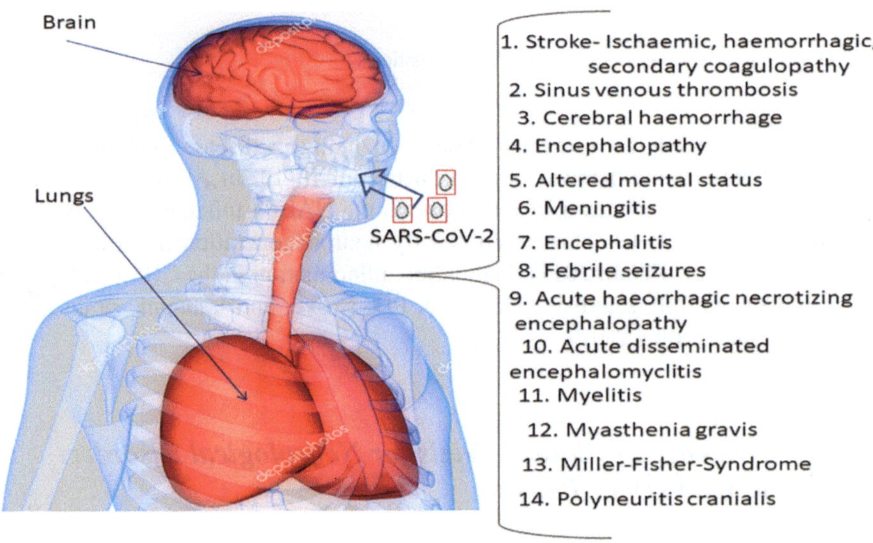


**Fig. 11.3** Neurological disorders reported occurring with COVID-19

ritonavir, favipiravir, and azithromycin) were tested for penetrating capacity in a well-tolerated brain, and hydroxychloroquine and baricitinib were found to be most likely to penetrate the brain (Richardson et al. 2020). COVID-19-infected patients' MRI studies confirm abnormalities in the cerebral parenchyma (Nuzzo et al. 2021).

### 11.4.2 Herbs for Neurological Disorders

Traditionally, *Bacopa monnieri* Linn. is used as a biomedicine for treating neurological disorders and psychiatric disorders (Russo and Borrelli 2005). No clinical studies have been conducted on the effects of *Bacopa monnieri* on the improvement of memory in healthy adults and older adults with age-related memory problems. *Bacopa monnieri* and its bioactive components have shown neuroprotective properties in reducing ROS, neuroinflammation, aggregation, inhibiting amyloid-β and improving cognitive abilities (Dubey and Chinnathambi 2019; Abdul Manap et al. 2019). The treatment of Alzheimer's disease has been reported in a few studies, but no studies have examined *Bacopa monnieri* alone in patients with depression (Brimson et al. 2021). Scientists have supported *Mucuna pruriens* as a treatment for PD, but the most common Parkinson's drug is Leva-dopa, which is abundant in *Mucuna pruriens* (Fothergill-Misbah et al. 2020; Cassani et al. 2016; Cilia et al. 2018; Khazdair et al. 2021). *Mucuna pruriens* treated Parkinson's patients show positive effects (https://clinicaltrials.gov/ct2/show/NCT02680977).

*Withania somnifera* is a neuroprotective herb. Withanone extracted from *Withania somnifera* inhibits amyloid β-42 and modulates proinflammatory cytokines (Pandey et al. 2018; Dar et al. 2017; Kuboyama et al. 2014; Lopresti and Smith 2021). The main phytocompounds in *Withania somnifera* are sitoindoside VII–X, withaferin A, withanosides IV, withanols, withanolide A, withanolide B, anaferine, beta-sitosterol, and withanolide D. These compounds are effective in treating anxiety disorders, Alzheimer's, Parkinson's, schizophrenia, Huntington's disease, dyslexia, depression, autism, addiction, amyotrophic lateral sclerosis, attention deficit hyperactivity disorder, and bipolar disorders (Zahiruddin et al. 2020). Table 11.2 shows the herbs commonly used in neurological disorders with their biomarkers for immune response to diseases.

A recent survey shows that 61.9% of herbal treatments and 25.3% of herbal-drug combinations are used by COVID-19 patients with mental illness or psychiatric issues (Alonso-Castro et al. 2021). Study outcomes have not been published for herbal treatments for neurological disorders in COVID-19 patients. An ongoing clinical trial on "Functional Assessment of Ashwagandha Root Extract during Weight Loss" evaluates the effect of Ashwagandha root extract capsule on nervous system disease (ClinicalTrials.gov identifier, NCT03112824) (https://www.clinicaltrials.gov/ct2/show/NCT03112824). Nuclear factor erythroid 2-related factor 2 (Nrf2) is a dominant biomarker for stress and associated neurological disorders. The combination of drugs (carnosol, withaferin A, and luteolin), rosemary, ashwagandha, and sophora japonica is undergoing a clinical trial (https://clinicaltrials.gov/ct2/show/NCT04638387). The study reported neuroprotective effects of ethanolic extract of *Acorus calamus* (Shukla et al. 2002). In Ayurveda, roots and rhizomes of the plant are used for the treatment of epilepsy, neurosis, insomnia, and other diseases (Vohora et al. 1990; Martis et al. 1991). An *Acorus calamus* leaf extract modulates the characteristics of interleukin-8 (IL-8), (IL-6), RNA protein level, and interferon regulatory factor 3 (IRF3), as well as nuclear

**Table 11.2** Herbs used in neurological disorder and their biomarker for the immune response against diseases

| S. No. | Herbs | Important phytochemicals | Biomarkers related to immune response | Neurological effects |
|---|---|---|---|---|
| 1 | *Acorus calamus* (Bach) | Galloylquinic acid, Bacopaside III and Bacopaside, α-asarone and β-asarone (trans and cis forms, respectively, of 2,4,5-trimethoxy1- propenyl benzene) acorenone, iso-acorone, (Z)-sesquilavandulol and dehydroxy isocalamendiol acofuran, a sesquiterpenoid, coumarine, saponin, asamyl alcohol, methyl eugenol, calamenol, α- pinene, acorine, acoretine, lucenin, acoradin (Yende et al. 2008) | Nitric oxide, interleukin-2 (IL-2), and tumor necrosis factor-a (TNF-a) (Mehrotra et al. 2003) | CNS depressant, anticonvulsant, anti-spasmodic, hypolipidemic, immuno-suppressive, anti-inflammatory, cryoprotective, and antioxidant (Rajput et al. 2014) |
| 2 | *Asparagus racemosus* (Shatavari) | Shatavarins, sarsasapogenin, quercetin, rutin (quercetin 3-O-rutinoside) and hyperoside (quercetin 3-O-galactoside), quercetin 3-glucuronide, ferulic, caffeic, and chlorogenic acids (Srivastava et al. 2018) | Significant increase of $CD^{3+}$ and $CD^{4}$/$CD^{8+}$, percentages effect on T cell activation. Significant upregulation of Th1 (IL-2, IFN-g) and Th2 (IL-4) cytokines suggesting its mixed Th1/Th2 adjuvant activity (Gautam et al. 2009) | Antidepressant activity (Meena et al. 2011) |
| 3 | *Bacopa monnieri* (Brahmi) | Dammarane-type triterpenoid saponins with jujubogenin and pseudojujubogenin as the aglycones including bacosides A1-A3, bacopa saponins A-G and bacopasides I–V (Shahid et al. 2017) | TNF-α and IL-6 from activated N9 microglial cells, caspase 1 and 3, matrix metalloproteinase-3 (Nemetchek et al. 2017) | Alzheimer's and Spinocerebellar ataxia (SCA) (Jeyasri et al. 2020) |
| 4 | *Blumea balsamifera* (Ngai camphor) | l-borneol, 10-epi-γ-eudesmol, γ-eudesmol, β-eudesmol, α-eudesmol, limonene, l-camphor, palmitic acid, and d-camphor (Pang et al. 2014) | LPS-induced RAW264.7 cells by decreasing activity of TLR4-NF-κB signaling pathway and inhibiting NLRP3 inflammasome activation (Liao et al. 2021) | Improve drugs delivery to the blood–brain barrier in the central nervous system (Zhang et al. 2017b) |

| | | | | |
|---|---|---|---|---|
| 5 | *Cannabis* (Bhang) | $\Delta^9$-tetrahydrocannabinol, $\Delta^8$-tetrahydrocannabinol, cannabinol, and cannabidiol (Ashton 2001) | IFN-γ producing CD8+ T cells, glycogen synthase kinase 3β (GSK-3β) hyperphosphorylation caused by Aβ (Nichols et al. 2021; Li et al. 2020) | Epilepsy and Alzheimer's disease (Hosami et al. 2021) |
| 6 | *Centella asiatica* (Mandookparni or Gotu kola) | triterpenoid saponins, asiaticoside, asiatic acid, madecassic acid, aglycones, saponins-asiaticoside B, brahminoside, centelloside, indcentelloside, thankuni side, isothankuniside, brahamic acid, betulic acid and alkaloid hydrocotyline, madecassic acid, glucose, rhamnose, sitosterol, Stigmasterol (Samy and Chow 2011) | Reduces amyloid-β pathology and activates NRF2-regulated antioxidant response pathway (Zweig et al. 2021) | Memory enhancement, Alzheimer's (Boondam et al. 2019) |
| 7 | *Convolvulus prostratus* (Shankha pushpi) | Microphyllic acid, kaempferol, kaempferol-3-glucoside, 3,4 dihydroxycinnamic acid, scopoletin, convolamine, convolidine, sitosterols, 2-butanone, pentanoic acid, cinnamic acid, silane, decanoic acid, 2-pentanol, ascorbic acid, 10-bromodecanoic acid, tridecane, phthalic acid, eicosane, octatriacontyl pentafluoropropionate, 1-octadecene sulphonyl chloride, squalene, pyrimidine, heneicosane, 1,2-benzenedicarboxylic acid, cyclononasiloxane, octadecamethyl, nonacosane, sulfurous acid pentadecyl 2-propyl ester, straight chain hydrocarbons (C22–C33), fatty acids (C14–C28), fatty alcohols (C24– C32), vitamin E, and cyclononasiloxane (Salehi et al. 2020) | Regulate the neuronal factor and proteins by inhibiting acetyl-cholinesterase and regulate the neuronal cascades, cytokine- IL-6, MCP-1, ICAM-1 (Balkrishna et al. 2020) | Enhance memory or as a brain and nerve tonic, Alzheimer's disease (Semwal et al. 2021) |

(continued)

**Table 11.2** (continued)

| S. No. | Herbs | Important phytochemicals | Biomarkers related to immune response | Neurological effects |
|---|---|---|---|---|
| 8 | *Curcuma longa* (Turmeric) | Demethoxycurcumin (2, DMC), bisdemethoxycurcumin (3, BDMC) and cyclocurcumin (4, CYC), tetrahydrocurcumin (5, THC) (Randino et al. 2016) | PI3K/Akt, NF-κB, TNF-α, modulating tau oligomers (Mathew and Hsu 2018; Panaro et al. 2020) | Alzheimer agents (Cascio et al. 2019) |
| 9 | *Mucuna pruriens* (Velvet beans) | 1–3,4-dihydroxyphenylalanine, pentadecanoic acid, 14-methyl-, methyl ester, dodecanoic acid, 9, 12-octadecadienoic acid (Z, Z)-methyl ester, 9,12-octadecadienoic acid, and 2-myristynoyl-glycinamide (Saikarthik et al. 2017; Lampariello et al. 2012) | (TNF-α), (IL-1β), IL-6, and enzymes such as nicotinamide adenine dinucleotide phosphate (NADPH) oxidase and cyclooxygenase-2 (COX-2) (Rai et al. 2017) | Parkinson's disease (Cilia et al. 2017) |
| 10 | *Nyctanthes-arbor tristis Linn.* (Har Singar or Parijat) | 4-hydroxy hexahydro benzofuran-7-one, 6β-hydroxyloganin, and arbortristoside, diterpenoids, nyctanthin, nyctanthoside, rengyolone, astragalin, flavonoids, anthocyanins, d-mannitol, essential oils, glycosides, carotenoids (Parekh and Soni 2020; Bharshiv et al. 2016) | Splenocytes proliferation and increased production of cytokines, especially IL-2 and IL-6 (Bharshiv et al. 2016) | Anti-epileptic and anxiolytic (Srikalyani and Ilango 2009) |
| 11 | *Ocimum tenuiflorum* (Tulsi) | Oleanolic acid, ursolic acid, rosmarinic acid, eugenol, carvacrol, linalool, B-caryophyllene (Singh and Chaudhuri 2018) | Increased Natural Killer (NK) and T-helper cells (Jamshidi and Cohen 2017) | Anti-stress, loss of memory, Alzheimer's disease (AD), Parkinson's disease (PD), Huntington's disease (HD). Oxidative stress and mitochondrial dysfunction (Singh and Chaudhuri 2018; Jamshidi and Cohen 2017) |

| | | | | |
|---|---|---|---|---|
| 12 | *Phyllanthus emblica* (Ambla) | Ascorbic acid (vitamin C), gallic acids, amlaic acid, arginine, aspartic acid, astragallin, β-carotene, β-sitosterol, chebulagic acid, chebulic acid, chebulaginic acid, chebulinic acid, corilagic acid, corilagin, cysteine, ellagic acid, emblicol, gibberellins, glutamic acid, glycine, histidine, isoleucine, kaempferol, leucodelphinidin, methionine, phenylalanine, phyllantidine, phyllemblic acid, quercetin, riboflavin, rutin, thiamin, threonine, tryptophan, tyrosine, valine, zeatin (Habib-ur-Rehman et al. 2007) | Epithelial barriers, phagocytes (neutrophils, macrophages), B- and T-lymphocytes, modulate cytokine production, decrease histamine levels, reduced proinflammatory cytokine, (TNF-α) and (IL-1β) levels, and appreciably upregulate the anti-inflammatory cytokine (IL-10) concentration (Chatterjee et al. 2010) | Memory deficit, Alzheimer's disease (Ashwlayan and Singh 2011; Biswas et al. 2015) |
| 13 | *Swertia chirata* (Chirayita) | Secoiridoid bitters, triterpenoids, xanthones, and alkaloids. Swertiamarin, gentiopicroside, amarogentin, and amaroswerin, amarogentin (Karan et al. 1999) | Inflammatory mediators by blocking the expression of COX-2 and phosphorylation of Akt, IKK-β, MAPK, and NF-κB, activation in LPS-stimulated macrophages (Hu et al. 2019) | Antidepressant properties, CNS stimulant, Parkinson's disease (Shikha et al. 2010) |
| 14 | *Vitex negundo* (Nirgundi) | Benzoic acid 3-hydroxy, Ledol, 9,12,15-octadecatrienoic acid, and Vitamin E, 4HPyran-4-one, 2,3-dihydro-3,5-dihydroxy-6-methyl caryophyllene, and n-hexadecanoic acid (Kumar et al. 2010) | Suppressed expressions of inflammatory Th1/Th2 cytokines, chemokines, endopeptidases (MMPs)s, oxidative effector enzyme (iNOS), adhesion molecules, IL-4/IFN-γ release with simultaneous enhancement in levels of IL-10, IFN-γ, MUC3, and enhanced phosphorylation of NF-κB, PI3K, Akt and p38MAPKs and downregulated AMPK (Tirpude et al. 2021) | Anti-stress (Tirpude et al. 2021) |

(continued)

**Table 11.2** (continued)

| S. No. | Herbs | Important phytochemicals | Biomarkers related to immune response | Neurological effects |
|---|---|---|---|---|
| 15 | *Withania somnifera* (Ashwagandha) | Withanolides, withaferin A, alkaloids, steroidal lactones, tropine, cuscohygrine, rutin, hyperoside, quercitrin, myricetin, quercetin, kaempferol, rhamnetin, gallic acid, caffeic acid, chlorogenic acid, ferulic acid, p-coumaric acid, p-hydroxybenzoic acids, caffeine, theobromine, theophylline, harmine, harmane, yohimbine, berberine, noscapine, and papaverine (Tripathi et al. 2020; Tiwari et al. 2015) | TNF-α-induced increases in CCL5 while attenuating the increase in CCL2 expression and NF-κB activation increase in dendrite growth markers, KLK8, and MAP2 (Grunz-Borgmann et al. 2015; Konar et al. 2019) | Alzheimer's disease (Konar et al. 2019) |

factor κB (NF-κB) activation (Kim et al. 2009; Sharma et al. 2020). Vitamin E and C intake are inversely associated with the risk of PD (Hantikainen et al. 2021). *Phyllanthus emblica* is an excellent source of vitamin C. Research indicates that *Phyllanthus emblica* reduces kainic acid-induced elevations of TNF-α in the brain, improving cognitive function levels in the brain and improving the cognitive deficit (Gaire and Subedi 2014). In neurodegenerative disorders, *Phyllanthus emblica* has proved to be beneficial due to its antioxidant, cholesterol-lowering, and anti-inflammatory properties (Husain et al. 2019; Uddin et al. 2016).

Borneol is a natural compound found in *Blumea balsamifera* that acts as blood–brain barrier permeability in the central nervous system. Borneol is an effective agent that can improve drug delivery to the brain: neurological function scores (NFS) and the cerebral infarction area (Kulkarni et al. 2021; Chen et al. 2019). Thus, *Blumea balsamifera* has the potential to provide neuroprotection for a variety of neurological disorders.

*Asparagus racemosus* ethanolic extract has been shown to enhance brain-derived neurotrophic factor (BDNF), as well as estrogen receptors (ERs) (Lalert et al. 2018). In molecular docking study involving *Asparagus racemosus (Wild.),* SARS-CoV-2, the NSP15 Endoribonuclease and Spike Receptor-binding domain was identified. In terms of their effectiveness against the spike receptor-binding domain and NSP15 endoribonuclease, asparoside-C, asparoside-D, and asparoside-F are the most effective (Chikhale et al. 2020). *Asparagus racemosus* phytocompounds exhibit adaptogenic, neuroprotective, antioxidant, anti-inflammatory, and nootropic properties without causing side effects. It may be a possible treatment for neurological disorders such as stress, anxiety, depression, and epilepsy (Majumdar et al. 2021).

*Cannabis sativa L* is a medicinal herb used in treating a variety of diseases and neurological disorders (Mechoulam 2019). The main neuroprotective ingredients found in cannabis are Δ9-tetrahydrocannabinol (THC) and cannabidiol (CBD). The neuroprotective properties of cannabinoids have led to them being referred to as the "Aspirin of the 21st century" (Baker et al. 2003). A hypothesis study on cannabidiol against SARS-CoV-2 proposed that it exerts immunomodulatory and anti-inflammatory effects, as well as decreasing uncontrolled cytokine production responsible for acute lung injury (Esposito et al. 2020). Cannabidiol has been suggested as a support drug against the COVID-19 pandemic with precaution (Paland et al. 2021; Anil et al. 2021).

*Convolvulus prostratus* is commonly known for its neuro-modulating properties. A number of phytocompounds from *Convolvulus prostratus*, including 4′-methoxy kaempferol, 7-methoxy quercetin, convolamine, scopoletin, and hydroxy-cinnamic acid, are responsible for neuroprotective effects in the brain (Alshoushan et al. 2021). An extract of *Convolvulus prostratus* modulates IL-1b, IL-6, TNF-a, alanine transaminase, and aspartate transaminase. Ayurveda shows that *Convolvulus prostratus* improves anxiety, irritability, inability to relax, lack of concentration, disturbed sleep, loss of memory, palpitations, headaches, dry mouth, upset stomach, and restlessness (Belwal et al. 2020).

*Swertia chirata* has traditionally been used in India to boost appetite and as a febrifuge (Liu et al. 2017). The herb's methanol extract is reported to have antidiabetic activity because it contains mangiferin, amarogentin, amaroswerin, sweroside, and swertiamarin as active constituents (Suryawanshi et al. 2009). Chirata sweat is known to be effective in the treatment of hepatitis, liver disorders, inflammation, chronic fever, malaria, anemia, GIT disorders, hypertension, mental disorders, and diabetes (Kumar and Van Staden 2016). The therapeutic potential of bellidifolin from *Swertia chirayita* has been explored for the treatment of inflammatory-mediated immune deficiency (Hu et al. 2019).

Traditional Indian medicine uses *Vitex negundo Linn.,* for treating respiratory and inflammatory disorders. Vitegnoside extracted from *Vitex negundo* inhibited p38 MAPK/MK2, JNK/c-Jun, and downstream NF-κB inflammatory transduction. In a study on the effect of vitegnoside from *Vitex negundo* on Alzheimer's disease, it was found that vitegnoside promoted neuroprotection by enhancing cell viability, maintaining cytomembrane integrity, and maintaining nuclear homogeneity (Wang et al. 2019).

Physiochemical analyses have demonstrated that crude extracts from *Nyctanthes arbor-tristis* have potential anti-inflammatory, anti-malarial, anti-viral, and immunostimulatory properties (Agrawal and Pal 2013). Researchers note that aqueous leaf extracts from *Nyctanthes arbor-tristis* are the most potent immunomodulators (Bhatia and Kaur 2001).

The herb *Centella asiatica* is well-known for its use in treating mental disorders in India and China. It is believed that triterpene derivatives like asiatic acid, madecassic acid, asiaticoside, madecassoside, and Brahmic acid provide positive effects on the aging brain (Samy and Chow 2011; Orhan 2012). Ethanolic extracts have been found to increase neural development (nerve growth factor (NGF) in human SH-SY5Y cell lines) (Soumyanath et al. 2005). This herb was traditionally used in memory enhancers and for treating Alzheimer's disease (Orhan et al. 2013).

*Curcuma longa* has been studied for its antidepressant properties. *Curcuma longa* contains curcumin, which has been found to be effective against neurological disorders such as Alzheimer's and multiple sclerosis (Witkin and Li 2013; Witkin et al. 2013). There is evidence that curcumin can effectively treat fearful or traumatic memories when used alone or in combination with existing medications (Monsey et al. 2015). The curcumin compound interacts directly and indirectly with various transcription factors such as nuclear factor kappa B (NF-κB), activator protein 1 (AP-1), β-catenin, and signal transducer and activator of transcription (STAT) proteins (Witkin et al. 2013; Liu et al. 2014; Shishodia et al. 2007). Additionally, it can act as a partial agonist of peroxisome proliferator-activated receptor-γ (PPAR-γ), a ligand-activated transcription factor involved in neurological disorders and anti-inflammatory signaling pathways. The nuclear receptor PPAR-γ affects metabolism, reproduction, and immune response (Bernardo et al. 2021). In addition to curcumin, other compounds in turmeric include digalloyl-hexoside, caffeic acid, hexoside, curdione, coumaric acid, caffeic acid, sinapic acid, quercetin-3-D-galactoside, casuarinin, bisdemethoxycurcumin, curcuminol, demethoxycurcumin, isorhamnetin, valoneic acid, and curcumin. Their therapeutic value varies according to the disease (Sabir et al. 2021; Maithili KarpagaSelvi et al. 2020). Furthermore, there is scientific evidence that curcumin may have a bioactive role against SARS-

CoV-2 infection as well (Babaei et al. 2020; Zahedipour et al. 2020; Rattis et al. 2021).

*Ocimum tenuiflorum* is a traditional herb used to prevent and treat the common cold, cough, headache, fever, influenza, sore throat, bronchitis, malarial fever, and migraine headaches (Joshi et al. 2017; Iqbal Chowdhury et al. 2020). *Ocimum tenuiflorum* oil contains 50.4% eugenol, ursolic acid, caryophyllene, linalool, and 1,8-cineole (Bhavya et al. 2018). Ocimumoside A and B are used for treating neurological disorders (Singh and Chaudhuri 2018). There is strong evidence that the daily administration of 10 g of an aqueous extract of fresh tulsi leaves for 2 weeks results in symptomatic improvement in patients with acute hepatitis viral infection (Jamshidi and Cohen 2017).

## 11.5 Conclusions

During the COVID-19 pandemic, the demand for medicinal plants was found to increase significantly. Among them, herbs such as *Curcuma longa, Ocimum tenuiflorum,* and *Withania somnifera* are gaining popularity as immune boosters against COVID-19 infection. Theoretically, SARS-CoV-2 causes inflammation in the body, which is controlled primarily by adaptive immune responses. The SARS-CoV-2 enters the cell through the ACE-2 receptor and is recognized by the toll-like receptor 7 (TLR-R) present in the endosome (Poulas et al. 2020). As a result, alpha interferon, TNF-alpha, as well as IL-12 and IL-6, are produced, and $CD8^{+}$ cytotoxic T cells are formed, and through $CD4^{+}$ helper T cells, antigen-specific B cells and antibodies are produced (Ahmadpoor and Rostaing 2020). Antibodies against unknown or known pathogens are produced by B cell receptors (BCRs), which recognize the pathogenic attack, counter-invading pathogens, and provide long-term protection against them (Nielsen and Boyd 2018). The immune system serves as the body's complete defense against pathogens. It is now clear that inborn medicinal herbs promote the immune system through various pathways against infection and inflammation. Many studies support the study objective, but there are still some gaps. Medicinal plants have the potential to treat neurological disorders at an affordable price in COVID-19 or post-COVID-19 patients. It is essential for a clear understanding of comprehensibility to conduct clinical trials on herbs and their secondary metabolisms, as well as their effects on various neurological disorders. Biomarkers, neuroradiological findings, and pathobiological mechanisms need to be re-examined to find the most suitable drugs. The interaction between humans and herbs, and the mechanism of action, should also be re-examined. The therapeutic potential of herbs for neurological disorders requires further mechanistic studies using modern scientific techniques and approaches. It is imperative to conduct further animal studies and clinical trials to explore how these herbs may help alleviate neurological disorders in COVID-19 patients.

**Consent for Publication** Not applicable.

**Conflict of Interest** The authors declare no conflict of interest.

**Funding** None.

**Acknowledgments** None.

## References

Abdul Manap AS, Vijayabalan S, Madhavan P, Chia YY, Arya A, Wong EH, Rizwan F, Bindal U, Koshy S (2019) Bacopa monnieri, a neuroprotective lead in Alzheimer disease: a review on its properties, mechanisms of action, and preclinical and clinical studies. Drug Target Insights 13: 1177392819866412. https://doi.org/10.1177/1177392819866412

Acharya A, Kevadiya BD, Gendelman HE, Byrareddy SN (2020) SARS-CoV-2 infection leads to neurological dysfunction. J Neuroimmune Pharmacol 15:167–173

Agrawal J, Pal A (2013) Nyctanthes arbor-tristis Linn—A critical ethnopharmacological review. J Ethnopharmacol 146(3):645–658

Ahmadpoor P, Rostaing L (2020) Why the immune system fails to mount an adaptive immune response to a Covid-19 infections. Transpl Int 33(7):824–825

Akhondzadeh S (2007) Herbal medicines in the treatment of psychiatric and neurological disorders. In: Low-cost approaches to promote physical and mental health. Springer, New York, NY, pp 119–138

Alonso-Castro AJ, Ruiz-Padilla AJ, Ortiz-Cortes M, Carranza E, Ramírez-Morales MA, Escutia-Gutiérrez R, Ruiz-Noa Y, Zapata-Morales JR (2021) Self-treatment and adverse reactions with herbal products for treating symptoms associated with anxiety and depression in adults from the central-western region of Mexico during the Covid-19 pandemic. J Ethnopharmacol 272: 113952

Alshoushan NE, Labyad NS, Taher YM, Mezogi JS (2021) Phytochemical screening and structure elucidation of isolated compounds from convolvulus dorycnium plant originated from west of Libya. Mediterran J Pharm Pharmaceut Sci 1(1):18–22

Amezcua-Guerra LM, Del Valle L, González-Pacheco H, Springall R, Márquez-Velasco R, Massó F, Brianza-Padilla M, Manzur-Sandoval D, González-Flores J, García-Ávila C, Juárez-Vicuña Y (2022) The prognostic importance of the angiotensin II/angiotensin-(1–7) ratio in patients with SARS-CoV-2 infection. Ther Adv Respir Dis 16:17534666221122544. https://doi.org/10.1177/17534666221122544

Anil SM, Shalev N, Vinayaka AC, Nadarajan S, Namdar D, Belausov E, Shoval I, Mani KA, Mechrez G, Koltai H (2021) Cannabis compounds exhibit anti-inflammatory activity in vitro in COVID-19-related inflammation in lung epithelial cells and pro-inflammatory activity in macrophages. Sci Rep 11(1):1–4

Ashton CH (2001) Pharmacology and effects of cannabis: a brief review. Br J Psychiatry 178(2): 101–106

Ashwlayan VD, Singh RA (2011) Reversal effect of Phyllanthus emblica (Euphorbiaceae) Rasayana on memory deficits in mice. Int J Appl Pharm 3:10–15

Azizi Jalilian F, Keshavarz G, Khazaei S, Nezamdoost M, Hashemi SH, Mamani M, Ansari N, Amini R, Khalkhali A, Keshavarz A, Ayubi E (2022) The effects of nutrition bio-shield superfood powder on immune system function: a clinical trial study among patients with COVID-19. Front Immunol 13:919402. https://doi.org/10.3389/fimmu.2022.919402

Babaei F, Nassiri-Asl M, Hosseinzadeh H (2020) Curcumin (a constituent of turmeric): new treatment option against COVID-19. Food Sci Nutr 8(10):5215–5227

Baker D, Pryce G, Giovannoni G, Thompson AJ (2003) The therapeutic potential of cannabis. Lancet Neurol 2(5):291–298
Balkrishna A, Thakur P, Varshney A (2020) Phytochemical profile, pharmacological attributes and medicinal properties of convolvulus prostratus–A cognitive enhancer herb for the management of neurodegenerative etiologies. Front Pharmacol 11:171
Barp A, Velardo D, Ciscato P, Sansone VA, Lunetta C (2021) Anti-HMGCR myopathy misdiagnosed as motor neuron disease and complicated with COVID-19 infection. Neurol Sci 42(5):1679–1682
Belwal T, Nabavi SM, Nabavi SF, Dehpour AR, Shirooie S (eds) (2020) Naturally occurring chemicals against Alzheimer's disease. Academic Press
Bernardo A, Plumitallo C, De Nuccio C, Visentin S, Minghetti L (2021) Curcumin promotes oligodendrocyte differentiation and their protection against TNF-α through the activation of the nuclear receptor PPAR-γ. Sci Rep 11(1):1–3
Bharshiv CK, Garg SK, Bhatia AK (2016 Jul) Immunomodulatory activity of aqueous extract of Nyctanthes arbor- tristis flowers with particular reference to splenocytes proliferation and cytokines induction. Indian J Pharmacol 48(4):412
Bhatia A, Kaur J (2001) Nyctanthes arbor-tristis leaves extract as Antagonizer of immunotoxic effects of chemical pesticides (experimental study). Int J Environ Stud 58(2):197–215
Bhavya ML, Chandu AG, Devi SS (2018) Ocimum tenuiflorum oil, a potential insecticide against rice weevil with anti-acetylcholinesterase activity. Ind Crop Prod 126:434–439
Biswas K, Islam A, Sharmin T, Biswas PK (2015) In-vitro cholinesterase inhibitory activity of dry fruit extract of Phyllanthus emblica relevant to the treatment of Alzheimer's disease. J Phytopharmacol 4:5–8. https://citeseerx.ist.psu.edu/viewdoc/download?doi=10.1.1.1080.8123&rep=rep1&type=pdf. Accessed dated 20 Mar 2021
Boni MF, Lemey P, Jiang X, Lam TT, Perry BW, Castoe TA, Rambaut A, Robertson DL (2020) Evolutionary origins of the SARS-CoV-2 sarbecovirus lineage responsible for the COVID-19 pandemic. Nat Microbiol 5(11):1408–1417
Boondam Y, Songvut P, Tantisira MH, Tapechum S, Tilokskulchai K, Pakaprot N (2019) Inverted U-shaped response of a standardized extract of Centella asiatica (ECa 233) on memory enhancement. Sci Rep 9(1):1. https://doi.org/10.1038/s41598-019-44867-z
Brief P (2020) COVID-19 and the need for action on mental health. World Health Organization
Brimson JM, Brimson S, Prasanth MI, Thitilertdecha P, Malar DS, Tencomnao T (2021) The effectiveness of Bacopa monnieri (Linn.) Wettst. as a nootropic, neuroprotective, or antidepressant supplement: analysis of the available clinical data. Sci Rep 11(1):1. https://doi.org/10.1038/s41598-020-80045-2
Brown EE, Kumar S, Rajji TK, Pollock BG, Mulsant BH (2020) Anticipating and mitigating the impact of the COVID-19 pandemic on Alzheimer's disease and related dementias. Am J Geriatr Psychiatry 28(7):712–721
Cascio FL, Puangmalai N, Ellsworth A, Bucchieri F, Pace A, Piccionello AP, Kayed R (2019) Toxic tau oligomers modulated by novel curcumin derivatives. Sci Rep 9(1):1–6. https://doi.org/10.1038/s41598-019-55419-w
Cassani E, Cilia R, Laguna J, Barichella M, Contin M, Cereda E, Isaias IU, Sparvoli F, Akpalu A, Budu KO, Scarpa MT (2016) Mucuna pruriens for Parkinson's disease: low-cost preparation method, laboratory measures and pharmacokinetics profile. J Neurol Sci 15(365):175–180
Chatterjee A, Chattopadhyay S, Bandyopadhyay SK (2010) Biphasic effect of Phyllanthus emblica L. extract on NSAID-induced ulcer: an antioxidative trail weaved with immunomodulatory effect. Evid Based Complement Alternat Med 2011:146808. https://doi.org/10.1155/2011/146808
Chaumont H, Etienne P, Roze E, Couratier C, Roger PM, Lannuzel A (2020) Acute meningoencephalitis in a patient with COVID-19. Rev Neurol 176(6):519–521
Chen ZX, Xu QQ, Shan CS, Shi YH, Wang Y, Chang RC, Zheng GQ (2019) Borneol for regulating the permeability of the blood-brain barrier in experimental ischemic stroke: preclinical evidence and possible mechanism. Oxidative Med Cell Longev 2019:2936737

Chikhale RV, Sinha SK, Patil RB, Prasad SK, Shakya A, Gurav N, Prasad R, Dhaswadikar SR, Wanjari M, Gurav SS (2020) In-silico investigation of phytochemicals from Asparagus racemosus as plausible antiviral agent in COVID-19. J Biomol Struct Dyn:1–5. https://doi.org/10.1080/07391102.2020.1784289

Cilia R, Laguna J, Cassani E, Cereda E, Pozzi NG, Isaias IU, Contin M, Barichella M, Pezzoli G (2017) Mucuna pruriens in Parkinson's disease: a double-blind, randomized, controlled, crossover study. Neurology 89(5):432–438

Cilia R, Laguna J, Cassani E, Cereda E, Raspini B, Barichella M, Pezzoli G (2018) Daily intake of Mucuna pruriens in advanced Parkinson's disease: a 16-week, noninferiority, randomized, crossover, pilot study. Parkinsonism Relat Disord 49:60–66

Dar NJ, Bhat JA, Satti NK, Sharma PR, Hamid A, Ahmad M (2017) Withanone, an active constituent from Withania somnifera, affords protection against NMDA-induced excitotoxicity in neuron-like cells. Mol Neurobiol 54(7):5061–5073

Ding Y, He X (1986) Traditional Chinese herbs in treatment of neurological and neurosurgical disorders. Can J Neurol Sci 13(3):210–213

Dubey T, Chinnathambi S (2019) Brahmi (Bacopa monnieri): an ayurvedic herb against the Alzheimer's disease. Arch Biochem Biophys 676:108153. https://doi.org/10.1016/j.abb.2019.108153

Ellul MA, Benjamin L, Singh B, Lant S, Michael BD, Easton A, Kneen R, Defres S, Sejvar J, Solomon T (2020) Neurological associations of COVID-19. Lancet Neurol 19(9):767–783

Elrayess MA, Zedan HT, Alattar RA, Abusriwil H, Al-Ruweidi MK, Almuraikhy S, Parengal J, Alhariri B, Yassine HM, Hssain AA, Nair A (2022) Soluble ACE2 and angiotensin II levels are modulated in hypertensive COVID-19 patients treated with different antihypertension drugs. Blood Press 31(1):80–90. https://doi.org/10.1080/08037051.2022.2055530

Esposito G, Pesce M, Seguella L, Sanseverino W, Lu J, Corpetti C, Sarnelli G (2020) The potential of cannabidiol in the COVID-19 pandemic. Br J Pharmacol 177(21):4967–4970

Fothergill-Misbah N, Maroo H, Cham M, Pezzoli G, Walker R, Cilia R (2020) Could Mucuna pruriens be the answer to Parkinson's disease management in sub-Saharan Africa and other low-income countries worldwide? Parkinsonism Relat Disord 73:3–7

Fotuhi M, Mian A, Meysami S, Raji CA (2020) Neurobiology of COVID-19. J Alzheimers Dis 76(1):3–19

Fuhrmann M, Bittner T, Jung CK, Burgold S, Page RM, Mitteregger G, Haass C, LaFerla FM, Kretzschmar H, Herms J (2010) Microglial Cx3cr1 knockout prevents neuron loss in a mouse model of Alzheimer's disease. Nat Neurosci 13(4):411–413

Gaire BP, Subedi L (2014) Phytochemistry, pharmacology and medicinal properties of Phyllanthus emblica Linn. Chin J Integr Med:1–8. https://doi.org/10.1007/s11655-014-1984-2

Galassi G, Marchioni A (2021) Myasthenia gravis at the crossroad of COVID-19: focus on immunological and respiratory interplay. Acta Neurologica Belgica 121:633–642

Gautam M, Saha S, Bani S, Kaul A, Mishra S, Patil D, Satti NK, Suri KA, Gairola S, Suresh K, Jadhav S (2009) Immunomodulatory activity of Asparagus racemosus on systemic Th1/Th2 immunity: implications for immunoadjuvant potential. J Ethnopharmacol 121(2):241–247

Ghosh R, De K, Roy D, Mandal A, Biswas S, Biswas S, Sengupta S, Naga D, Ghosh M, Benito-León J (2021) A case of area postrema variant of neuromyelitis optica spectrum disorder following SARS-CoV-2 infection. J Neuroimmunol 350:577439

Gonzalez-Latapi P, Fearon C, Fasano A, Lang AE (2021) Parkinson's disease and COVID-19: do we need to be more patient? Mov Disord 36(2):277

Grunz-Borgmann E, Mossine V, Fritsche K, Parrish AR (2015) Ashwagandha attenuates TNF-α-and LPS-induced NF-κB activation and CCL2 and CCL5 gene expression in NRK-52E cells. BMC Complement Altern Med 15(1):1–8. https://doi.org/10.1186/s12906-015-0958-z

Guisado-Vasco P, Cano-Megías M, Rodríguez-López M, de-Luna-Boquera IM, Carnevali-Ruiz D. (2020) COVID-19 and metabolic syndrome: NF-κB activation. Crossroads. Trends Endocrinol Metabol 31(11):802–803

Habib-ur-Rehman, Yasin KA, Choudhary MA, Khaliq N, Atta-ur-Rahman, Choudhary MI, Malik S (2007) Studies on the chemical constituents of Phyllanthus emblica. Nat Prod Res 21(9): 775–781

Hantikainen E, Lagerros YT, Ye W, Serafini M, Adami HO, Bellocco R, Bonn S (2021) Dietary antioxidants and the risk of Parkinson disease: the Swedish National March Cohort. Neurology 96(6):e895–e903. https://doi.org/10.1212/WNL.0000000000011373

Hartung HP, Aktas O (2020) COVID-19 and management of neuroimmunological disorders. Nat Rev Neurol 16(7):347–348

Hosami F, Manayi A, Salimi V, Khodakhah F, Nourbakhsh M, Nakstad B, Tavakoli-Yaraki M (2021) The pro- apoptosis effects of Echinacea purpurea and Cannabis sativa extracts in human lung cancer cells through caspase-dependent pathway. BMC Comple Med Therap 21(1):1–2

https://clinicaltrials.gov/ct2/show/NCT02680977. Accessed dated 15 May 2021

https://clinicaltrials.gov/ct2/show/NCT04638387. Accessed dated 20 May 2021

https://www.alz.co.uk/research/WorldAlzheimerReport2019.pdf. Accessed dated 13 May 2021

https://www.clinicaltrials.gov/ct2/show/NCT03112824. Accessed dated 15 May 2021

https://www.who.int/mental_health/neurology/neurological_disorders_report_web.pdf. Accessed dated 10 May 2021

Hu TY, Ju JM, Mo LH, Ma L, Hu WH, You RR, Chen XQ, Chen YY, Liu ZQ, Qiu SQ, Fan JT (2019) Anti- inflammation action of xanthones from Swertia chirayita by regulating COX-2/NF-κB/MAPKs/Akt signaling pathways in RAW 264.7 macrophage cells. Phytomedicine 55:214–221

Husain I, Zameer S, Madaan T, Minhaj A, Ahmad W, Iqubaal A, Ali A, Najmi AK (2019) Exploring the multifaceted neuroprotective actions of Emblica officinalis (Amla): a review. Metab Brain Dis 34(4):957–965

Iadecola C, Anrather J, Kamel H (2020) Effects of COVID-19 on the nervous system. Cell 183(1): 16–27

Iqbal Chowdhury I, Rahman MA, Hashem MA, Bhuiyan MM, Hajjar D, Alelwani W, Makki AA, Haque MA, Tangpong J, Bakhtiar MT (2020) Supplements of an aqueous combination of Justicia adhatoda and Ocimum tenuiflorum boost antioxidative effects and impede hyperlipidemia. Anim Models Experiment Med 3(2):140–151

Jamshidi N, Cohen MM (2017) The clinical efficacy and safety of Tulsi in humans: a systematic review of the literature. Evid Based Complement Alternat Med 2017:9217567. https://doi.org/10.1155/2017/9217567

Jeyasri R, Muthuramalingam P, Suba V, Ramesh M, Chen JT (2020) Bacopa monnieri and their bioactive compounds inferred multi-target treatment strategy for neurological diseases: a cheminformatics and system pharmacology approach. Biomol Ther 10(4):536

Johnson DA, Johnson JA (2015) Nrf2—a therapeutic target for the treatment of neurodegenerative diseases. Free Radic Biol Med 88:253–267

Joshi RK, Setzer WN, Da Silva JK (2017) Phytoconstituents, traditional medicinal uses and bioactivities of Tulsi (Ocimum sanctum Linn.): a review. Am J Essent Oil Nat Prod 5(1): 18–21. https://www.essencejournal.com/pdf/2017/vol5issue1/PartA/5-1-1-607.pdf

Karan M, Vasisht K, Handa SS (1999) Antihepatotoxic activity of Swertia chirata on carbon tetrachloride induced hepatotoxicity in rats. Phytother Res 13(1):24–30

Karuppan MK, Devadoss D, Nair M, Chand HS, Lakshmana MK (2021) SARS-CoV-2 infection in the central and peripheral nervous system-associated morbidities and their potential mechanism. Mol Neurobiol 58:2465–2480. https://doi.org/10.1007/s12035-020-02245-1

Katzenschlager R, Evans A, Manson A, Patsalos PN, Ratnaraj N, Watt H, Timmermann L, Van der Giessen R, Lees AJ (2004) Mucuna pruriens in Parkinson's disease: a double blind clinical and pharmacological study. J Neurol Neurosurg Psychiatry 75(12):1672–1677

Khazdair MR, Kianmehr M, Anaeigoudari A (2021) Effects of medicinal plants and flavonoids on Parkinson's disease: a review on basic and clinical evidences. Adv Pharmaceut Bull 11(2):224

Kim H, Han TH, Lee SG (2009) Anti-inflammatory activity of a water extract of Acorus calamus L. leaves on keratinocyte HaCaT cells. J Ethnopharmacol 122(1):149–156

Konar A, Gupta R, Shukla RK, Maloney B, Khanna VK, Wadhwa R, Lahiri DK, Thakur MK (2019) M1 muscarinic receptor is a key target of neuroprotection, neuroregeneration and memory recovery by i-Extract from Withania somnifera. Sci Rep 9(1):1–5. https://doi.org/10.1038/s41598-019-48238-6

Koralnik IJ, Tyler KL (2020) COVID-19: a global threat to the nervous system. Ann Neurol 88(1): 1–11

Kuboyama T, Tohda C, Komatsu K (2014) Effects of Ashwagandha (roots of Withania somnifera) on neurodegenerative diseases. Biol Pharm Bull 37(6):892–897

Kulkarni M, Sawant N, Kolapkar A, Huprikar A, Desai N (2021) Borneol: a promising monoterpenoid in enhancing drug delivery across various physiological barriers. AAPS PharmSciTech 22(4):1–7

Kumar V, Van Staden J (2016) A review of Swertia chirayita (Gentianaceae) as a traditional medicinal plant. Front Pharmacol 6:308

Kumar PP, Kumaravel S, Lalitha C (2010) Screening of antioxidant activity, total phenolics and GC-MS study of Vitex negundo. Afr J Biochem Res 4(7):191–195. https://academicjournals.org/journal/AJBR/article-abstract/743A94611519. Accessed dated 30 Apr 2021

Lalert L, Kruevaisayawan H, Amatyakul P, Ingkaninan K, Khongsombat O (2018) Neuroprotective effect of Asparagus racemosus root extract via the enhancement of brain-derived neurotrophic factor and estrogen receptor in ovariectomized rats. J Ethnopharmacol 225:336–341

Lam TT, Jia N, Zhang YW, Shum MH, Jiang JF, Zhu HC, Tong YG, Shi YX, Ni XB, Liao YS, Li WJ (2020) Identifying SARS-CoV-2-related coronaviruses in Malayan pangolins. Nature 583(7815):282–285

Lambrecq V, Hanin A, Munoz-Musat E, Chougar L, Gassama S, Delorme C, Cousyn L, Borden A, Damiano M, Frazzini V, Huberfeld G (2021) Association of clinical, biological, and brain magnetic resonance imaging findings with electroencephalographic findings for patients with COVID-19. JAMA Netw Open 4(3):e211489

Lampariello LR, Cortelazzo A, Guerranti R, Sticozzi C, Valacchi G (2012) The magic velvet bean of Mucuna pruriens. J Tradit Complement Med 2(4):331–339

Lampron A, ElAli A, Rivest S (2013) Innate immunity in the CNS: redefining the relationship between the CNS and its environment. Neuron 78(2):214–232

Lancet T. Mental health: time to invest in quality. Lancet (London, England). 2020;396(10257): 1045

Latinne A, Hu B, Olival KJ, Zhu G, Zhang L, Li H, Chmura AA, Field HE, Zambrana-Torrelio C, Epstein JH, Li B (2020) Origin and cross-species transmission of bat coronaviruses in China. Nat Commun 11(1):1–5

Leach M, MacGregor H, Scoones I, Wilkinson A (2021) Post-pandemic transformations: how and why COVID- 19 requires us to rethink development. World Dev 138:105233

Li H, Liu Y, Tian D, Tian L, Ju X, Qi L, Wang Y, Liang C (2020) Overview of cannabidiol (CBD) and its analogues: structures, biological activities, and neuroprotective mechanisms in epilepsy and Alzheimer's disease. Eur J Med Chem 192:112163

Liao J, Xie X, Wang W, Gao Y, Cai Y, Peng J, Li T, Yi Q, He C, Wang L (2021) Anti-inflammatory activity of essential oil from leaves of Blumea balsamifera (L.) DC through inhibiting TLR4/NF-kB signaling pathways and NLRP3 inflammasome activation in LPS-induced RAW264.7 macrophage cells. J Essent Oil Bear Plants 24(2):1–7. https://doi.org/10.1080/0972060X.2021.1912645

Liguori C, Pierantozzi M, Spanetta M, Sarmati L, Cesta N, Iannetta M, Ora J, Mina GG, Puxeddu E, Balbi O, Pezzuto G (2020) Subjective neurological symptoms frequently occur in patients with SARS-CoV2 infection. Brain Behav Immun 88:11–16

Liu ZJ, Liu HQ, Xiao C, Fan HZ, Huang Q, Liu YH, Wang Y (2014) Curcumin protects neurons against oxygen- glucose deprivation/reoxygenation-induced injury through activation of peroxisome proliferator-activated receptor-γ function. J Neurosci Res 92(11):1549–1559

Liu Y, Wang Y, Guo F, Zhan L, Mohr T, Cheng P, Huo N, Gu R, Pei D, Sun J, Tang L (2017) Deep sequencing and transcriptome analyses to identify genes involved in secoiridoid biosynthesis in the Tibetan medicinal plant Swertia mussotii. Sci Rep 7(1):1–4

Lopresti AL, Smith SJ (2021) Ashwagandha (Withania somnifera) for the treatment and enhancement of mental and physical conditions: a systematic review of human trials. J Herb Med 16: 100434

Maithili KarpagaSelvi N, Sridhar MG, Sripradha R (2020) Evaluation of free radical scavenging activities and phytochemical screening of Curcuma longa extracts. J Young Pharm 12(2):113

Majumdar S, Gupta S, Prajapati SK, Krishnamurthy S (2021) Neuro-nutraceutical potential of Asparagus racemosus: a review. Neurochem Int 145:105013

Makkar R, Behl T, Bungau S, Zengin G, Mehta V, Kumar A, Uddin M, Ashraf GM, Abdel-Daim MM, Arora S, Oancea R (2020) Nutraceuticals in neurological disorders. Int J Mol Sci 21(12): 4424

Martis G, Rao A, Karanth KS (1991) Neuropharmacological activity of Acorus calamus. Fitoterapia 42(4):331–336

Masters SL, O'Neill LA (2011) Disease-associated amyloid and misfolded protein aggregates activate the inflammasome. Trends Mol Med 17(5):276–282

Mathew D, Hsu WL (2018) Antiviral potential of curcumin. J Funct Foods 40:692–699

Mechoulam R (2019) The pharmacohistory of Cannabis sativa. In: Cannabinoids as therapeutic agents. Chapman and Hall/CRC, pp 1–20

Meena J, Ojha R, Muruganandam AV, Krishnamurthy S (2011) Asparagus racemosus competitively inhibits in vitro the acetylcholine and monoamine metabolizing enzymes. Neurosci Lett 503(1):6–9

Mehrotra S, Mishra KP, Maurya R, Srimal RC, Yadav VS, Pandey R, Singh VK (2003) Anticellular and immunosuppressive properties of ethanolic extract of Acorus calamus rhizome. Int Immunopharmacol 3(1):53–61

Mitra J, Kodavati M, Provasek VE, Rao KS, Mitra S, Hamilton DJ, Horner PJ, Vahidy FS, Britz GW, Kent TA, Hegde ML (2022) SARS-CoV-2 and the central nervous system: emerging insights into hemorrhage-associated neurological consequences and therapeutic considerations. Ageing Res Rev 80:101687. https://doi.org/10.1016/j.arr.2022.101687

Monsey MS, Gerhard DM, Boyle LM, Briones MA, Seligsohn MA, Schafe GE (2015) A diet enriched with curcumin impairs newly acquired and reactivated fear memories. Neuropsychopharmacology 40(5):1278–1288

Nalbandian A, Sehgal K, Gupta A, Madhavan MV, McGroder C, Stevens JS, Cook JR, Nordvig AS, Shalev D, Sehrawat TS, Ahluwalia N (2021) Post-acute COVID-19 syndrome. Nat Med 27(4):601–615

Nemetchek MD, Stierle AA, Stierle DB, Lurie DI (2017) The Ayurvedic plant Bacopa monnieri inhibits inflammatory pathways in the brain. J Ethnopharmacol 197:92–100

Nichols JM, Kummari E, Sherman J, Yang EJ, Dhital S, Gilfeather C, Yray G, Morgan T, Kaplan BL (2021) CBD suppression of EAE is correlated with early inhibition of splenic IFN-γ+ CD8+ T cells and modest inhibition of neuroinflammation. J Neuroimmune Pharmacol 16:346–362

Nielsen SC, Boyd SD (2018) Human adaptive immune receptor repertoire analysis—past, present, and future. Immunol Rev 284(1):9–23

Nuzzo D, Cambula G, Bacile I, Rizzo M, Galia M, Mangiapane P, Picone P, Giacomazza D, Scalisi L (2021) Long-term brain disorders in post Covid-19 neurological syndrome (PCNS) patient. Brain Sci 11(4):454

Orhan IE (2012) Centella asiatica (L.) Urban: from traditional medicine to modern medicine with neuroprotective potential. Evid Based Complement Alternat Med 2012:946259. https://doi.org/10.1155/2012/946259

Orhan IE, Atasu E, Senol FS, Ozturk N, Demirci B, Das K, Sekeroglu N (2013) Comparative studies on Turkish and Indian Centella asiatica (L.) Urban (gotu kola) samples for their enzyme inhibitory and antioxidant effects and phytochemical characterization. Ind Crop Prod 47:316–322

Paland N, Pechkovsky A, Aswad M, Hamza H, Popov T, Shahar E, Louria-Hayon I (2021) The immunopathology of COVID-19 and the cannabis paradigm. Front Immunol 12:327. https://doi.org/10.3389/fimmu.2021.631233

Panaro MA, Corrado A, Benameur T, Paolo CF, Cici D, Porro C (2020) The emerging role of curcumin in the modulation of TLR-4 signaling pathway: focus on neuroprotective and anti-rheumatic properties. Int J Mol Sci 21(7):2299

Pandey A, Bani S, Dutt P, Satti NK, Suri KA, Qazi GN (2018) Multifunctional neuroprotective effect of Withanone, a compound from Withania somnifera roots in alleviating cognitive dysfunction. Cytokine 102:211–221

Pang Y, Wang D, Fan Z, Chen X, Yu F, Hu X, Wang K, Yuan L (2014) Blumea balsamifera—A phytochemical and pharmacological review. Molecules 19(7):9453–9477

Parekh S, Soni A (2020) Nyctanthes arbor-tristis: comprehensive review on its pharmacological, antioxidant, and anticancer activities. J Appl Biol Biotechnol 8(01):95–104

Poulas K, Farsalinos K, Zanidis C (2020) Activation of TLR7 and innate immunity as an efficient method against COVID-19 pandemic: imiquimod as a potential therapy. Front Immunol 11: 1373

Rai SN, Birla H, Zahra W, Singh SS, Singh SP (2017) Immunomodulation of Parkinson's disease using Mucuna pruriens (Mp). J Chem Neuroanat 85:27–35

Rajput SB, Tonge MB, Karuppayil SM (2014) An overview on traditional uses and pharmacological profile of Acorus calamus Linn.(Sweet flag) and other Acorus species. Phytomedicine 21(3):268–276

Randino R, Grimaldi M, Persico M, De Santis A, Cini E, Cabri W, Riva A, D'Errico G, Fattorusso C, D'Ursi AM, Rodriquez M (2016) Investigating the neuroprotective effects of turmeric extract: structural interactions of β-amyloid peptide with single curcuminoids. Sci Rep 6(1):1–7. https://doi.org/10.1038/srep38846

Ransohoff RM, Brown MA (2012) Innate immunity in the central nervous system. J Clin Invest 122(4):1164–1171

Rattis BA, Ramos SG, Celes M (2021) Curcumin as a potential treatment for COVID-19. Front Pharmacol 12:1068

Reder AT, Centonze D, Naylor ML, Nagpal A, Rajbhandari R, Altin catal A, Kim M, Berdofe A, Radhakrishnan M, Jung E, Sandrock AW (2021) COVID-19 in patients with multiple sclerosis: associations with disease-modifying therapies. CNS Drugs 35(3):317–330

Richardson PJ, Ottaviani S, Prelle A, Stebbing J, Casalini G, Corbellino M (2020) CNS penetration of potential anti-COVID-19 drugs. J Neurol 267(7):1880–1882

Russo A, Borrelli F (2005) Bacopa monniera, a reputed nootropic plant: an overview. Phytomedicine 12(4):305–317

Sabir SM, Zeb A, Mahmood M, Abbas SR, Ahmad Z, Iqbal N (2021) Phytochemical analysis and biological activities of ethanolic extract of Curcuma longa rhizome. Braz J Biol 81(3):737–740. https://doi.org/10.1590/1519-6984.230628

Saikarthik J, Ilango S, Vijayakumar J, Vijayaraghavan R (2017) Phytochemical analysis of methanolic extract of seeds of Mucuna pruriens by gas chromatography mass spectrometry. Int J Pharm Sci Res 8(7):2916–2921. https://www.researchgate.net/profile/Saikarthik-Jayakumar/publication/336487249_PHYTOCHEMICAL_ANALYSIS_OF_METHANOLIC_EXTRACT_OF_SEEDS_OF_MUCUNA_PRURIENS_BY_GAS_CHROMATOGRAPHY_MASS_SPECTROMETRY/links/5da30cf1299bf116fea2e307/PHYTOCHEMICAL-ANALYSIS-OF-METHANOLIC-EXTRACT-OF-SEEDS-OF-MUCUNA-PRURIENS-BY-GAS-CHROMATOGRAPHY-MASS-SPECTROMETRY.pdf. Accessed on dated 17 June 2021

Salehi B, Krochmal-Marczak B, Skiba D, Patra JK, Das SK, Das G, Popović-Djordjević JB, Kostić AŽ, Anil Kumar NV, Tripathi A, Al-Snafi AE (2020) Convolvulus plant—A comprehensive review from phytochemical composition to pharmacy. Phytother Res 34(2):315–328

Samy RP, Chow VT (2011) Antimicrobial and phytochemical analysis of Centella asiatica (L.). Nat Preced. https://doi.org/10.1038/npre.2011.6033.1

Saunders JA, Estes KA, Kosloski LM, Allen HE, Dempsey KM, Torres-Russotto DR, Meza JL, Santamaria PM, Bertoni JM, Murman DL, Ali HH (2012) CD4+ regulatory and effector/memory T cell subsets profile motor dysfunction in Parkinson's disease. J Neuroimmune Pharmacol 7(4):927–938

Sellner J, Taba P, Öztürk S, Helbok R (2020) The need for neurologists in the care of COVID-19 patients. European. J Neurol 27(9):e31–e32

Semwal DK, Kumar A, Semwal RB, Andola HC (2021) Convolvulus prostratus. In: Naturally occurring chemicals against Alzheimer's disease. Academic Press, pp 409–424

Sen B (2020) Potentiality and possibility of Medicinal Plants on Ayurvedic Principle in prevention and treatment of COVID-19. J Ayurvedic Herb Med 6(2):100–107. http://www.ayurvedjournal.com/JAHM_202062_16.pdf. Accessed dated 15 June 2021

Shahid M, Subhan F, Ahmad N, Ullah I (2017) A bacosides containing Bacopa monnieri extract alleviates allodynia and hyperalgesia in the chronic constriction injury model of neuropathic pain in rats. BMC Complement Altern Med 17(1):1–2

Sharma V, Sharma R, Gautam DS, Kuca K, Nepovimova E, Martins N (2020) Role of Vacha (Acorus calamus Linn.) in neurological and metabolic disorders: evidence from ethnopharmacology, phytochemistry, pharmacology and clinical study. J Clin Med 9(4):1176

Shikha S, Nidhi M, Upama M (2010) Neurological studies of novel compounds from Swertia chirayita. J Chem Pharm Res 2(1):125–134. https://www.cabdirect.org/cabdirect/abstract/20103251947 Accessed dated 30 Apr 2021

Shishodia S, Singh T, Chaturvedi MM (2007) Modulation of transcription factors by curcumin. In: The molecular targets and therapeutic uses of curcumin in health and disease, pp 127–148

Shukla PK, Khanna VK, Ali MM, Maurya RR, Handa SS, Srimal RC (2002) Protective effect of Acorus calamus against acrylamide induced neurotoxicity. Phytother Res 16(3):256–260

Singh D, Chaudhuri PK (2018) A review on phytochemical and pharmacological properties of holy basil (Ocimum sanctum L.). Ind Crop Prod 118:367–382

Soumyanath A, Zhong YP, Yu X, Bourdette D, Koop DR, Gold SA, Gold BG (2005) Centella asiatica accelerates nerve regeneration upon oral administration and contains multiple active fractions increasing neurite elongation in-vitro. J Pharm Pharmacol 57(9):1221–1229

Srikalyani V, Ilango K (2009) Review on anti-epileptic and anxiolytic plants used in Manasamitra vatakam: an ayurvedic herbomineral formulation. Int J Pharma Bio Sci 10(2):71–78

Srivastava A, Singh V, Haque S, Pandey S, Mishra M, Jawed A, Shukla PK, Singh PK, Tripathi CK (2018) Response surface methodology-genetic algorithm based medium optimization, purification, and characterization of cholesterol oxidase from Streptomyces rimosus. Sci Rep 8(1):1–3. https://doi.org/10.1038/s41598-018-29241-9

Su X, Federoff HJ (2014) Immune responses in Parkinson's disease: interplay between central and peripheral immune systems. BioMed research international 2014:275178. https://doi.org/10.1155/2014/275178

Suryawanshi S, Asthana RK, Gupta RC (2009) Assessment of systemic interaction between Swertia chirata extract and its bioactive constituents in rabbits. Phytoth Res 23(7):1036–1038

Susilawathi NM, Tini K, Wijayanti IA, Rahmawati PL, Wardhana DP, Samatra DG, Sudewi AA (2021) Neurological manifestations of COVID-19: a clinical approach. Med J Indonesia. https://doi.org/10.13181/mji.rev.204821. Accessed dated 9 May 2021

Tang SW, Helmeste D, Leonard B (2021) Inflammatory neuropsychiatric disorders and COVID-19 neuroinflammation. Acta Neuropsychiatrica 30:1–3

Taquet M, Geddes JR, Husain M, Luciano S, Harrison PJ (2021) 6-month neurological and psychiatric outcomes in 236 379 survivors of COVID-19: a retrospective cohort study using electronic health records. Lancet Psychiatry 8(5):416–427

Tirpude NV, Sharma A, Joshi R, Kumari M, Acharya V, Vitex negundo Linn. (2021) Extract alleviates inflammatory aggravation and lung injury by modulating AMPK/PI3K/Akt/p38-NF-κB and TGF-β/Smad/Bcl2/caspase/LC3 cascade and macrophages activation in murine model of OVA-LPS induced allergic asthma. J Ethnopharmacol 271:113894. https://doi.org/10.1016/j.jep.2021.113894

Tiwari N, Mishra A, Bhatt G, Chaudhary A (2015) Evaluation of antistress potential of Negundin A from vitex negundo in acute stress induced mice. Europ J Med Plants:1–8

Tripathi MK, Singh P, Sharma S, Singh TP, Ethayathulla AS, Kaur P (2020) Identification of bioactive molecule from Withania somnifera (Ashwagandha) as SARS-CoV-2 main protease inhibitor. J Biomol Struct Dyn 39(15):5668–5681. https://doi.org/10.1080/07391102.2020.1790425

Tseng YH, Chen TH (2021) Care for patients with neuromuscular disorders in the COVID-19 pandemic era. Front Neurol 12:607790

Uddin MS, Al Mamun A, Hossain MS, Akter F, Iqbal MA, Asaduzzaman M (2016) Exploring the effect of Phyllanthus emblica L. on cognitive performance, brain antioxidant markers and acetylcholinesterase activity in rats: promising natural gift for the mitigation of Alzheimer's disease. Ann Neurosci 23(4):218–229

Vidrio AL, Nicolini H, Zarate CT, Castro TG, Rojop IJ, Magaña JM, López NM, Mendoza AD (2022) Association between SARS-CoV-2 infection and neuropsychiatric manifestations. COVID 2(9):1270–1286. https://doi.org/10.3390/covid2090094

Virhammar J, Nääs A, Fällmar D, Cunningham JL, Klang A, Ashton NJ, Jackmann S, Westman G, Frithiof R, Blennow K, Zetterberg H (2021) Biomarkers for central nervous system injury in cerebrospinal fluid are elevated in COVID-19 and associated with neurological symptoms and disease severity. Eur J Neurol 28(10):3324–3331. https://doi.org/10.1111/ene.14703

Vogrig A, Moritz CP, Camdessanché JP, Tholance Y, Antoine JC, Honnorat J, Gigli GL (2021) Unclear association between COVID-19 and Guillain-Barré syndrome. Brain 144(5):e45

Vohora SB, Shah SA, Dandiya PC (1990) Central nervous system studies on an ethanol extract of Acorus calamus rhizomes. J Ethnopharmacol 28(1):53–62

Wang Q, Jiang H, Wang L, Yi H, Li Z, Liu R (2019) Vitegnoside mitigates neuronal injury, mitochondrial apoptosis, and inflammation in an Alzheimer's disease cell model via the p38 MAPK/JNK pathway. J Alzheimers Dis 72(1):199–214

Wang H, Li T, Barbarino P, Gauthier S, Brodaty H, Molinuevo JL, Xie H, Sun Y, Yu E, Tang Y, Weidner W (2020) Dementia care during COVID-19. Lancet 395(10231):1190–1191

Witkin JM, Li X (2013) Curcumin, an active constituent of the ancient medicinal herb Curcuma longa L.: some uses and the establishment and biological basis of medical efficacy. CNS Neurol Disord Drug Targets 12(4):487–497

Witkin JM, Leucke S, Thompson LK, Lynch RA, Ding C, Heinz B, Catlow JT, Gleason SD, Li X (2013) Further evaluation of the neuropharmacological determinants of the antidepressant-like effects of curcumin. CNS Neurol Disord Drug Targets 12(4):498–505

Wu PF, Zhang Z, Wang F, Chen JG (2010) Natural compounds from traditional medicinal herbs in the treatment of cerebral ischemia/reperfusion injury. Acta Pharmacol Sin 31(12):1523–1531

Xiao K, Zhai J, Feng Y, Zhou N, Zhang X, Zou JJ, Li N, Guo Y, Li X, Shen X, Zhang Z (2020) Isolation of SARS-CoV-2-related coronavirus from Malayan pangolins. Nature 583(7815): 286–289

Yende S, Harle U, Rajgure D, Tuse T, Vyawahare N (2008) Pharmacological profile of Acorus calamus: an overview. Pharmacogn Rev 2(4):22–26. https://www.researchgate.net/profile/SubhashYende/publication/207178406_Pharmacological_profile_of_Acorus_calamus_An_Overview%27/links/568ba66a08ae1975839f89dd/Pharmacological-profile-of-Acorus-calamus-An-Overview.pdf. Accessed dated 10 Feb 2021

Younger DS (2021) Coronavirus 2019: clinical and neuropathological aspects. Curr Opin Rheumatol 33(1):49–57

Zahedipour F, Hosseini SA, Sathyapalan T, Majeed M, Jamialahmadi T, Al-Rasadi K, Banach M, Sahebkar A (2020) Potential effects of curcumin in the treatment of COVID-19 infection. Phytother Res 34(11):2911–2920

Zahiruddin S, Basist P, Parveen A, Parveen R, Khan W, Ahmad S (2020) Ashwagandha in brain disorders: a review of recent developments. J Ethnopharmacol 257:112876

Zhang Q, Lenardo MJ, Baltimore D (2017a) 30 years of NF-κB: a blossoming of relevance to human pathobiology. Cell 168(1–2):37–57

Zhang QL, Fu BM, Zhang ZJ (2017b) Borneol, a novel agent that improves central nervous system drug delivery by enhancing blood–brain barrier permeability. Drug Deliv 24(1):1037–1044

Zweig JA, Brandes MS, Brumbach BH, Caruso M, Wright KM, Quinn JF, Soumyanath A, Gray NE (2021) Prolonged treatment with Centella asiatica improves memory, reduces amyloid-β pathology, and activates NRF2-regulated antioxidant response pathway in 5xFAD mice. J Alzheimers Dis 81(4):1453–1468

# Chapter 12
# Glycyrrhizae Radix et Rhizoma (Gan Cao) for the Management of COVID-19

**Lin Ang, Eunhye Song, Hye Won Lee, and Myeong Soo Lee**

## 12.1 Background

### *12.1.1 General Description*

Glycyrrhizae Radix et Rhizoma is considered one of the most widely used medicinal plants for its health benefits. The Glycyrrhiza genus under the Fabaceae family has about 20 species that are mostly found in Asia, Europe, North America, and South America continents, and among them, 8 species are found in China (Li et al. 2020). The dry root and rhizome of the three recognized Glycyrrhiza species—*Glycyrrhiza uralensis* Fisch, *Glycyrrhiza glabra* L., and *Glycyrrhiza inflata* Batal—is referred to as Glycyrrhizae Radix et Rhizoma (glycyrrhiza), commonly known as licorice or liquorice, and also called *Gan Cao* in China. Traditional medicine uses glycyrrhiza widely to relieve coughs or to treat diverse illnesses such as hepatitis, influenza, and gastric ulcer (Guo et al. 2014; Pastorino et al. 2018). Glycyrrhiza has great economic potential, and its extract has been used widely as an ingredient in food, cosmetics, or tobacco (flavor) (Pastorino et al. 2018; Güçlü-Ustündağ and Mazza 2007; Isbrucker and Burdock 2006).

L. Ang · M. S. Lee (✉)
KM Science Research Division, Korea Institute of Oriental Medicine, Daejeon, South Korea
e-mail: anglin2808@kiom.re.kr

E. Song
Global Cooperation Center, Korea Institute of Oriental Medicine, Daejeon, South Korea
e-mail: esong@kiom.re.kr

H. W. Lee
KM Convergence Research, Korea Institute of Oriental Medicine, Daejeon, South Korea
e-mail: hwlee@kiom.re.kr

 
J.-T. Chen (ed.), *Ethnopharmacology and Drug Discovery for COVID-19: Anti-SARS-CoV-2 Agents from Herbal Medicines and Natural Products*,
https://doi.org/10.1007/978-981-99-3664-9_12

### *12.1.2 History of Pharmacological Use of Glycyrrhiza*

From the earliest days of recorded history, glycyrrhiza (primarily the species *Glycyrrhiza glabra* L., Leguminosae) has been used as a remedy. Historical records from various geographical regions and periods noted its extensive use (Armanini et al. 2002). The earliest reported use of glycyrrhiza for medicinal purposes dates back to ancient Assyrian, Egyptian, Chinese, and Indian cultures (Chopra and Chopra 1994; Thompson 1930). In addition, Greek sources indicate that glycyrrhiza was first used as a drug in Europe (Table 12.1) (Fiore et al. 2005). The plant's name is derived from two Greek words: *glukus* (sweet) and *rhiza* (root).

### *12.1.3 Comparison with Traditions of Glycyrrhiza Use outside Europe*

Ayurveda of the Indian subcontinent and the traditional medicine of China are the two major sources of traditional herbal medicinal knowledge outside of Europe. Glycyrrhiza is a common ingredient added to most herbal formulas in China. The oldest Chinese herbal classic, *Shennong Bencao Jing (The Divine Farmer's Classic of Materia Medica)*, which dates back to 200 BC, mentions the usage of glycyrrhiza. This work, according to mythology, is grounded on sources that date back to Emperor Shennong (around 2700 BC). Some of the described symptoms were widespread in Europe, including coughing, palpitations, pharyngitis, gastric pain, digestive tract ulcers, and sores. Additionally, the use of glycyrrhiza for drug and food poisoning intoxication is documented in the book (Fiore et al. 2005).

Ayurveda makes significant use of the latter glycyrrhiza recommendation. This medical knowledge includes phytotherapy, which was first formalized in India around 450 BC, although the oldest existing Sanskrit manuscript is from roughly 500 AD (Wujastyk 2003). In addition to its use in antidote mixes for diverse acute and chronic poisonings, *Glycyrrhiza glabra* has other uses that have been documented in Sanskrit, including voice improvement, viral respiratory tract infections, wound infections, surgical ear wounds, hemorrhaging wounds caused by bloodletting, and acute and chronic liver diseases such as hepatitis (Thyagarajan et al. 2002).

Antiviral effects are discussed in European, Indian, and Chinese documentations in the context of virally caused voice alterations in the larynx, viral infection-related coughs and pharyngitis, viral skin illnesses including condyloma and ulcers, and viral hepatitis.

**Table 12.1** History of use regarding glycyrrhiza

| Date | Country | Author | Documented information |
|---|---|---|---|
| 4th–3rd century BC | Greece | Theophrastus (Philosopher, botanist/4th–3rd century BC) | Suggested that the Greeks gained knowledge of the pharmacological applications of glycyrrhiza from the Scythians and called it "Scythian root" |
| 1st century AD | Italy | Pedanius Dioscorides (Greek physician and pharmacognosist in the Roman army/ca. 40–90 AD) | Listed glycyrrhiza among the 650 medicinal plants included in Medical Matters and categorized plants based on their medicinal therapeutic attributes |
| 1st century AD | Italy | Pliny the elder (23–79 AD) | Provided comprehensive information on glycyrrhiza in Natural History and recommended licorice for treating asthma, sore throat, mouth ulcer, and sterility |
| 1st century AD | Italy | Aulus Cornelius Celsus (around 25 BC–50 AD) | Reported the beneficial effects of glycyrrhiza |
| 1st century AD | Italy | Scribonius Largus (2–52 AD), | Reported the beneficial effects of glycyrrhiza |
| 2nd century AD | Italy | Claudius Galen (129–211 AD), | Reported the beneficial effects of glycyrrhiza |
| 5th century AD | Italy | Marcellus Empiricus (4th/5th century AD) | Reported the beneficial effects of glycyrrhiza |
| 5th century AD | Italy | Cassius Felix (5th century AD) | Reported the beneficial effects of glycyrrhiza |
| 6th–7th century AD | Spain | Saint Isidor (560–636 AD) | Addressed the origin of the term Glycyrrhiza in *Etymologies or Origins* |
| 8th–9th century AD | Italy | School of Salerno (8th–9th century AD) | Examined glycyrrhiza and its pharmacological properties in their work *Rule of Health* (*Regimen sanitatis)* |
| 10th–11th century AD | Iran | Ibn Sina/"Avicenna" (980–1037 AD) | Included glycyrrhiza in *The Canon of Avicenna*, which was considered an important recapitulation of medical knowledge and influenced Europe |
| 12th century AD | Germany | Hildegard von Bingen (1098–1179) | Compilation of a medical treatise, *Cause et cure*, integrated herbal knowledge including glycyrrhiza |
| 13th century AD | Syria | Abu Mohammed Abdallah ibn Ahmed al-Malaqi/"Ibn Al Baytar" (around 1197–1248) | Covered more than 1400 substances of plant, food, and drugs in his encyclopedia, *The Book of Medicinal and Nutritional Terms* |
| 14th century AD | Italy | *Jacobus Philippus* (1390–1400) | Documented the use of glycyrrhiza in *El Libro Agreg'a de Serapiom* |

(continued)

**Table 12.1** (continued)

| Date | Country | Author | Documented information |
|---|---|---|---|
| 15th–16th century AD | Germany | Leonhart Fuchs (1505–1566) | Created a botanical nomenclature and described glycyrrhiza |
| 16th century AD | Italy | Castore Durante (1529–1590) | Covered more than 900 herbal species in his work, *Herbario Nuovo (New Herbarium)*, ordered alphabetically according to common names, explained the applications and uses of herbs, which included glycyrrhiza |
| 17th century AD | United Kingdom | Nicholas Culpeper (1616–1654) | Documented medical uses of glycyrrhiza in *Complete Herbal* |
| 18th century AD | Italy | Guiseppe Donzelli | Described glycyrrhiza and explained its etymology as sweet root |
| 18th century AD | Sweden | Carl von Linne (1707–1778) | Classified plants into genera and species and discovered three distinct species among the genus Glycyrrhiza: *Glycyrrhiza glabra, Glycyrrhiza echinata*, and *Glycyrrhiza hirsute* |

## 12.2 Active Components of Glycyrrhizae Radix et Rhizoma

The triterpenoid saponin glycyrrhizin, which is also referred to as glycyrrhizic acid or glycyrrhizinic acid, is the primary active component in glycyrrhiza and is typically present in concentrations ranging from 6% and 10% (Li et al. 2020; Murray 2020). Flavonoids, such as isoliquiritigenin, liquiritigenin, isoliquiritin, glabrol, kumatakenin, and licochalcone A, B, and E, are also regarded as the glycyrrhiza's primary component. Glycyrrhiza has a broad range of pharmacological actions for its effects on the immune system, central nervous system, endocrine and cardiovascular systems, functions of the liver and kidneys, (Zhao et al. 2018; Rauchensteiner et al. 2005; Lin et al. 2005) as well as anti-inflammatory, anticancer, anti-virus, antibacterial, and so forth (Shon et al. 2004; Kim et al. 2004).

Glycyrrhiza has a significant amount of triterpenoid saponins (Table 12.2). In the study of active components of glycyrrhiza, 77 triterpene saponins were isolated. Various types of triterpenoid saponins can be found in each species of glycyrrhiza. The species *G. uralensis* isolated 50 oleanane-type pentacyclic triterpene saponins, *G. glabra* isolated 38, and *G. inflate* isolated only 13 saponins.

**Table 12.2** Types of triterpenoid saponins found in Glycyrrhizae Radix et Rhizoma

| Glycyrrhiza species | Components found |
|---|---|
| *G. uralensis* | Glycyrrhizin, Licoricesaponin G2, Araboglycyrrhizin, Uralsaponin B, Licoricesaponin B2, Licoricesaponin C2, Licoricesaponin H2, Licoricesaponin J2, Licoricesaponin K2, 18*β*-glycyrrhizic acid, Licoricesaponin M3 (uralsaponin T), Licoricesaponin N4, Uralsaponin V, 3*β*-*O*-*β*-D-glucuronopyranosyl-glycyrrhetinic acid, Licoricesaponin A3, Licoricesaponin E2, 22*β*-acetoxylglycyrrhizin, Uralsaponin D, 22*β*-acetoxyl-glycyrrhaldehyde, Licoricesaponin D3, Licoricesaponin F3, Licoricesaponin L3, 3*β*-*O*-[*β*-D-glucuronopyranosyl-(1 → 2)-*β*-D-glucuronopyranosyl]-glycyrretol, 3*β*-*O*-[*β*-D-glucuronopyranosyl-(1 → 2)-*β*-D-glucuronopyranosyl]-olean-9,12-diene-30-oic acid, Uralsaponin C, Uralsaponin E, Uralsaponin F, 3*β*-*O*-[*β*-D-glucuronopyranosyl-(1 → 2)-*β*-D-glucuronopyranosyl]-glycyrrhetic acid, Uralsaponin M, Uralsaponin N, Uralsaponin O, Uralsaponin P, Uralsaponin Q, Uralsaponin R, Uralsaponin S, Uralsaponin U, Uralsaponin W, Uralsaponin X, Uralsaponin Y, 3*β*-*O*-[*β*-D-glucuronopyranosyl-(1 → 2)-*β*-D-glucuronopyranosyl] -24-hydroxyglabrolide, Macedonoside E, 22*β*-acetyl-uralsaponin C, Glyuralsaponin A, Glyuralsaponin B, Glyuralsaponin C, Glyuralsaponin D, Glyuralsaponin E, Glyuralsaponin F, Glyuralsaponin G, Glyuralsaponin H |
| *G. glabra* | Glycyrrhizin, Licoricesaponin G2, Araboglycyrrhizin, Uralsaponin B, Licoricesaponin B2, Licoricesaponin C2, Licoricesaponin H2, Licoricesaponin J2, Licoricesaponin K2, 18*β*-glycyrrhizic acid, Licoricesaponin M3 (uralsaponin T), Licoricesaponin N4, Uralsaponin V, 3*β*-*O*-*β*-D-glucuronopyranosyl-glycyrrhetinic acid, Apioglycyrrhizin, Macedonoside A, Licoricesaponin O4, Licoricesaponin M1, Licoricesaponin M2, Licoricesaponin M3, Licoricesaponin M4, 30-hydroxyglycyrrhizin, Glycyrrhizin-20-methanoate, 24-hydroxyglucoglycyrrhizin, Rhaoglycyrrhizin, 11-deoxorhaoglycyrrhizin, Rhaoglucoglycyrrhizin, Rhaogalactoglycyrrhizin, 11-deoxo-20α-glycyrrhizin, 20α-galacturonoylglycyrrhizin, 20-α-rhaoglycyrrhizin, Glabasaponin A, Glabasaponin B, Glabasaponin C, Glabasaponin D, Glabasaponin E, Glabasaponin F, Glabasaponin G |
| *G. inflate* | Glycyrrhizin, Licoricesaponin G2, Araboglycyrrhizin, Licoricesaponin A3, Licoricesaponin E2, 22*β*-acetoxylglycyrrhizin, Uralsaponin D, 22*β*-acetoxyl-glycyrrhaldehyde, Apioglycyrrhizin, Macedonoside A, Licoricesaponin P2, licoricesaponin Q2, 24-hydroxy-licoricesaponin E2 |

## 12.3 Biological Activities

Glycyrrhiza has been found in a number of studies to have pharmacological benefits against cancer, viral infection, oxidative stress, inflammation, and immunoregulation (Li et al. 2019). Bioactive properties such as hepatoprotective, anti-inflammatory, antibacterial, antiviral, and anticancer effects are highly correlated to the herb's active components. Although most pharmacology research focused on glycyrrhizin and glycyrrhetinic acid, glycyrrhiza also contains a large number of flavonoids with substantial pharmacological effects on their own.

### 12.3.1 Anti-Inflammatory Properties

Glycyrrhiza possesses notable anti-inflammatory and anti-allergic properties (Kuroyanagi and Saito 1966; Cyong and Otsuka 1982) As glycyrrhizin or glycyrrhetinic acid attaches to glucocorticoid receptors, the majority of glycyrrhiza's anti-inflammatory effects have been related to cortisol effects. Despite so, many of the herb's activities inhibit or suppress cortisol (Kumagai et al. 1967a). Tryptophan oxygenase activation, hepatic cholesterol synthesis stimulation, hepatic glycogen synthesis stimulation, thymus atrophy inhibition, etc. are some of the antagonistic actions of cortisol. Glycyrrhizin, nevertheless, supports cortisol's regulation of the production of antibodies, the stress response, and inflammation. The glycyrrhiza's primary impact on glucocorticoid metabolism is probably brought about by the reduction of 5-β-reductase activity, which lengthens cortisol's half-life. Additionally, glycyrrhiza can enhance cortisol to convert it into cortisone (van Uum et al. 2002).

According to an animal study by Wang et al., it was found that glycyrrhizin reduced the inflammatory response caused by lipopolysaccharide (LPS) in mouse endometrial epithelial cells by blocking the TLR4 signaling pathway (Wang et al. 2017). In Akamatsu et al. in vitro study (Akamatsu et al. 1991), it was reported that glycyrrhizin prevented neutrophils from generating reactive oxygen species, a type of inflammatory mediator. In addition, glycyrrhizin may block the expression of high-mobility group protein B1 and the consequent generation of inflammatory cytokines in Sprague-Dawley rats after subarachnoid hemorrhage, as reported by Li et al. (2017). In Pang et al. study, inhibiting HMGB1 with glycyrrhizin reduced brain damage following diffuse axonal injury (DAI) in SD rats by acting as an anti-inflammatory (Pang et al. 2016). According to research by Li et al. using lipopolysaccharide (LPS)-induced inflammation mouse model, total saponins of glycyrrhiza may have an anti-inflammatory mechanism that involves inhibiting primary enzymes in the arachidonic acid metabolism pathway and reducing the release of inflammatory factors (Li et al. 2010).

### 12.3.2 Antiviral and Antimicrobial Properties

Studies have shown that glycyrrhizin and glycyrrhetinic acid induce interferon (Abe et al. 1982). Interferons attach to the surface of cells, where they trigger the production of intracellular proteins that stop viral DNA from being transcripted, resulting in antiviral effects.

Glycyrrhizin has a wide spectrum of antiviral activities and could be used as a natural antiviral agent (Moghimipour et al. 2015). It has been demonstrated to inhibit the growth of multiple DNA and RNA viruses in cell cultures, including severe acute respiratory syndrome coronavirus (SARS-CoV), human immunodeficiency virus (HIV), vaccinia, herpes simplex virus (HSV), Newcastle disease, Epstein-Barr, and vesicular stomatitis viruses, and to irreversibly inactivate HSV-1 (Cinatl et al. 2003; Sasaki et al. 2002; Lin 2003; Pompei et al. 1980). As it reduces the activity of the virus while simultaneously increasing the activity of the host cell, glycyrrhizin is an effective antiviral component against HIV, hepatitis B, coxsackievirus B3, enterovirus 71, HSV, and avian influenza A-H5N1 (Wang et al. 2015).

Upon administration of glycyrrhizin, the survival rate of mice with herpetic encephalitis increased approximately 2.5-fold, while HSV-1 replication was reduced in the control group (Sekizawa et al. 2001). Other glycyrrhiza components also exhibited immunomodulatory properties (Barfod et al. 2002). It has been demonstrated that glycyrrhizin blocks the replication of the SARS-CoV and varicella zoster virus in vitro (Baba and Shigeta 1987; Hoever et al. 2005). Wolkerstorfer et al. discovered that glycyrrhizin prevented the uptake of the influenza A virus into the cell (Wolkerstorfer et al. 2009). Glycyrrhizin also has a direct impact on anti-hepatitis B by influencing the extracellular secretion of hepatitis B surface antigen, reducing liver impairment in people with chronic hepatitis B, and eventually enhancing immunity. Additionally, HIV can be considerably slowed down in its growth by glycyrrhizin via activating the immune system (Sun et al. 2019).

Alcohol extracts from glycyrrhiza demonstrated in vitro antibacterial activity against *Streptococcus mutans*, *Haemophilus influenzae*, *Candida albicans, Bacillus subtilis, Staphylococcus aureus, Mycobacterium smegmatis*, *Moraxella catarrhalis*, and *Streptococcus pyogenes* (Fukai et al. 2002a, b; Tsukiyama et al. 2002; Mitscher et al. 1980). In general, the saponins have less antibacterial activity than the isoflavonoid components, which account for the majority of the antimicrobial activities.

### *12.3.3 Anticancer Properties*

The anticancer effects of glycyrrhiza components are extremely diverse with flavonoids and coumarins being the most active components (Wang and Nixon 2001). In an animal study (Takahashi et al. 2004), isoliquiritigenin has been found to reduce colon cancer in mice. It has been demonstrated that isoliquiritigenin dramatically reduces the growth of breast and prostate cancer cell lines (Kanazawa et al. 2003; Maggiolini et al. 2002). In addition, isoliquiritigenin inhibited leukopenia brought on by 5-fluorouracil treatment and greatly decreased pulmonary metastasis (Yamazaki et al. 2002). Licocoumarone, a coumarin compound in glycyrrhiza, was also found to cause cell apoptosis (Watanabe et al. 2002).

As demonstrated by Deng et al., glycyrrhizin significantly decreased the expression of thromboxane synthase (TxA2) and proliferating cell nuclear antigen, reversed liver and kidney damage in tumor-bearing mice, and may have this impact through inhibiting the TxA2 pathway (Deng et al. 2017). Additionally, dipotassium glycyrrhizinate also had anticancer effects on the glioblastoma cell lines by enhancing apoptosis and decreasing proliferation. The antitumor action of dipotassium glycyrrhizinate is correlated to NF-B suppression, where miR16 and miR146a are, respectively, mediated by IRAK2 and TRAF6 (Bonafé et al. 2019). The cytotoxic assay revealed no cytotoxic effect for glycyrrhizin, whereas their equivalent aglycones had shown cytotoxic activities against human cervical cancer cells and human breast adenocarcinoma cells (Zheng et al. 2010).

### *12.3.4 Hepatoprotective Properties*

There have been numerous in vitro and in vivo studies reporting on the potential mechanisms by which glycyrrhiza saponins present hepatoprotective actions. Through multitargeting therapeutic mechanisms such as anti-inflammation, anticancer, antioxidative stress, antisteatosis, antifibrosis, immunoregulation, and drug–drug interactions, glycyrrhizin has been shown to prevent liver disease and reduce drug-induced liver damage (Li et al. 2019).

In a study conducted using BALB/c mice, Tsuruoka et al. (Fukai et al. 2002a) demonstrated that glycyrrhizin could prevent the increase in alanine aminotransaminase and aspartate aminotransaminase while decreasing hepatocyte degeneration and inhibiting inducible nitric oxide synthase mRNA expression in the mice liver (Tsuruoka et al. 2009). Lee et al. observed that glycyrrhizin mitigated carbon tetrachloride ($CCl_4$)-induced liver damage in mice, most likely via downregulating proinflammatory mediators and inducing heme oxygenase-1 (Lee et al. 2007). According to Nakamura et al., glycyrrhizin inhibited soluble enzyme release from $CCl_4$-induced rat hepatocytes (Nakamura et al. 1985). Sato et al. discovered that glycyrrhizin could alter the expression of hepatitis B virus-related antigens on hepatocytes (Sato et al. 1996). Moreover, Lin et al. discovered that a three-day pretreatment with glycyrrhizin protected Sprague-Dawley rats from retrorsine-induced liver injury (Lin et al. 1999). Wistar rats are somewhat protected from ischemia-reperfusion-induced liver injury by glycyrrhizin (Nagai et al. 1992). Orazizadeh et al. also showed that glycyrrhizin successfully protects Wistar rats from $NTiO_2$-induced hepatotoxicity (Orazizadeh et al. 2014).

Other triterpenoid saponins found in glycyrrhiza also demonstrated hepatoprotective properties. Glyuralsaponins B and H exhibited antioxidant effects against cysteine-assisted $Fe^{2+}$-induced liver microsomal lipid peroxidation (Liu et al. 2019). In addition, galactosamine and carbon tetrachloride-induced liver damage could be prevented by glycyrrhetinic acid. The enzymatic action of nicotine adenine disphosphonucleotide also resulted in the inhibition of nonenzymatic lipid peroxidation and the formation of free radicals (Kiso et al. 1984).

### *12.3.5 Others*

Glycyrrhiza saponin monomers have a variety of additional pharmacological and physiological functions. Glycyrrhizin was found to have immunomodulatory, neuroprotective, and antioxidant properties (Ojha et al. 2016; Ming and Yin 2013;

Akman et al. 2015). Furthermore, glycyrrhizin may treat allergic rhinitis, by partly altering the Th1/Th2 balance by inhibiting the OX40 receptor and by boosting the activation of regulatory T cells (Fouladi et al. 2018).

Glycyrrhizin also has a different effect on estrogen metabolism by inhibiting estrogen action when the levels are too high and potentiating estrogen action when the levels are too low (Kumagai et al. 1967b). It has been demonstrated that glycyrrhetinic acid counteracts many of the estrogen's effects, especially exogenous estrogens (Kraus and Kaminskis 1969). Such estrogenic action is due to its isoflavonoid component which appears to have more estrogenic activity than the estrogen antagonism of glycyrrhetinic acid (Tamir et al. 2001). Notably, these same substances also slow down breast cancer cell growth (Maggiolini et al. 2002).

## 12.4 Clinical Applications of Glycyrrhizae Radix et Rhizoma

Glycyrrhiza is included frequently in traditional Eastern medicine, and it has been a staple of Western herbal medicine for decades. The most common clinical applications of glycyrrhiza containing glycyrrhizin are in the treatment of viral diseases (Murray 2020). Due to the diversity of the pharmacological effects, glycyrrhiza has also been used in many other diseases.

### *12.4.1 Obesity and Metabolic Syndrome*

Medications containing glycyrrhetinic acid may be useful in treating a variety of metabolic syndrome-related symptoms. In a pilot investigation (Armanini et al. 2003), normal-weight participants consuming glycyrrhetinic acid-containing commercial preparations reported a reduction in body fat mass. In another study (Fuhrman et al. 2002), moderately hypercholesterolemic patients supplemented with glycyrrhiza extract for 1 month decreased plasma susceptibility to oxidation, decreased systolic blood pressure, and increased resistance of plasma low-density lipoprotein.

In double-blind clinical research (Tominaga et al. 2009), overweight participants were randomly assigned to one of the four groups: low-dose licorice flavonoid oil (LFO), middle-dose LFO, high-dose LFO, and placebo. The participants in all LFO groups, except the placebo group, had considerably reduced body fat masses than at baseline, while the high-dose LFO group showed significant declines in body mass index and body weight.

Another study on LFO investigated its ability to improve the muscle mass of older people undergoing knee osteoarthritis rehabilitation and found that trunk muscular mass increased considerably more in the LFO group compared to those in the placebo group. Moreover, the percentage of body trunk fat and body fat in the LFO group was also lowered (Kinoshita et al. 2017).

### 12.4.2 Gastric and Duodenal Ulcers

Numerous clinical trials have demonstrated that deglycyrrhizinated licorice (DGL) is an effective antiulcer agent over the years. It has been demonstrated that DGL is a highly effective treatment for stomach ulcers (Turpie et al. 1969; Rees et al. 1979; Kassir 1985). In a clinical trial (Turpie et al. 1969), patients with stomach ulcers were given DGL or a placebo, and those receiving DGL experienced a considerably higher decrease in ulcer size with 44% of complete healing. DGL has also been proven to be more effective in comparison to cimetidine, ranitidine, and antacids in both the short-term treatment and maintenance of patients with peptic ulcers in several studies (Kassir 1985; Morgan et al. 1982, 1985). DGL is recommended to prevent gastric ulcers in patients requiring long-term treatment as it has been demonstrated in clinical studies to reduce the gastrointestinal bleeding caused by ulcerogenic medicines (Rees et al. 1979).

Duodenal ulcers can benefit from DGL as well. In previous research (Tewari and Wilson 1973), individuals with severe duodenal ulcers who received DGL treatment significantly improved, and no patients needed surgery after one year follow-up period. In another study (Kassir 1985), patients with verified chronic duodenal ulcers had their therapeutic effects of DGL compared to those of antacids or cimetidine. The healing rates were similar between the groups, although the DGL group saw fewer relapses than those who took cimetidine or antacids. These findings imply that DGL is a superior duodenal ulcer treatment, in addition to its protective properties and extremely low toxicity.

### 12.4.3 Human Immunodeficiency Virus and Acquired Immunodeficiency Syndrome

Preparations containing glycyrrhizin are promising in the treatment of acquired immunodeficiency syndrome (AIDS) and other human immunodeficiency virus (HIV)-related diseases. In double-blind clinical research on hemophiliac patients with HIV (Ikegami et al. 1993), none of the patients who received the glycyrrhizin progressed to immunological abnormalities or developed AIDS. The group without glycyrrhetinic acid, in contrast, saw a decrease in helper and total T-cell counts as well as antibody levels, and some patients even developed AIDS. In a different study (Ikegami et al. 1989), HIV-positive individuals without AIDS took glycyrrhizin for 1–2 years, and none of them developed AIDS-related symptoms. Glycyrrhizin causes an immediate improvement in immunological function in HIV-positive and

AIDS patients. Another study where glycyrrhizin was administered intravenously to symptom-free HIV-positive individuals reported that the patients' T-helper cell counts increased and their liver function improved (Mori et al. 1989). In another study where AIDS patients were administered glycyrrhizin intravenously for 30 days (Hattori et al. 1989), a decrease or absence of the P24 antigen indicating active disease was observed. These studies on HIV-positive and AIDS patients have produced favorable findings.

### *12.4.4 Gynecological Issues*

Another common application of glycyrrhiza is in the treatment of gynecological conditions, specifically premenstrual syndrome (PMS) and menopause. Given that glycyrrhizin has antiestrogenic actions and inhibits the breakdown of progesterone, administration of glycyrrhiza 2 weeks before the onset of menstruation may lessen PMS symptoms. Clinical research demonstrated that herbal combinations containing glycyrrhiza are also beneficial for dysmenorrhea (Tanaka 2003). The potential of glycyrrhiza isoflavones to impede serotonin reuptake suggests that they may also have antidepressant effects in PMS (Ofir et al. 2003).

### *12.4.5 Others*

Glycyrrhiza can be used to treat nearly all inflammatory and allergy conditions. In a double-blind clinical trial on postoperative sore throat (Agarwal et al. 2009), patients who had elective lumbar laminectomy were placed into two groups: gargling glycyrrhiza in water or water only before anesthesia. The occurrence and intensity of postoperative sore throat and post-extubation cough were lessened in the glycyrrhiza group compared to the water group.

Aphthous stomatitis is also a common ulcerative oral disease and DGL may promote healing effectively. In a study (Das et al. 1989), patients were directed to DGL mouthwash and most patients saw huge improvement within day 1 and complete recovery of the ulcers by day 3.

A review examined the potential use of the active components of glycyrrhiza in the treatment of rheumatoid arthritis via the cyclooxygenase (COX)-2/thromboxane A2 (TxA2) pathway (Huang et al. 2016). These active components have the potential to enhance the effects of non-steroidal anti-inflammatory drugs or disease-modifying antirheumatic drugs and also reduce their side effects by inhibiting the COX-2/TxA2 pathway, making them a potential add-on therapy for the treatment of rheumatoid arthritis.

## 12.5 Potential of Glycyrrhizae Radix et Rhizoma for the Management of COVID-19

The world is currently dealing with the coronavirus disease 2019 (COVID-19) pandemic, which is caused by severe acute respiratory syndrome coronavirus 2 (SARS-CoV-2). As glycyrrhiza is widely accessible and generally safe, it is frequently used in traditional Eastern medicine to treat COVID-19 infections. Glycyrrhizin was found to have the potential to lessen the severity of a COVID-19 infection by hindering the number of entry sites and providing an anti-inflammatory mechanism independent of angiotensin-converting enzyme 2 (ACE2) (Murck 2020).

### *12.5.1 ACE2 Receptor as SARS-CoV-2 Entry Receptor*

SARS-CoV-2 shares a high percentage of the genetic sequence of the SARS-CoV and has a similar cell entry receptor that is the angiotensin-converting enzyme II (ACE2) (Zhou et al. 2020). In the meantime, the SARS-CoV-2 infection shares many clinical resemblances with the SARS-CoV infection (Wang et al. 2020). Due to the similarities between SARS-CoV and SARS-CoV-2, the clinical experience and clinical research from SARS-CoV can be referred to treat SARS-CoV-2.

Recent studies suggest that SARS-CoV-2 and human ACE2 have strong correlations (Xu et al. 2020a; Wan et al. 2020). Letko et al. findings further validated that ACE2 is the receptor for SARS-CoV-2 as their findings found that the SARS-CoV-2 receptor-binding domain could only enter cells expressing ACE2 (Letko et al. 2020). Therefore, targeting ACE2 could be a hopeful potential method for preventing SARS-CoV-2 infection since virus entry depends on the host cell receptor.

### *12.5.2 Glycyrrhiza Anti-inflammatory Actions toward ACE2*

Glycyrrhiza may be a good candidate for treating COVID-19 and merits additional investigation in light of the experience and lessons learned from the fight against SARS. Glycyrrhizin, a triterpene saponin with a variety of biological actions and pharmacological effects, has been in vitro shown to be active against SARS-associated coronavirus (Cinatl et al. 2003). Glycyrrhizin is reported to work best when given before and after the viral adsorption phase as it can prevent viral attachment and penetration (Fujii et al. 2004). Additionally, glycyrrhizin can obstruct the replication and/or cytopathogenic effects of several respiratory viruses (Cinatl et al. 2003; Michaelis et al. 2011).

ACE2 plays a protective role for lung tissue by inhibiting the lipopolysaccharide (LPS)-induced toll-like receptor 4 (TLR4) pathway, where ACE2 overexpression could reduce the LPS-induced inflammation (Kaparianos and Argyropoulou 2011;

Ye and Liu 2020). Consequently, the downregulation of ACE2 may be considered concerning. Here, the immunomodulatory properties of glycyrrhizin are important for its antagonistic action of TLR4-dependent mechanisms, as glycyrrhetinic acid has a TLR4 antagonistic action that lowers inflammation in a variety of tissues, including the lung (Yu et al. 2005). Additionally, in an LPS model of inflammation, glycyrrhizin caused a decrease in TLR4 expression in the heart and lung, accompanied by a considerable decrease in the secretion of cytokines (Seo et al. 2017). Accordingly, in a mice study (Yu et al. 2005), glycyrrhizin protects against the acute respiratory distress syndrome (ARDS) caused by the TLR4 activator. In a mouse model of *Streptococcus* aureus infection (Yao and Sun 2019), intraperitoneal treatment of glycyrrhizin also significantly reduced the expression of inflammatory markers in the lungs. These findings are consistent with other studies where glycyrrhizin or glycyrrhetinic acid could inhibit inflammatory pathways via TLR4 (Thakur et al. 2017; Vitali et al. 2015). Overall, glycyrrhizin may induce anti-inflammatory actions by inhibiting TLR4 activities while regulating ACE2 (Ingraham et al. 2020).

### *12.5.3 Other Potential Roles of Glycyrrhiza in COVID-19 Infection*

Glycyrrhizin possesses cytokine-modulating activity and may potentially boost the immunological response (Xu et al. 2018). The proinflammatory cytokines are overproduced during a cytokine storm, which can worsen pathological damage in hosts (Liu et al. 2016). The severity of the disease was linked to the cytokine storm observed in critically ill COVID-19 patients (Huang et al. 2020). Therefore, it is crucially important to identify patients with cytokine storms and to intervene as soon as possible. During the SARS outbreak, glucocorticoids were frequently used in individuals with cytokine storm. Glucocorticoids have an anti-inflammatory impact and are prescribed to alleviate respiratory distress brought on by cytokine storms, but they also have an immunosuppressive effect that reduces the immune system's ability to remove viruses (Fujii et al. 2004). To avoid or lessen excessive cytokine storms in COVID-19 patients, glycyrrhizin is therefore anticipated to be administered in the early stages of the disease for a longer period with fewer side effects.

Antioxidants may be supportive in treating the cytokine storm brought on by infection since reactive oxygen species (ROS) influence inflammatory response (Liu et al. 2016). The buildup of intracellular ROS brought on by viral infection can be prevented by glycyrrhizin (Tong et al. 2020). Glycyrrhizin can inhibit the generation of ROS and inhibit the activation of signaling pathways which are known to be related to virus replication (Michaelis et al. 2011). Additionally, cytokine storms or prolonged inflammation due to SARS-CoV-2 can activate the complement and coagulation cascades, which may lead to the failure of many organs (Wang et al. 2020). Studies have shown that glycyrrhizin selectively inhibits thrombin

(Francischetti et al. 1997; Mendes-Silva et al. 2003). These findings suggest that glycyrrhizin may be effective against COVID-19 via multiple sites of action.

Conversely, bilateral diffuse alveolar damage with cellular fibromyxoid exudates can be seen in histological findings of COVID-19 (Xu et al. 2020b), where the impaired alveolar function may lead to hypoxia; therefore, inhibiting the hyperproduction of airway exudates is a promising strategy for preventing severe hypoxia. Notably, glycyrrhizin could also inhibit airway mucus hyperproduction (Nishimoto et al. 2010). Hence, glycyrrhizin may reduce hypoxia and alleviate clinical symptoms in individuals with COVID-19.

Additionally, glycyrrhizin could induce endogenous interferon which would prevent the transmission of viruses by preventing DNA and RNA viruses from replicating and by eradicating viral infections (Abe et al. 1982; Wang and Fish 2019). Hence, glycyrrhizin may also have an indirect effect in treating COVID-19. Besides, antiviral drugs for COVID-19 may also lead to drug-induced liver injury. In this context, glycyrrhizin could also play an adjunctive role in treating COVID-19 due to its hepatoprotective properties.

## 12.6 Conclusions

Glycyrrhiza may be a promising herb for the treatment of COVID-19. Over the years, total saponins or saponin monomers from glycyrrhiza have been explored and identified to have a variety of biological properties, including anti-inflammatory, antibacterial, antiviral, hepatoprotective, anticancer, and antioxidant actions. However, these pharmacological investigations on glycyrrhiza currently only include bioassays of a small number of saponin monomers. Additional research is required to examine the biological activity of additional triterpene saponins using in vitro and in vivo models. Due to the various pharmacological effect of glycyrrhiza, this herb is worth conducting more research to further validate its value in managing COVID-19, which is not necessarily to treat or prevent the disease, but to reduce the severity and symptomatology of COVID-19. Nevertheless, as the findings on biological actions drawn from in vivo and in vitro studies may not correspond with real-life clinical efficacy in patients, further assessments are needed to identify the optimal dosage and treatment regimen for the adequate application of glycyrrhiza in managing COVID-19.

## References

Abe N, Ebina T, Ishida N (1982) Interferon induction by glycyrrhizin and glycyrrhetinic acid in mice. Microbiol Immunol 26(6):535–539. https://doi.org/10.1111/j.1348-0421.1982.tb00207.x

Agarwal A, Gupta D, Yadav G, Goyal P, Singh PK, Singh U (2009) An evaluation of the efficacy of licorice gargle for attenuating postoperative sore throat: a prospective, randomized, single-blind study. Anesth Analg 109(1):77–81. https://doi.org/10.1213/ane.0b013e3181a6ad47

Akamatsu H, Komura J, Asada Y, Niwa Y (1991) Mechanism of anti-inflammatory action of glycyrrhizin: effect on neutrophil functions including reactive oxygen species generation. Planta Med 57(2):119–121. https://doi.org/10.1055/s-2006-960045

Akman T, Guven M, Aras AB, Ozkan A, Sen HM, Okuyucu A et al (2015) The neuroprotective effect of Glycyrrhizic acid on an experimental model of focal cerebral ischemia in rats. Inflammation 38(4):1581–1588. https://doi.org/10.1007/s10753-015-0133-1

Armanini D, Fiore C, Mattarello MJ, Bielenberg J, Palermo M (2002) History of the endocrine effects of licorice. Exp Clin Endocrinol Diabetes 110(6):257–261. https://doi.org/10.1055/s-2002-34587

Armanini D, De Palo CB, Mattarello MJ, Spinella P, Zaccaria M, Ermolao A et al (2003) Effect of licorice on the reduction of body fat mass in healthy subjects. J Endocrinol Investig 26(7): 646–650. https://doi.org/10.1007/bf03347023

Baba M, Shigeta S (1987) Antiviral activity of glycyrrhizin against varicella-zoster virus in vitro. Antivir Res 7(2):99–107. https://doi.org/10.1016/0166-3542(87)90025-8

Barfod L, Kemp K, Hansen M, Kharazmi A (2002) Chalcones from Chinese liquorice inhibit proliferation of T cells and production of cytokines. Int Immunopharmacol 2(4):545–555. https://doi.org/10.1016/s1567-5769(01)00202-8

Bonafé GA, Dos Santos JS, Ziegler JV, Umezawa K, Ribeiro ML, Rocha T et al (2019) Growth inhibitory effects of dipotassium glycyrrhizinate in glioblastoma cell lines by targeting microRNAs through the NF-κB signaling pathway. Front Cell Neurosci 13:216. https://doi.org/10.3389/fncel.2019.00216

Chopra RN, Chopra IC (1994) Indigenous drugs of India. Academic Publishers

Cinatl J, Morgenstern B, Bauer G, Chandra P, Rabenau H, Doerr HW (2003) Glycyrrhizin, an active component of liquorice roots, and replication of SARS-associated coronavirus. Lancet 361(9374):2045–2046. https://doi.org/10.1016/s0140-6736(03)13615-x

Cyong J, Otsuka Y (1982) A pharmacological study of the anti-inflammatory activity of Chinese herbs. A review. Acupunct Electrother Res 7(2–3):173–202. https://doi.org/10.3727/036012982816952116

Das SK, Das V, Gulati AK, Singh VP (1989) Deglycyrrhizinated liquorice in aphthous ulcers. J Assoc Physicians India 37(10):647

Deng QP, Wang MJ, Zeng X, Chen GG, Huang RY (2017) Effects of glycyrrhizin in a mouse model of lung adenocarcinoma. Cell Physiol Biochem 41(4):1383–1392. https://doi.org/10.1159/000467897

Fiore C, Eisenhut M, Ragazzi E, Zanchin G, Armanini D (2005) A history of the therapeutic use of liquorice in Europe. J Ethnopharmacol 99(3):317–324. https://doi.org/10.1016/j.jep.2005.04.015

Fouladi S, Masjedi M, Ghasemi R, M GH, Eskandari N. (2018) The in vitro impact of Glycyrrhizic acid on CD4+ T lymphocytes through OX40 receptor in the patients with allergic rhinitis. Inflammation 41(5):1690–1701. https://doi.org/10.1007/s10753-018-0813-8

Francischetti IM, Monteiro RQ, Guimarães JA (1997) Identification of glycyrrhizin as a thrombin inhibitor. Biochem Biophys Res Commun 235(1):259–263. https://doi.org/10.1006/bbrc.1997.6735

Fuhrman B, Volkova N, Kaplan M, Presser D, Attias J, Hayek T et al (2002) Antiatherosclerotic effects of licorice extract supplementation on hypercholesterolemic patients: increased resistance of LDL to atherogenic modifications, reduced plasma lipid levels, and decreased systolic blood pressure. Nutrition 18(3):268–273. https://doi.org/10.1016/s0899-9007(01)00753-5

Fujii T, Nakamura T, Iwamoto A (2004) Current concepts in SARS treatment. J Infect Chemother 10(1):1–7. https://doi.org/10.1007/s10156-003-0296-9

Fukai T, Marumo A, Kaitou K, Kanda T, Terada S, Nomura T (2002a) Anti-Helicobacter pylori flavonoids from licorice extract. Life Sci 71(12):1449–1463. https://doi.org/10.1016/s0024-3205(02)01864-7

Fukai T, Marumo A, Kaitou K, Kanda T, Terada S, Nomura T (2002b) Antimicrobial activity of licorice flavonoids against methicillin-resistant Staphylococcus aureus. Fitoterapia 73(6): 536–539. https://doi.org/10.1016/s0367-326x(02)00168-5

Güçlü-Ustündağ O, Mazza G (2007) Saponins: properties, applications and processing. Crit Rev Food Sci Nutr 47(3):231–258. https://doi.org/10.1080/10408390600698197

Guo J, Shang E, Zhao J, Fan X, Duan J, Qian D et al (2014) Data mining and frequency analysis for licorice as a "two-face" herb in Chinese formulae based on Chinese formulae database. Phytomedicine 21(11):1281–1286. https://doi.org/10.1016/j.phymed.2014.07.006

Hattori T, Ikematsu S, Koito A, Matsushita S, Maeda Y, Hada M et al (1989) Preliminary evidence for inhibitory effect of glycyrrhizin on HIV replication in patients with AIDS. Antivir Res 11(5–6):255–261. https://doi.org/10.1016/0166-3542(89)90035-1

Hoever G, Baltina L, Michaelis M, Kondratenko R, Baltina L, Tolstikov GA et al (2005) Antiviral activity of glycyrrhizic acid derivatives against SARS-coronavirus. J Med Chem 48(4): 1256–1259. https://doi.org/10.1021/jm0493008

Huang QC, Wang MJ, Chen XM, Yu WL, Chu YL, He XH et al (2016) Can active components of licorice, glycyrrhizin and glycyrrhetinic acid, lick rheumatoid arthritis? Oncotarget 7(2): 1193–1202. https://doi.org/10.18632/oncotarget.6200

Huang C, Wang Y, Li X, Ren L, Zhao J, Hu Y et al (2020) Clinical features of patients infected with 2019 novel coronavirus in Wuhan, China. Lancet 395(10223):497–506. https://doi.org/10.1016/s0140-6736(20)30183-5

Ikegami N, Akatani K, Imai M, Yoshioka K, Yano S (1993) Prophylactic effect of long-term oral administration of glycyrrhizin on AIDS development of asymptomatic patients. Int Conf AIDS. p 234

Ikegami N, Yoshioka K, Akatani K (1989) Clinical evaluation of glycyrrhizin on HIV-infected, asymptomatic hemophiliac patients in Japan. Int Conf AIDS, p 401

Ingraham NE, Lotfi-Emran S, Thielen BK, Techar K, Morris RS, Holtan SG et al (2020) Immunomodulation in COVID-19. Lancet Respir Med 8(6):544–546. https://doi.org/10.1016/s2213-2600(20)30226-5

Isbrucker RA, Burdock GA (2006) Risk and safety assessment on the consumption of licorice root (Glycyrrhiza sp.), its extract and powder as a food ingredient, with emphasis on the pharmacology and toxicology of glycyrrhizin. Regul Toxicol Pharmacol 46(3):167–192. https://doi.org/10.1016/j.yrtph.2006.06.002

Kanazawa M, Satomi Y, Mizutani Y, Ukimura O, Kawauchi A, Sakai T et al (2003) Isoliquiritigenin inhibits the growth of prostate cancer. Eur Urol 43(5):580–586. https://doi.org/10.1016/s0302-2838(03)00090-3

Kaparianos A, Argyropoulou E (2011) Local renin-angiotensin II systems, angiotensin-converting enzyme and its homologue ACE2: their potential role in the pathogenesis of chronic obstructive pulmonary diseases, pulmonary hypertension and acute respiratory distress syndrome. Curr Med Chem 18(23):3506–3515. https://doi.org/10.2174/092986711796642562

Kassir ZA (1985) Endoscopic controlled trial of four drug regimens in the treatment of chronic duodenal ulceration. Ir Med J 78(6):153–156

Kim SC, Byun SH, Yang CH, Kim CY, Kim JW, Kim SG (2004) Cytoprotective effects of Glycyrrhizae radix extract and its active component liquiritigenin against cadmium-induced toxicity (effects on bad translocation and cytochrome c-mediated PARP cleavage). Toxicology 197(3):239–251. https://doi.org/10.1016/j.tox.2004.01.010

Kinoshita T, Matsumoto A, Yoshino K, Furukawa S (2017) The effects of licorice flavonoid oil with respect to increasing muscle mass: a randomized, double-blind, placebo-controlled trial. J Sci Food Agric 97(8):2339–2345. https://doi.org/10.1002/jsfa.8044

Kiso Y, Tohkin M, Hikino H, Hattori M, Sakamoto T, Namba T (1984) Mechanism of antihepatotoxic activity of glycyrrhizin. I: effect on free radical generation and lipid peroxidation. Planta Med 50(4):298–302. https://doi.org/10.1055/s-2007-969714

Kraus SD, Kaminskis A (1969) The anti-estrogenic action of beta-glycyrrhetinic acid. Exp Med Surg 27(4):411–420

Kumagai A, Nanaboshi M, Asanuma Y, Yagura T, Nishino K (1967a) Effects of glycyrrhizin on thymolytic and immunosuppressive action of cortisone. Endocrinol Jpn 14(1):39–42. https://doi.org/10.1507/endocrj1954.14.39

Kumagai A, Nishino K, Shimomura A, Kin T, Yamamura Y (1967b) Effect of glycyrrhizin on estrogen action. Endocrinol Jpn 14(1):34–38. https://doi.org/10.1507/endocrj1954.14.34

Kuroyanagi T, Saito M (1966) Effect of prednisolone and glycyrrhizin on passive transfer in experimental allergic encephalomyelitis. Arerugi 15(2):67–74

Lee CH, Park SW, Kim YS, Kang SS, Kim JA, Lee SH et al (2007) Protective mechanism of glycyrrhizin on acute liver injury induced by carbon tetrachloride in mice. Biol Pharm Bull 30(10):1898–1904. https://doi.org/10.1248/bpb.30.1898

Letko M, Marzi A, Munster V (2020) Functional assessment of cell entry and receptor usage for SARS-CoV-2 and other lineage B betacoronaviruses. Nat Microbiol 5(4):562–569. https://doi.org/10.1038/s41564-020-0688-y

Li X, Qi Y, Cai R, Liu B, Song Y, Xie C (2010) Studies on the anti-inflammatory mechanism of total saponins of radix Glycyrrhiza in vitro. Chinese J Exp Tradit Med Formul 5:039

Li Y, Sun F, Jing Z, Wang X, Hua X, Wan L (2017) Glycyrrhizic acid exerts anti-inflammatory effect to improve cerebral vasospasm secondary to subarachnoid hemorrhage in a rat model. Neurol Res 39(8):727–732. https://doi.org/10.1080/01616412.2017.1316903

Li X, Sun R, Liu R (2019) Natural products in licorice for the therapy of liver diseases: progress and future opportunities. Pharmacol Res 144:210–226. https://doi.org/10.1016/j.phrs.2019.04.025

Li F, Liu B, Li T, Wu Q, Xu Z, Gu Y et al (2020) Review of constituents and biological activities of triterpene Saponins from Glycyrrhizae radix et Rhizoma and its Solubilization characteristics. Molecules 25(17). https://doi.org/10.3390/molecules25173904

Lin JC (2003) Mechanism of action of glycyrrhizic acid in inhibition of Epstein-Barr virus replication in vitro. Antivir Res 59(1):41–47. https://doi.org/10.1016/s0166-3542(03)00030-5

Lin G, Nnane IP, Cheng TY (1999) The effects of pretreatment with glycyrrhizin and glycyrrhetinic acid on the retrorsine-induced hepatotoxicity in rats. Toxicon 37(9):1259–1270. https://doi.org/10.1016/s0041-0101(98)00263-3

Lin ZJ, Qiu S-X, Wufuer A, Shum L (2005) Simultaneous determination of glycyrrhizin, a marker component in radix Glycyrrhizae, and its major metabolite glycyrrhetic acid in human plasma by LC–MS/MS. J Chromatogr B 814(2):201–207. https://doi.org/10.1016/j.jchromb.2004.10.026

Liu Q, Zhou YH, Yang ZQ (2016) The cytokine storm of severe influenza and development of immunomodulatory therapy. Cell Mol Immunol 13(1):3–10. https://doi.org/10.1038/cmi.2015.74

Liu YY, Yang YN, Feng ZM, Jiang JS, Zhang PC (2019) Eight new triterpenoid saponins with antioxidant activity from the roots of Glycyrrhiza uralensis Fisch. Fitoterapia 133:186–192. https://doi.org/10.1016/j.fitote.2019.01.014

Maggiolini M, Statti G, Vivacqua A, Gabriele S, Rago V, Loizzo M et al (2002) Estrogenic and antiproliferative activities of isoliquiritigenin in MCF7 breast cancer cells. J Steroid Biochem Mol Biol 82(4–5):315–322. https://doi.org/10.1016/s0960-0760(02)00230-3

Mendes-Silva W, Assafim M, Ruta B, Monteiro RQ, Guimarães JA, Zingali RB (2003) Antithrombotic effect of glycyrrhizin, a plant-derived thrombin inhibitor. Thromb Res 112(1–2):93–98. https://doi.org/10.1016/j.thromres.2003.10.014

Michaelis M, Geiler J, Naczk P, Sithisarn P, Leutz A, Doerr HW et al (2011) Glycyrrhizin exerts antioxidative effects in H5N1 influenza A virus-infected cells and inhibits virus replication and pro-inflammatory gene expression. PLoS One 6(5):e19705. https://doi.org/10.1371/journal.pone.0019705

Ming LJ, Yin AC (2013) Therapeutic effects of glycyrrhizic acid. Nat Prod Commun 8(3):415–418

Mitscher LA, Park YH, Clark D, Beal JL (1980) Antimicrobial agents from higher plants. Antimicrobial isoflavonoids and related substances from Glycyrrhiza glabra L. var. typica. J Nat Prod 43(2):259–269. https://doi.org/10.1021/np50008a004

Moghimipour E, Ameri A, Handali S, Ramezani Z, Azemi ME, Sadaghi-Nejad B (2015) In-vitro evaluation of antibacterial activity of Glycyrrhiza glabra and Acanthopyllum squarrusom total saponins. Rese J Pharmaceut Biol Chem Sci 6(5):764–768

Morgan AG, McAdam WA, Pacsoo C, Darnborough A (1982) Comparison between cimetidine and Caved-S in the treatment of gastric ulceration, and subsequent maintenance therapy. Gut 23(6): 545–551. https://doi.org/10.1136/gut.23.6.545

Morgan AG, Pacsoo C, McAdam WA (1985) Maintenance therapy: a two year comparison between Caved-S and cimetidine treatment in the prevention of symptomatic gastric ulcer recurrence. Gut 26(6):599–602. https://doi.org/10.1136/gut.26.6.599

Mori K, Sakai H, Suzuki S, Akutsu Y, Ishikawa M, Aihara M et al (1989) The present status in prophylaxis and treatment of HIV infected patients with hemophilia in Japan. Rinsho Byori 37(11):1200–1208

Murck H (2020) Symptomatic protective action of glycyrrhizin (licorice) in COVID-19 infection? Front Immunol 11:1239. https://doi.org/10.3389/fimmu.2020.01239

Murray MT (2020) Glycyrrhiza glabra (Licorice). In: Textbook of natural medicine, pp 641–7.e3. https://doi.org/10.1016/b978-0-323-43044-9.00085-6

Nagai T, Egashira T, Kudo Y, Yamanaka Y, Shimada T (1992) Attenuation of dysfunction in the ischemia-reperfused liver by glycyrrhizin. Jpn J Pharmacol 58(3):209–218. https://doi.org/10.1254/jjp.58.209

Nakamura T, Fujii T, Ichihara A (1985) Enzyme leakage due to change of membrane permeability of primary cultured rat hepatocytes treated with various hepatotoxins and its prevention by glycyrrhizin. Cell Biol Toxicol 1(4):285–295. https://doi.org/10.1007/bf00118193

Nishimoto Y, Hisatsune A, Katsuki H, Miyata T, Yokomizo K, Isohama Y (2010) Glycyrrhizin attenuates mucus production by inhibition of MUC5AC mRNA expression in vivo and in vitro. J Pharmacol Sci 113(1):76–83. https://doi.org/10.1254/jphs.09344fp

Ofir R, Tamir S, Khatib S, Vaya J (2003) Inhibition of serotonin re-uptake by licorice constituents. J Mol Neurosci 20(2):135–140. https://doi.org/10.1385/jmn:20:2:135

Ojha S, Javed H, Azimullah S, Abul Khair SB, Haque ME (2016) Glycyrrhizic acid attenuates neuroinflammation and oxidative stress in rotenone model of Parkinson's disease. Neurotox Res 29(2):275–287. https://doi.org/10.1007/s12640-015-9579-z

Orazizadeh M, Fakhredini F, Mansouri E, Khorsandi L (2014) Effect of glycyrrhizic acid on titanium dioxide nanoparticles-induced hepatotoxicity in rats. Chem Biol Interact 220:214–221. https://doi.org/10.1016/j.cbi.2014.07.001

Pang H, Huang T, Song J, Li D, Zhao Y, Ma X (2016) Inhibiting HMGB1 with Glycyrrhizic acid protects brain injury after DAI via its anti-inflammatory effect. Mediat Inflamm 2016:4569521. https://doi.org/10.1155/2016/4569521

Pastorino G, Cornara L, Soares S, Rodrigues F, Oliveira M (2018) Liquorice (Glycyrrhiza glabra): a phytochemical and pharmacological review. Phytother Res 32(12):2323–2339. https://doi.org/10.1002/ptr.6178

Pompei R, Pani A, Flore O, Marcialis MA, Loddo B (1980) Antiviral activity of glycyrrhizic acid. Experientia 36(3):304. https://doi.org/10.1007/bf01952290

Rauchensteiner F, Matsumura Y, Yamamoto Y, Yamaji S, Tani T (2005) Analysis and comparison of radix Glycyrrhizae (licorice) from Europe and China by capillary-zone electrophoresis (CZE). J Pharm Biomed Anal 38(4):594–600. https://doi.org/10.1016/j.jpba.2005.01.038

Rees WD, Rhodes J, Wright JE, Stamford LF, Bennett A (1979) Effect of deglycyrrhizinated liquorice on gastric mucosal damage by aspirin. Scand J Gastroenterol 14(5):605–607. https://doi.org/10.3109/00365527909181397

Sasaki H, Takei M, Kobayashi M, Pollard RB, Suzuki F (2002) Effect of glycyrrhizin, an active component of licorice roots, on HIV replication in cultures of peripheral blood mononuclear cells from HIV-seropositive patients. Pathobiology 70(4):229–236. https://doi.org/10.1159/000069334

Sato H, Goto W, Yamamura J, Kurokawa M, Kageyama S, Takahara T et al (1996) Therapeutic basis of glycyrrhizin on chronic hepatitis B. Antivir Res 30(2–3):171–177. https://doi.org/10.1016/0166-3542(96)00942-4

Sekizawa T, Yanagi K, Itoyama Y (2001) Glycyrrhizin increases survival of mice with herpes simplex encephalitis. Acta Virol 45(1):51–54

Seo EH, Song GY, Kwak BO, Oh CS, Lee SH, Kim SH (2017) Effects of glycyrrhizin on the differentiation of myeloid cells of the heart and lungs in lipopolysaccharide-induced septic mice. Shock 48(3):371–376. https://doi.org/10.1097/shk.0000000000000850

Shon Y-H, Lee H-S, Kim C-H, Lim J-K, Jeon B-H, Nam K-S (2004) Effect of herbal medicines on cytokine-induced cytotoxicity and major histocompatibility complex (MHC) class II antigen expression in rat thyroid cells. Biol Pharm Bull 27(3):371–374. https://doi.org/10.1248/bpb.27.371

Sun ZG, Zhao TT, Lu N, Yang YA, Zhu HL (2019) Research progress of glycyrrhizic acid on antiviral activity. Mini Rev Med Chem 19(10):826–832. https://doi.org/10.2174/1389557519666190119111125

Takahashi T, Takasuka N, Iigo M, Baba M, Nishino H, Tsuda H et al (2004) Isoliquiritigenin, a flavonoid from licorice, reduces prostaglandin E2 and nitric oxide, causes apoptosis, and suppresses aberrant crypt foci development. Cancer Sci 95(5):448–453. https://doi.org/10.1111/j.1349-7006.2004.tb03230.x

Tamir S, Eizenberg M, Somjen D, Izrael S, Vaya J (2001) Estrogen-like activity of glabrene and other constituents isolated from licorice root. J Steroid Biochem Mol Biol 78(3):291–298. https://doi.org/10.1016/s0960-0760(01)00093-0

Tanaka T (2003) A novel anti-dysmenorrhea therapy with cyclic administration of two Japanese herbal medicines. Clin Exp Obstet Gynecol 30(2–3):95–98

Tewari SN, Wilson AK (1973) Deglycyrrhizinated liquorice in duodenal ulcer. Practitioner 210(260):820–823

Thakur V, Nargis S, Gonzalez M, Pradhan S, Terreros D, Chattopadhyay M (2017) Role of glycyrrhizin in the reduction of inflammation in diabetic kidney disease. Nephron 137(2):137–147. https://doi.org/10.1159/000477820

Thompson RC (1930) Assyrian prescriptions for treating bruises or swellings. Am J Semit Lang Lit 47(Pt 1):1–25

Thyagarajan SP, Jayaram S, Gopalakrishnan V, Hari R, Jeyakumar P, Sripathi MS (2002) Herbal medicines for liver diseases in India. J Gastroenterol Hepatol 17(Suppl 3):S370–S376. https://doi.org/10.1046/j.1440-1746.17.s3.30.x

Tominaga Y, Nakagawa K, Mae T, Kitano M, Yokota S, Arai T et al (2009) Licorice flavonoid oil reduces total body fat and visceral fat in overweight subjects: a randomized, double-blind, placebo-controlled study. Obes Res Clin Pract 3(3):I–IV. https://doi.org/10.1016/j.orcp.2009.04.005

Tong T, Hu H, Zhou J, Deng S, Zhang X, Tang W et al (2020) Glycyrrhizic-acid-based carbon dots with high antiviral activity by multisite inhibition mechanisms. Small 16(13):e1906206. https://doi.org/10.1002/smll.201906206

Tsukiyama R, Katsura H, Tokuriki N, Kobayashi M (2002) Antibacterial activity of licochalcone A against spore-forming bacteria. Antimicrob Agents Chemother 46(5):1226–1230. https://doi.org/10.1128/aac.46.5.1226-1230.2002

Tsuruoka N, Abe K, Wake K, Takata M, Hatta A, Sato T et al (2009) Hepatic protection by glycyrrhizin and inhibition of iNOS expression in concanavalin A-induced liver injury in mice. Inflamm Res 58(9):593–599. https://doi.org/10.1007/s00011-009-0024-8

Turpie AG, Runcie J, Thomson TJ (1969) Clinical trial of deglycyrrhizinized liquorice in gastric ulcer. Gut 10(4):299–302. https://doi.org/10.1136/gut.10.4.299

van Uum SH, Walker BR, Hermus AR, Sweep CG, Smits P, de Leeuw PW et al (2002) Effect of glycyrrhetinic acid on 11 beta-hydroxysteroid dehydrogenase activity in normotensive and hypertensive subjects. Clin Sci (Lond) 102(2):203–211. https://doi.org/10.1042/cs20010194

Vitali R, Palone F, Pierdomenico M, Negroni A, Cucchiara S, Aloi M et al (2015) Dipotassium glycyrrhizate via HMGB1 or AMPK signaling suppresses oxidative stress during intestinal inflammation. Biochem Pharmacol 97(3):292–299. https://doi.org/10.1016/j.bcp.2015.07.039

Wan Y, Shang J, Graham R, Baric RS, Li F (2020) Receptor recognition by the novel coronavirus from Wuhan: an analysis based on decade-long structural studies of SARS coronavirus. J Virol 94(7). https://doi.org/10.1128/jvi.00127-20

Wang BX, Fish EN (2019) Global virus outbreaks: interferons as 1st responders. Semin Immunol 43:101300. https://doi.org/10.1016/j.smim.2019.101300

Wang ZY, Nixon DW (2001) Licorice and cancer. Nutr Cancer 39(1):1–11. https://doi.org/10.1207/S15327914nc391_1

Wang L, Yang R, Yuan B, Liu Y, Liu C (2015) The antiviral and antimicrobial activities of licorice, a widely-used Chinese herb. Acta Pharm Sin B 5(4):310–315. https://doi.org/10.1016/j.apsb.2015.05.005

Wang XR, Hao HG, Chu L (2017) Glycyrrhizin inhibits LPS-induced inflammatory mediator production in endometrial epithelial cells. Microb Pathog 109:110–113. https://doi.org/10.1016/j.micpath.2017.05.032

Wang D, Hu B, Hu C, Zhu F, Liu X, Zhang J et al (2020) Clinical characteristics of 138 hospitalized patients with 2019 novel coronavirus-infected pneumonia in Wuhan, China. JAMA 323(11):1061–1069. https://doi.org/10.1001/jama.2020.1585

Watanabe M, Hayakawa S, Isemura M, Kumazawa S, Nakayama T, Mori C et al (2002) Identification of licocoumarone as an apoptosis-inducing component in licorice. Biol Pharm Bull 25(10):1388–1390. https://doi.org/10.1248/bpb.25.1388

Wolkerstorfer A, Kurz H, Bachhofner N, Szolar OH (2009) Glycyrrhizin inhibits influenza A virus uptake into the cell. Antivir Res 83(2):171–178. https://doi.org/10.1016/j.antiviral.2009.04.012

Wujastyk D (2003) The roots of Ayurveda. Penguin

Xu X, Gong L, Wang B, Wu Y, Wang Y, Mei X et al (2018) Glycyrrhizin attenuates salmonella enterica Serovar typhimurium infection: new insights into its protective mechanism. Front Immunol 9:2321. https://doi.org/10.3389/fimmu.2018.02321

Xu X, Chen P, Wang J, Feng J, Zhou H, Li X et al (2020a) Evolution of the novel coronavirus from the ongoing Wuhan outbreak and modeling of its spike protein for risk of human transmission. Sci China Life Sci 63(3):457–460. https://doi.org/10.1007/s11427-020-1637-5

Xu Z, Shi L, Wang Y, Zhang J, Huang L, Zhang C et al (2020b) Pathological findings of COVID-19 associated with acute respiratory distress syndrome. Lancet Respir Med 8(4):420–422. https://doi.org/10.1016/s2213-2600(20)30076-x

Yamazaki S, Morita T, Endo H, Hamamoto T, Baba M, Joichi Y et al (2002) Isoliquiritigenin suppresses pulmonary metastasis of mouse renal cell carcinoma. Cancer Lett 183(1):23–30. https://doi.org/10.1016/s0304-3835(02)00113-1

Yao L, Sun T (2019) Glycyrrhizin administration ameliorates Streptococcus aureus-induced acute lung injury. Int Immunopharmacol 70:504–511. https://doi.org/10.1016/j.intimp.2019.02.046

Ye R, Liu Z (2020) ACE2 exhibits protective effects against LPS-induced acute lung injury in mice by inhibiting the LPS-TLR4 pathway. Exp Mol Pathol 113:104350. https://doi.org/10.1016/j.yexmp.2019.104350

Yu Z, Ohtaki Y, Kai K, Sasano T, Shimauchi H, Yokochi T et al (2005) Critical roles of platelets in lipopolysaccharide-induced lethality: effects of glycyrrhizin and possible strategy for acute respiratory distress syndrome. Int Immunopharmacol 5(3):571–580. https://doi.org/10.1016/j.intimp.2004.11.004

Zhao S, Yan X, Zhao Y, Wen J, Zhao Z, Liu H (2018) Dihydroisocoumarins from radix Glycyrrhizae. Chem Cent J 12(1):58. https://doi.org/10.1186/s13065-018-0427-0

Zheng YF, Qi LW, Cui XB, Peng GP, Peng YB, Ren MT et al (2010) Oleanane-type triterpene glucuronides from the roots of Glycyrrhiza uralensis Fischer. Planta Med 76(13):1457–1463. https://doi.org/10.1055/s-0029-1240907

Zhou P, Yang XL, Wang XG, Hu B, Zhang L, Zhang W et al (2020) A pneumonia outbreak associated with a new coronavirus of probable bat origin. Nature 579(7798):270–273. https://doi.org/10.1038/s41586-020-2012-7

# Chapter 13
# COVID-19-Induced Kidney Disease: Ethnopharmacological Intervention to Ameliorate Kidney Damage and Improve Kidney Function

**Angela E. Peter, V. Lakshmi Kalpana, Lakshmi Velaga, B. V. Sandeep, and Ganga Rao Battu**

## Abbreviations

| | |
|---|---|
| ACE2 | Angiotensin-converting enzyme 2 |
| ADME | Absorption, distribution, metabolism, excretion |
| AKI | Acute kidney injury |
| ALT | Alanine aminotransferase |
| AST | Aspartate aminotransferase |
| BUN | Blood urea nitrogen |
| CAT | Catalase |
| CBP | Complete blood picture |
| $CCl_4$ | Carbon tetrachloride |
| COVID-19 | Coronavirus disease 2019 |
| CRRT | Continuous renal replacement therapy |
| CXCL | C-X-C motif chemokine ligand |
| DPP4 | Dipeptidyl peptidase 4 |
| eGFR | Estimated glomerular filtration |
| ERK | Extracellular signal related-kinases |

A. E. Peter (✉) · B. V. Sandeep
Department of Biotechnology, College of Science and Technology, Andhra University, Visakhapatnam, India

V. L. Kalpana · L. Velaga
Department of Human Genetics, College of Science and Technology, Andhra University, Visakhapatnam, India

G. R. Battu
Andhra University College of Pharmaceutical Sciences, Andhra University, Visakhapatnam, India



J.-T. Chen (ed.), *Ethnopharmacology and Drug Discovery for COVID-19: Anti-SARS-CoV-2 Agents from Herbal Medicines and Natural Products*,
https://doi.org/10.1007/978-981-99-3664-9_13

| | |
|---|---|
| FGF | Fibroblast growth factor |
| G-CSF | Granulocyte colony-stimulating factor |
| GM-CSF | Granulocyte-macrophage colony-stimulating factor |
| GPx | Glutathione peroxidase |
| GSH | Glutathione |
| ICU | Intensive care unit |
| IFN | Interferon |
| IL | Interleukin |
| IP10 | Interferon gamma-induced protein 10 |
| JAK | Janus kinase |
| KIM-1 | Kidney injury molecule 1 |
| L-FABP | Liver-type fatty acid-binding protein |
| MAPK | Mitogen-activated protein kinases |
| MCP | Monocyte chemoattractant protein |
| MDA | Malondialdehyde |
| MIP | Macrophage inflammatory protein |
| MNC | Mononuclear cells |
| mTOR | Mammalian target of rapamycin |
| NF-κB | Nuclear factor kappa-light-chain-enhancer of activated B cells |
| PDGF | Platelet-derived growth factor |
| RBC | Red blood cell |
| RdRp | RNA-dependent RNA polymerase |
| RNA | Ribonucleic acid |
| SARS-CoV-2 | Severe acute respiratory syndrome coronavirus 2 |
| SOD | Superoxide dismutase |
| STAT | Signal transducers and activators of transcription |
| TCM | Traditional Chinese Medicine |
| TGF | Transforming growth factor |
| TLR | Toll-like receptors |
| TMPRSS2 | Transmembrane serine protease 2 |
| TNF | Tumor necrosis factor |
| VEGF | Vascular endothelial growth factor |

## 13.1 Introduction

The world saw an unprecedented event that began in the late December of 2019; a pandemic like never before; coronavirus disease 2019 (COVID-19). The first wave lasted till 2020, and the second, more fatal wave in 2021. Even now, many countries are still experiencing subsequent waves, India is now seeing a rise in cases and it is said to be the fourth wave. By late 2021, vaccines were rolled out and vaccination drives began. Notwithstanding, COVID-19 still possess a serious health threat.

The causative agent, severe acute respiratory syndrome coronavirus 2 (SARS-CoV-2) primarily infects the lungs, the respiratory system being the main target for COVID-19 disease. However, other organs and systems are known to be affected; the digestive system, renal system, cardiovascular system, and neural system, to name a few. Sometimes, SARS-CoV-2 directly infects the organs of these other systems. The inflammation produced in the body as a result of the SARS-CoV-2 infection is also responsible for negative effects on the other organ systems.

## 13.2 COVID-19 and Kidney Disease

The effect of COVID-19 disease on the renal system is undeniable, patients who have no history of kidney problems have been observed to develop kidney damage and reduction/loss of kidney function. A significant contributing feature is that a large group of severe COVID-19 patients also exhibit chronic co-existing conditions such as high blood pressure and diabetes, which greatly increase the risk of kidney disease (Coronavirus: kidney damage caused by COVID-19 2022). Studies have shown that kidney disease is prevalent in some COVID-19 patients at the time of admission itself, and many others develop acute kidney injury (AKI) during hospitalization (Cheng et al. 2020). According to the Kidney Disease Improving Global Outcome (KDIGO) criteria, the incidence of acute kidney injury in COVID-19 patients is 3–9% (Armaly et al. 2021). Reports of AKI in hospital settings, however, tell a different story; this rate of incidence varies and is often significantly higher. In hospital settings, AKI could be anywhere from 0% to 14.7%, the pooled incidence rate is 7%; and in ICU settings, this is higher still, 8.3–28.8%, pooled incidence rate being 19%, as evidenced by a meta-analysis of nine studies (Kaye et al. 2021). A study of 5994 hospitalized COVID-19 patients admitted to 13 academic and community hospitals in a metropolitan for about a month reported the occurrence of 36.6% AKI; 1993 patients suffered from AKI (Hirsch et al. 2020). A retrospective observational study of 3993 COVID-19 patients reported the incidence of 46% AKI; 1835 patients developed kidney injury (Chan et al. 2021). 23 out of 85 confirmed COVID-19 patients (27%) were found to suffer from AKI in one retrospective analysis (Batlle et al. 2020). In another prospective cohort study with 701 hospitalized COVID-19 patients, AKI was observed in 5.1% of patients (Armaly et al. 2021). Yet another observational retrospective study on 193 COVID-19 patients reported the occurrence of 23.3% AKI (Cheng et al. 2020). A single-center case series of 138 hospitalized patients the existence of 3.6% AKI among all patients (Allemailem et al. 2021). In another cohort, out of 63 COVID-19 patients, 32 patients (50.79%) had AKI (Wang et al. 2020). In a multi-centered retrospective observational study with 193 COVID-19 patients, among 128 non-severe COVID-19 patients, 12 (9%) had AKI. In the same study, among 65 severe COVID-19 patients, 43 (66%) suffered from AKI (Braun et al. 2020). A report of a study (single-centered, retrospective, observational study) of 710 critically ill COVID-19 patients, out of the 52 patients admitted into the ICU, 15 (29%) developed AKI (Li et al.

2020) A large cohort of 1099 COVID-19 patients reported a surprisingly low occurrence of AKI; 0.5% (Yang et al. 2020).

Several indications have been reported that point to altered kidney function and kidney injury during COVID-19 disease. Proteinuria, haematuria, altered CBP profiles, and reduction in estimated glomerular filtration (eGFR) have been reported in many COVID-19 patients (Armaly et al. 2021). The number of patients exhibiting proteinuria was varied, over a wide range of 26–63% in the different studies; 63% in 51 patients, 60% of 147 patients, 59% (on admission) of 193 patients, 53.9% of 193 patients, 43.9% (on admission) of 701 patients (Armaly et al. 2021; Kaye et al. 2021; Cheng et al. 2020; Braun et al. 2020; Guan et al. 2020; Gagliardi et al. 2020). Haematuria was also reported in COVID-19 patients, with different studies recording varying percent of incidence; 48% of 147 patients, 44% in 193 patients (on admission), 40.9% of 5449 patients, 26.7% out of 701 patients (on admission), 22.3% of 193 patients (Armaly et al. 2021; Kaye et al. 2021; Hirsch et al. 2020; Cheng et al. 2020; Braun et al. 2020). Studies revealed both proteinuria and haematuria to be independent risk factors of in-hospital mortality in COVID-19 patients. COVID-19 patients also exhibited elevated levels of blood urea nitrogen (BUN); in 31% of 193 patients, 14% of 193 patients, and 13% of 701 patients (Armaly et al. 2021; Li et al. 2020; Gagliardi et al. 2020). Along with the occurrence of proteinuria and haematuria, elevated blood urea nitrogen, serum creatinine, uric acid, and D-dimer were found to be significantly associated with mortality, as revealed with univariate Cox regression analysis of COVID-19 patients. Data from the same study also indicated an approximately 5.3 times increase in mortality risk in COVID-19 patients who developed AKI (Kaye et al. 2021; Meena et al. 2020). A dose-dependent relationship was observed between in-hospital mortality and the stage of AKI. The mortality risk in COVID-19 patients with stage 3 AKI was at least four times, a study on 701 patients reported (Cheng et al. 2020). In the study of 5994 COVID-19 patients, out of the 1993 patients who developed AKI during hospitalization, 35% (694) died; among these, 91% had stage 3 AKI, 64% stage 2 AKI, and 34% had stage 1 AKI (Hirsch et al. 2020).

One study reported the results of post-mortem autopsies conducted on six COVID-19 patients. The main damage in the kidney was found to be in the tubulointerstitium, which showed acute tubular necrosis of varying levels, luminal brush border sloughing, degeneration of vacuole, and lymphocyte infiltration to different proportions. In three of the cases, viral infection associated-syncytia was noted. The study also reported dilated capillary vessels without any severe glomerular injury, at the glomerular level (Diao et al. 2021). Another autopsy study of 26 COVID-19 patients revealed that some exhibited kidney damage. The changes identified through light microscopy include injury to the proximal tubule in association with loss of brush border, degeneration of vacuole (non-isometric), and necrosis (Su et al. 2020).

## 13.3 SARS-CoV-2 and the Kidneys

Some studies have reported the presence of the SARS-CoV-2 constituents—viral proteins and viral-like particles within the kidney. One study reported that in the tubular structures of the kidneys, the nucleocapsid protein and spike protein of SARS-CoV-2 were present, as revealed by immunohistochemical analysis. Both antigens were found to be restricted to ACE2-positive tubules; this was ascertained by immunofluorescence double staining. Viral RNA accumulation in the tubules was also found, as detected by in situ hybridization assay (Diao et al. 2021). Another study using electron microscopy revealed that in the podocytes and renal tubular epithelial cells, coronavirus-like particles having distinctive spikes were seen. The same study reported that the renal tubular epithelia were positive, on staining with SARS-CoV-2 nucleoprotein antibody (Su et al. 2020). Post-mortem analysis was carried out on 63 patients COVID-19 patients. Among the 63 patients, 32 had AKI. Out of these 32, the presence of RNA of SARS-CoV-2 was detected in the kidneys of 23 patients. This study also reported that this occurrence of SARS-CoV-2 RNA in the kidney was found to be associated with both, older age and a higher number of co-morbidities. Additionally, the presence of RNA in the kidney was significantly associated with a reduction in the survival time of patients. The same study also performed an extended analysis, SARS-CoV-2 was isolated from the autopsied kidneys of patients and was infected into cells in vitro. 48 hours post-cell infection, there was a 1000-fold increase in viral RNA, indicating the presence of highly infective virus from the kidneys of patients, even after death (Braun et al. 2020). A biopsy study of a COVID-19 patient who developed AKI was significant, electron microscopy revealed the presence of vacuoles with several spherical particles (50 to 110 nm), surrounded by spikes (9 to 10 nm) in podocyte cytoplasm. These particles could be viral inclusion bodies, that have been reported with SARS-CoV-2 infection (Kissling et al. 2020).

## 13.4 Mechanisms for Kidney Involvement in COVID-19

### 13.4.1 *Direct Cellular Entry*

To enter host cells, SARS-CoV-2 binds to its obligate receptor, angiotensin-converting enzyme II (ACE2). ACE2 is expressed on the cells in the lungs (alveolar type II cells), keratinocytes of the esophagus, epithelial cells of the stomach, cholangiocytes of the liver, colonocytes, small intestine and rectum, and also on the surface of kidney cells. The renal proximal tubular cells and podocytes express ACE2 receptors. It is interesting to note that ACE2 is more highly expressed in the brush border apical membrane of the proximal tubules than in the lung tissue. The ACE2 expression in the podocytes is considerably lower. In addition to the ACE2 receptor, the transmembrane protease serine 2 (TMPRSS2) of the host also plays a critical role in facilitating SARS-CoV-2 entry into the host cells. TMPRSS2 is involved in cleaving the viral S glycoprotein, aiding in the entry of SARS-CoV-2 into the cells. While the expression of TMPRSS2 in the proximal tubules in the kidney is relatively low, expression of the same by the distal tubules is more. Cathepsin B/L, cysteine proteases, DPP4, and glutamyl aminopeptidase expressed in the kidneys are suggested to be alternative viral S priming proteases for the TMPRSS2 in the proximal tubule (Ahmadian et al. 2021). It is hypothesized that SARS-CoV-2 enters the glomerular and arteriole capillaries, initially infecting the glomerular endothelial cells. Podocytes are subsequently infected, through the glomerular filtration barrier. SARS-CoV-2 eventually the proximal tubule surface and then infect the proximal tubular epithelium (Yin and Zhang 2020).

### *13.4.2 Role of ACE2 and Subsequent Injury*

Conversion of angiotensin II to Ang (1-7) is facilitated by the action of both soluble and insoluble ACE2. Ang (1-7) plays a significant role in controlling inflammation, vasoconstriction, and thrombosis. During acute lung injury, Ang II is increased, and during AKI, ACE2 is downregulated. This can cause the activation of the type I angiotensin receptor and a decrease in Ang (1-7) production. Both pulmonary inflammation and coagulation, which further aggravates AKI can result from the binding of angiotensin II to type I receptors. Entry of SARS-CoV-2 downregulates the expression of ACE2, consequently increasing the effects of Ang II and inhibiting the protective and anti-inflammatory of ACE2 in patients. COVID-19 patients with co-existing conditions such as cardiovascular diseases, diabetes, hypertension, and older age additionally exhibit ACE2 deficiency. In such patients, there can be a dysregulation between the protective angiotensin (1-7) and adverse AT1 roles in the renin-angiotensin system (RAS). ACE2 is also involved with the kallikrein–kinin system (KKS). An important member of this system, bradykinin has a key role to play in the process of inflammation. ACE2 brings about the hydrolysis and inactivation of [des-Arg973] BK (DABK), which is the active metabolite of bradykinin. COVID-19 can cause a decrease in the levels of ACE2, thus promoting KKS activation via BKB1R, the bradykinin B1 receptor. This results in increased recruitment of leukocytes and a surge of fluid in the lungs, aggravating the disease further and making resolution difficult (Batlle et al. 2020; Ahmadian et al. 2021).

### *13.4.3 Hyperinflammation and Dysregulated Immune Response Triggered Injury and Loss of Function*

It has been established that COVID-19 patients exhibit a dysregulated immune response, resulting in an extensive cytokine and chemokine release, also called a cytokine storm, that promotes the body to exist in a state of hyperinflammation. In patients with COVID-19, a significant increase in plasma levels of cytokines IL-6, IL-1β, TNF, IL-1RA, IFNγ, IL-7, IL-9, IL-10, G-CSF, GM-CSF, PDGF, FGF, VEGF, and chemokines IP10, CXCL8, MCP1, MIP1α, MIP1β can be observed, when compared with healthy individuals. This sort of dysfunctional immune response causes a failure in viral clearance. The ensuing continuous viral replication promotes the exaggerated inflammatory response and increased cytokine production. Other contributing factors include pyroptosis (an inflammatory form of programmed cell death) and ACE2 receptor mediated-increase in cytokines IL-6 and MCP1 (triggered by increased angiotensin II following the disruption of RAS) (Peter et al. 2021a). It is suggested that endothelial dysfunction, microcirculatory derangement, and tubular injury in the kidney are a consequence of the interaction of a large amount of circulating cytokines and chemokines with the kidney-resident cells (Soliman 2021). Different signaling pathways, JAK/STAT, TLR, NF-κB, mTOR, MAPK/ERK, and TGFβ are dysregulated that furthering the disease progression by promoting hypercytokinemia, hyperinflammation, thrombosis, and vasoconstriction. Hypercytokinemia and hyperinflammation induced by SARS-CoV-2 lead to the development of AKI in COVID-19 patients and can cause mortality (Ahmadian et al. 2021; Peter et al. 2021b).

### *13.4.4 Dysregulation of Complement System, Hypercoagulation and Vascular Consequences*

SARS-CoV-2 is also established to induce coagulopathy in the host. Increased clotting, disseminated intravascular coagulation, pulmonary infarction, and thrombosis were reported in severe COVID-19 patients. Irreversible kidney damage occurs as microvascular thrombosis and acute necrosis (both tubular and cortical), causing fibrinoid necrosis and glomerular ischemia. Dysfunction of the complement system has also been reported as a result of SARS-CoV-2 infection. In severe COVID-19, the dysregulated complement system promotes the disease state in the body. Widespread complement activation that is characterized by activation of C3, generation of C3a, fragment deposition of C3, and an increase in serum levels of C5a are observed in patients with severe COVID-19. All these events in turn stimulate histamine, leukotrienes, and prostaglandins, which eventually increase hypersensitivity symptoms, hypotension, hypoxia, flushing, and vasodilation. The complement and coagulation pathways stimulate each other, increasing disease severity (Ahmadian et al. 2021).

### 13.4.5 Organ Crosstalk

There are reports of the existence of crosstalk in ARDS, between the lung and kidney. ARDS can cause the development of AKI. The inflammatory and immune reactions during ARDS promote an increased release of circulating factors. Interaction of these with kidney-resident cells causes damage, resulting in AKI. Crosstalk between other organ systems and the kidney, like the cardiovascular system can also contribute to the development of AKI in COVID-19 patients. COVID-19-induced myocarditis leads to an impairment of cardiac output, compromised end-organ perfusion, and concomitant right-ventricular dysfunction, which ultimately causes diastolic dysfunction and venous congestion. This can be transmitted back to the kidney, resulting in kidney congestion, further compromising the process of perfusion. Additionally, hypoperfusion, hypotension, and renal vein congestion which are responsible for the reduction in the glomerular filtration rate (GFR) are associated with acute viral myocarditis and cytokine cardiomyopathy. Hypovolaemia, sepsis, and nephrotoxicity, known risk factors of cardiovascular comorbidity can also mediate AKI. This kind of dysfunction in one organ during infection leads to impairment in the other and increases disease severity (Ahmadian et al. 2021).

### 13.4.6 Other Factors

There are other factors also that come into play in patients with COVID-19 patients which can increase the risk of injury to the kidney. Increased severity of disease makes for longer in-hospital stays, ICU requirement, mechanical ventilation, hemodynamic instability, nephrotoxic drugs, and sepsis, factors that strongly contribute to kidney injury. In such patients, septic AKI acts synergistically with other mechanisms of kidney damage (Soliman 2021) (Fig. 13.1).

## 13.5 Addressing COVID-19 Related Kidney Injury

Case study data and various reports clearly show that there is a substantial number of patients exhibit kidney involvement as one of the disease manifestations of COVID-19. Reports also confirmed that AKI and reduced renal function are implicated in COVID-19 mortality rates, AKI is a predisposing factor to disease severity and in-hospital mortality. Consequently, successful management of COVID-19 involves effective treatment of COVID-19-induced kidney injury. As of now, there is still no specific treatment for COVID-19-induced AKI. Currently, the treatment of AKI in COVID-19 patients is just a supportive treatment procedure, the KDIGO supportive care guidelines are followed—nephrotoxic drugs are avoided, serum creatinine levels and urine output are monitored regularly and

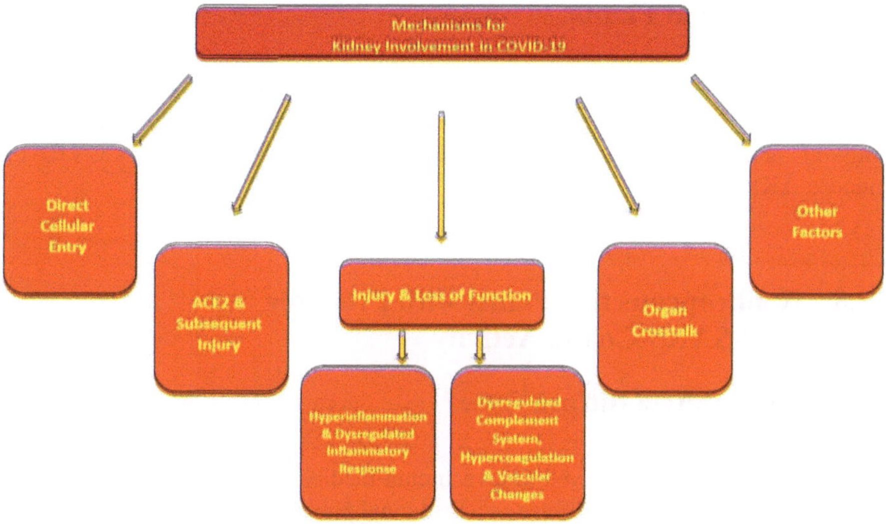


**Fig. 13.1** Factors contributing to the development of AKI during/post COVID-19

hemodynamic monitoring is carried out, if possible. Renal replacement therapy (RRT) is started early, to mitigate the damage and promote early resolution (Kaye et al. 2021).

## 13.6 The Lacuna

Although the occurrence of COVID-19-associated AKI is substantial and there is no doubt of the detrimental role it plays in the recovery of the patient, there is no specific treatment that is prescribed. At its best, the treatment is just medical management. Renal replacement therapy (RRT) and continuous renal replacement therapy (CRRT) are carried out, if necessary. However, the optimal time to initiate the RRT of CRRT on the patient has to be decided at the discretion of the physician. Before initiating the therapy, there should be careful consideration of fluid and volume overload management, and management of hyperkalemia and metabolic acidosis. Moreover, another factor to be considered is the continuous dependence on the therapy post-discharge. Some disadvantages of these therapies include the risk of transmission, especially in such a pandemic situation, the lack of resources like hospitals with fewer ICU beds, and the need for experienced personnel.

## 13.7 Medicinal Plants

Medicinal plants offer a viable solution for the lacuna presented above. Medicinal plants with established nephroprotective activity, and anti-inflammatory and antiviral activity can be suitable adjuvants in the therapy of COVID-19-induced AKI.

## 13.8 Nephroprotective Plants with Probable Anti-SARS-CoV-2 Activity

### *13.8.1* Boerhavia diffusa *Linn*

*Bierhavia diffusa* is a very well-known Indian medicinal plant, natively known as "*punarnava*." The use of *B. diffusa* as a medicinal plant is not restricted to the Indian subcontinent, it is also a native medicinal plant in different parts of Africa and South America. It is a perennial herb, and the whole plant, the leaves, and roots in particular have been utilized for hepatoprotective, gastrointestinal, respiratory, nephrological, and gynecological indications. Well known for the anti-inflammatory and immunomodulatory activity it possesses, *B. diffusa* is the major ingredient for over 35 different ayurvedic formulations. Several researchers have scientifically evaluated the pharmacological effects of the different parts of *B. diffusa* plant and have also successfully isolated bioactive compounds from it. In addition to ecdysteroids, flavonoids, flavonoid glycosides, lignans, purine nucleosides, steroids, and xanthones found in *B. diffusa*, a new class of isoflavonoids called rotenoids were isolated from its roots. Decoctions of the root, tender shoot, leaves, and the whole plant have been used for improving the renal system function. Decoction of the root and leaves have been used to treat inflammation and edema (*Boerhavia diffusa* n.d.; Mishra et al. 2014).

In a study conducted on male Wistar rats, gentamycin-induced AKI was produced. The efficacy of *B. diffusa* aqueous root extract in ameliorating kidney injury and improving kidney function was studied. High doses of gentamycin caused elevated levels of BUN, serum creatinine, and histopathological damage in rat models. Pre-treatment with 200 and 400 mg/kg/day *B. diffusa* extract, resulted in significantly lower serum creatinine and BUN levels, comparable to the standard treatment group (ALA-α-lipoic acid). AKI in gentamycin-treated rats was indicated by the presence of both, tubular necrosis and desquamation which involved more than half of the proximal tubules (median grade: 3). Significant difference was observed in the ALA and *B. diffusa* (200 mg/kg/day)-treated rats. In rats treated with ALA, prominent tubular epithelial necrosis and desquamation involving less than half of cortical tubules (median grade: 2) were observed. *B. diffusa* (200 mg/kg/day)-treated rats exhibited desquamation of tubular epithelial cells in small foci (median grade: 1) (Sawardekar and Patel 2015). Male Wistar albino rates were subject to kidney injury induced by acetaminophen. An increase in BUN, serum creatinine and structural damage; degeneration of epithelial cells, tubular necrosis, glomerular damage, and congestion indicated kidney injury in the animal models. Dose-dependent attenuation of elevation of both serum creatinine and BUN levels was observed on pre-treatment with *B. diffusa* aqueous extract (200 and 400 mg/kg/day). Moreover, pre-treatment with the aqueous extract of *B. diffusa* brought about a significant attenuation to renal cell damage, only minimal damage to the typical renal architecture was observed in these two groups (Pareta 2011).

An in silico study analyzed the efficacy of nine specific phytochemicals from *B. diffusa* to bind to and consequently inhibit SARS-CoV-2 main protease. Molecular docking analysis was carried out for boeravinone j, bioquercetin, biorobin, boerhavisterol, trans-caftaric acid, 2–3-4 beta-ecdysone, kaempferol, liriodendrin, quercetin. The least binding energies were obtained with biorobin; – 8.17 kcal/mol, bioquercetin; – 7.97 kcal/mol and boerhavisterol; – 6.77 kcal/mol. ADME studies were used to test the drug-likeness of biorobin, bioquercetin, and boerhavisterol. Among the three, the most suitable, following Lipinsky's rule was boerhavisterol. This suggests the potential therapeutic activity of *B. diffusa* for COVID-19 (Rutwick Surya and Praveen 2021).

### *13.8.2* Hygrophila auriculata *(Schumach.) Heine*

*Hygrophila auriculata* (K. Schum) Heine (syn.) *Asteracantha longifolia* Nees (syn) *Hygrophila spinosa* T. Anders. is an Indian medicinal plant well described in Ayurveda. A stout, erect herb found in wet places, native to Tropical and South Africa, the Indian subcontinent and Indo-China. The ashes, roots, and seeds of *H. auriculata* are widely used in traditional medicine to treat a wide range of ailments; inflammation, rheumatism, hepatic obstruction and jaundice, pain, gout, urinary infections, and edema (*Hygrophila auriculata* n.d.; Shanmugasundaram and Venkataraman 2006). The potency of the different plant parts of *H. auriculata* has been evaluated scientifically by different researchers. From the aerial parts of *H. auriculata*, lupeol, stigmasterol, and butelin have been isolated. Fatty acids have been reported to be present mainly in the seeds (Vijayakumar et al. 2006).

Gentamycin-induced kidney injury was produced in a group of study albino Wistar rats, and the efficacy of pre-treatment with *H. auriculata* methanol extract (200 and 400 mg/kg body weight; 8 days) was evaluated. Significant reduction in BUN, serum uric acid, and creatinine levels could be observed in the H. *auriculata*-pre-treated group. Animals in this group also exhibited recovery of the epithelial cell when compared to the non-treated group, which showed dilated tubules and degeneration of epithelial lining (Neharkar and Gaikwad 2019). Male Sprague-Dawley rats were treated with gentamycin to induce kidney injury for the study of the nephroprotective activity of the whole plant ethanolic extract of *H. auriculata*. Pre-treatment with *H. auriculata* ethanolic extract (250 mg/kg, po from ninth day to day 30) was found to cause a reduction in serum creatinine, 14 days post-treatment, and on the 30th day, serum creatinine was similar to healthy control rats. Gentamycin-induced increase in blood urea also was prevented, when measured on the 30th day, there was not much difference between the *H. auriculata* pre-treatment group and the healthy control group. Gentamycin was found to induce marked kidney damage, extensive proximal tubular necrosis, epithelial lining loss, edema; interstitial and perivascular and in the interstitium multiple focal collections of mononuclear cells. Definite changes in the glomeruli were observed. The group administered with 50 mg/kg ethanolic extract of *H. auriculata* exhibited areas of tubular degeneration and necrosis, perivascular edema, and tubulo-interstitial mononuclear cell infiltrates, throughout the cortex at different foci. The group treated with 250 mg/kg *H. auriculata* ethanolic extract showed much better outcomes, and varying degrees of regeneration with only slight degenerative changes. Small foci of mononuclear cell infiltration confined to the subcapsular area were scattered. The epithelial cells of the proximal convoluted tubules were almost intact. These findings confirm the therapeutic action and nephroprotective potential of *H. auriculata* (Kariyil et al. 2010).

Molecular docking analysis was carried out with two phytoconstituents of *H.* auriculata; apigenin 7-O-glucoside and cucurbitacin B. The binding of these to the N-domain and RBD domain was analyzed with Autodock Vina with Pyrx v0.8 platform, Pymol v2.5, Ligplot+ v2.2.4 and Discovery Studio Visualizer v21.1.0.20298. The phytochemicals of *H. auriculata* showed effective inhibition of the mutated N-Domain; the binding affinity for apigenin 7-O-glucoside was −6.5 kcal/mol and that of cucurbitacin B was −6.2 kcal/mol. Among the two, cucurbitacin B had a lower binding affinity. The compounds were studied in the Molinspiration server, ADMETlab 2.0 to analyze the adsorption, digestion, drug-likeness metabolism, excretion, lethal dosage, toxicity (ADMET) and toxicity (Bharathi et al. 2022). In another analysis of potent inhibitors of SARS-CoV-2 main proteases $3CL^{pro}$ (6 LU7) by in silico approach, luteolin 7- rutinoside from *H. auricualata* docked with $3CL^{pro}$, energy value exhibited −134.6 kcal/mol. This comprises of van der Waal interaction; −98.9 kcal/mol and hydrogen bonding interaction; −36.23 kcal/mol. This suggests that the individual phytoconstituents may be considered for formulations to effectively target SARS-CoV-2, thus treating COVID-19 (Vincent et al. 2020).

### *13.8.3* Phyllanthus niruri *L.*

*Phyllanthus niruri* is a perennial, tropical herb native to Tropical and Subtropical America (*Phyllanthus niruri* n.d.). It is a well-known traditional medicinal plant in South and South-East Asia, used for a wide range of indications such as diarrhea, dyspepsia, jaundice, genitourinary infections, and renal stones. In Ayurvedic and Unani medicine, the leaves and fruits are used in the treatment of gall stones and jaundice, and the Malay system of medicine is employed for treating cough and kidney diseases. *P. niruri* is used in the alleviation of liver injury secondary to various hepatotoxic agents in TCM (traditional Chinese Medicine). Extensive scientific research has been carried out to validate the ethnopharmacological activities of *P. niruri.* Several important bioactive compounds have been isolated and characterized from *P. niruri. P. niruri* is rich in alkaloids, coumarins, flavonoids, lignans, phenylpropanoids, tannins, and terpenes. Some bioactive compounds isolated and characterized from *P. niruri* include phyllanthin, nirtetralin, niranthin, nirurin, niruriflavone, nirphyllin, phyllnirurin (Lee et al. 2016).

Nephroprotective activity of *P. niruri* was studied using albino Wistar rats treated with gentamycin to produce kidney injury. Two doses of whole plant methanolic extract of *P. niruri* were used for the study, 200 mg/kg and 400 mg/kg, administered from the tenth to the 19th day of study. Treatment with gentamycin causes a decrease in body weight and urine volume and an increase in the weight of the kidney. Treatment with *P. niruri* protects from these effects, urine volume and kidney weight are almost the same in the control group and the *P. niruri* 400 mg/kg treatment group. Serum creatinine and protein, BUN are also significantly lower in the *P. niruri* treatment group. Gentamycin brought about glomerular congestion, and treatment with *P. niruri* showed moderate tubular degeneration and normal glomeruli and Bowman's capsule (Reddy et al. 2015).

The ability of two phytoconstituents of *P.* niruri; phyllanthin and hypophyllanthin to bind to spike glycoprotein (6LZG) and main protease (5R7Y) was determined using Molegro Virtual Docker 6.0. Both phytoconstituents possessed increased binding affinity to the COVID-19 inhibition sites when compared to their native ligand; 6LZG and 5R7Y. The rerank score of phyllanthin and hypophyllanthin was lower than 6LZG and 5R7Y, suggesting that they possess a greater interaction than the native ligands, both for spike glycoprotein (possible entry inhibitor) and main protease (potential translation and replication inhibitor) (Marhaeny et al. 2021).

### *13.8.4* Punica granatum *L.*

*Punica granatum,* commonly called pomegranate is a tree, native ranging from the Middle East region to the Himalayas in the Northern part of India. All over the Mediterranean region of Asia, Caucasus, northern Africa, and Europe, pomegranate was naturalized and its cultivation has been in process since ancient times. The fruit is a delicacy and is consumed for good taste and is also one of the few fruits advised to be eaten even during ill health. Pomegranate juice is also well consumed. Dried fruit is also used both, in foods and as a spice condiment (Lim 2012; *Punica granatum* n.d.). Pomegranate is known for its medicinal properties, it has been an inherent part of different traditional systems of medicine; Islamic, Persian,

Ayurveda, and Chinese. Traditionally, pomegranate is used for the treatment of diabetes, hypertension, peptic ulcer, different types of cancers, atherosclerosis, hyperlipidemia, and oral diseases (Ge et al. 2021). Extensive studies have been carried out on the parts of the pomegranate plant and numerous phytochemicals have been isolated and characterized. Some phytochemicals isolated are; from the roots; norpseudopelletierine, sedridine, hygrine and norhygrine, from the bark; punicotannic acid, punicalin and punicalagin, punicacorteins A, B, C, and D, from the leaves; punicafolin, granatins A and B, corilagin, strictinin, from the flowers; punicanolic acid, urolic and maslinic acids, ellagic acid, from the fruits; polyphenols, 35 dimers of flavanol-anthocyanin adducts, different ellagitannins, anthocyanin pigments and from the seeds; high contents of α-tocopherol and γ-tocopherol, estrogens, and glycosides have been reported (Lim 2012).

Kidney injury was induced in male Wistar rats using gentamycin. The nephroprotective activity of methanol *P. granatum* leaf extract was accessed, and three concentrations were studied, 100, 200, and 400 mg/kg *p.o. Kidney injury was indicated by elevated levels of* serum creatinine, and urea, degeneration of tubules alongside an increase in TNF-α, lipid peroxidation, and a decrease in the activity of antioxidant enzymes activity. Administration of methanolic extract of *P. granatum* along with gentamycin was found to protect the kidneys from the nephrotoxic effects of gentamicin. The best activity was found in the group of rats treated with the highest concentration of *P. granatum* extract, 400 mg/kg; indicated by a significant reduction in levels of serum creatinine and BUN and an increase in the concentration of serum albumin. Histopathological studies revealed substantial kidney injury on treatment with gentamycin; acute vacuolar degeneration, destruction and hyalinization of arterioles, tubular collapse, scarring and matrix-rich expansion of interstitium and tubule distortion, presence of mononuclear cells (MNCs) in the renal sections and edema. Treatment with *P. granatum* leaf extract was found to reduce the intensity of vacuolar degeneration, edema, and infiltration of MNC and also improve renal morphology. Substantial improvement of renal tubules and a reversal of many of the histopathological changes, evidenced by the preservation of tubular histology was observed in the 400 mg/kg methanol leaf extract treatment group. An enhancement of thickening in the basement membrane and glomerular size was also caused as a result of gentamycin treatment. Treatment with methanol extract at doses of 200 mg/kg and 400 mg/kg could effectively bring about a decrease in this gentamycin-induced increase of both, basement membrane thickening and glomerular size. Additionally, renal TNF-α levels were also altered. Gentamycin induced a considerable increase in the renal levels, but treatment with methanolic extract of *P. granatum* (at all three concentrations, 100, 200, and 400 mg/kg) reduced renal TNF-α levels substantially (Mestry et al. 2020). Another study employed carbon tetrachloride ($CCl_4$) to induce kidney injury in male CD4 albino mice and evaluated the nephroprotective effects of the *P. granatum* peel aqueous extract. Kidney damage was indicated by the elevated levels of serum creatinine and urea.

Histopathological evaluation showed significant degenerative changes, both in the glomerulus and Bowman's capsule with cells revealing evidence of necrosis and vacuolization in mice treated with $CCl_4$. Treatment with an aqueous extract of *P. granatum* peel in $CCl_4$-treated mice led to a significant decrease in serum urea and creatinine. Mice treated with *P. granatum* extract retained intact histological architecture in the kidney, with reduced damaged areas when compared to non-treated animals. In the *P. granatum* + $CCl_4$-treated group, some glomeruli seemed to be moderately congested, and a few tubules exhibited mildly swollen epithelial cells with normal nuclei. The renal tissue of the *P. granatum* + $CCl_4$-treated mice had a medium reduction in the content of collagen fiber as compared to the group treated with only CCl4. Reduced expression of Caspase-3 in the renal tissue was also noted in the animals treated with an aqueous extract of *P. granatum* as compared to non-treated animals. The study also analyzed the levels of mRNA expression of SOD (superoxide dismutase), catalase (CAT), and glutathione peroxidase (GPx). Substantial variation of gene expression levels of all the genes in the CCl4 and PPE + CCl4 treatment groups was observed. Treatment with *P. granatum* extract increased SOD and GPx expression, even slightly more that in the control animal group. CAT expression was increased in the *P. granatum*-treated groups but was notably lesser than in the control group (Emam et al. 2020).

The efficacy of the hydroethanolic extract of *P. granatum* as an antiviral drug against SARS-CoV-2 was studied. Three different concentrations of extract, ranging from 0.04 mg/mL to 1 mg/mL, were found to inhibit Spike and ACE2 interactions in a dose-dependent manner, up to 74%. Three individual polyphenols of *P. granatum*; punicalagin (PC), gallic acid (GA), and ellagic acid (EA) were tested to ascertain which had the most impact on the binding of Spike to ACE2. While GA did not have any effect on this binding, PC showed the most effect, with 49% inhibition, and EA with 36% inhibition. Further investigations studying the chemical interactions between the extract and Spike and extract and ACE2 were carried out. Though *P. granatum* extract bound to both proteins, the interaction with the Spike was tenfold strong as compared to that with ACE2. Gene expression analysis also revealed that *P. granatum* extract downregulates the expression of ACE2 and TMPRSS, the genes responsible for the virus access into the cells. Additionally, at a concentration of 0.2 mg/ml, *P. granatum* extract was able to inhibit the activity of one of the main proteins, 3CL protease, which is involved in viral replication by up to 80%. Among the individual phytoconstituents, PC exhibited the highest inhibition of enzymatic activity; 50% followed by EA; with 10% inhibition, and GA showed no inhibition of protease activity. This suggests that pomegranate juice is effectively reducing viral infection and replication and can be part of a formulation to treat COVID-19 (Tito et al. 2021).

### *13.8.5* Tribulus terrestris *L.*

*Tribulus terrestris* is an annual shrub, widely distributed in the Subtropical, Mediterranean, and desert climatic regions of the world, India, China, Mexico, Spain, southern USA, Mexico, and Bulgaria (Chhatre et al. 2014). It is commonly known as Devil's thorn and Puncture vine (Tribulus terrestris 2022). It has been used in folk medicine, extensively by Indian, Japanese, Bulgarian, Korean, and Chinese civilizations. It is known to possess a host of pharmacological properties; immunomodulatory, anti-inflammatory, anticariogenic, anticancer, diuretic, antidiabetic, hepatoprotective, hypolipidemic, cardiotonic, analgesic, antispasmodic, antiurolithic, aphrodisiac, absorption enhancing, antibacterial, anthelmintic, and larvicidal activities. Several important phytochemicals have been isolated from *T. terrestris;* protodioscin, neoprotodioscin, prototribestin from the aerial parts; tribufuriside D, tribufuroside E, tribulosaponin A, polianthoside D from the fruits (Ştefănescu et al. 2020).

Inbred male Balb/c mice were subject to kidney injury by the administration of cisplatin. Three doses of ethanolic extract of *T. terrestris;* 100, 300, and 500 mg/kg body weight were administered to assess the nephroprotective activity. While there were no significant differences in the body weight in the 300 and 500 mg/kg treatment groups, the animals in the 100 mg/kg treatment group showed a reduction in body weight. All three *T. terrestris*-treatment groups showed no significant difference in kidney weight, unlike the animals treated with cisplatin alone. Substantial kidney damage occurred with cisplatin treatment; cortical tubular necrosis, medullar tubular necrosis, and Bowman's space enlargement. Administration of *T. terrestris* extract brought about a dose-dependent reduction in kidney damage; with minimum damage seen in the 500 mg/kg treatment group, with scores closest to those of the control (Raoofi et al. 2015). Another study assessed the nephroprotective activity of the hydroalcoholic activity of *T. terrestris*, employing mercuric chloride to induce kidney injury. The study animals were male Wistar rats, and three doses of *T. terrestris* used; 100, 200, and 300 mg/kg/day for 7 days. Renal function was evaluated by checking BUN and serum creatinine levels. Dose-dependent reductions in the levels of serum creatinine and BUN were observed in the animals treated with *T. terrestris* extract; the highest reduction was observed with the 300 mg/kg dose, indicating improved renal function. A similar dose-dependent increase in GSH levels, GPx activity, inhibition of SOD activity, and suppression of renal MDA levels was noted on treatment with *T. terrestris* extract. AKI was effectively induced by cisplatin, as confirmed by the increase in KIM-1 and L-FABP levels. *T. terrestris* at lower concentrations; 100/200 mg/kg was ineffective in bringing about a significant reversal of the increased KIM-1 and L-FABP levels. *T. terrestris* extract at a higher concentration of 300 mg/kg was found to be efficacious in bringing about a substantial reduction in the levels of both KIM-1 and L-FABP. Histopathological studies revealed that treatment with *T. terrestris* resulted in 50–75% intact/viable tubular epithelium. Some necrosis could be observed, representing grade 1 protection using *T. terrestris*. This shows *T. terrestris* is effective in attenuating the nephrotoxic effects of mercuric chloride (Yadav et al. 2019).

The methanolic extract of the fruits of *T. terrestris* (different doses1, 10, 100, 1000 μM) exhibited substantial inhibition against the papain-like protease (PLpro). Six cinnamic amides isolated from the fruit of *T. terrestris* were assayed for their ability to inhibit SARS-CoV PLpro. All six compounds showed significant inhibitory activity, with IC50 values ranging between 15.8 and 70.1 μM. Terrestrimine was the most effective inhibitor, with an IC50 of 15.8 μM (Song et al. 2014).

### *13.8.6* Urtica dioica *L.*

The most common species of the Urticaceae family and one of the most studied medicinal plants worldwide, *Urtica dioica* is a unique herbaceous perineal flowering plant that has stinging hairs. It is widely distributed in the Indian Subcontinent, Europe, and a few regions in Africa. *U. dioica* is commonly called nettle; it is famous for its hairy, toothed leaves and for its sting. The stinging hair of the nettle has a bulbous tip, that breaks off leaving a sharp needle-like tube piercing the skin and injecting acetylcholine and histamine. This causes a burning and itching sensation that can sometimes last up to 12 hours (Dhouibi et al. 2020; Urtica dioica 2022). Traditionally, *U. dioica* is used to treat allergies, rheumatic conditions, urinary tract disorders, benign prostate hyperplasia, and as an anti-haemorrhagic agent. *U. dioica* is also used for cosmetic preparations; anti-dandruff and hair loss lotions. The leaves are eaten raw/blanched, steamed/fried and added to food items such as soups, sauces, pesto, quiches, purées. Formulations of *U. dioica* are also used to prepare homeopathic medications. Some of the pharmacological activities of *U. dioica* include anti-inflammatory, diuretic, anti-asthmatic, cardiovascular, astringent, depurative, anticancer, galactogogue, and stimulating effects. Phytosterols, lignans, isolectins, coumarins, triterpenoic acids, and monoterpendiols have been isolated from the roots of *U. dioica*. Rutin, quercetin-3-*O*-glucoside and kaempferol-3-*O*-glucoside were found in the leaves and stalks and rutinosyl flavonols, chlorogenic acid, and caffeoyl malic acid were reported from the leaves (Esposito et al. 2019).

AKI was induced in the study animals; male Wistar rats via the administration of gentamycin, for 8 days. The preventive effect of methanolic leaf extract of *U. dioica*, at a dose of 200 mg/kg/day, administered for 8 days was assessed in study animals. Kidney injury was indicated in the animals treated with gentamycin alone, they exhibited a substantial increase in BUN, plasma creatinine, urinary sodium excretion, fractional excretion of sodium and potassium, and MDA levels. Considerable renal damage was also found in the animals treated with gentamycin alone; tubular cells necrosis, production of intraluminal protein casts (grade 3), vacuolization of renal tubular cells (grade 4), increase in space of Bowman's capsule, decreased RBC count and dispersion of cells into the lumen of the tubule (grade 3). On the other hand, animals co-treated with methanol extract of *U. dioica* showed lower levels of

BUN and plasma creatinine, reduced excretion of urinary sodium, reduced fractional excretion of sodium and potassium and lower MDA levels. Improved renal blood flow, clearance of creatinine, urine osmolarity, and FRAP levels were also characteristic of the *U. dioica*-treated group. A substantial decrease in renal injury was also observed in *U. dioica*-treated animals. Decrease in cellular necrosis, decrement of the Bowman's space, increased number of RBCs in the glomeruli, and only grade 1 level of generation of protein casts within the lumen of the tubule and dispersion of cells into the lumen of the tubule (grade 1; $p < 0.001$) and grade 2 vacuolization of renal epithelial cells was observed (Hajihashemi et al. 2020). A similar study evaluating the nephroprotective activity of *U. dioica* was carried out in rabbits. Gentamycin was used to induce nephrotoxicity. Ethanolic of the whole *U. dioica* plant, at a dose of 200 mg/mL was used. Co-treatment with *U. dioica* extract was found to ameliorate the negative effects of gentamycin on the kidney. Serum creatinine, BUN, and MDA levels were lower in the co-treated group, values of BUN were almost equal to that of the control. Extensive kidney damage was seen in the gentamycin-treated group. Swelling of the renal corpuscle, increased urinary spaces, urinary pole destruction, degeneration of the epithelium of collecting tubules and Henle loop characterized by the presence of varying degrees of tubular epithelial vacuolar degeneration, diminished brush border, attenuation, loss of epithelial cellular detail with the abundant epithelium of proximal and distal convoluted tubules containing a massive cytoplasmic aggregation of hyaline droplet and slight interstitial hemorrhage. On the contrary, co-treatment with gentamycin and *U. dioica* showed almost normal kidney morphology. There were only a few focal hydropic regions of swelling of tubular epithelial cells, indicating the partial to complete protective role of *U. dioica* against the nephrotoxicity of gentamicin (Salih 2015).

SARS-CoV replication was inhibited by *U. dioica* agglutinin (UDA) in a dose-dependent manner in a study conducted in Vero 76 cells. Decreased yields of the virus (of Urbani strain) by 90% were also observed in UDA-treated Vero 76 cells. In vivo analysis was also carried out, by infecting BALB/c mouse models with SARS-CoV-1. UDA at doses of 5, 10, and 20 mg/kg/day were administered for 4 days. Treatment with UDA protected the mice against weight loss as a result of SARS-CoV, lower lung pathology scores, and protection from death. Further analysis revealed that UDA probably curbs infection with SARS-CoV by targeting the early stages of the replication cycle; virus adsorption or penetration. UDA also neutralizes the infectivity of the virus by possibly binding to the spike (S) glycoprotein of SARS-CoV. Binding studies suggested that the binding sites for UDA are plausibly the *N*-acetylglucosamine-like residues that are present on the glycosylated envelope glycoproteins, thus preventing the attachment of SARS-CoV to cells (Kumaki et al. 2011).

## *13.8.7* Vitis vinifera *L.*

Although *Vitis vinifera* distribution natively ranges from South Central and South-East Europe to Central Asia and North Iran, it has been introduced to several places around the globe; India, Korea, Afghanistan, Pakistan, Bangladesh, and some parts of Africa, Europe, and the USA. Commonly referred to as grape wine or parsley wine, it is a woody climber with coiled climbing tendrils (Vitis vinifera 2022). *V. vinifera* is one of the oldest domesticated plants, which is consumed for both, its nutritional and medicinal values. Since ancient times, it is known for its medicinal value, traditional systems of medicine around the world have chronicled the use of *V. vinifera* for different indications; Ayurveda for inflammation and cardiovascular diseases, in Malatya the fruit is suggested to aid the formation of blood, in Elazığ for the treatment of anemia, in Iran for ailments and memory problems, in Pakistan as a carminative, in Tuscany, the drink derived from the fruit is used as a digestive, a Persian physician chronicled its used for oral health and in Cyprus the alcohol marinade is used as a poultice, mouthwash, and liniment. Grape skin is a source of polyphenols, proanthocyanidins, flavan-3-ols, and resveratrol. The fruit is rich in antioxidant phenolics, anthocyanins, catechins, quercetin, resveratrol, procyanidins, epigallocatechin, and epigallocatechin-3-gallate (Ardid-Ruiz et al. 2020; Al-Warhi et al. 2022).

The potency of grape seed extract to protect against renal dysfunction was assessed in female albino rats. Dexamethasone was administered, which brought about kidney dysfunction, as evidenced by the high amounts of serum creatinine and uric acid. Two concentrations of aqueous grape seed extract were used for analysis; 200 and 400 mg/kg BW. A dose-dependent reduction in the levels of creatinine and uric acid was observed in the grape seed extract treatment groups. In fact, in the 400 mg/kg group, the levels of creatinine and uric acid were very similar to that of the control. Treatment with grape seed extract, especially at a dose of 400 mg/kg was able to effectively ameliorate the reduction in weight gain due to dexamethasone administration. The relative weight of the kidney also was improved in the grape seed treated group, when compared to the group that was administered only dexamethasone. This indicates that grape seed extract positively affects plausible renal alteration on the administration of medication like dexamethasone (Hasona et al. 2017). The nephroprotective activity of grape juice was evaluated in male Wistar rats that were fed on a high cholesterol diet. High cholesterol diet led to renal dysfunction, as evidenced by increased BUN, serum creatinine, and urea with altered morphology in the kidney; dilated proximal tubules with acidophilic casts in their lumina. Peritubular capillaries were also congested and dilated. The group of rats administered with grape juice exhibited lower levels of BUN, serum creatinine, and urea. Most tubules in this animal group appeared intact, only some were affected. Only a few peritubular capillaries were congested in the grape juice-treated group (Ali et al. 2015).

A study on the antiviral activity of *V. vinifera* leaf extract against SARS-CoV-2 by plaque reduction assay showed very promising results. Virus pre-treatment

**Fig. 13.2** Nephroprotective plants with antiviral activity (All plant images taken from Indian Biodiversity Portal)

(treatment before infection on the cell monolayer) and co-treatment (incubated with SARS-CoV-2 in addition to target cells) was found to yield the best results. Only a slight antiviral effect was noted when cells were treated first with extract and subsequently infected (cell pre-treatment), and when the extract was added after HSV-1 infection (post-treatment). This suggests that *V. vinifera* leaf extract acts on viral particles directly, and blocks interaction with the cell membrane. *V. vinifera* extract also strongly decreased the expression of S post-SARS-CoV-2 infection until a concentration of 10 μg/mL was reached, and it resumed being expressed in a dose-dependent manner at lower concentrations (Zannella et al. 2021).

The images of the plants with nephroprotective activity and probable anti-SARS-CoV-2 activity described above may be found in Fig. 13.2. Some more medicinal plants with nephroprotective activity and potential effect on COVID 19 (preventive&/treatment) are described in Table 13.1

## 13.9 Conclusion

One of the lethal manifestations of COVID-19 is the occurrence of acute kidney injury. AKI not only complicates and prolongs treatment but also accounts for poor prognosis and high mortality rates. Moreover, some patients who develop AKI may not regain renal function even post-discharge. This makes AKI a serious complication and warrants extensive study, in pursuit of suitable cures and efficient management systems. Medicinal plants open up fascinating horizons of treatment; nephroprotective plants have been in use via different indigenous systems of medicine. Medicinal plants constitute vital ingredients for formulations of different Ayurvedic and traditional Chinese medicines. Scientific evaluation has also validated the use of many of these plants and individual phytochemicals have been isolated and characterized. Some nephroprotective plants and phytochemicals also exhibit antiviral properties, in particular against SARS-CoV and SARS-CoV-2. The advantage of the use of medicinal plants is undisputable, with minimum side effects,

**Table 13.1** Medicinal plants with nephroprotective activity and potential COVID-19 effects (preventive&/treatment) (Mukherjee et al. 2022)

| S. No. | Medicinal plant | Mode of action | Reference |
|---|---|---|---|
| 1 | *Azadirachta indica* A.Juss. | Oral administration of crude aqueous leaf extract improved malaria-induced renal injury. Treatment with extract for 4 days brought levels of creatinine and BNU to near normalcy. Absence of any toxic effects on treatment<br>A pilot, double-blind, randomized controlled trial showed the prophylactic effect of neem capsules; it effectively decreased the risk of COVID-19 infection<br>Bark extract effectively inhibits in vitro infection and replication of SARS-CoV-2 and m-CoV-RSA59 and also reduces gene expression of envelope and nucleocapsid genes. In vivo, bark extract attenuates neuroinflammation and hepatitis by inhibiting viral replication and spread. Isolated fractions of NBE enriched in Nimbin isomers showed effective inhibition of m-CoV-RSA59 infection in vitro. In in silico analysis, NBE targets spike and RdRp of m-CoV and SARS-CoV-2 with great affinity | Somsak et al. (2015), Nesari et al. (2021), Sarkar et al. (2022) |
| 2 | *Magnolia officinalis* Rehder and E.H.Wilson | 95% ethanolic extract of the bark ameliorated kidney damage in a high-fat fed mouse model. Significant reduction in albumin-to-creatinine ratio, reduction in glomerular enlargement, and renal tubular epithelial damage over a 6 months treatment period. Reduced levels of TNF-α, plasminogen activator inhibitor in treated mice<br>Honokiol, a phytoconstituent from the bark of *Magnolia sps.* was found to inhibit the replication of SARS-CoV-2 in Vero E6 cells and human A549 cells. Honokiol was also effective in inhibiting newer variants of SARS-CoV-2 such as Omicron | Cui et al. (2013), Salgado-Benvindo et al. (2023) |
| 3 | *Syzygium aromaticum* (L.) Merr. and L.M. Perry | Ameliorative effect of clove bud hydroalcoholic extract on contrast-induced acute kidney injury established. A significant difference in degradation, flattening, necrosis of renal tubular cells and dilation of the tubular lumen in treated and non-treated groups | Nasri et al. (2020), Li et al. (2022) |

(continued)

**Table 13.1** (continued)

| S. No. | Medicinal plant | Mode of action | Reference |
|---|---|---|---|
| | | Water and ethanol extract of *Syzygium aromaticum* exhibited a dose-dependent suppression of SARS-CoV-2 spike protein binding to ACE2 and inhibition of ACE2 activity | |
| 4 | *Andrographis paniculata* (Burm. f.) Nees | In vitro treatment of human tubular epithelial cells (HK-2 cells) with andrographolide effectively averts renal tubular cell damage induced by high glucose. Additionally, andrographolide was also found to suppress mitochondrial dysfunction and NLRP3 inflammasome activation, thus protecting the progression of diabetic tubular injury and fibrosis<br>A significant inhibition in infectious virions production ($IC_{50}$ of 0.036 μg/mL and 0.034 μM respectively), determined by the plaque assay on treatment (postinfection) of andrographolide and *A. paniculata* in Calu-3 cells infected with SARS-CoV-2 was observed, indicating anti-SARS-CoV-2 activity | (Liu et al. 2021; Ngiamsuntorn et al. 2021) |
| 5 | *Tinospora cordifolia* (Willd.) Miers | Feeding diabetic rats with 5% bark of *T. cordifolia* effectively modulates changes in the kidney (increased glomerular area, glomerular filtration rate)<br>Saponarin, a phytoconstituent of *T. cordifolia* exhibited a very promising result in computational binding studies with the main protease of COVID-19, with a binding affinity of −8.75 kcal/mol.<br>The binding affinity of the experimental drug Remdesvir was found to be –8.65 kcal/mol | Joladarashi et al. (2012), Mulpuru and Mishra (2021) |
| 6 | *Cinnamomum zeylanicum* L. | Pre-treatment with an aqueous extract of cinnamon bark results in significant protection against kidney injury in acetaminophen-treated rats. Significant restoration of urea and creatinine levels and normalization of renal tissue architecture was observed on pre-treatment with cinnamon<br>Two phytoconstituents from *Cinnamomum zeylanicum*, cinnamyl acetate and caryophyllene oxide exhibited high binding affinities with the main protease of COVID-19, –5.38 | Hussain et al. (2019), Mulpuru and Mishra (2021) |

(continued)

**Table 13.1** (continued)

| S. No. | Medicinal plant | Mode of action | Reference |
|---|---|---|---|
| | | kcal/mol and –5.25 kcal/mol. The binding affinity of the experimental drug, Favipiravir was also determined, –5.76 kcal/mol | |
| 7 | *Piper nigrum* L. | Administration of ethanolic extract of fresh seeds (pre-treatment) was found to normalize levels of urea, creatinine, albumin, total protein, blood urea nitrogen levels and normalize histoarchitecture of kidney in study rats Computational studies revealed that piperine has a very high binding affinity (−7.0 kCal/mol) toward the RNA-binding pocket of the SARS-CoV-2 nucleocapsid | Gaikwad and Zodape (2019), Choudhary et al. (2020) |
| 8 | *Mucuna pruriens* (L.) DC. | Long-term treatment (90 days) with *M. pruriens* extract resulted in improved levels of serum ALT, urea and creatinine levels in study animals. Molecular docking using Biovia Discovery Studio was performed to study the binding of phytoconstituents from *M. pruriens*. High positive values of -CDocker energy and -CDocker interaction energy were observed with *M. pruriens* constituent, dopamine. This suggests that dopamine could possibly deactivate the nucleocapsid protein N-terminal RNA binding domain 6M3M enzyme, and consequently interrupting the life cycle of SARS-CoV-2 | Concessao et al. (2020), Mishra and Swain (2020) |
| 9 | *Allium sativum* L. | Treatment with ethanolic extract of garlic resulted in improved kidney function and as evidenced by improved parameters (urea, creatinine, blood urea nitrogen) to near-normal levels. A significant decrease (60–70%) of viral RNA and infectious viral particles in SARS-CoV-2 infected Vero E6 and Calu-3 cells treated with allicin was observed. Treatment with allicin in infected Calu-3 cells also brought about a reversal in several host pathways to levels similar to that of uninfected cells. Allicin also brought about a reduction in the abundance of structural proteins N, M, S, and ORF3 in the host–virus proteome | Anusuya et al. (2013), Mösbauer et al. (2021) |

(continued)

**Table 13.1** (continued)

| S. No. | Medicinal plant | Mode of action | Reference |
|---|---|---|---|
| 10 | *Chrysanthemum indicum* L. | Ethanolic extract was potent in both in vitro and in vivo studies, improving cell viability and renal function index, respectively. The herb was found to inhibit SARS-CoV-2 virus | Kim et al. (2015), Mukherjee et al. (2022) |
| 11 | *Zingiber officinale* Roscoe | Treatment with ginger restored renal function makers (creatinine clearance, reabsorption rate of urine albumin, creatinine, and urea levels), restoration and regeneration of glomeruli, renal tubules, and cells in cadmium-treated rats<br>6-gingerol, a phytochemical found in ginger exhibited high binding affinity (binding energy −2.8764 KJ/mol to −15.7591 KJ/mol) with several COVID-19 viral protein targets | Gabr et al. (2019), Rathinavel et al. (2020) |
| 12 | *Echinacea purpurea* L. Moench | Decrease in AST, blood urea nitrogen, and total and direct bilirubin in rats treated with hydroalcoholic extract of aerial parts of *E. purpurea*. The ability of *E. purpurea* to prevent SARS-CoV-2 infection was studied in a randomized, open, controlled, exploratory clinical study. In the test group, participants received *E. purpurea* extract (Echinaforce). 5 out of 60 individuals from this group contracted COVID-19 as opposed to 8 out of 60 in the control group. During viral infection, an increased dose of Echinaforce was given as treatment regime. This treatment was found to reduce overall viral load and there were fewer COVID-19-related hospitalizations when compared to the control group. | Rezaie et al. (2013), Kolev et al. (2022) |
| 13 | *Curcuma longa* L. | Protective effect against kidney injury and helps maintain normal kidney morphology in renal ischemic rat models<br>Effective neutralization of SARS-CoV-2 at subtoxic concentrations in Vero E6 and human Calu-3 cells was observed on treating with root extract of turmeric, dissolved turmeric capsule content and pure curcumin. Curcumin treatment brought about a substantial decrease in SARS-CoV-2 RNA levels in cell culture supernatants | Aggarwal and Harikumar (2009), Bormann et al. (2021) |

(continued)

**Table 13.1** (continued)

| S. No. | Medicinal plant | Mode of action | Reference |
|---|---|---|---|
| 14 | *Glycyrrhiza glabra* L. | The active compound of licorice, glycyrrhetinic acid reduces the frequency of hyperkalemia and serum potassium concentration in dialysis patients<br>The aqueous root extract of *G. glabra* showed a potent protective effect against COVID-19 in hamsters infected with SARS-CoV-2. Prophylactic treatment exhibited protection against loss in body weight, a decrease (35–40%) in viral load in the lungs, and reduced lung pathology. Reduced mRNA expression of proinflammatory cytokines and plasminogen activator-1 was also observed in vivo. In vitro, the extract reduced Th2 and Th17 differentiation and IL-4 and IL-17A production | Ferrari (2009), Rizvi at al. (2022) |

and offer high efficacy. This review aims to shed light on COVID-19-induced AKI and selected medicinal plants that can possess nephroprotective and antiviral activity. The use of these plants and their extracts is promising; they can be employed for the development of effective drugs with little or no side effects to combat COVID-19-induced AKI efficiently.

## References

Aggarwal BB, Harikumar KB (2009) Potential therapeutic effects of curcumin, the anti-inflammatory agent, against neurodegenerative, cardiovascular, pulmonary, metabolic, autoimmune and neoplastic diseases. Int J Biochem Cell Biol 41(1):40–59. https://doi.org/10.1016/j.biocel.2008.06.010. Epub 2008 Jul 9. PMID: 18662800; PMCID: PMC2637808

Ahmadian E, Hosseiniyan Khatibi SM, Razi Soofiyani S, Abediazar S, Shoja MM, Ardalan M, Zununi Vahed S (2021) Covid-19 and kidney injury: pathophysiology and molecular mechanisms. Rev Med Virol 31(3):e2176. https://doi.org/10.1002/rmv.2176. Epub 2020 Oct 6. PMID: 33022818; PMCID: PMC7646060

Ali S, Alahmadi A, Hamdy R, Huwait EA, Alansari A, Ayuob N (2015) Renoprotective effect of red grape (Vitis vinifera L.) juice and dark raisins against hypercholesterolemia-induced tubular renal affection in albino rats. Folia Morphol (Warsz) 78(1):91–100. https://doi.org/10.5603/FM.a2018.0069

Allemailem KS, Almatroudi A, Khan AA, Rahmani AH, Almarshad IS, Alekezem FS, Hassanein N, El-Kady AM (2021) Manifestations of renal system involvement in hospitalized patients with COVID-19 in Saudi Arabia. PLoS One 16(7):e0253036. https://doi.org/10.1371/journal.pone.0253036. PMID: 34264954; PMCID: PMC8282026

Al-Warhi T, Zahran EM, Selim S, Al-Sanea MM, Ghoneim MM, Maher SA, Mostafa YA, Alsenani F, Elrehany MA, Almuhayawi MS, Al Jaouni SK, Abdelmohsen UR, Elmaidomy

AH (2022) Antioxidant and wound healing potential of *Vitis vinifera* seeds supported by phytochemical characterization and docking studies. Antioxidants (Basel) 11(5):881. https://doi.org/10.3390/antiox11050881. PMID: 35624745; PMCID: PMC9137519

Anusuya N, Durgadevi P, Dhinek A, Mythily S (2013) Nephroprotective effect of ethanolic extract of garlic (*Allium sativum* L.) on cisplatin induced nephrotoxicity in male wistar rats. Asian J Pharm Clin Res 6:97–100

Ardid-Ruiz A, Harazin A, Barna L, Walter FR, Bladé C, Suárez M, Deli MA, Aragonès G (2020) The effects of Vitis vinifera L phenolic compounds on a blood-brain barrier culture model: expression of leptin receptors and protection against cytokine-induced damage. J Ethnopharmacol 247:112253. https://doi.org/10.1016/j.jep.2019.112253. Epub 2019 Sep 25. PMID: 31562952

Armaly Z, Kinaneh S, Skorecki K (2021) Renal manifestations of Covid-19: physiology and pathophysiology. J Clin Med 10(6):1216. https://doi.org/10.3390/jcm10061216. PMID: 33804075; PMCID: PMC8000200

Batlle D, Soler MJ, Sparks MA, Hiremath S, South AM, Welling PA, Swaminathan S (2020) COVID-19 and ACE2 in cardiovascular, lung, and kidney working group. Acute kidney injury in COVID-19: emerging evidence of a distinct pathophysiology. J Am Soc Nephrol 31(7): 1380–1383. https://doi.org/10.1681/ASN.2020040419. Epub 2020 May 4. PMID: 32366514; PMCID: PMC7350999

Bharathi M, Sivamaruthi BS, Kesika P, Thangaleela S, Chaiyasut C (2022) In silico screening of potential phytocompounds from several herbs against SARS-CoV-2 Indian Delta variant B.1.617.2 to inhibit the spike glycoprotein trimer. Appl Sci 12:665. https://doi.org/10.3390/app12020665

*Boerhavia diffusa* (n.d.) Plants of the world online. https://powo.science.kew.org/taxon/urn:lsid:ipni.org:names:326899-2. Accessed on the 29th June 2022

Bormann M, Alt M, Schipper L, van de Sand L, Le-Trilling VTK, Rink L, Heinen N, Madel RJ, Otte M, Wuensch K, Heilingloh CS, Mueller T, Dittmer U, Elsner C, Pfaender S, Trilling M, Witzke O, Krawczyk A (2021) Turmeric root and its bioactive ingredient curcumin effectively neutralize SARS-CoV-2 in vitro. Viruses 13(10):1914. https://doi.org/10.3390/v13101914. PMID: 34696344; PMCID: PMC8537626

Braun F, Lütgehetmann M, Pfefferle S, Wong MN, Carsten A, Lindenmeyer MT, Nörz D, Heinrich F, Meißner K, Wichmann D, Kluge S, Gross O, Pueschel K, Schröder AS, Edler C, Aepfelbacher M, Puelles VG, Huber TB (2020) SARS-CoV-2 renal tropism associates with acute kidney injury. Lancet 396(10251):597–598. https://doi.org/10.1016/S0140-6736(20)31759-1. Epub 2020 Aug 17. PMID: 32818439; PMCID: PMC7431179

Chan L, Chaudhary K, Saha A, Chauhan K, Vaid A, Zhao S, Paranjpe I, Somani S, Richter F, Miotto R, Lala A, Kia A, Timsina P, Li L, Freeman R, Chen R, Narula J, Just AC, Horowitz C, Fayad Z, Cordon-Cardo C, Schadt E, Levin MA, Reich DL, Fuster V, Murphy B, He JC, Charney AW, Böttinger EP, Glicksberg BS, Coca SG, Nadkarni GN, Mount Sinai COVID Informatics Center (MSCIC) (2021) AKI in hospitalized patients with COVID-19. J Am Soc Nephrol 32(1):151–160. https://doi.org/10.1681/ASN.2020050615. Epub 2020 Sep 3. PMID: 32883700; PMCID: PMC7894657

Cheng Y, Luo R, Wang K, Zhang M, Wang Z, Dong L, Li J, Yao Y, Ge S, Xu G (2020) Kidney disease is associated with in-hospital death of patients with COVID-19. Kidney Int 97(5): 829–838. https://doi.org/10.1016/j.kint.2020.03.005. Epub 2020 Mar 20. PMID: 32247631; PMCID: PMC7110296

Chhatre S, Nesari T, Somani G, Kanchan D, Sathaye S (2014) Phytopharmacological overview of Tribulus terrestris. Pharmacogn Rev 8(15):45–51. https://doi.org/10.4103/0973-7847.125530

Choudhary P, Chakdar H, Singh D, Selvaraj C, Singh SK, Kumar S, Saxena AK (2020) Computational studies reveal piperine, the predominant oleoresin of black pepper (Piper nigrum) as a potential inhibitor of SARS-CoV-2 (COVID-19). Curr Sci 119:1333–1342

Concessao P, Bairy LK, Raghavendra AP (2020) Protective effect of *Mucuna pruriens* against arsenic-induced liver and kidney dysfunction and neurobehavioral alterations in rats. Vet World.

13(8):1555–1566. https://doi.org/10.14202/vetworld.2020.1555-1566. Epub 2020 Aug 12. PMID: 33061227; PMCID: PMC7522945

Coronavirus: kidney damage caused by COVID-19. https://www.hopkinsmedicine.org/health/conditions-and-diseases/coronavirus/coronavirus-kidney-damage-caused-by-covid19#:~:text=Signs%20of%20kidney%20problems%20in,kidney%20injury%20may%20require%20dialysis. Accessed 25 June 2022

Cui W, Wang Y, Chen Q, Sun W, Cai L, Tan Y, Kim KS, Kim KH, Kim YH (2013) Magnolia extract (BL153) ameliorates kidney damage in a high fat diet-induced obesity mouse model. Oxid Med Cell Longev. 2013:367040. https://doi.org/10.1155/2013/367040. Epub 2013 Nov 28. PMID: 24381715; PMCID: PMC3863519

Dhouibi R, Affes H, Ben Salem M, Hammami S, Sahnoun Z, Zeghal KM, Ksouda K (2020) Screening of pharmacological uses of Urtica dioica and others benefits. Prog Biophys Mol Biol 150:67–77. https://doi.org/10.1016/j.pbiomolbio.2019.05.008. Epub 2019 Jun 1. PMID: 31163183

Diao B, Wang C, Wang R, Feng Z, Zhang J, Yang H, Tan Y, Wang H, Wang C, Liu L, Liu Y, Liu Y, Wang G, Yuan Z, Hou X, Ren L, Wu Y, Chen Y (2021) Human kidney is a target for novel severe acute respiratory syndrome coronavirus 2 infection. Nat Commun 12(1):2506. https://doi.org/10.1038/s41467-021-22781-1. PMID: 33947851; PMCID: PMC8096808

Emam NM, Anjum S, Okail HA, Ibrahim MAR, Ahmad T (2020) Pomegranate peel extract protects against carbon tetrachloride-induced nephrotoxicity in mice through increasing antioxidants status. Biomed Rep 13(3):13. https://doi.org/10.3892/br.2020.1320. Epub 2020 Jun 26. PMID: 32765852; PMCID: PMC7391291

Esposito S, Bianco A, Russo R, Di Maro A, Isernia C, Pedone PV (2019) Therapeutic perspectives of molecules from *Urtica dioica* extracts for cancer treatment. Molecules 24(15):2753. https://doi.org/10.3390/molecules24152753. PMID: 31362429; PMCID: PMC6695697

Ferrari P (2009) Licorice: a sweet alternative to prevent hyperkalemia in dialysis patients? Kidney Int 76(8):811–812. https://doi.org/10.1038/ki.2009.282. PMID: 19789539

Gabr SA, Alghadir AH, Ghoniem GA (2019) Biological activities of ginger against cadmium-induced renal toxicity. Saudi J Biol Sci 26(2):382–389. https://doi.org/10.1016/j.sjbs.2017.08.008. Epub 2017 Aug 18. PMID: 31485182; PMCID: PMC6717148

Gagliardi I, Patella G, Michael A, Serra R, Provenzano M, Andreucci M (2020) COVID-19 and the kidney: from epidemiology to clinical practice. J Clin Med 9(8):2506. https://doi.org/10.3390/jcm9082506. PMID: 32759645; PMCID: PMC7464116

Gaikwad V, Zodape G (2019) Ameliorative effect of *piper Nigrum* on ethionamide and Para-amino salicylic acid induced nephrotoxicity in Sprague-dawley rats. Saudi J Med Pharm Sci 5:448–455

Ge S, Duo L, Wang J, Gegenzhula, Yang J, Li Z, Tu Y (2021) A unique understanding of traditional medicine of pomegranate, Punica granatum L. and its current research status. J Ethnopharmacol 271:113877. https://doi.org/10.1016/j.jep.2021.113877. Epub 2021 Jan 27. PMID: 33515685

Guan W, Ni Z, Hu Y, Liang W, Ou C, He J (2020) Clinical characteristics of coronavirus disease 2019 in China. N Engl J Med. https://doi.org/10.1056/NEJMoa2002032

Hajihashemi S, Ahmadi M, Chehrei A, Ghanbari F (2020) Ameliorative effect of cotreatment with the methanolic leaf extract of *Urtica dioica* on acute kidney injury induced by gentamicin in rats. Avicenna J Phytomed 10(3):273–286. PMID: 32523882; PMCID: PMC7256274

Hasona NA, Alrashidi AA, Aldugieman TZ, Alshdokhi AM, Ahmed MQ (2017) Vitis vinifera extract ameliorate hepatic and renal dysfunction induced by dexamethasone in albino rats. Toxics 5(2):11. https://doi.org/10.3390/toxics5020011. PMID: 29051443; PMCID: PMC5606666

Hirsch JS, Ng JH, Ross DW, Sharma P, Shah HH, Barnett RL, Hazzan AD, Fishbane S, Jhaveri KD (2020) Northwell COVID-19 Research Consortium; Northwell Nephrology COVID-19 Research Consortium. Acute kidney injury in patients hospitalized with COVID-19. Kidney Int 98(1):209–218. https://doi.org/10.1016/j.kint.2020.05.006. Epub 2020 May 16. PMID: 32416116; PMCID: PMC7229463

Hussain Z, Khan JA, Arshad A, Asif P, Rashid H, Arshad MI (2019) Protective effects of Cinnamomum zeylanicum L. (Darchini) in acetaminophen-induced oxidative stress, hepatotoxicity and nephrotoxicity in mouse model. Biomed Pharmacother 109:2285–2292. https://doi.org/10.1016/j.biopha.2018.11.123. Epub 2018 Nov 28. PMID: 30551486

*Hygrophila auriculata* (n.d.) Plants of the world online. https://powo.science.kew.org/taxon/urn:lsid:ipni.org:names:49646-1. Accessed on the 29th June 2022

Joladarashi D, Chilkunda ND, Salimath PV (2012) Tinospora cordifolia consumption ameliorates changes in kidney chondroitin sulphate/dermatan sulphate in diabetic rats. J Nutr Sci 1:e7. https://doi.org/10.1017/jns.2012.6. PMID: 25191554; PMCID: PMC4153103

Kariyil B, Joy D, Mercey K (2010) Therapeutic effect of ethanolic extract of *Hygrophila spinosa* T. Anders on gentamicin-induced nephrotoxicity in rats. Indian J Exp Biol 48:911–917

Kaye AD, Okeagu CN, Tortorich G, Pham AD, Ly EI, Brondeel KC, Eng MR, Luedi MM, Urman RD, Cornett EM (2021) COVID-19 impact on the renal system: pathophysiology and clinical outcomes. Best Pract Res Clin Anaesthesiol 35(3):449–459. https://doi.org/10.1016/j.bpa.2021.02.004. Epub 2021 Mar 2. PMID: 34511232; PMCID: PMC7920815

Kim TW, Kim YJ, Park SR, Seo CS, Ha H, Shin HK, Jung JY (2015) Chrysanthemum indicum attenuates cisplatin-induced nephrotoxicity both in vivo and in vitro. Nat Prod Commun. 10(3): 397–402

Kissling S, Rotman S, Gerber C, Halfon M, Lamoth F, Comte D, Lhopitallier L, Sadallah S, Fakhouri F (2020) Collapsing glomerulopathy in a COVID-19 patient. Kidney Int 98(1): 228–231. https://doi.org/10.1016/j.kint.2020.04.006. Epub 2020 Apr 15. PMID: 32471639; PMCID: PMC7156952

Kolev E, Mircheva L, Edwards MR, Johnston SL, Kalinov K, Stange R, Gancitano G, Berghe WV, Kreft S (2022) Echinacea Purpurea For the Long-Term Prevention of Viral Respiratory Tract Infections During Covid-19 Pandemic: A Randomized, Open, Controlled, Exploratory Clinical Study. Front. Pharmacol. 13:856410. https://doi.org/10.3389/fphar.2022.856410

Kumaki Y, Wandersee MK, Smith AJ, Zhou Y, Simmons G, Nelson NM, Bailey KW, Vest ZG, Li JK, Chan PK, Smee DF, Barnard DL (2011) Inhibition of severe acute respiratory syndrome coronavirus replication in a lethal SARS-CoV BALB/c mouse model by stinging nettle lectin, Urtica dioica agglutinin. Antiviral Res 90(1):22–32. https://doi.org/10.1016/j.antiviral.2011.02.003. Epub 2011 Feb 19. PMID: 21338626; PMCID: PMC3085190

Lee NYS, Khoo WKS, Adnan MA, Mahalingam TP, Fernandez AR, Jeevaratnam K (2016) The pharmacological potential of *Phyllanthus niruri*. J Pharm Pharmacol 68(8):953–969. https://doi.org/10.1111/jphp.12565

Li Z, Wu M, Yao J, et al (2020) Caution on kidney dysfunctions of COVID-19 patients. medRxiv. https://doi.org/10.1101/2020.02.08.20021212

Li Y, Liu Z, Zeng M, El Kadiri A, Huang J, Kim A, He X, Sun J, Chen P, Wang TTY, Zhang Y, Gao B, Xie Z, Yu LL (2022). Chemical Compositions of Clove (Syzygium aromaticum (L.) Merr. & L.) Extracts and Their Potentials in Suppressing SARS-CoV-2 Spike Protein-ACE2 Binding, Inhibiting ACE2, and Scavenging Free Radicals. J Agric Food Chem. 70(45):14403–14413. https://doi.org/10.1021/acs.jafc.2c06300. Epub 2022 Nov 1. PMID: 36318658

Lim TK (2012) Punica granatum. In: Edible medicinal and non-medicinal plants, pp 136–194. https://doi.org/10.1007/978-94-007-5653-3_10. PMCID: PMC7122081

Liu W, Liang L, Zhang Q, Li Y, Yan S, Tang T, Ren Y, Mo J, Liu F, Chen X, Lan T (2021) Effects of andrographolide on renal tubulointersticial injury and fibrosis. Evidence of its mechanism of action Phytomedicine 91:153650. https://doi.org/10.1016/j.phymed.2021.153650. Epub 2021 Jul 9. PMID: 34332282

Marhaeny HD, Widyawaruyanti A, Widiandani T, Fuad Hafid A, Wahyuni TS (2021) Phyllanthin and hypophyllanthin, the isolated compounds of *Phyllanthus niruri* inhibit protein receptor of corona virus (COVID-19) through *in silico* approach. J Basic Clin Physiol Pharmacol 32(4): 809–815. https://doi.org/10.1515/jbcpp-2020-0473. PMID: 34214339.

Meena P, Bhargava V, Rana DS, Bhalla AK, Gupta A (2020) COVID-19 and the kidney: a matter of concern. Curr Med Res Pract 10(4):165–168. https://doi.org/10.1016/j.cmrp.2020.07.003. Epub 2020 Jul 18. PMID: 32839726; PMCID: PMC7367802.fhi

Mestry SN, Gawali NB, Pai SA, Gursahani MS, Dhodi JB, Munshi R, Juvekar AR (2020) Punica granatum improves renal function in gentamicin-induced nephropathy in rats via attenuation of oxidative stress. J Ayurveda Integr Med. 11(1):16–23. https://doi.org/10.1016/j.jaim.2017.09.006. Epub 2018 Mar 17. PMID: 29555255; PMCID: PMC7125389

Mishra and Swain (2020) In silico Analysis of Phytochemicals from Mucuna pruriens against COVID-19. Ind. J. Nat. Sci. 10(60):25457. https://media.cutm.ac.in/naacdvvqueries/3.4.5/screenshots-ofresearch- articles/1716.pdf

Mishra S, Aeri V, Gaur PK, Jachak SM (2014) Phytochemical, therapeutic, and ethnopharmacological overview for a traditionally important herb: Boerhavia diffusa Linn. Biomed Res Int 2014:808302. https://doi.org/10.1155/2014/808302. Epub 2014 May 14. PMID: 24949473; PMCID: PMC4053255

Mösbauer K, Fritsch VN, Adrian L, Bernhardt J, Gruhlke MCH, Slusarenko AJ, Niemeyer D, Antelmann H (2021) The effect of Allicin on the proteome of SARS-CoV-2 infected Calu-3 cells. Front Microbiol 12:746795. https://doi.org/10.3389/fmicb.2021.746795

Mukherjee PK, Efferth T, Das B, Kar A, Ghosh S, Singha S, Debnath P, Sharma N, Bhardwaj PK, Haldar PK (2022) Role of medicinal plants in inhibiting SARS-CoV-2 and in the management of post-COVID-19 complications. Phytomedicine 98:153930. https://doi.org/10.1016/j.phymed.2022.153930. Epub ahead of print. PMID: 35114450; PMCID: PMC8730822

Mulpuru V, Mishra N (2021) Computational identification of SARS-CoV-2 inhibitor in *Tinospora cordifolia.* Cinnamomum zeylanicum and Myristica fragrans VirusDis 32:511–517. https://doi.org/10.1007/s13337-021-00721-3

Nasri P, Hooshyar N, Yalameha B, Savari MA, Khosravifarsani M, Mardani S, Noorshargh P, Dehkordi AH, Nasri H, Golestaneh E, Bijad E, Ghasemi B, Nowrouzi N (2020) Ameliorative impact of hydroalcoholic extract of clove buds (*Syzygium aromaticum*) on contrast-induced acute kidney injury; an experimental histopathological study. J Nephropathol 9:1–5

Neharkar V, Gaikwad K (2019) Nephroprotective activity of *Hygrophila auriculata* leaves methanolic extract against drug-induced nephrotoxicity in rats. Pharma Sci Monit 10(3):82–89

Nesari TM, Bhardwaj A, ShriKrishna R, Ruknuddin G, Ghildiyal S, Das A, Pandey AK, Chaudhary N, Soman G, Barde M (2021) Neem (Azadirachta Indica A. Juss) capsules for prophylaxis of COVID-19 infection: a pilot, double-blind, randomized controlled trial. Altern Ther Health Med 27(S1):196–203. PMID: 33891569

Ngiamsuntorn K, Suksatu A, Pewkliang Y, Thongsri P, Kanjanasirirat P, Manopwisedjaroen S, Charoensutthivarakul S, Wongtrakoongate P, Pitiporn S, Chaopreecha J, Kongsomros S, Jearawuttanakul K, Wannalo W, Khemawoot P, Chutipongtanate S, Borwornpinyo S, Thitithanyanont A, Hongeng S (2021) Anti-SARS-CoV-2 activity of *Andrographis paniculata* extract and its major component Andrographolide in human lung epithelial cells and cytotoxicity evaluation in major organ cell representatives. J Nat Prod 84(4):1261–1270. https://doi.org/10.1021/acs.jnatprod.0c01324. Epub 2021 Apr 12. PMID: 33844528; PMCID: PMC8056600

Pareta SK (2011) Protective effects of *Boerhaavia diffusa* against acetaminophen induced nephrotoxicity in rats. Pharmacol Online 2:698–706

Peter AE, Sandeep BV, Rao BG, Kalpana VL (2021a) Calming the storm: natural Immunosuppressants as adjuvants to target the cytokine storm in COVID-19. Front Pharmacol 11:583777. https://doi.org/10.3389/fphar.2020.583777. PMID: 33708109; PMCID: PMC7941276

Peter AE, Sandeep BV, Rao BG, Kalpana VL (2021b) Nanotechnology to the rescue: treatment perspective for the immune dysregulation observed in COVID-19. Front Nanotechnol 3: 644023. https://doi.org/10.3389/fnano.2021.644023

*Phyllanthus niruri* (n.d.) Plants of the world online. https://powo.science.kew.org/taxon/urn:lsid:ipni.org:names:194900-2. Accessed on the 29th June 2022

*Punica granatum* (n.d.) Plants of the world online. https://powo.science.kew.org/taxon/urn:lsid:ipni.org:names:554129-1. Accessed on the 30th June 2022

Raoofi A, Khazaei M, Ghanbari A (2015) Protective effect of hydroalcoholic extract of tribulus terrestris on cisplatin induced renal tissue damage in male mice. Int J Prev Med 6:11. https://doi.org/10.4103/2008-7802.151817. PMID: 25789143; PMCID: PMC4362283

Rathinavel T, Palanisamy M, Palanisamy S, Subramanian A, Thangaswamy S (2020) Phytochemical 6-gingerol - a promising drug of choice for COVID-19. Int J Advw Sci Eng 6:1482–1489

Reddy GS, Raparla LP, Reddy GR, Charan DV (2015) Evaluation of Nephroprotective activity of the Methanolic extract of *Phyllanthus niruri* (family - Euphorbiaceae). Int J Pharm Phytopharmacological Res 4(5):276–280

Rezaie A, Fazlara A, Haghi Karamolah M, Shahriari A, Najaf Zadeh H, Pashmforosh M (2013) Effects of Echinacea purpurea on Hepatic and Renal Toxicity Induced by Diethylnitrosamine in Rats. Jundishapur J Nat Pharm Prod 8(2):60–64. Epub 2013 May 4. PMID: 24624189; PMCID: PMC3941908

Rizvi ZA, Babele P, Sadhu S, Madan U, Tripathy MR, Goswami S, Mani S, Kumar S, Awasthi A, Dikshit M (2022) Prophylactic treatment of *Glycyrrhiza glabra* mitigates COVID-19 pathology through inhibition of pro-inflammatory cytokines in the hamster model and NETosis. Front Immunol 13:945583. https://doi.org/10.3389/fimmu.2022.945583

Rutwick Surya U, Praveen N (2021) A molecular docking study of SARS-CoV-2 main protease against phytochemicals of *Boerhavia diffusa* Linn. For novel COVID-19 drug discovery. Virus 32(1):46–54. https://doi.org/10.1007/s13337-021-00683-6. Epub 2021 Mar 18. PMID: 33758772; PMCID: PMC7971947

Salih NA (2015) Effect of nettle (*Urtica dioica*) extract on gentamicin induced nephrotoxicity in male rabbits. Asian Pac. J. Trop. Biomed. 5(9):756–760. https://doi.org/10.1016/j.apjtb.2015.07.005

Salgado-Benvindo C, Leijs AA, Thaler M, Tas A, Arbiser JL, Snijder EJ, van Hemert MJ (2023). Honokiol Inhibits SARS-CoV-2 Replication in Cell Culture at a Post-Entry Step. Microbiol Spectr. 11(3):e0327322. https://doi.org/10.1128/spectrum.03273-22. Epub 2023 May 4. PMID: 37212560; PMCID: PMC10269499.

Sarkar L, Oko L, Gupta S, Bubak AN, Das B, Gupta P, Safiriyu AA, Singhal C, Neogi U, Bloom D, Banerjee A, Mahalingam R, Cohrs RJ, Koval M, Shindler KS, Pal D, Nagel M, Sarma JD (2022) Azadirachta indica A. Juss bark extract and its Nimbin isomers restrict β-coronaviral infection and replication. Virology 569:13–28. https://doi.org/10.1016/j.virol.2022.01.002. Epub 2022 Feb 15. PMID: 35219218; PMCID: PMC8844965

Sawardekar SB, Patel TC (2015) Evaluation of the effect of *Boerhavia diffusa* on gentamicin-induced nephrotoxicity in rats. J Ayurveda Integr Med 6(2):95–103. https://doi.org/10.4103/0975-9476.146545. PMID: 26166999; PMCID: PMC4484055

Shanmugasundaram P, Venkataraman S (2006) Hepatoprotective and antioxidant effects of Hygrophila auriculata (K. Schum) Heine Acanthaceae root extract. J Ethnopharmacol 104(1–2):124–128. https://doi.org/10.1016/j.jep.2005.08.058. Epub 2005 Oct 5. PMID: 16213687

Soliman NA (2021) COVID-19 infection and the kidneys: learning the lesson. J Infect Public Health 14(7):922–926. https://doi.org/10.1016/j.jiph.2021.05.010. Epub 2021 May 26. PMID: 34119846; PMCID: PMC8152202

Somsak V, Chachiyo S, Jaihan U, Nakinchat S (2015) Protective effect of aqueous crude extract of neem (Azadirachta indica) leaves on plasmodium berghei-induced renal damage in mice. J Trop Med 2015:961205. https://doi.org/10.1155/2015/961205. Epub 2015 Aug 25. PMID: 26379714; PMCID: PMC4562167

Song YH, Kim DW, Curtis-Long MJ, Yuk HJ, Wang Y, Zhuang N, Lee KH, Jeon KS, Park KH (2014) Papain-like protease (PLpro) inhibitory effects of cinnamic amides from Tribulus terrestris fruits. Biol Pharm Bull 37(6):1021–1028. https://doi.org/10.1248/bpb.b14-00026

Ştefănescu R, Tero-Vescan A, Negroiu A, Aurică E, Vari CE (2020) A comprehensive review of the phytochemical, pharmacological, and toxicological properties of *Tribulus terrestris* L. Biomol Ther 10(5):752. https://doi.org/10.3390/biom10050752. PMID: 32408715; PMCID: PMC7277861

Su H, Yang M, Wan C, Yi LX, Tang F, Zhu HY, Yi F, Yang HC, Fogo AB, Nie X, Zhang C (2020) Renal histopathological analysis of 26 postmortem findings of patients with COVID-19 in China. Kidney Int 98(1):219–227. https://doi.org/10.1016/j.kint.2020.04.003. Epub 2020 Apr 9. PMID: 32327202; PMCID: PMC7194105

Tito A, Colantuono A, Pirone L, Pedone E, Intartaglia D, Giamundo G, Conte I, Vitaglione P, Apone F (2021) Pomegranate Peel extract as an inhibitor of SARS-CoV-2 spike binding to human ACE2 receptor (*in vitro*): a promising source of novel antiviral drugs. Front Chem 9: 638187. https://doi.org/10.3389/fchem.2021.638187

*Tribulus terrestris*. Plants of the World Online. https://powo.science.kew.org/taxon/urn:lsid:ipni.org:names:873489-1. Accessed on the 30th June 2022

*Urtica dioica*. Plants of the World Online. https://powo.science.kew.org/taxon/urn:lsid:ipni.org:names:260630-2. Accessed on the 30th June 2022

Vijayakumar M, Govindarajan R, Rao GM, Rao CV, Shirwaikar A, Mehrotra S, Pushpangadan P (2006) Action of Hygrophila auriculata against streptozotocin-induced oxidative stress. J Ethnopharmacol 104(3):356–361. https://doi.org/10.1016/j.jep.2005.09.030. Epub 2005 Nov 9. PMID: 16289604

Vincent S, Arokiyaraj S, Saravanan M, Dhanraj M (2020) Molecular docking studies on the anti-viral effects of compounds from Kabasura Kudineer on SARS-CoV-2 $3CL^{pro}$. Front Mol Biosci 7:613401. https://doi.org/10.3389/fmolb.2020.613401

*Vitis vinifera*. Plants of the world online. https://powo.science.kew.org/taxon/urn:lsid:ipni.org:names:30478388-2. Accessed on the 30th June 2022.

Wang D, Hu B, Hu C, Zhu F, Liu X, Zhang J, Wang B, Xiang H, Cheng Z, Xiong Y, Zhao Y, Li Y, Wang X, Peng Z (2020) Clinical characteristics of 138 hospitalized patients with 2019 novel coronavirus-infected pneumonia in Wuhan, China. JAMA 323(11):1061–1069. https://doi.org/10.1001/jama.2020.1585. Erratum in: JAMA. 2021 Mar 16;325(11):1113. PMID: 32031570; PMCID: PMC7042881

Yadav HN, Sharma US, Singh S, Gupta YK (2019) Effect of *Tribulus terrestris* in mercuric chloride-induced renal accumulation of mercury and nephrotoxicity in rat. J Adv Pharm Technol Res 10(3):132–137. https://doi.org/10.4103/japtr.JAPTR_386_18. PMID: 31334096; PMCID: PMC6621347

Yang X, Yu Y, Xu J, Shu H, Xia J, Liu H, Wu Y, Zhang L, Yu Z, Fang M et al (2020) Clinical course and outcomes of critically ill patients with SARS-CoV-2 pneumonia in Wuhan, China: a single-centered, retrospective, observational study. Lancet Respir Med 8:475–481. https://doi.org/10.1016/S2213-2600(20)30079-5

Yin W, Zhang PL (2020) Infectious pathways of SARS-CoV-2 in renal tissue. J Nephropathol 9(4): e37. https://doi.org/10.34172/jnp.2020.37

Zannella C, Giugliano R, Chianese A, Buonocore C, Vitale GA, Sanna G, Sarno F, Manzin A, Nebbioso A, Termolino P, Altucci L, Galdiero M, de Pascale D, Franci G (2021) Antiviral activity of *Vitis vinifera* leaf extract against SARS-CoV-2 and HSV-1. Viruses 13(7):1263. https://doi.org/10.3390/v13071263. PMID: 34209556; PMCID: PMC8310055

# Chapter 14
# Phytochemicals and Nutraceuticals Targeting SARS-CoV-2: An In Silico Analysis

**Saranya Nallusamy, Caroline Ravikumar, Jayakanthan Mannu, Bharathi Nathan, Gnanam Ramasamy, Raveendran Muthurajan, and Mohankumar Subbarayalu**

## 14.1 Introduction

A new coronavirus was reported in December 2019, which is called severe acute respiratory syndrome coronavirus 2 (SARS-CoV-2), which causes coronavirus disease 2019 (COVID-19) (Ko et al. 2020). Since its emergence, it caused a menace to humankind and spread to nearly every country and area in the world (John Hopkins University 2020). COVID-19 was proclaimed a pandemic by the World Health Organization (WHO) as it threatened human life and caused huge economic losses throughout the world. Although immunization is a positive step, it might not completely guard against the illness; as a result, it will still be necessary to create potent COVID-19 treatment and prevention.

Traditional medicines which employ plant extracts, nutraceuticals, and other plant products remain the base for present-day drug discovery and development programs (Veeresham 2012). Plant phytochemicals, dietary supplements, and functional foods are nutraceuticals that also imbibe medicinal properties within them (Varzakas et al. 2016). Plethora reports are available for COVID-19-related research and the hunt for phytochemicals from varied sources grows exponentially.

---

**Supplementary Information** The online version contains supplementary material available at https://doi.org/10.1007/978-981-99-3664-9_14.

---

S. Nallusamy · C. Ravikumar · J. Mannu · B. Nathan · G. Ramasamy
Department of Plant Molecular Biology and Bioinformatics, Tamil Nadu Agricultural University, Coimbatore, India

R. Muthurajan · M. Subbarayalu (✉)
Department of Plant Biotechnology, Tamil Nadu Agricultural University, Coimbatore, India



J.-T. Chen (ed.), *Ethnopharmacology and Drug Discovery for COVID-19: Anti-SARS-CoV-2 Agents from Herbal Medicines and Natural Products*,
https://doi.org/10.1007/978-981-99-3664-9_14

Behl et al. (2021) have indicated the importance of plant-based functional foods in combating human viral diseases elaboratively. They have discussed the importance of flavonoids, chalcones, coumarins, phenolic compounds, etc., and plant sources such as ginger, garlic, moringa, mint leaves, etc., in alleviating the provoking of viral diseases.

A recent study on the application of Ayurvedic Kadha for the management and prevention of SARS-CoV-2 was reported by Maurya and Sharma et al. (2022). The role of phytochemicals found in the Kadha in controlling the viral infection and multiplication in the host cells was investigated using in silico approach. Most of the phytochemicals found in the herbs ashwagandha, black pepper, clove, tulsi, and giloy were effective in inhibiting COVID-19.

Qazi et al. (2021) have specifically studied the antiviral activity of phytochemicals from *Aeglemarmelos*, *Vetiveriazizanioides*, *Moringa oleifera*, and *Punica granatum* against protein targets such as 3CLpro and RNA-dependent RNA polymerase through docking and molecular dynamics simulation studies. They have provided lead molecules from the above functional food sources that may hinder viral replication and transcription.

Considering the medicinal significance of the phytochemical compounds against COVID-19 targets, in this chapter, we discuss the research on virtual screening of around 744 phytochemicals from 37 different plant species for specific COVID-19 targets such as SARS-CoVMpro, MERS-CoVMpro, SARS-CoV2 MPro, Spike glycoprotein, RNA-dependent RNA polymerase, NSP16, NSP10, NSP3 NSP15, and NSP9. Furthermore, in order to understand the effect of phytochemicals from Indian-based functional foods that are used day-to-day and their impact on SARS-CoV-2 targets, main protease, and spike glycoprotein have been investigated using the docking approach.

## 14.2 Methodology

The general methodology followed in exploring the inhibitory activity of phytochemicals on covid targets using in silico approach is shown in Fig. 14.1. A druggable target for SARS-CoV-2 is identified through a literature review or by using biological sequence databases. Small molecules or ligand molecules are curated from plant phytochemical databases like Dr. Duke's database (1992–2016)/Pubchem database (Kim et al. 2019). The experimentally determined tertiary structures of the proteins were retrieved Protein Data Bank (Berman et al. 2000). Identified targets and ligands are prepared in minimized energy conformation with proper bond angles and distances to improve the docking accuracy. The binding /active sites of the protein targets were identified using the CASTp server (Tian et al. 2018), which is the site the ligand molecule is set for docking. Tools like PyRx 0.8 (Dallakyan and Olson 2015) are used for virtual screening/docking of the phytochemicals using Auto dock/Auto vina module. Docked complexes are analyzed using Discovery Studio Visualizer (BIOVIA) (Accelrys Inc., San Diego, CA, USA).

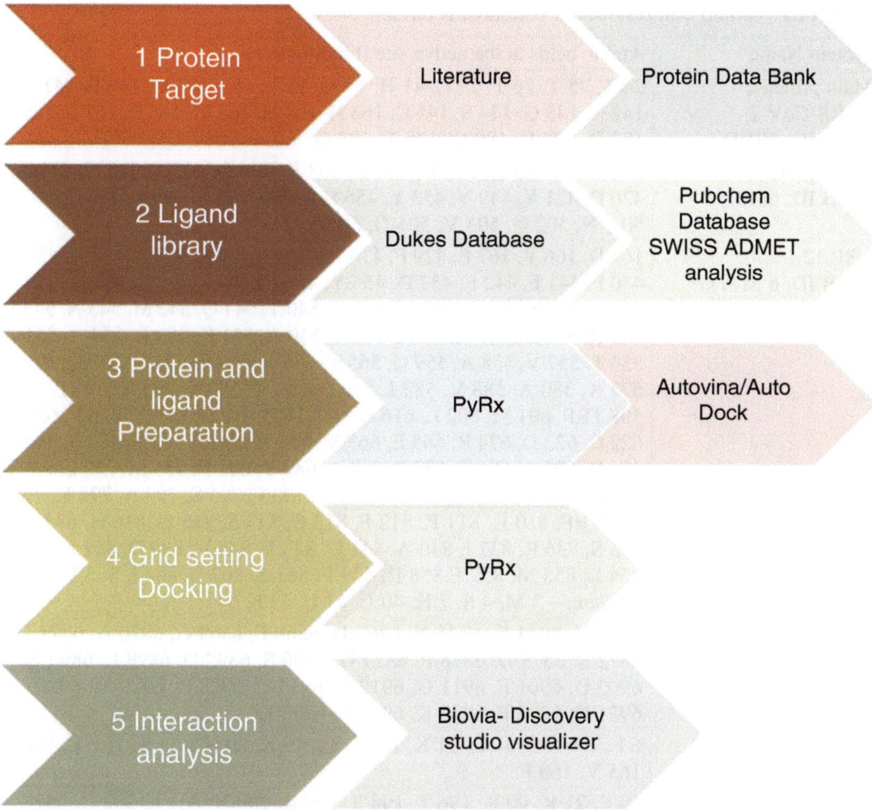


**Fig. 14.1** General methodology followed in the identification of phytochemicals that may inhibit SARS-CoV-2 targets using in silico approach

## 14.3 In Silico Screening of Medicinal Compounds Against CoVID-19 Targets

Our traditional medicines implementing Ayurveda, Siddha, and Unani aged around 5000 years ago remain the golden base behind the current drug discovery programs. Nallusamy et al. (2021) have performed a detailed study on phytochemicals to combat COVID-19 in the identification of novel molecules that could act as an inhibitor of SARS-CoV-2 targets. Auto Vina wizard available in PyRx (Dallakyan and Olson 2015) was used for screening 744 ligands (chemical compounds and bioactive compounds from medicinal herbs) against seven major protein targets. Active site residues were predicted using the CASTp server for the above-mentioned targets are given in Table 14.1. Phytochemicals from "Kabasurakudineer" (Siddha medicine prescribed for immunity booster, AYUSH Board of Government of India) were also included for ligand library preparation. Furthermore, a comparison of the medicinal compounds with that of already existing drugs such as

**Table 14.1** Amino acid residues of coronavirus binding sites

| Protein Name | Amino acids at the active site (binding site) |
|---|---|
| Main protease—SARS-CoV-2 (PDB ID: 5R81) | 24 T, 25 T, 26 T, 27 L, 41 H, 44 C, 45 T, 46 S, 49 M, 140 F, 141 L, 142 N, 143 G, 144 S, 145 C, 163 H, 164 H, 165 M, 166 E, 167 L, 168 P, 187 D, 188 R, 189 Q, 190 T, 192 Q |
| Spike protein (PDB ID: 6M0J) | *Chain E*—403 R, 405 D, 406 E, 408 R, 409 Q, 415 T, 416 G, 417 K, 420 D, 421 Y, 449 Y, 453 Y, 456 F, 493 Q, 494 S, 495 Y, 496 G, 497 F, 50 1 N, 502 G, 503 V, 504 G, 505 Y |
| NSP 12 (PDB ID: 6 M71) | 164 D, 166 V, 167 E, 429 F, 430 K, 431 E, 436 E, 437 L, 438 K, 439 H, 440 F, 441 F, 442 F, 452 D, 455 Y, 456 Y, 494 I, 496 N, 497 N, 499 D, 500 K, 501 S,503 G, 507 N, 511 K, 540 T, 541 Q, 542 M, 543 N, 544 L, 545 K, 546 Y, 547 A, 548 I, 549 S, 550 A, 551 K, 553 R, 554 A, 555 R, 556 T, 557 V, 558 A, 559 G, 565 T, 568 N, 569 R, 572 H, 573 Q, 576 L, 577 K, 580 A, 588 V, 589 I, 590 G, 591 T, 592 S, 593 K, 594 F, 598 TRP, 601 M, 602 L, 616 G, 617 TRP, 618 D, 619 Y, 620 P, 621 K, 622 C, 623 D, 624 R, 665 E, 667 V, 676 K, 680 T, 681 S, 682 S, 683 G, 684 D, 685 A, 686 T, 687 T, 688 A, 689 Y, 691 N, 756 M, 758 L, 759 S, 760 D, 761 D, 762 A, 763 V, 792 V, 793 F, 795 S, 797 A, 798 K, 79 C, 800 TRP, 810 H, 811 E, 812 F, 813 C, 814 S, 815 Q, 816 H, 833 D, 835 S, 836 R, 837 I, 840 A, 841 G, 843 F, 844 V, 845 D, 847 I, 848 V, 854 L, 855 M, 857 E,858 R, 859 F, 861 S, 862 L, 864 I, 865 D<br>*C Chain*—3 M, 4 S, 7 K, 40 L, 41 L, 43 K |
| NSP10 - NSP16 (PDB ID: 6W4H) | 6841 N, 6844 K, 6845 Y, 6867 H, 6868 F, 6869 G, 6870 A, 6871 G, 6872 S, 6873 D, 6878 P, 6879 G, 6896 S, 6897 D, 6898 L, 6899 N, 6900 D, 6901 F, 6911 G, 6912 D, 6913 C, 6928 D, 6929 M, 6930 Y, 6931 D, 6932 P, 6933 K, 6947 F, 6968 K |
| NSP3 (PDB ID: 6 W02) | 6 F, 7 S, 8 G, 10 L, 11 K, 12 L,18 I, 19 K, 20 N, 158 K, 161 Y, 162 D, 165 V, 168 F |
| NSP15 (PDB ID: 6 W01) | 69 E, 71 K, 90 K, 196 T, 198 S, 199 R, 200 N, 201 L, 202 Q, 252 L, 255 L, 259 F, 266 L, 268 D, 272 M, 273 D. 274 S, 275 T, 277 K, 279 Y, 295 V, 296 I, 297 D |
| NSP9 (PDB ID: 6W4B) | 13 M, 33 Y, 38 G, 39 G, 40 R, 42 V, 57F, 58 P, 59 K, 60 S, 66 I, 68 T |

hydroxychloroquine, chloroquine, and ivermectin was performed to quench the drug-likeness behavior of the molecules against SARS-CoV-2 targets.

Out of the 605 phytochemicals from 37 plant species, 33 phytochemicals from 22 plant species were reported as the best inhibitors with higher binding affinities against seven targets of SARS-CoV-2. Among the above-mentioned 22 plant species, four plants were found to be constituents of "Kabasurakudineer."

Plantwise comparison of phytochemicals that may inhibit SARS-CoV-2 targets using a docking approach highlighted the following plants, *Vitex negundo*, *Solanum nigrum*, *Pedalium murex*, *Azadirachta indica*, *Terminalia chebula*, *Cissus quadrangularis*, *Clerodendrum serratum*, and *Ocimum basilicum* as predominant entities among the chosen 37 plants. Moreover, phytochemicals such as amentoflavone, catechin-o-gallate, agasthisflavone, and chlorogenin showed higher binding affinity against multiple targets of CoVID-19.

Amentoflavone reported docking scores of −9.3 kcal/mol, −8.3 kcal/mol, −7.4 kcal/mol, −8.5 kcal/mol, and −8.2 kcal/mol for the SARS-CoV-2 targets such as NSP 12, NSP 9, NSP 3, NSP10-NSP16, and spike protein, respectively. Biological activities such as antitumor, antiviral, anti-inflammatory, antioxidative, antidiabetic, and antifungal have been observed for amentoflavone (Yu et al. 2017). Also, it was found to have antisenescence activity in the cardiovascular and central nervous systems (Park et al. 2011).

In the case of agathisflavone, binding affinity values of −8.9 kcal/mol, −8.2 kcal/mol, −8.2 kcal/mol, and −6.6 kcal/mol were predicted on docking with targets NSP 12, SARS-CoV-2 main protease, spike protein, and NSP 3, respectively. Figure 14.2 provides the interaction pattern of RdRp and Spike glycoprotein with ivermectin, agathisflavone, and amentoflavone.

Catechins are the class of compounds that are generally reported in antiviral studies due to their property of increased affinity toward cellular membranes (Kaihatsu et al. 2018). Catechin-o-gallate showed significant binding interactions with the affinity values of −7.3 kcal/mol, −6.6 kcal/mol, −7.9 kcal/mol, and −8.5 kcal/mol against spike protein, NSP3, NSP 12 (RNA-dependent RNA polymerase), and NSP 15, respectively. Chlorogenin which is a constituent of *Solanum torvum* also showed higher binding affinity against multiple targets such as NSP9 (−8.2 kcal/mol), MPro (−7.7 kcal/mol), and NSP16-NSP10 (−7.6 kcal mol), respectively.

## 14.4 Docking Studies on Food Phytochemicals against SARS-CoV-2 Main Protease (MPro) and Spike Glycoprotein

A total of 63 food crops with 4462 phytochemicals were screened against SARS-CoV-2 MPro and spike glycoprotein to understand the antiviral property of day-to-day food phytochemicals that can help to withstand the adverse effect of COVID-19. For this study, chosen crops were categorized into seven types vegetables, fruits, supplements, spices, pulses, millet, and herbs mainly based on their properties and usage. Details of the number of plants taken under each category are provided in Table 14.2. Information on the phytochemicals, PubChem id, and the plant part are given in Supplementary Table 14.S1.

### *14.4.1 Docking Experiment and Drug-Like Molecule Prediction*

In silico screening of phytochemicals for SARS-CoV-2 antiviral activity was performed using PyRx (0.8) AutoDock module (Dallakyan and Olson 2015). Except

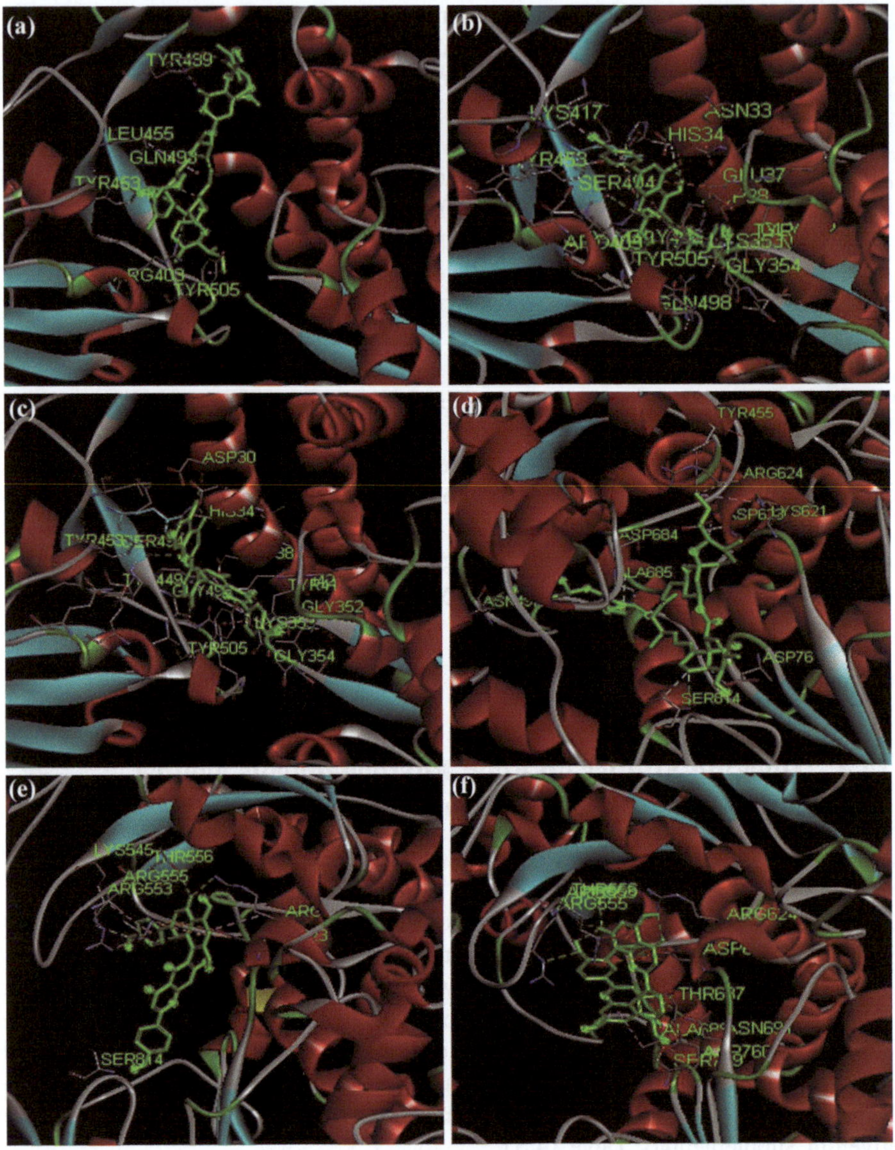


**Fig. 14.2** Docked complex structures showing the interaction of spike protein with that of (**a**) ivermectin, (**b**) agasthisflavone, (**c**) amentoflavone and NSP12 with that of (**d**) ivermectin, (**e**) agasthisflvaone, (**f**) amentoflavone

for the docking protocol, all the methods are similar to the previously discussed article by Nallusamy et al. (2021). Details of food phytochemicals were collected from Duke's database and their structures were obtained from the PubChem database. The protein structure was prepared by adding hydrogen atoms by Autodock to

**Table 14.2** Information on the number of plants in each category that was used for phytochemical curation is provided

| S. No | Name | Number of plants |
|---|---|---|
| 1 | Vegetables | 15 |
| 2 | Fruits | 11 |
| 3 | Supplement | 3 |
| 4 | Spices | 7 |
| 5 | Pulses | 4 |
| 6 | Millets | 3 |
| 7 | Herbs | 20 |
| | Total | 63 |

make macromolecule modulation and energy minimization was carried out using SWISS PDB Viewer. The ligand was prepared to the lowest energy conformation using the uff force field available in PyRx software. PyRx (0.8) AutoDock module was used for the generation of pdbqt files of ligands and proteins. Docking was performed using Autodock wizard with the parameters which include Genetic Algorithm runs (100), number of individuals in the population (150), Maximum number of generations (27000), and rate of crossover (0.8). A threshold of −5.0 kcal/ mol was set to streamline the process of identification of antiviral compounds. In the present study, compounds are considered active when their binding affinity value is equal to or greater than −5.0 kcal/mol. All the analyses were performed based on this threshold value.

Fig. 14.3 provides the details of the numbers of phytochemicals curated under each category and the number of phytochemicals that score binding affinity greater than or equal to −5.0 kcal/mol against the spike glycoprotein and MPro. Tables 14.3 and 14.4 provide the details of the docking score and the compounds that are predicted as lead-like inhibitor molecules against SARS-CoV-2 MPro and spike protein. The best three compounds from each category are discussed elaborately in the present chapter for both the target proteins.

#### 14.4.1.1 Vegetables

Oleanolic acid (*Cyprus rotundus*), Gibberellin-a-3 (garlic), and Tocopherols (drum stick) were identified with best binding affinity scores of −11.4 kcal/mol, −9.4 kcal/ mol, and − 7.8 kcal/mol, respectively, for the target spike glycoprotein. In the case of MPro, oleanolic acid, gibberellin-a-3, and rotundifoline (mint leaves) were reported as inhibitors with the highest binding affinity of (−12.7 kcal/mol), − 11.9 kcal/mol, and − 9.4 kcal/mol, respectively.

Oleanolic Acid: *Cyprus rotundus*

The antiviral activity of essential oils from *Cyprus rotundus* has a proven role against hepatitis B virus (HBV) and also has hepatoprotective characteristics (Wu and

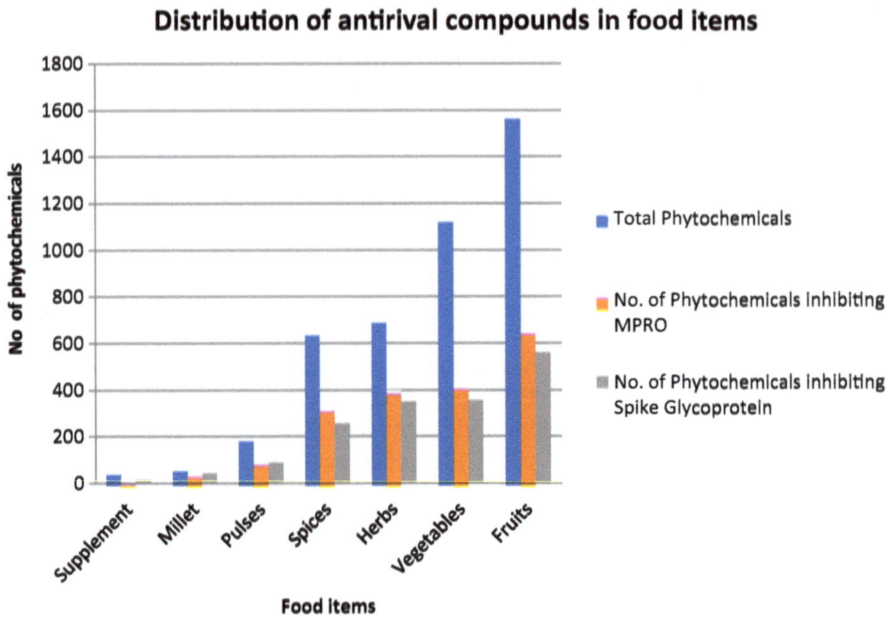


**Fig. 14.3** Details of phytochemicals studied under each food category and their inhibition potential statistics based on docking score are given

McGoogan 2020; Khwaza et al. 2018). Animal and test-tube studies have displayed the antiviral nature of oleanolic acid against herpes viruses, HIV influenza, and hepatitis, which is a component of *C. rotundus*.

Giberellin-a-3- Garlic

Garlic (*Allium sativum* L.) is a commonly consumed food and has been used as medicine against a number of virus diseases such as HIV, HSV 2, pneumonia, and rotavirus, etc. (Rouf et al. 2020; Bayan et al. 2014).

Tocopherols- Leaf- Drum Stick

*Moringa olefiera* is one of the most important traditional plants with medicinal value where most of its parts are edible using which many Indian dishes are prepared. Apart from its antiviral property, it is used to treat many medicinal properties like anemia, headache, malaria, diabetes, tuberculosis, etc. (Biswas et al. 2020; Siddiqui et al. 2022). Moreover, tocopherol has been reported for its drug-likeness property against SARS-CoV-2 spike glycoprotein using docking and simulation approaches by Siddiqui et al. (2022).

**Table 14.3** Screening of top three phytochemicals for each food category against SARS-CoV-2 spike protein

| Phytocompound | Pubchem id | Plant parts | Plant | Binding affinity Kcal/mol |
|---|---|---|---|---|
| *Vegetables* | | | | |
| Oleanolic acid | 10,494 | Tuber, bulb | *Cyprus rotundus,* | −11.4 |
| Gibberellin-a-3 | 52,921,570 | Bulb | *Allium sativum* | −9.4 |
| Tocopherols | 14,986 | Leaf | *Moringa oleifera* | −7.8 |
| *Spices* | | | | |
| Crataegolic acid | 73,659 | Flower | *Syzygiumaromaticum* | −12 |
| Zingiberol | 5,317,270 | Rhizome | *Zingiber officinale* | −8 |
| Gamma-eudesmol | 6,432,005 | Rhizome | *Zingiber officinale* | −8 |
| Sesquiterpenes | 53,790,872 | Leaf | *Cinnamomum verum* | −7.9 |
| *Millet* | | | | |
| Violanthin | 448,438 | Finger millet | Finger millet | −7.6 |
| Narigenin | 442,428 | Finger millet | Finger millet | −7.5 |
| Orientin | 5,281,675 | Finger millet | Finger millet | −7 |
| Flavonoids | 53,787,266 | Pearl millet | Pearl millet | −7 |
| *Pulses* | | | | |
| Delta-tocopherol | 92,094 | Seed | *Arachis hypogaea* | −7.8 |
| Vitamin E | 14,985 | Seed | *Arachis hypogaea* | −7.7 |
| Cajanone | 325,518 | Plant | *Cajanus cajan* | −7.5 |
| *Herbs* | | | | |
| Maslinic acid | 73,659 | Leaf | *Eucalyptus globulus* | −12 |
| Triterpenoid | 451,674 | Leaves | *Abutilon indicum* | −11.6 |
| Oleanolic acid | 10,494 | Leaf | *Eucalyptus globulus* | −11.4 |
| Oleanolic acid | 10,494 | Leaf | *Ocimum basilicum* | −11.4 |
| Oleanolic acid | 10,494 | Leaf | *Ocimum tenuiflorum* | −11.4 |
| *Fruits* | | | | |
| Corilagin | 73,568 | Fruit | *Phyllanthus emblica* | −12 |
| Citrusin-ii | 102,082,877 | Fruit | *Citrus sinensis* | −11.8 |
| Oleanolic acid | 10,494 | Pericarp | *Vitis vinifera* | −11.4 |
| *Supplements* | | | | |
| Coronene | 9115 | Ghee | Ghee | −7.4 |
| Riboflavin | 493,570 | Plant | Spirulina | −7.1 |
| Epoxygedunin | 6,708,724 | Ghee | Ghee | −7.1 |

Rotundifoline: Mint Leaves

Rotundifoline is a novel compound with antiviral activity obtained from the present analysis of food phytochemicals.

**Table 14.4** Screening of top three phytochemicals for each food category against SARS-CoV-2 MPro

| Phytocompound | Pubchem id | Plant part | Plant | Binding affinity kcal/mol |
|---|---|---|---|---|
| *Vegetable* | | | | |
| Oleanolic acid | 10,494 | Tuber, bulb | *Cyprus rotundus* | −12.7 |
| Gibberellin-a-3 | 52,921,570 | Bulb | *Allium sativum* | −11.9 |
| Rotundifoline | 139,055,911 | Leaves | *Mentha piperita L* Mint leaves | −9.4 |
| *Spices* | | | | |
| Crataegolic acid | 73,659 | Flower | *Syzygiumaromaticum* | −12.3 |
| Sesquiterpenes | 53,790,872 | Leaf | *Cinnamomum verum* | −8.7 |
| Sesamin | 72,307 | Flower, root | *Piper longum* | −8.3 |
| *Millet* | | | | |
| Vitexin | 5,280,441 | Panicle | Finger millet | −8.2 |
| Isoorientin | 114,776 | Panicle | Finger millet | −7.8 |
| Isovitexin | 162,350 | Panicle | Finger millet | −7.4 |
| *Pulses* | | | | |
| Alpha-amyrin | 73,170 | Root | *Cajanus cajan* | −8.3 |
| Vitamin e | 14,985 | Seed | *Arachis hypogaea* | −7.8 |
| 5,7,2′-trihydroxyisoflavone | 44,257,268 | Root | *Cajanus cajan* | −7.6 |
| *Herbs* | | | | |
| Nimbinene | 44,715,635 | Leaf | *Azadirachta indica* | −13.1 |
| Nimbandiol | 157,277 | Leaf | *Azadirachta indica* | −12.9 |
| Triterpenoid | 451,674 | Leaves | *Abutilon indicum* | −12.8 |
| *Fruits* | | | | |
| Citrusin-ii | 102,082,877 | Fruit | *Citrus sinensis* | −15.1 |
| Corilagin | 73,568 | Fruit | *Phyllanthus emblica* | −14.2 |
| Limonexic-acid | 5,321,283 | Fruit | *Citrus sinensis* | −13.3 |
| *Supplement* | | | | |
| Coronene | 9115 | Ghee | Ghee | −7.4 |
| Epoxygedunin | 6,708,724 | Ghee | Ghee | −7.1 |
| Riboflavin | 493,570 | Plant | Spirulina | −7.1 |
| Spathulenol | 92,231 | Honey | Honey | −5.8 |

#### 14.4.1.2 Spices

Among the spices category, crataegolic acid (*Syzygiumaromaticum)* scored good binding affinity of −12.0 kcal/mol and − 12.3 kcal/mol (Fig. 14.4b) for spike glycoprotein and MPro, respectively. Other compounds such as zingiberol (*Zingiber officinale)*, gamma-eudesmol (*Zingiber officinale)*, and Sesquiterpenes (Cinnamon) were reported for the spike glycoprotein. Sesquiterpenes (−8.7 kcal/mol) was also reported as an inhibitor for the MPro target, scoring high among the top three

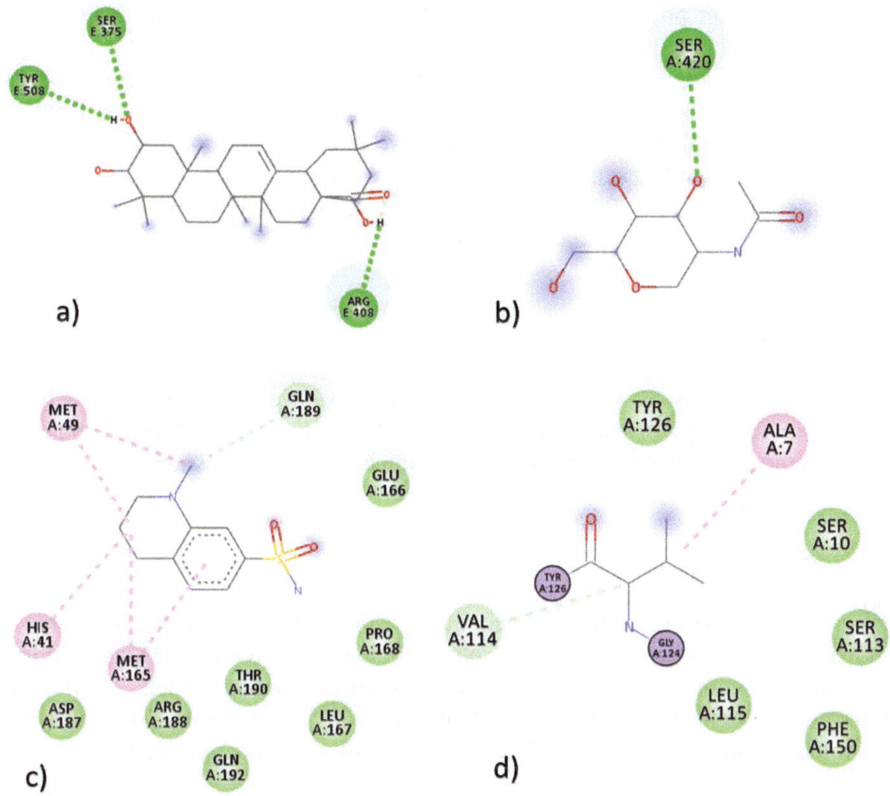


**Fig. 14.4** Docked complex structures of spike protein with (**a**) Corilagin, (**b**) Crataegolic acid (**c**) Citrusin-ii (**d**) Corilagin

inhibitors. Sesamin from *Piper Longum* was the third-highest inhibitor among the screened compounds.

### Crataegolic Acid-*Syzgiumaromaticum*

*Syzygium aromaticum* L (Clove) is a well known dietary spice that has been used for several years in traditional medicine for many disorders such as respiratory ailments, antiviral, and anti-inflammatory properties (Vicidomini et al. 2021).

### *Zingiber officinale*

*Zingiber officinale* Roscoe (Ginger) is an important spice in many food items and remains the constituent of phenolic compounds, terpenes, polysaccharides, lipids, organic acids, and raw fibers (Mao et al. 2019; San Chang et al. 2013). Ginger also has the important property of smoothing the airway muscle for easier breathing (Townsend et al. 2013).

Zingiberol (Baliga et al. 2011) remains the main constituent of ginger rhizome aroma and is identified as a potent compound against spike glycoprotein. Furthermore, another compound from ginger, namely, gamma-eudesmol was observed to have the highest binding affinity against spike glycoprotein which was also reported by Tran et al. (My et al. 2020) reported as one of the active inhibitors for SARS-CoV-2 MPro through computational studies.

Sesquiterpenes-Cinnamon

Cinnamon which is an essential oil was found to have strong antiviral activity against a wide range of pathogenic viruses (Allahverdiyev et al. 2004). Cinnamon contains the phytochemical sesquiterpenes which are major class terpenes.

*Piper longum*

*Piper longum* L. (Long pepper) is recognized as one of the important constituents of indigenous medicines. It has varied pharmacological activities such as antiviral, anticancer, immunomodulatory, cardioprotective, antimicrobial, antifungal, bioavailability-enhancing and antioxidant (Choudhary and Singh 2018). Sesamin from long pepper has reported −8.3 kcal/mol binding affinity against MPro.

#### 14.4.1.3 Millets

A total of 63 phytochemicals from three millets, namely proso millet, finger millet, and pearl millet were taken for screening the anti-SARS-CoV-2 inhibitors. Phytochemicals from finger millet (violanthin, narigenin, orientin, vitexin, isoorientin, and isovitexin) and pearl millet (Flavonoid) have the highest score against spike glycoprotein and MPro targets. Major phytochemicals from finger millet have been reported as top binding inhibitors of covid targets. Minor cereals such as finger millets and pearl millets are highly enriched with phytochemicals, vitamins, calcium, dietary fiber (Hassan et al. 2021), and polyphenol compounds. Additionally, it possesses antidiabetic, antitumorigenic, atherosclerogenic effects, antioxidant, and antimicrobial properties (Devi et al. 2014).

Violanthin was reported as a SARS-CoV-2 MPro inhibitor (Vincent et al. 2020) that may have a role in reducing shortness of breath, sore throat, etc. Naringenin is an important flavonoid known for its biological activities such as antioxidant, anti-inflammatory, and antiviral properties (Tsiani and Den Hartogh 2019). Orientin and isoorientin were reported for several activities including antioxidant, antiaging, antiviral, antibacterial, anti-inflammation, vasodilatation and cardioprotective, antiadipogenesis, antinociceptive, radiation protective, neuroprotective, and antidepressant-like effects (Lam et al. 2016). Vitexin and isovitexin were used in conventional Chinese medicines, where they have many properties such as antioxidant, anticancer, anti-inflammatory, antihyperalgesic, and neuroprotective effects (He et al. 2016).

#### 14.4.1.4 Pulses

Four pulses such as ground nut, sprouted green gram, black gram, and pigeon pea containing 191 phytochemicals were virtually screened for their anti-covid properties. Peanut (Delta-tocopherol, Vitamin E) and pigeon pea (Cajanone, alpha-amyrin, 5,7,2′-trihydroxyisoflavone) were the two pulses that reported the highest binding affinity phytochemicals with the chosen SARS-CoV 2 targets, respectively.

*Oil seed - Arachis hypogaea* (Peanut)

*Arachis hypogaea L.* is an important commercial crop worldwide that has been used for the production of oil and it serves as an important source of protein. Its pharmacological attributes lie in heart disease, body weight management, anticancer, antiviral, and antidiabetes properties (Harrod et al. 2022). Delta-tocopherol is well known for its medicinal value such as anticancer, lysosomal storage disorders, and anti-inflammatory properties. Its derivative D-α-tocopherol polyethylene glycol succinate (TPGS) (Makau et al. 2018) was identified as a potential antiviral compound against SARS-CoV-2 and β-coronaviruses that showed strong synergy with remdesivir through in silico screening and in vitro viral inhibition assay studies. A fat-soluble antioxidant, vitamin E stops the production of reactive oxygen species when fat undergoes oxidation. Furthermore, the role of vitamin E in the prevention of chronic diseases is being investigated (Huang et al. 2007).

*Cajanus cajan* (Pigeon Pea)

Pigeon pea is used in Chinese and Brazilian traditional medicine known for their nutritional and medicinal properties such as antioxidant, antidiabetic, antimicrobial, DNA damage protective, and xanthine oxidase inhibitory properties and to treat cough (Mathew et al. 2017). This study has reported the antiviral properties of three compounds mentioned above which are constituents of the pigeon pea against SARS-CoV-2 targets. Cajanone is one of the well-studied antimycobacterial compounds (Coronado-Aceves et al. 2017) and is predicted to have antiviral properties in the present study. Alpha-amyrin () possesses antioxidant, anti-inflammatory, antiarthritis, and antiviral properties. 5,7,2′-trihydroxyisoflavone has shown its relevance to an anti-covid property in the cited literature (Sainz-Cort and Heeroma 2020).

#### 14.4.1.5 Herbs

Maslinic acid/Crataegolic acid (−12.0 kcal/mol), Triterpeniod (−11.6 kcal/mol), and Oleanolic acid (−11.4 kcal/mol) are the top three phytochemicals reported for the target spike glycoprotein. *Eucalyptus globulus* (Martin 2005), *Abutilon indicum* (Abat et al. 2017), *Ocimum basilicum* (Shahrajabian et al. 2020), and *Ocimum tenuiflorum* (Prakash and Gupta 2005) are medicinal plants that are well known for their pharmacological activities have reported the above-mentioned phytochemicals variedly.

For the MPro target, nimbinene (13.1 kcal/mol), nimbandiol (−12.9 kcal/mol), and triterpenoid (−12.8 kcal/mol) were predicted as inhibitors. Nimbinene and nimbandiol are phytoconstitutents of *Azadirachta indica* (Alzohairy 2016), which is a plant of traditional medicinal value used generation over generations.

#### 14.4.1.6 Fruits

Orange (Gonçalves et al. 2017) has a proven record of antiviral activity, and this study has found two compounds, namely citrusin-ii and limonexic acid (only MPro) were predicted as inhibitors of SARS-CoV-2 targets. Citrusin -ii (Fig. 14.4c) was predicted to have the highest binding score of −15.1 kcal/mol among all the screened plant metabolites of various food categories against the target MPRO. Limonexic acid (−13.3 kcal/mol for spike protein), which is a limonoid (Tundis et al. 2014), has shown anticancer, antibacterial, antifungal, antimalarial, and antiviral activities.

Oleanolic acid from grapes was listed in the top three predicted inhibitors of spike glycoprotein. Grapes juice was reported to have antiviral properties (Cliver and Kostenbader Jr 1979) in a number of studies. Among the 417 phytochemicals investigated in grapes, 174 and 193 phytochemicals have shown binding affinity equal to or more than −5.0 kcal/mol for spike and MPro targets. Among all the other fruits investigated in the present study, grapes had the maximum phytochemicals and maximum predicted inhibitors within the cutoff value of −5.0 kcal/mol.

Corilagin, which is a phytoconstituent of gooseberry (an Indian traditional medicinal plant) and an anti-SARS-CoV-2 inhibitor (Yang et al. 2021) exhibited a binding score of −12.0 kcal/mol and −14.2 kcal/mol for MPro and spike glycoprotein (Fig. 14.4a, d), respectively.

#### 14.4.1.7 Supplements

In supplementary food items phytochemical investigation, epoxygedunin, riboflavin and spathulenol from ghee, spirulina, and honey were predicted as inhibitors against MPro, spike glycoprotein. A smaller set of 49 phytochemicals were studied under this category. Regarding the properties of the reported phytochemicals, spathulenol (do Nascimento et al. 2018) has antioxidant, anti-inflammatory, antiproliferative, and antimycobacterial properties in earlier studies. Riboflavin, which is a B vitamin, is known for its antioxidant, antiaging, anti-inflammatory, antinociceptive, and anticancer properties (Suwannasom et al. 2020). Epoxygedunin (James and Thomas 2019) was observed to have anticancer properties from earlier reported studies.

Plants that possess a higher number of antiviral molecules have been analyzed in which only the top ten plants are discussed. Grape (193 and 174), Pepper (83 and 74), Orange (72 and 72), Ginger (71 and 63), Carrot (67 and 71), Onion (62 and 56), Eucalyptus (61 and 63), Fenugreek (56 and 59), and *Piper longum* (53 and 49) are the plants reported to have a higher number of phytochemicals that inhibits both MPro and spike glycoprotein protein targets. Banana and papaya were observed to have 49 phytochemicals each with a binding affinity score of more than −5.0 kcal/mol against spike glycoprotein alone, whereas wheatgrass reported 58 phytochemicals showing an antiviral property specific to MPro. Fruits, vegetables, spices, and herbs are the categories of the above-mentioned plants which possess a higher number of anti-covid phytochemicals compared to pulses, supplements, and millet. Information on the total number of phytochemicals and antiviral activity based on the docking score are given in the Supplementary Fig. 14.S1.

Network illustrates (Fig. 14.5a) the compounds that may inhibit spike glycoprotein and MPro. Interestingly, eleven compounds such as coronene, riboflavin, sesquiterpenes, giberellin-a3, corilagin, crategolic acid, oleanolic acid, triterpenol, epoxygedunin, citrusin-ii, and vitamin E have shown highest binding affinity against both spike glycoprotein and Mpro targets.

To understand the distribution of phytochemicals with antiviral properties among the chosen 63 plants, the percentage of antiviral molecules was computed for each plant based on the ratio of a total number of compounds and a total number of compounds with a docking score equal to or greater than −5.0 kcal/mol. A total number of compounds that have reported binding affinity equal to or greater than −5.0 kcal/mol are considered antiviral molecules. Figure 14.5b shows the presence of the highest number of phytochemicals in the millets and the least number in the supplement food items such as ghee, honey, and spirulina. The second and third highest were observed with herbs and pulses, respectively. This statistic provides knowledge on the phytochemicals present in the particular food and their antiviral efficiency on comparison within seven food categories taken for this analysis.

## 14.5 Conclusion

Virtual screening of thousands of food nutraceuticals and phytochemicals resulted in the prediction of potent inhibitors from daily consumed food items. Although the present work is based on predictions, they provide us the clue to explore the phytochemicals that can be further taken for drug testing and clinical trials. Naturally occurring phytochemicals have a good impact on human health with fewer hazards/side effects in comparison to synthetic chemicals. Plants contain vital

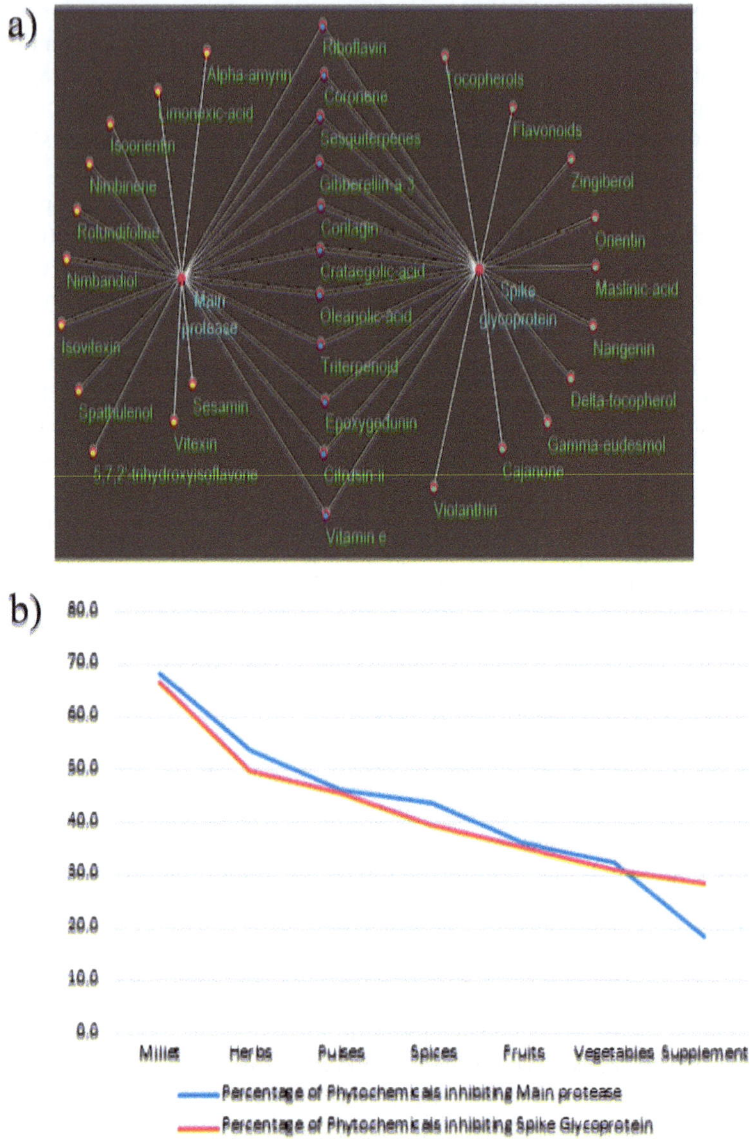


**Fig. 14.5** (**a**) Network diagram of predominant inhibitors having the highest docking score unique and common to main protease and spike glycoprotein, respectively (**b**) Percentage value forecasting the impact of different food categories and their antiviral property

phytochemicals that emphasize healthy food habits and lifestyles to combat severe viral diseases such as COVID. It is essential to consume all food items in different categories in different combinations that may have a synergistic effect to inhibit the function of viral targets. Some of the food items are neglected totally from our dining due to the changing lifestyle. In this line, millets, which are high nutrient enriched small grains have drawn the attention of the globe has invoked the announcement of "International millet year in 2023" to create awareness among the public. Food remains the hub of most of the antiviral molecules and it enlightens the way of healthy living by combating several diseases and symptoms.

## References

Abat JK, Kumar S, Mohanty A (2017) Ethnomedicinal, phytochemical and ethnopharmacological aspects of four medicinal plants of Malvaceae used in Indian traditional medicines: a review. Medicines 4(4):75

Allahverdiyev A, Duran N, Ozguven M, Koltas S (2004) Antiviral activity of the volatile oils of Melissa officinalis L. against herpes simplex virus type-2. Phytomedicine 11(7–8):657–661

Alzohairy MA (2016) Therapeutics role of Azadirachta indica (neem) and their active constituents in diseases prevention and treatment. Evid Based Complement Alternat Med 2016:7382506

Baliga MS, Haniadka R, Pereira MM, D'Souza JJ, Pallaty PL, Bhat HP, Popuri S (2011) Update on the chemopreventive effects of ginger and its phytochemicals. Crit Rev Food Sci Nutr 51(6): 499–523

Bayan L, Koulivand PH, Gorji A (2014) Garlic: a review of potential therapeutic effects. Avicen J Phytomed 4(1):1

Behl T, Rocchetti G, Chadha S, Zengin G, Bungau S, Kumar A, Mehta V, Uddin MS, Khullar G, Setia D, Arora S, Sinan KI, Ak G, Putnik P, Gallo M, Montesano D (2021) Phytochemicals from plant foods as potential source of antiviral agents: an overview. Pharmaceuticals 14(4):381

Berman HM, Westbrook J, Feng Z, Gilliland G, Bhat TN, Weissig H, Shindyalov IN, Bourne PE (2000) The protein data bank. Nucleic Acids Res 28(1):235–242

Biswas D, Nandy S, Mukherjee A, Pandey DK, Dey A (2020) Moringa oleifera Lam. And derived phytochemicals as promising antiviral agents: a review. S Afr J Bot 129:272–282

Choudhary N, Singh V (2018) A census of P. longum's phytochemicals and their network pharmacological evaluation for identifying novel drug-like molecules against various diseases, with a special focus on neurological disorders. PLoS One 13(1):e0191006

Cliver DO, Kostenbader KD Jr (1979) Antiviral effectiveness of grape juice. J Food Prot 42(2): 100–104

Coronado-Aceves EW, Gigliarelli G, Garibay-Escobar A, Zepeda RER, Curini M, Cervantes JL, Espitia-Pinzón CII, Superchi S, Vergura S, Marcotullio MC (2017, 206) New Isoflavonoids from the extract of Rhynchosiaprecatoria (Humb. &Bonpl. ex Willd.) DC. and their antimycobacterial activity. J Ethnopharmacol:92–100

Dallakyan S, Olson AJ (2015) Small-molecule library screening by docking with PyRx. In: Chemical biology. Humana Press, New York, NY, pp 243–250

Devi PB, Vijayabharathi R, Sathyabama S, Malleshi NG, Priyadarisini VB (2014) Health benefits of finger millet (Eleusine coracana L.) polyphenols and dietary fiber: a review. J Food Sci Technol 51(6):1021–1040

do Nascimento KF, Moreira FMF, Santos JA, Kassuya CAL, Croda JHR, Cardoso CAL, do Carmo Vieira M, Ruiz ALTG, Foglio MA, de Carvalho JE, Formagio ASN (2018) Antioxidant, anti-inflammatory, antiproliferative and antimycobacterial activities of the essential oil of Psidium guineense Sw. and spathulenol. J Ethnopharmacol 210:351–358

Gonçalves D, Lima C, Ferreira P, Costa P, Costa A, Figueiredo W, Cesar T (2017) Orange juice as dietary source of antioxidants for patients with hepatitis C under antiviral therapy. Food Nutr Res 61(1):1296675

Harrod KS, Pacl HT, Tipper JL, Ahmad S, Ahmad A, Holder G, Petit C, Green T, Steyn AJC, Might M (2022) Water soluble tocopherol derivatives inhibit the SARS-CoV-2 RNA-dependent RNA

polymerase. In: B107. Molecular mechanisms of SARS-CoV-2, influenza, and other lung infections. American Thoracic Society, pp A3610–A3610
Hassan ZM, Sebola NA, Mabelebele M (2021) The nutritional use of millet grain for food and feed: a review. Agric Food Sec 10(1):1–14
He M, Min JW, Kong WL, He XH, Li JX, Peng BW (2016) A review on the pharmacological effects of vitexin and isovitexin. Fitoterapia 115:74–85
Huang HY, Caballero B, Chang S, Alberg AJ, Semba RD, Schneyer C, Bass EB (2007) Multivitamin/mineral supplements and prevention of chronic disease: executive summary. Am J Clin Nutr 85(1):265S–268S
James JIMCY, Thomas JIBU (2019) Anticancer activity of microalgae extract on human cancer cell line (MG-63). Asian J Pharm Clin Res 12(1):139–142
John Hopkins University (2020) Coronavirus resource center
Kaihatsu K, Yamabe M, Ebara Y (2018) Antiviral mechanism of action of epigallocatechin-3-O-gallate and its fatty acid esters. Molecules 23(10):2475
Khwaza V, Oyedeji OO, Aderibigbe BA (2018) Antiviral activities of oleanolic acid and its analogues. Molecules 23(9):2300
Kim S, Chen J, Cheng T, Gindulyte A, He J, He S, Li Q, Shoemaker BA, Thiessen PA, Yu B, Zaslavsky L, Bolton EE (2019) PubChem 2019 update: improved access to chemical data. Nucleic Acids Res 47(D1):D1102–D1109
Ko WC, Rolain JM, Lee NY, Chen PL, Huang CT, Lee PI, Hsueh PR (2020) Arguments in favour of remdesivir for treating SARS-CoV-2 infections. Int J Antimicrob Agents 55(4):105933
Lam KY, Ling APK, Koh RY, Wong YP, Say YH (2016) A review on medicinal properties of orientin. Adv Pharmacol Sci 2016:4104595
Makau JN, Watanabe K, Mohammed MM, Nishida N (2018) Antiviral activity of peanut (Arachis hypogaea L.) skin extract against human influenza viruses. J Med Food 21(8):777–784
Mao QQ, Xu XY, Cao SY, Gan RY, Corke H, Beta T, Li HB (2019) Bioactive compounds and bioactivities of ginger (Zingiber officinale roscoe). Foods 8(6):185
Martin YC (2005) A bioavailability score. J Med Chem 48(9):3164–3170
Mathew D, Manila TM, Divyasree P, Sandhya Rajan VTK (2017) Therapeutic molecules for multiple human diseases identified from pigeon pea (*Cajanus cajan* L. Millsp.) through GC–MS and molecular docking. Food Sci Hum Well 6(4):202–216
Maurya DK, Sharma D (2022) Evaluation of traditional ayurvedic Kadha for prevention and management of the novel Coronavirus (SARS-CoV-2) using in silico approach. J Biomol Struct Dyn 40(9):3949–3964
My TTA, Loan HTP, Hai NTT, Hieu LT, Hoa TT, Thuy BTP, Quang DT, Triet NT, Anh TTV, Dieu NTX, Nhung NTA (2020) Evaluation of the inhibitory activities of COVID-19 of Melaleuca cajuputi oil using docking simulation. Chem Select 5(21):6312–6320
Nallusamy S, Mannu J, Ravikumar C, Angamuthu K, Nathan B, Nachimuthu K, Ramasamy G, Muthurajan R, Subbarayalu M, Neelakandan K (2021) Exploring phytochemicals of traditional medicinal plants exhibiting inhibitory activity against main protease, Spike glycoprotein, RNA-dependent RNA polymerase and non-structural proteins of SARS-CoV-2 through virtual screening. Front Pharmacol 12:1704
Park NH, Lee CW, Bae JH, Na YJ (2011) Protective effects of amentoflavone on Lamin A-dependent UVB-induced nuclear aberration in normal human fibroblasts. Bioorg Med Chem Lett 21(21):6482–6484
Prakash PAGN, Gupta N (2005) Therapeutic uses of Ocimum sanctum Linn (Tulsi) with a note on eugenol and its pharmacological actions: a short review. Indian J Physiol Pharmacol 49(2):125
Qazi S, Das S, Khuntia BK, Sharma V, Sharma S, Sharma G, Raza K (2021) In silico molecular docking and molecular dynamic simulation analysis of phytochemicals from Indian foods as potential inhibitors of SARS-CoV-2 RdRp and 3CLpro. Nat Prod Commun 16(9): 1934578X211031707

Rouf R, Uddin SJ, Sarker DK, Islam MT, Ali ES, Shilpi JA, Nahar L, Tiralongo E, Sarker SD (2020) Antiviral potential of garlic (Allium sativum) and its organosulfur compounds: a systematic update of pre-clinical and clinical data. Trends Food Sci Technol 104:219–234

Sainz-Cort A, Heeroma JH (2020) The interaction between the endocannabinoid system and the renin angiotensin system and its potential implication for COVID-19 infection. J Cannab Res 2(1):1–4

San Chang J, Wang KC, Yeh CF, Shieh DE, Chiang LC (2013) Fresh ginger (Zingiber officinale) has anti-viral activity against human respiratory syncytial virus in human respiratory tract cell lines. J Ethnopharmacol 145(1):146–151

Shahrajabian MH, Sun W, Cheng Q (2020) Chemical components and pharmacological benefits of Basil (Ocimumbasilicum): a review. Int J Food Prop 23(1):1961–1970

Siddiqui S, Upadhyay S, Ahmad R, Barkat M, Jamal A, Alothaim AS, Hassan M, Rahman MA, Arshad M, Ahamad T, Khan MF, Ahmad J (2022) Interaction of bioactive compounds of Moringa oleifera leaves with SARS-CoV-2 proteins to combat COVID-19 pathogenesis: a phytochemical and in silico analysis. Appl Biochem Biotechnol 194(12):5918–5944

Suwannasom N, Kao I, Pruß A, Georgieva R, Bäumler H (2020) Riboflavin: the health benefits of a forgotten natural vitamin. Int J Mol Sci 21(3):950

Tian W, Chen C, Lei X, Zhao J, Liang J (2018) CASTp 3.0: computed atlas of surface topography of proteins. Nucleic Acids Res 46(W1):W363–W367

Townsend EA, Siviski ME, Zhang Y, Xu C, Hoonjan B, Emala CW (2013) Effects of ginger and its constituents on airway smooth muscle relaxation and calcium regulation. Am J Respir Cell Mol Biol 48(2):157–163

Tsiani E, Den Hartogh DJ (2019) Antidiabetic properties of naringenin: a citrus fruit polyphenol. Biomol Ther 9(3):1–21

Tundis R, Loizzo MR, Menichini F (2014) An overview on chemical aspects and potential health benefits of limonoids and their derivatives. Crit Rev Food Sci Nutr 54(2):225–250

Varzakas T, Zakynthinos G, Verpoort F (2016) Plant food residues as a source of nutraceuticals and functional foods. Foods 5(4):88

Veeresham C (2012) Natural products derived from plants as a source of drugs. J Adv Pharm Technol Res 3(4):200

Vicidomini C, Roviello V, Roviello GN (2021) Molecular basis of the therapeutical potential of clove (Syzygiumaromaticum L.) and clues to its anti-COVID-19 utility. Molecules 26(7):1880

Vincent S, Arokiyaraj S, Saravanan M, Dhanraj M (2020) Molecular docking studies on the anti-viral effects of compounds from kabasura kudineer on SARS-CoV-2 3CLpro. Front Mol Biosci 7:613401

Wu Z, McGoogan JM (2020) Characteristics of and important lessons from the coronavirus disease 2019 (COVID-19) outbreak in China: summary of a report of 72 314 cases from the Chinese Center for Disease Control and Prevention. JAMA 323(13):1239–1242

Yang LJ, Chen RH, Hamdoun S, Coghi P, Ng JP, Zhang DW, Guo X, Xia C, Law BYK, Wong VKW (2021) Corilagin prevents SARS-CoV-2 infection by targeting RBD-ACE2 binding. Phytomedicine 87:153591

Yu S, Yan H, Zhang L, Shan M, Chen P, Ding A, Li SFY (2017) A review on the phytochemistry, pharmacology, and pharmacokinetics of amentoflavone, a naturally-occurring biflavonoid. Molecules 22(2):299

# Chapter 15
# Therapeutic and Prophylactic Effects of Plant Derivatives Against SARS CoV-2

**Vinotha Viswanathan and Vaseeharan Baskaralingam**

## 15.1 Introduction

The single-stranded enveloped RNA virus family is known as coronavirus which are belonging to the Coronaviridae family. Coronavirus has caused respiratory tract, gastrointestinal tract, and neurological diseases (Wu et al. 2020; Bhavaniramya et al. 2022b). As stated by WHO (World Health Organization), COVID-19 was caused by SARS-CoV-2 (severe acute respiratory syndrome coronavirus-2) which cause severe health risk worldwide. The family of coronaviridae has Letovirinae and Orthocoronavirinae as subfamilies. Among, Orthocoronavirinae consist of α, β, γ, and δ coronaviruses. Amidst, α and β viruses infect mammals while the γ and δ variants affect the avians and are also opportunistic to mammals. Herein, SARS-CoV-1 and 2 belong to the β variants (Cui et al. 2019; Amanat and Krammer 2020; Harrison et al. 2020). SARS-CoV-2 (Fig. 15.1) mainly causes acute respiratory distress (ARDS), pneumonia, renal failure, and finally death. Fever and cough are the common symptoms of SARS-CoV-2 (Fig. 15.2a). However, asymptomatic conditions have also been recorded recently (Wang et al. 2020; Chan et al. 2020). The mode of transmission of SARS-CoV-2 is highly through respiratory droplets, person-to-person or any surfaces contact, and oral transmission from fecal (Fig. 15.2b) (Otter et al. 2016; Harrison et al. 2020).

During the pandemic condition, SARS-CoV-2 was recognized as a major public health emergency by the unit of WHO as it is been a key threat to the human population. Moreover, it influences the global economy. The SARS-CoV-2 infection is been a threatened disease globally due to its easy and rapid transmission between human populations through droplets (Srivastava and Saxena 2020; Ghosh et al.

V. Viswanathan · V. Baskaralingam (✉)
Biomaterials and Biotechnology in Animal Health Lab, Department of Animal Health and Management, Alagappa University, Karaikudi, Tamil Nadu, India

 
J.-T. Chen (ed.), *Ethnopharmacology and Drug Discovery for COVID-19: Anti-SARS-CoV-2 Agents from Herbal Medicines and Natural Products*,
https://doi.org/10.1007/978-981-99-3664-9_15

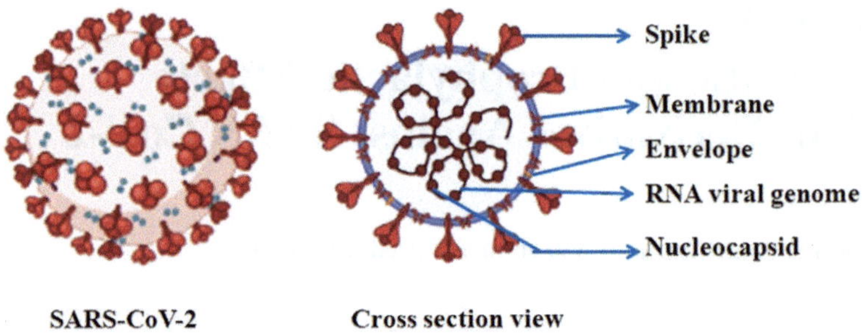


**Fig. 15.1** Structure of Severe Acute Respiratory Syndrome Corona Virus-2 (SARS-CoV-2)

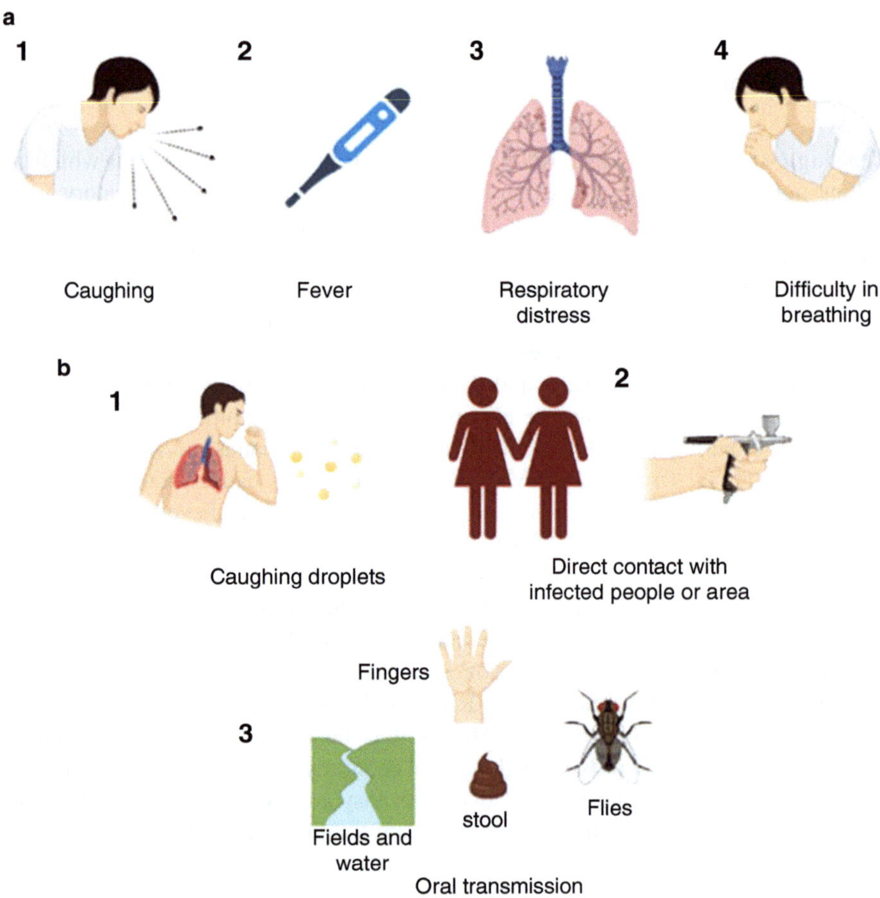


**Fig. 15.2** (**a**) Symptoms and (**b**) Transmission of SARS-CoV-2 infection

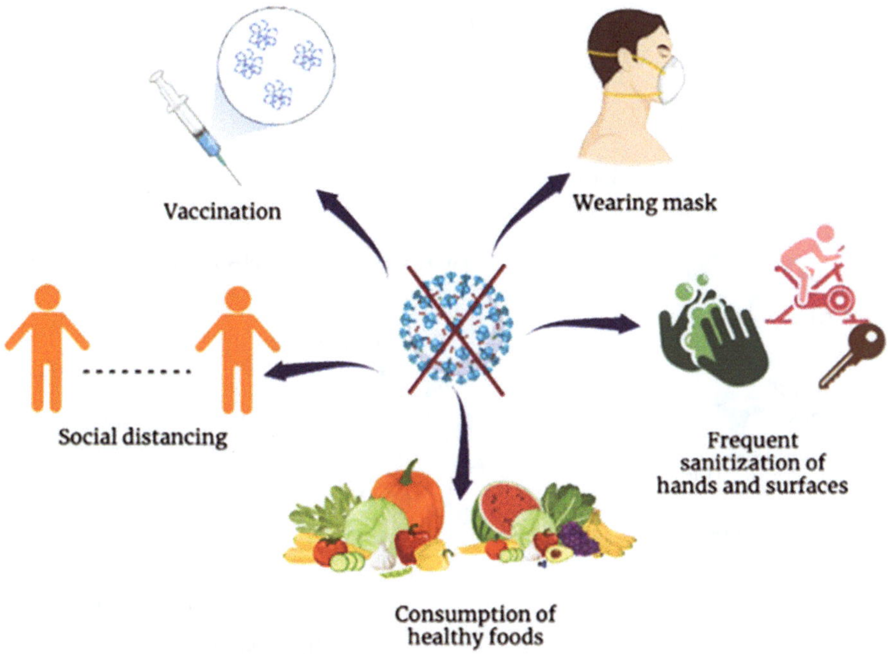


**Fig. 15.3** Prevention measures of SARS-CoV-2 infection

2021). Predominantly, vaccination against COVID-19 viruses at frequent intervals helps to prevent or reduce the risk of transmission and susceptibility in the host. Moreover, the practice of frequent hand washing, using mouth covering, keeping at least 6 feet distance between people, isolation of infected people, and consumption of nutrient-rich food could enhance immunity, and avoidance of dispensable gathering might reduce the risk of COVID-19 (Fig. 15.3) (Srivastava and Saxena 2020; Prather et al. 2020; Zolezzi et al. 2021; Centers for Disease Control and Prevention 2021). Still, no approved medications are available to fight against COVID-19. Hence, the only way to defeat SARS-CoV-2 is to find a drug to fight against them at their specific receptors involved in the replication process. Nowadays, many researchers are focused on the development of a product that could target the specific receptors [i.e., angiotensin-converting enzyme II (ACE2), spike protein, RNA-dependent RNA polymerase (RdRp), nonstructural protein (Fig. 15.4)] (Elfiky 2020; Abdelli et al. 2021; Babadaei et al. 2021; Velusamy et al. 2021; Bhavaniramya et al. 2022a). Recently, Senthil Kumar et al. (2021) reported that antcins isolated from *Antrodia* species significantly restrain the ACE2 receptor.

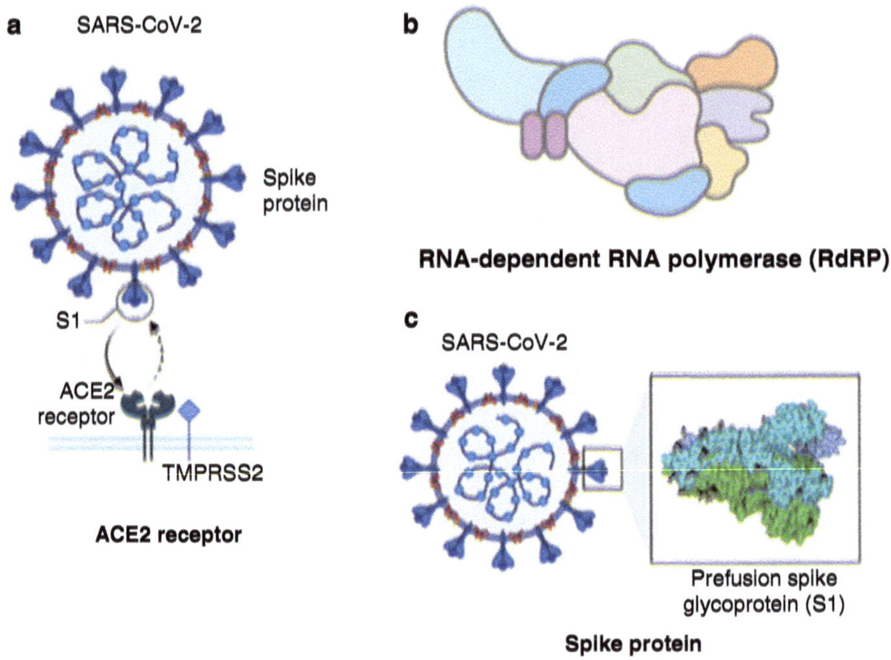


**Fig. 15.4** SARS-CoV-2 recognition receptors (**a**) Angiotensin-converting enzyme II (ACE2), (**b**) RNA-dependent RNA polymerase (RdRp), and (**c**) Spike protein

## 15.2 Therapeutic Plants

Therapeutic plants are the ultimate agents who always participated in the day-to-day life of human beings. In the form of food, medicine, prophylactic agent, and traditional cultures the plants are taken place in the daily routine. Moreover, therapeutic plants are rich in secondary metabolites including flavonoids, phenolics, polyphenols, tannins, saponins, and alkaloids possess antibacterial, antifungal, antioxidant, anti-analgesic, anti-inflammatory, and especially antiviral activity. Typically, various parts of the plants including leaves, roots, stem or barks, flowers, and seeds were extremely utilized in the antiviral activity from ancient times (Fig. 15.5). For instance, *Scutellaria edelbergii* extract was reported for its antibacterial, anti-inflammatory, and anti-analgesic activity in Swiss albino mice (Shah et al. 2021); anti-diabetic and anti-cancer activity of *Cuminum cyminum* was also reported earlier (Mnif and Aifa 2015). Similarly, *Aloe vera* species was found to be effective in antibacterial, antiviral, and anti-cancer activities. In addition, *Aloe vera* could regulate the immune system and are a hepatoprotective agent (Gao et al. 2019). On the other side, a combination of essential oil from three different medicinal plants exhibits prophylaxis against respiratory syncytial virus (RSV) (Tseliou et al. 2019). Some therapeutic plants and their derivatives reported for antiviral properties are listed in Table 15.1.

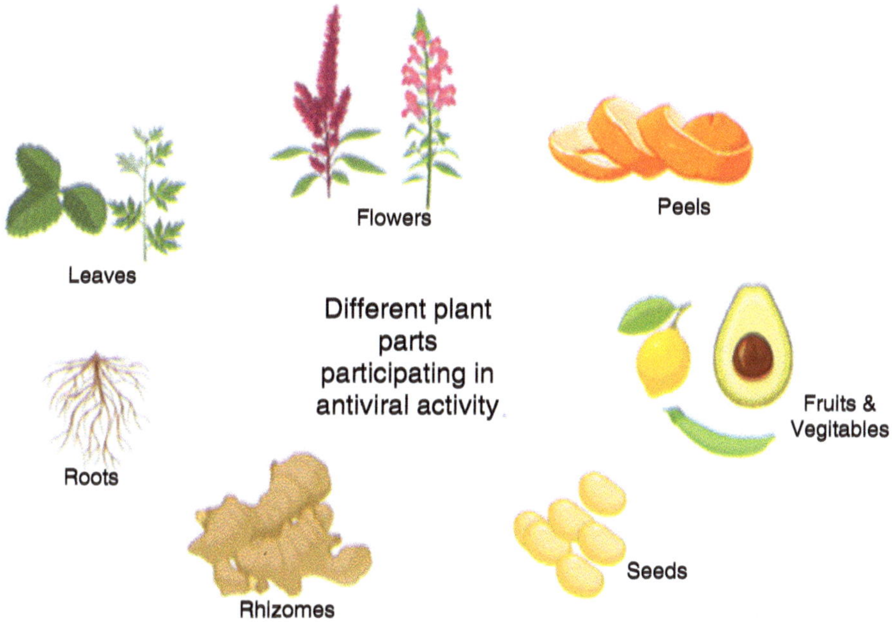


**Fig. 15.5** Various parts of therapeutic plants involved in antiviral activity

## 15.3 The Therapeutic Effect of Plant Derivatives Against SARS-CoV-2

As stated above, therapeutic plants and their derivatives have the potential to control viral infections from ancient times. Therapeutic plant derivatives also expressed prophylactic effects against COVID-19 (Divya et al. 2020). Recently, Shree et al. (2022) reported the phyto-compounds Withanoside V and Somniferine derived from Ashwagantha, Tinocordiside derived from Giloy, and Vicenin, Isorientin 4'-O-glucoside 2"-O-p-hydroxybenzoagte and Ursolic acid from Tulsi expressed high affinity toward the virus particle. Similarly, the botanical derivatives from Amla, Bhumi Amla, and Giloy are proficiently engaging the protease binding point in COVID-19 (Murugesan et al. 2021). Identical in silico studies on the relationship between plant derivatives and the viral particle have been listed in Table 15.2. Also, the chemical structure of some therapeutic plant derivatives that can fight against SARS-CoV-2 has been illustrated in Fig. 15.6.

**Table 15.1** Therapeutic plants and their derivatives reported for antiviral properties

| Sl. No | Name of the plant | Part used | Nature of the sample | Plant derivatives if any | Viral infection | Reference |
|---|---|---|---|---|---|---|
| 1 | *Nepeta nepetella* | | Aqueous extract | – | HSV1 | Abad et al. (2000) |
| 2 | *Dittrichia viscosa* | – | Aqueous extract | – | VSV, poliovirus type 1 | Abad et al. (2000) |
| 3 | *Desmodium canadense*, *Mentha piperita*, and *Thymus vulgaris* | – | Aqueous extract | – | Avian infectious bronchitis virus | Lelešius et al. (2019) |
| 4 | *Morus alba var. alba*, *Morus alba var. rosea, and Morus rubra* | Leaves and bark | Aqueous and methanol extract | – | Human coronavirus (HCoV 229E), and single-stranded RNA viral strains | Thabti et al. (2020) |
| 5 | *Veronica persica Poir.* | – | Methanol extract | – | HSVs 1 and 2 | Sharifi-Rad et al. (2018) |
| 6 | *Antidesma bunius*, *Caseria graveolens*, *Cleistanthus patulus*, *Colebrookea oppositifolia*, *Hypericum gaitii*, *Millettia extensa*, *Protium serratum*, *Rubus ellipticus*, and *Ventilago maderaspatana* | – | Aqueous extract | – | Enterovirus-71 | Panda et al. (2017) |
| 7 | *Euphorbia coopire* and *Morus alba* | Flowers and leaves | Chloroform and aqueous methanol extract | 7-galloyl catechin, Gallic acid, Kaempferol 3-O-β-(6″-*O*-galloyl)-glucopyranoside, quercetin 3-*O*-*β*-(6″-*O*-galloyl)-glucopyranoside, and Kaempferol | HSV-1 | El-Toumy et al. (2018) |
| 8 | *Baccharis gaudichaudiana* | Aerial parts | Organic extract | – | HSV-1 | Visintini Jaime et al. (2013) |

| | | | | | | |
|---|---|---|---|---|---|---|
| 9 | *Solanum nigrum* | Seeds | Methanol and chloroform extract | – | Hepatitis C virus | Javed et al. (2011) |
| 10 | *Acacia nilotica* | – | Acetone and methanol extract | – | Hepatitis C virus | Rehman et al. (2011) |
| 11 | *Zingiber officinale* | Rhizomes | Aqueous extract | – | Chikungunya virus (CHIKV) | Kaushik et al. (2020) |
| 12 | *Andrographis paniculata* and *Momordica charantia* | – | Methanol extract | – | Dengue virus serotype 1 (DENV-1) | Tang et al. (2012) |
| 13 | *Trachyspermum ammi* | Seeds | Essential oil | – | Japanese encephalitis virus (JEV) | Roy et al. (2015) |
| 14 | – | – | – | Baicalein | Dengue virus serotype 2 (DENV-2) | Zandi et al. (2012) |
| 15 | *Zingiber officinale* | Rhizome (fresh) | Hot water extract | – | Human respiratory syncytial virus (HRSV) | San Chang et al. (2013) |
| 16 | *Phyllanthus emblica* | Branches and leaves | – | Pentagalloylglucose (PGG) | Influenza A virus | Liu et al. (2011) |
| 17 | *Agrimonia pilosa* | Whole plant | Ethanol extract | – | Human influenza viruses (H1N1 and H3N2 influenza A&B subtypes) | Shin et al. (2010) |
| 18 | *Phyllanthus* spp. | – | Aqueous and methanol extract | – | DENV-2 | Lee et al. (2013) |
| 19 | *Scutia buxifolia* | Leaves and stem bark | Ethanol extract | – | HSV-1 | Boligon et al. (2013) |

(continued)

**Table 15.1** (continued)

| Sl. No | Name of the plant | Part used | Nature of the sample | Plant derivatives if any | Viral infection | Reference |
|---|---|---|---|---|---|---|
| 20 | *Lagerstroemia speciosa* | Leaves | – | Ellagic acid | Human rhinoviruses (HRVs) | Park et al. (2014) |
| 21 | *Azardirachta indica* | Bark | Aqueous extract | – | HSV-1 | Tiwari et al. (2010) |
| 22 | *Bletilla striata* | Rhizome | Ethanol extract | Diphenanthrenes | Influenza A virus | Shi et al. (2017) |
| 23 | *Nepeta nuda* | Aerial parts | Chloroform extract | – | HSV-1 and 2 | Todorov et al. (2015) |
| 24 | *Byrsonima verbascifolia*, *Myracrodruon urundeuva*, *Eugenia dysenterica* and *Hymenaea courbaril* | Leaves | Ethanol extract | – | Rotavirus SA11 | Cecílio et al. (2012) |
| 25 | *Radix isatidis* | Root and leaves | Aqueous extract | – | Scbrood virus (SBV) | Sun et al. (2021) |
| 26 | *Acorus calamus*, *Cymbopogon citrates*, and *Myristica fatua* | – | Methanol extract | – | DENV-2 | Rosmalena et al. (2019) |
| 27 | *Embelia ribes* | Fruits | Ethyl acetate | – | Influenza virus A | Hossan et al. (2018) |
| 28 | *Syzygium jambos* | Leaves | Ethanol | – | HSV-1 | Twilley et al. (2017) |
| 29 | *Artocarpus heterophyllus* | Leaves | Dichloromethane | – | Hepatitis C virus (HCV) | Hafid et al. (2017) |

| | | | | | | |
|---|---|---|---|---|---|---|
| 30 | *Portulaca oleracea* | Aerial part | Aqueous extract | – | Influenza A viruses | Li et al. (2019) |
| 31 | *Cissampelos pariera* | – | Alcoholic extract | – | DENV – 1 to 4 | Sood et al. (2015) |
| 32 | *Ocimum sanctum* and *Acacia arabica* | Leaves | Crude extract | Terpenoid (Ocimum) and polyphenol (acacia) | Avian influenza (AI) H9N2 | Ghoke et al. (2018) |
| 33 | *Heracleum lehmannianum*, *H. ponticum*, *H. aconitifolium* | Root | Essential oil | – | Influenza virus A and B | Tkachenko (2007) |
| 34 | *Radix isatidis* | – | – | Clemastanin B, Epigoitrin, Phenylpropanoids portion and Phenylpropanoids, alkaloids, and organic acid combination | Influenza virus | Ping et al. (2016) |
| 35 | *Coridothymus capitatus*, *Origanum dictamnus*, and *Salvia fruticosa* | – | Combination of essential oils | – | Influenza A/H1N1 virus, influenza B and human rhinovirus 14 (HRV14) | Tseliou et al. (2019) |
| 36 | *Lavandula angustifolia* and *Salvia officinalis* | – | Essential oil | – | Avian influenza H5N1 virus | Abou Baker et al. (2021) |

**Table 15.2** In silico studies on the relationship between plant derivatives and SARS-CoV-2

| Sl. No | Name of the plant | Plant derivatives | SARS-CoV-2 target site/Mode of interaction | Reference |
|---|---|---|---|---|
| 1 | *Curcuma longa* | Cyclocurcumin | SARS-CoV-2 (Mpro)/formation of two hydrogen bonds with amino acid residues of COVID-19 Mpro | Rajagopal et al. (2020) |
| 2 | *Withania somnifera* | Withanolide and Withanone | SARS-CoV-2 (Mpro)/quercetin-3-rutinoside-7-glucoside | Kushwaha et al. (2021) |
| 3 | *Curcuma longa* | Cuminoid and Tetrahydroxycurcumin | SARS-CoV-2 (Mpro) /hydrogen bonding | Ghosh et al. (2021) |
| 4 | *Sesamum indicum* | Sesami, Sesaminol, and Sesamolin | SARS-CoV-2 (Mpro)/formation of two hydrogen bonds with amino acid residues | Natesh et al. (2021a) |
| 5 | *Ferula asafoetida* | Farnesiferol B | SARS-CoV-2 (Mpro)/amino acid residues and indicated H-bond | Natesh et al. (2021a) |
| 6 | *Camellia sinensis* | Allocatechin-3-gallate | SARS-CoV-2 (Mpro)/amino acid residues and indicated H-bond | Ghosh et al. (2021) |
| 7 | *Heracleum candicans* | Candibirin H, Candibirin G | SARS-CoV-2 3CLpro and PLpro/hydrogen bonding | Natesh et al. (2021b) |
| 8 | *Acacia nilotica* | Catechin 5-O-gallate, Kaempferol 3-glucoside 7-rhamnoside, Kaempferol7,4′-diglucoside | SARS-CoV-2 3CLpro and PLpro, SARS-CoV-2 spike protein/hydrogen bonding | Natesh et al. (2021b) |
| 9 | *Hypericum perforatum* L. | Hypericin, Pseudohypericin, 6,6′-biapigenin | SARS-CoV-2 3CLpro, SARS-CoV-2 spike protein and PLpro/hydrogen bonding | Natesh et al. (2021b) |
| 10 | *Tribulus terrestris* L. | Tribuloside, Tribulosin | SARS-CoV-2 spike protein/hydrogen bonding | Natesh et al. (2021b) |
| 11 | – | Deacetylnomilin, Ichangin, nomilin, and β-amyrin (Terpinoids) | SARS-CoV-2 spike and Mpro/hydrogen bonding | Giofrè et al. (2021) |

(continued)

**Table 15.2** (continued)

| Sl. No | Name of the plant | Plant derivatives | SARS-CoV-2 target site/Mode of interaction | Reference |
|---|---|---|---|---|
| 12 | *Oxalis pes-caprae* | Caeruleanone A | SARS-CoV-2 (Mpro)/hydrophobic interaction | Gul et al. (2022) |
| 13 | *Ocimum menthiifolium* | Apigenin-7-O-rutinoside, Prunin, and Acaciin | nsp16/10 complex of SARS-CoV-2 | Albohy et al. (2022) |
| 14 | *Rhus succedanea* | Amentoflavone and Agathisflavone (as biflavonoid) | SARS-CoV-2 (Mpro)/interact with catalytic residues | Lokhande et al. (2020) |
| 15 | *Camellia sinensis* | Oolonghomobisflavan-A (polyphenol) | SARS-CoV-2 (Mpro) | Bhardwaj et al. (2021) |
| 16 | *Camellia sinensis* | Epigallocatechin gallate, Gallocatechin-3-gallate, and Epicatechin gallate | SARS-CoV-2 (Mpro)/interact with catalytic residues | Ghosh et al. (2021) |
| 17 | *Nigella sativa* | Campesterol, Cycloeucalenol, alpha-spinasterol and Beta-sitosterol | N-terminal RNA binding domain and Papain-like protease of SARS-CoV-2 | Siddiqui et al. (2022) |
| 18 | *Silybum marianum*, *Withania somnifera*, *Tinospora cordifolia*, and *Aloe barbadensis* | Silybin, Withaferin A, Cordioside, and Catechin and quercetin | SARS-CoV-2 | Pandit and Latha (2020) |
| 19 | *Boesenbergia rotunda* | Panduratin A | SARS-CoV-2 | Kanjanasirirat et al. (2020) |
| 20 | *Echinacea purpurea* | Echinaforce | SARS-CoV-2 | Signer et al. (2020) |

## 15.4 Prophylactic Effect of Plant Derivatives

It is said, “prevention is better than cure” which means prevention is a safe way to stay away from any problem, especially infections or diseases, and also this practice shields us from the severity of the problem. From ancient times, the therapeutic plants and the phyto-compounds derived from the therapeutic plants are well-known prophylactic agents as well as curative agents. Prophylaxis of SARS-CoV-2 has predominantly depending on the improvement of immunity. Some plants are a rich source of immune boosters. Specifically, the consumption of *Phyllanthus emblica* (Amla) enhances the immune system and protects from SARS-CoV-2 infection because Amla is rich in vitamin C (Kapoor et al. 2020). Magzoub (2020) suggested that the consumption of ginger (*Zingiber officinale*) juice (12 mg/250 mL) can diminish the vulnerability to SARS-CoV-2 by increasing the amount of IgM.

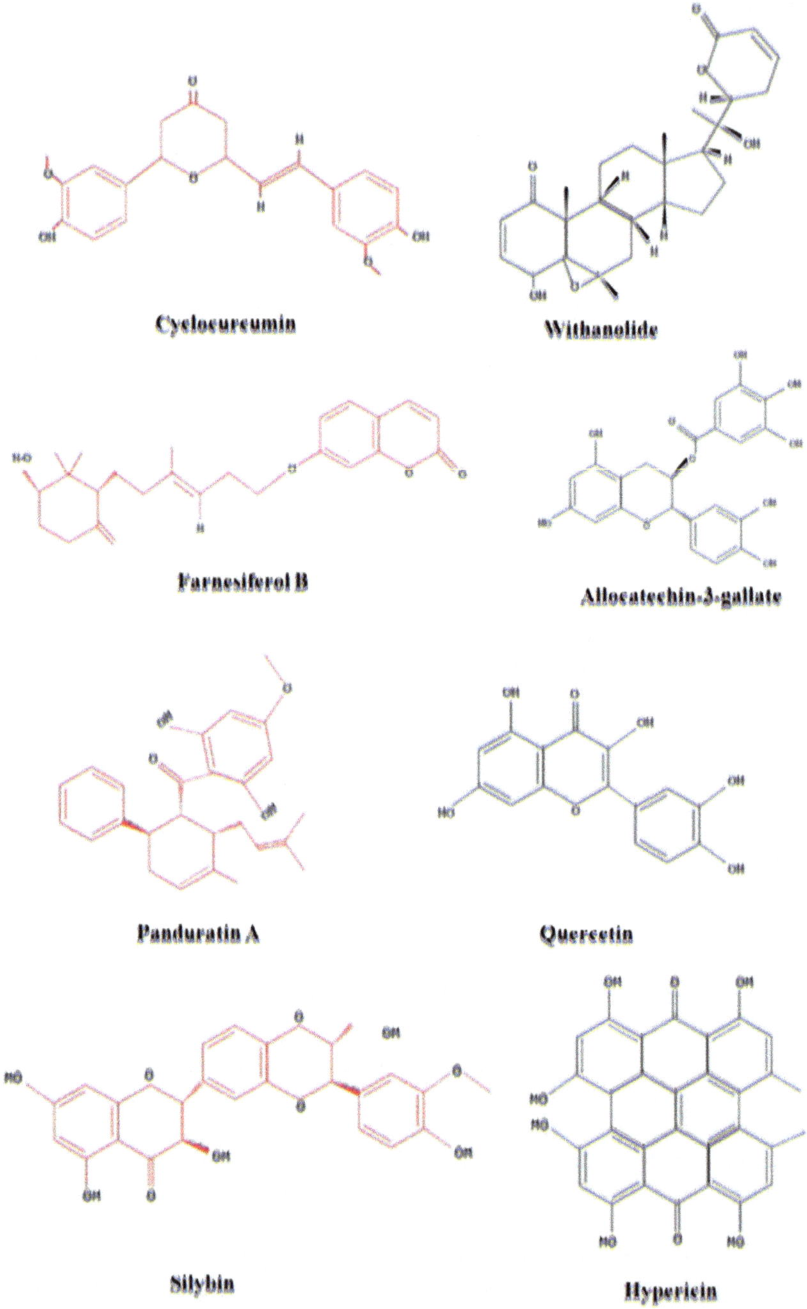


**Fig. 15.6** Therapeutic plant derivatives involved in antiviral activity against SARS-CoV-2

Predominantly, the AEC2 receptor facilitates the entry of the virus inside the host; hence it is recognized as a significant receptor. Therefore, the inhibition of the AEC2 receptor is greatly preferable for prophylactic activity (Senthil Kumar et al. 2021). Previously, some in silico studies revealed that the polyphenols isolated from pomegranate peels have high affinity toward the Mpro region of SARS-CoV-2 suggesting the binding probability with ACE2, also the affinity was more stable which offers prevention of infection (Rakshit et al. 2021). Alike, pre- and post-administration of Oleandrin a phytochemical isolated from *Nerium oleander* restrain the replication of viral particles and reduce the risk of infection (Plante et al. 2020). Limited studies are available on the experiment of the prophylactic effect of plant derivatives. Hence, there is an open forum and also an immense need to experiment with the prophylactic effect of plant derivatives to develop effective alternatives to existing vaccines.

## 15.5 Conclusion and Prospective

From the ancient period, plants and their derivatives played a vital role in the control of many pathogenic microorganisms, especially viruses. Moreover, the therapeutic plants and their derivatives were effectively involved in the prophylactic activity. Numerous studies revealed that extracts, essential oils, and secondary metabolites from therapeutic plants could efficiently control the viral replication process. Particularly, some plant derivatives, including phenols, saponins, flavonoids, etc., could interrupt specific receptor recognition. Many in silico studies suggest the utilization of therapeutic plant derivatives to control the infection. Though, the insufficient reports on the experimental pieces of evidence on the relationship between the therapeutic plant derivatives and the SARS-CoV-2 virus have been the major limitation. Due to this reason, the development of therapeutic plant-mediated drugs/vaccine development was getting delayed. Therefore, more in vitro and clinical trials on the prophylactic and antiviral efficiency of therapeutic plant derivatives against SARS-CoV-2 infection were needed and it might be facilitated the development of target-specific and highly safe alternatives.

**Acknowledgments** First author [V. Vinotha - IF170397] sincerely acknowledges the financial support from the Department of Science and Technology, New Delhi, India under INSPIRE fellowship scheme.

## References

Abad MJ, Guerra JA, Bermejo P, Irurzun A, Carrasco L (2000 Dec) Search for antiviral activity in higher plant extracts. Phytother Res 14(8):604–607

Abdelli I, Hassani F, Bekkel Brikci S, Ghalem S (2021) *In silico* study the inhibition of angiotensin converting enzyme 2 receptor of COVID-19 by *Ammoides verticillata* components harvested from Western Algeria. J Biomol Struct Dyn 39(9):3263–3276

Abou Baker DH, Amarowicz R, Kandeil A, Ali MA, Ibrahim EA (2021) Antiviral activity of *Lavandula angustifolia* L. and *Salvia officinalis* L. essential oils against avian influenza H5N1 virus. J Agric Food Res 4:100135

Albohy A, Zahran EM, Abdelmohsen UR, Salem MA, Al-Warhi T, Al-Sanea MM, Abelyan N, Khalil HE, Desoukey SY, Fouad MA, Kamel MS (2022) Multitarget *in silico* studies of *Ocimum menthiifolium*, family Lamiaceae against SARS-CoV-2 supported by molecular dynamics simulation. J Biomol Struct Dyn 40(9):4062–4072

Amanat F, Krammer F (2020) SARS-CoV-2 vaccines: status report. Immunity 52(4):583–589

Babadaei MM, Hasan A, Vahdani Y, Bloukh SH, Sharifi M, Kachooei E, Haghighat S, Falahati M (2021) Development of remdesivir repositioning as a nucleotide analog against COVID-19 RNA dependent RNA polymerase. J Biomol Struct Dyn 39(10):3771–3779

Bhardwaj VK, Singh R, Sharma J, Rajendran V, Purohit R, Kumar S (2021) Identification of bioactive molecules from tea plant as SARS-CoV-2 main protease inhibitors. J Biomol Struct Dyn 39(10):3449–3458

Bhavaniramya S, Ramar V, Vishnupriya S, Palaniappan R, Sibiya A, Vaseeharan B (2022a) Comprehensive analysis of SARS-COV-2 drug targets and pharmacological aspects in treating the COVID-19. Curr Mol Pharmacol 15(2):393–417

Bhavaniramya S, Sibiya A, Alothaim AS, Al Othaim A, Ramar V, Veluchamy A, Manikandan P, Vaseeharan B (2022b) Evaluating the structural and immune mechanism of Interleukin-6 for the investigation of goat Milk peptides as potential treatments for COVID-19. J King Saud Univ Sci 34(4):101924

Boligon AA, Kubiça TF, Mario DN, de Brum TF, Piana M, Weiblen R, Lovato L, Alves SH, Santos RC, dos Santos Alves CF, Athayde ML (2013) Antimicrobial and antiviral activity-guided fractionation from *Scutia buxifolia* Reissek extracts. Acta Physiol Plant 35(7):2229–2239

Cecílio AB, de Faria DB, de Carvalho Oliveira P, Caldas S, de Oliveira DA, Sobral ME, Duarte MG, de Souza Moreira CP, Silva CG, de Almeida VL (2012) Screening of Brazilian medicinal plants for antiviral activity against rotavirus. J Ethnopharmacol 141(3):975–981

Centers for Disease Control and Prevention (2021) How to protect yourself and others. Natl Cent Immun Respir Dis(NCIRD), Div Viral Dis. Available at https://www.cdc.gov/coronavirus/2019-ncov/vaccines/fully-vaccinated.html

Chan JF, Yuan S, Kok KH, To KK, Chu H, Yang J, Xing F, Liu J, Yip CC, Poon RW, Tsoi HW (2020) A familial cluster of pneumonia associated with the 2019 novel coronavirus indicating person-to-person transmission: a study of a family cluster. Lancet 395(10223):514–523

Cui J, Li F, Shi ZL (2019) Origin and evolution of pathogenic coronaviruses. Nat Rev Microbiol 17(3):181–192

Divya M, Vijayakumar S, Chen J, Vaseeharan B, Durán-Lara EF (2020) South Indian medicinal plants can combat deadly viruses along with COVID-19?-a review. Microb Pathog 148:104277

Elfiky AA (2020) Anti-HCV, nucleotide inhibitors, repurposing against COVID-19. Life Sci 248: 117477

El-Toumy SA, Salib JY, El-Kashak WA, Marty C, Bedoux G, Bourgougnon N (2018) Antiviral effect of polyphenol rich plant extracts on herpes simplex virus type 1. Food Sci Human Wellness 7(1):91–101

Gao Y, Kuok KI, Jin Y, Wang R (2019) Biomedical applications of *Aloe vera*. Crit Rev Food Sci Nutr 59(Suppl 1):S244–S256

Ghoke SS, Sood R, Kumar N, Pateriya AK, Bhatia S, Mishra A, Dixit R, Singh VK, Desai DN, Kulkarni DD, Dimri U (2018) Evaluation of antiviral activity of *Ocimum sanctum* and *Acacia arabica* leaves extracts against H9N2 virus using embryonated chicken egg model. BMC Complement Altern Med 18(1):1–10

Ghosh R, Chakraborty A, Biswas A, Chowdhuri S (2021) Evaluation of green tea polyphenols as novel corona virus (SARS CoV-2) main protease (Mpro) inhibitors–an *in silico* docking and molecular dynamics simulation study. J Biomol Struct Dyn 39(12):4362–4374

Giofrè SV, Napoli E, Iraci N, Speciale A, Cimino F, Muscarà C, Molonia MS, Ruberto G, Saija A (2021) Interaction of selected terpenoids with two SARS-CoV-2 key therapeutic targets: an *in silico* study through molecular docking and dynamics simulations. Comput Biol Med 134: 104538

Gul F, Khan I, Iqbal J, Abbasi BA, Shahbaz A, Capasso R, Amaro-Estrada I, Jardan YA, Cossio-Bayugar R, Mahmood T (2022) Phytochemistry, biological activities and *in silico* molecular docking studies of *Oxalis pes-caprae* L. compounds against SARS-CoV-2. J King Saud Uni Sci 34(6):102136

Hafid AF, Aoki-Utsubo C, Permanasari AA, Adianti M, Tumewu L, Widyawaruyanti A, Wahyuningsih SP, Wahyuni TS, Lusida MI, Hotta H (2017) Antiviral activity of the dichloromethane extracts from *Artocarpus heterophyllus* leaves against hepatitis C virus. Asian Pac J Trop Biomed 7(7):633–639

Harrison AG, Lin T, Wang P (2020) Mechanisms of SARS-CoV-2 transmission and pathogenesis. Trends Immunol 41(12):1100–1115

Hossan MS, Fatima A, Rahmatullah M, Khoo TJ, Nissapatorn V, Galochkina AV, Slita AV, Shtro AA, Nikolaeva Y, Zarubaev VV, Wiart C (2018) Antiviral activity of *Embelia ribes* Burm. f. against influenza virus *in vitro*. Arch Virol 163(8):2121–2131

Javed T, Ashfaq UA, Riaz S, Rehman S, Riazuddin S (2011) *In-vitro* antiviral activity of *Solanum nigrum* against hepatitis C virus. Virol J 8(1):1–7

Kanjanasirirat P, Suksatu A, Manopwisedjaroen S, Munyoo B, Tuchinda P, Jearawuttanakul K, Seemakhan S, Charoensutthivarakul S, Wongtrakoongate P, Rangkasenee N, Pitiporn S (2020) High-content screening of Thai medicinal plants reveals *Boesenbergia rotunda* extract and its component Panduratin A as anti-SARS-CoV-2 agents. Sci Rep 10(1):1–2

Kapoor MP, Suzuki K, Derek T, Ozeki M, Okubo T (2020) Clinical evaluation of *Emblica officinalis* Gatertn (Amla) in healthy human subjects: health benefits and safety results from a randomized, double-blind, crossover placebo-controlled study. Contemp Clin Trials Commun 17:100499

Kaushik S, Jangra G, Kundu V, Yadav JP, Kaushik S (2020) Anti-viral activity of *Zingiber officinale* (Ginger) ingredients against the Chikungunya virus. Virus 31(3):270–276

Kushwaha PP, Singh AK, Prajapati KS, Shuaib M, Gupta S, Kumar S (2021) Phytochemicals present in Indian ginseng possess potential to inhibit SARS-CoV-2 virulence: a molecular docking and MD simulation study. Microb Pathog 157:104954

Lee SH, Tang YQ, Rathkrishnan A, Wang SM, Ong KC, Manikam R, Payne BJ, Jaganath IB, Sekaran SD (2013) Effects of cocktail of four local Malaysian medicinal plants (*Phyllanthus* spp.) against dengue virus 2. BMC Complement Altern Med 13(1):1–3

Lelešius R, Karpovaitė A, Mickienė R, Drevinskas T, Tiso N, Ragažinskienė O, Kubilienė L, Maruška A, Šalomskas A (2019) *In vitro* antiviral activity of fifteen plant extracts against avian infectious bronchitis virus. BMC Vet Res 15(1):1

Li YH, Lai CY, Su MC, Cheng JC, Chang YS (2019) Antiviral activity of *Portulaca oleracea* L. against influenza a viruses. J Ethnopharmacol 241:112013

Liu G, Xiong S, Xiang YF, Guo CW, Ge F, Yang CR, Zhang YJ, Wang YF, Kitazato K (2011) Antiviral activity and possible mechanisms of action of pentagalloylglucose (PGG) against influenza A virus. Arch Virol 156(8):1359–1369

Lokhande K, Nawani N, Venkateswara SK, Pawar S (2020) Biflavonoids from *Rhus succedanea* as probable natural inhibitors against SARS-CoV-2: a molecular docking and molecular dynamics approach. J Biomol Struct Dyn:1–3

Magzoub M (2020) Life style guideline of ginger (Zingiber officinale) as prophylaxis and treatment for coronaviruses (SARS-CoV-2) infection (COVID-19). Saudi J Biomed Res 5(6):125–127

Mnif S, Aifa S (2015) Cumin (*Cuminum cyminum* L.) from traditional uses to potential biomedical applications. Chem Biodivers 12(5):733–742

Murugesan S, Kottekad S, Crasta I, Sreevathsan S, Usharani D, Perumal MK, Mudliar SN (2021) Targeting COVID-19 (SARS-CoV-2) main protease through active phytocompounds of ayurvedic medicinal plants–*Emblica officinalis* (Amla), *Phyllanthus niruri* Linn.(Bhumi Amla) and *Tinospora cordifolia* (Giloy)–a molecular docking and simulation study. Comput Biol Med 136: 104683

Natesh J, Mondal P, Penta D, Salam AA, Meeran SM (2021a) Culinary spice bioactives as potential therapeutics against SARS-CoV-2: computational investigation. Comput Biol Med 128:104102

Natesh J, Mondal P, Kaur B, Salam AA, Kasilingam S, Meeran SM (2021b) Promising phytochemicals of traditional Himalayan medicinal plants against putative replication and transmission targets of SARS-CoV-2 by computational investigation. Comput Biol Med 133:104383

Otter JA, Donskey C, Yezli S, Douthwaite S, Goldenberg S, Weber DJ (2016) Transmission of SARS and MERS coronaviruses and influenza virus in healthcare settings: the possible role of dry surface contamination. J Hosp Infect 92(3):235–250

Panda SK, Padhi L, Leyssen P, Liu M, Neyts J, Luyten W (2017) Antimicrobial, anthelmintic, and antiviral activity of plants traditionally used for treating infectious disease in the Similipal Biosphere Reserve, Odisha, India. Front Pharmacol 8:658

Pandit M, Latha N (2020). In silico studies reveal potential antiviral activity of phytochemicals from medicinal plants for the treatment of COVID-19 infection

Park SW, Kwon MJ, Yoo JY, Choi HJ, Ahn YJ (2014) Antiviral activity and possible mode of action of ellagic acid identified in *Lagerstroemia speciosa* leaves toward human rhinoviruses. BMC Complement Altern Med 14(1):1–8

Ping X, Weiyang Y, Jianwei C, Xiang L (2016) Antiviral activities against influenza virus (FM1) of bioactive fractions and representative compounds extracted from Banlangen (*Radix Isatidis*). J Tradit Chin Med 36(3):369–376

Plante KS, Plante JA, Fernandez D, Mirchandani D, Bopp N, Aguilar PV, Sastry KJ, Newman RA, Weaver SC (2020) Prophylactic and therapeutic inhibition of *in vitro* SARS-CoV-2 replication by oleandrin. bioRxiv

Prather KA, Wang CC, Schooley RT (2020) Reducing transmission of SARS-CoV-2. Science 368(6498):1422–1424

Rajagopal K, Varakumar P, Baliwada A, Byran G (2020) Activity of phytochemical constituents of *Curcuma longa* (turmeric) and *Andrographis paniculata* against coronavirus (COVID-19): an *in silico* approach. Fut J Pharm Sci 6(1):1–10

Rakshit M, Muduli S, Srivastav PP, Mishra S (2021) Pomegranate peel polyphenols prophylaxis against SARS-CoV-2 main protease by *in-silico* docking and molecular dynamics study. J Biomol Struct Dyn:1–5

Rehman S, Ashfaq UA, Riaz S, Javed T, Riazuddin S (2011) Antiviral activity of *Acacia nilotica* against hepatitis C virus in liver infected cells. Virol J 8(1):1–6

Rosmalena R, Elya B, Dewi BE, Fithriyah F, Desti H, Angelina M, Hanafi M, Lotulung PD, Prasasty VD, Seto D (2019) The antiviral effect of indonesian medicinal plant extracts against dengue virus *in vitro* and *in silico*. Pathogens 8(2):85

Roy S, Chaurvedi P, Chowdhary A (2015) Evaluation of antiviral activity of essential oil of *Trachyspermum Ammi* against Japanese encephalitis virus. Pharm Res 7(3):263

San Chang J, Wang KC, Yeh CF, Shieh DE, Chiang LC (2013) Fresh ginger (*Zingiber officinale*) has anti-viral activity against human respiratory syncytial virus in human respiratory tract cell lines. J Ethnopharmacol 145(1):146–151

Senthil Kumar KJ, Gokila Vani M, Hsieh HW, Lin CC, Wang SY (2021) Antcins from *Antrodia cinnamomea* and *Antrodia salmonea* inhibit angiotensin-converting enzyme 2 (ACE2) in epithelial cells: can be potential candidates for the development of SARS-CoV-2 prophylactic agents. Plan Theory 10(8):1736

Shah M, Murad W, Ur Rehman N, Halim SA, Ahmed M, Rehman H, Zahoor M, Mubin S, Khan A, Nassan MA, Batiha GE (2021) Biomedical applications of *Scutellaria edelbergii* Rech. f.: *In vitro* and *in vivo* approach. Molecules 26(12):3740

Sharifi-Rad J, Iriti M, Setzer WN, Sharifi-Rad M, Roointan A, Salehi B (2018) Antiviral activity of *Veronica persica* Poir. on herpes virus infection. Cell Mol Biol 64(8):11–17

Shi Y, Zhang B, Lu Y, Qian C, Feng Y, Fang L, Ding Z, Cheng D (2017) Antiviral activity of phenanthrenes from the medicinal plant *Bletilla striata* against influenza A virus. BMC Complement Altern Med 17(1):1–4

Shin WJ, Lee KH, Park MH, Seong BL (2010) Broad-spectrum antiviral effect of *Agrimonia pilosa* extract on influenza viruses. Microbiol Immunol 54(1):11–19

Shree P, Mishra P, Selvaraj C, Singh SK, Chaube R, Garg N, Tripathi YB (2022) Targeting COVID-19 (SARS-CoV-2) main protease through active phytochemicals of ayurvedic medicinal plants–*Withania somnifera* (Ashwagandha), *Tinospora cordifolia* (Giloy) and *Ocimum sanctum* (Tulsi)–a molecular docking study. J Biomol Struct Dyn 40(1):190–203

Siddiqui S, Upadhyay S, Ahmad R, Gupta A, Srivastava A, Trivedi A, Husain I, Ahmad B, Ahamed M, Khan MA (2022) Virtual screening of phytoconstituents from miracle herb *Nigella sativa* targeting nucleocapsid protein and papain-like protease of SARS-CoV-2 for COVID-19 treatment. J Biomol Struct Dyn 40(9):3928–3948

Signer J, Jonsdottir HR, Albrich WC, Strasser M, Züst R, Ryter S, Ackermann-Gäumann R, Lenz N, Siegrist D, Suter A, Schoop R (2020) *In vitro* virucidal activity of Echinaforce®, an *Echinacea purpurea* preparation, against coronaviruses, including common cold coronavirus 229E and SARS-CoV-2. Virol J 17(1):1–1

Sood R, Raut R, Tyagi P, Pareek PK, Barman TK, Singhal S, Shirumalla RK, Kanoje V, Subbarayan R, Rajerethinam R, Sharma N (2015) *Cissampelos pareira* Linn: natural source of potent antiviral activity against all four dengue virus serotypes. PLoS Negl Trop Dis 9(12): e0004255

Srivastava N, Saxena SK (2020) Prevention and control strategies for SARS-CoV-2 infection. In: Coronavirus disease 2019 (COVID-19). Springer, Singapore, pp 127–140

Sun L, Zhang X, Xu S, Hou C, Xu J, Zhao D, Chen Y (2021) Antiviral activities of a medicinal plant extract against sacbrood virus in honeybees. Virol J 18(1):1–10

Tang LI, Ling AP, Koh RY, Chye SM, Voon KG (2012) Screening of anti-dengue activity in methanolic extracts of medicinal plants. BMC Complement Altern Med 12(1):1–10

Thabti I, Albert Q, Philippot S, Dupire F, Westerhuis B, Fontanay S, Risler A, Kassab T, Elfalleh W, Aferchichi A, Varbanov M (2020) Advances on antiviral activity of *Morus* spp. plant extracts: human coronavirus and virus-related respiratory tract infections in the spotlight. Molecules 25(8):1876

Tiwari V, Darmani NA, Yue BY, Shukla D (2010) *In vitro* antiviral activity of neem (*Azardirachta indica* L.) bark extract against herpes simplex virus type-1 infection. Phytother Res 24(8): 1132–1140

Tkachenko KG (2007) Antiviral activity of the essential oils of some *Heracleum* L. species. J Herbs Spices Med Plants 12(3):1–2

Todorov D, Shishkova K, Dragolova D, Hinkov A, Kapchina-Toteva V, Shishkov S (2015) Antiviral activity of medicinal plant *Nepeta nuda*. Biotechnol Biotechnolog Equip 29(Supp 1):S39–S43

Tseliou M, Pirintsos SA, Lionis C, Castanas E, Sourvinos G (2019) Antiviral effect of an essential oil combination derived from three aromatic plants (*Coridothymus capitatus* (L.) Rchb. f., *Origanum dictamnus* L. and *Salvia fruticosa* Mill.) against viruses causing infections of the upper respiratory tract. J Herb Med 17:100288

Twilley D, Langhansová L, Palaniswamy D, Lall N (2017) Evaluation of traditionally used medicinal plants for anticancer, antioxidant, anti-inflammatory and anti-viral (HPV-1) activity. S Afr J Bot 112:494–500

Velusamy P, Kiruba K, Su CH, Arun V, Anbu P, Gopinath SC, Vaseeharan B (2021) SARS-CoV-2 spike protein: site-specific breakpoints for the development of COVID-19 vaccines. J King Saud Univ Sci 33(8):101648

Visintini Jaime MF, Redko F, Muschietti LV, Campos RH, Martino VS, Cavallaro LV (2013) *In vitro* antiviral activity of plant extracts from Asteraceae medicinal plants. Virol J 10(1):1

Wang D, Hu B, Hu C, Zhu F, Liu X, Zhang J, Wang B, Xiang H, Cheng Z, Xiong Y, Zhao Y (2020) Clinical characteristics of 138 hospitalized patients with 2019 novel coronavirus–infected pneumonia in Wuhan, China. JAMA 323(11):1061–1069

Wu D, Wu T, Liu Q, Yang Z (2020) The SARS-CoV-2 outbreak: what we know. Int J Infect Dis 94: 44–48

Zandi K, Teoh BT, Sam SS, Wong PF, Mustafa MR, AbuBakar S (2012) Novel antiviral activity of baicalein against dengue virus. BMC Complement Altern Med 12(1):1–9

Zolezzi M, Paravattil B, El-Gaili T (2021) Using motivational interviewing techniques to inform decision-making for COVID-19 vaccination. Int J Clin Pharm 43(6):1728–1734

# Chapter 16
# Therapeutic Potential of Essential Oils against SARS-CoV-2 Infection

**Muhammad Adil, Pragya Tiwari, and Jen-Tsung Chen**

## 16.1 Introduction

The present era has witnessed a tremendous global health crisis and rising mortality rates due to the ongoing coronavirus disease 2019 (COVID-19) pandemic. The SARS-CoV-2 (a new strain of coronavirus), primarily originated in China, gradually dissiminated across the world and created a major health disaster to the global economy. However, this expanding health catastrophe was effectively tackled by the high-throughput advances in medical science, via the development of various COVID-19 vaccines (Alshrari et al. 2022). In the last few years, a new variant of coronavirus has emerged and controlled with effective medications/vaccines. Moreover, another variant, B.1.167 has been currently detected in many countries including India, Canada and the United Kingdom (Sanyaolu et al. 2021). With a fast mutation and multiplication characteristics, the emergence of a more deadly and infectious viral strain (Malabadi et al. 2021) cannot be ruled out, however, with effectual healthcare and bio-based therapeutics, these challenges can be addressed and countered to a major extent.

M. Adil (✉)
Pharmacology and Toxicology Section, University of Veterinary and Animal Sciences, Lahore, Jhang Campus, Jhang, Pakistan
e-mail: muhammad.adil@uvas.edu.pk

P. Tiwari (✉)
Department of Biotechnology, Yeungnam University, Gyeongsan, Gyeongbuk, Republic of Korea
e-mail: pragyatiwari@ynu.ac.kr

J.-T. Chen
Department of Life Sciences, National University of Kaohsiung, Kaohsiung, Taiwan
e-mail: jentsung@nuk.edu.tw



J.-T. Chen (ed.), *Ethnopharmacology and Drug Discovery for COVID-19: Anti-SARS-CoV-2 Agents from Herbal Medicines and Natural Products*,
https://doi.org/10.1007/978-981-99-3664-9_16

Herbal medicines have been traditionally used for the treatment of human diseases and are gaining recognition as effective therapeutics in the current era. Bioactive components and extracts from multiple plants have demonstrated pharmacological effects in the treatment of multiple diseases including diabetes, cancer and neurological diseases, either used individually or in combination therapies. The volatile secondary metabolites from plants, known as essential oils, are classified into distinct chemical classes as hydrocarbons (sesquiterpenes, monoterpenes, diterpenes, etc.), oxygenated hydrocarbons (terpenoids comprising phenols, aldehydes, alcohols, etc.), non-terpene compounds (eugenol, cinnamaldehyde, etc.), and other compounds (Schnitzler et al. 2011; Bora et al. 2020). Essential oils are widely used in our daily life as aromatherapeutic agents, air fresheners, perfumes and aromatic waters. Besides, essential oils also comprise an integral part of folk and traditional medicine, and exhibit multiple bioactivities including antioxidant, antimicrobial, immunomodulatory, sedative and antirheumatic properties (Brahmi et al. 2016; Dhifi et al. 2016; Bora et al. 2020). While essential oils are produced by more than 17,000 plant species, classified into diverse plant families, and extracted from multiple plant parts, namely flowers, leaves, fruits and roots, primarily an aromatic ingredient of specialized plant tissues (Schnitzler et al. 2011), with aromatic essential oils forms the basis of aromatherapy.

Plant-derived essential oils constitute a complex mixture of phytochemicals, originating from phenylpropanoids, monoterpene and sesquiterpene classes (Ma and Yao 2020; Tshibangu et al. 2020) and possess wider applications in pharmaceutical, perfume and soap industries. Regarded as a prospective strategy for tacking SARS-CoV-2 in recent times, essential oils are effective in restricting the growth of diverse pathogens, on account of their volatile components and chemical composition (Swamy et al. 2016). The use of essential oils highlights several advantages, including high lipophilicity, rapid action, small size, virucidal properties at low concentrations, high vapor pressure, low cost of preparation, and prospects as a prophylactic agent in the initial stages of viral pneumonia (Elsebai and Albalawi 2022). A key example is eucalyptol, with its vital physicochemical properties and pharmacological functions, which reveals its prospects as a drug in the prevention and treatment of COVID-19. Moreover, other major examples of essential oils as effective therapeutics against SARS-CoV-2 comprise the eucalyptus oil (jensenone and eucalyptol/1,8-cineol), garlic oil (garlicin, allitridin, ajoene), aromatic herbs (essential oils of *Laurus nobilis*, *Juniperus oxycedrus* ssp. oxycedrus), and other essential oils with potent antiviral activities (Elsebai and Albalawi 2022).

In a quest toward the discovery of potent therapeutics against SARS-CoV-2, essential oils show antiviral activity, owing to their lipophilicity, disruption of the viral membrane and viral disintegration (Elsebai and Albalawi 2022). Moreover, many bioactive components of crude essential oils act on different parts of the virus (entry in the cell, transcription, virus assembly and translation), and are gaining popularity as attractive bio-based therapeutics against SARS-CoV-2. Studies have suggested that essential oils and their bioactive molecules can be used as adjuvants for viral disease treatment, resulting in milder disease forms and improving the overall health status of patients (Damiescu et al. 2022). Consequently, randomized clinical trials are necessary for standardizing the bioactive constituents and determining the efficacy of potent active compounds, of essential oils against SARS-CoV-2.

## 16.2 Physicochemical and Pharmacokinetic Characteristics of Essential Oils

Some essential oils, such as eucalyptol, have distinct physicochemical properties, causing their presence in the lungs via exhalation and exerting their pharmacological action in low doses (Usachev et al. 2013). While asymptomatic patients are crucial in the diseas transmission, eucalyptol exerts virucidal effects both on oral administration in the lower respiratory tract and inhalation in the upper respiratory tract, thus preventing the virus from spreading to other body organs (Wölfel et al. 2020). Many essential oils are commercially marketed as antimicrobials and are important alternatives to chemical agents, specifically in the mileau of rising microbial resistance to conventional drugs (Gutierrez et al. 2008). Studies have documented the effects of essential oils from *Anthemis hyalina*, *Nigella sativa* and *Citrus sinensis* on coronavirus replication and TRP gene expression (Ulasli et al. 2014).

The initial progress of SARS-CoV-2 occurs via targeting lungs and airways, the primary site of infection, therefore, it is suggested that essential oils should be administered via inhalation, causing a direct action in the lungs and other parts of respiratory tract. This administration route of essential oils provides a better option to hamper the binding interactions between SARS-CoV-2 spike proteins and their ACE-2 receptors, located in the parenchymatous lung cells (Elsebai and Albalawi 2022). In addition, lung-based delivery of potent essential oils with anti-SARS-CoV-2 function constitutes a remarkable approach for achieving the desired outcome (Dent et al. 2013). Another key advantage of inhaled drugs with reference to essential oils comprises their low molecular weights and high vapor pressures, therefore these substances are exhaled from the lungs and lead to non-specific anti-inflammatory action, further improving their pharmacological function (Dent et al. 2013). These characteristics enable the eucalyptol to persist in the lungs and cause its local virucidal effect (Sharifi-Rad et al. 2017).

With the pharmacological effects of essential oils and their constituents well reported, the application of essential oils against viral diseases is widely explored in the current era. Essential oils are capable of penetrating and subsequently rupturing the viral cell membrane by their sufficient lipophilic potential (Elsebai and Albalawi 2022). The metabolism of essential oils proceeds via multiple degradative pathways and functions of different enzymes in vivo. It is equally important to determine the chemical profiles of essential oils and their respective constituents. The safety profiles of essential oils can be elucidated by the biotransformation of essential oils in phase I metabolism (reduction, oxidation and hydrolysis etc.) and phase II metabolism (Al-Harrasi et al. 2020). In addition, studies have reported the pharmacodynamics of multiple essential oils during in vitro conditions however, the availability of essential oils in different organs has been unknown. This necessitates a requirement to explicate the absorption, distribution, metabolism and excretion of essential oils for understanding the link between in vivo and in vitro studies. Moreover, it is also imperative to comprehend the bioavailability as well as the pharmacokinetic properties of essential oils and their constituents (Al-Harrasi et al. 2020).

## 16.3 Therapeutic Basis of Essential Oils with Reference to SARS-CoV-2 Infection

Apart from the symptomatic treatment in terms of anti-inflammatory, anticoagulant and antipyretic agents, monoclonal antibodies and antiviral drugs such as monlupiravir and remdesivir have been previously used against COVID-19 infection with varying success rates (Elsebai and Albalawi 2022). Essential oils exhibit diverse pharmacological spectra characterized by anticancer, antioxidant, anti-inflammatory, insecticide and antimicrobial effects (Galvan et al. 2021). Bioactive substances of various essential oils such as tea tree oil, thyme oil and eucalyptus oil are used as an adjunct to conventional therapeutic agents against flu and common cold, whereas, those of *Allium sativum*, *Echinacea purpurea*, *Echinacea angustifolia* and *Zingiber officinale* are useful for their immunostimulant potential (Damiescu et al. 2022). Essential oils from anise (*Illicium verum*), eucalyptus, sandalwood (*S. album*), chamomile (*Matricaria recutita*), and others have shown promising antiviral effects against HSV-1 and/or HSV-2 (Schnitzler et al. 2001), essential oils from *Artemisia glabella* (Seidakhmetova et al. 2002), *Origanum acutidens* (Sökmen et al. 2004), *Oenanthe crocata* (Bonsignore et al. 2004), against influenza viruses, essential oils from *Eucalyptus globulus* (eucalyptus), *Ocimum basilicum album* (tropical basil), *Lavandula latifolia* (lavender), and others against HSV-1, highlight the potential efficacy of essential oils against multiple viruses.

Essential oils have been extensively used in folk medicine on account of their miscellaneous pharmacological effects including antimicrobial, immunomodulatory, antioxidant, anti-inflammatory, mucolytic and antirheumatic properties (Brahmi et al. 2016; Dhifi et al. 2016). Moreover, spoilage bacteria and food-borne pathogens are highly susceptible to the combinations of essential oils (Gutierrez et al. 2008; Mutlu-Ingok and Karbancioglu-Guler 2017). The antiviral efficacy of essential oils has been documented against a wide range of pathologically important viruses including yellow fever virus, human and avian influenza viruses, human herpes viruses (HSV-1 and HSV-2), Zika virus, avian influenza A virus ($H_1N_1$), influenza A virus ($H_5N_1$), human immunodeficiency virus (HIV), and SARS-COV-2 (Astani et al. 2010; Schnitzler et al. 2011; Ma and Yao 2020; Rouf et al. 2020). The effectiveness of essential oils against enveloped single-stranded positive RNA viruses has been validated through in vitro assays as well as experimental studies using animal models (Wilkin et al. 2020). Likewise, both in vitro and in vivo studies have revealed the additive pharmacological effects of oseltamivir and germacrone combination against the influenza virus (Liao et al. 2013). Moreover, the hsp1 was successfully targeted by a synergistic combination of acyclovir and 4., piperitenone oxide which is the active ingredient of *Mentha suaveolens* essential oil (Civitelli et al. 2014). Concurrent use of *Melissa officinalis* essential oil improved the antiviral effect of oseltamivir against avian influenza A virus (Pourghanbari et al. 2016). Improved clinical outcomes characterized by reduced lung pathological index, pulmonary cytokines expression and viral titers, whereas, enhanced survival rate have been associated with the synergistic pharmacological benefit resulting from the co-administration of oseltamivir and eucalyptol (Lai et al. 2017).

Immunomodulatory, mucolytic and anti-inflammatory effects are also useful to combat viral respiratory diseases including the SARS-COV-2 infection. Consequently, the use of essential oils is gradually increasing in medicine, cosmetics, pharmaceutics, food and other industries. Simply targeting the SARS-COV-2 using specific antiviral drugs may not provide a complete solution to COVID-19 infection. Therefore, essential oils characterized by anti-inflammatory, antioxidant, antimicrobial and immunomodulatory effects can properly relieve the clinical manifestations of COVID-19 infection by suppressing the pulmonary inflammatory cascade and preventing secondary bacterial infection (Wilkin et al. 2020). Based upon their high volatility and ease of inhalational administration, essential oils can be effectively used to subside sore throat, coughing, expectoration, nasal congestion and rhinorrhea (Horváth and Ács 2015; Valussi et al. 2021). The optimal antimicrobial efficacy of essential oils entails their application for disinfecting closed areas wherein enhanced proliferation of SARS-COV-2 is projected.

## 16.4 Major Bioactive Constituents and Pharmacological Mechanisms of Essential Oils Against SARS-CoV-2

Essential oils and their purified bioactive substances may represent a relatively less explored source of useful antiviral drugs for add-on therapy of viral infections (Wilkin et al. 2020). Whole essential oils are usually preferred over purified constituents by virtue of multiple pharmacological actions, reduced probability of microbial resistance and synergistic effects (Popa et al. 2020). Antifungal, antibacterial and antiviral activities have been documented for many essential oils and their derivative compounds (Swamy et al. 2016; Wińska et al. 2019). Sesquiterpenes and monoterpene hydrocarbons have been implicated in triggering the potential antiviral activity of essential oils (Sobrinho et al. 2021). Different pharmacological mechanisms of essential oils have been shown in Fig. 16.1. The synergistic constituents of essential oils in combination therapies, improve the efficacy of other antiviral drugs and alleviate COVID-19 symptoms (Da Silva et al. 2020). Various steps of viral replication such as cell entry, assembly, transcription and translation

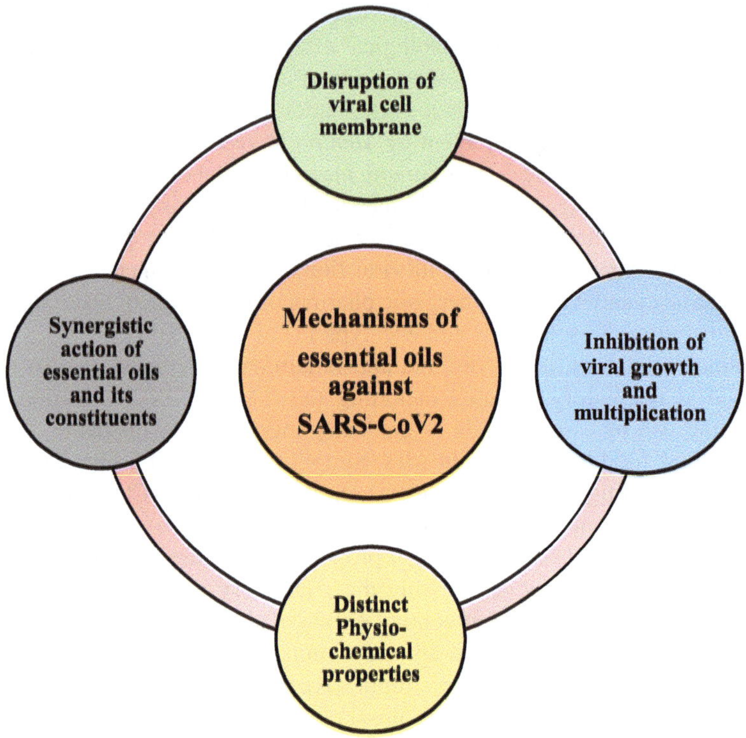


**Fig. 16.1** Different mechanisms of action of essential oils against SARS-CoV-2

can be impaired by several bioactive ingredients of crude essential oils (Elsebai and Albalawi 2022). Several proteins of SARS-COV-2 have been effectively targeted using different bioactive components of essential oils including carvone, thymoquinone, camphene, limonin, carvacrol, eucalyptol and thymol (Neto et al. 2021). The majority of tested substances primarily targeted the spike protein of SARS-COV-2 that has been associated with the cell entry of virus following its attachment to specific ACE-2 receptors (Neto et al. 2021).

Molecular docking analysis revealed the synergistic inhibitory actions caused by 17 different compounds of garlic essential oil on Mpro and ACE-2 proteins of SARS-COV-2 (Thuy et al. 2020). ACE-2 antagonism has also been attributed to the principal ingredient of *Ammoides verticillata* essential oil, called isothymol (Abdelli et al. 2021). Moreover, the bioactive substance called zerumbone exhibited a high affinity for spike protein-human cell ACE-2 receptor complex (Neto et al. 2021; Sharbidre et al. 2021). Endoribonuclease which is essential for the replication of SARS-COV-2 and restricts the host defense mechanism can be efficiently inhibited using curion (Hackbart et al. 2020). Another substance known as caryophyllene demonstrated a high affinity for non-structural RNA-binding protein-9 RP1A as well as spike protein, spro (Neto et al. 2021). Likewise, the bioactive components of South Asian plant, *Melaleuca cajuputi*, including α-eudesmol, β-selinenol, γ-eudesmol, cineol, linalool, guaiol and terpineol displayed a synergistic inhibitory effect on viral replication via interacting with ACE-2 receptor (My et al. 2020). Cysteine protease enzyme having a pivotal role in the viral replication process was successfully targeted by spatholinol, himachalol and eudismol (Neto et al. 2021). Several major ingredients of different essential oils such as pulegone, menthol, camphene, cinnamaldehyde, anethole, geraniol, thymol and carvacrol inhibited viral replication by targeting the S1 element of the spike protein (Kulkarni et al. 2020). Menthol, eugenol, carvacrol and cinnamaldehyde suppressed the synthesis of proinflammatory cytokines and recruitment of macrophages in the bronchoalveolar fluid (Wani et al. 2021). Essential oils of juniper, lemon and geranium are also known for their inhibitory effects on ACE-2 receptor of SARS-COV-2 (Kumar et al. 2020). Ethanolic preparation consisting of methyl salicylate, eucalyptol, thymol and menthol (Listerine®) demonstrated significant antiviral effect during an in vitro study (Stathis et al. 2021). Moreover, in silico evidence is available regarding the blockage of S1 unit of viral spike protein by mouth rinses comprising of pinocarveol, caryophyllene, myrtenol and carvacrol (Yadalam et al. 2021).

Different bioactive compounds, including γ-terpinene, p-cymene, limonene, thymol and isothymol, derived from the Algerian plant, *Ammoides verticillate*, were screened for their in silico ACE-2 inhibitory actions. Isothymol was found as the most effective ACE-2 receptor antagonist with its binding affinity equivalent to that of chloroquine and captopril (Abdelli et al. 2021). The diverse flora of the Amazon region has also been examined for potential antiviral efficacy against SARS-CoV-2 using the molecular docking technique. Several of the tested compounds such as humulene epoxide II, hedycaryol, guaia-6,9-diene, germacrene A, azulenol, aristochene, amorpha-4,9-dien-2-ol, allo-aromadendrene epoxide, 14-hydroxy-muurolene, α-muurolene, α-calacorene, α-cadinene, α-amorphene and (E)-α-atlantone have reflected antiviral effects (Amparo et al. 2021). Buriti oil obtained from *Mauritia flexuosa* and containing α-carotene, 9-cis-β-carotene and 13-cis-β-carotene has been implicated in interacting with the viral peptidase enzyme (Costa et al. 2021). Similarly, a palm tree known as Coco de Mer (*Lodoicea maldivica*) comprising monoterpenes, bicyclogermacrene and β-caryophyllene has been linked with antimicrobial potential (Giuliani et al. 2020). Faster recovery and suppression of clinical manifestations were documented in COVID-19 patients following the twice utilization of *Nigella sativa* oil capsules (500 mg) for 10 days (Koshak et al. 2021). Based upon their preliminary antiviral actions, the essential oils of *Melaleuca alternifolia* (tea tree) and *Laurus nobilis* are also considered as potential anti-COVID agents, requiring further pharmacological screening (Loizzo et al. 2008; Schnitzler et al. 2011).

## 16.5 Conclusion and Future Recommendations

Essential oils obtained from medicinal plants have been traditionally used in several parts of the world for different pharmaceutical benefits. Accordingly, the pharmacotherapeutic significance of aromatherapy seems inevitable in the milieu of SARS-COV-2 infection. Different parts of cultivated or wild plants including leaves, flowers, bark, root or fruits are used for the extraction of essential oils (Damiescu et al. 2022). Standardized preparations of certain essential oils are commercially available in the form of nasal sprays, gels, creams and capsules, and are widely used for the treatment of different viral infections (Damiescu et al. 2022). Several beneficial aspects of essential oils including high lipophilicity, smaller molecular size, rapid onset of action, wide safety margin, high vapor pressure, efficient virucidal activity, ease of synthesis and formulation, convenient oral administration, and lack of first-pass metabolism reflect their therapeutic and prophylactic significance in conjunction with the typical antiviral drugs for the treatment of viral pneumonia (Elsebai and Albalawi 2022). Essential oils can be used as an adjunct to conventional antiviral agents in countries with deprived healthcare facilities and limited availability of effective vaccines (Damiescu et al. 2022). Nevertheless, highly concentrated formulations of some essential oils can give rise to detrimental effects such as vomiting, nausea, dizziness and epileptic seizures upon oral administration (Damiescu et al. 2022). Lack of standardized dosage forms, direct cytotoxic reactions and poor solubility are some of the drawbacks that limit the widespread application of some essential oils for therapeutic purpose (Damiescu et al. 2022). Besides, some essential oils can also lead to tissue irritation or a burning sensation on the cutaneous/mucosal surface. However, the nano-emulsion formulations have considerably minimized the likelihood of undesirable effects and enhanced the therapeutic utility of essential oils (Damiescu et al. 2022).

Several commercially available essential oils have been linked with efficient antimicrobial effects and thereby necessitate further investigation of their pharmacological potential to combat the extensively rising microbial resistance (Gutierrez et al. 2008; Wińska et al. 2019; Kakhki et al. 2020). Eucalyptol is considered a promising candidate drug against SARS-COV-2 infection on account of its suitable physicochemical profile and diverse pharmacological effects (Elsebai and Albalawi 2022). The currently available in silico data regarding the antiviral efficacy of essential oils against SARS-COV-2 should be further validated using in vitro assays, animal models and clinical studies (Neto et al. 2021). Apart from the confirmation of antiviral efficacy at experimental and clinical levels, the safety profile of essential oils and their purified bioactive products also require in-depth evaluation through toxicity studies. Randomized clinical studies are required for determining the efficacy and safety of standardized essential oil formulations (Damiescu et al. 2022). Keeping in view the antioxidant, immunomodulatory and anti-inflammatory properties of essential oils against SARS-COV-2 infection, their pharmacological actions and therapeutic effects should be further explored through experimental and clinical studies (Elsebai and Albalawi 2022).

## References


Abdelli I, Hassani F, Bekkel Brikci S, Ghalem S (2021) In silico study the inhibition of angiotensin converting enzyme 2 receptor of COVID-19 by Ammoides verticillata components harvested from Western Algeria. J Biomol Struct Dyn 39(9):3263–3276

Al-Harrasi A, Bhatia S, Ahmed MM, Anwer K, Sharma P (2020) Pharmacokinetics of essentials oils. In: Role of essential oils in the management of COVID-19. CRC Press, pp 359–370

Alshrari AS, Hudu SA, Imran M, Asdaq SMB, Ali AM, Rabbani SI (2022) Innovations and development of COVID-19 vaccines: a patent review. J Infect Public Health 15(1):123–131

Amparo TR, Seibert JB, Silveira BM, Costa FSF, Almeida TC, Braga SFP, da Silva GN, Dos Santos ODH, de Souza GHB (2021) Brazilian essential oils as source for the discovery of new anti-COVID-19 drug: a review guided by in silico study. Phytochem Rev 20(5):1013–1032

Astani A, Reichling J, Schnitzler P (2010) Comparative study on the antiviral activity of selected monoterpenes derived from essential oils. Phytother Res 24(5):673–679

Bonsignore L, Casu L, Loy G, Deidda D, Genco F (2004) Analysis of the essential oil of Oenanthe crocata L. and its biological activity. J Essent Oil Res 16(3):266–269

Bora H, Kamle M, Mahato DK, Tiwari P, Kumar P (2020) Citrus essential oils (CEOs) and their applications in food: an overview. Plan Theory 9(3):357

Brahmi F, Achat S, Guendouze-Bouchefa N, Benazzouz-Smail L, Elsebai MF, Madani K (2016) Recent advances in the identification and the study of composition and activities of medicinal plants. J Coast Life Med 4:983–999

Civitelli L, Panella S, Marcocci ME, De Petris A, Garzoli S, Pepi F, Vavala E, Ragno R, Nencioni L, Palamara AT (2014) In vitro inhibition of herpes simplex virus type 1 replication by Mentha suaveolens essential oil and its main component piperitenone oxide. Phytomedicine 21(6):857–865

Costa AN, de Sá ÉR, Bezerra RD, Souza JL, Lima FC (2021) Constituents of Buriti oil (Mauritia flexuosa L.) like inhibitors of the SARS-coronavirus main peptidase: an investigation by docking and molecular dynamics. J Biomol Struct Dyn 39(13):4610–4617

Da Silva JKR, Figueiredo PLB, Byler KG, Setzer WN (2020) Essential oils as antiviral agents, potential of essential oils to treat SARS-CoV-2 infection: an in-silico investigation. Int J Mol Sci 21(10):3426

Damiescu R, Lee DY, Efferth T (2022) Can essential oils provide an alternative adjuvant therapy for COVID-19 infections and pain management at the same time? Pharmaceuticals 15(11):1387

Dent AG, Sutedja TG, Zimmerman PV (2013) Exhaled breath analysis for lung cancer. J Thorac Dis 5(Suppl 5):S540

Dhifi W, Bellili S, Jazi S, Bahloul N, Mnif W (2016) Essential oils' chemical characterization and investigation of some biological activities: a critical review. Medicines (Basel, Switzerland) 3(4):E25–E25

Elsebai M, Albalawi M (2022) Essential oils and COVID-19. Molecules (Basel, Switzerland) 27(22):7893–7893

Galvan D, Effting L, Torres Neto L, Conte-Junior CA (2021) An overview of research of essential oils by self-organizing maps: a novel approach for meta-analysis study. Compr Rev Food Sci Food Saf 20(4):3136–3163

Giuliani C, Pieraccini G, Santilli C, Tani C, Bottoni M, Schiff S, Fico G, Papini A, Falsini S (2020) Anatomical investigation and GC/MS analysis of 'Coco de Mer', Lodoicea maldivica (Arecaceae). Chem Biodivers 17(11):e2000707

Gutierrez J, Barry-Ryan C, Bourke P (2008) The antimicrobial efficacy of plant essential oil combinations and interactions with food ingredients. Int J Food Microbiol 124(1):91–97

Hackbart M, Deng X, Baker SC (2020) Coronavirus endoribonuclease targets viral polyuridine sequences to evade activating host sensors. Proc Natl Acad Sci 117(14):8094–8103

Horváth G, Ács K (2015) Essential oils in the treatment of respiratory tract diseases highlighting their role in bacterial infections and their anti-inflammatory action: a review. Flavour Fragr J 30(5):331–341

Kakhki MT, Sedaghat N, Mohsenzadeh M (2020) Chemical composition, antioxidative, antibacterial, and time-kill activities of some selected plant essential oils against foodborne pathogenic and spoilage organisms. In: Veterinary research forum, vol 4. Veterinary Research Forum, pp 339–346

Koshak AE, Koshak EA, Mobeireek AF, Badawi MA, Wali SO, Malibary HM, Atwah AF, Alhamdan MM, Almalki RA, Madani TA (2021) Nigella sativa for the treatment of COVID-19: an open-label randomized controlled clinical trial. Complement Ther Med 61:102769

Kulkarni SA, Nagarajan SK, Ramesh V, Palaniyandi V, Selvam SP, Madhavan T (2020) Computational evaluation of major components from plant essential oils as potent inhibitors of SARS-CoV-2 spike protein. J Mol Struct 1221:128823

Kumar KS, Vani MG, Wang C-S, Chen C-C, Yu-Chien C, Li-Ping L, Huang C-H, Lai C-S, Sheng-Yang W (2020) Geranium and lemon essential oils and their active compounds downregulate angiotensin-converting enzyme 2 (ACE2), a SARS-CoV-2 spike receptor-binding domain, in epithelial cells. Plants 9(6):770

Lai Y-N, Li Y, Fu L-C, Zhao F, Liu N, Zhang F-X, Xu P-P (2017) Combinations of 1, 8-cineol and oseltamivir for the treatment of influenza virus A (H3N2) infection in mice. J Med Virol 89(7):1158–1167

Liao Q, Qian Z, Liu R, An L, Chen X (2013) Germacrone inhibits early stages of influenza virus infection. Antivir Res 100(3):578–588

Loizzo MR, Saab AM, Tundis R, Statti GA, Menichini F, Lampronti I, Gambari R, Cinatl J, Doerr HW (2008) Phytochemical analysis and in vitro antiviral activities of the essential oils of seven Lebanon species. Chem Biodivers 5(3):461–470

Ma L, Yao L (2020) Antiviral effects of plant-derived essential oils and their components: an updated review. Molecules (Basel, Switzerland) 25(11):2627

Malabadi RB, Kolkar KP, Meti NT, Chalannavar RK (2021) Role of botanical essential oils as a therapy for controlling coronavirus (SARS-CoV-2) disease (Covid-19). Int J Res Scient Innov 8(4):105–118

Mutlu-Ingok A, Karbancioglu-Guler F (2017) Cardamom, cumin, and dill weed essential oils: chemical compositions, antimicrobial activities, and mechanisms of action against campylobacter spp. Molecules 22(7):1191

My TTA, Loan HTP, Hai NTT, Hieu LT, Hoa TT, Thuy BTP, Quang DT, Triet NT, Anh TTV, Dieu NTX (2020) Evaluation of the inhibitory activities of COVID-19 of Melaleuca cajuputi oil using docking simulation. Chem Select 5(21):6312–6320

Neto LT, Monteiro MLG, Galvan D, Conte-Junior CA (2021) An evaluation of the potential of essential oils against SARS-CoV-2 from in silico studies through the systematic review using a Chemometric approach. Pharmaceuticals (Basel, Switzerland) 14(11):1138

Popa M, Măruțescu L, Oprea E, Bleotu C, Kamerzan C, Chifiriuc MC, Pircalabioru GG (2020) In vitro evaluation of the antimicrobial and immunomodulatory activity of culinary herb essential oils as potential Perioceutics. Antibiotics (Basel, Switzerland) 9(7):428

Pourghanbari G, Nili H, Moattari A, Mohammadi A, Iraji A (2016) Antiviral activity of the oseltamivir and Melissa officinalis L. essential oil against avian influenza A virus (H9N2). Virus Dis 27:170–178

Rouf R, Uddin SJ, Sarker DK, Islam MT, Ali ES, Shilpi JA, Nahar L, Tiralongo E, Sarker SD (2020) Antiviral potential of garlic (Allium sativum) and its organosulfur compounds: a systematic update of pre-clinical and clinical data. Trends Food Sci Technol 104:219–234

Sanyaolu A, Okorie C, Marinkovic A, Haider N, Abbasi AF, Jaferi U, Prakash S, Balendra V (2021) The emerging SARS-CoV-2 variants of concern. Therapeut Adv Infect Dis 8: 20499361211024372

Schnitzler P, Schön K, Reichling J (2001) Antiviral activity of Australian tea tree oil and eucalyptus oil against herpes simplex virus in cell culture. Pharmazie 56(4):343–347

Schnitzler P, Astani A, Reichling J (2011) Antiviral effects of plant-derived essential oils and pure oil components. In: Lipids and essential oils as antimicrobial agents, pp 239–254

Seidakhmetova R, Beisenbaeva A, Atazhanova G, Suleimenov E, Pak R, Kulyyasov A, Adekenov S (2002) Chemical composition and biological activity of the essential oil from Artemisia glabella. Pharm Chem J 36(3):135–138

Sharbidre A, Dhage P, Duggal H, Meshram R (2021) In silico investigation of Tridax procumbens phyto-constituents against SARS-CoV-2 infection. Biointerface Res Appl Chem 11:12120–12148

Sharifi-Rad J, Salehi B, Schnitzler P, Ayatollahi S, Kobarfard F, Fathi M, Eisazadeh M, Sharifi-Rad M (2017) Susceptibility of herpes simplex virus type 1 to monoterpenes thymol, carvacrol, p-cymene and essential oils of Sinapis arvensis L., Lallemantia royleana Benth. And Pulicaria vulgaris Gaertn. Cell Mol Biol 63(8):42–47

Sobrinho ACN, de Morais SM, Marinho MM, de Souza NV, Lima DM (2021) Antiviral activity on the Zika virus and larvicidal activity on the Aedes spp. of Lippia alba essential oil and β-caryophyllene. Ind Crop Prod 162:113281

Sökmen M, Serkedjieva J, Daferera D, Gulluce M, Polissiou M, Tepe B, Akpulat HA, Sahin F, Sokmen A (2004) In vitro antioxidant, antimicrobial, and antiviral activities of the essential oil and various extracts from herbal parts and callus cultures of Origanum acutidens. J Agric Food Chem 52(11):3309–3312

Stathis C, Victoria N, Loomis K, Nguyen SA, Eggers M, Septimus E, Safdar N (2021) Review of the use of nasal and oral antiseptics during a global pandemic. Future Microbiol 16(2):119–130

Swamy MK, Akhtar MS, Sinniah UR (2016) Antimicrobial properties of plant essential oils against human pathogens and their mode of action: an updated review. Evid Based Complement Alternat Med 2016:3012462

Thuy BTP, My TTA, Hai NTT, Hieu LT, Hoa TT, Thi Phuong Loan H, Triet NT, Anh TTV, Quy PT, Tat PV (2020) Investigation into SARS-CoV-2 resistance of compounds in garlic essential oil. ACS omega 5(14):8312–8320

Tshibangu DS, Matondo A, Lengbiye EM, Inkoto CL, Ngoyi EM, Kabengele CN, Bongo GN, Gbolo BZ, Kilembe JT, Mwanangombo DT (2020) Possible effect of aromatic plants and essential oils against COVID-19: review of their antiviral activity. J Complemen Alternat Med Res 11(1):10–22

Ulasli M, Gurses SA, Bayraktar R, Yumrutas O, Oztuzcu S, Igci M, Igci YZ, Cakmak EA, Arslan A (2014) The effects of Nigella sativa (Ns), Anthemis hyalina (Ah) and Citrus sinensis (Cs) extracts on the replication of coronavirus and the expression of TRP genes family. Mol Biol Rep 41:1703–1711

Usachev EV, Pyankov OV, Usacheva OV, Agranovski IE (2013) Antiviral activity of tea tree and eucalyptus oil aerosol and vapour. J Aerosol Sci 59:22–30

Valussi M, Antonelli M, Donelli D, Firenzuoli F (2021) Appropriate use of essential oils and their components in the management of upper respiratory tract symptoms in patients with COVID-19. J Herb Med 28:100451

Wani AR, Yadav K, Khursheed A, Rather MA (2021) An updated and comprehensive review of the antiviral potential of essential oils and their chemical constituents with special focus on their mechanism of action against various influenza and coronaviruses. Microb Pathog 152:104620

Wilkin PJ, Al-Yozbaki M, George A, Gupta GK, Wilson CM (2020) The undiscovered potential of essential oils for treating SARS-CoV-2 (COVID-19). Curr Pharm Des 26(41):5261–5277

Wińska K, Mączka W, Łyczko J, Grabarczyk M, Czubaszek A, Szumny A (2019) Essential oils as antimicrobial agents-myth or real alternative? Molecules (Basel, Switzerland) 24(11):2130

Wölfel R, Corman VM, Guggemos W, Seilmaier M, Zange S, Müller MA, Niemeyer D, Jones TC, Vollmar P, Rothe C (2020) Virological assessment of hospitalized patients with COVID-2019. Nature 581(7809):465–469

Yadalam PK, Varatharajan K, Rajapandian K, Chopra P, Arumuganainar D, Nagarathnam T, Sohn H, Madhavan T (2021) Antiviral essential oil components against SARS-CoV-2 in pre-procedural mouth rinses for dental settings during COVID-19: a computational study. Front Chem 9:642026

# Chapter 17
# Antiviral Properties of South Indian Plants Against SARS-CoV-2

**Kathirvel Bharathi, Ajith Sivasangar Latha, Arumugam Jananisri, Venkataramanaravi Bavyataa, Boopalan Rajan, Balasubramanian Balamuralikrishnan, Mariadhas Valan Arasu, Naif Abdullah Al-Dhabi, Beulah Catharine, and Arumugam Vijaya Anand**

## Abbreviations

| | |
|---|---|
| ACE2 | Angiotensin-converting enzyme 2 |
| H1N1 | Influenza |
| HBV and HCV | Hepatitis B and C virus |
| HIV | Human immunodeficiency viruses |
| $M^{pro}$ | Main protease |
| NSPs | Non-structural proteins |
| $PL^{pro}$ | Papain-like protease |
| SARS-CoV-2 | Severe acute respiratory syndrome |
| TNF-α | Tumor necrosis factor-α |

K. Bharathi (✉) · A. Sivasangar Latha · A. Jananisri · V. Bavyataa · A. V. Anand (✉)
Department of Human Genetics and Molecular Biology, Bharathiar University, Coimbatore, Tamil Nadu, India
e-mail: avahgmb@buc.edu.in

B. Rajan
Department of Biochemistry and Biotechnology, Annamalai University, Chidambaram, India

B. Balamuralikrishnan
Department of Food Science and Biotechnology, Sejong University, Seoul, South Korea

M. V. Arasu · N. A. Al-Dhabi
Department of Botany and Microbiology, College of Science, King Saud University, Riyadh, Saudi Arabia

B. Catharine
Department of Biochemistry and Biotechnology, Avinashilingam Institute for Home Science and Higher Education for Women is a Women's Deemed University, Coimbatore, Tamil Nadu, India

 
J.-T. Chen (ed.), *Ethnopharmacology and Drug Discovery for COVID-19: Anti-SARS-CoV-2 Agents from Herbal Medicines and Natural Products*,
https://doi.org/10.1007/978-981-99-3664-9_17

## 17.1 Introduction

Coronavirus disease 2019 (COVID-19) is a pandemic disease that originated in Wuhan in China in December 2019 and was disclosed as a pandemic by World Health Organization in March 2020. COVID-19 is caused by severe acute respiratory syndrome coronavirus 2 (SARS-CoV-2) and it belongs to the family of Coronaviridae. Studies have highlighted that coronaviruses caused severe acute respiratory syndrome (SARS) in 2002 and Middle East respiratory disease (MERS) in 2012 (Mlozi 2022). In the past years, 40 million people were affected by COVID-19 and six million deaths were recorded across the globe. The widespread of COVID-19 is mainly due to a lack of immunity in the host because of its novel nature. COVID-19 is a life-threatening disease that has become a huge socio-economic burden (Shanmugam et al. 2020; Kuchi Bhotla et al. 2020; Kuchi Bhotla et al. 2021a, b; Pushparaj et al. 2022; Meyyazhagan et al. 2022; Mukherjee et al. 2022).

Nucleocapsid, spike, membrane, and envelope are the four structural proteins present in the virions of COVID-19. The assembly and release of viral particles depend on the viral envelope. The spike, membrane, and envelope proteins are responsible for the development of the viral envelope, whereas the RNA genome is held by a nucleocapsid (Yashvardhini et al. 2021). The spike glycoprotein along with the TMPRSS2 aids the penetration of the virus to the Angiotensin-converting enzyme 2 (ACE2). The spike (S) protein is split into two subunits S1 and S2 by furin. The ACE2 receptor is present in various parts of the body including the heart, lungs, kidney, intestine, testis, and blood vessels. The S2 subunit is further cleaved by the action of lysosomal cathepsin or by TMPRSS2 (Boopathi et al. 2021; Jackson et al. 2022).

The viral genome is released into the cytosol of the host cell after being uncoated from the endosome. Translation of the viral genome takes place, generating viral replicase proteins from open reading frames 1a and 1b. These replicase proteins are cleaved into non-structural proteins (NSPs), such as Papain-like protease ($PL^{pro}$) and main protease ($M^{pro}$). $M^{pro}$ facilitates the cleavage of non-structural proteins 12 (RNA-dependent RNA polymerase, or RdRp) and $PL^{pro}$ is involved in the processing of Nsp1–3. The rearrangement of the replicase proteins promotes the replication of the viral genome and sub-genomic RNAs and also helps to translate the structural and accessory proteins needed for the construction of new viral particles (Luan et al. 2020; Harrison et al. 2020).

In the lungs, the host immune defense mechanism produces cytokines in response to inflammation, activating neutrophils, dendritic cells, and macrophages. Neutrophils, which are a type of white blood cell, can cause lung injury when they are released in large numbers. The immune response also activates T cells, which are a type of adaptive immune cell. The production of large amounts of cytokines, like interleukins and tumor necrosis factor-α (TNF-α), in response to the inflammation can lead to a cytokine storm, which can cause cell death in the alveoli and interfere with oxygen supply. This can lead to respiratory failure and other serious complications (Channappanavar and Perlman 2017; Tomar et al. 2020). Antiviral drugs such as molnupiravir, remdesivir, paxlovid, hydroxychloroquine, and chloroquine (antimalarial), and lopinavir and ritonavir (antihuman immunodeficiency viruses, HIV) are prescribed for the treatment of COVID-19 (Chen et al. 2020).

Due to the lack of proper allopathic medication and vaccination, there is a need for an alternative approach. In ancient times, each country had a traditional medicine system that played a vital role in health care. Now the World Health Organization also has encouraged the usage of plant-based products for the treatment and management of COVID-19. The Indian traditional system of medicine is 4000 years old and is practiced throughout generations by all cultures. India is listed as one of the 12 plant diversity centers in the world. In South India, the Eastern Ghats and Western Ghats are the two significant geographic zones. The medicinal plants and their formulation from Ayurveda, Siddha, Unani, Homeopathy, and Naturopathy had a huge contribution to the management of COVID-19 (Sivasankari et al. 2013; Chopra et al. 2021).

In part, the Southern part of the Siddha medicinal system is more prevalent (Nargis Begum et al. 2009; Vijayakumar et al. 2015, 2019; Manikandan and Anand 2016; Sattanathan et al. 2020; Vijaya Anand et al. 2020). The five basic elements (Panchabootham), six tastes (Arusuvai), and three senses of humor (Mukkuttram) form the fundamental principle of Siddha medicine (Thas 2008). The clinical investigation of disease is based on the eight examinations using eyes (Vizhi), color (Niram), tongue (Naa), speech (Mozhi), pulse diagnosis (Naadi), sensation (Sparisam), urine (Moothiram), and stool (Malam) (Kannan et al. 2022). Diabetic ulcers, leprosy, eczema, vitiligo, pompholyx, alopecia, and warts are treated using Siddha medicine. Other than these herbal formulations like Nilavembu kudineer had shown efficiency against Chikungunya and Dengue virus. During COVID-19, a Siddha formulation of Kaba Sura Kudineer was approved by the Ministry of Ayush as a COVID-19 drug to improve immunity (Bala 2021).

In the current study, the literature was obtained from authenticated databases such as Springer, Nature, Biomed Central, Research Gate, ScienceDirect, Elsevier, Wiley, and PubMed for relevant content. COVID-19, SARS-CoV-2, South Indian plants, origin, pharmacological activity, ethnopharmacology, virucidal, and antiviral effects were used as the keywords to search the articles. Around 105 plants present in South India were selected from the database and scrutinized based on their antiviral property and their origin in South India. The present study is focused on documenting the important pharmacological activity, antiviral and anti-SARS-CoV-2 effect of plants present in South India.

## 17.2 Pharmacological and Ethnopharmacological Use

In the present study, antiviral and anti-SARS-CoV-2 properties of 22 (*Andrographis paniculata*, *Azadirachta indica*, *Tinospora cordifolia*, *Terminalia chebula*, *Phyllanthus amarus*, *Phyllanthus emblica*, *Aegel marmelos*, *Cinnamomum verum*, *Euphorbia hirta*, *Terminalia arjuna*, *Moringa oleifera*, *Syzygium cumini*, *Cyperus rotundus*, *Solanum trilobatum*, *Murraya koenigii*, *Cassia fistula*, *Nerium oleander*, *Gymnema sylvestre*, *Cassia alata*, *Avicennia marina*, *Clerodendrum inerme*, and *Catharanthus roseus*) South Indian have been summarized. The phytocompounds isolated from medicinal plants and herbs, plant products, and formulations have shown effects against various human viruses including Influenza A virus and its subtypes, Hepatitis B and C virus (HBV, HCV), Polioviruses, Chikungunya virus, Zika virus, Herpes simplex virus type 1 (HSV-1), Dengue virus, and HIV, and viruses that affect the plants (tobacco mosaic virus and sunn-hemp rosette virus) and livestock (Bombyx mori nucleopolyhedrovirus). The pharmacological activity of South Indian plants is given in Tables 17.1 and 17.2 contains the ethnopharmacological uses of South Indian plants concerning treating COVID-19 (Arumugam et al. 2020; Poochi et al. 2020; Anand et al. 2021; Kuchi Bhotla et al. 2021a, b; Chandra Manivannan et al. 2022).

## 17.3 South Indian Plants with Antiviral Activity

The strong anti-HIV activity was exhibited by the aerial parts (aqueous extract) of the *Andrographis paniculata*. Similarly, andrographolide derived from ethanolic extract (whole plant) exhibited activity against HIV by preventing the activity of cyclin-dependent kinase and resulting in the dysfunction of the cell cycle of HIV. In vitro analysis showed that the methanolic extract of *Andrographis paniculata* has considerable activity against the Dengue virus (Hossain et al. 2014). Jadhav and Karuppayil (2021) studied andrographolide and its derivatives found in *Andrographis paniculata* for antiviral effects against HSV-1. The findings of the study denoted that pre-treated 3,14,19-triacetylandrographolide, 14-acetyl-3, 19-isopropylideneandrographolide, and 14-acetylandrographolide were the potential in suppressing the infection and complete inhibition of the viral multiplication was achieved by post-treatment.

In in vitro studies, ethanolic extract of *Andrographis paniculata* (leaves) presented antiretroviral activity against Simian retrovirus-2 (Kumar et al. 2021a, b). The andrographolide and its derivatives were found to be effective against the Influenza A virus. Cai et al. (2015) demonstrated that 14-deoxy-11,12-dehydroandrographolide compounds extracted from *Andrographis paniculata* acted against the Influenza A virus by constraining the viral replication. Another compound 14-a-lipoyl andrographolide also presented an antiviral effect against Influenza A and its subtypes by restricting the viral hemagglutinin and binding (Intharuksa et al. 2022).

HuH7 cells infected with the Hepatitis C virus were treated with a methanolic fraction of *Phyllanthus amarus*. The treatment with *Phyllanthus amarus* diminished the replication of viral RNA of the Hepatitis C virus (Mao et al. 2016). HIV replication was limited by *Phyllanthus amarus* administration in in vitro studies (Notka et al. 2004). Ghosh et al. (2022) reported that water and alcoholic extracts of *Phyllanthus amarus* significantly decreased the replication of HIV in HeLa cells.

**Table 17.1** Pharmacological activity of South Indian plants

| Name of plant and family | Common name | Origin | Pharmacological activity | Reference |
|---|---|---|---|---|
| *Aegle marmelos* and Rutaceae | Bael or Wood apple | Native to **India**, parts of South Asia | Antioxidant, hepatoprotective, antidiabetic/ antiobesity, cardioprotective, antipyretic/anti-inflammatory, antimicrobial, anticancer, antiulcer, and antiviral | Dutta et al. (2014) |
| *Andrographis paniculata* and Acanthaceae | Nilavembu | Taiwan, Mainland China, and **India**. Asia, Sri Lanka, Thailand, and Vietnam | Anti-inflammatory, common cold, hepatoprotective, antihyperglycemic, antihyperlipidemic, antiparasitic, antimicrobial, cardiovascular, anticancer, immunomodulatory, and antiviral | Kumar et al. (2021a, b) |
| *Avicennia marina* and Acanthaceae | Gray mangrove or white mangrove | China, **India**, Pakistan, and Egypt | Antioxidant, antidiabetic, antimicrobial, antiviral, anthelmintic, antimalarial, anticancer, anti-inflammatory, and sexual activity | ElDohaji et al. (2020) |
| *Azadirachta indica* and Meliaceae | Neem | Myanmar (earlier Burma) or Upper Myanmar, **India** Pakistan Asia, Africa, and Saudi Arabia | Antiviral, antimalarial, antifungal, antibacterial, antiarthritic, anti-inflammatory, antipyretic, antigastric, spermicidal, diuretic, antifungal, and immunomodulatory | Moin et al. (2021) |
| *Cassia alata* and Caesalpiniaceae | Candle bush | **India,** Africa, and Asia | Anticancer, antimalarial, anthelmintic, anti-inflammatory, antioxidant, antibacterial, antifungal, hepatoprotective, cardioprotective, broncho-relaxant, antiviral, antidiabetic, and antihyperlipidemic | Dewi et al. (2019) |

(continued)

**Table 17.1** (continued)

| Name of plant and family | Common name | Origin | Pharmacological activity | Reference |
|---|---|---|---|---|
| *Cassia fistula* and Fabaceae | Golden shower | Asia, Pakistan, **India,** Sri Lanka Bangladesh, Myanmar, and Thailand | Antitussive, leukotriene inhibition, clastogenic effect, antipyretic, antioxidant, laxative, anti-inflammatory, wound healing, hepatoprotective, antifungal, antibacterial, hypocholesterolemic and hypoglycemic, antiparasitic, hypolipidemic, anti-fertility, antileishmaniasis, and antiviral | Danish et al. (2011) |
| *Catharanthus roseus* and Lauraceae | Madagascar periwinkle | **India,** China, and Sri Lanka | Antidiabetic, anticancer and cytotoxic, larvicidal and pupicidal, antimicrobial, antioxidant, and antiviral | Kumar et al. (2022) |
| *Cinnamomum verum* and Lauraceae | Ceylon cinnamon | Sri Lanka, **India**, Asia, China, Indonesia, Burma, the Caribbean, Australia, Africa, and Madagascar | Antimicrobial, antioxidant, anti-inflammatory, anticancer, antidiabetic, wound healing, anti-HIV, anti-Parkinson, antianxiety, antidepressant, and antiviral | Pathak and Sharma (2021) |
| *Clerodendrum inerme* and Verbenaceae | Indian privet | **India,** Northern Africa, Asia, Egypt, and Madagascar | Antioxidant, antimicrobial, anti-inflammatory, antimalarial, antiviral, antidiabetic, anticancer, and analgesic | Kar et al. (2014) |
| *Cyperus rotundus* and Cyperaceae | Korai, Korai kilangu, Muthakasu, Koraipullu | **India**, Africa, Southern Asia, and Southern, and central Europe | Analgesic, antiandrogenic, antimicrobial, cytotoxic, anticonvulsant, antidiabetic, antidiarrheal, antigenotoxic, anti-inflammatory antilipidemic and cardioprotective, antiplatelet, antiuropathogenic, | Peerzada et al. (2015) |

(continued)

**Table 17.1** (continued)

| Name of plant and family | Common name | Origin | Pharmacological activity | Reference |
|---|---|---|---|---|
| | | | hepatoprotective, lactogenic, antidepressant, antiviral, neuroprotective, and inotropic | |
| *Euphorbia hirta* and Euphorbiaceae | Asthma-plant | **India**, Sri Lanka Asia, Madagascar Europe, and North America | Antiviral, antiallergic, antibacterial, antidiabetic, antidiarrheal, antioxidant, antitumor, anxiolytic and sedative, and diuretic | Khursheed and Jain (2022) |
| *Gymnema sylvestre* and Apocynaceae | Gurmar | **India**, Sri Lanka, Africa, Australia, Vietnam, and Malaysia | Antidiabetic, antiarthritic, dental caries, antibiotic and antimicrobial, anti-inflammatory, anticancer, cytotoxic, antihyperlipidemic, immunostimulatory, hepatoprotective, antiviral, wound healing, and ethnobotanical | Tiwari et al. (2014) |
| *Moringa oleifera* and Moringaceae | Moringa | **India**, Kenia, Ethiopia and Somalia, Northern **India**, Greeks, and Romans | Antioxidant, anticancer, anti-inflammatory, immunomodulatory, hypoglycemic, antiviral, hypolipidemic, hepatic, and kidney protective | Mahato et al. (2022) |
| *Murraya koenigii* and Rutaceae | Curry Leaf | **India**, Sri Lanka, Malaysia, and South Africa | Antibacterial, antifungal, antiprotozoal, antitrichomonal, immunomodulation, antioxidant, nephroprotective, antipyretic, anthelmintic, antiulcer, antidiarrheal, cardioprotective, antiosteoporotic, antiviral, inotropic, mosquitocidal and larvicidal, dental caries, and hypoglycemic | Goel et al. (2020) |

(continued)

**Table 17.1** (continued)

| Name of plant and family | Common name | Origin | Pharmacological activity | Reference |
|---|---|---|---|---|
| *Nerium oleander* and Apocynaceae | Oleander | **Indo**-Pakistan subcontinent | Anticancer, antimicrobial, cardiotonic, antidiabetic, anti-inflammatory, antinociceptive, larvicidal, antiviral, hepatoprotective, and antioxidant | Ebrahimi et al. (2018) |
| *Phyllanthus amarus* and Phyllanthaceae | Indian gooseberry | **India**, American, African, Madagascar Australasian, and Asian | Antiamnesic, antibacterial, anticancer, antidiarrhoeal, gastroprotective and antiulcer, antifungal analgesic, anti-inflammatory, antiallodynic, antihematogenic, antinociceptive, antioxidant, antiplasmodial, antiviral, aphrodisiac, contraceptive, diuretic, antihypertensive, hepatoprotective, hypoglycemic and hypocholesterolemic, immunomodulatory, nephroprotective, radioprotective, and spasmolytic | Patel et al. (2011) |
| *Solanum trilobatum* and Solanaceae | Purple fruited pea eggplant | The southern region of **India** along with some other countries like Srilanka, Thailand, and Vietnam | The antimicrobial, hemolytic, protective effect, anti-inflammatory oviposition, deterrent and skin repellent, antioxidant, antibacterial, antiviral, antidiabetic, and immunomodulatory | Ranjith et al. (2010) |
| *Syzygium cumini* and Myrtaceae | Malabar plum | **India**, South Asia, Burma, Bangladesh, Pakistan, Nepal, Indonesia, and Sri Lanka | Antioxidant, anti-inflammatory, anticancer, radioprotection hyperlipidemia and cardioprotective, antidiabetic, antiviral, gastroprotective, | Ayyanar and Subash-Babu (2012) |

(continued)

**Table 17.1** (continued)

| Name of plant and family | Common name | Origin | Pharmacological activity | Reference |
|---|---|---|---|---|
| | | | antidiarrheal, and antimicrobial | |
| *Terminalia arjuna* and Combretaceae | Arjun | **India**, Burma, Mauritius, and Sri Lanka | Coronary flow, cardiac hemodynamics, antioxidant, cardioprotective, hypolipidemic, antiviral, antiatherogenic, and blood pressure | Amalraj and Gopi (2017) |
| *Terminalia chebula* and Combretaceae | Harida, Harada, Haritaki and Harar | Asia, Bangladesh, **India**, China, Nepal, Pakistan, Vietnam, Thailand, Sri Lanka, Africa, Myanmar, Afghanistan, Brazil, and Iran | Anti-inflammatory and anti-arthritic, antioxidative, hepatoprotective, antidiabetic and antihyperglycemic, anticancer, hypolipidemic and hypo-cholesterolemic, cardioprotective, neuroprotective, gastroprotective, anticonvulsant, molluscicidal, nephroprotective, radioprotective, wound healing, antiaging, cytoprotective, antispermatogenic, antidiarrheal and antimotility, antipsychiatry, antiviral, antifungal, antiparasitic, and antibacterial | Bag et al. (2013) |
| *Tinospora cordifolia* and Menispermaceae | Guduchi | **India**, Sri Lanka, China, Thailand, Malaysia, Bangladesh, and Africa | Antidiarrhoeal, anthem, antiviral, antimicrobial, antidiarrhoeal, anticancer, anthelmintic, and antidiarrhoeal | Choudhary et al. (2013) |
| *Phyllantus emblica* and Phyllanthaceae | Amla | Asia, **India**, Nepal, Bangladesh, Pakistan, Bhutan, Sri Lanka, and China | Immunomodulatory, analgesic, antipyretic, Anti-inflammatory, anticancer, antiproliferative, antimutagenic, | Saini et al. (2022) |

(continued)

**Table 17.1** (continued)

| Name of plant and family | Common name | Origin | Pharmacological activity | Reference |
|---|---|---|---|---|
| | | | diabetes, and associated problems, antimicrobial, neuroprotective, antiviral, antimalarial, and antioxidant | |

The extracts also had activity against the reverse transcriptase, protease, and integrase enzymes of HIV. Corilagin and ellagitannins geraniin isolated compounds of *Phyllanthus amarus* are considered as the mediators of antiviral effect.

The water extract (leaves) of *Azadirachta indica* has presented inhibition against of replication of a wide range of viruses including Measles, Chikungunya, Dengue virus, and Vaccinia in in vitro and in vivo analysis (Parida et al. 2002, Subapriya and Nagini 2005). The methanolic extract (leaves) of *Azadirachta indica* extract also presented a virucidal effect against Coxsackievirus virus B4. On pre-treatment with bark extract of *Azadirachta indica* actively prevented the entry of HSV-1 (Alzohairy 2016). The methanolic extract of *Avicennia marina* was investigated for anti-HSV-1 and HIV effects. The findings revealed that *Avicennia marina* has a strong inhibition effect against HSV-1 by blocking the viral entry, but has not shown activity against HIV (Namazi et al. 2013). The methanolic and iridoid glycoside derivatives of *Avicennia marina* diminished the multiplication of the virus (Behbahani 2014). ElDohaji et al. (2020) suggested that glycerin crude extract of *Avicennia marina* has antiviral properties against Poliovirus and HSV-1.

CIP-29, a protein that induces systemic antiviral resistance derived from *Clerodendrum inerme*, has been utilized to develop systemic resistance to the Sunn-hemp rosette virus in *Cyamopsis tetragonoloba* (Prasad et al. 2014). Adhikari et al. (2021) demonstrated that *Clerodendrum inerme* and *Gymnema sylvestre* had the potential to decline the mouse coronavirus and HSV infection. In vitro analysis of methanolic extract *Gymnema sylvestre* successfully prevented HIV and HBV infections (Subashini and Rajendran 2015). The alatains A and alatains B phytoconstituents of *Cassia alata* have shown activity against the plant tobacco mosaic virus (Zhou et al. 2020). Other compounds such as alataindolein A, B, C, and D and alatachromone A have also shown antitobacco mosaic virus (Yang et al. 2022).

Moringa, a phytocompound derived from *Moringa oleifera* (seeds), was found to have antiviral properties against the Influenza (H1N1) subtype by suppressing viral plaque formation and hemolysis in RAW264.7 cells that were infected with the H1N1 virus, according to Xiong et al. (2021). The water extract of *Moringa oleifera* has the potential against the Newcastle disease virus (Tolba et al. 2022). The ethanolic and water extracts of *Moringa oleifera* are composed of high anthraquinone and flavonoids. *Moringa oleifera* has shown inhibitory activity against various viruses like Polioviruses, Cytomegalovirus, and HIV (Hamza et al. 2021). Eleven

**Table 17.2** Ethnopharmacological uses of South Indian Plants

| Name of the plants | References |
|---|---|
| *Aegle marmelos*<br>**Leaf:** To treat abscesses, ulcers, vomiting, backache, weakness of the heart, cuts, blood sugars, acute bronchitis, dropsy, and diarrhea injuries caused by animals and beriberi. Ophthalmic infections and asthmatic complaints, cold, cough, and other respiratory complaints, hair growth, and wound healing<br>**Root:** To treat fish poison, fevers, increases seminal fluid volume, remedy for heart palpitation and melancholia, hypoglycemia, gastric troubles, dog bite, antiamoebic, heart disorders, and rheumatism<br>**Flowers** are used as a tonic for the intestine, stomach, antidiabetic, antidysenteric, local anesthetic, epilepsy, and diaphoretic<br>**Fruit:** Treat diarrhea, burn cases, gastric troubles, dysentery, laxative, constipation, digestive, tonic, ulcer, brain and heart tonic, gonorrhea, intestinal parasites, lower blood pressure, epilepsy, and cure intestinal parasites, like *Ascaris lumbricoides* and *Entamoeba histolytica*<br>**Bark:** Fish poison and fever treatment | Dutta et al. (2014) |
| *Andrographis Paniculata*<br>**Leaf:** Antioedema activity, hepatoprotective effect, antiviral/antiretroviral, antioxidant, antimicrobial, antifertility, anti-inflammatory, antidiabetic, and anticancer<br>**Stem:** Anticancer, antimicrobial, and antioxidant<br>**Root:** Antimicrobial<br>**Whole plants:** Analgesic | Kumar et al. (2021a, b) |
| *Avicennia marina*<br>**Leaf:** Anticancer (India and Iran), antioxidant (China), simplex virus 2 (HSV-2) (Iran), weak radical scavenging (China), antiviral (Indonesia), and antibacterial (Indonesia)<br>**Root:** Antiglycation (Pakistan)<br>**Fruit:** Antioxidant (China) | ElDohaji et al. (2020) |
| *Azadirachta indica*<br>**Leaf:** Antifungal, antisemitic, antihelminthic, anticlotting agent, antitumor, antituberculosis, antiviral, antiseptic, cosmetics, contraceptive, insecticides, fertilizers, insect repellents, and mosquito and nematicides<br>**Flower:** Stimulant, analgesic, elimination of intestinal worms, and phlegm and bile suppression<br>**Fruit:** Leprosy and intestinal worms, yield oil and cake an antibacterial effect, relieves piles, urinary disorders, intestinal worms, phlegm, epistaxis, diabetes, eye problem, leprosy, and wounds<br>**Barks:** Antifungal, antiallergenic, antitumor, antiprotozoal, alternative, curative of fever, and analgesic<br>**Deodorant:**<br>**1. Gum:** Effective against skin diseases such as scabies, ringworms, ulcers, and wounds<br>**2. Twigs:** Oral deodorant, toothache reliever, and tooth cleaners | Moin et al. (2021) |
| *Cassia alata*<br>**Leaf:** Gonorrhea, heart failure, abdominal pains, edema and purgative, ringworm, skin problems, impetigo, favus, and other mycoses, psoriasis, syphilis sores, chronic lichen planus, herpes, shingles, scabies, | Dewi et al. (2019) |

(continued)

**Table 17.2** (continued)

| Name of the plants | References |
|---|---|
| rash and itching, eczema, stomach pain during pregnancy and leaf decoctions, hemorrhoids, dysentery, jaundice, blood in the urine (schistosomiasis, gonorrhea), hernia, headache, and one-sided weakness or paralysis<br>**Stem:** Antiasthma, aperients, improve the visual activity and diuretic agent. (Chinese medicine) | |
| *Cassia fistula*<br>**Leaf:** The leaves extract is used as a laxative, used externally as an emollient, reduced mutagenicity in *E. coli*, in insect bites, a poultice is used for chilblains, rheumatism, facial paralysis, swelling, laxative properties, antiperiodic, the leaves are used in piles, jaundice, also externally skin eruptions, rheumatism ulcers, eczema, ringworms, insect bites, and pustules<br>**Stem:** Prescribed as an astringent, febrifuge, strong purgative, tonic, fever<br>**Root:** Chest pain, migraine, blood dysentery, joint pain, blood sugar level up to 30%, chest pain, swellings, amenorrhea, fever, retained excretions and biliousness, and heart diseases<br>**Root bark:** Exhibits cardiac disorders, anti-inflammatory activity, rheumatic conditions, biliousness, wounds, hemorrhages, boils, and various skin diseases and ulcers<br>**Flower:** Purgative, febrifugal, biliousness, and astringent. Antifertility activity in female albino rats. The heated pods are applied to swellings on the neck due to cold<br>**Fruits:** Snakebite, cathartic, skin diseases, asthma, diabetes, abortifacient, antipyretic, lessens inflammation and heat of the body, demulcent, throat troubles, useful in chest complaints, diseases of the eye and gripping, liver complaints, colic, constipation, urinary disorders, and chlorosis<br>**Seed:** Carminative, laxative, improves the appetite, antipyretic activity, cooling, biliousness, jaundice, swollen throat, skin disease, and hypoglycemic activity<br>**Bark**: Antidysentric properties, tonic, skin complaints, the powder or decoction of the bark is administered in jaundice, leprosy, heart diseases, and syphilis | Danish et al. (2011) |
| *Catharanthus roseus*<br>**Leaf:** Toothache, menorrhagia, blood circulation, and memory loss (India). Diabetes mellitus, amenorrhea (Philippines), Reducing toothaches, soothing insect bites, regulating blood sugar, healing wounds, cardiovascular diseases, and improving circulation and memory (Sri Lanka). Diabetes (England) and Vomitive (France). Diabetes; combat primary inertia in childbirth; (Dominica). Menorrhagia and diabetes (Australia), Anemia, Ulcers, inflammation (Myanmar), Traditionally used for diabetes (Spain), Astringent, diuretic, cough remedy (China), Diabetes, rheumatism (Mozambique), Indigestion, dyspepsia (Mauritius), Menorrhagia, rheumatism (Africa) and Toothache, fever, skin diseases, diuretic (India)<br>**Stem:** Diabetes (West Bengal), Dysmenorrheal (Indonesia)<br>**Root:** Menorrhagia, tonic, stomachic (India), abortion, emmenagogue, dysmenorrhea with scanty flow abortion (Philippines), diabetes | Kumar et al. (2022) |

(continued)

**Table 17.2** (continued)

| Name of the plants | References |
|---|---|
| (Mexico), vermifuge, purgative, hemostatic, depurative and toothache remedy (France). hypertensive, febrifuge (Mozambique) vomitive, vermifuge, purgative, hemostatic toothache remedies, and depurative (Madagascar). fevers, malaria (Brazil), stomach problems (Kenya), venereal disease (Venda), and febrifuge (Australia)<br>**Fruit:** Eyewash for the eyes of infants (Rodriguez Islands & Jamaica), asthma, tuberculosis, and flatulence (Bahamas) | |
| *Cinnamomum verum*<br>**Leaf:** To treat asthma, rheumatism, cephalalgia, febrifuge, inflammatory diseases, and ophthalmia (China, India, Japan, Korea, and Thailand).<br>**Root:** To treat venereal infections, coughs, elephantiasis, skin diseases, tropical burns, rheumatism, febrifuge, vermifuge, malaria, antihypertensive, sedative properties, and antidiabetic. | Pathak and Sharma (2021) |
| *Clerodendrum*<br>**Leaf:** To treat food flavoring agents (Asia). Leukorrhea, puerperal fever, and eczema<br>**Stem:** Rheumatic pain, dermatitis, body ache, bile, dysentery, and stomachache<br>**Root:** Kidney stone removal, cholera, tonsillitis, vomiting, colic, flatulence in children, bone fracture, jaundice, malaria<br>**Flower:** Blood dysentery<br>**Fruit:** Bronchitis<br>**Whole plant:** Bone fracture, Colic, cholera, vomiting, diarrhea, dermatitis, flatulence in children, dysentery, menstrual irregularities, dysmenorrhea, painful menstruation, emmenagogue, high fever, bronchitis, epilepsy, dysentery, galactagogue, gynecological disorder, indigestion, malaria, urinary disorders, constipation, intermittent fevers, kidney-stone removal, leprosy, itching, anthelmintic, dysentery, leukorrhea, lowering blood glucose level, puerperal fever, bile, rheumatic pain, body ache, fever, scorpion bite, skin diseases/blisters, jaundice, snake-bite, stomach, and bowel disorder, tonsillitis, urinary problems, diaphoretic, wound healing, stimulant, astringent, diuretic, tonic, vermifuge, demulcent, stomachic, and emmenagogue (India, Pakistan, and Bangladesh) | Kar et al. (2014) |
| *Euphorbia hirta*<br>**Leaf:** Antifungal, antibacterial activities, warmed and rubbed on itchy soles, surma to cure eye sores and lower eyelids<br>Stem: Gastrointestinal disorders (dysentery, diarrhea, intestinal parasitosis, etc.), bronchial and respiratory diseases (bronchitis, hay fever, asthma, etc.), and conjunctivitis. Tonic properties, hypotensive, anxiolytic, analgesic, antipyretic, anti-inflammatory activities, and anxiolytic<br>**Root:** Anticancer activity, prostaglandins I2, E2, and D2, inhibits aflatoxin contamination, e incognita, snake bites, and decoction | Peerzada et al. (2015) |
| *Gymnema sylvestre*<br>**Leaf:** Antisweet tested, diabetes mellitus, stomachic, and diuretic. Decreases gluconeogenic enzyme and pathological changes, snakebite, stimulating insulin, inhibition glucose absorption, hypolipidemic, | Khursheed and Jain (2022) |

(continued)

Table 17.2 (continued)

| Name of the plants | References |
|---|---|
| antiatherosclerotic, and anti-inflammatory. Inhibit glucose absorption, antimicrobial, diabetes, obesity, dianex herbal formulation, diabecon, regulated–induced hyperglycemia, and gustatory sensation<br>**Stem:** Cardiotonic, diuretic, digestive, stimulant, laxative, uterine tonic, and stomachic<br>**Root:** Cardiotonic, diuretic, digestive, stimulant, laxative, uterine tonic, and stomachic | |
| *Moringa oleifera*<br>**Leaf:** Antioxidant, antibacterial, antifungal, anti-inflammatory, cardioprotective, immunomodulatory, neuroprotective, and anticancer<br>**Stem:** Antioxidant, antifungal, and anticancer<br>**Root:** Antioxidants<br>**Fruit:** Antioxidants and anticancer.<br>**Bark:** Antioxidant<br>**Seed:** Antifungal and anticancer | Tiwari et al. (2014) |
| *Murraya koenigii*<br>**Leaf:** Stomachic, febrifuge, purgative, antihelminthic, antianemic, antiulcer, analgesic, antiamnesic, antinociceptive, cooling, itching, and anti-inflammation. Hair growth to cure eruption and bruises, vomiting, night blindness, hypercholesterolemia lightning, and bites of poisonous animals<br>**Stem:** Datum for cleaning, and strengthening gums and teeth<br>**Root:** Analgesic, antihelminthic, inflammation, cooling agent and itching, leucoderma, kidney pain, and blood disorders<br>**Flower:** Astringent<br>**Whole plant:** Leucoderma, diabetes mellitus, kidney pain, body aches, and vomiting<br>**Bark:** Datum for cleaning, and strengthening gums and teeth | Mahato et al. (2022) |
| *Nerium oleander*<br>**Leaf:** To the treatment of microbial and fungal diseases, expectorant, heart tonic, diuretic, emetic, and diaphoretic. Scabies and to reduce swellings. Snakes and other venomous bites, cure eye diseases. Nerium is emetic, cardiotonic, diaphoretic, cardiotonic, expectorant, sternutatory, diuretic, anticancer, antibacterial, and antifungal<br>**Root:** Anticancer, skin complaints, herpes, and also ringworm infection<br>**Fruit:** Treatment of microbial and fungal diseases, emetic, cardiotonic, diaphoretic, cardiotonic, expectorant, sternutatory, diuretic, anticancer, antibacterial, and antifungal. The hydroalcoholic and aqueous extract of the flowers is antinociceptive and cardiotonic<br>**Bark:** Treatment of microbial and fungal diseases, expectorant, heart tonic, diuretic, emetic, and diaphoretic. Athletic, febrifuge, and intermittent fever. Oil prepared from the root bark is used in the treatment of leprosy and skin diseases of a scaly Nature<br>**Seeds** are used as a purgative in rheumatism and dropsy<br>**Whole plant:** Anticancer, warts, against corns, carcinoma, cancerous ulcers, hard tumors, or ulcerating | Goel et al. (2020) |
| *Phyllanthus amarus*<br>**Leaf:** Diabetics, jaundice, gonorrhea and syphilis, malaria, skin | Ebrahimi et al. (2018) |

(continued)

Table 17.2 (continued)

| Name of the plants | References |
|---|---|
| diseases, menstrual, chronic dysentery, syphilis, stomachache, kidney problems, rashes, and itches<br>**Root:** Jaundice<br>**Whole plant:** Migraine, gonorrhea and syphilis, jaundice, malaria, skin diseases, dyspepsia, liver disease, moderate constipation, anorexia, irritable bowel syndrome, chronic colitis, aphrodisiac, urinary tract infection, diabetes, dysentery, cleaning uterus, stomach ache, health promotion, laxative, and disease prevention | |
| *Solanum trilobatum*<br>**Leaf:** Cure snake poison and increase male fertility. Antimicrobial activity and antimitotic against fungi and bacteria and anticancer activity<br>**Stem:** Antimitotic, anti-inflammatory, and antiulcerogenic<br>**Root:** Treat acute and chronic bronchitis, tuberculosis, and treating cough<br>**Flower:** Antibacterial<br>**Fruit:** Antibacterial and treats cough | Patel et al. (2011) |
| *Syzygium cumini*<br>**Leaf:** Strengthen the gums and teeth, dysentery with bloody discharge, and diarrhea in children<br>**Fruit** Removes bad smell from the mouth, astringent to the bowels, stomachic, biliousness, diuretic, antidiabetic, and astringent. Astringent in chronic diarrhea and efficient<br>**Bark:** Digestive, anthelmintic, astringent to the bowels, bronchitis, treatment of sore throat, thirst, asthma, dysentery, biliousness, blood purifier, and ulcers<br>**Seed:** Good for diabetes and astringent to the bowels. Treat cough, cold, skin problems, fever, rashes, and the throat, mouth genitourinary tract ulcers, and intestines (Tamil Nadu) | Ranjith et al. (2010) |
| *Terminalia arjuna*<br>**Leaf:** Earache, heart ailments (Malabar tribe and Kerala)<br>**Fruit:** Treat wounds and blood in the urine<br>**Bark**: To treat demulcent, astringent, cardiotonic, expectorant, antidysenteric, and styptic, useful in fracture, urinary astringent, leukorrhea, ulcers, anemia, diabetes, cirrhosis, and cardiomyopathy. To kill worms in teeth and cure headaches | Ayyanar and Subash-Babu (2012) |
| *Terminalia chebula*<br>Fruit: Mild laxative, tonic, stomachic, antispasmodic, alterative, hemorrhoids, ophthalmia, bleeding gums, dental caries, anti-inflammatory, ulcerated oral cavity, purifying, and healing capacity for wounds, analgesics, sore throat, oral ulcers, bleeding and ulceration in gums, and dentifrice in loose gums<br>Increase appetite, digestive aid, liver stimulant, stomachic, gastrointestinal prokinetic agent, chronic diarrhea, nervous weakness, nervous irritability, promotes the receiving power of the five senses<br>It is adjuvant in hemorrhages due to its astringent nature and good for chronic cough, chorizo, sore throat as well as asthma. Also, it is useful in renal calculi, dysuria, retention of urine, and skin disorders with discharges like allergies, urticaria, and other erythematous disorders | Amalraj and Gopi (2017) |

(continued)

**Table 17.2** (continued)

| Name of the plants | References |
|---|---|
| *Terminalia chebula*<br>**Fruit:** Reduces swelling, hastens, healing, and cleanses wounds, ulcers, skin disorders, and erysipelas, anti-inflammatory, haritaki prevents problems of the throat and stomatitis<br>**Bark:** Used as a diuretic and tonic for the heart. | Bag et al. (2013) |
| *Tinospora cordifolia*<br>**Leaf:** Fever, cholera, rheumatism, snake-bite, vermifuge, syphilitic sores, wash for sore eyes, internal inflammation, wound itching, reduce thirst, increase appetite, jaundice, rheumatism, relieve body pain leaf, abdominal pain, paralysis, leprosy, and skin disease<br>**Stem:** Snake-bite, rheumatism, vermifuge, wound itching, wash for sore eyes and syphilitic sores, internal inflammation, increase appetite, thirst, appetizer, antipyretic, stomachic, bitter tonic, mucous congætion, ears problems, treatment of stomachache, parasites, jaundice, treat fevers caused by smallpox, cholera, and malaria. Diabetes, hypertension, postpartum remedy, and muscle pain (Murut community), garo, non-garo, and prevention of intestinal disorders (Bangladesh)<br>**Root:** Snake-bite, cholera, rheumatism, vermifuge, syphilitic sores, wash for sore eyes, internal inflammation, thirst, increase appetite, and wound itching | Choudhary et al. (2013) |
| *Phyllantus emblica*<br>**Leaf:** Antimalarial<br>**Root:** Eczema (China), hemostatic and astringent (Nepali). Aperient, refrigerant, and febrile condition<br>**Flower:** Refrigerant and aperient<br>**Fruit:** Laxative, diuretic, and inflammation of the eye<br>**Bark:** Refrigerant, aperient, and gonorrhea | Saini et al. (2022) |

compounds extracted from the methanolic extract of *Moringa oleifera* (seeds) were then evaluated against the H1NI virus. The results revealed that glucomoringin, moringa A, and vitexin had virucidal effects against the H1N1 virus (Xiong et al. 2022).

Somu et al. (2019) tested 14 fractions isolated from *Aegle marmelos* (leaves extract) for their antiviral property. Seselin exhibited significant effectiveness against Bombyx mori nucleopolyhedrovirus present in larvae of silkworms. Kaur et al. (2020) suggested *Aegle marmelos* has the strongest efficacy against HIV. Wahab et al. (2020) tested the efficiency of the methanolic extract of *Catharanthus roseus* on HSV-1. The secondary metabolites terpenoid, alkaloid, and saponin expressed anti-HSV-1 effect by decreasing plaque formation in both pre- and post-treatment. In vitro studies by Xu et al. (2015) in a hepatoblastoma cell line transfected with Hepatitis B disclosed that among the 37 sesquiterpenoids extracted from methanolic extract of *Cyperus rotundus* (roots), eudesmane-type sesquiterpenoids expressed good antiviral effect against Hepatitis B.

Ogbole et al. (2021) have reported that *Euphorbia hirta* extract (leaves) has a protective action against Enterovirus C99, Coxsackievirus A13, and Coxsackievirus

A20. The hydromethanolic extract of *Euphorbia hirta* showed activity against the Simian immunodeficiency viruses, HIV-1, and HIV-2 by decreasing the viral replication (Gyuris et al. 2009). The virucidal property of casuarinin derived from *Terminalia arjuna* (bark) reduced the cell growth, plaque formation, and virality of HSV-2 (Cheng et al. 2002). In silico analysis, compounds present in *Terminalia arjuna* were tested for antiviral effects against the Zika virus. The outcome of the results revealed that tannic acid from *Terminalia arjuna* can control the Zika virus (Priya et al. 2018). Singh et al. (2021) described that *Cinnamomum verum* extract had efficacy against the HIV protease.

The aqueous extract of *Nerium oleander* was tested against various viruses such as Reovirus type-1, Vesicular stomatitis virus, Yellow fever virus, HSV, HIV-1, and Poliovirus type-1. *Nerium oleander* was found to be constructive in controlling Polio virus type-1 by decreasing the formation of plaque (Sanna et al. 2021). Mononuclear cells infected with HIV were used in detecting the performance of *Nerium oleander* against the virus. The extract reduced the effect of the virus by diminishing the envelope protein, GP-120 expression, and replication of HIV (Singh et al. 2013). Silver nanoparticles synthesized from *Tinospora cordifolia* have expressed antiviral activity against Chikungunya virus-infected Vero cells (Sharma et al. 2019).

The methanolic extract of the *Terminalia chebula* (fruits) presented notable inhibitory activity (El-Mekkawy et al. 1995). Kurokawa et al. (1995) indicated that the extract of *Terminalia chebula* and acyclovir together can provide better protection by attenuating the infection of HSV-1. Chebulinic acid and chebulagic acid extracted from ethanolic extract of *Terminalia chebula* (fruits) were purified and examined against HSV-2. During post-viral infection in Vero cells, the phytochemical displayed a significant virucidal effect against HSV-2 by preventing viral attachment, penetration, and plaque formation. The activity of phytochemicals was found to be greater than the activity exhibited by the standard drug acyclovir (Kesharwani et al. 2017). Another research involving fruits of *Terminalia chebula* (galloyl moiety) exhibited notable effects against HIV-1 (Ahn et al. 2002).

The punicalagin and chebulagic acid isolated from dried fruits of *Terminalia chebula* showed remarkable activity against HSV-1 in lung cells by suppressing the transmission and binding of viral particles (Lin et al. 2011). The phytocompounds of *Terminalia chebula* (fruits) were assessed for anti-HCV activity two tannins (chebumeinin A and chebumeinin B) showed positive effects on the reduction of HCV (Ajala et al. 2014). Oyuntsetseg et al. (2014) reported that subtypes of the influenza virus were decreased by the treatment with an aqueous extract of *Terminalia chebula* (seeds). Mishra et al. (2018) suggested that *Terminalia chebula* has anti-HIV-1 and HSV-2 properties. Likewise, the chebulinic acid derived from flowers of *Terminalia chebula* exhibited an inhibitory effect against HSV-2 and Dengue virus by blocking the entry of the virus. However, the chebulinic acid was not effective against Chikungunya virus infection (Thomas et al. 2022).

The phytoconstituents like putranjivain A, digallic acid, 1-*O*-galloyl-*β*-D-glucose, kaempferol-3-*O*-*β*-D-glucoside, 1,6-di-*O*-galloyl-*β*-D-glucose, quercetin-3-*O*-*β*-D-glucoside found in methanolic extract of *Phyllanthus emblica* presented anti-HIV property (El-Mekkawy et al. 1995). Coxsackievirus virus B3 was combated by the

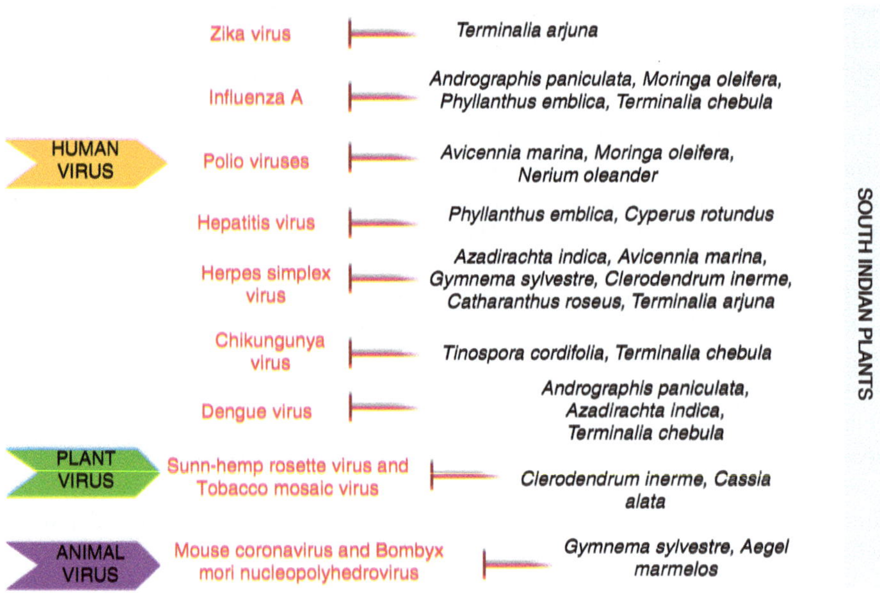


**Fig. 17.1** Antiviral activity of South Indian plants

phyllaemblicin B compound derived from *Phyllanthus emblica* in both animal models and cell-based analysis (Wang et al. 2009). Phyllaemblic acid methyl, phyllaemblicin B, and phyllaemblicin C also affected Coxsackievirus virus B3 (Liu et al. 2009). 1,2,4,6-tetra-*O*-galloyl-$\beta$-D-glucose (phenolic compound) extracted from *Phyllanthus emblica* exhibited activity against the HBV and HSV-1 (Xiang et al. 2010, 2011). Phyllaemblicins G1–G8 and glochicoccin D, phyllaemblicin F, and phyllaemblic acid were tested for antihepatitis B properties. The findings indicate that phyllaemblicins G6–G8 have antihepatitis B activity (Lv et al. 2014).

Lv et al. (2015) revealed that glochicoccinoside D had an effect against the influenza A virus. The gel formulated from ethanolic extract of *Phyllanthus emblica* was found to have the potential of decreasing HSV-2 and HIV-1 (Mishra et al. 2018). Triphala formulation containing *Phyllanthus emblica* has repressed the action of the Dengue virus in Vero cells (Panya et al. 2021). 1,2,3,4,6-penta-*O*-galloyl-$\beta$-D-glucose isolated from the various parts (leaves, barks, and roots) of *Phyllanthus emblica* demonstrated virucidal effect on Influenza virus in MDCK cells. Glochicoccinoside D and phyllaemblicin B from roots of *Phyllanthus emblica* also showed an effect against the subtype (H3N2) of the Influenza virus (Saifulazmi et al. 2022) (Fig. 17.1).

## 17.4 Anti-SARS-COV-2 Activity of South Indian Plants

Calu-3 lung cells affected with the SARS-CoV-2 virus were used for testing the ability of *Andrographis paniculata* and its derived andrographolide. The results showed that *Andrographis paniculata* could suppress the generation of virions (Sa-Ngiamsuntorn et al. 2021). Andrographolide and andrographidine C of *Andrographis paniculata* presented higher binding toward the targets. Especially the andrographidine C had a stronger binding affinity to phosphatidylinositol-4,5-bisphosphate 3-kinase catalytic subunit gamma and ACE receptors (Xie et al. 2021). SARS-CoV-2 induces host response by inducing 36 genes causing cytokine storm. The study reported that andrographolide can restrict the cytokine storm by suppressing the NFkB1 pathway (Rehan et al. 2021). Recent research by Das et al. (2022) also demonstrated that andrographolide has the efficacy to attenuate the cytokine storm by regulating interleukin 6 and TLR4-MD2. Andrographolide also could interact with various viral entry points like cathepsin L, TMPRSS-2, furin, and ACE2.

Latha et al. (2022) investigated the anti-SARS-CoV-2 activity of ethanolic extracts of *Withania somnifera*, *Centella asiatica*, *Calophyllum inophyllum*, *Zanthoxylum piperitum*, and *Andrographis paniculata* on nucleoprotein and envelope genes of SARS-CoV-2 virus. The results showed that in in vitro analysis, *Centella asiatica*, *Andrographis paniculata*, and *Withania somnifera* displayed activity against both genes. Among them, *Andrographis paniculata* had a higher activity than the standard drug (remdesivir). The compounds isolated from *Centella asiatica* (kaempferol), *Withania somnifera* (coagulin G), and *Andrograhis paniculata* (andrographidine C) have also shown better inhibitory effects against the spike protein, $M^{pro}$, and RdRp, respectively, in in silico method.

Wanaratna et al. (2022) examined the efficiency of the *Andrograhis paniculata* extract in individuals affected with mild SARS-CoV-2 infection. *Andrograhis paniculata* extract treatment showed betterment in levels of C-reactive protein without adverse effects. Tanwettiyanont et al. (2022) reported administration of *Andrograhis paniculata* did not show a significant reduction in pneumonia in individuals with mild SARS-CoV-2. Intharuksa et al. (2022) stated that *Andrograhis paniculata* has the capability to reduce the production of the virion and also can bind with ACE and $3CL^{pro}$ receptors.

Borkotoky and Banerjee (2021) conducted a study to examine the competency of the 70 compounds derived from *Azadirachta indica* against the proteins (envelope and membrane) of SARS-CoV-2 which are useful in the assembly of viruses. Nimbolin A (−11.2 kcal/mol) and nimocin (−10.2 kcal/mol) from *Azadirachta indica* showed a good binding affinity to envelope and membrane protein respectively. Desacetylgedunin present in seeds of *Azadirachta indica* was found to have significant binding with the papain-like protease which was revealed in in silico studies conducted by Baildya et al. (2021). Shadrack et al. (2021) reported that the virus entry site (spike-receptor binding domain (RBD)-ACE2) was obstructed by quercetin, azadirachtin H, and margocin isolated from *Azadirachta indica*.

Azadirachtin was effective the inhibiting $M^{pro}$, $PL^{pro}$, and RdRp proteins. In vivo studies, oral treatment of nimbin and its isomers derived inhibited viral production and dissemination of m-CoV-RSA59 and SARS-CoV-2 (Foka et al. 2022).

Sagar and Kumar (2020) evaluated the phytoconstituents present in *Tinospora cordifolia* against the major targets (RBD, RdRp, surface glycoprotein, and $M^{pro}$) important for the attachment and multiplication of the SARS-CoV-2 virus. The outcome of the study demonstrated that berberine, magnoflorine, tinocordiside, and isocolumbin showed higher binding scores against all four targets. Both $M^{pro}$ and surface glycoprotein were potentially inhibited by isocolumbin and tinocordiside. Another in silico study by Chowdhury (2021) revealed that berberine present in *Tinospora cordifolia* can bind with $3CL^{pro}$ which is essential during the replication of the virus.

The phytocompounds present in *Tinospora cordifolia* were analyzed for anti-SARS-CoV-2 activity. Among 11 compounds scrutinized, xanosporic acid, tembetarine, cardiofolioside B, berberine, and tinosponone were found to be potential inhibitors of $3CL^{pro}$ and tinosponone had the top binding affinity to $3CL^{pro}$ (Krupanidhi et al. 2021). Mulpuru and Mishra (2021) reported that saponarin found in *Tinospora cordifolia* was effective in arresting the $M^{pro}$ of the SARS-CoV-2 virus. Jena et al. (2021) investigated the effectiveness of phytocompounds of *Tinospora cordifolia* against the spike protein of SARS-CoV-2. The results suggested that several compounds present in *Tinospora cordifolia* have the ability to counteract the attachment of spike protein with the ACE receptor.

Sarkar et al. (2022) studied 16 compounds of *Terminalia chebula* for their anti-SARS-CoV-2 activity. The observations revealed that daucosterol, 1,3,6-trigalloyl glucose, and β-sitosterol compounds had a higher inhibitory effect against all eight ($M^{pro}$, $PL^{pro}$, non-structural protein (NSP) 10, helicase, NSP9 RNA binding protein, NSP15 endoribonuclease, RBD, and RdRp) targets. Arjunin, chebulinic acid, chebulanin, chebulagic acid, punicalin, 1,6 di-*O*-galloyl-D-glucose, 1,2,3,4,6 penta galloyl β-Dglucose, galloyl glucose, ellagitanin, terflavin C, terflavin D, casuarinin, and corilagin in *Terminalia chebula* were tested antiviral property against SARS-CoV-2. The findings revealed that the compounds of *Terminalia chebula* can be used as a therapeutic alternative for the treatment of COVID infection (Naik et al. 2021). Ghosh et al. (2022) evaluated 22 various phytochemicals present in the *Terminalia chebula* against the SARS-CoV-2 virus. The findings reported that sterols and triterpenoids have strong binding against the $M^{pro}$, whereas daucosterol exhibited stronger affinity.

Based on the literature, 190 phytochemicals present in *Phyllanthus amarus* were selected and docked against the $M^{pro}$ and $3CL^{pro}$. Out of 190, 16 phytochemicals showed a good binding affinity with scores between −8.9 and −9.6 kcal/mol. Among them, quercetin-3-*O*-glucuronide and myricitrin topped the binding score (Murthy et al. 2021). Murugesan et al. (2021) recommended that astragalin, apigenin-6-*C*-glucosyl7-Oglucoside, amritoside, and pectolinarin present in *Phyllanthus amarus*, *Tinospora cordifolia*, and *Phyllanthus niruri* has an antiviral effect against the SARS-CoV-2 virus. Sree et al. (2022) demonstrated that *Phyllanthus amarus* has the potential to combat SARS-CoV-2 by acting against the $M^{pro}$. The phytoconstituents of *Phyllanthus amarus* displayed better binding energy than the reference (lopinavir).

Chikhale et al. (2021) reported that chlorogenic acid, myricetin, and quercitrin found in the *Phyllanthus emblica* have the potential to act against the protein targets of the SARS-CoV-2 virus. Pandey et al. (2021) established the importance of phytochemicals present in *Phyllanthus emblica* and *Aegel marmelos* through in silico analysis. Three compounds from *Phyllanthus emblica* and two compounds from *Aegel marmelos* were analyzed and the results exhibited that (2S)-eriodictyol 7-*O*-(6″-*O*-galloyl)-*β*-D glucopyranoside from *Phyllanthus emblica* had good binding score against the RdRp of SARS-CoV-2 virus. Varnasseri et al. (2022) suggested that tea prepared using *Phyllanthus emblica* lowered the length of stay and improved the C-reactive protein and oxygen saturation level.

In silico studies of phytochemicals present in cinnamon were examined by Prasanth et al. (2021). Pavetannin C1 and tenufolin presented a higher correlation with both $M^{Pro}$ and spike protein. Other compounds such as 6-cinnamtannin-B1, proanthocyanidin-A2, glucopyranosylprocyanidin B1, kaempferol 3-*α*-*L*-arabinofuranoside-7- rhamnoside also showed active binding to spike and $M^{pro}$ protein. Yakhchali et al. (2021) recommended the usage of *Cinnamomum verum* in the treatment of SARS-CoV-2 because *Cinnamomum verum* is a well-known antiobstructive drug that provides great relief for lung obstruction.

Khursheed and Jain (2022) reported that methanolic extracts (leaves and roots) of *Euphorbia hirta* can inhibit the ACE. A computational approach was used by Parmar et al. (2022) to check the anti-SARS-CoV-2 effect of *Euphorbia hirta*. Myricetin was efficient in forbidding the $M^{pro}$ and 3-o-*α*-rhamnoside, isoquercitrin, and quercetin were found to be efficient in constraining the RdRp. Rutin and euphrobianin showed significant activity against both RdRp and $M^{pro}$. Around 170 phytocomounds present in *Euphorbia hirta* were scrutinized using molecular docking. The results demonstrated that trigallic acid (−9.2 kcal/mol), trigalloylquinic acid (−9.0 kcal/mol), quercetin-3-*O*-rhamnoside (−9.0 kcal/mol), myricetin-3-*O*-rhamnoside (−9.0 kcal/mol), and pedunculagin II (−8.9 kcal/mol) higher affinity against the SARS-CoV- $M^{pro}$ (Cayona and Creencia 2022).

Arjunetin present in ethanolic extract of *Terminalia arjuna* was tested against SARS-CoV-2 using docking analysis. The findings established that arjunetin showed antiviral activity by acting against the RdRp, $3CL^{pro}$, and $PL^{pro}$, and the binding of arjunetin was greater than the lopinavir and remdesivir control drugs (Arumugam et al. 2022). Wanarase et al. (2022) reported that N-desmethyl sildenafil and triterpenoid present in *Terminalia arjuna* could suppress the multiplication of the SARS-CoV-2 virus by inhibiting the $M^{pro}$.

Thirty-three bioactive compounds of ethanolic extract *Moringa oleifera* (fruit) were examined for their antiviral properties against SARS-CoV-2 infection. The outcome of the study indicated that the phytochemicals of *Moringa oleifera* were efficient in binding with spike and ACE receptors (Siddiqui et al. 2022a). Sen et al. (2022) demonstrated that flavonoids (kaempferol, isorhamnetin, and apigenin) present in *Moringa oleifera* have activity against the $M^{pro}$. Sivani et al. (2021) proposed that isorhamnetin-3-*O*-rutinoside and rutin antiviral compounds of *Moringa oleifera* may have activity against the SARS-CoV-2 virus. The ethanolic extract of *Moringa oleifera* (leaf) was subjected to gas chromatography-mass spectroscopy. The β-sitosterol and β-tocopherol compounds exhibited good interaction with spike-RBD-ACE2 and spike glycoprotein, respectively (Siddiqui et al. 2022a, b). Aini et al. (2022) screened the phytoconstituents of *Moringa oleifera* and *Syzygium cumini* for the anti-SARS-CoV-2 effect. The observation showed that myricetin from *Moringa oleifera* and ellagic acid from *Syzygium cumini* were the best candidates for a therapeutic approach against the SARS-CoV-2 virus.

Kumar et al. (2021a, b) proposed that two secondary metabolites (stigmasta-5,22-dien-3-ol and β-amyrin) in *Cyperus rotundus* have the ability to attenuate the $M^{pro}$. The plant-specific compounds of *Solanum trilobatum* including α-solanine, solanidine, and solasodine studied for their effect against SARS-CoV-2. The findings of the study revealed that α-solanine (−203.0 kcal/mol) had higher binding energy against the $M^{pro}$. Other two compounds solasodine (−101.6 kcal/mol) and solanidine (−69.0 kcal/mol) also exhibited good binding affinity to the target (Anandakumar et al. 2020). Molecular docking methods were used by Wadanambi et al. (2022) to analyze the carbazole alkaloids (koenine, o-methylmurrayamine A, girinimbine, and mukonicine) found in *Murraya koenigii*. Mukonicine, o-methylmurrayamine A, koenine, and girinimbine All the phytochemicals significantly prevented the viral replication of SARS-CoV-2 by suppressing $M^{pro}$ activation.

Dawood et al. (2021) focused on anthraquinone extracted from ethanolic extract of *Cassia fistula* (pods) to investigate the antiviral property. The anthraquinone performed well in the antioxidant assay. In silico analysis disclosed that anthraquinone (aloe-emodin and rhein) can restrict the $M^{pro}$. The efficacy of *Nerium oleander* and its derivative oleandrin were measured for their antiviral activity in Vero cells. The administration of oleandrin to Vero cells infected during pre- and post-infection presented a decline in the viral production of SARS-CoV-2 (Plante et al. 2021). Subramani et al. (2020) suggested that active phytoconstituents, especially the gymnemic acids of *Gymnema sylvestre* can reduce the replication of the virus by binding to the $M^{pro}$ site. Ernawati et al. (2021) assessed the phytoconstituents of *Dendrophthoe petandra* and *Cassia alata* for antiviral activity. The authors identified that quercetin, aloe-emodin, emodin, and kaempferol from *Dendrophthoe petandra* and *Cassia alata* have the potential to act as an inhibitor for SARS-CoV-2 infection.

A new steroid versisterol from *Avicennia marina* showed strong activity against the $3CL^{pro}$ in in silico analysis (Elsbaey et al. 2022). Kar et al. (2021) used molecular docking methods to evaluate the anti-SRAS-CoV-2 property of 12 different species of *Clerodendrum* spp. $M^{pro}$ enzyme, RdRp, and spike protein were fixed as the target sites. The compounds stigmasterol, taraxerol, and friedelin were securitized from *Clerodendrum* spp. The finding depicted that taraxerol had a better binding affinity to the target sites. The vindolininol, apparicine, and 12-chlorotabersonine of *Catharanthus roseus* were selected from the database based on their structural similarity with chloroquine and hydroxychloroquine drugs. The compounds were also checked for their pharmacokinetics and drug-likeness (Lipinski's) properties.

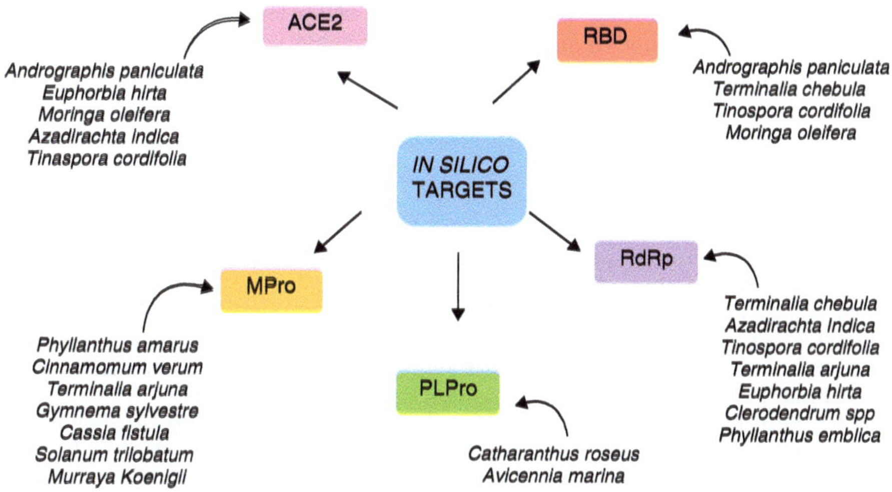


**Fig. 17.2** South Indian plant effects against major in silico targets of COVID-19

The outcome of the study demonstrated that compounds have a strong binding score to the $PL^{pro}$ and spike protein (Kalaria and Patel 2020). Figure 17.2 represents the South Indian plant effect against major in silico targets of COVID-19.

## 17.5 Conclusion

The outbreak of SARS-CoV-2 has become a global health crisis due to the unavailability of proper medication and vaccines. In the current scenario, an alternative or complementary treatment or medicine is mandatory to control the spread of infection. To date only, antimalarial and anti-HIV drugs are prescribed to individuals affected with COVID-19. India is the second largest populated country. It is hectic to handle a pandemic like COVID-19. Till present, around 4.5 million people were affected by COVID-19 with a death rate of 5 lakhs in India. The World Health Organization has recommended the utilization of plant-based medicine and products to tackle the virus. India holds a heritage of having traditional medicinal systems such as Ayurveda, Siddha, and Unani. Siddha medicine has been practiced in the Southern parts of India.

The current chapter has concentrated on documenting the antiviral and anti-SARS-CoV-2 effects of plants that are indigenous to South India. Many commonly available and day-to-day consumed plants and herbs including drumstick, gooseberry, neem, java plum, and green chiretta have shown activity against SARS-CoV-2. The secondary metabolites and plant extracts have exhibited better anti-SARS-CoV-2 effects against the pivotal targets of SARS-CoV-2 like $M^{pro}$, $PL^{pro}$, RdRp, spike protein, and ACE2 receptors in in silico analysis. Thus, studying the action of secondary metabolites, formulations, and plant-based products on these targets in employing in vivo and in vitro analysis could pay a deep insight into the discovery of novel therapeutic approaches to combat COVID-19.

## References


Adhikari B, Marasini BP, Rayamajhee B, Bhattarai BR, Lamichhane G, Khadayat K, Adhikari A, Khanal S, Parajuli N (2021) Potential roles of medicinal plants for the treatment of viral diseases focusing on COVID-19: a review. Phytother Res 35(3):1298–1312

Ahn MJ, Kim CY, Lee JS, Kim TG, Kim SH, Lee CK, Lee BB, Shin CG, Huh H, Kim J (2002) Inhibition of HIV-1 integrase by galloyl glucoses from *Terminalia chebula* and flavonol glycoside gallates from *Euphorbia pekinensis*. Planta Med 68(5):457–459

Aini NS, Kharisma VD, Widyananda MH, Murtadlo AA, Probojati RT, Turista DD, Tamam MB, Jakhmola V, Sari DP, Albari MT, Pernamasari D (2022) *In silico* screening of bioactive compounds from *Syzygium cumini* L. and *Moringa oleifera* L. against SARS-CoV-2 *via* tetra inhibitors. Pharm J 14(4):267

Ajala OS, Jukov A, Ma CM (2014) Hepatitis C virus inhibitory hydrolysable tannins from the fruits of *Terminalia chebula*. Fitoterapia 99:117–123

Alzohairy MA (2016) Therapeutics role of *Azadirachta indica* (neem) and their active constituents in diseases prevention and treatment. Evid Based Complement Alternat Med 2016:7382506

Amalraj A, Gopi S (2017) Medicinal properties of *Terminalia arjuna* (Roxb.) Wight & Arn.: a review. J Tradit Complement Med 7(1):65–78

Anand AV, Balamuralikrishnan B, Kaviya M, Bharathi K, Parithathvi A, Arun M, Senthilkumar N, Velayuthaprabhu S, Saradhadevi M, Al-Dhabi NA, Arasu MV, Yatoo MI, Tiwari R, Dhama K (2021) Medicinal plants, phytochemicals, and herbs to combat viral pathogens including SARS-CoV-2. Molecules 26(6):1775

Anandakumar S, Kannan D, Wilson E, Narayanan KB, Suresh G, Kanakavalli K, Manoharan MT (2020) Potential phytopharmaceutical constituents of *Solanum trilobatum* L. as significant inhibitors against COVID-19: robust-binding mode of inhibition by molecular docking, PASS-aid bioactivity and ADMET investigations 2020

Arumugam VA, Sangeetha T, Zareena F, Pavithra R, Ann M, Sanjeev A, Sunantha B, Meyyazhagan A, Yatoo MI, Sharun K, Tiwari R, Pandey MK (2020) COVID-19 and the world with co-morbidities of heart disease, hypertension and diabetes. J Pure Appl Microbiol 14(3):1623–1638

Arumugam GS, Sen A, Dash SS, Mitra K, Doble M, Rajaraman G, Gummadi SN (2022) Arjunetin as a promising drug candidate against SARS-CoV-2: molecular dynamics simulation studies. J Biomol Struct Dyn 40(22):12358–12379

Ayyanar M, Subash-Babu P (2012) *Syzygium cumini* (L.) Skeels: a review of its phytochemical constituents and traditional uses. Asian Pac J Trop Biomed 2(3):240–246

Bag A, Bhattacharyya SK, Chattopadhyay RR (2013) The development of *Terminalia chebula* Retz. (*Combretaceae*) in clinical research. Asian Pac J Trop Biomed 3(3):244–252

Baildya N, Khan AA, Ghosh NN, Dutta T, Chattopadhyay AP (2021) Screening of potential drug from *Azadirachta Indica* (Neem) extracts for SARS-CoV-2: an insight from molecular docking and MD-simulation studies. J Mol Struct 1227:129390

Bala D (2021) Attitudes, beliefs, and self-use of Kabasura Kudineer among urban and rural population in Tamil Nadu, India: a comparative cross-sectional study. J Fam Med Prim Care 10(1):158–166

Behbahani M (2014) Evaluation of anti-HIV-1 activity of a new iridoid glycoside isolated from *Avicenna marina*, *in vitro*. Int Immunopharmacol 23(1):262–266

Boopathi S, Poma AB, Kolandaivel P (2021) Novel 2019 coronavirus structure, mechanism of action, antiviral drug promises and rule out against its treatment. J Biomol Struct Dyn 39(9): 3409–3418

Borkotoky S, Banerjee M (2021) A computational prediction of SARS-CoV-2 structural protein inhibitors from *Azadirachta indica* (neem). J Biomol Struct Dyn 39(11):4111–4121

Cai W, Li Y, Chen S, Wang M, Zhang A, Zhou H, Chen H, Jin M (2015) 14-Deoxy-11, 12-dehydroandrographolide exerts anti-influenza A virus activity and inhibits replication of H5N1 virus by restraining nuclear export of viral ribonucleoprotein complexes. Antivir Res 118: 82–92

Cayona R, Creencia E (2022) Phytochemicals of *Euphorbia hirta* L. and their inhibitory potential against SARS-CoV-2 main protease. Front Mol Biosci 8:801401

Chandra Manivannan A, Malaisamy A, Eswaran M, Meyyazhagan A, Arumugam VA, Rengasamy KRR, Balasubramanian B, Liu WC (2022) Evaluation of clove phytochemicals as potential antiviral drug candidates targeting SARS-CoV-2 main protease: computational docking, molecular dynamics simulation, and pharmacokinetic profiling. Front Mol Biosci 9:918101

Channappanavar R, Perlman S (2017) Pathogenic human coronavirus infections: causes and consequences of cytokine storm and immunopathology. Semin Immunopathol 39(5):529–539

Chen PL, Lee NY, Cia CT, Ko WC, Hsueh PR (2020) A review of treatment of Coronavirus disease 2019 (COVID-19): therapeutic repurposing and unmet clinical needs. Front Pharmacol 11: 584956

Cheng HY, Lin CC, Lin TC (2002) Antiherpes simplex virus type 2 activity of casuarinin from the bark of *Terminalia arjuna* Linn. Antivir Res 55(3):447–455

Chikhale RV, Sinha SK, Khanal P, Gurav NS, Ayyanar M, Prasad SK, Wanjari MM, Patil RB, Gurav SS (2021) Computational and network pharmacology studies of *Phyllanthus emblica* to tackle SARS-CoV-2. Phytomed Plus 1(3):100095

Chopra D, Bhandari B, Dwivedi S (2021) Beneficial role of Indian medicinal plants in COVID-19. MGM J Med Sci 8(2):166

Choudhary N, Siddiqui MB, Azmat S, Khatoon S (2013) *Tinospora cordifolia*: ethnobotany, phytopharmacology and phytochemistry aspects. Int J Pharm Sci Res 4(3):891

Chowdhury P (2021) *In silico* investigation of phytoconstituents from Indian medicinal herb '*Tinospora cordifolia* (giloy)' against SARS-CoV-2 (COVID-19) by molecular dynamics approach. J Biomol Struct Dyn 39(17):6792–6809

Danish M, Singh P, Mishra G, Srivastava S, Jha KK, Khosa RL (2011) *Cassia fistula* Linn. (Amulthus)-An important medicinal plant: a review of its traditional uses, phytochemistry and pharmacological properties. J Nat Prod Plant Resour 1(1):101–118

Das BS, Das NC, Swain SS, Mukherjee S, Bhattacharya D (2022) Andrographolide induces anti-SARS-CoV-2 response through host-directed mechanism: an *in silico* study. Futur Virol 17:651. https://doi.org/10.2217/fvl-2021-0171

Dawood YA, Fadal SA, Almousawi UM, Al-Hejjaj MY (2021) Antioxidant activity and *in-silico* study of anthraquinone glycosides extracted from *Cassia fistula* against the main protease (7BUY) in SARS-CoV-2. Biomed Pharmacol J 14(3):1669–1675

Dewi R, Firza Y, Ali M, Saleih F, Suede A, Abdul S (2019) A review on *Cassi alata*: pharmacological, traditional and medicinal aspects. Australian Herbal Insight 2(1):5

Dutta A, Lal N, Naaz M, Ghosh A, Verma R (2014) Ethnological and ethnomedicinal importance of *Aegle marmelos* (L.) Corr (Bael) among indigenous people of India. Am J Ethnomed 1(5): 290–312

Ebrahimi F, Ghorbani Nohooji M, Miri SM (2018) Agronomic and pharmacological aspects of *Nerium oleander*: an important medicinal plant. In: The first national congress and international fair of medicinal plants and strategies for Persian medicine that affect diabetes, pp 9–11

ElDohaji LM, Hamoda AM, Hamdy R, Soliman SSM (2020) Avicennia marina a natural reservoir of phytopharmaceuticals: curative power and platform of medicines. J Ethnopharmacol 263: 113179

El-Mekkawy S, Meselhy MR, Kusumoto IT, Kadota S, Hattori M, Namba T (1995) Inhibitory effects of Egyptian folk medicines on human immunodeficiency virus (HIV) reverse transcriptase. Chem Pharm Bull (Tokyo) 43(4):641–648

Elsbaey M, Ibrahim MA, Hegazy ME (2022) Versisterol, a new endophytic steroid with 3CL protease inhibitory activity from *Avicennia marina* (Forssk.) Vierh. RSC Advances 12(20): 12583–12589

Ernawati T, Angelina M, Dewi RT, Fajriah S. Investigation of the potential COVID-19 inhibitor from *Cassia alata* and *Dendrophthoe petandra* using computational approach. InIOP conference series: materials science and engineering, IOP Publishing, 2021; 1011(1):012017)

Foka FET, Manamela N, Mufamadi SM, Mufhandu HT (2022) Potential of *Azadirachta indica* as a capping agent for antiviral nanoparticles against SARS-CoV-2. Biomed Res Int 2022:5714035

Ghosh R, Badavath VN, Chowdhuri S, Sen A (2022) Identification of alkaloids from *Terminalia chebula* as potent SARS-CoV-2 main protease inhibitors: an *in silico perspective*. Chem Select 7(14):e202200055

Goel A, Sharma A, Kulshrestha S (2020) A phytopharmacological review on *Murraya koenigii*: an important medicinal plant. Inventi Rapid Ethnopharmacol 62:113–119

Gyuris A, Szlávik L, Minárovits J, Vasas A, Molnár J, Hohmann J (2009) Antiviral activities of extracts of *Euphorbia hirta* L. against HIV-1, HIV-2 and SIVmac251. In Vivo 23(3):429–432

Hamza M, Ali A, Khan S, Ahmed S, Attique Z, Ur Rehman S, Khan A, Ali H, Rizwan M, Munir A, Khan AM, Siddique F, Mehmood A, Nouroz F, Khan S (2021) nCOV-19 peptides mass fingerprinting identification, binding, and blocking of inhibitors flavonoids and anthraquinone of *Moringa oleifera* and hydroxychloroquine. J Biomol Struct Dyn 39(11):4089–4099

Harrison AG, Lin T, Wang P (2020) Mechanisms of SARS-CoV-2 transmission and pathogenesis. Trends Immunol 41(12):1100–1115

Hossain MS, Urbi Z, Sule A, Hafizur Rahman KM (2014) *Andrographis paniculate* (Burm. f.) Wall.ex Nees: a review of ethnobotany, phytochemistry, and pharmacology. Sci World J 2014: 274905

Intharuksa A, Arunotayanun W, Yooin W (2022) Sirisa-Ard P.A comprehensive review of *Andrographis paniculata* (Burm. f.) nees and its constituents as potential lead compounds for COVID-19 drug discovery. Molecules 27(14):4479

Jackson CB, Farzan M, Chen B, Choe H (2022) Mechanisms of SARS-CoV-2 entry into cells. Nat Rev Mol Cell Biol 23(1):13–20

Jadhav AK, Karuppayil SM (2021) *Andrographis paniculata* (Burm. F) Wall ex Nees: antiviral properties. Phytother Res 35(10):5365–5373

Jena S, Munusami P, Mm B, Chanda K (2021) Computationally approached inhibition potential of *Tinospora cordifolia* towards COVID-19 targets. Virus Dis 32(1):65–77

Kalaria RK, Patel HK (2020) Naturally occurring phytochemical as inhibitors from *Catharanthus roseus*: an *in silico* approaches for drug development against COVID-19 2020

Kannan M, Sathiyarajeswaran P, Sasikumar D, Geetha A, Mohanapriya M, Vinod NP, Manickam P, Kanakavalli K, Parthibhan P, Pitchiah Kumar M, Kannan R, Sivaraman G (2022) Safety and efficacy of a Siddha medicine fixed regimen for the treatment of asymptomatic and mild COVID-19 patients. J Ayurveda Integrat Med 13(3):100589

Kar P, Goyal AK, Das AP, Sen A (2014) Antioxidant and pharmaceutical potential of *Clerodendrum* L.: an overview. Int J Green Pharm (IJGP) 8(4):210

Kar P, Sharma NR, Singh B, Sen A, Roy A (2021) Natural compounds from *Clerodendrum spp.* as possible therapeutic candidates against SARS-CoV-2: an *in silico* investigation. J Biomol Struct Dyn 39(13):4774–4785

Kaur R, Sharma P, Gupta GK, Ntie-Kang F, Kumar D (2020) Structure-activity-relationship and mechanistic insights for anti-HIV natural products. Molecules 25(9):2070

Kesharwani A, Polachira SK, Nair R, Agarwal A, Mishra NN, Gupta SK (2017) Anti-HSV-2 activity of *Terminalia chebula* Retz extract and its constituents, chebulagic and chebulinic acids. BMC Complement Altern Med 17(1):110

Khursheed A, Jain V (2022) *Euphorbia hirta* as a gold mine of high-value phytochemicals: a comprehensive review of its pharmacological activities and possible role against SARS-CoV-2. Biomed Res Therap 9(2):4930–4949

Krupanidhi S, Abraham Peele K, Venkateswarulu TC, Ayyagari VS, Nazneen Bobby M, John Babu D, Venkata Narayana A, Aishwarya G (2021) Screening of phytochemical compounds of *Tinospora cordifolia* for their inhibitory activity on SARS-CoV-2: an *in silico* study. J Biomol Struct Dyn 39(15):5799–5803

Kuchi Bhotla H, Kaul T, Balasubramanian B, Easwaran M, Arumugam VA, Pappusamy M, Muthupandian S, Meyyazhagan A (2020) Platelets to surrogate lung inflammation in COVID-19 patients. Med Hypotheses 143:110098

Kuchi Bhotla H, Balasubramanian B, Arumugam VA, Pushparaj K, Murugesh E, Rathinasamy B (2021a) Insinuating cocktailed components in biocompatible-nanoparticles could act as an impressive neo-adjuvant strategy to combat COVID-19. Nat Resour Hum Health:3–7

Kuchi Bhotla H, Balasubramanian B, Meyyazhagan A, Pushparaj K, Easwaran M, Pappusamy M, Alwin Robert A, Arumugam VA, Tsibizova V, Msaad Alfalih A, Aljowaie RM, Saravanan M, Di Renzo GC (2021b) Opportunistic mycoses in COVID-19 patients/survivors: epidemic inside a pandemic. J Infect Public Health 14(11):1720–1726

Kumar S, Singh B, Bajpai V (2021a) *Andrographis paniculata* (Burm. f.) Nees: traditional uses, phytochemistry, pharmacological properties and quality control/quality assurance. J Ethnopharmacol 275:114054

Kumar SB, Krishna S, Pradeep S, Mathews DE, Pattabiraman R, Murahari M, Murthy TK (2021b) Screening of natural compounds from *Cyperus rotundus* Linn against SARS-CoV-2 main protease (Mpro): an integrated computational approach. Comput Biol Med 134:104524

Kumar S, Singh B, Singh R (2022) *Catharanthus roseus* (L.) G. Don: a review of its ethnobotany, phytochemistry, ethnopharmacology and toxicities. J Ethnopharmacol 284:114647

Kurokawa M, Nagasaka K, Hirabayashi T, Uyama S, Sato H, Kageyama T, Kadota S (1995) Efficacy of traditional herbal medicines in combination with acyclovir against herpes simplex virus type 1 infection *in vitro* and *in vivo*. Antivir Res 27(1–2):19–37

Latha D, Hrishikesh D, Shiban G, Chandrashekar C, Bharath BR (2022) In silico, in vitro screening of plant extracts for anti-SARS-CoV-2 activity and evaluation of their acute and sub-acute toxicity. Phytomed Plus 2(2):100233

Lin LT, Chen TY, Chung CY, Noyce RS, Grindley TB, McCormick C, Lin TC, Wang GH, Lin CC, Richardson CD (2011) Hydrolyzable tannins (chebulagic acid and punicalagin) target viral glycoprotein-glycosaminoglycan interactions to inhibit herpes simplex virus 1 entry and cell-to-cell spread. J Virol 85(9):4386–4398

Liu Q, Wang YF, Chen RJ, Zhang MY, Wang YF, Yang CR, Zhang YJ (2009) Anti-coxsackie virus B3 norsesquiterpenoids from the roots of *Phyllanthus emblica*. J Nat Prod 72(5):969–972

Luan B, Huynh T, Cheng X, Lan G, Wang HR (2020) Targeting proteases for treating COVID-19. J Proteome Res 19(11):4316–4326

Lv JJ, Wang YF, Zhang JM, Yu S, Wang D, Zhu HT, Cheng RR, Yang CR, Xu M, Zhang YJ (2014) Anti-hepatitis B virus activities and absolute configurations of sesquiterpenoid glycosides from *Phyllanthus emblica*. Org Biomol Chem 12(43):8764–8774

Lv JJ, Yu S, Xin Y, Cheng RR, Zhu HT, Wang D, Yang CR, Xu M, Zhang YJ (2015) Anti-viral and cytotoxic norbisabolane sesquiterpenoid glycosides from *Phyllanthus emblica* and their absolute configurations. Phytochemistry 117:123–134

Mahato DK, Kargwal R, Kamle M, Sharma B, Pandhi S, Mishra S, Kumar P (2022) Ethnopharmacological properties and nutraceutical potential of *Moringa oleifera*. Phytomed Plus 2(1):100168

Manikandan R, Anand AV (2016) Evaluation of antioxidant activity of Psidium guajava Linn. in streptozotocin–induced diabetic rats. Free Radical Antioxid 6(1):72–76

Mao X, Wu LF, Guo HL, Chen WJ, Cui YP, Qi Q, Li S, Liang WY, Yang GH, Shao YY, Zhu D, She GM, You Y, Zhang LZ (2016) The genus Phyllanthus: an ethnopharmacological, phytochemical, and pharmacological review. Evid Based Complement Alternat Med 2016:7584952

Meyyazhagan A, Pushparaj K, Balasubramanian B, Kuchi Bhotla H, Pappusamy M, Arumugam VA, Easwaran M, Pottail L, Mani P, Tsibizova V, Di Renzo GC (2022) COVID-19 in pregnant women and children: insights on clinical manifestations, complexities, and pathogenesis. Int J Gynaecol Obstet 156(2):216–224

Mishra NN, Kesharwani A, Agarwal A, Polachira SK, Nair R, Gupta SK (2018) Herbal gel formulation developed for anti-human immunodeficiency virus (HIV)-1 activity also inhibits *in vitro* HSV-2 infection. Viruses 10(11):580

Mlozi SH (2022) The role of natural products from medicinal plants against COVID-19: traditional medicine practice in Tanzania. Heliyon 8(6):e09739

Moin MS, Siddiqui JI, Alam MA, Khatoon F, Khan S, Minhajuddin A (2021) Ethnomedicinal potential of widely used plant *Azadirachta indica* A. Juss: a comprehensive review. J Phytopharmacol 10:456

Mukherjee PK, Efferth T, Das B, Kar A, Ghosh S, Singha S, Debnath P, Sharma N, Bhardwaj P, Haldar PK (2022) Role of medicinal plants in inhibiting SARS-CoV-2 and in the management of post-COVID-19 complications. Phytomedicine 2022:153930

Mulpuru V, Mishra N (2021) Computational Identification of SARS-CoV-2 Inhibitor in *Tinospora cordifolia*, *Cinnamomum zeylanicum* and *Myristica fragrans*. Virus Dis 32(3):511–517

Murthy TK, Joshi T, Gunnan S, Kulkarni N, Priyanka V, Kumar SB, Gowrishankar BS (2021) *In silico* analysis of *Phyllanthus amarus* phytochemicals as potent drugs against SARS-CoV-2 main protease. Curr Res Green Sustain Chem 4:100159

Murugesan S, Kottekad S, Crasta I, Sreevathsan S, Usharani D, Perumal MK, Mudliar SN (2021) Targeting COVID-19 (SARS-CoV-2) main protease through active phytocompounds of ayurvedic medicinal plants - *Emblica officinalis* (Amla), *Phyllanthus niruri* Linn.(Bhumi Amla) and *Tinospora cordifolia* (Giloy) - A molecular docking and simulation study. Comput Biol Med 136:104683

Naik KS, Sahithya MD, Kumar SC, Al-Fahad D, Srinath M (2021) Identification of potent bioactive compounds from Terminalia chebula (Retz.) targeting multiple receptors of SARS-CoV-2 through *in silico* approach: king of Ayurveda against COVID-19. Eur J Pharm Med Res 8(12):209–217

Namazi R, Zabihollahi R, Behbahani M, Rezaei A (2013) Inhibitory activity of *Avicennia marina*, a medicinal plant in Persian folk medicine, against HIV and HSV. Iran J Pharm Res 12(2): 435–443

Nargis Begum T, Muhammad Ilyas MH, Burkanudeen A, Kalavathy S, Vijaya Anand A, Sampathkumar P, Jaswanth A (2009) Hypoglycemic and antihyperlipidemic activity of ethanolic leaf extract of *Azima tetracantha* Lam. on alloxan-induced diabetic rats. *Journal of Cell and Tissue*. Research 9(1):1681–1685

Notka F, Meier G, Wagner R (2004) Concerted inhibitory activities of *Phyllanthus amarus* on HIV replication *in vitro* and *exvivo*. Antivir Res 64(2):93–102

Ogbole OO, Akinleye TE, Nkumah AO, Awogun AO, Attah AF, Adewumi MO, Adeniji AJ (2021) *In vitro* antiviral activity of peptide-rich extracts from seven Nigerian plants against three non-Polio Enterovirus species C serotypes. Virol J 18(1):161u

Oyuntsetseg N, Khasnatinov MA, Molor-Erdene P, Oyunbileg J, Liapunov AV, Danchinova GA, Oldokh S, Baigalmaa J, Chimedragchaa C (2014) Evaluation of direct antiviral activity of the Deva-5 herb formulation and extracts of five Asian plants against Influenza A virus H3N8. BMC Complement Altern Med 14:235

Pandey K, Lokhande KB, Swamy KV, Nagar S, Dake M (2021) *In silico* exploration of phytoconstituents from *Phyllanthus emblica* and *Aegle marmelos* as potential therapeutics against SARS-CoV-2 RdRp. Bioinform Biol Insights 15:11779322211027403

Panya A, Jantakee K, Punwong S, Thongyim S, Kaewkod T, Yenchitsomanus PT, Tragoolpua Y, Pandith H (2021) Triphala in traditional ayurvedic medicine inhibits dengue virus infection in Huh7 hepatoma cells. Pharmaceuticals (Basel) 14(12):1236

Parida MM, Upadhyay C, Pandya G, Jana AM (2002) Inhibitory potential of neem (*Azadirachta indica Juss*) leaves on dengue virus type-2 replication. J Ethnopharmacol 79(2):273–278

Parmar G, Shah A, Shah S, Seth AK (2022) Identification of bioactive phytoconstituents from the plant *Euphorbia hirta* as potential inhibitor of SARS-CoV-2: an *in silico* approach. Biointerface Res Appl Chem 12:1385–1396

Patel JR, Tripathi P, Sharma V, Chauhan NS, Dixit VK (2011) *Phyllanthus amarus*: ethnomedicinal uses, phytochemistry and pharmacology: a review. J Ethnopharmacol 138(2):286–313

Pathak R, Sharma H (2021) A review on medicinal uses of *Cinnamomum verum* (Cinnamon). J Drug Deliv Therapeut 11(6-S):161–166

Peerzada AM, Ali HH, Naeem M, Latif M, Bukhari AH, Tanveer A (2015) *Cyperus rotundus* L.: traditional uses, phytochemistry, and pharmacological activities. J Ethnopharmacol 174:540–560

Plante KS, Dwivedi V, Plante JA, Fernandez D, Mirchandani D, Bopp N, Aguilar PV, Park JG, Tamayo PP, Delgado J, Shivanna V (2021) Antiviral activity of oleandrin and a defined extract of *Nerium oleander* against SARS-CoV-2. Biomed Pharmacother 138:111457

Poochi SP, Easwaran M, Balasubramanian B, Anbuselvam M, Meyyazhagan A, Park S, Bhotla HK, Anbuselvam J, Arumugam VA, Keshavarao S, Kanniyappan GV, Pappusamy M, Kaul T (2020) Employing bioactive compounds derived from *Ipomoea obscura* (L.) to evaluate potential inhibitor for SARS-CoV-2 main protease and ACE2 protein. Food Front 1:168–179

Prasad V, Mishra SK, Srivastava S, Srivastava A (2014) A virus inhibitory protein isolated from Cyamopsis tetragonoloba (L.) Taub. upon induction of systemic antiviral resistance shares partial amino acid sequence homology with a lectin. Plant Cell Rep 33(9):1467–1478

Prasanth DSNBK, Murahari M, Chandramohan V, Panda SP, Atmakuri LR, Guntupalli C (2021) *In silico* identification of potential inhibitors from *Cinnamon* against main protease and spike glycoprotein of SARS CoV-2. J Biomol Struct Dyn 39(13):4618–4632

Priya S, Kumar NS, Hemalatha S (2018) Antiviral phytocompounds target envelope protein to control Zika virus. Comput Biol Chem 77:402–412

Pushparaj K, Kuchi Bhotla H, Arumugam VA, Pappusamy M, Easwaran M, Liu WC, Issara U, Rengasamy KRR, Meyyazhagan A, Balasubramanian B (2022) Mucormycosis (black fungus) ensuing COVID-19 and comorbidity meets - Magnifying global pandemic grieve and catastrophe begins. Sci Total Environ 805:150355

Ranjith MS, Ranjitsingh AJA, Shankar SG, Vijayalaksmi GS, Deepa K, Babu K, Sidhu HS (2010) *Solanum trilobatum* in the management of atopy: through inhibition of mast cell degranulation and moderation of release of interleukins. Pharm Res 2(1):10

Rehan M, Ahmed F, Howladar SM, Refai MY, Baeissa HM, Zughaibi TA, Kedwa KM, Jamal MS (2021) A computational approach identified andrographolide as a potential drug for suppressing COVID-19-induced cytokine storm. Front Immunol 12:648250

Sagar V, Kumar AH (2020) Efficacy of natural compounds from *Tinospora cordifolia* against SARS-CoV-2 protease, surface glycoprotein and RNA polymerase. Biol Eng Med Sci Rep 6(1): 06–08

Saifulazmi NF, Rohani ER, Harun S, Bunawan H, Hamezah HS, Nor Muhammad NA, Azizan KA, Ahmed QU, Fakurazi S, Mediani A, Sarian MN (2022) A review with updated perspectives on the antiviral potentials of traditional medicinal plants and their prospects in antiviral therapy. Life (Basel) 12(8):1287

Saini R, Sharma N, Oladeji OS, Sourirajan A, Dev K, Zengin G, Kumar V (2022) Traditional uses, bioactive composition, pharmacology, and toxicology of *Phyllanthus emblica* fruits: a comprehensive review. J Ethnopharmacol 282:114570

Sa-Ngiamsuntorn K, Suksatu A, Pewkliang Y, Thongsri P, Kanjanasirirat P, Manopwisedjaroen S (2021) Anti-SARS-CoV-2 activity of *Andrographis paniculata* extract and its major component andrographolide in human lung epithelial cells and cytotoxicity evaluation in major organ cell representatives. J Nat Prod 84(4):1261–1270

Sanna G, Madeddu S, Serra A, Collu D, Efferth T, Hakkim FL, Rashan L (2021) Anti-poliovirus activity of *Nerium oleander* aqueous extract. Nat Prod Res 35(4):633–636

Sarkar A, Agarwal R, Bandyopadhyay B (2022) Molecular docking studies of phytochemicals from *Terminalia chebula* for identification of potential multi-target inhibitors of SARS-CoV-2 proteins. J Ayurv Integrat Med 13(2):100557

Sattanathan G, Palanisamy T, Padmapriy S, Arumugam VA, Park S, Kim IH, Balasubramanian B (2020) Influences of dietary inclusion of algae *Chaetomorpha aerea* enhanced growth performance, immunity, haematological response and disease resistance of Labeo rohita challenged with *Aeromonas hydrophila*. Aquacult Rep 2020(17):100353

Sen D, Bhaumik S, Debnath P, Debnath S (2022) Potentiality of *Moringa oleifera* against SARS-CoV-2: identified by a rational computer aided drug design method. J Biomol Struct Dyn 40(16):7517–7534

Shadrack DM, Vuai SA, Sahini MG, Onoka I (2021) *In silico* study of the inhibition of SARS-COV-2 viral cell entry by neem tree extracts. RSC Adv 11(43):26524–26533

Shanmugam R, Thangavelu S, Fathah Z, Yatoo MI, Tiwari R, Pandey MK, Dhama J, Chandra R, Malik Y, Dhama K, Sha R, Chaicumpa W, Shanmugam V, Arumugam AV (2020) SARS-CoV-2/COVID-19 pandemic-an update. J Exp Biol Agric Sci 8(1):S219–S245

Sharma V, Kaushik S, Pandit P, Dhull D, Yadav JP, Kaushik S (2019) Green synthesis of silver nanoparticles from medicinal plants and evaluation of their antiviral potential against Chikungunya virus. Appl Microbiol Biotechnol 103(2):881–891

Siddiqui S, Ahmad R, Alaidarous M, Zia Q, Ahmad Mir S, Alshehri B, Srivastava A, Trivedi A (2022a) Phytoconstituents from *Moringa oleifera* fruits target ACE2 and open spike glycoprotein to combat SARS-CoV-2: an integrative phytochemical and computational approach. J Food Biochem 2022:e14062

Siddiqui S, Upadhyay S, Ahmad R, Barkat MA, Jamal A, Alothaim AS, Hassan MZ, Rahman MA, Arshad M, Ahamad T, Khan MF, Shankar H, Ali M, Kaleem S, Ahmad J (2022b) Interaction of bioactive compounds of *Moringa oleifera* leaves with SARS-CoV-2 proteins to combat COVID-19 pathogenesis: a phytochemical and *in silico* analysis. Appl Biochem Biotechnol 194(12):5918–5944

Singh S, Shenoy S, Nehete PN, Yang P, Nehete B, Fontenot D, Yang G, Newman RA, Sastry KJ (2013) *Nerium oleander* derived cardiac glycoside oleandrin is a novel inhibitor of HIV infectivity. Fitoterapia 84:32–39

Singh N, Rao AS, Nandal A, Kumar S, Yadav SS, Ganaie SA, Narasimhan B (2021) Phytochemical and pharmacological review of *Cinnamomum verum* J. Presl-a versatile spice used in food and nutrition. Food Chem 338:127773

Sivani BM, Venkatesh P, Murthy TK, Kumar SB (2021) *In silico* screening of antiviral compounds from *Moringa oleifera* for inhibition of SARS-CoV-2 main protease. Curr Res Green Sustain Chem 4:100202

Sivasankari B, Pitchaimani S, Anandharaj M (2013) A study on traditional medicinal plants of Uthapuram, Madurai District, Tamilnadu, South India. Asian Pac J Trop Biomed 3(12):975–979

Somu C, Karuppiah H, Sundaram J (2019) Antiviral activity of seselin from *Aegle marmelos* against nuclear Polyhedrosis virus infection in the larvae of silkworm, Bombyx mori. J Ethnopharmacol 245:112155

Sree MP, Yogan GP, Gopal V (2022) Inhibitory potentials of phytoconstituents of *Phyllanthus amarus* against severe acute respiratory syndrome-corona virus-2 main protease: a computational aided approach. Indian J Pharm Sci 84(5):1203–1209

Subapriya R, Nagini S (2005) Medicinal properties of neem leaves: a review. Curr Med Chem Anticancer Agents 5(2):149–146

Subashini MS, Rajendran P (2015) In vitro screening of anti HBV and anti HIV properties of Gymnema sylvestre R. Br leaves from Kolli Hills, Tamilnadu, India. Int J Curr Microbiol App Sci 4:542–547

Subramani SK, Gupta Y, Manish M, Prasad GB (2020) *Gymnema Sylvestre* a potential inhibitor of COVID-19 main protease by MD simulation study 2020

Tanwettiyanont J, Piriyachananusorn N, Sangsoi L, Boonsong B, Sunpapoa C, Tanamatayarat P, Na-Ek N, Kanchanasurakit S (2022) Use of *Andrographis paniculata* (Burm.f.) Wall.ex Nees and risk of pneumonia in hospitalised patients with mild coronavirus disease 2019: a retrospective cohort study. Front Med 9:947373

Thas JJ (2008) Siddha medicine--background and principles and the application for skin diseases. Clin Dermatol 26(1):62–78

Thomas N, Patil P, Sharma A, Kumar S, Singh VK, Alagarasu K, Parashar D, Tapryal S (2022) Studies on the antiviral activity of chebulinic acid against dengue and chikungunya viruses and *in silico* investigation of its mechanism of inhibition. Sci Rep 12(1):10397

Tiwari P, Mishra BN, Sangwan NS (2014) Phytochemical and pharmacological properties of *Gymnema sylvestre*: an important medicinal plant. Biomed Res Int 2014:830285

Tolba HMN, Elmaaty AA, Farag GK, Mansou DA, Elakkad HA (2022) Immunological effect of *Moringa oleifera* leaf extract on vaccinated and non-vaccinated Hubbard chickens experimentally infected with Newcastle virus. Saudi J Biol Sci 29(1):420–426

Tomar B, Anders HJ, Desai J, Mulay SR (2020) Neutrophils and neutrophil extracellular traps drive necroinflammation in COVID-19. Cell 9(6):1383

Varnasseri M, Siahpoosh A, Hoseinynejad K, Amini F, Karamian M, Yad MJY, Cheraghian B, Khosravi AD (2022) The effects of add-on therapy of *Phyllanthus Emblica* (Amla) on laboratory confirmed COVID-19 cases: a randomized, double-blind, controlled trial. Complement Ther Med 65:102808

Vijaya Anand A, Velayuthaprabhu S, Rengarajan RL, Sampathkumar P, Radhakrishnan R (2020) Bioactive compounds of Guava (L.). In: Bioactive compounds in underutilized fruits and nuts. Springer, Cham, pp 503–527

Vijayakumar K, Anand AV, Manikandan R (2015) *In vitro* antioxidant activity of ethanolic extract of *Psidium guajava* leaves. Int J Res Stud Biosci 3(5):145–149

Vijayakumar K, Rengarajan RL, Radhakrishnan R, Mathew S, Qadri I, Vijaya Anand A (2019) *Psidium guajava* leaf extracts and their quercetin protect HepG2 cell lines against $CCL_4$ induced cytotoxicity. Indian J Clin Biochem 34(3):324–329

Wadanambi PM, Jayathilaka N, Seneviratne KN (2022) A computational study of carbazole alkaloids from *Murraya koenigii* as potential SARS-CoV-2 main protease inhibitors. Appl Biochem Biotechnol 195:1–24

Wahab NZA, Azizul A, Ibrahim N (2020) Phytochemistry, cytotoxicity and antiviral activity of *Catharanthus roseus*. Ira J Microbiol 12(5):460–465

Wanarase S, Chavan S, Sharma S (2022) A molecular docking study: targeting COVID-19 (SARS-CoV-2) main protease using active phytocompounds from *Terminalia arjuna*. Innov J Med Sci 10(5):9–14

Wanaratna K, Leethong P, Inchai N, Chueawiang W, Sriraksa P, Tabmee A, Sirinavin S (2022) Efficacy and safety of *Andrographis paniculata* extract in patients with mild COVID-19: a randomized controlled trial. Arch Int Med Res 5(3):423–427

Wang YF, Wang XY, Ren Z, Qian CW, Li YC, Kaio K, Wang QD, Zhang Y, Zheng LY, Jiang JH, Yang CR, Liu Q, Zhang YJ, Wang YF (2009) Phyllaemblicin B inhibits Coxsackie virus B3 induced apoptosis and myocarditis. Antivir Res 84(2):150–158

Xiang YF, Ju HQ, Li S, Zhang YJ, Yang CR, Wang YF (2010) Effects of 1,2,4,6-tetra-O-galloyl-β-D-glucose from *P. emblica* on HBsAg and HBeAg secretion in HepG2.2.15 cell culture. Virol Sin 25(5):375–380

Xiang Y, Pei Y, Qu C, Lai Z, Ren Z, Yang K, Xiong S, Zhang Y, Yang C, Wang D, Liu Q, Kitazato K, Wang Y (2011) *In vitro* anti-herpes simplex virus activity of 1,2,4,6-tetra-O-galloyl-β-D-glucose from *Phyllanthus emblica* L. (*Euphorbiaceae*). Phytother Res 25(7): 975–982

Xie R, Lin Z, Zhong C, Li S, Chen B, Wu Y, Huang L, Yao H, Shi P, Huang J (2021) Deciphering the potential anti-COVID-19 active ingredients in *Andrographis paniculata* (Burm. F.) Nees by combination of network pharmacology, molecular docking, and molecular dynamics. R Soc Chem Adv 11(58):36511–36517

Xiong Y, Rajoka MSR, Mehwish HM, Zhang M, Liang N, Li C, He Z (2021) Virucidal activity of moringa A from *Moringa oleifera* seeds against Influenza A viruses by regulating TFEB. Int Immunopharmacol 95:107561

Xiong Y, Riaz Rajoka MS, Zhang M, He Z (2022) Isolation and identification of two new compounds from the seeds of *Moringa oleifera* and their antiviral and anti-inflammatory activities. Nat Prod Res 36(4):974–983

Xu HB, Ma YB, Huang XY, Geng CA, Wang H, Zhao Y, Yang TH, Chen XL, Yang CY, Zhang XM, Chen JJ (2015) Bioactivity-guided isolation of anti-hepatitis B virus active sesquiterpenoids from the traditional Chinese medicine: rhizomes of *Cyperus rotundus*. J Ethnopharmacol 171:131–140

Yakhchali M, Taghipour Z, Mirabzadeh Ardakani M, Alizadeh Vaghasloo M, Vazirian M, Sadrai S (2021) Cinnamon and its possible impact on COVID-19: the viewpoint of traditional and conventional medicine. Biomed Pharmacother 143:112221

Yang PS, Dai JM, Gu XJ, Xiong W, Huang DQ, Qiu SY, Zheng JN, Li Y, Yang FX, Zhou M (2022) Indole alkaloids and chromones from the stem bark of *Cassia alata* and their antiviral activities. Molecules 27(10):3129

Yashvardhini N, Samiksha S, Jha DK (2021) Pharmacological intervention of various Indian medicinal plants in combating COVID-19 infection. Biomed Res Ther 8(7):4461–4475

Zhou M, Liu C, Yang J, Niu D, Dong M, Yang G, Li X, Hu Q (2020) Alatains A and B, unique hetero-dimeric polyphenols from *Cassia alata* and their anti-Tobacco mosaic virus activity. Fitoterapia 147:104763

# Chapter 18
# Immune-Boosting Plants Used in Turkish Folk Medicine and Their Potential Against COVID-19

**Seyid Ahmet Sargin and Yunus Dogan**

## 18.1 Introduction

The coronavirus disease 2019 (COVID-19) outbreak, which emerged in Wuhan, China in late 2019, is still the first contagious epidemic the world has encountered after the Spanish Flu in terms of scope and size. The virus, namely severe acute respiratory syndrome coronavirus-2 (SARS-CoV-2) with its enveloped and single-stranded RNA content, rapidly expanded into a global dimension, with rising serious mortality rates after it turned into a regional epidemic (TAS 2020). The first verified COVID-19 case in Turkey was detected on March 11. As of 10 May 2021, 5,016,141 confirmed cases and 42,746 deaths have been reported to the World Health Organization (WHO 2021). To combat this rapidly emerging threat, prevention and treatment protocols were established and some social restrictions were imposed in a very short time (Konakci et al. 2020).

The immune system is a complex system that forms the defense mechanism against diseases in a living thing, recognizes and destroys pathogens and tumor cells, and protects the body from foreign and harmful substances (Scudiero et al. 2021). Two types of immunities, innate and adaptive, play a powerful role in increased susceptibility to infections. The likelihood of being diseased depends on the effectiveness of the immune response. Therefore, low immunity results in a reduced ability to fight the pathogen and become more susceptible to disease. Most of the deaths associated with coronavirus are due to the suppressed immune system of the human body that is unable to fight against it, rather than the damage caused by

S. A. Sargin (✉)
Faculty of Education, Alanya Alaaddin Keybubat University, Antalya, Turkey

Y. Dogan
Buca Faculty of Education, Dokuz Eylul University, Izmir, Turkey
e-mail: yunus.dogan@deu.edu.tr



J.-T. Chen (ed.), *Ethnopharmacology and Drug Discovery for COVID-19: Anti-SARS-CoV-2 Agents from Herbal Medicines and Natural Products*,
https://doi.org/10.1007/978-981-99-3664-9_18

the virus (Bhatta 2020). When the immune response is low, weak, or damaged, infections such as coronavirus attack people, especially young and older people. It also becomes an open invitation for other diseases such as diabetes, heart disease, or cancer (Arshad et al. 2020; WHO 2020). Since it is not yet a registered drug or vaccine 100% effective against COVID-19 and all its mutants or variants, the immune system is still the best defense we have. As long as the immune system is functioning normally, infections such as COVID-19 may go undetected (Chowdhury et al. 2020). It has been found that patients suffering from infectious and non-infectious diseases of the lungs are at greater risk of this viral infection due to their low immune system (Opitz et al. 2010; Khanal et al. 2020). Therefore, strengthening immunity (natural body system) can make a major contribution as a prophylactic measure against multiple pathogenic conditions as well as maintaining optimum health (Khanal et al. 2020; Nicholson 2016).

The most common symptoms are fever, headache, cough, difficulty breathing, and diarrhea. In some cases or different variants of the virus, these symptoms may also remain silent. Or, on the contrary, it might be presented as severe pneumonia causing shortness of breath, kidney failure, or even death. After the virus enters the respiratory mucosa cells of another host through droplets that are released by coughing and sneezing (or close-range speech), it releases its genetic code to initiate viral replication, then uses the following three mechanisms to circumvent the immune response: The virus implements various mechanisms to circumvent the immune response. In the first step, it inhibits the rapid expression of interferon type 1 (IFN-1), known as the "initial alarm," which modulates immune cells to the "antiviral state" in the case of the system encountering the virus. Second, it blocks IFN-1's existing signals in the system through inhibition of the phosphorylation of STAT-1 (signal converter and transcription activator 1). Its third mechanism is the dysfunctioning of the immune system through excessive and prolonged production of IFN-1 by plasmacytoid dendritic cells (pDCs). The process causes an overproduction of activated neutrophils and macrophages, resulting in lung immunopathology (e.g. acute respiratory distress syndrome) (Taghizadeh-Hesary and Akbari 2020). This, in turn, activates a rapid reaction of innate immunity and results in the release of cytokines in large amounts, which play a key role in determining the extent of the infection. They damage beneficial cells and organs in the body, including defense cells, thus causing destruction and leading to death (Scudiero et al. 2021). In severe cases, a severe immune response is observed with deep lung exudate production, which limits ventilation and can subsequently trigger an uncontrolled inflammatory response, acute respiratory distress syndrome (ARDS), and septic shock, following bilateral pneumonia (Galmarini et al. 2020). These two complications are the primary reasons for hospitalization in intensive care and mortality for COVID-19 in patients with a history of smoking and comorbidities (Kakodkar et al. 2020). Chronic obstructive pulmonary disease, including chronic inflammation of the upper respiratory tract (URI), hypertension, diabetes, malignant tumor, coronary heart disease, cerebrovascular disease, and chronic kidney disorders are among the most common of these clinical stories (Scudiero et al. 2021; Chowdhury et al. 2020).

Due to their ability to easily mutate their genetic material and gain resistance, developing a broad-spectrum antiviral drug/vaccine that is completely safe and has the potential of 100% effectiveness and it reaching the market level is difficult and quite time-consuming (Oyston and Robinson 2012). Moreover, synthetically developed antiviral drugs sometimes have harmful or even fatal side effects that can cause significant health problems (Ghildiyal et al. 2020). Indeed, a newly developed popular vaccine that received market approval a few months ago created clots as a side effect and caused deaths in vaccinated people. Within a week, its use was suspended by about 18 countries worldwide (Tobaiqy et al. 2021). Therefore, herbs and plant-based herbal remedies seem promising in reducing these side effects because to their capacity to have strong antiviral activity (Ghildiyal et al. 2020; Jahan and Onay 2020).

A healthy and proper diet can strengthen the immune system and prevent the development of diseases and immune depression. For example, antioxidants can prevent or repair damage caused by harmful agents such as viruses and free radicals. The most powerful antioxidant in the human body is glutathione (GSH). If its level in the body increases, there is a simultaneous decrease in cytokine factors such as interleukin-6 (IL-6) and tumor necrosis factor-α (TNF-α), which cause to worsen the disease (pro-inflammatory), thereby helping the patient recover (Scudiero et al. 2021). Especially medicinal plants, due to their rich content of minerals and vitamins, regularly contribute to the healthy development of the general intestinal microbiome, which constitutes 85% of the immune system and keeps the immune system active in the fight against infections. Highlights of immune-enhancing herbs are garlic, black cumin, and licorice (Arshad et al. 2020; WHO 2020). They produce many important phytochemicals through their secondary metabolism and act as a defense against stress caused by environmental triggers and pathogens. In fact WHO's announcement that 80% of the world's population trusts them for treatment highlights this fact (Ekor 2014; Khan et al. 2021).

Herbal preparations/medicines are always defined as a therapeutic regimen that, rather than consisting of a single compound that interacts with a single target, is a concerted pharmacological intervention of several compounds that interact with multiple targets (Wang et al. 2012). The results showed that not only do they modulate pathways related to strengthening the immune system, but also modulate multiple pathways that contribute to the progression of multiple disease pathogenesis that would add beneficial effects on specific issues such as hypertension and diabetes patients (Khanal et al. 2020). After the first SARS-CoV epidemic, it has been shown in many studies that plant extracts and phytochemicals produced from plants affect the virus with different mechanisms. These consist of direct antiviral activity, immune stimulator, inflammation modulator, and symptom management. Direct antiviral activity includes inhibition of early replication through viral entry inhibition and late replication inhibition (Khan et al. 2021).

The most likely mechanism for immune enhancers is to trigger humoral and cellular immune responses (Vaghasiya et al. 2010). For example, it has been found that aqueous and methanol extracts of basil leaves and seed oil increase the immune response by increasing T-helper and natural killer cells, lymphocyte count, phagocytic activity, neutrophil count, antibody titer, and the like (Jamshidi and Cohen 2017; Singh et al. 2017). Similar positive effects of a wide variety of flavonoids found naturally in plants, such as quercetin, naringin, hesperetin, and catechin, against severe acute respiratory syndrome coronavirus (SARS-CoV) have been reported (Khanna et al. 2020). Researchers such as Terali et al. (2020) and Bibi et al. (2020) also elucidated the inhibitory action pathway against the ACE2 enzyme receptor, which enables COVID-19 to enter host cells. Hence, Utomo et al. (2020) reported that curcumin isolated from turmeric (*Curcuma longa*) inhibits SARS-CoV-2 protease, spike glycoprotein-RBD, and PD-ACE2 receptors, based on their molecular insertion studies. As a result of their in silico studies, Jakhmola Mani et al. (2020) stated that active ingredients such as a-hederin, thymohydroquinone, and thymoquinone, obtained from the extract of *Nigella sativa* seeds, efficiently bind and block ACE2 and reduce hypoxia and inflammation caused by oxidative stress, strengthening immunity and can greatly assist in the fight against COVID-19. Also, ACE2 inhibitors have been identified in various species such as *Allium sativum, Cerasus avium, Berberis integerrima, Alcea digitata, Rubia tinctorum, Peganum harmala*, etc. (Heidary et al. 2020). It has also been disclosed that herbal blockers which can inhibit the "pathogenic" kinase (PAK1), which plays a major role in the infection of many viruses, can support the immune system and serve as potential therapeutic agents against COVID-19 (Asih et al. 2021). Also, curcumin rhizome (Nemati et al. 2020) and tea-shaped extract of *Schinus molle* (pink peppercorn) fruits (Ji et al. 2019) contribute to the blocking of PAK1, strengthening the immune system against COVID-19 and other viral infections, and its essential anti-inflammatory formulations have been reported. Conversely, Shaghaghi (2020) showed in molecular dynamic simulations with terpenoid compounds, including thymoquinone extracted from *Nigella sativa*, that thymoquinone can reduce the likelihood of infection of SARS-CoV-2 (Elfiky 2020). Therefore, it is envisaged that medicinal plants may be effective in the prevention and treatment of COVID-19, either through direct use or their isolated compounds. However, many more advanced studies are needed for optimum effective dosage formulation and usage descriptions.

The following can be given as examples of similar studies conducted in Turkey: In an in silico study conducted by Adem et al. (2020), phytochemicals such as rutin, apiin, hesperidin, diosmin, and diacetylcurcumin were identified as potent COVID-19-main-protease (Mpro) inhibitors. Yilmaz et al. (2020) reported that black mulberry (*Morus nigra*) syrup increased serum lysozyme, myeloperoxidase, superoxide dismutase, and catalase activities and expression levels of immune-related genes, increasing innate immune parameters and antioxidant-related gene expression responses. In another study, Bilen et al. (2020) revealed that the aqueous methanolic extract of lemon balm (*Melissa officinalis*) has antioxidant and cell-mediated-immune system-stimulating effects by increasing lysozyme, pepsin, and trypsin activities. Thus, they highlighted that these herbal use recipes can play a vital role in reducing the symptoms of COVID-19 infection and preventing the further spread of such pandemics.

The COVID-19 pandemic is perhaps the most dangerous and deadly epidemic in human history, due to its rapid infection ability from person to person and its high incidence of mortality. The worldwide spreading rate and the concern that vaccines will be insufficient increase the need for new and natural medicine resources that will stimulate and strengthen our immunity without causing complications such as coagulation and side effects. However, well-designed, further-clinical studies are needed to demonstrate the potential efficacy of local herbal diets and preparations that have the potential to increase immunity against SARS-CoV-2 infection (Tewari et al. 2020). Turkey has a potential worth of ore for researchers and industry who are interested in this issue because it has the richest flora of Europe and the Middle East. The country's official pharmacopoeia includes contributions from 37 countries, including the European Pharmacopoeia. The number of monographs with rich content prepared to standardize the use of commonly used herbs is increasing day by day (Şekeroğlu and Gezici 2020).

While the previous study exhibited a national-plant list that promises a potential for anti-influenza activity in Turkey, in this study, the most comprehensive national-wide inventory of important medicinal plants that make the immune system quite active and strong is presented. In addition, the repercussions of these plants in the traditional medicine works of neighboring and nearby countries and the experimental studies conducted worldwide for the above-mentioned activities were examined, and the results were presented in tables and graphs. The taxa, which have not been the subject of any experimental research, constitute a unique reserve for new investigators in the field of natural medicine.

## 18.2 Materials and Methods

### *18.2.1 Gathering Data*

The study has been conducted in three different stages by using only English and Turkish terms and by searching electronic databases respected by scientific circles such as Web of Science, Scopus, ScienceDirect, PubMed, ProQuest, Medline, HighWire Press, Cochrane Library, Google Scholar, and EBSCO. As a first step, an inventory roster of herbs consumed as immune system stimulants, enhancers, or boosters in traditional Turkish medicine was displayed (Table 18.1). To reach more point-by-point data, master and doctoral dissertations carried out throughout the country (Fig. 18.1) were also embodied in the study. To this end, approximately 1000 national works were determined between January 1999 and December 2020, and a consensus reached only over 99 (Table 18.2). The therapeutic plants traditionally utilized by the nation's citizens for centuries were meticulously distinguished from those works according to the ensuing gauges and introduced in (Table 18.1). The equivalents of those plants in peer works conducted in nearby nations were examined in the stage after that. Finally, the implications of the empirical works in the global literature on immune-boosting and other similar

**Table 18.1** The list of plant taxa used for immune-boosting and related activities in Turkish folk medicine

| Plant name | Family | C/W/E | English name | Part(s) | Preparation(s) | Report(s) |
|---|---|---|---|---|---|---|
| *Achillea millefolium* L. | Compositae | W | Common yarrow | Leaves, flowers | Infusion | Kalafatçılar and Kalafatçılar (2010), Baytop (1999), Özer et al. (2005), Akan and Bakır-Sade (2015), Everest and Ozturk (2005) |
| ***Achillea nobilis*** **L.** | Compositae | W | Noble yarrow | Aerial parts, flowers | Infusion | Çekin et al. (2016), Sargin et al. (2015), Nacakcı and Dutkuner (2015), Everest and Ozturk (2005) |
| *Allium cepa* L. | Amaryllidaceae | C | Onion | Bulbs, leaves | Eaten raw, boiling, juice with some honey | Jahan and Onay (2020), Köse (2019), Özcan et al. (2020), Uçar et al. (2020), Yeşilada (2012), Ege and Elmastaş (2020), Uzun and Kaya (2016), Çakılcıoğlu and Türkoğlu (2007), Ekşi et al. (2020) |
| *Allium sativum* L. | Amaryllidaceae | C | Garlic | Leaves, bulbs, flowers | Eaten raw, cooked, tincture (prepared with the bulbs, lemon, and vinegar) | Jahan and Onay (2020), Kalafatçılar and Kalafatçılar (2010), Köse (2019), Özcan et al. (2020), Kavaklı et al. (2020b), Uçar et al. (2020), Yeşilada (2012), Ege and Elmastaş (2020), Uzun and Kaya (2016), Çakılcıoğlu and Türkoğlu (2007), Ekşi et al. (2020) |

| | | | | | | |
|---|---|---|---|---|---|---|
| ***Althaea officinalis* L.** | Malvaceae | C/W | Common marsh | Buds, flowers | Infusion | Kalafatçılar and Kalafatçılar (2010), Baytop (1999), Ötnü and Akan (2020) |
| *Apium graveolens* L. | Apiaceae | C | Celery | Leaves, seeds | Decoction, mixture (eaten after milling with cloves and seeds of anise, turnip, radish, and mixed with honey) | Baytop (1999), Özcan et al. (2020), Sargin (2013), Uçar et al. (2020) |
| ***Asparagus acutifolius* L.** | Asparagaceae | C/W | Wild asparagus | Aerial parts | Infusion | Çekin et al. (2016), Ege and Elmastaş (2020), Demirci-Kayıran (2019), Ozturk et al. (2017b) |
| ***Astragalus* spp.** | Leguminosae | W | Milkvetches | Aerial parts, roots, young shoots | Infusion, eaten raw after peeling, infusion after dried and pulverized, mixture (obtained by drying and pulverizing the fresh shoots and mixing with honey), tincture (prepared with lemon and vinegar) | Kalafatçılar and Kalafatçılar (2010), Sargin (2013), Uçar et al. (2020), Tuzlacı and Doğan (2010a), Arı et al. (2015), Akbulut et al. (2017), Doğan (2008), Korkmaz and Karakurt (2014), Polat et al. (2013) |
| *Avena sativa* L. | Poaceae | C/W | Oat, common oat | Fruits, seeds | Pastry, meal | Baytop (1999), Özcan et al. (2020), Çekin et al. (2016), Yeşilada (2012) |
| *Brassica oleracea* L. | Brassicaceae | C | Cabbage, flowering cabbage, kale | Aerial parts, leaves | Decoction, meal, salad | Özcan et al. (2020), Gül and Seçkin-Dinler (2016), Gökçe (2011), Akbulut et al. (2017) |
| *Brassica rapa* L. | Brassicaceae | C | Turnip | Tubers, roots, leaves, seeds | Juice, boiling-meal, mixture (eaten after milling with cloves and seeds of anise, celery, radish and mixed with honey) | Özcan et al. (2020), Sargin (2013), Yeşilada (2012), Gökçe (2011) |

(continued)

**Table 18.1** (continued)

| Plant name | Family | C/W/E | English name | Part(s) | Preparation(s) | Report(s) |
|---|---|---|---|---|---|---|
| *Calendula officinalis* L. | Compositae | C/W | Pot marigold, ruddles | Flowers | Infusion | Kalafatçılar and Kalafatçılar (2010), Çekin et al. (2016), Onbaşlı and Dal (2020) |
| *Camellia sinensis* (L.) Kuntze | Theaceae | C | Green tea | Leaves | Infusion | Genç (2010), Özcan et al. (2020), Yeşilada (2012), Doğan and Doğan (2020), Korkmaz et al. (2016) |
| *Capparis spinosa* L. | Capparaceae | W | Caper bush, flinders rose | Fruits, buds, roots, barks | Infusion, decoction | Baytop (1999), Genç (2010), Çekin et al. (2016) |
| *Capsicum annuum* L. | Solanaceae | C | Ornamental pepper | Fruits | Eaten raw, spice, decoction, meal | Kavaklı et al. (2020b), Uzun and Koca (2020), Sargin and Büyükcengiz (2019) |
| *Castanea sativa* mill. | Fagaceae | C/W | Chestnut, sweet chestnut | Fruits, seeds, flowers | Decoction, boiling, infusion | Köse (2019), Kavaklı et al. (2020a) |
| ***Centaurea hyalolepis* Boiss.** | Compositae | W | Yellow star-thistle, knapweed | Flowers | Restorative paste (with honey after crashing) | Kılıç (2019), Altundag and Ozturk (2011), Tuzlacı and Doğan (2010a), Doğan (2008) |
| ***Ceratonia siliqua* L.** | Leguminosae | C/W | Carob | Fruits | Eaten raw, boiling, molasses | Kurt and Karaoğul (2018), Biçer (2020), Doğan and Doğan (2020), Sargin and Büyükcengiz (2019) |
| *Cichorium intybus* L. | Compositae | W | Blue daisy, blue dandelion, blue sailors, blue weed | Aerial parts, roots | Infusion, decoction | Kılıç (2019), Şenkardeş (2014), Çekin et al. (2016), Özer et al. (2005), Ozturk et al. (2017b) |
| ***Cistus creticus* L.** | Cistaceae | W | Pink rock-rose, hoary rock-rose | Leaves | Infusion | Şekeroğlu and Gezici (2020), Baytop (1999), Güneş (2010), Ege and Elmastaş |

| | | | | | | |
|---|---|---|---|---|---|---|
| | | | | | | (2020), Yeşilada et al. (2019), Doğan and Doğan (2020) |
| ***Cistus laurifolius* L.** | Cistaceae | W | Laurel-leaved rock-rose | Leaves | Infusion | Şekeroğlu and Gezici (2020), Ege and Elmastaş (2020), Yeşilada et al. (2019), Doğan and Doğan (2020) |
| ***Cistus salviifolius* L.** | Cistaceae | W | Sage-leaved rock-rose | Leaves | Infusion | Şekeroğlu and Gezici (2020), Baytop (1999), Ege and Elmastaş (2020), Yeşilada et al. (2019), Doğan and Doğan (2020) |
| *Citrus* spp. | Rutaceae | C/W | Oranges, lemons, grapefruits, pomelos, limes | Pericarps, fruits | Decoction, jam, eaten raw, juice, marmalade, dropped in teas and soups, gargle | Jahan and Onay (2020), Şekeroğlu and Gezici (2020), Baytop (1999), Köse (2019), Genç (2010), Özcan et al. (2020), Yeşilada (2012), Akan and Bakır-Sade (2015), Güneş (2010) |
| ***Cota austriaca* (Jacq.) Sch.Bip.** | Compositae | W | Corn chamomile | Capitulum, flowers, aerial parts | Infusion | Kılıç (2019), Genç (2010), Şenkardeş (2014), Alkış (2019), Doğan (2014), Tetik et al. (2013), Kaval et al. (2014) |
| *Crataegus monogyna* Jacq. | Rosaceae | C/W | Hawthorn, may | Flowers, fruits, seeds | Infusion, eaten raw, infusion, eaten raw, decoction | Şenkardeş (2014), Kavaklı et al. (2020b), Olgun (2019), Paksoy et al. (2016) |
| ***Crataegus orientalis* pall. Ex M.Bieb.** | Rosaceae | C/W | Oriental hawthorn | Flowers, fruits, seeds | Infusion, eaten raw, infusion, eaten raw, decoction | Şenkardeş (2014), Kavaklı et al. (2020b), Gençay (2007) |
| ***Daucus carota* L.** | Apiaceae | C/W | Wild carrot, bird's nest | Seeds | Spice after milling | Özcan et al. (2020), Akan and Bakır-Sade (2015) |

(continued)

**Table 18.1** (continued)

| Plant name | Family | C/W/E | English name | Part(s) | Preparation(s) | Report(s) |
|---|---|---|---|---|---|---|
| ***Drosera rotundifolia* L.** | Droseraceae | C/W | Round-leaved sundew, common sundew | Flowers | Infusion | Kalafatçılar and Kalafatçılar (2010), Baytop (1999), Genç (2010) |
| ***Echinophora tenuifolia* subsp. *sibthorpiana* (Guss.) Tutin** | Apiaceae | C/W | Turkish çördük, tarhana | Aerial parts | Soup, meal (after dried and pulverized) | Sargin (2013), Özer et al. (2005), Bulut et al. (2014) |
| *Elaeagnus angustifolia* L. | Elaeagnaceae | C/W | Silverberry | Fruits, leaves, flowers | Eaten raw, infusion | Şenkardeş (2014), Biçer (2020), Ozturk et al. (2017a) |
| *Equisetum arvense* L. | Equisetaceae | W | Field horsetail, common horsetail | Aerial parts | Infusion, foot bath | Kalafatçılar and Kalafatçılar (2010), Şanlı (2006), Everest and Ozturk (2005), Korkmaz et al. (2016) |
| ***Eriobotrya japonica* (Thunb.) Lindl.** | Rosaceae | C/W | Loquat | Leaves | Infusion mix (with *Cydonia* leaves and *Tilia* flowers) | Baytop (1999), Hürkul and Köroğlu (2019), Sargin (2015) |
| ***Eruca vesicaria* (L.) Cav.** | Brassicaceae | C/W | Rocket, garden rocket | Leaves | Eaten raw, salad | Deniz et al. (2010), Akan and Bakır-Sade (2015), Kavaklı et al. (2020a) |
| ***Eupatorium cannabinum* L.** | Compositae | W | Hemp-agrimony, holy rope | Aerial parts | Infusion | Kalafatçılar and Kalafatçılar (2010), Baytop (1999), Genç (2010), Mumcu and Korkmaz (2018) |
| ***Ferula orientalis* L.** | Apiaceae | W | – | Whole parts | Decoction, infusion after dried and pulverized | Ötnü and Akan (2020), Altundag and Ozturk (2011), Tuzlacı and Doğan (2010a), Bulut et al. (2014), Doğan (2008) |

| | | | | | | |
|---|---|---|---|---|---|---|
| ***Ficus carica* L.** | Moraceae | C/W | Common fig | Fruits | Eaten raw or dried with/without nuts | Gençay (2007), Bulut (2008), Arasan (2014) |
| *Fragaria vesca* L. | Rosaceae | C/W | Strawberry, wild strawberry | Fruits | Jam, syrup | Özcan et al. (2020), Sargin (2013), Uçar et al. (2020), Gül and Seçkin-Dinler (2016) |
| *Glycyrrhiza glabra* L. | Leguminosae | C/W | Licorice, licorice | Leaves, roots | Infusion after pulverizing, syrup | Şekeroğlu and Gezici (2020), Kılıç (2019), Kavaklı et al. (2020b), Uçar et al. (2020), Ötnü and Akan (2020), Özer et al. (2005), Akan and Bakır-Sade (2015), Tütenocaklı (2014), Yeşil et al. (2019), Mumcu and Korkmaz (2018), Akbaş and Akçakaya (2020), Küpeli-Akkol et al. (2020) |
| *Helianthus tuberosus* L. | Compositae | C/W | Sunroot, sunchoke, earth apple | Tubers | Eaten raw, salad, soup, meal | Baytop (1999), Özçelik et al. (2016) |
| *Hordeum vulgare* L. | Poaceae | C/W | Barley | Fruits, seeds | Decoction, additive (the flour is obtained after drying, and grinding is added to the meals) | Olgun (2019), Özdemir-Nath (2016), Doğan (2014) |
| *Hypericum perforatum* L. | Hypericaceae | C/W | St. John's wort | Flowering branches, aerial parts | Infusion after drying, infusion mix (with the leaves of *Urtica* spp. and *Rubus* spp) | Köse (2019), Gül and Seçkin-Dinler (2016), Özdemir-Nath (2016), Yeşilyurt et al. (2017a, b), Nacakcı and Dutkuner (2015), Sargin and Büyükcengiz (2019) |

(continued)

**Table 18.1** (continued)

| Plant name | Family | C/W/E | English name | Part(s) | Preparation(s) | Report(s) |
|---|---|---|---|---|---|---|
| ***Hyssopus officinalis* L.** | Lamiaceae | W | Hyssop | Flowers, flowering branches, leaves | Infusion | Jahan and Onay (2020), Çekin et al. (2016), Özer et al. (2005) |
| *Juglans regia L.* | Juglandaceae | C/W | Walnut, common walnut | Leaves, pericarp, fruits | Infusion, mixture (eaten with honey and dried fig) | Kalafatçılar and Kalafatçılar (2010), Baytop (1999), Kavaklı et al. (2020b), Nacakcı and Dutkuner (2015) |
| ***Juniperus oxycedrus* L.** | Cupressaceae | W | Cade, cade juniper, prickly juniper | Fruits, seeds, leaves, tars, cones, gall | Decoction, infusion, distillation | Şenkardeş (2014), Deniz et al. (2010), Şanlı (2006), Akgul et al. (2018), Kökçü et al. (2015), Tuzlacı and Sadıkoğlu (2007) |
| ***Lavandula stoechas* L.** | Lamiaceae | C/W | Spanish lavender, topped lavender | Leaves, flowering branches | Infusion | Ötnü and Akan (2020), Güneş (2010), Bulut (2008), Nacakcı and Dutkuner (2015), Bozyel and Merdamert-Bozyel (2020), Bulut and Tuzlacı (2009) |
| ***Lepidium sativum* L.** | Brassicaceae | C/W | Garden cress | Aerial parts, seeds | Infusion, restorative paste (prepared with honey) | Baytop (1999), Deniz et al. (2010), Kavaklı et al. (2020a), Çiçek (2019), Ongan (2018) |
| ***Linum usitatissimum* L.** | Linaceae | C/W | Flax | Seeds, fixed oils | Eaten raw, drunk | Kalafatçılar and Kalafatçılar (2010), Köse (2019), Özcan et al. (2020), Kavaklı et al. (2020b) |
| *Malus domestica* Borkh. | Rosaceae | C | Apple | Fruits | Eaten raw, juice, decoction (with pear peels after dried) | Sargin (2013), Uçar et al. (2020), Olgun (2019) |

| | | | | | | |
|---|---|---|---|---|---|---|
| ***Malus sylvestris* (L.) mill.** | Rosaceae | C/W | European crab apple | Fruits | Eaten raw, juice, decoction (with pear peels after dried) | Uçar et al. (2020), Olgun (2019), Günbatan et al. (2016), Doğan (2014) |
| ***Malva neglecta* Wallr.** | Malvaceae | W | Cheese plant, dwarf mallow | Aerial parts | Infusion, decoction | Arasan (2014), Bulut and Tuzlacı (2013), Doğan (2014), Şenkardeş (2014), Olgun (2019) |
| ***Malva nicaeensis* all.** | Malvaceae | W | Bull mallow, French mallow | Leaves | Decoction with the aerial parts of *Urtica dioica.* | Gençay (2007), Bulut (2008), Kızılarslan (2008), Uzun and Kaya (2016) |
| *Malva sylvestris* (L.) mill. | Malvaceae | W | Large-flowered mallow, high mallow | Leaves, aerial parts | Infusion, meal (prepared with rice, radish, onion, and butter). Another meal is prepared with the leaves, *Papaver rhoeas* petals, *Urtica dioica,* and *Spinacia oleracea* leaves | Köse (2019), Sargin (2013), Özer et al. (2005), Özdemir-Nath (2016), Bulut (2008), Kızılarslan (2008), Tuzlacı and Doğan (2010b) |
| *Matricaria chamomilla* L. | Compositae | W | Chamomile, German chamomile | Aerial parts, flowering branches, flowers | Infusion | Kalafatçılar and Kalafatçılar (2010), Genç (2010), Özcan et al. (2020), Uçar et al. (2020), Özer et al. (2005), Sargin et al. (2015), Tuzlacı and Sadıkoğlu (2007), Polat et al. (2015) |
| *Melissa officinalis* L. | Lamiaceae | C/W | Lemon balm | Aerial parts | Infusion after drying, infusion mix (with sage leaves, elder fruits, black pepper, and ginger) | Uçar et al. (2020), Özer et al. (2005), Ege and Elmastaş (2020), Yeşilada et al. (2019), Demirci-Kayıran (2019), Güneş (2017), Özhatay et al. (2009), Diken (2009) |

(continued)

**Table 18.1** (continued)

| Plant name | Family | C/W/E | English name | Part(s) | Preparation(s) | Report(s) |
|---|---|---|---|---|---|---|
| ***Mentha longifolia* (L.) L.** | Lamiaceae | W | Horse mint | Aerial parts | Eaten raw, infusion | Jahan and Onay (2020), Kılıç (2019, 2016), Şenkardeş (2014), Özer et al. (2005), Olgun (2019), Kurt and Karaoğul (2018), Yeşilyurt et al. (2017b), Arı et al. (2015) |
| *Mentha pulegium* L. | Lamiaceae | W | Pennyroyal, pennyrile, squaw mint | Leaves | Infusion after drying, infusion mix (with the leaves of *Petroselinum crispum* and *Origanum vulgare*) | Jahan and Onay (2020), Köse (2019), Kurt and Karaoğul (2018), Yeşilyurt et al. (2017b) |
| ***Mentha spicata* L.** | Lamiaceae | W | Spearmint | Aerial parts, leaves | Infusion, spice (after powdering) | Jahan and Onay (2020), Köse (2019), Deniz et al. (2010), Kurt and Karaoğul (2018), Yeşilyurt et al. (2017b), Güneş (2010, 2017) |
| *Mentha x piperita* L. | Lamiaceae | C/W | Peppermint | Leaves, aerial parts | Eaten raw, salad, spice, syrup, infusion mix-1 (with lemon juice), infusion mix-2 (with the leaves of *Petroselinum crispum* and*Origanum vulgare*) | Jahan and Onay (2020), Kalafatçılar and Kalafatçılar (2010), Kılıç (2019), Genç (2010), Şenkardeş (2014), Uçar et al. (2020), Özer et al. (2005), Gül and Seçkin-Dinler (2016), Kurt and Karaoğul (2018), Yeşilyurt et al. (2017a, b), Uzun and Koca (2020), Akay (2019), Bozyel and Merdamert-Bozyel (2020) |

| | | | | | | |
|---|---|---|---|---|---|---|
| *Mespilus germanica* L. | Rosaceae | C/W | Medlar, medlar tree | Leaves, fruits | Infusion, eaten raw | Köse (2019), Gül and Seçkin-Dinler (2016), Uzun and Koca (2020) |
| *Morus alba* L. | Moraceae | C/W | White mulberry | Fruits, leaves | Syrup, molasses, churchkhela, infusion | Şenkardeş (2014), Çekin et al. (2016), Olgun (2019), Özdemir-Nath (2016), Gençay (2007), Çiçek (2019), Ege and Elmastaş (2020), Doğan (2014) |
| *Morus nigra* L. | Moraceae | C/W | Black mulberry, blackberry | Fruits, leaves | Molasses, syrup, infusion | Gül and Seçkin-Dinler (2016), Bulut (2008), Ege and Elmastaş (2020) |
| *Myrtus communis* L. | Myrtaceae | C/W | Common myrtle or true myrtle | Fruits, leaves | Eaten raw, infusion | Akan and Bakır-Sade (2015), Güneş (2010), Korkmaz et al. (2016) |
| *Nasturtium officinale* R.Br. | Brassicaceae | W | Watercress, yellowcress | Leaves, flowers | Infusion, salad, meal | Kılıç (2019), Şanlı (2006), Gençay (2007), Kocabaş et al. (2017) |
| ***Nigella arvensis* L.** | Ranunculaceae | C/W | Wild fennel, field fennel flower | Flowers, seeds | Infusion after dried, restorative paste (with honey after crushed), spices | Kılıç (2019), Deniz et al. (2010), Güneş (2017), Arı et al. (2015), Kılıç et al. (2020) |
| *Nigella damascena* L. | Ranunculaceae | C/W | Love-in-a-mist, ragged lady | Seeds, essential oils | Infusion, eaten raw | Genç (2010), Çekin et al. (2016), Özçelik et al. (2016) |
| *Nigella sativa* L. | Ranunculaceae | C/W | Black cumin | Seeds, essential oils | Eating raw, infusion after crashing, spices | Kalafatçılar and Kalafatçılar (2010), Baytop (1999), Köse (2019), Uçar et al. (2020), Ötnü and Akan (2020), Deniz et al. (2010), Kurt and Karaoğul (2018), Biçer |

(continued)

**Table 18.1** (continued)

| Plant name | Family | C/W/E | English name | Part(s) | Preparation(s) | Report(s) |
|---|---|---|---|---|---|---|
| | | | | | | (2020), Arı et al. (2015), Arıtuluk (2010) |
| *Olea europaea* L. | Oleaceae | C/W | Olive | Leaves, fruits, fixed oils | Infusion, cataplasm with one tablespoon molasses, tarhana, and flour, drunk one table-spoon or added to salads | Şekeroğlu and Gezici (2020), Köse (2019), Özcan et al. (2020), Ötnü and Akan (2020), Yeşilada et al. (2019) |
| ***Origanum onites* L.** | Lamiaceae | W | Pot marjoram, Cretan oregano | Aerial parts with flowers, flowering branches | Foot bath, infusion-1 (with sage leaves), infusion-2 (with anise and mint leaves) | Deniz et al. (2010), Kalafatçılar and Kalafatçılar (2010), Nacakcı and Dutkuner (2015), Bozyel and Merdamert-Bozyel (2020), Şekeroğlu and Gezici (2020) |
| *Origanum vulgare* L. | Lamiaceae | W | Ornamental oregano | Aerial parts with/without flowers, flowering branches | Foot bath, infusion 1 (with sage leaves), infusion 2 (with anise and mint leaves) | Şekeroğlu and Gezici (2020), Deniz et al. (2010), Özer et al. (2005), Bozyel and Merdamert-Bozyel (2020) |
| ***Papaver rhoeas* L.** | Papaveraceae | W | Flanders poppy | Flowers, petals | Infusion with/without *Urtica dioica* leaves, meal (prepared with the leaves of *Spinacia oleracea* and *Malva* spp. or *Urtica dioica*) | Özdemir-Nath (2016), Gençay (2007), Kızılarslan (2008), Özhatay et al. (2009), Tütenocaklı (2014), Kültür (2007) |
| *Petroselinum crispum* (mill.) fuss | Apiaceae | C/W | Parsley | Leaves, aerial parts | Eaten raw, juice (with lemon), salad, infusion with/without the leaves of *Mentha spp* and*Origanum* spp. | Şenkardeş (2014), Uçar et al. (2020), Ötnü and Akan (2020), Olgun (2019), Güneş (2010), Kavaklı et al. (2020a), Çiçek (2019) |
| ***Phoenix dactylifera* L.** | Arecaceae | C/W | Date, date palm | Fruits | Eaten raw | Baytop (1999), Genç (2010), Kavaklı et al. (2020a), Biçer (2020) |

| | | | | | | |
|---|---|---|---|---|---|---|
| ***Pinus brutia* Ten.** | Pinaceae | C/W | Calabrian pine, Turkish pine | Barks, shoots, pollens, cones, fruits, essential oils | Decoction, eaten with honey. Pickle, prepared with seeds, cones, lemons, garlic, and water. Medicinal bath, frankincense | Jahan and Onay (2020), Kalafatçılar and Kalafatçılar (2010), Genç (2010), Sargin (2013), Gürbüz et al. (2019) |
| ***Pinus nigra* J.F. Arnold** | Pinaceae | W | Black pine | Resins, tars, essential oils | Decoction, medicinal bath, frankincense | Jahan and Onay (2020), Kalafatçılar and Kalafatçılar (2010), Genç (2010), Sargin (2013), Gürbüz et al. (2019) |
| ***Pinus pinea* L.** | Pinaceae | C/W | Stone pine, umbrella pine, parasol pine | Fruits, pollens, essential oils | Eaten with molasses, honey, dried figs, or grapes | Jahan and Onay (2020), Genç (2010), Ötnü and Akan (2020), Gürbüz et al. (2019) |
| *Pinus sylvestris* L. | Pinaceae | W | Redwood, Scots fir | Buds, resins, cones, essential oils | Decoction, medicinal bath, frankincense | Jahan and Onay (2020), Kalafatçılar and Kalafatçılar (2010), Genç (2010), Gürbüz et al. (2019) |
| *Portulaca oleracea* L. | Polygonaceae | W | Common purslane, fatweed | Aerial parts | Eating raw, boiling, roasting | Kılıç (2019), Köse (2019), Ötnü and Akan (2020), Arasan (2014), Kocabaş et al. (2017) |
| *Prunus avium* (L.) L. | Rosaceae | C/W | Sweet cherry | Fruits stalks | Infusion, restorative paste (prepared with a cherry stalk and rye after drying and pulverizing and sweetening with honey or molasses) | Sargin (2013), Sargin et al. (2015), Özdemir-Nath (2016), Gürbüz et al. (2019), Hürkul and Köroğlu (2019), Korkmaz et al. (2016) |
| ***Prunus cerasifera* Ehrh.** | Rosaceae | C/W | Cherry plum | Fruits | Eaten raw, infusion, decoction | Çiçek (2019), Arı et al. (2015), Hürkul and Köroğlu (2019) |
| ***Prunus domestica* L.** | Rosaceae | C/W | Plum, European plum | Fruits | Eaten raw, compote | Ötnü and Akan (2020), Kavaklı et al. (2020a) |

(continued)

**Table 18.1** (continued)

| Plant name | Family | C/W/E | English name | Part(s) | Preparation(s) | Report(s) |
|---|---|---|---|---|---|---|
| ***Prunus spinosa* L.** | Rosaceae | W | Sloe, blackthorn | Fruits | Eaten raw, decoction | Şanlı (2006), Özdemir-Nath (2016), Yeşilyurt et al. (2017b), Özhatay et al. (2009) |
| *Punica granatum* L. | Lythraceae | C/W | Pomegranate | Fruits, flowers, seeds, pericarps | Eaten raw, juice, infusion, decoction | Baytop (1999), Köse (2019), Sargin (2013), Kavaklı et al. (2020b), Olgun (2019), Kurt and Karaoğul (2018), Özdemir-Nath (2016), Gençay (2007), Güneş (2010), Çiçek (2019), Arasan (2014), Tuzlacı and Sadıkoğlu (2007), Kilic and Bagci (2013), Korkmaz and Karakurt (2014), Ozturk et al. (2017a), Polat et al. (2011), Şahin (2019) |
| *Pyrus communis* L. | Rosaceae | C/W | European pear, common pear | Fruits | Eaten raw, decoction with the dried peel of the apple after dried | Güneş (2017), Akgul et al. (2018), Doğan (2014) |
| *Rhus coriaria* L. | Anacardiaceae | C/W | Tanner's sumach, Sicilian sumac | Leaves, fruits | Infusion, spice | Uçar et al. (2020), Nacakcı and Dutkuner (2015), Doğan and Doğan (2020), Özçelik et al. (2016) |
| ***Rosa* × *damascena* Herm.** | Rosaceae | C/W | Rose, damask rose | Flowers | Infusion, jam, syrup, juice, pastry, vinegar | Baytop (1999), Kavaklı et al. (2020b), Çekin et al. (2016) |
| *Rosa canina* L. | Rosaceae | C/W | Dog rose, briar rose, common briar | Fruits, leaves, flowers, petals, roots, stems | Eaten raw, infusion after drying and pulverizing, infusion mix (with lemon balm, | Şanlı (2006), Kültür (2007), Sarıkan (2007), Bulut (2008), Doğan (2008), |

| | | | | | | |
|---|---|---|---|---|---|---|
| | | | | | elderflower, plantaginis leaf, and a little honey or stevia) decoction, jam, Marmalate | Kızılarslan (2008), Özhatay et al. (2009), Deniz et al. (2010), Kalafatçılar and Kalafatçılar (2010), Kılıç (2019), Köse (2019), Şenkardeş (2014), Çekin et al. (2016), Yeşilada (2012), Sargin et al. (2015), Gül and Seçkin-Dinler (2016), Olgun (2019), Yeşilyurt et al. (2017b), Çiçek (2019), Gürbüz et al. (2019), Altundag and Ozturk (2011), Tuzlacı and Doğan (2010a), Uzun and Kaya (2016), Yeşil et al. (2019), Bulut et al. (2017), Günbatan et al. (2016), Mumcu and Korkmaz (2018), Paksoy et al. (2016), Arıtuluk (2010), Erarslan et al. (2020), Kavaklı et al. (2020a, b), Uçar et al. (2020), Ötnü and Akan (2020) |
| ***Rosa hemisphaerica*** **Herrm.** | Rosaceae | W | Sulphur rose | Fruits | Eaten raw, decoction | Şenkardeş (2014), Kavaklı et al. (2020b), Uzun and Kaya (2016) |
| *Rosmarinus officinalis* L. | Lamiaceae | C/W | Rosemary | Leaves, fruits, young shoots | Infusion, syrup, jam, eaten raw | Genç (2010), Kavaklı et al. (2020b), Ötnü and Akan (2020), Yeşilada (2012), Deniz et al. (2010), Güneş (2010), Şahin (2019) |

(continued)

**Table 18.1** (continued)

| Plant name | Family | C/W/E | English name | Part(s) | Preparation(s) | Report(s) |
|---|---|---|---|---|---|---|
| ***Rubus caesius* L.** | Rosaceae | C/W | Dewberry | Fruits, leaves, roots | Syrup, jam, eaten raw, infusion after drying and pulverizing, infusion mix (with the leaves of *Urtica* spp. and the aerial parts of *Hypericum perforatum*) | Baytop (1999), Özcan et al. (2020), Kurt and Karaoğul (2018), Çiçek (2019), Paksoy et al. (2016) |
| ***Rubus canescens* DC.** | Rosaceae | W | Woolly blackberry | Leaves, fruits, young shoots | Infusion, syrup, jam, Marmalate, eaten raw | Kalafatçılar and Kalafatçılar (2010), Özcan et al. (2020), Gül and Seçkin-Dinler (2016), Şanlı (2006), Özhatay et al. (2009), Kültür (2007), Tuzlacı and Doğan (2010b) |
| ***Rubus sanctus* Schreb.** | Rosaceae | W | Holy bramble | Fruits, roots, flowers, leaves | Syrup, jam, eaten raw, decoction, infusion, jam, infusion after drying, infusion mix (with the leaves of *Urtica* spp. and the aerial parts of *Hypericum perforatum*) | Kalafatçılar and Kalafatçılar (2010), Kılıç (2019), Özcan et al. (2020), Gül and Seçkin-Dinler (2016), Şanlı (2006), Özdemir-Nath (2016), Bulut (2008), Kızılarslan (2008), Akay (2019), Hürkul and Köroğlu (2019), Tuzlacı and Doğan (2010b) |
| ***Salvia absconditiflora* Greuter & Burdet** | Lamiaceae | E | – | Aerial parts | Infusion | Kılıç (2019), Şenkardeş (2014), Sargin (2013), Sargin et al. (2015), Yeşilada et al. (2019) |
| *Salvia officinalis* L. | Lamiaceae | C/W | Culinary sage, golden sage, garden sage | Aerial parts | Infusion after drying, infusion mix-1 (with/without the flowers or fruits of *Sambucus* | Kalafatçılar and Kalafatçılar (2010), Genç (2010), Kavaklı et al. (2020a, b), |

| | | | | | | |
|---|---|---|---|---|---|---|
| | | | | | *nigra*, the leaves of *Melissa officinalis*, black pepper, and ginger). Infusion mix-2 (with flowers of *Tilia* and *teucrium*, *Urtica* and *Thymus* leave with honey) | Uçar et al. (2020), Yeşilada (2012), Akan and Bakır-Sade (2015), Yeşilada et al. (2019), Uzun and Koca (2020) |
| ***Salvia palaestina* Benth.** | Lamiaceae | W | Palestinian sage | Aerial parts | Infusion | Kılıç (2019), Yeşilada et al. (2019), Kocabaş et al. (2017) |
| ***Salvia tomentosa* Mill.** | Lamiaceae | W | Balsamic sage | Aerial parts | Infusion after drying, infusion mix (with flowers of *Tilia argentea* and *teucrium polium* or *Urtica dioica* and *Thymus longicaulis* leaves with honey) | Sargin (2013), Deniz et al. (2010), Sargin et al. (2015), Yeşilada et al. (2019), Alkış (2019) |
| ***Salvia verticillata* subsp. *amasiaca* (Freyn & Bornm.) Bornm.** | Lamiaceae | W | Lilac sage, whorled clary | Aerial parts | Infusion | Köse (2019), Altundag and Ozturk (2011), Doğan (2014) |
| *Sambucus ebulus* L. | Adoxaceae | W | European dwarf elder | Fruits, roots, seeds | Eaten raw, decoction, infusion after drying and pulverizing, infusion mix (with sage, Melissa leaves, black pepper, and ginger) | Baytop (1999), Yeşilada et al. (2019), Ozturk et al. (2017a), Öztürk and Ölçücü (2011) |
| *Sambucus nigra* L. | Adoxaceae | W | Elderberry, European elder | Leaves, flowers, fruits, barks | Infusion after drying and pulverizing, infusion mix-1 (with lemon balm, plantaginis leaf, rosehip fruit powder, and a little honey or stevia). Infusion mix-2 (with sage, Melissa leaves, black pepper, and ginger) | Kalafatçılar and Kalafatçılar (2010), Ötnü and Akan (2020), Yeşilada (2012), Şanlı (2006), Ege and Elmastaş (2020), Yeşilada et al. (2019), Akbaş and Akçakaya (2020), Kökçü et al. (2015) |

(continued)

**Table 18.1** (continued)

| Plant name | Family | C/W/E | English name | Part(s) | Preparation(s) | Report(s) |
|---|---|---|---|---|---|---|
| ***Satureja cuneifolia* Ten.** | Lamiaceae | W | Apulian savory | Aerial parts | A glass of its infusion is drunk twice a day | Sargin (2013), Sargin et al. (2015), Güneş (2017) |
| ***Satureja hortensis* L.** | Lamiaceae | W | Summer savory | Aerial parts, flowering branches, essential oil | Infusion, spice, lotion | Olgun (2019), Altundag and Ozturk (2011), Tuzlacı and Doğan (2010a), Doğan (2014), (2008), Tetik et al. (2013), Kılıç (2016), Polat et al. (2013) |
| ***Satureja thymbra* L.** | Lamiaceae | W | Thyme-leaved savory | Aerial parts, flowering branches, essential oil | Infusion, spice, lotion | Sargin et al. (2015), Nacakcı and Dutkuner (2015), Sargin (2015) |
| *Sesamum indicum* L. | Pedaliaceae | C/W | Sesame | Seeds | Mixture (crushed and mixed with boiled grape juice), eaten raw | Baytop (1999), Ötnü and Akan (2020), Güneş (2017) |
| *Silybum marianum* (L.) Gaertn. | Compositae | W | Milk thistle, Marian thistle | Seeds, stems, fruits | Eaten raw after peeling, infusion | Baytop (1999), Kılıç (2019), Sargin et al. (2015), Akbaş and Akçakaya (2020) |
| *Solanum lycopersicum* L. | Solanaceae | C | Tomato | Fruits | Eaten raw or after boiling, soup, salad | Köse (2019), Özcan et al. (2020), Sargin (2013), Uçar et al. (2020), Gökçe (2011) |
| *Spinacia oleracea* L. | Amaranthaceae | C | Spinach | Leaves | Boiling, salad, pastry material. Mixture-meal (prepared with *Papaver rhoeas* petals and *Malva* spp. or *Urtica dioica* leaves) | Özcan et al. (2020), Sargin (2013), Gençay (2007), Uzun and Kaya (2016) |

| | | | | | | |
|---|---|---|---|---|---|---|
| *Taraxacum* spp. | Compositae | W | Dandelions | Flowers, roots | Infusion, meal, mixture-meal (with *Allium cepa, Allium sativum, Solanum lycopersicum*, and *Capsicum annuum*), decoction, tincture (prepared with lemon and vinegar) | Kılıç (2019), Genç (2010), Gül and Seçkin-Dinler (2016), Kurt and Karaoğul (2018), Özdemir-Nath (2016) |
| ***Thymbra capitata* (L.) Cav.** | Lamiaceae | W | Spanish oregano, cone-head thyme | Aerial parts, flowering branches, essential oil | Infusion, lotion, spice | Şekeroğlu and Gezici (2020), Kalafatçılar and Kalafatçılar (2010), Sargin et al. (2015), Doğan and Doğan (2020) |
| ***Thymbra spicata* L.** | Lamiaceae | W | Thyme spiked | Aerial parts, flowering branches, essential oil | Infusion, lotion, spice, foot bath, bath with/without Matricaria | Şekeroğlu and Gezici (2020), Kalafatçılar and Kalafatçılar (2010), Kılıç (2019), Sargin et al. (2015), Akan and Bakır-Sade (2015), Güneş (2010), Doğan and Doğan (2020) |
| ***Thymbra sintenisii* Bornm. & Azn.** | Lamiaceae | W | – | Aerial parts, flowers | Infusion, lotion, spice | Şekeroğlu and Gezici (2020), Kalafatçılar and Kalafatçılar (2010), Kılıç (2019), Doğan and Doğan (2020) |
| ***Thymus fallax* Fisch. & C.A.Mey.** | Lamiaceae | W | – | Aerial parts, flowering branches, whole parts | Infusion, foot bath, bath with/without Matricaria | Jahan and Onay (2020), Şekeroğlu and Gezici (2020), Gül and Seçkin-Dinler (2016), Olgun (2019) |
| ***Thymus longicaulis* C.Presl** | Lamiaceae | W | Creeping thyme | Flowering branches, aerial parts | Infusion | Jahan and Onay (2020), Şekeroğlu and Gezici (2020), Gül and Seçkin-Dinler (2016), Günbatan et al. (2016) |

(continued)

**Table 18.1** (continued)

| Plant name | Family | C/W/E | English name | Part(s) | Preparation(s) | Report(s) |
|---|---|---|---|---|---|---|
| ***Thymus praecox* Opiz** | Lamiaceae | W | – | Aerial parts, flowering branches, whole parts | Infusion, foot bath, bath with/without Matricaria | Jahan and Onay (2020), Şekeroğlu and Gezici (2020), Gül and Seçkin-Dinler (2016), Günbatan et al. (2016), Alkış (2019) |
| ***Tilia cordata* Mill.** | Malvaceae | W | Bast, small-leaved linden | Leaves, fruits, flowers, bracts | Decoction mix (with cinnamon and cloves) | Kalafatçılar and Kalafatçılar (2010), Şenkardeş (2014), Kavaklı et al. (2020a, b), Yeşilyurt et al. (2017b), Gökçe (2011) |
| ***Tilia platyphyllos* Scop.** | Malvaceae | W | Broad-leaved lime | Flowers, bracts | Decoction with cinnamon and cloves, infusion after drying, infusion mix (with *Salvia tomentosa* leaves), cold maceration | Kalafatçılar and Kalafatçılar (2010), Kavaklı et al. (2020b), Gökçe (2011), Bulut and Tuzlacı (2013), Bulut et al. (2017) |
| *Tilia tomentosa* Moench | Malvaceae | C/W | European white lime, silver lime, silver linden | Leaves, flowers, fruits, barks, bracts, roots | Cold maceration, infusion after drying, infusion mix (with *Salvia tomentosa* leaves) | Kavaklı et al. (2020b), Çekin et al. (2016), Kurt and Karaoğul (2018), Gürbüz et al. (2019), Gökçe (2011), Akay (2019), Bulut et al. (2017) |
| *Tribulus terrestris* L. | Zygophyllaceae | W | Caltrop, devil's-thorn | Fruits | Infusion after dried | Baytop (1999), Kılıç (2019), Şenkardeş (2014), Deniz et al. (2010), Nacakcı and Dutkuner (2015), Demirci-Kayıran (2019), Özçelik et al. (2016), Kökçü et al. (2015) |

| | | | | | | |
|---|---|---|---|---|---|---|
| *Urtica dioica* L. | Urticaceae | W | Stinging nettle, perennial nettle, tall nettle, common nettle | Aerial parts (without flowering), leaves, roots, seeds | Infusion after drying, infusion mix (with/without the petals of *Papaver rhoeas* or the leaves of *Rubus* spp. and the aerial parts of *Hypericum perforatum*), the roots are eaten raw after preparing a mixture (after milling with mustard seeds and mixing with honey). Mixture-meal (with the leaves, *Papaver rhoeas* petals, and *Malva* spp. and *Spinacia oleracea* leaves). Seeds are eaten with honey | Jahan and Onay (2020), Kalafatçılar and Kalafatçılar (2010), Kılıç (2019), Köse (2019), Genç (2010), Şenkardeş (2014), Çekin et al. (2016), Yeşilada (2012), Deniz et al. (2010), Özer et al. (2005), Olgun (2019), Şanlı (2006), Özdemir-Nath (2016), Yeşilyurt et al. (2017a, b), Gençay (2007), Çiçek (2019), Uzun and Koca (2020), Demirci-Kayıran (2019), Everest and Ozturk (2005), Kızılarslan (2008), Uzun and Kaya (2016), Ongan (2018), Özçelik et al. (2016), Tütenocaklı (2014), Bulut and Tuzlacı (2009, 2013), Mumcu and Korkmaz (2018), Paksoy et al. (2016), Sarıkan (2007), Akgul et al. (2018), Doğan (2014), Kılıç et al. (2020), Tetik et al. (2013), Diken (2009), Erarslan et al. (2020), Kaval et al. (2014), Onbaşlı and Dal (2020), Polat et al. (2015) |
| ***Urtica pilulifera* L.** | Urticaceae | W | Roman nettle | Aerial parts, leaves, seeds | Salad, pastry, infusion, restorative paste (prepared with ginger, cinnamon, | Kalafatçılar and Kalafatçılar (2010), Genç (2010), Yeşilada (2012), Sargin et al. |

(continued)

Table 18.1 (continued)

| Plant name | Family | C/W/E | English name | Part(s) | Preparation(s) | Report(s) |
|---|---|---|---|---|---|---|
| | | | | | turmeric, and *Nigella sativa* seeds after milling and mixing with honey) | (2015), Şanlı (2006), Demirci-Kayıran (2019), Everest and Ozturk (2005), Ongan (2018) |
| ***Urtica urens*** **L.** | Urticaceae | W | Small nettle | Aerial parts | Infusion after dried | Kalafatçılar and Kalafatçılar (2010), Kılıç (2019), Köse (2019), Şenkardeş (2014), Yeşilada (2012), Özer et al. (2005), Yeşilyurt et al. (2017b), Bulut (2008), Demirci-Kayıran (2019), Everest and Ozturk (2005), Ongan (2018), Tütenocaklı (2014), Bulut and Tuzlacı (2009, 2013) |
| ***Vaccinium arctostaphylos*** **L.** | Ericaceae | W | Caucasian whortleberry | Fruits, leaves | Eaten raw, infusion | Kalafatçılar and Kalafatçılar (2010), Baytop (1999), Genç (2010), Kubat (2019) |
| ***Vaccinium myrtillus*** **L.** | Ericaceae | W | Bilberry, wineberry, whortleberry | Fruits, leaves | Eaten raw/dried, jam infusion | Kalafatçılar and Kalafatçılar (2010), Baytop (1999), Köse (2019), Özcan et al. (2020), Yeşilada (2012), Şanlı (2006), Kubat (2019) |
| ***Viburnum opulus*** **L.** | Adoxaceae | W | Guelder-rose | Fruits | Eaten raw, juice | Köse (2019), Kurt and Karaoğul (2018), Ongan (2018), Tütenocaklı (2014) |
| *Viscum album* L. | Santalaceae | W | European mistle-toe, common mistletoe | Fruits, leaves, stems, aerial parts | Infusion, decoction, tincture (prepared with lemon and vinegar) | Kalafatçılar and Kalafatçılar (2010), Baytop (1999), Genç (2010), Uzun and Koca |

| | | | | | | |
|---|---|---|---|---|---|---|
| | | | | | | (2020), Bulut et al. (2017), Ozturk et al. (2017a), Öztürk and Ölçücü (2011), Polat et al. (2011) |
| *Vitis labrusca* L. | Vitaceae | C/W | Fox grape, concord grape | Fruits, seeds | Eaten raw, eaten with honey after dried and pulverized | Kalafatçılar and Kalafatçılar (2010), Biçer (2020), Kubat (2019), Sarıkan (2007) |
| ***Vitis vinifera* L.** | **Vitaceae** | **C** | **Grapevine** | **Fruits, seeds** | **Eaten raw or dried, cataplasm (with tarhana flour), molasses, churchkhela** | **Kalafatçılar and Kalafatçılar (2010), Kılıç (2019), Köse (2019), Özcan et al. (2020), Sargin (2013), Şanlı (2006), Biçer (2020), Güneş (2017), Sarıkan (2007), Kılıç et al. (2020)** |
| ***Withania somnifera* (L.) Dunal** | **Solanaceae** | **W** | **Ashwagandha, Indian ginseng** | **Roots** | **Infusion after dried and pulverized** | **Kalafatçılar and Kalafatçılar (2010), Kubat (2019), Küpeli-Akkol et al. (2020)** |
| ***Ziziphora clinopodioides* lam.** | **Lamiaceae** | **W** | **Blue mint bush** | **Flowering branches, aerial parts** | **Infusion, spices** | **Sargin (2013), Sargin et al. (2015), Altundag and Ozturk (2011), Tuzlacı and Doğan (2010a)** |
| ***Ziziphus jujuba* mill.** | **Rhamnaceae** | **C/W** | **Jujube, red date** | **Fruits** | **Eaten raw or after dried, decoction mix (prepared with pear, ginger, and lemon), tincture (prepared with lemon and vinegar)** | **Çekin et al. (2016), Uzun and Koca (2020), Arıtuluk (2010)** |

*C* Cultured plants, *W* wild plans, *E* endemic plants, *CW* cultured and wild plants. **Boldly highlighted taxa** (which are 60 in total and their immune-boosting and other similar activity effects have not been investigated experimentally yet)

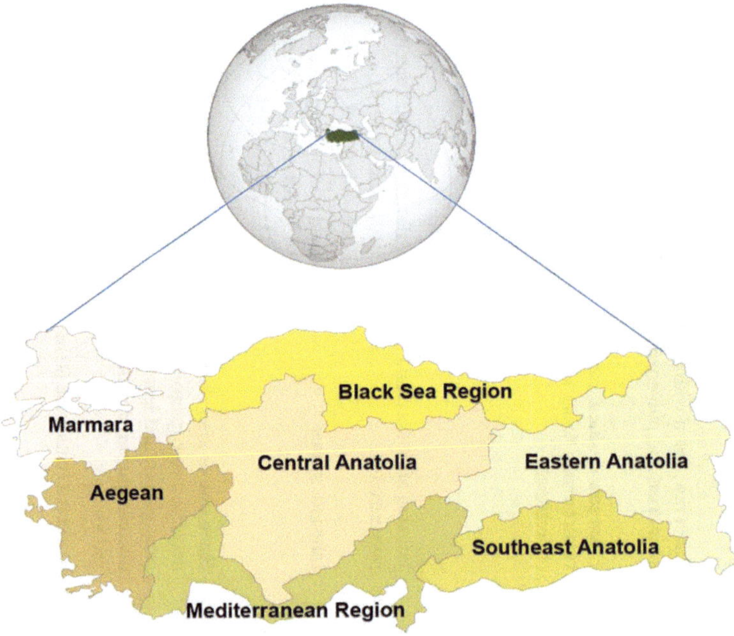


**Fig. 18.1** Regional map of Turkey

activities, such as immunostimulant, immunomodulatory, immune-enhancing, and immunotherapeutic are discussed, and the results are exhibited in (Table 18.3).

### 18.2.2 Evaluating Data

For the primary and moment portion research, the works identified to be within the scope of the study were minutely analyzed, contrasted, and preferred with respect to the following criteria:

1. The works ought to be conducted within the borders of Turkey.
2. It must be presented in an ethnobotanical, ethnomedicinal, or ethnopharmacological framework.
3. The author must have an academic title or citation records from other scientific works must be available.
4. The plant taxa mentioned in the work should be given in the study together with their scientific and local names.

The third screening study, which compares the taxa determined as a result of the first screening with the experimental studies published in the world literature, was made based on the following criteria:

**Table 18.2** Distribution of ninety-nine works selected from ethnomedical studies conducted in Turkey by region

| Region | Work | Citation | Citation % | Selected Studies |
|---|---|---|---|---|
| Eastern Anatolia | 15 | 53 | 7.5 | Olgun (2019), Altundag and Ozturk (2011), Tuzlacı and Doğan (2010a), Arasan (2014), Doğan (2008, 2014), Korkmaz et al. (2016), Tetik et al. (2013), Çakılcıoğlu and Türkoğlu (2007), Kaval et al. (2014), Kılıç (2016), Kilic and Bagci (2013), Öztürk and Ölçücü (2011), Polat et al. (2011, 2013) |
| Marmara | 12 | 68 | 9.6 | Özdemir-Nath (2016), Yeşilyurt et al. (2017b), Bulut (2008), Güneş (2017), Kızılarslan (2008), Akay (2019), Özhatay et al. (2009), Tütenocaklı (2014), Bulut and Tuzlacı (2009), Kökçü et al. (2015), Kültür (2007), Tuzlacı and Doğan (2010b) |
| Mediterranean | 12 | 66 | 9.3 | Şanlı (2006), Güneş (2010), Nacakcı and Dutkuner (2015), Uzun and Koca (2020), Demirci-Kayıran (2019), Everest and Ozturk (2005), Özçelik et al. (2016), Alkış (2019), Kocabaş et al. (2017), Sargin and Büyükcengiz (2019), Ozturk et al. (2017b), Sargin (2015) |
| Black Sea | 10 | 85 | 12.0 | Köse (2019), Kavaklı et al. (2020b), Gül and Seçkin-Dinler (2016), Kurt and Karaoğul (2018), Gürbüz et al. (2019), Kubat (2019), Mumcu and Korkmaz (2018), Yeşilyurt et al. (2017a), Korkmaz and Karakurt (2014), Polat et al. (2015) |
| Southeastern Anatolia | 10 | 78 | 11.0 | Kılıç (2019), Ötnü and Akan (2020), Akan and Bakır-Sade (2015), Gençay (2007), Çiçek (2019), Yeşil et al. (2019), Akgul et al. (2018), Arıtuluk (2010), Kılıç et al. (2020), Ozturk et al. (2017a) |
| Aegean | 9 | 66 | 9.3 | Sargin (2013), Deniz et al. (2010), Sargin et al. (2015), Arı et al. (2015), Bulut and Tuzlacı (2013), Bulut et al. (2017), Sarıkan (2007), Tuzlacı and Sadıkoğlu (2007), Şahin (2019) |
| Central Anatolia | 7 | 49 | 6.9 | Şenkardeş (2014), Kavaklı et al. (2020a), Uzun and Kaya (2016), Ongan (2018), Günbatan et al. (2016), Paksoy et al. (2016), Akbulut et al. (2017) |
| All regions | 24 | 246 | 34.4 | Jahan and Onay (2020), Şekeroğlu and Gezici (2020), Kalafatçılar and Kalafatçılar (2010), Baytop (1999), Genç (2010), Özcan et al. (2020), Çekin et al. (2016), Uçar et al. (2020), Yeşilada (2012), Özer et al. (2005), Ege and Elmastaş (2020), Yeşilada et al. (2019), Biçer (2020), Doğan and Doğan (2020), Gökçe (2011), Bozyel and Merdamert-Bozyel (2020), Hürkul and Köroğlu (2019), Akbaş and Akçakaya (2020), Bulut et al. (2014), Diken (2009), Ekşi et al. (2020), Erarslan et al. (2020), Küpeli-Akkol et al. (2020), Onbaşlı and Dal (2020) |
| Total | 99 | 711 | | |

**Table 18.3** Worldwide immune-boosting and other similar activity research results of the taxa detected in the study

| Plant species | Active compounds effective on immune response(s) (and used parts) | Type of activity | Type of immune response | Mechanism of action | References |
|---|---|---|---|---|---|
| *Achillea millefolium* L. | Not specified (aerial parts) | Immune-boosting | Humoral | By reducing the levels of serum lipids and boosting the immune response. Thus, it is proposed that can be used as an antibiotic alternative | Yakhkeshi et al. (2012) |
| *Allium cepa* L. | Agglutinin (bulbs) | Immunomodulatory | The cell-mediated | By increasing the production of nitric oxide, production of pro-inflammatory cytokines, enhancing the proliferation of murine thymocytes, and elevating the expression levels of cytokines | Prasanna and Venkatesh (2015) |
| *Allium sativum* L. | Not specified (bulbs) | Immune-boosting | The cell-mediated | By increasing CD4 (cluster of differentiation 4) T lymphocytes and total WBC (white blood cells) synergistically with *Allium cepa* | Mirabeau and Samson (2012) |
| *Apium graveolens* L. | Not specified (leaves) | Immunostimulant | The cell-mediated | By increasing total white and red blood counts, hemoglobin content, and hematocrit value | Mohamed et al. (2018) |
| *Avena sativa* L. | β-Glucan (seeds) | Immunomodulatory | The cell-mediated | By increasing cytokine levels | Rösch et al. (2016) |
| *Brassica oleracea* L. | Not specified (aerial parts) | Immune-boosting, immunostimulant | Humoral and the cell-mediated | By showing significant improvement in phagocytic and bactericidal activity | Pisuttharachai et al. (2020) |
| *Brassica rapa* L. | Not specified (roots) | Immunostimulant | The cell-mediated | By stimulating their endocytic/pinocytosis activity and releasing nitric oxide, tumor necrosis factor-alpha (TNF-α), and interleukin-6 cytokines | Wufuer et al. (2020) |

| | | | | | |
|---|---|---|---|---|---|
| *Calendula officinalis* L. | Not specified (flowers) | Immunostimulant | The cell-mediated | By increasing the production of nitric oxide on a level comparable with that of lipopolysaccharides | Danilets et al. (2010) |
| *Camellia sinensis* (L.) Kuntze | Strictinin, catechin (leaves) | Immunostimulant, immunomodulatory | The cell-mediated | By using the phagocytic activity of macrophage-like cells as an indicator of immune function activation | Monobe et al. (2008) |
| *Capparis spinosa* L. | Quercetin and kaempferol glycosides (buds) | Immunomodulatory | The cell-mediated | Inhibiting the extracellular virus release upregulates their production of IL-12, IFN-γ, and TNF-α. It should thus be employed for the treatment of immunocompromised hosts | Arena et al. (2008) |
| *Capsicum annuum* L. | Not specified (fruits) | Immunomodulatory | Humoral and the cell-mediated | Oral administration with capsicum extract might activate the CD4+ T cells leading to cytokine production as well as CD19+ B cells in Peyer's patches | Park et al. (2010) |
| *Castanea sativa* mill. | Not specified (fruits) | Immunostimulant | The cell-mediated | Through the skin mucus and serum immune parameters were significantly increased in fish fed with CSP compared to the control | Van Doan et al. (2020) |
| *Cichorium intybus* L. | Oxycoumarins, hydroxycinnamic acids, and flavonoids (aerial parts) | Immunomodulatory | Humoral and the cell-mediated | By reducing the suppressive azathioprine effect on the cell-mediated immune response, antibody response, and phagocytosis with macrophages | Saybel et al. (2020) |
| *Citrus* spp. | Auraptene (fruits) | Immunomodulatory | Humoral and the cell-mediated | By activating splenocytes and lymphocytes from mesenteric lymph nodes and increasing IL-4, interferon-gamma (IFNγ), and TNF-α production | Nishimoto et al. (2012) |
| *Crataegus monogyna* Jacq. | Not specified (leaves) | Immunomodulatory | Humoral and the cell-mediated | By increasing the total number, percentage, and absolute count of T and B splenocytes | Lis et al. (2020) |

(continued)

**Table 18.3** (continued)

| Plant species | Active compounds effective on immune response(s) (and used parts) | Type of activity | Type of immune response | Mechanism of action | References |
|---|---|---|---|---|---|
| *Elaeagnus angustifolia* L. | Not specified (fruits) | Immunostimulant | Innate, humoral, and the cell-mediated | By increasing the spleen and thymus index of mice, increasing the phagocytosis of the mononuclear phagocyte system, and stimulating hemolysin formation | Lian et al. (2009) |
| *Equisetum arvense L.* | Not specified (aerial parts) | Immune-enhancing | Innate, humoral, and the cell-mediated | By enhancement of cytokine (IL-2, INF-γ) production in Th1 cells | Yukitake and Yamamoto (2011) |
| *Fragaria vesca* L. | Gamma-aminobutyric acid (fruits) | Immunomodulatory | Innate, humoral, and the cell-mediated | By increasing IL-10 and IFN-γ in mice | Cataldo et al. (2020) |
| *Glycyrrhiza glabra* L. | Not specified (roots) | Immunomodulatory | Humoral and the cell-mediated | By increasing leukocytes, phagocytic activities, and antibody-secreting cells of the mouse spleen in combination with zinc | Mazumder et al. (2012) |
| *Helianthus tuberosus* L. | Inulin (tubers) | Immunostimulant, immunomodulatory | Humoral and the cell-mediated | By increasing total immunoglobulin and lysozyme activity | Tiengtam et al. (2017) |
| *Hordeum vulgare* L. | β-Glucan (seeds) | Immunomodulatory | The cell-mediated | By increasing cytokine levels | Rösch et al. (2016) |
| *Hypericum perforatum* L. | Germacrene D, β-caryophyllene, α-humulene (essential oils from leaves and flowers) | Immunomodulatory | Innate | By inhibiting neutrophil $Ca^{2+}$ mobilization, chemotaxis, and reactive oxygen species (ROS) production | Schepetkin et al. (2020) |

| | | | | | |
|---|---|---|---|---|---|
| *Juglans regia L.* | Not specified (fruits) | Immunomodulatory | Innate, humoral, and the cell-mediated | By promoting macrophage phagocytosis and natural killer activity, elevating the percentages of CD3+ and CD4, upregulating the production of antibody IgM, IgA, IgG, and SIgA, and regulating cytokines, Th1 produced | Mao et al. (2020) |
| *Malus domestica* Borkh. | Catechin, procyanidin B, quercetin-3-galactoside, and quercetin xyloside (fruit peels) | Immunostimulant | Innate and the cell-mediated | By increasing the lysozyme and phagocytic activity of macrophages | Riaz et al. (2020) |
| *Malva sylvestris* (L.) mill. | Not specified (leaves) | Immunomodulatory | The cell-mediated | By acting as macrophage and T helper-1 (Th-1) activators | El Ghaoui et al. (2008) |
| *Matricaria chamomilla* L. | Not specified (flowering branches) | Immunomodulatory | The cell-mediated | Through activating immunoregulation cells of peripheral blood and increasing the sensitivity of effector cells to helper signals | Uteshev et al. (1999) |
| *Melissa officinalis* L. | Not specified (aerial parts) | Immunostimulant | The cell-mediated | By increasing lipase and lysozyme activity, and elevating pepsin and trypsin | Bilen et al. (2020) |
| *Mentha pulegium* L. | Not specified (aerial parts) | Immune-enhancing, immunostimulant | Innate | By increasing phagocytosis, lysozyme, respiratory burst, and total white and red blood cells (WBC/RBC) | Akbary et al. (2015) |
| *Mentha x piperita* L. | Not specified (aerial parts) | Immunostimulant | Humoral and the cell-mediated | By increasing erythrocytes, leucocytes, hematocrit, hemoglobin, phagocytic activity, lysozyme, anti-protease, and anti-bactericidal activities, as well as elevating serum protein and globulin levels | Talpur (2014) |
| *Mespilus germanica* L. | Not specified (leaves) | Immunostimulant | Humoral | By increasing the immunoglobulin levels and interleukin 8 levels as well as elevating the level of lysozyme activity | Hoseinifar et al. (2017) |

(continued)

**Table 18.3** (continued)

| Plant species | Active compounds effective on immune response(s) (and used parts) | Type of activity | Type of immune response | Mechanism of action | References |
|---|---|---|---|---|---|
| *Morus alba* L. | 2-Formyl-5-(hydroxymethyl)-1H-pyrrole-1-butanoic acid and 2-formyl-5-(methoxymethyl)-1H-pyrrole-1-butanoic acid (fruits) | Immunostimulant | The cell-mediated | By the enhancement of nitric oxide, TNF-α, and IL-12 production and the stimulation of phagocytic activity | Kim et al. (2013) |
| *Morus nigra L.* | Not specified (fruits) | Immunomodulatory | Innate | By increasing the activities of serum lysozyme, myeloperoxidase, superoxide dismutase, and catalase and increasing the expression levels of immune-related genes (interleukin 1, beta, tumor necrosis factor, immunoglobulin M, interferon-gamma and heat shock protein 70) in the spleen and antioxidant-related genes | Yilmaz et al. (2020) |
| *Myrtus communis* L. | Not specified (leaves) | Immunostimulant | Innate | By increasing the number of erythrocytes (RBCs), leukocytes (WBCs), hematocrit, and lysozyme activity | Taee et al. (2017) |
| *Nasturtium officinale* R.Br. | Not specified (leaves) | Immunomodulatory, immunostimulant | Humoral and the cell-mediated | By enhancing hemoglobin (Hb), mean corpuscular hemoglobin concentration (MCHC), lysozyme and complement activities, as well as total protein and globulin levels | Asadi et al. (2012) |
| *Nigella damascena* L. | Damascenine and β-Elemene (seed-essential oils) | Immunomodulatory | The cell-mediated | By inhibiting the release of interleukin 1 beta (IL-1β) and interleukin 8 (IL-8), decreasing matrix metallopeptidase 9 (MMP-9) production | Sieniawska et al. (2019) |

| | | | | | |
|---|---|---|---|---|---|
| *Nigella sativa* L. | Thymoquinone (seed-essential oils) | Immunomodulatory, immunotherapeutic | Innate and the cell-mediated | By possessing reproducible antioxidant effects through enhancing the oxidant scavenger system and augmenting the T cell- and natural killer cell-mediated immune responses | Salem (2005) |
| *Olea europaea* L. | Oleuropein (leaves) | Immunomodulatory | The cell-mediated | By inducing and maintaining high proliferation in lymphocytes | Randon and Attard (2007) |
| *Origanum vulgare* L. | Not specified (aerial parts) | Immunostimulant | Innate and the cell-mediated | By increasing serum total protein, albumin, globulin, respiratory burst activity, phagocytic activity, and serum lysozyme activity | Pourmoghim et al. (2015) |
| *Petroselinum crispum* (mill.) fuss | Not specified (aerial part-essential oils) | Immunomodulatory | The cell-mediated | Through its inhibitory effect on PHA-stimulated splenocytes which occurred due to the production of cytokines such as IFN-γ and IL-2, which are vital for T-cell proliferation and the signaling pathways | Karimi et al. (2012) |
| *Phoenix dactylifera* L. | Chlorogenic acid, caffeic acid, pelargonin, ferulic acid, trypsin (fruits) | Immunomodulatory | Humoral and the cell-mediated | By stimulating FN-γ mRNA expression significantly in mouse Peyer's patch cell cultures and increasing the number of IFN-γ + CD4+ CD49b + and IL-12 + CD11b + cells | Karasawa et al. (2011) |
| *Pinus sylvestris* L. | Not specified (cones) | Immune-enhancing | Humoral | By affecting the expansion of T cells (both CD4 and CD8) during the proliferative phase | Bradley et al. (2014) |
| *Portulaca oleracea* L. | Not specified (aerial parts) | Immunostimulant, immunomodulatory | Innate, humoral, and the cell-mediated | By enhancing splenocyte proliferation and upregulated inflammatory cytokines (IL-2, IL-12, TNF-α, and IFN-γ) and natural killer (NK) cell activity in vitro | Park et al. (2019) |

(continued)

**Table 18.3** (continued)

| Plant species | Active compounds effective on immune response(s) (and used parts) | Type of activity | Type of immune response | Mechanism of action | References |
|---|---|---|---|---|---|
| *Prunus avium* (L.) L. | Not specified (fruits) | Immunomodulatory | The cell-mediated | By inducing the NO release from RAW264.7 cells and the expression of several immune-related molecular (TNFα, IL6, IL10, GCSF, iNOS, COX-2) | Cao et al. (2018) |
| *Punica granatum* L. | Not specified (pericarps) | Immunomodulatory | Humoral and the cell-mediated | By increasing antibody density to typhoid-H antigen and enhancing the inhibition of leukocyte migration | Ross et al. (2001) |
| *Pyrus communis* L. | Not specified (fruits) | Immunostimulant | The cell-mediated | By increasing phagocytic activities | Pavithra et al. (2020) |
| *Rhus coriaria* L. | Not specified (fruits) | Immune-enhancing | Innate, humoral, and the cell-mediated | By increasing the number of leukocyte (WBC) and erythrocyte (RBC), lymphocyte, monocyte, and neutrophil value as well as serum lysozyme and TNF-α and IL-1b | Slavov et al. (2013) |
| *Rosa × damascena* Herrm. | Not specified (petals) | Immunomodulatory | Humoral and the cell-mediated | Through Peyer's patch cells and enhanced production of IL-6 from murine macrophages | Saaby et al. (2011) |
| *Rosa canina* L. | Oleanolic acid, betulinic acid, and ursolic acid (fruits) | Immunomodulatory | Humoral and the cell-mediated | By inhibiting the lipopolysaccharide-induced interleukin-6 release | Al Sheyab et al. (2012) |
| *Rosmarinus officinalis* L. | Not specified (leaves) | Immunomodulatory | Humoral | By increasing IgM (immunoglobulin M) response at all antibody titers and proliferation of spleen cells | Capek and Hribalova (2004) |
| *Salvia officinalis* L. | 3-O-methyl-galactose (aerial parts) | Immunomodulatory | The cell-mediated | By inducing the proliferation of rat thymocytes | Gharaei et al. (2020) |

| | | | | | |
|---|---|---|---|---|---|
| *Sambucus ebulus* L. | Not specified (leaves) | Immunostimulant | The cell-mediated | By increasing the levels of IFN-γ, IL-4, and nitric oxide and switching the host immune responses toward the Th1 response | Heidari-Kharaji et al. (2019) |
| *Sambucus nigra* L. | Cyanidin 3-glucoside and quercetin (fruits) | Immunomodulatory | The cell-mediated | By blocking viral glycoproteins as well as indirect effects by increased expression of IL-6, IL-8, and TNF | Torabian et al. (2019) |
| *Satureja hortensis* L. | Not specified (essential oils) | Immunomodulatory | Innate | Through a significant enhancement in serum lysozyme for the innate immune response of angelfish | Ghafari-Farsani et al. (2018) |
| *Sesamum indicum* L. | Not specified (seeds) | Immunomodulatory | Innate, humoral, and the cell-mediated | By increasing splenocyte proliferation, balancing the secretion of Th1 and Th2 cytokines, inhibiting the secretion of key pro-inflammatory mediators (IL-6, TNFα, and nitric oxide), and increasing the cytotoxic activity of natural killer (NK) cells | Majdalawieh et al. (2020) |
| *Silybum marianum* (L.) Gaertn. | Silymarin (seeds) | Immunomodulatory | Innate and the cell-mediated | Through the enhancement of hematological and immunological parameters including lysozyme and complement activities, total protein, and globulin levels | Ahmadi et al. (2012) |
| *Solanum lycopersicum* L. | Not specified (fruits) | Immune-enhancing | Humoral and the cell-mediated | By increasing phagocytosis, chemotaxis, and levels of immunoglobulins (TIg, IgG, IgM, and IgA) | Elwan et al. (2019) |
| *Spinacia oleracea* L. | Not specified (leaves) | Immunostimulant, immune-enhancing | Innate | Through a substantial improvement of serum protein profile, lysozyme activity, hematocrit percent, and serum superoxide dismutase enzyme (SOD) | Ayoub and Khames (2019) |

(continued)

**Table 18.3** (continued)

| Plant species | Active compounds effective on immune response(s) (and used parts) | Type of activity | Type of immune response | Mechanism of action | References |
|---|---|---|---|---|---|
| *Taraxacum* spp. | Taraxasterol (not specified) | Immunomodulatory | The cell-mediated | By increasing hepatic glutathione (GSH) and superoxide dismutase (SOD) production, inhibiting the release of pro-inflammatory cytokines tumor necrosis factor-α (TNF-α), interleukin-6 (IL-6), IL-1β, interferon-γ (IFN-γ) and IL-4 | Sang et al. (2019) |
| *Thymbra spicata* L. | Not specified (aerial parts) | Immunomodulatory | The cell-mediated | By increasing the decreased level of the IL-6 cytokine in response to TNF-α | Mendi et al. (2017) |
| *Tilia tomentosa* Moench | Not specified (flowers) | Immunostimulant | Innate | By increasing lysozyme activity and respiratory burst activity | Almabrok et al. (2018) |
| *Tribulus terrestris* L. | Saponins (fruits) | Immunomodulatory | The cell-mediated | By increasing phagocytic activity | Tilwari et al. (2011) |
| *Urtica dioica* L. | Quercetin-3-O-rutinoside, kaempherol-3-O-rutinoside, and isorhamnetin-3-O-glucoside (leaves and roots) | Immunostimulant | The cell-mediated | By chemotactic effects in 4, 8, and 16 μg doses and showing high intra-cellular killing activity | Akbay et al. (2003) |
| *Viscum album* L. | Not specified (aerial parts) | Immune-boosting | The cell-mediated | By increasing phagocytic activity and candidacidal activity of neutrophils and stimulating the levels of CD4 + CD25+ and CD8 + CD25+ T cells | Fidan et al. (2008) |
| *Vitis labrusca* L. | Arabino-3, 6-galactan (pericarp) | Immunostimulant | Innate and the cell-mediated | Results of comparisons between the taxa detected in the study and research on immune-stimulating and related activities carried out globally | Kim et al. (2010) |

| | | | | | |
|---|---|---|---|---|---|
| *Vitis vinifera* L. | Not specified (seeds) | Immunostimulant | Innate | By increasing nitroblue tetrazolium (NBT) activity, myeloperoxidase (MPO) activity, phagocytic activity, and lysozyme activity | Al-Atbee and Al-Niaeem (2021) |
| *Withania somnifera* (L.) Dunal | Not specified (roots) | Immunostimulant | The cell-mediated | By increasing nitric oxide (NO) production by macrophages | Iuvone et al. (2003) |
| *Ziziphora clinopodioides* lam. | Not specified (aerial parts) | Immune-boosting | The cell-mediated | By improving the immune system and blood biochemical parameters of laying hens with *Melissa officinalis* and *Tanacetum balsamita* | Nobakht et al. (2012) |
| *Ziziphus jujuba* mill. | Not specified (fruits) | Immunomodulatory, immunostimulant | Innate, humoral, and the cell-mediated | By increasing spleen and thymus indices, promoting serum hemolysin formation, and enhancing the phagocytic activity of macrophages | Zou et al. (2018) |

1. It should be an experimental study (in vitro or in vivo).
2. The scientific name of the plant used should be included in the title of the work or the text.
3. One of the words "immune-boosting, immunostimulant, immunomodulator, immune-enhancing, or immunotherapeutic" should be included in the title.
4. The method of obtaining the compound or extracts used from the plant and the mechanism of action determined should be included in the text.
5. For each taxon, only one experimental study should be selected that meets the above criteria. However, in the title, studies with active compounds are preferred.

### *18.2.3 Data Arrangement*

Accurate-scientific nomenclature with international validity of the plant taxa identified in national surveys was performed using "Turkey Plant List" (Güner et al. 2012), "The Plant List" (2021), and "International Plant Names Index" (2021) and exhibited in (Table 18.1) after arranging in alphabetical order.

The same table contains scientific names, information about pharmaceutical uses, and citation details. The English names of taxa have been added to the table using the following databases or search engines: Springer Link (https://link.springer.com/article), Encyclopedia of Life (https://eol.org), and USDA Plants (https://plants.sc.egov.usda.gov/java). The common names of plants native to Turkey and those that are considered to be a part of the "Irano-Turanian element," however, cannot be verified because they are only known in Turkey and its near surroundings and may not yet be known in English-speaking nations. The results of the research on whether the taxa listed in (Table 18.1) have been the subject of experimental studies in the world literature on their immune-boosting and other similar activities are presented in (Table 18.3). The same table also includes the scientific names of the taxa that match, active compounds (if any), parts used, kind of activity, type of immunological response, method of action, and associated citations.

### *18.2.4 Comparative Analysis*

After determining the definitive list of plants that have the potential to stimulate or strengthen the immune system in Turkish traditional medicine, a second scanning work has been performed to comport with the resemblance percentage with peer works carried out in nearby countries, and the outcomes are submitted in (Table 18.4). Since there are very few taxa related to the immune system in the identified works, we tried to select as many works as possible from each country. However, the negative effects of repeated taxa on the total number and percentage have been avoided. For example; if taxa related to the immune system were mentioned in only three of the studies conducted in one country, only different

**Table 18.4** Taxon similarity percentages in similar studies conducted in neighboring and nearby countries

| Countries | Regions | Total taxa used for immunity | Similar taxa # | Similarity % | References |
|---|---|---|---|---|---|
| Albania | All | 14 | 9 | 64.3 | Pieroni (2008), Pieroni et al. (2005, 2014a, b), Pieroni and Quave (2014), Quave and Pieroni (2014), Tomasini and Theilade (2019) |
| Cyprus | All | 23 | 14 | 60.9 | Karousou and Deirmentzoglou (2011), Lardos et al. (2011), Özkum et al. (2013) |
| Bulgaria | All | 16 | 9 | 56.3 | Ivancheva and Stantcheva (2000), Kozuharova et al. (2013), Koleva et al. (2015) |
| Palestine | All | 31 | 16 | 51.6 | Abu-Rabia (2005), Ali-Shtayeh et al. (2011), Eid and Jaradat (2020), Jaradat (2005) |
| Bosnia and Herzegovina | All | 22 | 11 | 50.0 | Redžić (2007), Šaric Kundalic et al. (2015) |
| Iran | Mashhad, Azerbaijanprovince | 32 | 16 | 50.0 | Amiri and Joharchi (2013), Latifian and Arslanoğlu (2018), Miraldi et al. (2001), Naghibi et al. (2005) |
| Kosovo | Anadrini region, Albanian Alps, and South Kosovo | 15 | 7 | 46.7 | Mullalija et al. (2021), Mustafa et al. (2012) |
| Macedonia | All | 15 | 7 | 46.7 | Gjorgieva-Ackova et al. (2011), Pieroni et al. (2013), Rexhepi et al. (2013) |
| Nakhchivan | All | 26 | 12 | 46.2 | Ibadullayeva et al. (2015), Novruzova et al. (2015), Ozturk and Hakeem (2018) |
| Iraq | Northern Iraq | 22 | 10 | 45.5 | Ahmed (2016), Ahmad and Askari (2015), Kawarty et al. (2020), Mati and de Boer (2011) |

(continued)

**Table 18.4** (continued)

| Countries | Regions | Total taxa used for immunity | Similar taxa # | Similarity % | References |
|---|---|---|---|---|---|
| Croatia | Northeastern Istria, Knin area, and Pannonian region | 16 | 7 | 43.8 | Varga et al. (2019), Vitasović-Kosić et al. (2017), Žuna-Pfeiffer et al. (2020) |
| Azerbaijan | All | 7 | 3 | 42.9 | Alakbarov (2001) |
| Armenia | All | 12 | 5 | 41.7 | Nanagulyan et al. (2020) |
| Syria | Aleppo | 12 | 5 | 41.7 | Alachkar et al. (2011), Rihawy et al. (2010) |
| Georgia | All | 10 | 4 | 40.0 | Akhalkatsi et al. (2018), Bussmann et al. (2016a, b), Martkoplishvili and Kvavadze (2015) |
| Greece | Thessaloniki (northern) | 68 | 27 | 39.7 | Hanlidou et al. (2004), Tsioutsiou et al. (2019) |
| Serbia | Eastern and Southeastern | 63 | 24 | 38.1 | Janaćković et al. (2019), Matejić et al. (2020), Popović et al. (2012), Zlatković et al. (2014), Živković et al. (2020) |
| Ukraine | Southwestern | 35 | 13 | 37.1 | Mattalia et al. (2020a, b), Sõukand and Pieroni (2016), Stryamets et al. (2015) |
| Lebanon | All | 11 | 4 | 36.4 | Arnold et al. (2015), Jeambey et al. (2009), Marouf et al. (2015), Redwan et al. (2017) |
| Jordan | All | 26 | 9 | 34.6 | Aburjai et al. (2007), Alzweiri et al. (2011), Assaf et al. (2013), Al-Quran (2014), Lev and Amar (2002) |
| Romania | Transylvania, Dobruja, and West Romania | 36 | 12 | 33.3 | Daraban et al. (2013), Papp et al. (2014), Petran et al. (2020), Pieroni et al. (2012, 2015) |
| Montenegro | Northern Montenegro | 13 | 4 | 30.8 | Menković et al. (2011, 2014) |
| Israel | All | 30 | 9 | 30.0 | Lev and Amar (2000), Said et al. (2002) |

(continued)

**Table 18.4** (continued)

| Countries | Regions | Total taxa used for immunity | Similar taxa # | Similarity % | References |
|---|---|---|---|---|---|
| Russia | All | 15 | 4 | 26.7 | Shikov et al. (2017), Spiridonov et al. (2005) |

taxa in these three works were included in the study. If *Melissa officinalis* was repeated for the same activity in all three, its number was considered to be 1, and it was not allowed to affect the total number as 3. The table includes country, region, the total number of taxa used for immunity, the number of similar taxa, the percentage of similarity, and references to related studies. Countries are listed in descending order of similarity percentages.

## 18.3 Results and Discussion

Medicinal herbs are a 1000-year-old source in a variety of traditional herbal medicine systems, from raw uses in the form of direct preparations, such as infusions, decoctions, maceration, and pastes, to the extraction of key compounds. In fact, their secondary metabolism is a phytochemical treasure with promising results for all of humanity. Therefore, as trials for testing vaccines continue, traditional herbal remedies in the form of decoctions, tea, or powders require increased clinical testing and research to alleviate symptoms (Khan et al. 2021). As in many parts of the world, people living in Turkey, especially in rural areas, have rich traditional knowledge and experience about the utilization of therapeutic plants in the avoiding and handling of assorted illnesses, but also open to serious clinical evaluation studies (Sargin 2021).

### *18.3.1 Regional Analysis*

The dispersion of 99 works with respect to the regions where they were acquired was as follows: 15 in Eastern Anatolia (15.2%), 12 in the Mediterranean and Marmara (12.1%), 10 in Southeastern Anatolia and Black Sea (10.1%), 9 in the Aegean (9.1%), 7 in Central Anatolia (7.1%), and 24 general studies across all regions (24.2%). The reason why ethnomedicinal works related to immunity come mostly from Eastern Anatolia, Mediterranean and Marmara regions may be because separate regional studies have been carried out for each region spread over a wide area. After all, these regions contain very different topographies.

The sectional dissolution of 711 total reports received was as follows: Black Sea: 85 (12.1%), Southeastern Anatolia: 78 (11.0%), Marmara: 68 (9.6%), Mediterranean and Aegean: 66 (9.3%), Eastern Anatolia: 53 (7.5%), Central Anatolia: 49 (6.9%), and general studies covering all regions: 246 (34.4%). The fact that the obtained taxon references come mostly from the Black Sea region may be because these regions contain locations with very high biodiversity and endemism such as Bolu, Düzce, and Artvin (Güner et al. 2012). In addition, the traditional high-plateau life, which is scattered far away from the city and district centers, and the abundant rainy climate in these regions, may have prompted the local people to use more varieties of plants (Terzioğlu et al. 2015).

The following can be said as the reason for the lowest percentages in Central Anatolia: First of all, the capital Ankara and Eskişehir, known as the city of universities, are located within the borders of the region. Therefore, widespread urbanization might be providing local people with easy access to healthcare services (Keser and Kurt 2016) and reduce the tendency to plant. Secondly, it can be shown that there are not many options for a medicinal plant even for people who are far from healthcare, due to the dominance of the plateau landform in the region, the low annual rainfall percentage (Apaydin et al. 2011), and the much lower level of plant biodiversity than other regions (Kenar et al. 2020).

### *18.3.2 Data Analysis*

The bibliographies exploited for the choice of 128 plant taxa that are used traditionally for immunity enhancement consist of 67 articles, 23 theses, seven books, and two bulletins. These plant taxa are most commonly Lamiaceae (25 taxa, 19.5%), Rosaceae (18 taxa, 14.1%), Compositae (11 taxa, 8.6%), Malvaceae (7 taxa, 5.5%), Apiaceae and Brassicaceae (5 taxa, 3.9%), and other families (57 taxa, 44.5%). The rationale behind the Lamiaceae family is the foremost favored in Turkish folk medicine can be demonstrated as the habit of preparing many conventional immune-boosting preparations such as lotion, medical baths, spice, infusion mix, and syrup due to the family that contains the highest dosage of essential oils (Bozyel and Merdamert-Bozyel 2020) and experience of receiving good results (Başer et al. 2006) in immunostimulation and flu treatment. In addition, 44.2% of species belonging to the Lamiaceae family, 65.2% of species belonging to *Origanum* genus, 52.6% of species belonging to *Thymus* genus, and 28% of species belonging to *Satureja* genus are endemic in Turkey (Bozdemir 2019). Therefore, it is not coincidental that this family comes first in presenting the richest variety of taxa to our study.

In the works carried out in different regions of Turkey, the most common genera are *Citrus* and *Salvia* (5 taxa, 3.9%) and *Mentha, Pinus, Prunus*, and *Taraxacum* (4 taxa, 3.1%). It is not surprising that *Citrus*, *Salvia*, and *Mentha* appeared in the first place. They are grown both naturally and culturally (for healing and ornamental purposes) in park gardens, balconies, and pots in almost every part of the country. *Citrus* species are actually an exotic genus settled in the flora of Turkey. It is seen on all roadsides and gardens, especially in the southern regions (Baytop 1999; Başer et al. 2006). Moreover, the *C. limon* species is sold as a "panacea!" or "Lemon for tea and soup!" in almost all local markets of the country. It is also always in high demand for direct use or mixtures. In particular, the infusion mixture prepared from *Mentha x piperita* (peppermint) leaves and *C. limon* peel can be said to be the most widely used traditional preparation in the country to increase body resistance against winter diseases (Sargin 2021; Tuzlacı and Eryaşar-Aymaz 2001). The other three genera, *Pinus, Prunus*, and *Taraxacum*, are also naturally distributed in a wide area in the country (Baytop 1999, 2007). *Prunus* and *Pinus* fruits and *Taraxacum* leaves have been traditionally consumed for centuries for healing and food to strengthen the body against diseases (Sargin and Büyükcengiz 2019; Ozturk et al. 2017a; Dogan et al. 2004).

Among the determined plants, 61 were wild (47.7%), 55 were wild and cultural (43.0%), 11 were cultivated (8.6%), and one (*Salvia absconditiflora*) was endemic (0.9%). In Table 18.1, these details are displayed in a column; cultivated "C," endemic "E," wild taxa as "W," and cultivated & wild "CW." The majority of plant parts are consumed as fruits (39.8%), leaves (33.6%), flowers, flowering branches, petals and capitula (32.8%), aerial parts (32.0%), seeds and cones (19.5%), roots, rhizomes and bulbs (13.3%), essential oils, fixed oils, resins, and tars (12.5%), whole parts, stems, barks and shoots (10.9%), and other parts (buds, bracts, pericarps, fruit stalks, galls, and pollens) (10.2%). Those parts were mostly used as infusion and infusion mix (80.5%), eating raw or after milling (35.2%), decoction, decoction mix, boiling (31.3%), jam, syrup, marmalate, molasses (21.9%), meal, roast, soup, pastry, cooked (18.0%), pickle, vinegar, tincture, maceration (14.8%), external uses such as foot bath, medicinal bath lotion and frankincense (14.1%), restorative paste, spices drop and additives (11.7%), churchkhela and mixtures (11.7%), drunk, juice, gargle (10.9%), and salad (7.8%). In terms of utilization parts, *Juniperus oxycedrus, Pinus brutia, Rosa canina*, and *Tilia tomentosa* (6 parts, 4, 7%), while the taxa with the most usage types are *Rubus sanctus* (8 types, 6.3%), *Citrus* spp. (7 types, 5.5%), *Mentha x piperita, Rosa x damascena, R. canina* and *Thymbra spicata* (6 types, 4.7%) are the taxa with the greatest abundance (Table 18.1).

### *18.3.3 Comparison of the Findings with Research from Surrounding Nations*

The number of taxa that match studies from surrounding and neighboring nations out of the 128 that we specified in our research is 81; the similarity rate (from 81×100/128) was computed as 63.3%. 18 plants, including *Urtica dioica* (with 11 citations

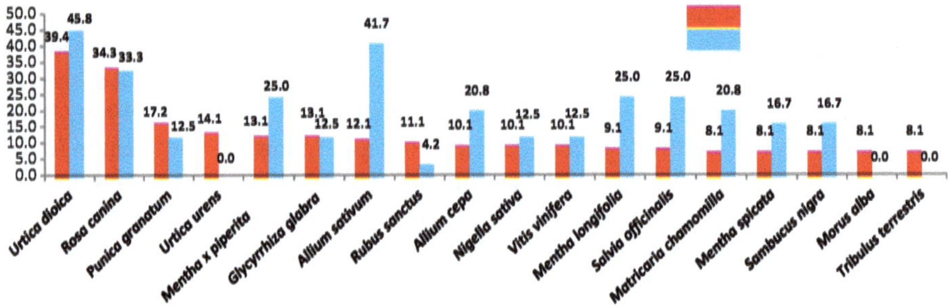


**Fig. 18.2** The most reported plants are in Turkey and neighboring countries

and 45.8%), *Achillea millefolium* and *Allium sativum* (with 10 citations and 41.7%), and *Juglans regia, Rosa canina,* and *Taraxacum* spp. (8 citations and 33.3%) (Fig. 18.2), have been found as the most reported plants. These herbs may have gained popularity because they have been used for ages in Turkish traditional medicine and are more long-lasting and effective at boosting immunity. Especially, the emergence of *Urtica* spp. as the most used plant as an immune booster in 11 countries (45.8%), other than Turkey, may be because it has proven itself in the traditional medicine of many cultures rather than the fact that its distribution area is wide and easily accessible. The similarity of 63.3% between the findings of investigations carried out in 24 close and neighboring nations may confirm the powerful effectiveness of these taxa (Table 18.4). 47 plants such as *Urtica urens, Morus alba,* and *Tribulus terrestris* do not resemble at all (Fig. 18.2), may be because they are not naturally widespread in these countries because of geographical or climatic differences, or even if they do, they are not traditionally preferred for immunity.

Nevertheless, the maximum similarity was detected with Albania (64.3%), Cyprus (60.9%), and Bulgaria (56.3%), while the minimum one was with Russia (26.7%), Israel (30.0%), and Montenegro (30.8%). In countries with high similarity, there may be similarities in climate, vegetation, and landforms, as well as cultural and plant use habits from the Ottoman period. Differences in religious practices, social-cultural norms, flora, and topography may be the root causes of the poor similarity with Russia, Israel, and Montenegro (Table 18.4).

### *18.3.4 Evaluation of Worldwide Works in Comparison*

As a result of the comparison screening of experimental studies conducted worldwide under the title of activities such as immunostimulant, immunomodulator, immune-enhancing, or immunotherapeutic, reflections of only 68 (53.1%) of 128 taxa were detected. 60 taxa (46.9%), which have not been investigated yet, are among the valuable data of this study and are shown in bold in Table 18.1. The percentage distribution of the activities mentioned in the titles of the studies is as follows: Immunomodulatory (47.1%), immunostimulant (27.9%), immunomodulatory and immunostimulant (7.4%), immune-enhancing (5.9%), immune-boosting (5.9%), immune-enhancing and immunostimulant (1.5%), immune-boosting and immunostimulant (1.5%), and immunomodulatory and immunotherapeutic (1.5%). For only 23 taxa (18.0%) out of 68 matching taxa, the results were given with the active ingredients, while the remaining 45 taxa (35.2%) were not specified. The prominent among 39 active substances are quercetin derivatives (such as quercetin xyloside, quercetin-3-galactoside, and quercetin-3-O-rutinoside) (10.3%), β-glucan, and catechin (5.1%), respectively (Table 18.3).

The action mechanisms in the studies identified are listed according to the immune responses on which they were based: Cell-mediated (42.6%), humoral and cell-mediated (22.1%), innate only (11.8%), innate, humoral, and cell-mediated (11.8%), humoral only (5.9%), and innate and cell-mediated (5.9%), respectively. According to these three basic immune responses, it has been determined that the mechanisms were carried out in the following eight different groups: By increasing spleen and thymus indices/serum immune parameters/total white and red blood cells (WBC/RBC) (20.3%), by increasing pinocytosis, phagocytic and bactericidal activity/acting as macrophage and T helper-1 (Th-1) activators/switching the host immune responses toward Th1 response (20.3%), by increasing interleukin cytokine levels/elevating the levels of tumor necrosis factor-alpha (TNF-α) and interferon-gamma (IFNγ)/elevating the expression levels of cytokines/activating the CD4+ T cells leading to cytokine production (18.0%), by activating the CD19+ B cells in Peyer's patches/increasing total immunoglobulin/reducing the suppressive azathioprine effect on the cell-mediated immune response, and antibody response (13.5%), by promoting serum hemolysin formation and increasing lysozyme, myeloperoxidase, superoxide dismutase catalase, and superoxide dismutase activities (12.8%), by increasing production of nitric oxide (NO) on a level comparable with that of lipopolysaccharides/reducing the levels of serum lipids/inhibiting the lipopolysaccharide-induced interleukin-6 release (6.8%), by increasing respiratory burst activity/reactive oxygen species (ROS) production/increasing the hematocrit level and red blood cell count/hemoglobin content and hematocrit value (4.5%), and by increasing the cytotoxic activity of natural killer (NK) cells (3.8%), respectively (Table 18.3).

When the plant parts used for 68 taxa (53.1%) that were determined to match in the comparison of Turkey and the world literature were examined, a concordance of 94.1% was obtained. This result may have proved the accuracy and reliability of part selection and application methods of plants used as immunostimulants and enhancers in Turkish folk medicine.

### 18.3.5 Comparative Analysis of Effective Substances

The taxa with the largest quantities of potent substances that have been demonstrated to have immune-boosting and other associated benefits include *Phoenix dactylifera* (5 chemicals, 12.8%, out of 39), *Malus domestica* (4; 10.25%), *Cichorium intybus,*

*Hypericum perforatum, Urtica dioica*, and *Rosa canina* (3; 7.7%) (Table 18.3). *Phoenix dactylifera* (date palm) is widespread in a vast area from North Africa to the Middle East, parts of Central and South America, and Southern Europe for both its production and consumption of its fruit for food and healing purposes (Baytop 1999; Karasawa et al. 2011). The reason why it stands out in terms of active ingredient richness in this study may be that it is rich in fiber, mineral, and other nutritional sources, as well as its antibacterial, antifungal, and immunostimulant properties, as a widely effective folk remedy in preventing various infectious diseases and increasing body resistance (Qadir et al. 2020). As in the geographical area mentioned above, the consumption of dried fruit becomes widespread in Turkey, especially during the holy month of Ramadan. It is known that while fasting, the body compensates for the mineral and energy loss, thus keeping the immune system active (Baytop 1999, 2007; Kavaklı et al. 2020a). The study by Karasawa et al. (Karasawa et al. 2011) provides additional evidence for this finding. They investigated the immunomodulatory effects of hot water extract of ripe fruits compared to prunes and fig fruits in mice and declared that the polyphenols and polysaccharides they isolated stimulated the cellular immune system in mice. Therefore, this finding may be a shred of evidence that this taxon might be more effective than others in terms of immunomodulatory activity.

As a result of this screening, quercetin and its derivatives came to the fore with their immunomodulatory and immunostimulant activities and it was determined that they were isolated from the taxa of *Capparis spinosa, Malus domestica, Urtica dioica*, and *Sambucus nigra* (Akbay et al. 2003; Arena et al. 2008; Torabian et al. 2019; Riaz et al. 2020) (Tablsssse 18.3). The fact that quercetin and its derivatives appear to be the most prevalent effective substances may not be a coincidence. These chemicals are the best ones for boosting or activating immunity, according to numerous experimental types of research. A few of these are given below:

Colunga-Biancatelli et al. (Colunga-Biancatelli et al. 2020) provided evidence that the co-administration of vitamin C and quercetin shows a synergistic antiviral effect and enhances its efficacy due to its antiviral and immunomodulatory properties and the capacity of ascorbate to recycle quercetin. Wang et al. (Wang et al. 2020) increased the immunostimulant properties, antioxidant index, and disease resistance of zebrafish with the optimal quercetin level. Singh et al. (Singh et al. 2017) reported for the first time the in vivo immunostimulatory activity of quercetin in ovalbumin-vaccinated Balb/c mice. Their administration of quercetin showed the dominance of the Th2 immune response by increasing IgG1 antibody titers, as well as increased infiltration of CD11c + dendritic cells in the mouse peritoneum and the production of LPS-activated IL-1β and nitric oxide (NO) by peritoneal macrophages. These outcomes may validate that quercetin and its derivates are ideal aspirants for a functional nutraceutical or phytotherapy by stimulating or positively modulating immunity. Additionally, many taxa commonly employed in Turkish traditional medicine, *Avena sativa, Camellia sinensis, Hordeum vulgare*, and *Malus domestica*, have active compounds such as β-glucan and catechin. Some studies (Özcan et al. 2020; Yeşilada 2012; Ege and Elmastaş 2020; Riaz et al. 2020; Monobe et al. 2008; Rösch et al. 2016) pointed out that beta-glucan and catechin act through one or more of the eight groups of action mechanisms mentioned above, and that these molecules could potentially give positive results in the treatment of COVID-19 in terms of immune-stimulating or accelerating activities.

### *18.3.6 Exotic Plants*

14 medicinal exotic herbs, such as *Alpinia officinarum* (lesser galangal), *Cinchona pubescens* (quina), *Echinacea angustifolia* (coneflower), *Hibiscus sabdariffa* (hibiscus), *Malpighia emarginata* (acerola cherry), and *Vaccinium macrocarpon* (cranberry) are used for stimulating or boosting the immune system in Turkish folk medicine and traded in public markets and herbal shops. Scientific names, families, English names, used parts, preparations, native lands, and related references of these plants are introduced in Table 18.5.

The native to Southeast Asia *Citrus* species and the native to North Africa *Phoenix dactylifera* (date palm) listed in (Table 18.1) are essentially exotic taxa that were subsequently incorporated into the native flora. After being grown for centuries along the coastlines of the Mediterranean and Aegean, on highways, in parks, and gardens, they have shown a natural expansion. Other plants that have adapted to the flora during the Ottoman era are *Solanum lycopersicum* (tomato, South America), *Morus alba* and *M. nigra* (mulberry, China), *Camellia sinensis* (tea plant, native to South and Southeast Asia), *Withania somnifera* (Indian ginseng), *Capsicum annuum* (pepper, South and Central America), *Prunus cerasifera* (myrobalan plum, Japan), and *Ziziphus jujuba* (jujube, Southeast Asia) in the same table (Kaya et al. 2018; Polat and Selvi 2020).

## 18.4 Conclusion

Suffering from a global epidemic of COVID-19 mutants and variants, the whole world has turned its attention and hope to natural treatments that can keep the immune system active and strong. Recent studies of some medicinal plant preparations hold great prospects that they may be potential candidates to develop effective and safe immune-boosting drugs against this relentless pandemic. Initiatives in this regard will primarily be hope for millions of people with weakened immune systems due to old age and pre-existing medical problems, waiting for treatment in quarantine at home or in intensive care in a hospital. In particular, recent research on phytochemicals like quercetin, beta-glucan, and catechin makes us feel a little more optimistic about the future.

Thanks to a strong immune system, a person with good body health can easily get rid of severe acute respiratory syndrome coronavirus 2 (SARS-CoV-2) infections without any complications. It is known that the side effects of existing synthetic drugs and the necessity of taking a certain dose while in the state of infection make

**Table 18.5** Exotic plants traditionally used for immune-boosting and other similar activities in Turkey

| Sc. Names | Families | English names | Parts | Preparations | Native lands | References |
|---|---|---|---|---|---|---|
| *Alpinia officinarum* hance | Zingiberaceae | Lesser galangal | Rhizomes | Spice, decoction, or infusion (after pulverizing) | Southeast Asia | Ötnü and Akan (2020), Gökçe (2011), Akbaş and Akçakaya (2020), Korkmaz and Karakurt (2014) |
| *Cinchona pubescens* Vahl | Rubiaceae | Red cinchona, quina | Barks | Decoction or infusion (after pulverizing), syrup, or liqueur prepared after pulverizing | Central and South America | Jahan and Onay (2020), Kalafatçılar and Kalafatçılar (2010), Baytop (1999), Genç (2010), Kavaklı et al. (2020b), Akbaş and Akçakaya (2020) |
| *Cinnamomum verum* J.Presl | Lauraceae | Cinnamon, true cinnamon tree | Barks | Decoction with the pulverized bark, linden leaves, and cloves. Restorative paste (prepared with the dried and pulverized bark, black pepper fruits, ginger rhizomes, nettle seeds, turmeric rhizomes, and *Nigella sativa* seeds after milling and mixing with honey) | South and Southeast Asia | Kalafatçılar and Kalafatçılar (2010), Genç (2010), Şenkardeş (2014), Kavaklı et al. (2020a, b), Şanlı (2006), Yeşilyurt et al. (2017b), Everest and Ozturk (2005), Gökçe (2011), Bulut and Tuzlacı (2013), Bulut et al. (2017) |
| *Curcuma longa* L. | Zingiberaceae | Turmeric | Rhizomes | Restorative paste (prepared with the dried and pulverized rhizomes, ginger, cinnamon, and the seeds of nettle and *Nigella sativa* after milling and mixing with honey) | Indian subcontinent and Southeast Asia | Jahan and Onay (2020), Şekeroğlu and Gezici (2020), Genç (2010), Uçar et al. (2020), Yeşilada (2012), Şanlı (2006), Ege and Elmastaş (2020), Demirci-Kayıran (2019), Doğan and Doğan (2020), Everest and Ozturk (2005), Ongan |

| | | | | | | |
|---|---|---|---|---|---|---|
| | | | | | | (2018), Akbaş and Akçakaya (2020) |
| *Echinacea angustifolia* DC. | Compositae | Narrow-leaf coneflower | Roots, rhizomes, flowers | Decoction or infusion (after pulverizing) | North America | Kalafatçılar and Kalafatçılar (2010), Genç (2010), Yeşilada (2012), Ege and Elmastaş (2020), Doğan and Doğan (2020), Kavaklı et al. (2020b), Uçar et al. (2020) |
| *Hibiscus sabdariffa* L. | Malvaceae | Roselle, carcade, hibiscus | Leaves | Infusion | West and East Africa and Southeast Asia including Northeastern India. | Kalafatçılar and Kalafatçılar (2010), Baytop (1999), Ötnü and Akan (2020), Yeşilada (2012), Akan and Bakır-Sade (2015), Arıtuluk (2010) |
| *Malpighia emarginata* DC. | Malpighiaceae | Acerola cherry | Fruits | Decoction or infusion (after pulverizing) | Central and South America | Genç (2010), Ötnü and Akan (2020), Yeşilada (2012) |
| *Panax ginseng* C.A.Mey. | Araliaceae | Asian ginseng | Roots | Decoction or infusion (after pulverizing), paste (with honey) | Russian Far East (outer Manchuria), Northeast China, and the Korean Peninsula | Jahan and Onay (2020), Kalafatçılar and Kalafatçılar (2010), Baytop (1999), Genç (2010), Kavaklı et al. (2020b), Ötnü and Akan (2020), Yeşilada (2012), Kurt and Karaoğul (2018), Gökçe (2011), Ongan (2018), Kökçü et al. (2015), Korkmaz and Karakurt (2014), Onbaşlı and Dal (2020) |
| *Piper nigrum* L. | Piperaceae | Black pepper | Unripe fruits | Infusion mix (with the dried and pulverized fruits and ginger rhizomes, melissa leaves, sage | India | Baytop (1999), Uçar et al. (2020), Özer et al. (2005), Ege and Elmastaş (2020), |

(continued)

**Table 18.5** (continued)

| Sc. Names | Families | English names | Parts | Preparations | Native lands | References |
|---|---|---|---|---|---|---|
| | | | | leaves, and elder fruits), restorative paste1 (all of them above are eaten by mixing with honey), restorative paste2 (prepared with the dried and pulverized fruits and ginger rhizomes, cinnamon, turmeric, nettle seeds, and *Nigella sativa* seeds after milling and mixing with honey) | | Yeşilada et al. (2019), Demirci-Kayıran (2019), Doğan and Doğan (2020), Güneş (2017), Özhatay et al. (2009), Diken (2009) |
| *Rheum officinale* Baill. | Polygonaceae | Chinese rhubarb | Roots, rhizomes | Infusion after pulverizing, syrup, paste (with honey) | China | Jahan and Onay (2020), Kalafatçılar and Kalafatçılar (2010), Baytop (1999), Genç (2010) |
| *Syzygium aromaticum* (L.) Merr. & L. M.Perry | Myrtaceae | Cloves | Flower buds, essential oil | Pastille, infusion, frankincense, decoction mix (with cinnamon and linden leaves), mixture (eaten after milling with seeds of celery and anise, turnip, radish, and mixed with honey) | Maluku Islands | Baytop (1999), Özcan et al. (2020), Sargin (2013), Uçar et al. (2020) |
| *Terminalia citrina* Roxb. Ex Fleming | Combretaceae | – | Fruits, seeds | Paste (prepared with honey after dried and pulverized) | East India | Baytop (1999), Ötnü and Akan (2020), Kökçü et al. (2015) |
| *Vaccinium macrocarpon* Aiton | Ericaceae | Cranberry | Fruits | Eaten after dried, juice, decoction | North America | Genç (2010), Ötnü and Akan (2020), Yeşilada (2012) |

| | | | | | | |
|---|---|---|---|---|---|---|
| *Zingiber officinale* roscoe | Zingiberaceae | Ginger | Rhizomes | Infusion mix (with the dried and pulverized rhizomes, melissa leaves, sage leaves, elder fruits, and black pepper), restorative paste1 (all of them above are eaten by mixing with honey), restorative paste2 (prepared with the dried and pulverized rhizomes, cinnamon, turmeric, nettle seeds, and *Nigella sativa* seeds after milling and mixing with honey) | South Asia | Baytop (1999), Uçar et al. (2020), Özer et al. (2005), Ege and Elmastaş (2020), Yeşilada et al. (2019), Demirci-Kayıran (2019), Doğan and Doğan (2020), Güneş (2017), Özhatay et al. (2009), Akbaş and Akçakaya (2020), Diken (2009) |

vital organs, such as the kidney and liver, even riskier (Marra et al. 2020; Vitiello et al. 2021). In fact, new cases are added every day to the reports that dozens of people who had newly developed vaccines lost their lives due to unpredictable coagulation. Countries that were lining up to be able to supply the relevant vaccines a while ago, announced that they, therefore, ban the use of these vaccines one after another (Evenett et al. 2021). At this point, medicinal plants might almost be the rescuers of humanity, for they have been providing us with the successful results observed in immune system boosting and similar activities for centuries, provided that they are routinely utilized following the guidelines laid out in their pharmacopeias. However, there is also a misperception that herbal medicines are completely safe and have no side effects. In addition, complications that may occur in the case of the use of existing drugs are another issue that should be taken into consideration. For this reason, we strongly recommend that people and establishments that keep and sell fresh or dried medicinal herbs in public markets and shops are inspected more strictly by local governments and that they are sold to the public following international norms. In these matters, the directives, instructions, and practices of the American Food and Drug Administration (FDA) and the European Medicines Agency (EMA) can be taken as a model and the existing system can be revised. We firmly advocate the urgent need to inform the public in more detail by recording the herbal mixtures and prescriptions of traditional empirical practices with proven reliability and effectiveness after the necessary clinical tests, revealing their ingredients and active compounds through experimental studies, and updating their knowledge in their pharmacopeias. We sincerely believe that more effective results in public enlightenment will be attained if non-governmental organizations, scientific communities, and official authorities/institutions organize some sort of enlightenment seminars with the participation of local residents on plants that can be used safely against COVID-19 and similar epidemics.

Considering that Turkey has the richest flora in Europe and the Middle East and its invaluable ethnomedical history, it presents a treasure-worthy potential for those who want to do serious research on this subject. In fact, this study presents a nationally useful inventory of immune-enhancing plants that are regularly used in traditional Turkish medicine, not only against COVID-19 but for the prevention and treatment of epidemic diseases in general. We present 68 (53.1%) plants whose effective immunostimulating and strengthening properties have been confirmed in experimental studies in the world literature as ready-made material for the relevant pharmacological sectors. In addition to the 60 (46.9%) taxa that have not been searched yet, which are shown in bold in Table 18.1, the first 18 plants (Fig. 18.2), which received more citations than others about immunity enhancement in our research and were also prominent in neighboring country studies, we strongly recommend that these should urgently be subjected to clinical trials firstly. In this way, we will feel peaceful and happy if we can be instrumental in the survival of humanity from the current and future epidemics and pandemics with minimal damage, alleviation of patients' agonies, and the development of a healthier and happier generation.

## References


Abu-Rabia A (2005) Herbs as a food and medicine source in Palestine. Asian Pac J Cancer Prev 6: 404–407

Aburjai T, Hudaib M, Tayyem R, Yousef M, Qishawi M (2007) Ethnopharmacological survey of medicinal herbs in Jordan, the Ajloun Heights region. J Ethnopharmacol 110(2):294–304. https://doi.org/10.1016/j.jep.2006.09.031

Adem S, Eyupoglu V, Sarfraz I, Rasul A, Ali M (2020) Identification of potent COVID-19 main protease (Mpro) inhibitors from natural polyphenols: an in silico strategy unveils a hope against CORONA. Preprints 46(3):1–16. https://doi.org/10.20944/preprints202003.0333.v1

Ahmad SA, Askari AA (2015) Ethnobotany of the Hawraman region of Kurdistan Iraq. Harv Pap Bot 20(1):85–89. https://doi.org/10.3100/hpib.v20iss1.2015.n8

Ahmadi K, Banaee M, Vosoghei AR, Mirvaghefei AR, Ataeimehr B (2012) Evaluation of the immunomodulatory effects of silymarin extract (*Silybum marianum*) on some immune parameters of rainbow trout, *Oncorhynchus mykiss* (Actinopterygii: Salmoniformes: Salmonidae). Acta Ichthyol Piscat 42(2):113–120. https://doi.org/10.3750/AIP2011.42.2.04

Ahmed HM (2016) Ethnopharmacobotanical study on the medicinal plants used by herbalists in Sulaymaniyah Province, Kurdistan. Iraq J Ethnobiol Ethnomed 12:8. https://doi.org/10.1186/s13002-016-0081-3

Akan H, Bakır-Sade Y (2015) Kâhta (Adıyaman) merkezi ve Narince köyü'nün etnobotanik açıdan araştırılması. Bitlis Eren Univ J Sci Technol 4(2):219–248. https://doi.org/10.17798/beufen.47724

Akay D (2019) Medicinal plants using as folk medicine for respiratory system diseases in Edirne. Master thesis. Trakya Univ Sci Ins Edirne

Akbary P, Ghareghani Poor M, Fereidouni MS (2015) Effects of Zataria multiflora Boiss and *Mentha pulegium* L extracts on phagocytosis, lysozyme, respiratory burst and blood cells of rainbow trout (Oncorhynchus mykiss Walbaum). J Vet Res 70(4):447–454. https://doi.org/10.22059/JVR.2016.56466

Akbaş MN, Akçakaya A (2020) COVID-19 and phytotherapy. Bezmialem Sci 8(4):428–437

Akbay P, Basaran AA, Undeger U, Basaran N (2003) In vitro immunomodulatory activity of flavonoid glycosides from *Urtica dioica* L. Phytother Res 17:34–37. https://doi.org/10.1002/ptr.1068

Akbulut S, Özkan ZC, Kalankan G (2017) Some plants and their effects used in traditional treatment of diseases at Çorum Province in Turkey. Int J Second Metab 4(3, special issue 2): 330–339. https://doi.org/10.21448/ijsm.373823

Akgul A, Akgul A, Senol SG, Yildirim H, Secmen O, Dogan Y (2018) An ethnobotanical study in Midyat (Turkey), a city on the silk road where cultures meet. J Ethnobiol Ethnomed 14:12. https://doi.org/10.1186/s13002-017-0201-8

Akhalkatsi M, Goloshvili T, Gvaladze G (2018) Rare and endangered medicinal plants of Georgia. Int J Res Stud Sci Eng 5(3):16–25

Al Sheyab FM, Abuharfeil N, Salloum L, Hani RB, Awad DS (2012) The effect of rosemary (*Rosmarinus officinalis*. L) plant extracts on the immune response and lipid profile in mice. J Biol Life Sci 3(1):2157–6076. https://doi.org/10.5296/jbls.v3i1.906

Alachkar A, Jaddouh A, Elsheikh MS, Bilia AR, Vincieri FF (2011) Traditional medicine in Syria: folk medicine in Aleppo governorate. Nat Prod Commun 6(1):79–84. https://doi.org/10.1177/1934578X1100600119

Alakbarov FU (2001) Medicinal plants used in medieval Azerbaijan phytotherapy. J Herb Pharmacother 1(3):35–49. https://doi.org/10.1080/J157v01n03_04

Al-Atbee IA, Al-Niaeem KS (2021) Efficiency of grape (*Vitis vinifera*) seed oil on nonspecific immune response and histopathological effects in common carp Cyprinus carpio challenged with *Pseudomonas aeruginosa*. Biol Appl Environ Res 5(1):130–140. https://doi.org/10.51304/baer.2021.5.1.130

Ali-Shtayeh MS, Jamous RM, Jamous RM (2011) Herbal preparation use by patients suffering from cancer in Palestine. Complement Ther Clin Pract 17(4):235–240. https://doi.org/10.1016/j.ctcp.2011.06.002

Alkış AN (2019) Şanlıurfa'daki aktarlarda zayıflamak amacıyla kullanılan bitkilerin fitoterapi yönünden araştırılması. Master Thesis. Harran University, Science Institute, Şanlıurfa

Almabrok AA, Amhamed ID, Mohamed GA, Bilen S, Altief TAS (2018) Effect of *Tilia tomentosa* methanolic extract on growth performance, digestive enzyme activity, immune system and haematological indices of common carp (*Cyprinus carpio*). J Mar Sci Technol 7(1):12–20

Al-Quran S (2014) Used ethnobotany of medicinal plants by inhabitants of Al-Mafraq, Jordan. Arnaldoa 21(1):119–126

Altundag E, Ozturk M (2011) Ethnomedicinal studies on the plant resources of East Anatolia. Turkey Proc Soc Behav Sci 19:756–777. https://doi.org/10.1016/j.sbspro.2011.05.195

Alzweiri M, Al Sarhan A, Mansi K, Hudaib M, Aburjai T (2011) Ethnopharmacological survey of medicinal herbs in Jordan, the northern Badia region. J Ethnopharmacol 137(1):27–35. https://doi.org/10.1016/j.jep.2011.02.007

Amiri MS, Joharchi MR (2013) Ethnobotanical investigation of traditional medicinal plants commercialized in the markets of Mashhad. Iran Avicenna J Phytomed 3(3):254–271

Apaydin H, Anli AS, Ozturk F (2011) Evaluation of topographical and geographical effects on some climatic parameters in the central anatolia region of Turkey. Int J Climatol 31(9): 1264–1279. https://doi.org/10.1002/joc.2154

Arasan Ş (2014) Plants used in folk medicine and their usage areas in Savur (Mardin) region. Master thesis. Yüzüncüyıl University, Science Institute, Van

Arena A, Bisignano G, Pavone B, Tomaino A, Bonina FP, Saija A, Cristani M, D'Arrigo M, Trombetta D (2008) Antiviral and immunomodulatory effect of a lyophilized extract of *Capparis spinosa* L. buds. Phytother Res 22(3):313–317. https://doi.org/10.1002/ptr.2313

Arı S, Temel M, Kargıoğlu M, Konuk M (2015) Ethnobotanical survey of plants used in Afyonkarahisar-Turkey. J Ethnobiol Ethnomed 11(1):1–15. https://doi.org/10.1186/s13002-015-0067-6

Arıtuluk ZC (2010) Tefenni (Burdur) ilçesinin florası ve halk ilaçları. MSc Thesis. Hacettepe University, Health Sciences Institute, Ankara

Arnold N, Baydoun S, Chalak L, Raus T (2015) A contribution to the flora and ethnobotanical knowledge of mount Hermon, Lebanon. Fl Medit 25:13–55

Arshad MS, Khan U, Sadiq A, Khalid W, Hussain M, Yasmeen A, Asghar Z, Rehana H (2020) Coronavirus disease (COVID-19) and immunity booster green foods: a mini-review. Food Sci Nutr 8(8):3971–3976. https://doi.org/10.1002/fsn3.1719

Asadi MS, Mirvaghefei AR, Nematollahi MA, Banaee M, Ahmadi K (2012) Effects of watercress (nasturtium nasturtium) extract on selected immunological parameters of rainbow trout (Oncorhynchus mykiss). Open Vet J 2(1):32–39

Asih SC, Irdiani R, Sahlan M, Nasikin M (2021) Molecular docking study between 3 thai medicinal plants compounds and COVID-19 therapeutic protein targets: Sars-cov-2 main protease, ace-2, and pak-1. Int J Appl Pharm 13(special issue 2):41–48; https://scholar.ui.ac.id/en/publications/molecular-docking-study-between-3-thai-medicinal-plants-compounds

Assaf AM, Haddadin RN, Aldouri NA, Alabbassi R, Mashallah S, Mohammad M, Bustanji Y (2013) Anti-cancer, anti-inflammatory and anti-microbial activities of plant extracts used against hematological tumors in traditional medicine of Jordan. J Ethnopharmacol 145(3): 728–736. https://doi.org/10.1016/j.jep.2012.11.039

Ayoub HF, Khames DK (2019) Effect of spinach (*Spinacia oleracea*) leaves powder and vitamin e on growth performance and non-specific immune response of Nile tilapia (*Oreochromis niloticus*). Abbassa Int J Aqua 12(1):65–84

Başer KHC, Tümen G, Malyer H, Kırımer N (2006) Plants used for common cold in Turkey. In: Proceedings of the IVth international congress of ethnobotany (ICEB 2005), pp 133–137

Baytop T (1999) Türkiye'de bitkiler ile tedavi: Geçmişten Bugüne. Nobel Tıp Kitabevleri, İstanbul

Baytop T (2007) Türkçe bitkı adlari sözlüğü. Türk Tarih Kurumu yayınları

Bhatta A (2020) Choice of food: a preventive measure during COVID-19 outbreak. Eurasian J Med Sci 2(1):88–92. https://doi.org/10.46405/ejms.v2i1.49

Bibi N, Gul S, Ali J, Kamal MA (2020) Viroinformatics approach to explore the inhibitory mechanism of existing drugs repurposed to fight against COVID-19. Eur J Pharmacol 885: 173496. https://doi.org/10.1016/j.ejphar.2020.173496

Biçer A (2020) A study on foods that boost the immune system during the COVID-19 pandemic. NATURENGS Special issue:28–37

Bilen S, Altief TAS, Özdemir KY, Salem MOA, Terzi E, Güney K (2020) Effect of lemon balm (*Melissa officinalis*) extract on growth performance, digestive and antioxidant enzyme activities, and immune responses in rainbow trout (*Oncorhynchus mykiss*). Fish Physiol Biochem 46(1): 471–481. https://doi.org/10.1007/s10695-019-00737-z

Bozdemir Ç (2019) Türkiye'de yetişen kekik türleri, ekonomik önemi ve kullanım alanları. Yüzüncü Yıl Univ J Agr Sci 29(3):583–594. https://doi.org/10.29133/yyutbd.511777

Bozyel ME, Merdamert-Bozyel E (2020) Ethnomedicinal uses of genus Lavandula (Lamiaceae) in Turkish traditional medicine. Int J Acad 4(2):5–16

Bradley WG, Holm KN, Tanaka A (2014) An orally active immune adjuvant prepared from cones of Pinus sylvestris, enhances the proliferative phase of a primary T cell response. BMC Complement Altern Med 14(1):1–11. https://doi.org/10.1186/1472-6882-14-163

Bulut GE (2008) Ethnobotanical researches in Bayramiç (Çanakkale) region. PhD Thesis. Marmara Univ. Health Sci. Ins, Istanbul

Bulut G, Tuzlacı E (2009) Folk medicinal plants of Bayramiç (Çanakkale-Turkey). J Fac Pharm İst Univ 40:87–99

Bulut G, Tuzlacı E (2013) An ethnobotanical study of medicinal plants in Turgutlu (Manisa–Turkey). J Ethnopharmacol 149:633–647. https://doi.org/10.1016/j.jep.2013.07.016

Bulut G, Tuzlacı E, Doğan A, Şenkardeş İ, Doğan A, Şenkardeş İ (2014) An ethnopharmacological review on the Turkish Apiaceae species. J Fac Pharm İst Univ 44(2):163–179

Bulut G, Haznedaroğlu MZ, Doğan A, Koyu H, Tuzlacı E (2017) An ethnobotanical study of medicinal plants in Acıpayam (Denizli-Turkey). J Herb Med 10:64–81. https://doi.org/10.1016/j.hermed.2017.08.001

Bussmann RW, Zambrana NYP, Sikharulidze S, Kikvidze Z, Kikodze D, Tchelidze D, Khutsishvili M, Katevan B, Hart RE (2016a) A comparative ethnobotany of Khevsureti, Samtskhe-Javakheti, Tusheti, Svaneti, and Racha-Lechkhumi, Republic of Georgia (Sakartvelo). Caucasus J Ethnobiol Ethnomed 12:43. https://doi.org/10.1186/s13002-016-0110-2

Bussmann RW, Paniagua Zambrana NY, Sikharulidze S, Kikvidze Z, Kikodze D, Tchelidze D, Katevan B, Hart RE (2016b) Medicinal and food plants of Svaneti and Lechkhumi, Sakartvelo (Republic of Georgia). Caucasus Med Aromat Plants 5(266):2167–0412. https://doi.org/10.4172/2167-0412.1000266

Çakılcıoğlu U, Türkoğlu İ (2007) Plants and fruits used for cholesterol treatment by the folk in Elaziğ. Phytol Balc 13(2):239–245

Cao J, Tang D, Wang Y, Li X, Hong L, Sun C (2018) Characteristics and immune-enhancing activity of pectic polysaccharides from sweet cherry (*Prunus avium*). Food Chem 254:47–54. https://doi.org/10.1016/j.foodchem.2018.01.145

Capek P, Hribalova V (2004) Water-soluble polysaccharides from *Salvia officinalis* L. possessing immunomodulatory activity. Phytochemistry 65(13):1983–1992. https://doi.org/10.1016/j.phytochem.2004.05.020

Cataldo PG, Villena J, Elean M, de Giori GS, Saavedra L, Hebert EM (2020) Immunomodulatory properties of a γ-aminobutyric acid-enriched strawberry juice produced by Levilactobacillus brevis CRL 2013. Front Microbiol 11:610016

Çekin MD, Alpınar K, Küçük İ (2016) Zeytinburnu Tıbbi Bitkiler Bahçesi. İstanbul: Zeytinburnu Belediyesi Kültür Yayınları 35:1–6

Chowdhury MA, Hossain N, Kashem MA, Shahid MA, Alam A (2020) Immune response in COVID-19: A review. J Infect Public Health 13(11):1619–1629. https://doi.org/10.1016/j.jiph.2020.07.001

Çiçek İ (2019) Çermik ilçesi ve köylerinin (Diyarbakır) etnobotanik özellikleri. Master Thesis. Bingöl Univ. Sci. Ins, Bingöl

Colunga-Biancatelli RML, Berrill M, Catravas JD, Marik PE (2020) Quercetin and vitamin C: an experimental, synergistic therapy for the prevention and treatment of SARS-CoV-2 related disease (COVID-19). Front Immunol 11:1451. https://doi.org/10.3389/fimmu.2020.01451

Danilets MG, Bel'skiĭ I, Gur'Ev AM, Belousov MV, Bel'Skaia NV, Trofimova ES, Uchasova EG, Alhmedzhanov RR, Ligacheva AA, Lusbov MS, Agefonov VI (2010) Effect of plant polysaccharides on TH1-dependent immune response: screening investigation. Eksp Klin Farmakol 73(6):19–22

Daraban IN, Arsene GG, Turcus V, Ardelean A (2013) Assessment on bioeconomical potential for medicinal plants in salty meadows from the Aradului plain (W. Romania). Studia Universitatis" Vasile Goldis" Arad Seria Stiintele Vietii 23(1):71

Demirci-Kayıran SD (2019) A research on the present uses of the medicinal plants in De Materia medica written by Dioscorides in eastern Mediterranean region. Lokman Hekim Derg 9(2): 189–202. https://doi.org/10.31020/mutftd.519382

Deniz L, Serteser A, Kargıoğlu M (2010) Uşak Üniversitesi ve yakın çevresindeki bazı bitkilerin mahalli adları ve etnobotanik özellikleri. AKÜ Fen Ve Mühendislik Bilimleri Dergisi 10(1): 57–72

Diken ME (2009) Antioxidant contents of some medicinal herbs. Master thesis. Balıkesir Univ. Sci. Ins, Balıkesir

Doğan A (2008) Ovacık (Tunceli) yöresinin geleneksel halk ilacı olarak kullanılan bitkileri. Master thesis. Marmara Univ Health Sci Ins, Istanbul

Doğan A (2014) Ethnobotanical researches in Pertek (Tunceli) region. PhD Thesis. Marmara University, Health Science Institute, Istanbul

Doğan Y, Doğan S (2020) Koronavirüs Pandemisi ve Türkiye'de Bitkisel Üretime Etkisi. Artuklu Kaime Uluslararası İktisadi ve İdari Araştırmalar Dergisi 3(1):56–78

Dogan Y, Baslar S, Ay G, Mert HH (2004) The use of wild edible plants in western and central anatolia (Turkey). Econ Bot 58:684–690. https://doi.org/10.1663/0013-0001(2004)058[0684:TUOWEP]2.0.CO;2

Ege M, Elmastaş M (2020) Antiviral Etkili Fitoterapötikler: Tıbbi Bitkiler ve Fitokimyasallar. J Integr Anatol Med 1(2):5–20

Eid AM, Jaradat N (2020) Public knowledge, attitude, and practice on herbal remedies used during pregnancy and lactation in West Bank Palestine. Front Pharmacol 11:46. https://doi.org/10.3389/fphar.2020.00046

Ekor M (2014) The growing use of herbal medicines: issues relating to adverse reactions and challenges in monitoring safety. Front Pharmacol 4:177. https://doi.org/10.3389/fphar.2013.00177

Ekşi G, Özkan AMG, Koyuncu M (2020) Garlic and onions: an eastern tale. J Ethnopharmacol 253: 112675. https://doi.org/10.1016/j.jep.2020.112675

El Ghaoui WB, Ghanem EB, Chedid LA, Abdelnoor AM (2008) The effects of Alcea rosea L., Malva sylvestris L. and Salvia libanotica L. water extracts on the production of anti-egg albumin antibodies, interleukin-4, gamma interferon and interleukin-12 in BALB/c mice. Phytother Res 22(12):1599–1604. https://doi.org/10.1002/ptr.2530

Elfiky AA (2020) Natural products may interfere with SARS-CoV-2 attachment to the host cell. J Biomol Struct Dyn 39(9):3194. https://doi.org/10.1080/07391102.2020.1761881

Elwan HA, Elnesr SS, Mohany M, Al-Rejaie SS (2019) The effects of dietary tomato powder (*Solanum lycopersicum* L.) supplementation on the haematological, immunological, serum biochemical and antioxidant parameters of growing rabbits. J Anim Physiol Anim Nutr 103(2):534–546. https://doi.org/10.1111/jpn.13054

Erarslan ZB, Ecevit-Genç G, Kültür Ş (2020) Medicinal plants traditionally used to treat skin diseases in Turkey–eczema, psoriasis, vitiligo. Ankara Üniversitesi Eczacılık Fakültesi Dergisi 44(1):137–166. https://doi.org/10.33483/jfpau.586114

Evenett SJ, Hoekman B, Rocha N, Ruta M (2021) COVID-19 vaccine challenges: what have we learned so far and what remains to be done? Health Policy 125(5):553–567. https://doi.org/10.1016/j.healthpol.2021.03.013

Everest A, Ozturk E (2005) Focusing on the ethnobotanical uses of plants in Mersin and Adana provinces (Turkey). J Ethnobiol Ethnomed 1(1):1–6. https://doi.org/10.1186/1746-4269-1-6

Fidan I, Ozkan S, Gurbuz I, Yesilyurt E, Erdal B, Yolbakan S, Imir T (2008) The efficiency of *Viscum album* ssp. *album* and *Hypericum perforatum* on human immune cells in vitro. Immunopharmacol Immunotoxicol 30(3):519–528. https://doi.org/10.1080/08923970802135286

Galmarini V, Ferrara F, Tosco P, Russo A, Molinari G, Capelli R (2020) Coronavirus, the immune system and vitamin D. Lo Scalpello J 34:140–144. https://doi.org/10.36149/0390-5276-161

Genç L (2010) Tıbbi ve Aromatik Bitkilerin Kullanım Alanları ve Etiği. Anadolu University Web-Ofset, Eskişehir

Gençay A (2007) Cizre (Şırnak)'nin etnobotanik özellikleri. Master Thesis. Yüzüncüyıl University, Science Institute, Van

Ghafari-Farsani H, Gerami MH, Farsani MN, Rashidiyan G, Mehdipour N, Ghanad M, Faggio C (2018) Effect of different levels of essential oils (*Satureja hortensis*) in diet on improvement growth, blood biochemical and immunity of angelfish (*Pterophyllum scalare* Schultze, 1823). Nat Prod Res 2018:1–6. https://doi.org/10.1080/14786419.2018.1434635

Gharaei A, Shafie M, Mirdar Harijani J, Hasanein P, Arshadi A (2020) Immune responses and haematological parameters changes of rainbow trout (*Oncorhynchus mykiss*) under effects of dietary administration of sumac (*Rhus coriaria* L.). J Agric Sci Technol 22(1):173–186

Ghildiyal R, Prakash V, Chaudhary VK, Gupta V, Gabrani R (2020) Phytochemicals as antiviral agents: recent updates. In: Swamy M (ed) Plant-derived bioactives. Springer, Singapore. https://doi.org/10.1007/978-981-15-1761-7_12

Gjorgieva-Ackova D, Kadifkova-Panovska T, Baceva K, Stafilov T (2011) Metalic trace elements in medicinal plants from Macedonia. Middle East J Sci Res 7(1):109–114

Gökçe NP (2011) Doğanın mucizesi şifalı bitkiler. Sis yayıncılık, İstanbul

Gül V, Seçkin-Dinler B (2016) Some medical and aromatic plants growing naturally in Kumru region (Ordu). Süleyman Demirel Univ Ziraat Fak 11(1):146–156

Günbatan T, Gürbüz İ, Özkan AMG (2016) The current status of ethnopharmacobotanical knowledge in Çamlıdere (Ankara, Turkey). Turk J Bot 40(3):241–249

Güner A, Aslan S, Ekim T, Vural M, Babaç MT (2012) Türkiye Bitkileri Listesi (Damarlı Bitkiler). Nezahat Gökyiğit Botanical Garden and Flora Research Association Publication, Istanbul, pp 47–83

Güneş S (2010) Karaisalı (Adana) ve köylerinde halkın kullandığı doğal bitkilerin etnobotanik yönden araştırılması. Master Thesis. Niğde University, Science Institute, Niğde

Güneş F (2017) Medicinal plants used in the Uzunköprü district of Edirne. Turkey Acta Soc Bot Pol 86(4):3565. https://doi.org/10.5586/asbp.3565

Gürbüz İ, Özkan AMG, Akaydin G, Salihoğlu E, Günbatan T, Demirci F, Yeşilada E (2019) Folk medicine in Düzce Province (Turkey). Turk J Bot 43(6):769–784. https://doi.org/10.3906/bot-1905-13

Hanlidou E, Karousou R, Kleftoyanni V, Kokkini S (2004) The herbal market of Thessaloniki (N Greece) and its relation to the ethnobotanical tradition. J Ethnopharmacol 91(2–3):281–299. https://doi.org/10.1016/j.jep.2004.01.007

Heidari-Kharaji M, Fallah-Omrani V, Badirzadeh A, Mohammadi-Ghalehbin B, Nilforoushzadeh MA, Masoori L, Montakhab-Yeganeh H, Zare M (2019) *Sambucus ebulus* extract stimulates cellular responses in cutaneous leishmaniasis. Parasite Immunol 41(1):e12605. https://doi.org/10.1111/pim.12605

Heidary F, Varnaseri M, Gharebaghi R (2020) The potential use of Persian herbal medicines against COVID-19 through angiotensin-converting enzyme 2. Arch Clin Infect Dis 15(COVID-19): e102838. https://doi.org/10.5812/archcid.102838

Hoseinifar SH, Zou HK, Miandare HK, Van Doan H, Romano N, Dadar M (2017) Enrichment of common carp (Cyprinus carpio) diet with medlar (Mespilus germanica) leaf extract: effects on skin mucosal immunity and growth performance. Fish Shellfish Immunol 67:346–352. https://doi.org/10.1016/j.fsi.2017.06.023

Hürkul MM, Köroğlu A (2019) Etnobotanik Bir Derleme: Amygdaloideae (Rosaceae) Alt Familyası. Fabad J Pharm Sci 44(1):35–46

Ibadullayeva S, Gasimov H, Gahramanova M, Zulfugarova P, Novruzova L (2015) Medico-ethnobotanical inventory (liver and gallbladder ducts illnesses) of Nakhchivan AR. Azerbaijan Int J Sci 4(6):80–88

International Plant Names Index. (2021).http://www.ipni.org. Accessed 28 Apr 2021

Iuvone T, Esposito G, Capasso F, Izzo AA (2003) Induction of nitric oxide synthase expression by *Withania somnifera* in macrophages. Life Sci 72(14):1617–1625. https://doi.org/10.1016/S0024-3205(02)02472-4

Ivancheva S, Stantcheva B (2000) Ethnobotanical inventory of medicinal plants in Bulgaria. J Ethnopharmacol 69(2):165–172. https://doi.org/10.1016/S0378-8741(99)00129-4

Jahan I, Onay A (2020) Potentials of plant-based substance to inhabit and probable cure for the COVID-19. Turk J Biol 44(3):228–241. https://doi.org/10.3906/biy-2005-114

Jakhmola-Mani R, Sehgal N, Dogra N, Saxena S, Pande Katare D (2020) Deciphering underlying mechanism of Sars-CoV-2 infection in humans and revealing the therapeutic potential of bioactive constituents from Nigella sativa to combat COVID-19: in-silico study. J Biomol Struct Dyn 40(6):2417. https://doi.org/10.1080/07391102.2020.1839560

Jamshidi N, Cohen MM (2017) The clinical efficacy and safety of Tulsi in humans: a systematic review of the literature. Evid Based Complement Alternat Med 13:9217567. https://doi.org/10.1155/2017/9217567

Janaćković P, Gavrilović M, Savić J, Marin PD, Stevanović ZD (2019) Traditional knowledge on plant use from Negotin Krajina (eastern Serbia): an ethnobotanical study. Indian J Tradit Knowl 18(1):25–33

Jaradat N (2005) Ethnopharmacological survey of natural products in Palestine. An Najah Univ J Res 19:13–67

Jeambey Z, Johns T, Talhouk S, Batal M (2009) Perceived health and medicinal properties of six species of wild edible plants in North-East Lebanon. Public Health Nutr 12(10):1902–1911. https://doi.org/10.1017/S1368980009004832

Ji Y, Li S, Ho CT (2019) Chemical composition, sensory properties and application of Sichuan pepper (*zanthoxylum* genus). Food Sci Human Well 8(2):115–125. https://doi.org/10.1016/j.fshw.2019.03.008

Kakodkar P, Kaka N, Baig MN (2020) A comprehensive literature review on the clinical presentation, and management of the pandemic coronavirus disease 2019 (COVID-19). Cureus 12(4): e7560. https://doi.org/10.7759/cureus.7560

Kalafatçılar ÖA, Kalafatçılar Aİ (2010) Bitkiler ve Sağlık. Sidas Yayınları, İzmir

Karasawa K, Uzuhashi Y, Hirota M, Otani H (2011) A matured fruit extract of date palm tree (*Phoenix dactylifera* L.) stimulates the cellular immune system in mice. J Agric Food Chem 59(20):11287–11293. https://doi.org/10.1021/jf2029225

Karimi MH, Ebadi P, Amirghofran Z (2012) Parsley and immunomodulation. Expert Rev Clin Immunol 8(4):295–297. https://doi.org/10.1586/eci.12.12

Karousou R, Deirmentzoglou S (2011) The herbal market of Cyprus: traditional links and cultural exchanges. J Ethnopharmacol 133(1):191–203. https://doi.org/10.1016/j.jep.2010.09.034

Kavaklı SA, Özgür K, Ugurlu E (2020a) Eskişehir halk pazarlarında satılan bazı tıbbi-aromatik bitkiler ve kullanım alanları. Ağaç ve Orman 1(1):1–11

Kavaklı SA, Yaylacı ŞG, Ugurlu E (2020b) Kastamonu Aktarlarında Satılan Bazı Tıbbi Aromatik Bitkiler ve Kullanım Alanları. Iğdır univ fen bilim enst derg 10(4):2962–2974. https://doi.org/10.21597/jist.739021

Kaval I, Behçet L, Cakilcioglu U (2014) Ethnobotanical study on medicinal plants in Geçitli and its surrounding (Hakkari-Turkey). J Ethnopharmacol 155(1):171–184. https://doi.org/10.1016/j.jep.2014.05.014

Kawarty AMMA, Behçet L, Çakılcıoğlu U (2020) An ethnobotanical survey of medicinal plants in Ballakayati (Erbil, North Iraq). Turk J Bot 44(3):345–357. https://doi.org/10.3906/bot-1910-39

Kaya LG, Kaynakci-Elinc Z, Yucedag C, Cetin M (2018) Environmental outdoor plant preferences: a practical approach for choosing outdoor plants in urban or suburban residential areas in Antalya, Turkey. Fresenius Environ Bull 27(12):7945–7952

Kenar N, Sekerciler F, Coban S (2020) The phytosociology, ecology, and plant diversity of new plant communities in central Anatolia (Turkey). Hacquetia 19(1):1–22

Keser A, Kurt E (2016) The contributions of waqfs and urban development to the settlement of public health services: an overview to Turkish administrative history of Anatolian Seljuks. IOSR-JHSS 21(10):66–79. https://doi.org/10.9790/0837-2110126679

Khan T, Khan MA, Ullah N, Nadhman A (2021) Therapeutic potential of medicinal plants against COVID-19: the role of antiviral medicinal metabolites. Biocatal Agric Biotechnol 31:101890. https://doi.org/10.1016/j.bcab.2020.101890

Khanal P, Duyu T, Patil BM, Dey YN, Pasha I, Wanjari M, Gurav SS, Maity A (2020) Network pharmacology of AYUSH recommended immune-boosting medicinal plants against COVID-19. J Ayurveda Integr Med 13(1):100374. https://doi.org/10.1016/j.jaim.2020.11.004

Khanna K, Kohli SK, Kaur R, Bhardwaj A, Bhardwaj V, Ohri P, Sharma A, Ahmad A, Bhardwaj R, Ahmad P (2020) Herbal immune-boosters: substantial warriors of pandemic COVID-19 battle. Phytomedicine 85:153361. https://doi.org/10.1016/j.phymed.2020.153361

Kılıç Ö (2016) An ethnobotanical survey from Bingol (Turkey). RAJAR 2(10):685–691. https://doi.org/10.18535/rajar/v2i10.05

Kılıç M (2019) An ethnobotanical research on plants grown in Artuklu (Mardin) region. PhD Thesis. Manisa Celal Bayar Univ. Sci. Ins. Manisa

Kilic Ö, Bagci E (2013) An ethnobotanical survey of some medicinal plants in Keban (Elazığ—Turkey). J Med Plants Res 7(23):1675–1684. https://doi.org/10.5897/JMPR2013.4451

Kılıç M, Yıldız K, Kılıç FM (2020) Traditional uses of medicinal plants in Artuklu. Turkey Hum Ecol 48:619. https://doi.org/10.1007/s10745-020-00180-2

Kim D, Lee J, Kim KJ, Hong HC, Shin KS, Yu KW (2010) Macrophage stimulating polysaccharide purified from peels of grape (*Vitis labrusca*). Food Sci Biotechnol 19(2):479–486. https://doi.org/10.1007/s10068-010-0067-1

Kim SB, Chang BY, Jo YH, Lee SH, Han SB, Hwang BY, Kim SY, Lee MK (2013) Macrophage activating activity of pyrrole alkaloids from *Morus alba* fruits. J Ethnopharmacol 145(1): 393–396. https://doi.org/10.1016/j.jep.2012.11.007

Kızılarslan Ç (2008) İzmit Körfezi'nin güney kesiminde etnobotanik bir araştırma. Master Thesis. Istanbul University, Health Science Institute, Istanbul

Kocabaş YZ, Erol A, Aktolun O (2017) Medicinal plants of flora of KSU avsar campus (Kahramanmaras) and surrounding areas. Aksaray J Sci Müh 1(2):110–120; http://asujse.aksaray.edu.tr/en/download/article-file/323447.

Kökçü B, Esen O, Uysal İ (2015) Medicinal plants sold in Çanakkale/Turkey city center herbalists. Biodicon 8(3):80–91

Koleva V, Dragoeva A, Nanova Z, Koynova T, Dashev G (2015) An ethnobotanical study on current status of some medicinal plants used in Bulgaria. Int J Curr Microbiol App Sci 4(4): 297–305

Konakci G, Uran BNO, Erkin O (2020) In the Turkish news: coronavirus and "alternative & complementary" medicine methods. Complement Ther Med 53:102545. https://doi.org/10.1016/j.ctim.2020.102545

Korkmaz M, Karakurt E (2014) Kelkit (Gümüşhane) aktarlarında satılan tıbbi bitkilerin etnobotanik özellikleri. Süleyman Demirel univ fen bilim enst derg 18(3):60–80
Korkmaz M, Karakuş S, Özçelik H, Selvi S (2016) An ethnobotanical study on medicinal plants in Erzincan. Turkey Indian J Tradit Know 15(2):192–202
Köse M (2019) Ethnobotanical features of Güneysu (Rize) district. Master Thesis. Artvin Çoruh Univ. Sci. Ins. Artvin
Kozuharova E, Lebanova H, Getov I, Benbassat N, Napier J (2013) Descriptive study of contemporary status of the traditional knowledge on medicinal plants in Bulgaria. Afr J Pharm Pharmacol 7(5):185–198. https://doi.org/10.5897/AJPP12.871
Kubat Ö (2019) Salarha vadisi (Rize) tıbbi ve aromatik bitkileri. Master thesis. Artvin Çoruh Univ. sci. Ins. Artvin
Kültür Ş (2007) Medicinal plants used in Kırklareli province (Turkey). J Ethnopharmacol 111(2): 341–364. https://doi.org/10.1016/j.jep.2006.11.035
Küpeli-Akkol EK, Karpuz B, Sobarzo-Sánchez E, Khan H (2020) A phytopharmacological overview of medicinal plants used for prophylactic and treatment of colitis. Food Chem Toxicol 144: 111628. https://doi.org/10.1016/j.fct.2020.111628
Kurt P, Karaoğul E (2018) Bartın'da Aktarlarda Satılan Tıbbi Aromatik Bitkiler ve Ülkemizdeki Pazar Payları. Bartın Orman Fakültesi Dergisi 20(1):73–80. https://doi.org/10.24011/barofd.400016
Lardos A, Prieto-Garcia J, Heinrich M (2011) Resins and gums in historical iatrosophia texts from Cyprus–a botanical and medico-pharmacological approach. Front Pharmacol 2:32. https://doi.org/10.3389/fphar.2011.00032
Latifian E, Arslanoğlu ŞF (2018) Traditional medicinal plants of Azerbaijan province of Iran. Agric Sci 9:157–170
Lev E, Amar Z (2000) Ethnopharmacological survey of traditional drugs sold in Israel at the end of the 20th century. J Ethnopharmacol 72(1–2):191–205. https://doi.org/10.1016/S0378-8741(00)00230-0
Lev E, Amar Z (2002) Ethnopharmacological survey of traditional drugs sold in the kingdom of Jordan. J Ethnopharmacol 82(2–3):131–145. https://doi.org/10.1016/S0378-8741(02)00182-4
Lian Y, Chen H, Li B, Yang J, Yuan J (2009) Immunoloregulation effect of *Elaeagnus angustifolia* L. polysaccharides in immunosuppressive mice. J Anhui Agric Sci 37:7481–7482
Lis M, Szczypka M, Suszko-Pawłowska A, Sokół-Łętowska A, Kucharska A, Obmińska-Mrukowicz B (2020) Hawthorn (*Crataegus monogyna*) phenolic extract modulates lymphocyte subsets and humoral immune response in mice. Planta Med 86(2):160–168. https://doi.org/10.1055/a-1045-5437
Majdalawieh AF, Farraj JF, Carr RI (2020) *Sesamum indicum* (sesame) enhances NK anti-cancer activity, modulates Th1/Th2 balance, and suppresses macrophage inflammatory response. Asian Pac J Trop Med 10(7):316–324. https://doi.org/10.4103/2221-1691.284946
Mao R, Wu L, Zhu N, Liu X, Hao Y, Liu R, Du Q, Li Y (2020) Immunomodulatory effects of walnut (*Juglans regia* L.) oligopeptides on innate and adaptive immune responses in mice. J Funct Foods 73:104068. https://doi.org/10.1016/j.jff.2020.104068
Marouf M, Batal M, Moledor S, Talhouk SN (2015) Exploring the practice of traditional wild plant collection in Lebanon. Food Cult Soc 18(3):355–378. https://doi.org/10.1080/15528014.2015.1043103
Marra F, Smolders EJ, El-Sherif O, Boyle A, Davidson K, Sommerville AJ, Marzolini C, Siccardi M, Burger D, Gibbons S, Khoo S, Back D (2020) Recommendations for dosing of repurposed COVID-19 medications in patients with renal and hepatic impairment. Drugs R D 21:9–27. https://doi.org/10.1007/s40268-020-00333-0
Martkoplishvili I, Kvavadze E (2015) Some popular medicinal plants and diseases of the upper Palaeolithic in Western Georgia. J Ethnopharmacol 166:42–52. https://doi.org/10.1016/j.jep.2015.03.003

Matejić JS, Stefanović N, Ivković M, Živanović N, Marin PD, Džamić AM (2020) Traditional uses of autochthonous medicinal and ritual plants and other remedies for health in Eastern and South-eastern Serbia. J Ethnopharmacol 261:113186. https://doi.org/10.1016/j.jep.2020.113186

Mati E, de Boer H (2011) Ethnobotany and trade of medicinal plants in the Qaysari market, Kurdish autonomous region. Iraq J Ethnopharmacol 133(2):490–510. https://doi.org/10.1016/j.jep.2010.10.023

Mattalia G, Stryamets N, Pieroni A, Sõukand R (2020a) Knowledge transmission patterns at the border: ethnobotany of Hutsuls living in the Carpathian Mountains of Bukovina (SW Ukraine and NE Romania). J Ethnobiol Ethnomed 16:41. https://doi.org/10.1186/s13002-020-00391-3

Mattalia G, Stryamets N, Grygorovych A, Pieroni A, Sõukand R (2020b) Borders as crossroads: the diverging routes of herbal knowledge of Romanians living on the Romanian and Ukrainian sides of Bukovina. Front Pharmacol 11:598390

Mazumder PM, Pattnayak S, Parvani H, Sasmal D, Rathinavelusamy P (2012) Evaluation of immunomodulatory activity of Glycyrrhiza glabra L roots in combination with zing. Asian Pac J Trop Biomed 2(1):S15–S20. https://doi.org/10.1016/S2221-1691(12)60122-1

Mendi A, Yağcı BG, Kızıloğlu M, Saraç N, Uğur A, Yılmaz D, Uçkan D (2017) *Thymbra spicata* var. *intricata* induces mesenchymal stem cell proliferation and osteogenic differentiation. Braz Arch Biol Technol 60:e17160391. https://doi.org/10.1590/1678-4324-2017160391

Menković N, Šavikin K, Tasić S, Zdunić G, Stešević D, Milosavljević S, Vincek D (2011) Ethnobotanical study on traditional uses of wild medicinal plants in Prokletije Mountains (Montenegro). J Ethnopharmacol 133(1):97–107. https://doi.org/10.1016/j.jep.2010.09.008

Menković N, Šavikin K, Zdunić G, Milosavljević S, Živković J (2014) Medicinal plants in northern Montenegro: traditional knowledge, quality, and resources. In: Ethnobotany and biocultural diversities in the Balkans. Springer, New York, NY, pp 197–228. https://doi.org/10.1007/978-1-4939-1492-0_11

Mirabeau TY, Samson ES (2012) Effect of Allium cepa and Allium sativum on some immunological cells in rats. Afr J Tradit Complement Altern Med 9(3):374–379. https://doi.org/10.4314/ajtcam.v9i3.11

Miraldi E, Ferri S, Mostaghimi V (2001) Botanical drugs and preparations in the traditional medicine of West Azerbaijan (Iran). J Ethnopharmacol 75(2–3):77–87. https://doi.org/10.1016/S0378-8741(00)00381-0

Mohamed GA, Amhamed ID, Almabrok AA, Barka ABA, Bilen S, Elbeshti RT (2018) Effect of celery (Apium graveolens) extract on the growth, haematology, immune response and digestive enzyme activity of common carp (Cyprinus carpio). J Mar Sci Technol 7(2):51–59. https://doi.org/10.33714/masteb.457721

Monobe M, Ema K, Kato F, Maeda-Yamamoto M (2008) Immunostimulating activity of a crude polysaccharide derived from green tea (Camellia sinensis) extract. J Agric Food Chem 56(4):1423–1427. https://doi.org/10.1021/jf073127h

Mullalija B, Mustafa B, Hajdari A, Quave CL, Pieroni A (2021) Ethnobotany of rural and urban Albanians and Serbs in the Anadrini region. Kosovo Genet Resour Crop Evol 68:1825–1848. https://doi.org/10.1007/s10722-020-01099-9

Mumcu Ü, Korkmaz H (2018) Ethnobotanical uses of alien and native plant species of Yeşilırmak Delta, Samsun. Turkey Acta Biologica Turcica 31(3):102–113

Mustafa B, Hajdari A, Krasniqi F, Hoxha E, Ademi H, Quave CL, Pieroni A (2012) Medical ethnobotany of the Albanian alps in Kosovo. J Ethnobiol Ethnomed 8:6. https://doi.org/10.1186/1746-4269-8-6

Nacakcı FM, Dutkuner İ (2015) Kumluca (Antalya)'da etnobotanik bir çalışma. Türkiye Ormancılık Dergisi 19(2):113–119. https://doi.org/10.18182/tjf.421970

Naghibi F, Mosadegh M, Mohammadi MS, Ghorbani AB (2005) Labiatae family in folk medicine in Iran: from ethnobotany to pharmacology. Iran J Pharm Sci 4(2):63–79

Nanagulyan S, Zakaryan N, Kartashyan N, Piwowarczyk R, Łuczaj Ł (2020) Wild plants and fungi sold in the markets of Yerevan (Armenia). J Ethnobiol Ethnomed 16:26. https://doi.org/10.1186/s13002-020-00375-3

Nemati M, Asl ER, Pouya FD, Rasmi Y (2020) Curcumin, an inhibitor of PAK1, potential treatment for COVID-19. J Infect 3(2):1–3

Nicholson LB (2016) The immune system. Essays Biochem 60(3):275–301. https://doi.org/10.1042/EBC20160017

Nishimoto S, Muranaka A, Nishi K, Kadota A, Sugahara T (2012) Immunomodulatory effects of citrus fruit auraptene in vitro and in vivo. J Funct Foods 4(4):883–890. https://doi.org/10.1016/j.jff.2012.06.005

Nobakht A, Mansoub NH, Nezhady MAM (2012) Effect of *Melissa officinalis* L., *Tanacetum balsamita* L. and *Ziziphora clinopodioides* L. on performance, blood biochemical and immunity parameters of laying hens. Asian J Anim Vet Adv 7(1):74–79. https://doi.org/10.3923/ajava.2012.74.79

Novruzova LA, Maharramov SH, Ibadullayeva SJ (2015) Management of gastroenterostomy illness with herbs in veterinary practice in Nakhchivan Autonomous Republic (Azerbaijan). Int J Vet Sci 4(4):224–226

Olgun Ş (2019) Ethnobotanical characteristics of Arıcak (Elazığ) district. Master Thesis Bingöl Univ Sci Inst Bingöl

Onbaşlı D, Dal A (2020) Türk Farmakopesindeki Tıbbi Bitkilerinin İncelenmesi. Erciyes Univ Sağlık Bil Fak Derg 7(1):22–36

Ongan D (2018) Kayseri'de yaşayan yetişkin bireylerde bitkisel ürün kullanım alışkanlığı. Sağlık Bilimleri Dergisi 27(2):125–131

Opitz B, van Laak V, Eitel J, Suttorp N (2010) Innate immune recognition in infectious and noninfectious diseases of the lung. Am J Respir Crit Care Med 181(12):1294–1309. https://doi.org/10.1164/rccm.200909-1427SO

Ötnü H, Akan H (2020) Plants sold for phytotherapy in pharmacies and herbalists of Şanlıurfa. KSU J Agric Nat 23(4):947–965. https://doi.org/10.18016/ksutarimdoga.vi.688167

Oyston P, Robinson K (2012) The current challenges for vaccine development. J Med Microbiol 61(7):889–894. https://doi.org/10.1099/jmm.0.039180-0

Özcan M, Arslan D, Ünver A (2020) Fonksiyonel Gıdalar Ve Fitokimyasallar. Akademik Gıda 1(5):40–45

Özçelik H, Çinbilgel İ, Muca B, Tavuç İ, Koca A, Bebekli Ö (2016) Burdur ili bitki envanteri (ekonomik, nadir ve endemik bitkileri). Burdur Belediyesi, Sistem Ofset ve Matb, Ankara

Özdemir-Nath E (2016) Ethnobotanical researches in Savaştepe and kepsut (Balıkesir) regions. PhD Thesis. Istanbul Univ. Health Sci. Ins, Istanbul

Özer Z, Elibüyük IÖ, Önen H, Elibüyük EA (2005) Sağlıklı bir yaşamdır yabancı otlar: Türk mutfak kültürü üzerine araştırmalar. In: Toygar K, Toygar NB (eds) Türk halk kültürünü araştırma ve tanıtma vakfı, vol 12, pp 113–194

Özhatay N, Akalın E, Genç GE, Kültür Ş (2009) Ethnomedicinal uses of the wild vascular plants from European Turkey (Turkish Thrace). IV Balkan Botanical Congress, Sofia, pp 613–623

Özkum D, Ömrüm A, Toklu HZ (2013) Herbal medicine use among diabetes mellitus patients in Northern Cyprus. J Med Plant Res 7(22):1652–1664. https://doi.org/10.5897/JMPR12.1207

Ozturk M, Hakeem KR (2018) Plant and human health, volume 1: ethnobotany and physiology. Springer Nature, Cham

Öztürk F, Ölçücü C (2011) Ethnobotanical features of some plants in the district of Şemdinli (Hakkari-Turkey). Int J Acad Res 3(1):117–121

Ozturk M, Altay V, Gucel S, Altundag E (2017a) Chapter 5: plant diversity of the drylands in Southeast Anatolia-Turkey: role in human health and food security. In: Ansari AA, Gill SS (eds) Plant biodiversity: monitoring, assessment and conservation. CABI, Wallingford, pp 83–124

Ozturk M, Altay V, Gonenç TM (2017b) Herbal from high mountains in the East Mediterranean. In: Bhojraj S et al (eds) Drug discovery from herbs—approaches and applications, chapter 24, centre for science technology of the non-aligned and other developing countries. DAYA Publishing House, New Delhi, pp 327–367

Paksoy MY, Selvi S, Savran A (2016) Ethnopharmacological survey of medicinal plants in Ulukışla (Niğde-Turkey). J Herb Med 6(1):42–48. https://doi.org/10.1016/j.hermed.2015.04.003

Papp N, Birkás-Frendl K, Bencsik T, Stranczinger S, Czégényi D (2014) Survey of traditional beliefs in the Hungarian Csángó and Székely ethnomedicine in Transylvania. Romania Rev Bras Farmacogn 24(2):141–152. https://doi.org/10.1016/j.bjp.2014.03.005

Park MY, Kim DH, Jin MR (2010) Immunomodulatory effects of orally administrated capsicum extract on Peyer's patches. J Physiol Pathol Korean Med 24(3):446–451

Park YM, Lee HY, Kang YG, Park SH, Lee BG, Park YJ, Oh HG, Moon DI, Kim YP, Park DS, Lee HM, Kim OJ, Yang HJ, Lee YR (2019) Immune-enhancing effects of *Portulaca oleracea* L.–based complex extract in cyclophosphamide-induced splenocytes and immunosuppressed rats. Food Agric Immunol 30(1):13–24. https://doi.org/10.1080/09540105.2018.1540552

Pavithra K, Vijayadurga G, Dinesh D, Ramasubramanian V, Nadu T (2020) Effect of muskmelon, *Cucumis melo* and pears, *Pyrus communis* extract on growth, haematology, immune response and disease resistance in Labeo rohita against *Pseudomonas aeruginosa*. J Emerg Technol 7(11):976–789

Petran M, Dragos D, Gilca M (2020) Historical ethnobotanical review of medicinal plants used to treat children diseases in Romania (1860s–1970s). J Ethnobiol Ethnomed 16:15. https://doi.org/10.1186/s13002-020-00364-6

Pieroni A (2008) Local plant resources in the ethnobotany of Theth, a village in the northern Albanian alps. Genet Resour Crop Evol 55:1197–1214. https://doi.org/10.1007/s10722-008-9320-3

Pieroni A, Quave CL (2014) Ethnobotany and biocultural diversities in the Balkans: perspectives on sustainable rural development and reconciliation. Springer, New York/Heidelberg

Pieroni A, Dibra B, Grishaj G, Grishaj I, Maçai SG (2005) Traditional phytotherapy of the Albanians of Lepushe, northern Albanian alps. Fitoterapia 76(3–4):379–399. https://doi.org/10.1016/j.fitote.2005.03.015

Pieroni A, Quave CL, Giusti ME, Papp N (2012) "We are Italians!": the hybrid ethnobotany of a venetian diaspora in eastern Romania. Hum Ecol 40:435–451. https://doi.org/10.1007/s10745-012-9493-4

Pieroni A, Rexhepi B, Nedelcheva A, Hajdari A, Mustafa B, Kolosova V, Cianfaglione K, Quave CL (2013) One century later: the folk botanical knowledge of the last remaining Albanians of the upper Reka Valley, mount Korab. Western Macedonia J Ethnobiol Ethnomed 9:22. https://doi.org/10.1186/1746-4269-9-22

Pieroni A, Nedelcheva A, Hajdari A, Mustafa B, Scaltriti B, Cianfaglione K, Quave CL (2014a) Local knowledge on plants and domestic remedies in the mountain villages of Peshkopia (Eastern Albania). J Mt Sci 11:180–193. https://doi.org/10.1007/s11629-013-2651-3

Pieroni A, Cianfaglione K, Nedelcheva A, Hajdari A, Mustafa B, Quave CL (2014b) Resilience at the border: traditional botanical knowledge among Macedonians and Albanians living in Gollobordo. Eastern Albania J Ethnobiol Ethnomed 10:31. https://doi.org/10.1186/1746-4269-10-31

Pieroni A, Nedelcheva A, Dogan Y (2015) Local knowledge of medicinal plants and wild food plants among Tatars and Romanians in Dobruja (South-East Romania). Genet Resour Crop Evol 62:605–620. https://doi.org/10.1007/s10722-014-0185-3

Pisuttharachai D, Nalinanon W, Areechon N (2020) Effects of dietary supplementation with broccoli sprouts (Brassica oleracea) on the hematology and immunological responses of Nile tilapia (Oreochromis niloticus). J Fish Environ 44(1):65–75

Polat R, Selvi S (2020) Turizm Yerleşim Alanlarında Peyzajda Kullanılan Bitkilerin Çeşitliliği: Edremit Körfezi (Balıkesir) Örneği. Tr J Nature Sci 9(Special issue):62–72

Polat R, Satıl F, Çakılcıoğlu U (2011) Medicinal plants and their use properties of sold in herbal market in Bingöl (Turkey) district. Biodicon 4(3):25–35

Polat R, Cakilcioglu U, Satıl F (2013) Traditional uses of medicinal plants in Solhan (Bingöl—Turkey). J Ethnopharmacol 148(3):951–963. https://doi.org/10.1016/j.jep.2013.05.050

Polat R, Cakilcioglu U, Kaltalioğlu K, Ulusan MD, Türkmen Z (2015) An ethnobotanical study on medicinal plants in Espiye and its surrounding (Giresun-Turkey). J Ethnopharmacol 163:1–11. https://doi.org/10.1016/j.jep.2015.01.008

Popović Z, Smiljanić M, Matić R, Kostić M, Nikić P, Bojović S (2012) Phytotherapeutical plants from the Deliblato Sands (Serbia): traditional pharmacopoeia and implications for conservation. Indian J Tradit Knowl 11(3):385–400

Pourmoghim H, Haghighi M, Rohani MS (2015) Effect of dietary inclusion of *Origanum vulgare* extract on non-specific immune responses and hematological parameters of rainbow trout (*Oncorhynchus mykiss*). Bull Env Pharmacol Life Sci 4:33–39

Prasanna VK, Venkatesh YP (2015) Characterization of onion lectin (*Allium cepa* agglutinin) as an immunomodulatory protein inducing Th1-type immune response in vitro. Int Immunopharmacol 26(2):304–313. https://doi.org/10.1016/j.intimp.2015.04.009

Qadir A, Shakeel F, Ali A, Faiyazuddin M (2020) Phytotherapeutic potential and pharmaceutical impact of *Phoenix dactylifera* (date palm): current research and future prospects. J Food Sci Technol 57:1191–1204. https://doi.org/10.1007/s13197-019-04096-8

Quave CL, Pieroni A (2014) Fermented foods for food security and food sovereignty in the Balkans: a case study of the Gorani people of Northeastern Albania. J Ethnobiol 34(1):28–43. https://doi.org/10.2993/0278-0771-34.1.28

Randon AM, Attard E (2007) The in vitro immunomodulatory activity of oleuropein, a secoiridoid glycoside from *Olea europaea* L. Nat Prod Commun 2(5):515–519. https://doi.org/10.1177/1934578X0700200501

Redwan Z, Baydoun SA, Nasser H, Arnold-Apostolides N (2017) Socio-economic importance of some high value wild medicinal plants from Lebanon. In: International symposium on survey of uses of plant genetic resources to the benefit of local populations, vol 1267, pp 247–252. https://doi.org/10.17660/ActaHortic.2020.1267.37

Redžić SS (2007) The ecological aspect of ethnobotany and ethnopharmacology of population in Bosnia and Herzegovina. Coll Antropol 31(3):869–890

Rexhepi B, Mustafa B, Hajdari A, Rushidi-Rexhepi J, Quave CL, Pieroni A (2013) Traditional medicinal plant knowledge among Albanians, Macedonians and Gorani in the Sharr Mountains (Republic of Macedonia). Genet Resour Crop Evol 60:2055–2080. https://doi.org/10.1007/s10722-013-9974-3

Riaz A, Yousaf K, Abdin M, Jabbar S, Abid M, Muhammad M (2020) Extraction optimization, purification and immunostimulatory activity in vitro of polyphenols from apple (*Malus domestica*) peel. Sains Malays 49(7):1553–1566. https://doi.org/10.17576/jsm-2020-4907-08

Rihawy MS, Bakraji EH, Aref S, Shaban R (2010) Elemental investigation of Syrian medicinal plants using PIXE analysis. Nucl Instrum Meth B 268(17–18):2790–2793. https://doi.org/10.1016/j.nimb.2010.06.034

Rösch C, Meijerink M, Delahaije RJ, Taverne N, Gruppen H, Wells JM, Schols HA (2016) Immunomodulatory properties of oat and barley β-glucan populations on bone marrow derived dendritic cells. J Funct Foods 26:279–289. https://doi.org/10.1016/j.jff.2016.07.023

Ross RG, Selvasubramanian S, Jayasundar S (2001) Immunomodulatory activity of *Punica granatum* in rabbits—a preliminary study. J Ethnopharmacol 78(1):85–87. https://doi.org/10.1016/S0378-8741(01)00287-2

Saaby L, Jäger AK, Moesby L, Hansen EW, Christensen SB (2011) Isolation of immunomodulatory triterpene acids from a standardized rose hip powder (*Rosa canina* L.). Phytother Res 25(2):195–201. https://doi.org/10.1002/ptr.3241

Şahin B (2019) Ethnobotany in Kemalpaşa (İzmir) district. Master thesis. Ege Univ. Sci. Ins, Izmir

Said O, Khalil K, Fulder S, Azaizeh H (2002) Ethnopharmacological survey of medicinal herbs in Israel, the Golan Heights and the West Bank region. J Ethnopharmacol 83(3):251–265. https://doi.org/10.1016/S0378-8741(02)00253-2

Salem ML (2005) Immunomodulatory and therapeutic properties of the Nigella sativa L. seed. Int Immunopharmacol 5(13–14):1749–1770. https://doi.org/10.1016/j.intimp.2005.06.008

Sang R, Yu Y, Ge B, Xu L, Wang Z, Zhang X (2019) Taraxasterol from *taraxacum* prevents concanavalin A-induced acute hepatic injury in mice via modulating TLRs/NF-κB and Bax/Bc1-2 signalling pathways. Artif Cells Nanomed Biotechnol 47(1):3929–3937. https://doi.org/10.1080/21691401.2019.1671433

Şanlı BZ (2006) Bursa ve çevresinden toplanan ve ticareti yapılan bazı ekonomik bitkiler. Master Thesis. Uludağ University, Science Institute, Bursa

Sargin SA (2013) Alaşehir ve çevresinde (Manisa) tarımsal biyoçeşitlilik ve etnobotanik araştırmaları. PhD Thesis. Balıkesir University, Science Institute, Balıkesir

Sargin SA (2015) Ethnobotanical survey of medicinal plants in Bozyazı district of Mersin. Turkey J Ethnopharmacol 173:105–126. https://doi.org/10.1016/j.jep.2015.07.009

Sargin SA (2021) Potential anti-influenza effective plants used in Turkish folk medicine: a review. J Ethnopharmacol 265:113319. https://doi.org/10.1016/j.jep.2020.113319

Sargin SA, Büyükcengiz M (2019) Plants used in ethnomedicinal practices in Gulnar district of Mersin. Turkey J Herb Med 15:1–18. https://doi.org/10.1016/j.hermed.2018.06.003

Sargin SA, Selvi S, López V (2015) Ethnomedicinal plants of Sarigöl district (Manisa). Turkey J Ethnopharmacol 171:64–84. https://doi.org/10.1016/j.jep.2015.05.031

Šaric Kundalic B, Ahmedbegovic A, Cilovic E, Ademovic Z, Kerleta-Tuzovic V, Izic B (2015) Ethnobotanical study of traditionally used plants in human therapy of Trestenica and Tulovici, North East Bosnia and Herzegovina. Pharmacia 18(2):221–234

Sarıkan I (2007) Kazdağları yöresinin geleneksel ilaçlarının saptanması. Master thesis. Ege Univ. Health Sci. Ins. Izmir

Saybel OL, Rendyuk TD, Dargaeva TD, Nikolaev SM, Khobrakova VB (2020) Phenolic compounds and immunomodulating activity of chicory (Cichorium intybus L.). Extract Pharmacogn J 12(5):1104–1107. https://doi.org/10.5530/pj.2020.12.156

Schepetkin IA, Özek G, Özek T, Kirpotina LN, Khlebnikov AI, Quinn MT (2020) Chemical composition and immunomodulatory activity of *Hypericum perforatum* essential oils. Biomol Ther 10(6):916. https://doi.org/10.3390/biom10060916

Scudiero O, Lombardo B, Brancaccio M, Mennitti C, Cesaro A, Fimiani F, Gentile L, Moscarella E, Amodio F, Ranieri A, Gragnano F, Laneri S, Mazzaccara C, Di Micco P, Caiazza M, D'Alicandro G, Limongelli G, Calabrò P, Pero R, Frisso G (2021) Exercise, immune system, nutrition, respiratory and cardiovascular diseases during COVID-19: a complex combination. Int J Environ Res Public Health 18(3):1–20. https://doi.org/10.3390/ijerph18030904

Şekeroğlu N, Gezici S (2020) Koronavirüs Pandemisi ve Türkiye'nin Bazı Şifalı Bitkileri. Anatol Clin 25(Special Issue 1):163–182. https://doi.org/10.21673/anadoluklin.724210

Şenkardeş İ (2014) Ethnobotanical researches in the southern districts of Nevşehir (Acıgöl, Derinkuyu, Gülşehir, Nevşehir-Merkez, Ürgüp). PhD Thesis. Marmara Univ. Health Sci. Ins. Istanbul

Shaghaghi N (2020) Molecular docking study of novel COVID-19 protease with low risk Terpenoides compounds of plants. ChemRxiv

Shikov AN, Tsitsilin AN, Pozharitskaya ON, Makarov VG, Heinrich M (2017) Traditional and current food use of wild plants listed in the Russian pharmacopoeia. Front Pharmacol 8:841. https://doi.org/10.3389/fphar.2017.00841

Sieniawska E, Michel P, Mroczek T, Granica S, Skalicka-Woźniak K (2019) *Nigella damascena* L. essential oil and its main constituents, damascenine and β-elemene modulate inflammatory response of human neutrophils ex vivo. Food Chem Toxicol 125:161–169. https://doi.org/10.1016/j.fct.2018.12.057

Singh D, Tanwar H, Jayashankar B, Sharma J, Murthy S, Chanda S, Singh SB, Ganju L (2017) Quercetin exhibits adjuvant activity by enhancing Th2 immune response in ovalbumin immunized mice. Biomed Pharmacother 90:354–360. https://doi.org/10.1016/j.biopha.2017.03.067

Slavov A, Kiyohara H, Yamada H (2013) Immunomodulating pectic polysaccharides from waste rose petals of *Rosa damascena* mill. Int J Biol Macromol 59:192–200. https://doi.org/10.1016/j.ijbiomac.2013.04.054

Sõukand R, Pieroni A (2016) The importance of a border: medical, veterinary, and wild food ethnobotany of the Hutsuls living on the Romanian and Ukrainian sides of Bukovina. J Ethnopharmacol 185:17–40. https://doi.org/10.1016/j.jep.2016.03.009

Spiridonov NA, Konovalov DA, Arkhipov VV (2005) Cytotoxicity of some Russian ethnomedicinal plants and plant compounds. Phytother Res 19(5):428–432. https://doi.org/10.1002/ptr.1616

Stryamets N, Elbakidze M, Ceuterick M, Angelstam P, Axelsson R (2015) From economic survival to recreation: contemporary uses of wild food and medicine in rural Sweden, Ukraine and NW Russia. J Ethnobiol Ethnomed 11:53. https://doi.org/10.1186/s13002-015-0036-0

Taee HM, Hajimoradloo A, Hoseinifar SH, Ahmadvand H (2017) The effects of dietary Myrtle (*Myrtus communis* L.) supplementations on growth performance and some innate immune responses in rainbow trout (Oncorhynchus mykiss). Int J Aquat Biol 5(4):252–259. https://doi.org/10.22034/ijab.v5i4.331

Taghizadeh-Hesary F, Akbari H (2020) The powerful immune system against powerful COVID-19: a hypothesis. Med Hypotheses 140:109762. https://doi.org/10.1016/j.mehy.2020.109762

Talpur AD (2014) *Mentha piperita* (peppermint) as feed additive enhanced growth performance, survival, immune response and disease resistance of Asian seabass, Lates calcarifer (Bloch) against Vibrio harveyi infection. Aquaculture 420:71–78. https://doi.org/10.1016/j.aquaculture.2013.10.039

TAS, 2020. Turkish academy of sciences: the assessment report on COVID-19 global outbreak; http://www.tuba.gov.tr/files/yayinlar/raporlar/T%C3%9CBA%20Assessment%20Report%20on%20COVID-19%20Global%20Outbreak.pdf. Accessed 11 Apr 2021

Terali K, Baddal B, Gülcan HO (2020) Prioritizing potential ACE2 inhibitors in the COVID-19 pandemic: insights from a molecular mechanics-assisted structure-based virtual screening experiment. J Mol Graph Model 100:107697. https://doi.org/10.1016/j.jmgm.2020.107697

Terzioğlu S, Tüfekçioğlu A, Küçük M (2015) Vegetation and plant diversity of high-altitude mountains in eastern Karadeniz (Black Sea) region of Turkey and climate change interactions. In: Öztürk M, Hakeem K, Faridah-Hanum I, Efe R (eds) Climate change impacts on high-altitude ecosystems. Springer, Cham. https://doi.org/10.1007/978-3-319-12859-7_15

Tetik F, Civelek S, Cakilcioglu U (2013) Traditional uses of some medicinal plants in Malatya (Turkey). J Ethnopharmacol 146(1):331–346. https://doi.org/10.1016/j.jep.2012.12.054

Tewari S, David J, David B (2020) A critical review on immune-boosting therapeutic diet against coronavirus (COVID-19). J Sci Technol 5:43–49. https://doi.org/10.46243/jst.2020.v5.i5.pp43-49

The Plant List (2021).; https://www.theplantlist.org/. Accessed 27 Apr 2021

Tiengtam N, Paengkoum P, Sirivoharn S, Phonsiri K, Boonanuntanasarn S (2017) The effects of dietary inulin and Jerusalem artichoke (*Helianthus tuberosus*) tuber on the growth performance, haematological, blood chemical and immune parameters of Nile tilapia (*Oreochromis niloticus*) fingerlings. Aquac Res 48(10):5280–5288. https://doi.org/10.1111/are.13341

Tilwari A, Devi PU, Prabha B, Shukla NP (2011) Immunomodulatory effect of fractions of saponins from *Tribulus Terrestris* on non-specific immunity using in vitro phagocytosis. Int J Drug Dev Res 1(4):202–207

Tobaiqy M, Elkout H, MacLure K (2021) Analysis of thrombotic adverse reactions of COVID-19 AstraZeneca vaccine reported to EudraVigilance database. medRxiv 9(4):393. https://doi.org/10.1101/2021.03.19.21253980

Tomasini S, Theilade I (2019) Local knowledge of past and present uses of medicinal plants in Prespa National Park. Albania Econ Bot 73:217–232. https://doi.org/10.1007/s12231-019-09454-3

Torabian G, Valtchev P, Adil Q, Dehghani F (2019) Anti-influenza activity of elderberry (*Sambucus nigra*). J Funct Foods 54:353–360. https://doi.org/10.1016/j.jff.2019.01.031

Tsioutsiou EE, Giordani P, Hanlidou E, Biagi M, De Feo V, Cornara L (2019) Ethnobotanical study of medicinal plants used in Central Macedonia, Greece. Evid Based Complement Alternat Med 2019:4513792. https://doi.org/10.1155/2019/4513792

Tütenocaklı T (2014) Yenice (Çanakkale) ve çevresinde tarımsal bitki biyoçeşitliliği ve etnobotanik araştırmalar. PhD Thesis. Çanakkale Onsekiz Mart University, Science Institute, Çanakkale

Tuzlacı E, Doğan A (2010a) Turkish folk medicinal plants, IX: Ovacık (Tunceli). Marmara. Pharm J 14:136–143. https://doi.org/10.12991/201014449

Tuzlacı E, Doğan A (2010b) Turkish folk medicinal plants, IX: Ovacık (Tunceli). Marmara Pharm J 14:136–143. https://doi.org/10.12991/201014449

Tuzlacı E, Eryaşar-Aymaz P (2001) Turkish folk medicinal plants, part IV: Gönen (Balıkesir). Fitoterapia 72(4):323–343. https://doi.org/10.1016/S0367-326X(00)00277-X

Tuzlacı E, Sadıkoğlu E (2007) Turkish folk medicinal plants, part VI: Koçarlı (Aydın). J Fac Pharm 39:25–37

Uçar D, Kanat T, Müslümanoğlu AY, Kayaycı MZ (2020) Coronavirus ve Fitoterapi. Bütünleyici ve Anadolu Tıbbı Dergisi 1(2):49–57

Uteshev BS, Laskova IL, Afanas' ev, V.A. (1999) The immunomodulating activity of the heteropolysaccharides from German chamomile (Matricaria chamomilla) during air and immersion cooling. Eksp Klin Farmakol 62(6):52–55

Utomo RY, Meiyanto E (2020) Revealing the potency of citrus and galangal constituents to halt SARS-CoV-2 infection. Preprints 1:2020030214. https://doi.org/10.20944/preprints202003.0214.v1

Uzun M, Kaya A (2016) Ethnobotanical research of medicinal plants in Mihalgazi (Eskişehir). Turkey Pharm Biol 54(12):2922–2932. https://doi.org/10.1080/13880209.2016.1194863

Uzun SP, Koca C (2020) Ethnobotanical survey of medicinal plants traded in herbal markets of Kahramanmaraş. Plant Divers 42(6):443–454. https://doi.org/10.1016/j.pld.2020.12.003

Vaghasiya J, Datani M, Nandkumar K, Malaviya S, Jivani N (2010) Comparative evaluation of alcoholic and aqueous extracts of *Ocimum sanctum* for immunomodulatory activity. Int J Pharm Biol Res 1(1):25–29

Van Doan H, Hoseinifar SH, Hung TQ, Lumsangkul C, Jaturasitha S, El-Haroun E, Paolucci M (2020) Dietary inclusion of chestnut (*Castanea sativa*) polyphenols to Nile tilapia reared in biofloc technology: impacts on growth, immunity, and disease resistance against *Streptococcus agalactiae*. Fish Shellfish Immunol 105:319–326. https://doi.org/10.1016/j.fsi.2020.07.010

Varga F, Šolić I, Dujaković MJ, Łuczaj Ł, Grdiša M (2019) The first contribution to the ethnobotany of inland Dalmatia: medicinal and wild food plants of the Knin area. Croatia Acta Soc Bot Pol 88(2):3622

Vitasović-Kosić I, Juračak J, Łuczaj Ł (2017) Using Ellenberg-Pignatti values to estimate habitat preferences of wild food and medicinal plants: an example from northeastern Istria (Croatia). J Ethnobiol Ethnomed 13:31. https://doi.org/10.1186/s13002-017-0159-6

Vitiello A, La Porta R, D'Aiuto V, Ferrara F (2021) The risks of liver injury in COVID-19 patients and pharmacological management to reduce or prevent the damage induced. Egypt Liver J 11(1):1–6. https://doi.org/10.1186/s43066-021-00082-y

Wang Y, Fan X, Qu H, Gao X, Cheng Y (2012) Strategies and techniques for multi-component drug design from medicinal herbs and traditional Chinese medicine. Curr Top Med Chem 12(12): 1356–1362. https://doi.org/10.2174/156802612801319034

Wang J, Zhang C, Zhang J, Xie J, Yang L, Xing Y, Li Z (2020) The effects of quercetin on immunity, antioxidant indices, and disease resistance in zebrafish (Danio rerio). Fish Physiol Biochem 46(2):759–770. https://doi.org/10.1007/s10695-019-00750-2

WHO, 2020. Water, sanitation, hygiene and waste management for COVID-19: technical brief; https://apps.who.int/iris/bitstream/handle/10665/331305/WHO-2019-NcOV-IPC_WASH-2020.1-eng.pdf. Accessed 27 Mar 2021

WHO, 2021. Turkey: WHO coronavirus disease (COVID-19) dashboard with vaccination data; https://COVID-19.who.int/region/euro/country/tr. Accessed 10 May 2021

Wufuer R, Bai J, Liu Z, Zhou K, Taoerdahong H (2020) Biological activity of Brassica rapa L. polysaccharides on RAW264. 7 macrophages and on tumor cells. Bioorg Med Chem 28(7): 115330. https://doi.org/10.1016/j.bmc.2020.115330

Yakhkeshi S, Rahimi S, Hemati MH (2012) Effects of yarrow (*Achillea millefolium* L.), antibiotic and probiotic on performance, immune response, serum lipids and microbial population of broilers. J Agric Sci Technol 14(4):799–810

Yeşil Y, Çelik M, Yılmaz B (2019) Wild edible plants in Yeşilli (Mardin-Turkey), a multicultural area. J Ethnobiol Ethnomed 15:52. https://doi.org/10.1186/s13002-019-0327-y

Yeşilada E (2012) İyileştiren bitkiler. Hayykitap, Istanbul

Yeşilada E, Kantarcı G, Öztekin İ, Ortancıl Ö, Mudun A, Zahmacıoğlu O, Öz M, Göktuna Z, Badıllı-Demirbaş Ş, Aklar-Çörekçi A (2019) Geleneksel ve Tamamlayıcı Tıp YÜGETAT Bülteni. Yeditepe Üniversitesi Hastaneleri, Istanbul

Yeşilyurt EB, Şimşek I, Akaydin G, Yeşilada E (2017a) An ethnobotanical survey in selected districts of the black Sea region (Turkey). Turk J Bot 41:47–62. https://doi.org/10.3906/bot-1606-12

Yeşilyurt EB, Şimşek I, Tuncel T, Akaydın G, Yeşilada E (2017b) Plants used as folk medicine in some settlements of the Marmara region. Marmara Pharm J 21:132–148. https://doi.org/10.12991/marupj.259891

Yilmaz S, Ergün S, Yigit M, Yilmaz E, Ahmadifar E (2020) Dietary supplementation of black mulberry (*Morus nigra*) syrup improves the growth performance, innate immune response, antioxidant status, gene expression responses, and disease resistance of Nile tilapia (*Oreochromis niloticus*). Fish Shellfish Immunol 107:211–217. https://doi.org/10.1016/j.fsi.2020.09.041

Yukitake J, Yamamoto Y (2011) Enhancement of cytokine (IL-2, INF-γ) production in Th1 cells by crude protein extract of Equisetum arvense LINNE. J Anal Bio Sci 34(5):339–344

Živković J, Ilić M, Šavikin K, Zdunić G, Ilić A, Stojković D (2020) Traditional use of medicinal plants in South-eastern Serbia (Pčinja District): ethnopharmacological investigation on the current status and comparison with half a century old data. Front Pharmacol 11:1020. https://doi.org/10.3389/fphar.2020.01020

Zlatković BK, Bogosavljević SS, Radivojević AR, Pavlović MA (2014) Traditional use of the native medicinal plant resource of Mt. Rtanj (eastern Serbia): ethnobotanical evaluation and comparison. J Ethnopharmacol 151:704–713. https://doi.org/10.1016/j.jep.2013.11.037

Zou M, Chen Y, Sun-Waterhouse D, Zhang Y, Li F (2018) Immunomodulatory acidic polysaccharides from *Zizyphus jujuba* cv. Huizao: insights into their chemical characteristics and modes of action. Food Chem 258:35–42. https://doi.org/10.1016/j.foodchem.2018.03.052

Žuna-Pfeiffer T, Krstin L, Špoljarić Maronić D, Hmura M, Eržić I, Bek N, Stević F (2020) An ethnobotanical survey of useful wild plants in the North-Eastern part of Croatia (Pannonian region). Pl Biosyst 154(4):463–473. https://doi.org/10.1080/11263504.2019.1635222

# Chapter 19
# A Comparison Study of Medicinal Plants Used Against SARS-CoV-2 and Those Recommended Against Malaria in Africa

**Abdirahman Elmi, Ahmed Said Mohamed, Saredo Said, and Remy Bationo**

## 19.1 Introduction

African medicinal plants have been used for thousands of years, and many of them are well-known for their health benefits and therapeutic values. They have been listed either in complete catalogs mentioning the scientific and local names, the family, the genus, and the traditional use or either through scientific articles presenting the phytochemical and biological aspects of a plant for a given biological activity. African plants are well known not only for their food resources but also for their therapeutic values. When epidemics or pandemics appear, the use of traditional medicine may have a high demand because they are natural and less expensive remedies.

In this chapter, first, the disease with a long history for humankind in Africa was discussed based on the literature of several decades. This is a very dangerous and deadly disease, normally, fewer than 5-year-old children and pregnant women are the most affected groups according to the record. It has been reported that malaria leads to more than 400,000 death per year, 92% in Africa. However, in 2018, compared to the year 2000, nearly 600,000 lives were saved due to the success of investments in fighting this disease. Nevertheless, the parasites can develop resistance to the treatments. Hence, future research needs to explore some more effective treatments and in the meantime should consider accessibility, feasibility, and cost.

A. Elmi (✉) · A. S. Mohamed · S. Said
Centre d'Etudes et de Recherche de Djibouti, Medicinal Research Institute, IRM-CERD, Djibouti City, Djibouti

R. Bationo
Laboratoire de Chimie Organique et de Physique appliquées, Université Joseph KI-ZERBO, Ouaga, Burkina Faso

 
J.-T. Chen (ed.), *Ethnopharmacology and Drug Discovery for COVID-19: Anti-SARS-CoV-2 Agents from Herbal Medicines and Natural Products*,
https://doi.org/10.1007/978-981-99-3664-9_19

In addition to malaria, the global pandemic of coronavirus disease 2019 (COVID-19) arose in 2019 and subsequently in turn caused a serious mortality rate in Africa in the past years. In this scenario, timely vaccine and drug development is a great challenge due to the increasing number of variants found. Interestingly, a symptomatic resemblance of two diseases, i.e., COVID-19 vs. malaria, was found despite the large difference in classification, i.e., viral vs. parasitic. Subsequently, antimalarial agents or antimalarial candidates were evaluated for their potential for combating severe acute respiratory syndrome coronavirus 2 (SARS-CoV-2).

This chapter aims in the first part to summarize the ethnobotanical research works for SARS-CoV-2 in Africa. Then, it focuses on phytochemical analyses and the biological/clinical test of the plants with anti-SARS-CoV-2 effects. Finally, the similarities and differences between COVID-19 and malaria were comprehensively discussed.

## 19.2 Ethnobotany Report in Africa on Anti-Coronavirus Plants

The African continent was greatly affected by COVID-19, which is a deadly disease caused by a new type of coronavirus that is highly contagious. Unfortunately, in 2019, this disease had no treatment in the form of medicine or vaccines. Since the emergence of the SARS-CoV-2 pandemic, the global scientific community has continued to explore a cure to combat the pandemic. In the past years, many laboratories and/or pharmaceutical industries have been looking for existing molecule structures that can respond biologically to this disease or in particular also new plant-based molecules to overcome this pandemic (Llivisaca-Contreras et al. 2021; Costanzo et al. 2020). Herbal medicine research is very important and has a high potential to discover medicines for the management of COVID-19 (Elmi et al. 2020a, 2021; Lim et al. 2021; Jomah et al. 2020). Several studies have been reported and have found promising remedies (Fouedjou et al. 2021; Najem et al. 2022). Ethnobotanical studies against SARS-CoV-2 are generally carried out focusing on plants commonly used in Africa to relieve cough, fever, fatigue, and respiratory problem which are also the main and most common symptoms of COVID-19.

In 2020, Vroh conducted a study of plant diversity in traditional medicine for key symptoms of COVID-19 in sub-Saharan Africa (Vroh 2020). The result of this work identified 99 species use against frequent symptoms of COVID-19 in sub-Saharan African countries (Burkina Faso, Madagascar, Togo, Gabon, Cameroon, Benin, Senegal, Ivory Coast, and South Africa, Mozambique, Ghana, Niger, Mali, Ethiopia, Democratic Republic of Congo, Central African Republic). Fourteen of these species mentioned (*Terminalia macroptera, Terminalia catappa, Citrus limon, Anogeissus leiocarpus, Zea mays, Eucalyptus camaldulensis, Artemisia afra, Calantica cerasifolia, Lantana camara, Ocimum gratissimum, Zingiber officinale, Lippia javanica, Scoparia dulcis,* and *Crossopteryx febrifuga*) are used at the same time for the treatment of two out of the three common symptoms and no species is used for the treatment of the three symptoms at the same time. By priority of use, 5 plants (*Zingiber officinale, Lippia javanica, Ocimum gratissimum, Citrus limon,* and *Artemisia afra*) are very common because of their diverse and common uses in traditional medicine in sub-Saharan Africa.

In the next year, a broader study was carried out by Beressa et al. in 2021, bringing together a total of 36 studies grouping the usefulness of African medicinal plants against viral infections. Among 328 screened plants showing antiviral activities, 10 plants were found to possess anti-SARS-CoV-2 activity. This work proposed a source of certain African medicinal plants, with antiviral effects that could potentially be a remedy against COVID-19 in the future (Beressa et al. 2021).

Adeleye et al. in 2021 have also reported some information focusing on the ethnomedicine identification of plants present in Africa as a means against SARS-CoV-2. This work highlighted 15 plants: *Abrus precatorius* (Iwu 2013), *Achyranthes aspera* (Sharma and Chaudhary 2015), *Allium sativum* (Thuy et al. 2020; Tijani et al. 2019; Asowata-Ayodele et al. 2016), *Annona muricata* (Roger et al. 2015; Coria-Téllez et al. 2018), *Artemisia afra* (More et al. 2012), *Azadirachta indica* (Xuan et al. 2004; Kumar and Navaratnam 2013; Borkotoky and Banerjee 2021), *Cryptolepis sanguinolenta* (Odoh and Akwuaka 2012; Osafo et al. 2017), *Curcuma longa* (Ayati et al. 2019; Taoheed et al. 2017), *Euphorbia hirta* (Jeje TO et al. 2016), *Garcinia kola* (Seanego and Ndip 2012; Ekene and Erhirhie 2014), *Glycyrrhiza glabra* (Murck 2020; El-Saber Batiha et al. 2020), *Moringa oleifera* (Bhattacharya et al. 2018; Shaji 2020), *Nigella sativa* (Koshak and Koshak 2020; Rahman 2020), *Psidium guajava* (Gutiérrez et al. 2008; Laksmiani et al. 2020), *and Zingiber officinale* (Mao et al. 2019; Prasad et al. 2020) from different African countries (Adeleye et al. 2021).

Most of these plants have shown effective or mean activity against SARS-CoV-2 to relieve the intensity of symptoms of this disease. For example, Oyebamiji et al. and Trivedi et al. in 2020 showed that *Annona muricata* seeds can be used as an inhibitor against SARS-CoV-2 (Trivedi et al. 2020; Oyebamiji et al. 2020). Also, *Curcuma longa* is a plant with potential anti-SARS-CoV-2 effects, reported both by Wen *et al.*, in 2007 (Wen et al. 2007) and Zahedipour and Coll, in 2020 (Zahedipour et al. 2020). In addition, an interesting study was also carried out in 2020 by Lin and Ying, showing the good anti-inflammatory activity of curcumin which can be used as a treatment for pneumonia in patients affected by COVID-19 infection (Liu and Ying 2020).

As respiratory perturbations are also one of the major symptoms of SARS-CoV-2. Agbor *et al.* in 2021 selected ethnobotanical scientific articles on respiratory diseases in Africa over the past 10 years (2011–2021). This review has been the subject of 14 scientific articles on different respiratory diseases which is a track to identify future anti-SARS-CoV-2 plants. It has been listed at around 143 plants belonging to 60 families, whose species of the family *Fabaceae* and *Lamiaceae* are the most cited (18 citations each). The plants that have been identified are used for about 15 diseases and/or symptoms related to the respiratory system. Among the 143 plants listed, *Ocimum gratissimum L.* (4 citations), *Entandrophragma cylindricum (Sprague), Scyphocephalium ochocoa Warb, Rubia cordifolia,* and *Allium sativum L.* (3 citations each) are most popular. Based on this review, *Ocimum gratissimum* is more used against respiratory disorders and can be useful for the local management of COVID-19 (Agbor and Ndjib 2021).

Traditional medicine is also practiced in Maghreb countries (North Africa) to treat infectious diseases (Chebaibi et al. 2022). During the pandemic, ethnobotanical studies are carried out to identify natural substances against COVID-19. An ethnobotanical study was carried out in Fez city (Morocco) in 2021 (Benkhaira et al. 2021). Benkhaira *et al.* listed around 49 medicinal plants belonging to 28 botanical families, used against SRAS-CoV-2. Considering PUV (Plant Use Value), there are 4 plants used a lot, namely *Syzygium aromaticum* (0.46), *Thymus vulgaris* (0.46), *Eucalyptus globulus* (0.4), and *Artemisia vulgaris* (0.36) (Trotter and Logan 1986).

In Algeria, a recent ethnobotanical study was likewise carried out in 2022 in the region of Bejaia (Northeast of Algeria). Brahmi *et al.* counted about 23 medicinal plants (*Allium sativum, Allium cepa, Pistacia lentiscus, Pimpinella anisum, Artemisia herba-alba* (Boughendjioua et al. 2022), *Matricaria chamomilla, Lavandula stoechas, Rosmarinus officinalis, Ocimum basilicum, Melissa officinalis, Mentha spicata, Origanum vulgare, Thymus vulgaris, Salvia officinalis, Cinnamomum verum, Eucalyptus globules, Syzygium aromaticum, Olea europaea, Pinus halepensis, Citrus x limon, Citrus sinensis, Aloysia citriodora, and Zingiber officinalis*) belonging to 12 families (Amaryllidaceae, Anacardiaceae, Apiaceae, Asteraceae, Lamiaceae, Lauraceae, Myrtaceae, Oleaceae, Pinaceae, Rutaceae, Verbenaceae, and Zingiberaceae) were listed in the Bejaia localities against COVID-19 infection. This work was done through data analysis using relative citation frequency (RFC), family importance value (FIV), and plant part value (PPV) (Orch et al. 2020). Among these plants, *Aloysia citriodora* (RFC = 0.248), *Mentha spicata* (RFC = 0.145), *Citrus limon* (RFC = 0.135), *Thymus vulgaris* (RFC = 0.118), *Zingiber officinalis* (RFC = 0.09), *Artemisia herba-alba* (RFC = 0.065), and *Eucalyptus Globules* (RFC = 0.063) are used. The leaves of these plants which are used (65%) in the form of infusion (43.6%) are administered orally (95.03%) to treat and relieve certain symptoms of COVID-19 (Brahmi et al. 2022).

Another study was conducted in Algeria, particularly in Seraidi, a region rich in medicinal plants. An ethnobotanical survey was evaluated to meet the health needs, of antiviral plants of interest over the pandemic, which is also a disease of viral origin. This survey listed 32 species (*Origanum vulgare, Thymus capitatus, Lavandula angustifolia, Thymus vulgaris, Artemisia herba-alba, Zingiber officinale, Curcuma longa, Glycyrrhiza glabra, Cinnamomum zeylanicum, Eucalyptus globules, Syzygium aromaticum, Allium sativum, Pistacia lentiscus, Allium cepa, Olea europaea, Verbena officinalis, Citrus sinensis, Citrus limon, Mentha spicata, Cupressus arizonica, Melissa officinalis, Pulmonaria officinalis, Salvia officinalis, Artemisia absinthium, Ziziphus lotus, Nigella sativa, Pinus sylvestris, Myristica fragrans, Urtica dioica, Origanum majorana, Brassica rapa,* and *Arisarum vulgare*) to belong to 20 botanical families, of which the Lamiaceae is the most represented family in terms of use as an antiviral (Salima et al. 2022). The species identified in the Seraidi region were mentioned also in the online survey conducted by Hamdani and Houari in 2020, carried out on 500 people, including 49% suffering from COVID-19, in different regions of northern Algeria (Hamdani and Houari 2020).

Since 2019, the search for a miracle plant or phytoconstituents to help combat this new global SARS-CoV-2 pandemic has continued to grow. The African continent has largely contributed to the search for relevant means through ethnobotanical studies to combat this pandemic using traditional medicinal knowledge and its terrestrial forest wealth (Tegen et al. 2021; Belmouhoub et al. 2021).

Currently, no medicine can treat SARS-CoV-2 except vaccines, and still not effective, due to the virus variance of SARS-CoV-2. The search for plant-derived therapeutics for COVID-19 is necessary for Africa. Therefore, the encryptions of African medicinal plants, used to treat different viral infections are a priority for possible evaluations against SARS-CoV-2.

## 19.3 Phytocompounds from Anti-Coronavirus Plants

Secondary metabolites are molecules that are not directly involved in the growth process of living organisms, unlike primary metabolites (carbohydrates, amines, and lipids). However, their role is not secondary to the survival of the producing organisms. They are produced during the degradation of primary molecules and have various functions, in particular protection against aggressors (bacteria, viruses, etc.). Several categories of specialized metabolites are classically distinguished according to their biochemical nature and their biosynthetic origin. They can be divided into three main families such as polyphenols, nitrogen compounds, and terpene-sterols. The different types of these compounds have been studied in the fight against COVID-19 (Figs. 19.1, 19.2, and 19.3).

### *19.3.1 Polyphenols*

Preliminary research suggests that polyphenols, a type of antioxidant found in many plants, may help protect against the virus that causes COVID-19. In vitro, kaempferol inhibits coronaviruses by blocking the 3a viral channels implicated in viral proliferation (Schwarz et al. 2014). Quercetin, epigallocatechin gallate, and gallocatechin gallate showed very promising effects against SARS-CoV 3CL (pro) with a weak half inhibition concentration of 73, 73, and 47 μM, respectively (Nguyen et al. 2012). A recent study has shown that quercetin can block the normal operations of the SARS-CoV-2 viral proteins inside human cells around 85%. This value rises to 93% if this compound is combined with vitamin D (Glinsky 2020). Other phenolic compounds showed more potent against papain-like protease (PLpro) than against 3-chymotrypsin-like protease (3CLpro). The $IC_{50}$ of fourteen compounds (broussochalcone B, broussochalcone A, 4-hydroxyisolonchocarpin,

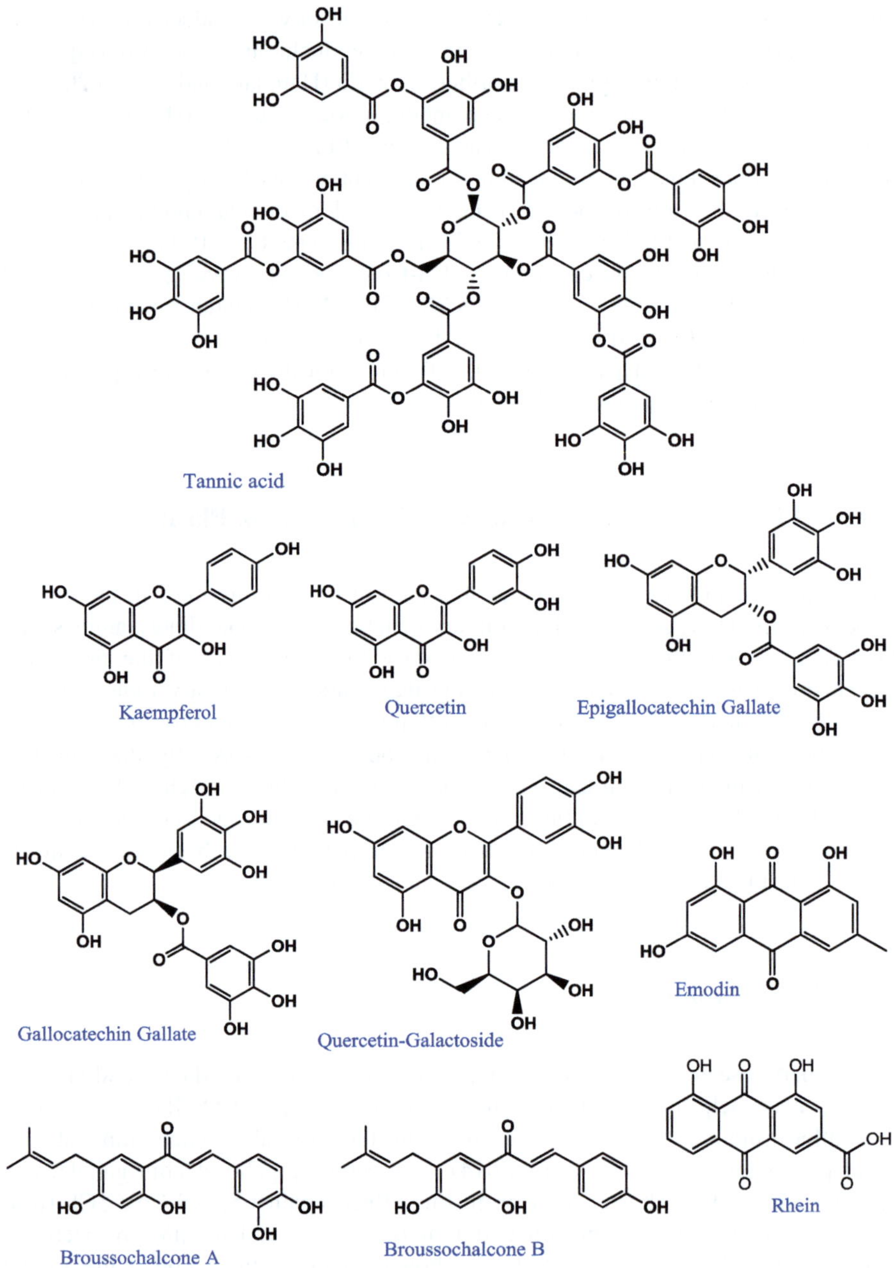


**Fig. 19.1** Structure of polyphenols tested against coronavirus

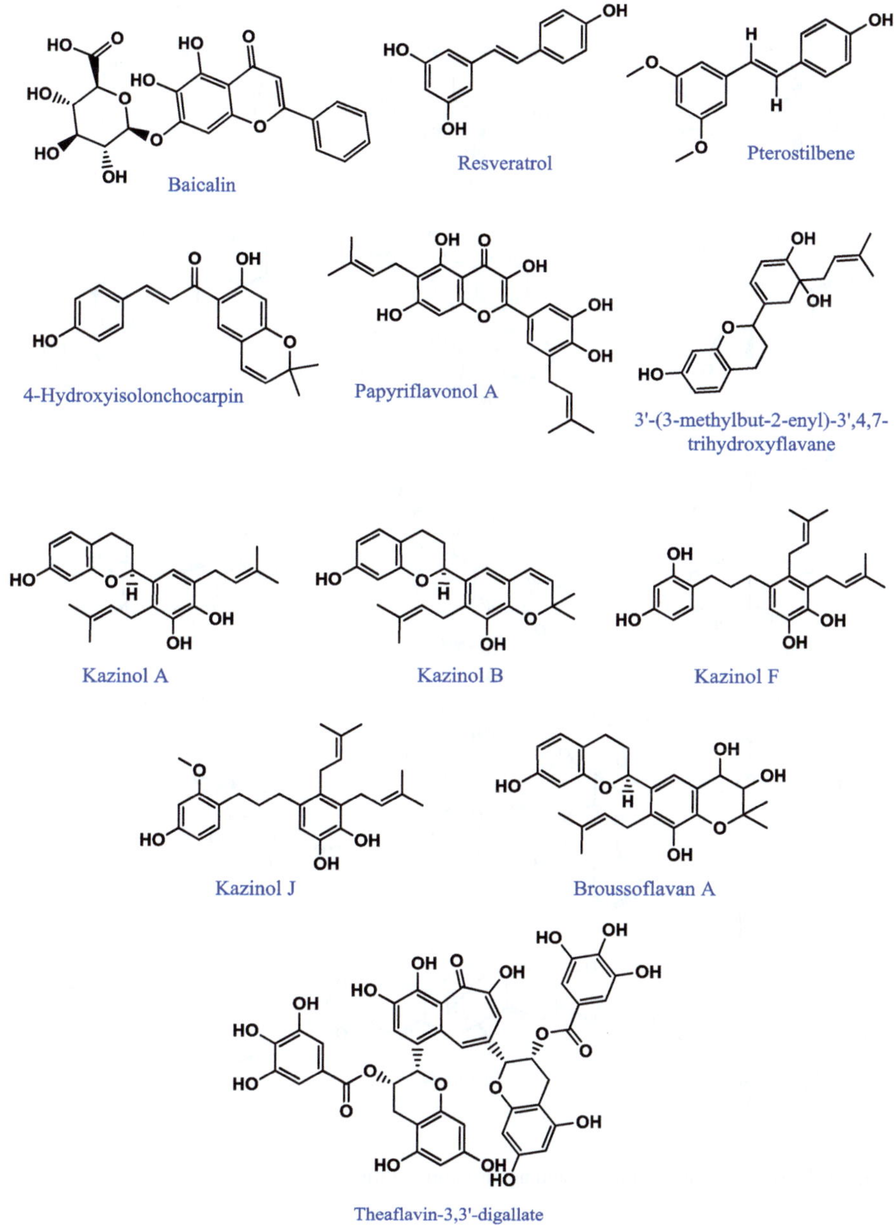


**Fig. 19.1** (continued)

kaempferol, papyriflavonol A, 3′-(3-methylbut-2-enyl)-3′,4,7-trihydroxyflavane, kazinol A, kazinol B, kazinol J, broussoflavan A, kazinol F, quercetin, and quercetin-β-galactoside) ranged from 30.2 to 233.3 μM and 3.7 to 66.2 μM against SARS-CoV 3CLpro and SARS-CoV PLpro, respectively (Park et al. 2017). An in

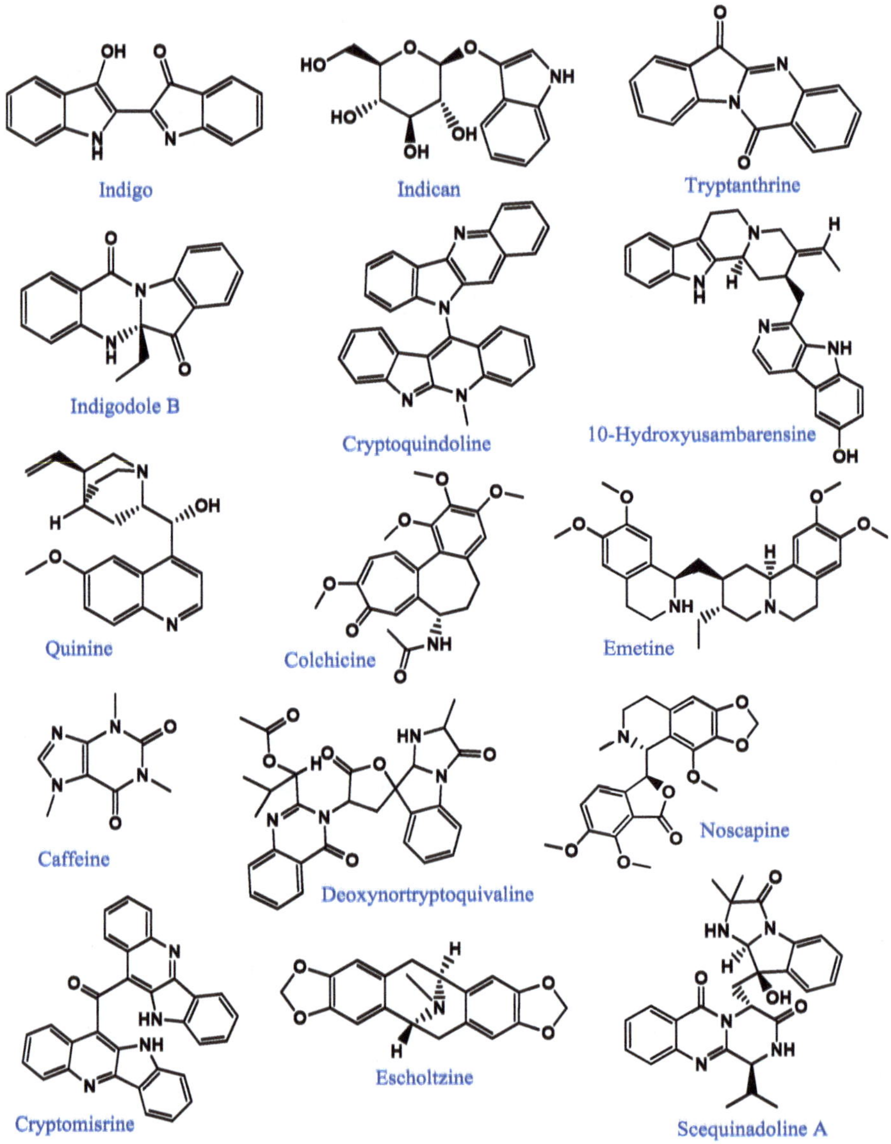


**Fig. 19.2** Structure of alkaloids tested against coronavirus

silico study showed that the flavonoids from Djiboutian medicinal plants (including rutin, catechin, and kaempferol) have better binding energy, between −9.1 and −7.25 kcal/mol, than the reference remdesivir, and hydroxychloroquine with −7.194 and − 5.816 kcal/mol, respectively, at the active site SARS-CoV-2 Mp (Elmi et al. 2020b).

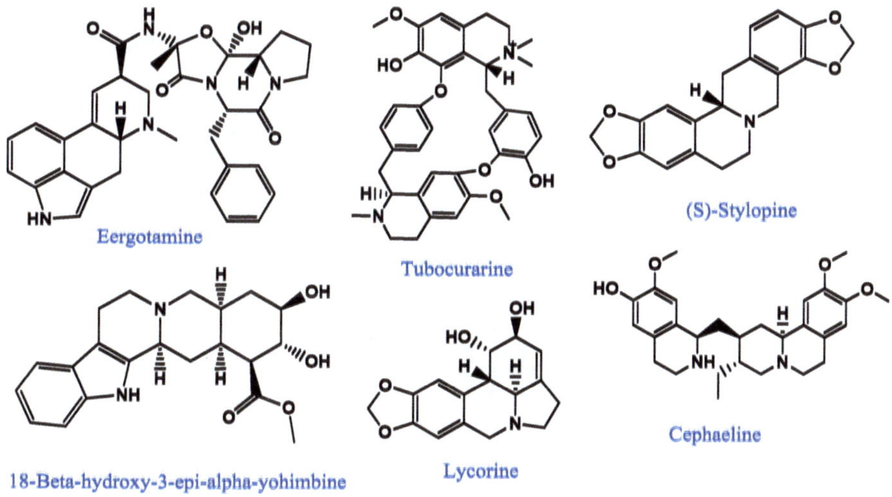


**Fig. 19.2** (continued)

Catechin derivatives interact vigorously with one or both the amino acids, His41 and Cys145, from SARS-CoV-2-Mpro, a key component of this viral reproduction (Ghosh et al. 2021). Two anthraquinones, emodin, and rhein, common to the genre aloe prevent landing S-protein on ACE-II (Ho et al. 2007). Tannic acid, theaflavin-3,3′-digallate, and baicalin were demonstrated as effective viral 3CLPro inhibitors (Chen et al. 2005; Su et al. 2020). Also, resveratrol and pterostilbene are promising antiviral compounds to treat SARS-CoV-2 infection in in vitro tests using African green monkey kidney cells (ter Ellen et al. 2020).

### *19.3.2 Alkaloids*

Indigo and Indicant isolated from *Satis indigotica* root showed efficiency in blocking the cleavage processing of the 3CL$^{pro}$ (Lin et al. 2005) Tryptanthrin and indigodole B (5a *R*-ethyl tryptanthrin) present in *Strobilanthes cusia leaf* exhibited potent antiviral activity against HCoV-NL63 replication, particularly by blocking viral RNA genome synthesis (Tsai et al. 2020). In silico test 10-hydroxyusambarensin and cryptoquindoline, found in different African medicinal plants are capable to inhibit the 3CL$^{pro}$ with a high score to both old and emerging SARS-CoV (Gyebi et al. 2021). Quinine, currently prescribed in Africa for malaria, has a good interaction with the Lys353 residue localized on the ACE2 receptor (Gonzalez et al. 2022), and with to NSP12 target (Sumitha et al. 2020).

The most common target in the case of alkaloids is SARS-CoV-2 Mp, with examples including 10-hydroxyusambarensine, caffeine, emetine, deoxy-nortryptoquivaline, ergotamine ; noscapine, cryptomisrin, escholtzine, 18-hidroxy-3-epi-alpha-yohimbine, (*S*)-stylopine, scequinadoline A and tubocurarine (Gyebi

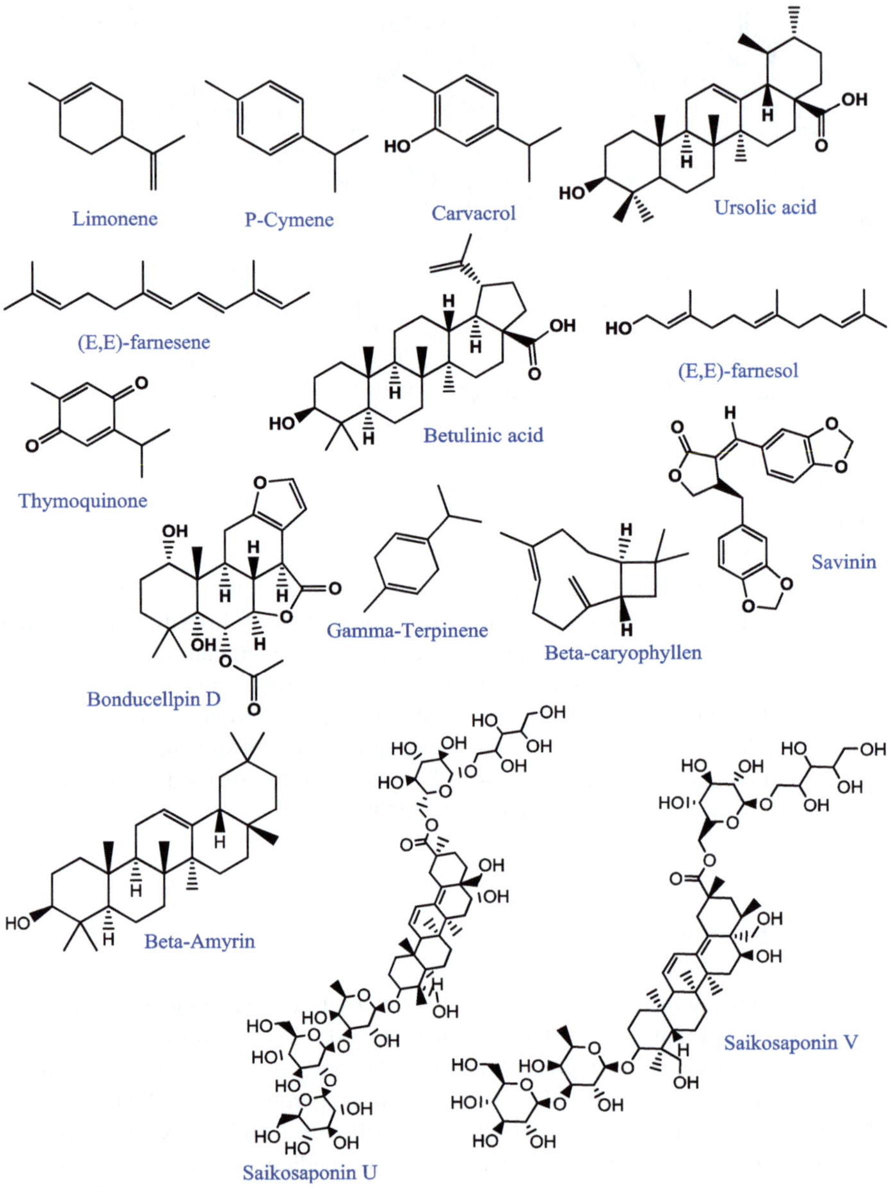


**Fig. 19.3** Structure of alkaloids tested against coronavirus

et al. 2021, 2022; Borquaye et al. 2020; Das et al. 2021; Elzupir 2022; Singh and Florez 2020; Garg and Roy 2020; Gurung et al. 2020; Ismail et al. 2021; Kumar et al. 2022; Qiao et al. 2020; Quimque et al. 2021; Sisakht et al. 2021). In in vitro tests, emetine, lycorine, and cephaeline alkaloids were evaluated as inhibitors of viral protein synthesis to have cross-interactions with human ribosomes on different catalytic sites (Ren et al. 2022). Finally, colchicine is the alkaloid that has had the largest clinical trials in different countries. Indeed, in Greece it is valued for its significantly enhanced time to treatment (Deftereos et al. 2020), in Brazil about the lower need for hospitalization maintenance (Lopes et al. 2021), in Russia, it is observed a brief and statistically significant decrease, and normalization of CRP (C-reactive protein) (Mareev et al. 2021), in Canada, USA, and Italia, a lower incidence of death or hospital admission than placebo (Tardif et al. 2021; Brunetti et al. 2020; Manenti et al. 2021). However, unfortunately, there are no studies in Africa till present.

### *19.3.3 Terpenoids*

Limonene, *P*-cymene, and γ-terpinene extracted from *Ammoides verticillata* (Desf.) Briq., collected in western Algeria, have a good inhibition on the ACE 2 receptor of SARS-CoV-2 (Abdelli et al. 2021). Different terpenic compounds reported significant interaction on sites or receptors of SRAS-CoV such as betulinic acid and savinin (Wen et al. 2007), saikosaponin U and saikosaponin V (Sinha et al. 2021), carvacrol, ursolic acid (Sampangi-Ramaiah et al. 2020), bonducellpin D (Gurung et al. 2020), β-amyrin (Borkotoky and Banerjee 2021), β-caryophyllene (Narkhede et al. 2020), thymoquinone (Bouchentouf and Missoum 2020), (*E, E*)-farnesol and (*E, E*)-α/β-farnesene (da Silva et al. 2020).

## 19.4 Bioassay and Clinical Evidence

In Africa where the use of plants is very present, a rapid recourse to these resources has been observed. In emergencies, plants that have been used for thousands of years can be taken without worrying about their toxicity as opposed to newly synthesized molecules. However, due to poor scientific infrastructure, there are few confirmatory bioassays or clinical studies. Because of the similarity of the symptoms between COVID-19 and malaria, *Artemisia* sp. has been one of the most investigated plants in plant care. Extracts of this plant (aqueous and alcoholic) as well as its main molecule (artemisinin) and its derivatives (artesunate and artemether) were tested in vitro against SARS-CoV-2. The originality of this study is to test the effectiveness of these preparations or pure compounds in pre-treatment (pt) and treatment (t) after infection (Zhou et al. 2021). In both cases (pt and t), the alcoholic extract has an effective concentration of about 50% smaller than the aqueous extract at 173 μg/mL (pt)/142 μg/mL (t) and 390 μg/mL (pt)/260 μg/mL (t), respectively. Overall, artesunate had the best activity with $EC_{50}$ of 12 μg/mL (pt) and 7 μg/mL (t). The latter molecule is a known antimalarial drug in pharmacy and this reinforces the possible link between malaria and COVID-19.

Another example is “covid organics,” a preparation of *artemisia annua* in hot water promoted by the Malagasy government and distributed in several African countries. Its use was debated at the beginning of the epidemic due to the lack of tangible evidence of its efficacy, and the WHO was late in agreeing to evaluate it. Since then multiple studies have been carried out. An in vitro evaluation shows an inhibitory action of this drink with an effective half concentration ($EC_{50}$) of 7.73% of raw drink and with a selectivity index of 5.28 (Nie et al. 2021). Some clinical studies of natural products and their derivatives against COVID-19 are summarized in Table 19.1.

A team of Malagasy researchers with technical support from WHO conducted an efficacy and safety evaluation of artemisinin and 1,8-cineole capsule for COVID-19 patients (Rakotosaona et al. 2022). The study was a phase 3 randomized, double-blind, placebo-controlled trial involving more than 1500 patients. The median time to recovery was 14 and 21 days with the capsule or placebo treatment, respectively. According to this clinical study, none of the patients treated with CVO + C developed the severe form of this disease, and all the vital functions of their organism are preserved (kidney and liver, among others). A similar study of artemisinin combined with piperaquine showed a faster clearance of SARS-CoV-2 than control patients (Li et al. 2021). Also, different studies were performed in which pharmaceutical preparations containing artemisia plant, artemisinin, and/or its derivatives about the virological clearance (NCT04502342, ChiCTR2000033049, and PACTR202103601407640), shorter time to clinical improvement (NCT04382040, NCT04374019, and CTRI/2020/09/028044), and to avoid hospitalizations and oxygen use (NCT04530617).

Other plants and essential oils that have targeted the symptoms (NCT04501965, NCT04980573, NCT04981314, and NCT04705753) and sanitary deterioration (NCT04400890 and NCT04851821) were studied.

In Nigeria, a new compound, harpagide 5-O-β-D-glucopyranoside from *Clerodendrum volubile* leaf has a promising ability to suppress T-cell proliferation (Erukainure et al. 2022). The latter is involved in the activation of cytokine release syndrome, and studies have indicated that targeting T-cell proliferation may prevent the severe form of COVID-19 (Zhou et al. 2020a).

Three extracts of Egyptian medicinal plants (*Psiadia punctulata, Aframomum melegueta,* and *Nigella sativa*) and their isolated compounds achieved high inhibition rates between 31.5–66.4% and 63.21–73.8%, respectively, for the extracts and compounds against the major proteases of SARS-CoV-2 (Abdallah et al. 2022).

In Madagascar, (−)-6-epi-artemisinin isolated from *Saldinia proboscidea, a typical source, showed interesting* in vitro *activity against SARS-CoV-2* (Randrianarivo et al. 2021).

**Table 19.1** Some clinical studies of natural products and their derivatives against COVID-19

| Trial code | Preparations | Participants | Outcomes measure | Status |
|---|---|---|---|---|
| NCT04382040 | Artemisinin+ curcumin+ frankincense (=Boswellia) | 50 | Time to clinical improvement | Completed |
| NCT04502342 | Artesunate +amodiaquine | 30 | Virological clearance | Unknown |
| NCT04374019 | artesunate + Artemisia annua | 13 | Clinical deterioration | Terminated |
| NCT04530617 | Artemisia annua | 246 | Rate of hospitalizations and oxygen use | Terminated |
| CTRI/2020/09/028044 | Artemisinin | 120 | Accelerated the recovery of patients | Terminated |
| PACTR202103601407640 | Artemisinin and 1,8-cineole | 276 | With complete clearance of the SARS-CoV-2 virus at day 28 | Terminated |
| ChiCTR2000033049 | Artemisinin + piperaquine | 41 | Shorten the time to reach undetectable SARS-CoV-2 | Terminated |
| NCT04501965 | Quinquina-stevia/ azithromycin | 231 | Virologic clearance and COVID-19 symptoms development | Unknown |
| NCT04400890 | Resveratrol/ vitamin D3 | 100 | Hospitalization rates for COVID-19 | Terminated |
| NCT04980573 | Citrus essential oil | 47 | Fatigue symptom | Terminated |
| NCT04981314 | Echinacea purpurea | 230 | Fever | Recruiting |
| NCT04851821 | Quercetin | 80 | Avoid emergency/ disappearance of symptoms | Completed |
| NCT04705753 | Essential oil mixture | 20 | Number of patients with symptom resolution in 14 days max | Completed |
| NCT04997395 | Cbd and thc | 30 | Long covid symptoms (pain, fatigue, anxiety, oxygen saturation, etc.) attenuation | Recruitment |
| NCT03944447 | Medical cannabis | 200.000 | COVID-19 infection rates in cannabis users and symptoms attenuation | Recruitment |

(continued)

**Table 19.1** (continued)

| Trial code | Preparations | Participants | Outcomes measure | Status |
|---|---|---|---|---|
| NCT05228626 | Zanthoxylum and Warburgia sp | 584 | Clinical recovery in 8 days | Not yet recruiting |

## 19.5 Malaria and COVID-19: Common Features and Their Differences

Malaria is a parasitic disease involving the genus *Plasmodium*. It is also transmitted and spread to humans by the bite of an infected female mosquito of the species Anopheles (Talapko et al. 2019). In 2020, the WHO annual report estimated the number of deaths at 627,000 in the world, the majority in Africa (WHO 2021). The emerging disease, COVID-19, was caused by SARS-CoV-2 and coronavirus disease 2019 (COVID-19), which is a serious threat to global health (Fisher and Heymann 2020). The COVID-19 outbreak started in Wuhan, a city in China's Hubei province, at the end of December 2019 and since it has spread rapidly all over the world and caused an outbreak (Sharma et al. 2021). Then WHO declared the COVID-19 outbreak a pandemic (World Health Organization 2020). Although malaria and COVID-19 may have several similar performances, clinical diagnosis of both diseases is based on patient symptoms and physical examination findings (Fig. 19.4).

The responsible agents of these diseases are viruses (COVID-19) and parasites (malaria), respectively, living organisms that are found all around us. The rapid spread of SARS-CoV-2 throughout the world is due, among other things, to its mode of transmission (direct human-to-human), whereas the parasite responsible for malaria needs an intermediary, the mosquito. The pendency of the COVID has accelerated the search for preventive and curative vaccines. Currently, about ten vaccines have been licensed by the WHO or state agencies, while the first malaria vaccine is currently only used in a limited number of countries for children (Bejon et al. 2008; Liu and Ye 2022).

### 19.5.1 *The Role of ACE2 in Malaria and COVID-19*

SARS-CoV-2 interacts with angiotensin-converting enzyme 2 (ACE2) as a receptor for entry into human cells, resulting in the downregulation of ACE2 in infected individuals (Yan et al. 2020; Sommerstein et al. 2020). ACE2 is an enzyme that is present on the cell surface of various organs in the body (heart, lungs, kidneys, etc.) (Zhou et al. 2020b). The amount of ACE2 and its action is crucial in identifying the severity of COVID-19. ACE2 plays a major role in the renin-angiotensin system (RAS) which antagonizes the action of angiotensin-converting enzyme (ACE) by converting angiotensin II to angiotensin 1–7. Its counteraction regulates blood

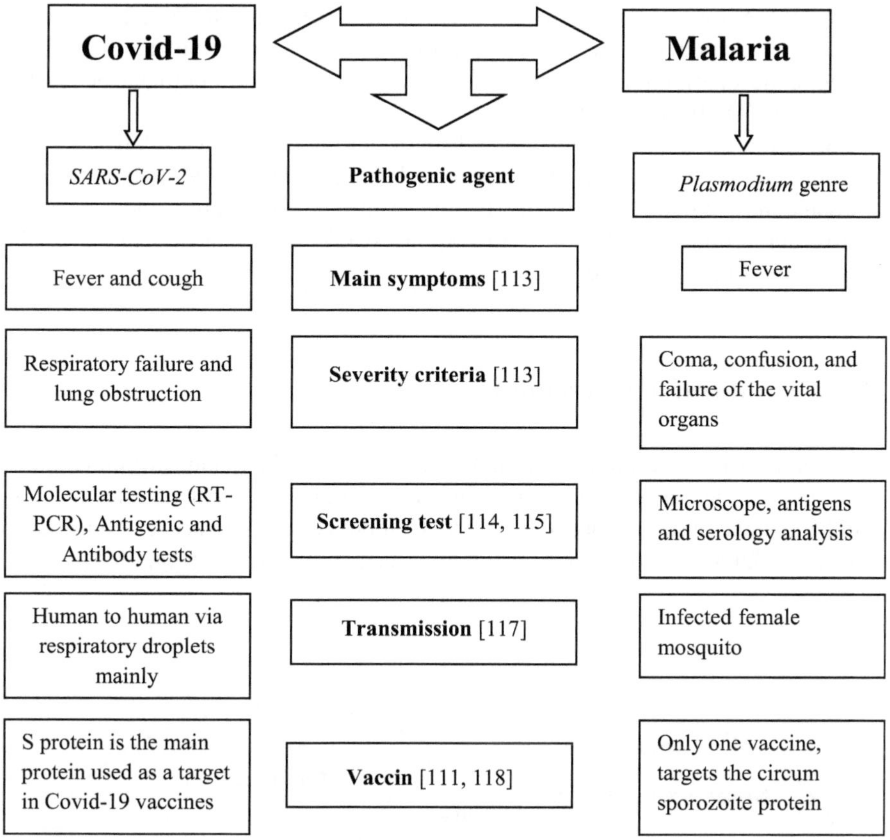


**Fig. 19.4** Description of a comparison between COVID-19 and malaria

pressure by dilating blood vessels in a healthy adult (Stewart et al. 2008; Donoghue et al. 2000).

Besides some genetic variants of ACE2 are characterized by genetic deletion/insertion polymorphism (D/I). This characteristic is a good indicator of the severity of COVID-19 (Gemmati and Tisato 2020). Indeed, this polymorphism is related to an increase in Ang II concentration (Gemmati and Tisato 2020; Gemmati et al. 2020) and consequently leads to a negative disease evolution (Verma et al. 2021). In contrast, in malaria, high Ang II is beneficial because it has a parasiticidal effect and allows normal blood flow (Saraiva et al. 2011).

### *19.5.2 Impact the Country of the Two Diseases*

The worldwide distribution of malaria and COVID-19 infections shows an inverse impact. Indeed, in highly malaria-endemic countries, including the African continent, the impact (contagion and death) is low compared to Western countries that have eradicated malaria. Different hypotheses have been formulated, such as the demographic distribution of a predominantly young African continent or the low logistics of COVID-19 detection.

But within the continent, there is a disparity in the impact of COVID-19 which has done more damage in the North and South, while the malaria-endemic sub-Saharan region is relatively unscathed. Therefore the role of hydroxychloroquine (HCQ) and chloroquine (CQ) used against malaria in this region should be considered.

### *19.5.3 Natural Therapeutics*

In this non-exhaustive literature search, we have listed 31 botanical families most used against malaria and COVID-19 in Africa. Eight botanical families are both popular against these two diseases, namely Combretaceae, Asteraceae, Annonaceae, Rutaceae, Meliaceae, Euphorbiaceae, Lamiaceae, and Fabaceae (Fig. 19.5).

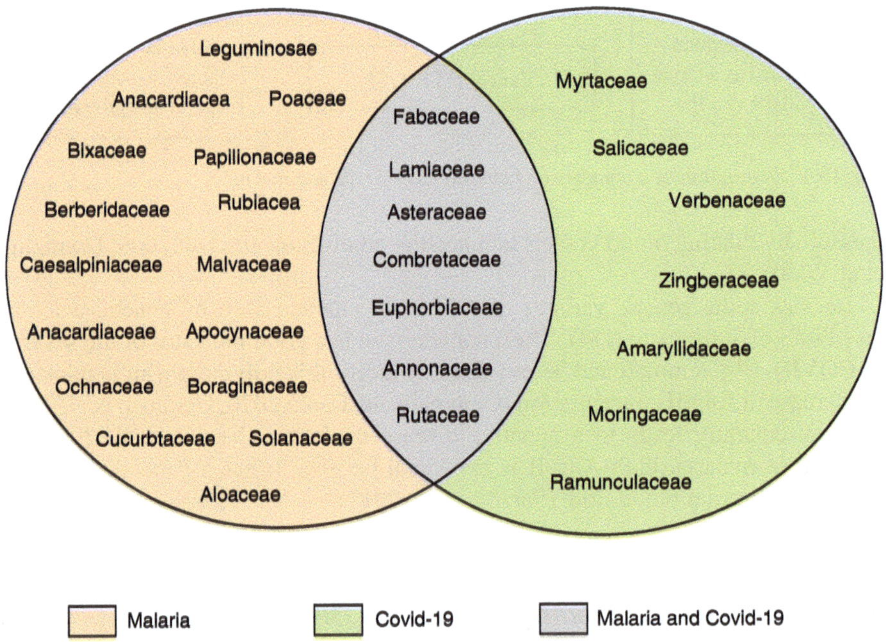


**Fig. 19.5** Antimalarial major botanic family and anti-covid major botanic family plants (Asase et al. 2005; Diarra et al. 2015; Dike et al. 2012; Manya et al. 2020; Manuel et al. 2020; Opio et al. 2018; Oladeji et al. 2020; Yetein et al. 2013; Odugbemi et al. 2007)

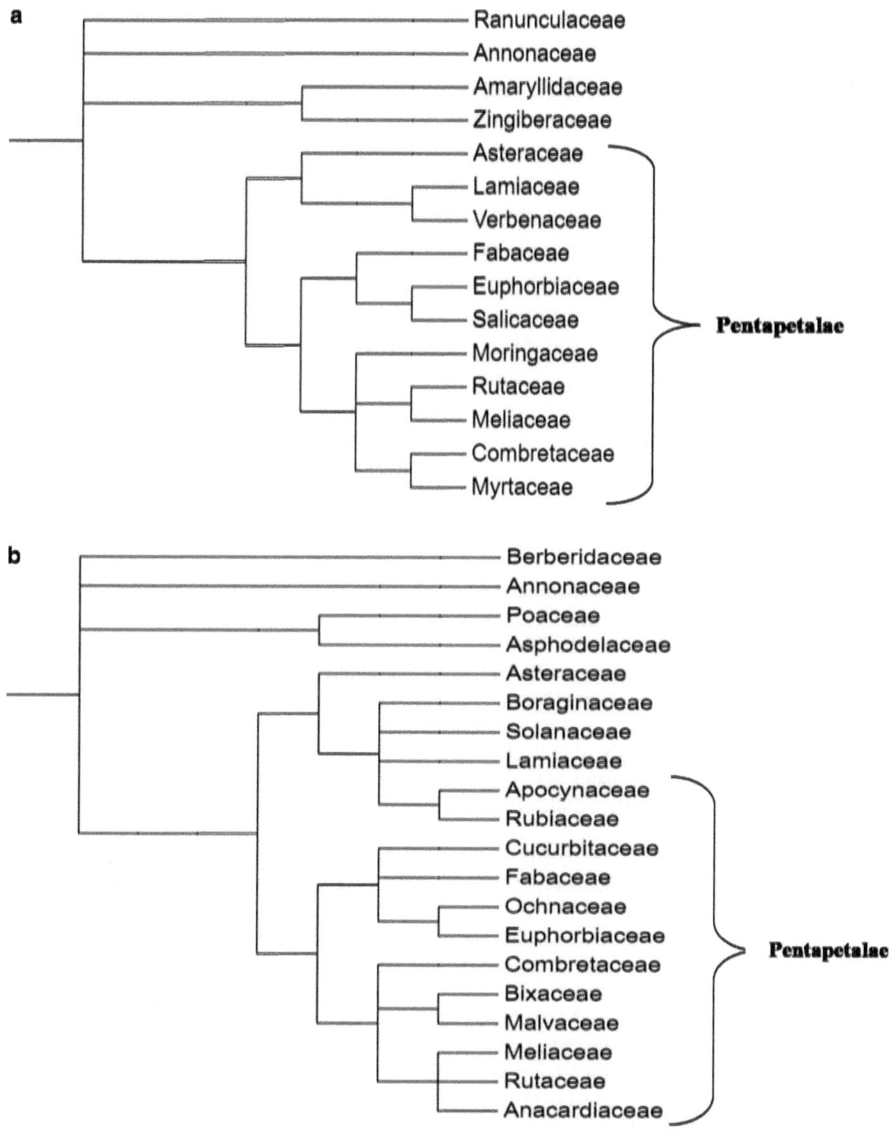


**Fig. 19.6** Phylogenic organization (done by PhyloT software) of the plant families potentially used against COVID-19 (**a**) and malaria (**b**) in Africa (made with phyloT Sofware (Letunic and Bork 2021))

Most of the families listed belong to the Pentapetalae group (Fig. 19.6). The latter is the most diverse group of extant flowering plants and has a characteristic type of flower made up of whorls of five pieces each (Chanderbali et al. 2017).

## 19.6 Conclusion

SARS-CoV-2 has largely affected Africa even though the impact has been smaller than on other continents. Because of the high use of medicinal plants in Africa, various studies have been interested in identifying them and documenting their efficacy on patients infected by the coronavirus. The plants frequently cited are those known to be used against major symptoms of the disease such as fever, cough, and difficulty in breathing, among others. The strong similarity of symptoms with malaria has directed the search for remedies for plants and antimalarial drugs. The identification of active compounds against this new virus is rare as the targeting process usually takes a long time. However, apart from treating the symptoms, there is a high demand to explore drugs that combat protein targets of this virus.

## References


Abdallah HM, El-Halawany AM, Darwish KM et al (2022) Bio-guided isolation of SARS-CoV-2 main protease inhibitors from medicinal plants: in vitro assay and molecular dynamics. Plants (Basel) 11:1914. https://doi.org/10.3390/plants11151914

Abdelli I, Hassani F, Bekkel Brikci S, Ghalem S (2021) In silico study the inhibition of angiotensin converting enzyme 2 receptor of COVID-19 by Ammoides verticillata components harvested from Western Algeria. J Biomol Struct Dyn 39:3263–3276. https://doi.org/10.1080/07391102.2020.1763199

Adeleye OA, Femi-Oyewo MN, Bamiro OA et al (2021) Ethnomedicinal herbs in African traditional medicine with potential activity for the prevention, treatment, and management of coronavirus disease 2019. Future J Pharm Sci 7:72. https://doi.org/10.1186/s43094-021-00223-5

Agbor GA, Ndjib R (2021) Systematic review of plants used against respiratory diseases related to COVID-19 in Africa. J Drug Deliv Ther 11:141–153. https://doi.org/10.22270/jddt.v11i4-S.4957

Asase A, Oteng-Yeboah AA, Odamtten GT, Simmonds MSJ (2005) Ethnobotanical study of some Ghanaian anti-malarial plants. J Ethnopharmacol 99:273–279. https://doi.org/10.1016/j.jep.2005.02.020

Asowata-Ayodele AM, Afolayan AJ, Otunola GA (2016) Ethnobotanical survey of culinary herbs and spices used in the traditional medicinal system of Nkonkobe municipality, eastern cape, South Africa. S Afr J Bot 104:69–75. https://doi.org/10.1016/j.sajb.2016.01.001

Ayati Z, Ramezani M, Amiri MS et al (2019) Ethnobotany, phytochemistry and traditional uses of curcuma spp. and pharmacological profile of two important species (C. longa and C. zedoaria): a review. Curr Pharm Des 25:871–935. https://doi.org/10.2174/1381612825666190402163940

Bejon P, Lusingu J, Olotu A et al (2008) Efficacy of RTS,S/AS01E vaccine against malaria in children 5 to 17 months of age. N Engl J Med 359:2521–2532. https://doi.org/10.1056/NEJMoa0807381

Belmouhoub M, Aberkane B, Bey MB (2021) Ethnopharmacological survey on medicinal plants used by Algerian population to prevent SARS-CoV-2 infection. Ethnobot Res Appl 22:1–13

Benkhaira N, Koraichi SI, Fikri-Benbrahim K (2021) Ethnobotanical survey on plants used by traditional healers to fight against COVID-19 in fez city, Northern Morocco. Ethnobot Res Appl 21:1–18

Beressa TB, Deyno S, Mtewa AG et al (2021) Potential benefits of antiviral African medicinal plants in the management of viral infections: systematic review. Front Pharmacol 12:682794

Bhattacharya A, Tiwari P, Sahu PK, Kumar S (2018) A review of the phytochemical and pharmacological characteristics of Moringa oleifera. J Pharm Bioallied Sci 10:181. https://doi.org/10.4103/JPBS.JPBS_126_18

Borkotoky S, Banerjee M (2021) A computational prediction of SARS-CoV-2 structural protein inhibitors from Azadirachta indica (neem). J Biomol Struct Dyn 39:4111–4121. https://doi.org/10.1080/07391102.2020.1774419

Borquaye LS, Gasu EN, Ampomah GB et al (2020) Alkaloids from *Cryptolepis sanguinolenta* as potential inhibitors of SARS-CoV-2 viral proteins: an *in silico* study. Biomed Res Int 2020: e5324560. https://doi.org/10.1155/2020/5324560

Bouchentouf S, Missoum N (2020) Identification of compounds from nigella sativa as new potential inhibitors of 2019 novel Coronavirus (COVID-19): molecular docking study. Chemistry

Boughendjioua H, Aiouaz S, Medkour L, et al (2022) Chemical composition (a new chemotype) of Algerian Mugwort (Artemisia Herba-Alba Asso) essential oil and his possible potential against COVID-19

Brahmi F, Iblhoulen Y, Issaadi H et al (2022) Ethnobotanical survey of medicinal plants of Bejaia localities from Algeria to prevent and treat coronavirus (COVID-19) infection shortened title: phytomedicine to manage COVID-19 pandemic. Adv Tradit Med (ADTM). https://doi.org/10.1007/s13596-022-00649-z

Brunetti L, Diawara O, Tsai A et al (2020) Colchicine to weather the cytokine storm in hospitalized patients with COVID-19. J Clin Med 9:2961. https://doi.org/10.3390/jcm9092961

Chanderbali AS, Berger BA, Howarth DG et al (2017) Evolution of floral diversity: genomics, genes and gamma. Philos Trans R Soc B Biol Sci 372:20150509. https://doi.org/10.1098/rstb.2015.0509

Chebaibi M, Bousta D, Bourhia M et al (2022) Ethnobotanical study of medicinal plants used against COVID-19. Evid Based Complement Alternat Med 2022:e2085297. https://doi.org/10.1155/2022/2085297

Chen C-N, Lin CPC, Huang K-K et al (2005) (NaN/NaN/NaN) inhibition of SARS-CoV 3C-like protease activity by theaflavin-3,3′-digallate (TF3). Evid Based Complement Alternat Med 2: 209–215. https://doi.org/10.1093/ecam/neh081

Coria-Téllez AV, Montalvo-Gónzalez E, Yahia EM, Obledo-Vázquez EN (2018) Annona muricata: a comprehensive review on its traditional medicinal uses, phytochemicals, pharmacological activities, mechanisms of action and toxicity. Arab J Chem 11:662–691. https://doi.org/10.1016/j.arabjc.2016.01.004

Costanzo M, Giglio MARD, Roviello GN (2020) SARS-CoV-2: recent reports on antiviral therapies based on lopinavir/ritonavir, darunavir/umifenovir, hydroxychloroquine, remdesivir, favipiravir and other drugs for the treatment of the new coronavirus. Curr Med Chem 27: 4536–4541. https://doi.org/10.2174/0929867327666200416131117

da Silva JKR, Figueiredo PLB, Byler KG, Setzer WN (2020) Essential oils as antiviral agents, potential of essential oils to treat SARS-CoV-2 infection: an in-silico investigation. Int J Mol Sci 21:3426. https://doi.org/10.3390/ijms21103426

Das S, Sarmah S, Lyndem S, Singha Roy A (2021) An investigation into the identification of potential inhibitors of SARS-CoV-2 main protease using molecular docking study. J Biomol Struct Dyn 39:3347–3357. https://doi.org/10.1080/07391102.2020.1763201

Deftereos SG, Giannopoulos G, Vrachatis DA et al (2020) Effect of colchicine vs standard care on cardiac and inflammatory biomarkers and clinical outcomes in patients hospitalized with coronavirus disease 2019: the GRECCO-19 randomized clinical trial. JAMA Netw Open 3: e2013136. https://doi.org/10.1001/jamanetworkopen.2020.13136

Diarra N, van't Klooster C, Togola A et al (2015) Ethnobotanical study of plants used against malaria in Sélingué subdistrict, Mali. J Ethnopharmacol 166:352–360. https://doi.org/10.1016/j.jep.2015.02.054

Dike IP, Obembe OO, Adebiyi FE (2012) Ethnobotanical survey for potential anti-malarial plants in South-Western Nigeria. J Ethnopharmacol 144:618–626. https://doi.org/10.1016/j.jep.2012.10.002

Donoghue M, Hsieh F, Baronas E et al (2000) A novel angiotensin-converting enzyme–related carboxypeptidase (ACE2) converts angiotensin I to angiotensin 1–9. Circ Res 87:e1–e9. https://doi.org/10.1161/01.RES.87.5.e1

Ekene ENE, Erhirhie EO (2014) Garcinia kola: a review of its ethnomedicinal, chemical and pharmacological properties. Int J Cur Res Rev 6:1–7

Elmi A, Sayem SA-J, Ahmed M, Abdoul-Latif F (2020a) Natural compounds from Djiboutian medicinal plants as inhibitors of covid-19 by in silico investigations. Int J Curr Pharm Res 1:52–57. https://doi.org/10.22159/ijcpr.2020v12i4.39051

Elmi A, Sayem SA-J, Ahmed M, Abdoul-Latif F (2020b) Natural compounds from Djiboutian medicinal plants as inhibitors of COVID-19 by in silico investigations. Int J Curr Pharm Sci 1: 52–57. https://doi.org/10.22159/ijcpr.2020v12i4.39051

Elmi A, Mohamed AS, Siddiqui N et al (2021) Identification of potential inhibitors of SARS-CoV-2 from Artemisia annua compounds by in silico evaluation and their density functional theory (DFT). J Drug Deliv Ther 11:71–82. https://doi.org/10.22270/jddt.v11i1-s.4702

El-Saber Batiha G, Magdy Beshbishy A, El-Mleeh A et al (2020) Traditional uses, bioactive chemical constituents, and pharmacological and toxicological activities of Glycyrrhiza glabra L. (Fabaceae). Biomolecules 10:352. https://doi.org/10.3390/biom10030352

Elzupir AO (2022) Caffeine and caffeine-containing pharmaceuticals as promising inhibitors for 3-chymotrypsin-like protease of SARS-CoV-2. J Biomol Struct Dyn 40:2113–2120. https://doi.org/10.1080/07391102.2020.1835732

Erukainure OL, Atolani O, Muhammad A et al (2022) Targeting the initiation and termination codons of SARS-CoV-2 spike protein as possible therapy against COVID-19: the role of novel harpagide 5-O-β-D-glucopyranoside from *Clerodendrum volubile* P Beauv. (Labiatae). J Biomol Struct Dyn 40:2475–2488. https://doi.org/10.1080/07391102.2020.1840439

Fisher D, Heymann D (2020) Q&A: the novel coronavirus outbreak causing COVID-19. BMC Med 18:57. https://doi.org/10.1186/s12916-020-01533-w

Fouedjou RT, Chtita S, Bakhouch M et al (2021) Cameroonian medicinal plants as potential candidates of SARS-CoV-2 inhibitors. J Biomol Struct Dyn 1:1–15. https://doi.org/10.1080/07391102.2021.1914170

Garg S, Roy A (2020) In silico analysis of selected alkaloids against main protease (Mpro) of SARS-CoV-2. Chem Biol Interact 332:109309. https://doi.org/10.1016/j.cbi.2020.109309

Gemmati D, Tisato V (2020) Genetic hypothesis and pharmacogenetics side of renin-angiotensin-system in COVID-19. Genes 11:1044. https://doi.org/10.3390/genes11091044

Gemmati D, Bramanti B, Serino ML et al (2020) COVID-19 and individual genetic susceptibility/receptivity: role of ACE1/ACE2 genes, immunity, inflammation and coagulation. Might the double X-chromosome in females be protective against SARS-CoV-2 compared to the single X-chromosome in males? Int J Mol Sci 21:3474. https://doi.org/10.3390/ijms21103474

Ghosh R, Chakraborty A, Biswas A, Chowdhuri S (2021) Evaluation of green tea polyphenols as novel corona virus (SARS CoV-2) main protease (Mpro) inhibitors—an in silico docking and molecular dynamics simulation study. J Biomol Struct Dyn 39:4362–4374. https://doi.org/10.1080/07391102.2020.1779818

Glinsky GV (2020) Tripartite combination of candidate pandemic mitigation agents: vitamin D, quercetin, and Estradiol manifest properties of medicinal agents for targeted mitigation of the COVID-19 pandemic defined by genomics-guided tracing of SARS-CoV-2 targets in human cells. Biomedicine 8:129. https://doi.org/10.3390/biomedicines8050129

Gonzalez BL, de Oliveira NC, Ritter MR et al (2022) The naturally-derived alkaloids as a potential treatment for COVID-19: a scoping review. Phytother Res 36:2686–2709. https://doi.org/10.1002/ptr.7442

Gurung AB, Ali MA, Lee J et al (2020) Unravelling lead antiviral phytochemicals for the inhibition of SARS-CoV-2 Mpro enzyme through in silico approach. Life Sci 255:117831. https://doi.org/10.1016/j.lfs.2020.117831

Gutiérrez RMP, Mitchell S, Solis RV (2008) Psidium guajava: a review of its traditional uses, phytochemistry and pharmacology. J Ethnopharmacol 117:1–27. https://doi.org/10.1016/j.jep.2008.01.025

Gyebi GA, Ogunro OB, Adegunloye AP et al (2021) Potential inhibitors of coronavirus 3-chymotrypsin-like protease (3CLpro): an in silico screening of alkaloids and terpenoids from African medicinal plants. J Biomol Struct Dyn 39:3396–3408. https://doi.org/10.1080/07391102.2020.1764868

Gyebi GA, Adegunloye AP, Ibrahim IM et al (2022) Prevention of SARS-CoV-2 cell entry: insight from in silico interaction of drug-like alkaloids with spike glycoprotein, human ACE2, and TMPRSS2. J Biomol Struct Dyn 40:2121–2145. https://doi.org/10.1080/07391102.2020.1835726

Hamdani FZ, Houari N (2020) Phytothérapie et COVID-19. Une étude fondée sur une enquête dans le nord de l'Algérie. Phytothérapie 18:248–254. https://doi.org/10.3166/phyto-2020-0241

Ho T-Y, Wu S-L, Chen J-C et al (2007) Emodin blocks the SARS coronavirus spike protein and angiotensin-converting enzyme 2 interaction. Antivir Res 74:92–101. https://doi.org/10.1016/j.antiviral.2006.04.014

Ismail EMOA, Shantier SW, Mohammed MS et al (2021) Quinoline and quinazoline alkaloids against COVID-19: an *in silico* multitarget approach. J Chem 2021:e3613268. https://doi.org/10.1155/2021/3613268

Iwu MM (2013) Handbook of African medicinal plants, 2nd edn. CRC Press, Boca Raton

Jeje TO, Ibraheem O, Brai BI, Ibukun EO (2016) Pharmacological potential of asthma weed ( Euphorbia hirta ) extract toward eradication of plasmodium berghei in infected albino. Int J Toxicol Pharmacol Res 8:130–137

Jomah S, Asdaq SMB, Al-Yamani MJ (2020) Clinical efficacy of antivirals against novel coronavirus (COVID-19): a review. J Infect Public Health 13:1187–1195. https://doi.org/10.1016/j.jiph.2020.07.013

Koshak DAE, Koshak PEA (2020) Nigella sativa L as a potential phytotherapy for coronavirus disease 2019: a mini review of in silico studies. Curr Ther Res 93:100602. https://doi.org/10.1016/j.curtheres.2020.100602

Kumar VS, Navaratnam V (2013) Neem (Azadirachta indica): prehistory to contemporary medicinal uses to humankind. Asian Pac J Trop Biomed 3:505–514. https://doi.org/10.1016/S2221-1691(13)60105-7

Kumar N, Awasthi A, Kumari A et al (2022) Antitussive noscapine and antiviral drug conjugates as arsenal against COVID-19: a comprehensive chemoinformatics analysis. J Biomol Struct Dyn 40:101–116. https://doi.org/10.1080/07391102.2020.1808072

Laksmiani NPL, Larasanty LPF, Santika AAGJ et al (2020) Active compounds activity from the medicinal plants against SARS-CoV-2 using in silico assay. Biomed Pharm J 13:873–881

Letunic I, Bork P (2021) Interactive tree of life (iTOL) v5: an online tool for phylogenetic tree display and annotation. Nucleic Acids Res 49:W293–W296. https://doi.org/10.1093/nar/gkab301

Li G, Yuan M, Li H et al (2021) Safety and efficacy of artemisinin-piperaquine for treatment of COVID-19: an open-label, non-randomised and controlled trial. Int J Antimicrob Agents 57:106216. https://doi.org/10.1016/j.ijantimicag.2020.106216

Lim XY, Teh BP, Tan TYC (2021) Medicinal plants in COVID-19: potential and limitations. Front Pharmacol 12:611408

Lin C-W, Tsai F-J, Tsai C-H et al (2005) Anti-SARS coronavirus 3C-like protease effects of Isatis indigotica root and plant-derived phenolic compounds. Antivir Res 68:36–42. https://doi.org/10.1016/j.antiviral.2005.07.002

Liu Y, Ye Q (2022) Safety and efficacy of the common vaccines against COVID-19. Vaccine 10:513. https://doi.org/10.3390/vaccines10040513

Liu Z, Ying Y (2020) The inhibitory effect of curcumin on virus-induced cytokine storm and its potential use in the associated severe pneumonia. Front Cell Dev Biol 8:479

Llivisaca-Contreras SA, Naranjo-Morán J, Pino-Acosta A et al (2021) Plants and natural products with activity against various types of coronaviruses: a review with focus on SARS-CoV-2. Molecules 26:4099. https://doi.org/10.3390/molecules26134099

Lopes MI, Bonjorno LP, Giannini MC et al (2021) Beneficial effects of colchicine for moderate to severe COVID-19: a randomised, double-blinded, placebo-controlled clinical trial. RMD Open 7:e001455. https://doi.org/10.1136/rmdopen-2020-001455

Manenti L, Maggiore U, Fiaccadori E et al (2021) Reduced mortality in COVID-19 patients treated with colchicine: results from a retrospective, observational study. PLoS One 16:e0248276. https://doi.org/10.1371/journal.pone.0248276

Manuel L, Bechel A, Noormahomed EV et al (2020) Ethnobotanical study of plants used by the traditional healers to treat malaria in Mogovolas district, Northern Mozambique. Heliyon 6: e05746. https://doi.org/10.1016/j.heliyon.2020.e05746

Manya MH, Keymeulen F, Ngezahayo J et al (2020) Antimalarial herbal remedies of Bukavu and Uvira areas in DR Congo: an ethnobotanical survey. J Ethnopharmacol 249:112422. https://doi.org/10.1016/j.jep.2019.112422

Mao Q-Q, Xu X-Y, Cao S-Y et al (2019) Bioactive compounds and bioactivities of ginger (Zingiber officinale roscoe). Foods 8:185. https://doi.org/10.3390/foods8060185

Mareev VY, Orlova YA, Plisyk AG et al (2021) Proactive anti-inflammatory therapy with colchicine in the treatment of advanced stages of new coronavirus infection. The first results of the COLORIT study. Kardiologiia 61:15–27. https://doi.org/10.18087/cardio.2021.2.n1560

More G, Lall N, Hussein A, Tshikalange TE (2012) Antimicrobial constituents of Artemisia afra Jacq. Ex Willd. Against periodontal pathogens. Evid Based Complement Alternat Med 2012: e252758. https://doi.org/10.1155/2012/252758

Murck H (2020) Symptomatic protective action of glycyrrhizin (Licorice) in COVID-19 infection? Front Immunol 11:1239

Najem M, Ibijbijen J, Nassiri L (2022) Phytotherapy in response to COVID-19 and risks of intoxication: a field study in the city of Meknes (Morocco). J Pharm Pharm Res 10:357–386. https://doi.org/10.56499/jppres21.1257_10.3.357

Narkhede RR, Pise AV, Cheke RS, Shinde SD (2020) Recognition of natural products as potential inhibitors of COVID-19 Main protease (Mpro): in-silico evidences. Nat Prod Bioprospect 10: 297–306. https://doi.org/10.1007/s13659-020-00253-1

Nguyen TTH, Woo H-J, Kang H-K et al (2012) Flavonoid-mediated inhibition of SARS coronavirus 3C-like protease expressed in Pichia pastoris. Biotechnol Lett 34:831–838. https://doi.org/10.1007/s10529-011-0845-8

Nie C, Trimpert J, Moon S et al (2021) In vitro efficacy of Artemisia extracts against SARS-CoV-2. Virol J 18:182. https://doi.org/10.1186/s12985-021-01651-8

Odoh UE, Akwuaka CI (2012) Pharmacognostic profile of root of Cryptolepis sanguinolenta (lindl.) Schlechter. Pharm J 4:40–44. https://doi.org/10.5530/pj.2012.28.8

Odugbemi TO, Akinsulire OR, Aibinu IE, Fabeku PO (2007) Medicinal plants useful for malaria therapy in Okeigbo, Ondo state, Southwest Nigeria. Afr J Tradit Complement Altern Med 4: 191–198

Oladeji OS, Oluyori AP, Bankole DT, Afolabi TY (2020) Natural products as sources of antimalarial drugs: ethnobotanical and ethnopharmacological studies. Scientifica 2020:1–22. https://doi.org/10.1155/2020/7076139

Opio D, Andama E, Kureh G (2018) Ethnobotanical survey of antimalarial plants in areas of: Abukamola, Angeta, Oculokori and Omarari of Alebtong District in Northern Uganda. EJMP 21:1–14. https://doi.org/10.9734/EJMP/2017/38043

Orch H, Benkhnigue O, Zidane L, Douira A (2020) Treatment of urinary treatment of urolithiasis: ethnobotanical study of plants used by the population bordering the forest of IzarèneLithiasis: ethnobotanical study of plants used by the population bordering the forest of Izarène. Ethnobot Res Appl 19:1–15

Osafo N, Mensah KB, Yeboah OK (2017) Phytochemical and pharmacological review of Cryptolepis sanguinolenta (Lindl.) Schlechter. Adv Pharmacol Pharmacol Sci 2017:e3026370. https://doi.org/10.1155/2017/3026370

Oyebamiji AK, Oladipo EK, Olotu TM et al (2020) In-vitro investigation on selected compounds in Annona Muricata seed: a potential SARS-CoV nsp12 polymerase inhibitors down regulating 2019-nCoV. Int J Trad Nat Med 10:13–25

Park J-Y, Yuk HJ, Ryu HW et al (2017) Evaluation of polyphenols from Broussonetia papyrifera as coronavirus protease inhibitors. J Enzyme Inhib Med Chem 32:504–512. https://doi.org/10.1080/14756366.2016.1265519

Prasad A, Muthamilarasan M, Prasad M (2020) Synergistic antiviral effects against SARS-CoV-2 by plant-based molecules. Plant Cell Rep 39:1109–1114. https://doi.org/10.1007/s00299-020-02560-w

Qiao Z, Zhang H, Ji H-F, Chen Q (2020) Computational view toward the inhibition of SARS-CoV-2 spike glycoprotein and the 3CL protease. Computation 8:53. https://doi.org/10.3390/computation8020053

Quimque MTJ, Notarte KIR, Fernandez RAT et al (2021) Virtual screening-driven drug discovery of SARS-CoV2 enzyme inhibitors targeting viral attachment, replication, post-translational modification and host immunity evasion infection mechanisms. J Biomol Struct Dyn 39: 4316–4333. https://doi.org/10.1080/07391102.2020.1776639

Rahman MT (2020) Potential benefits of combination of Nigella sativa and Zn supplements to treat COVID-19. J Herb Med 23:100382. https://doi.org/10.1016/j.hermed.2020.100382

Rakotosaona R, Mioramalal SA et al (2022) Efficacy and safety of CVO PLUS CURATIF capsules, Malagasy improved traditional medication for treating COVID-19: a randomized, double- blind, placebo-controlled trial. Arch Clin Biomed Res 06:1. https://doi.org/10.26502/acbr.50170295

Randrianarivo S, Rasolohery C, Rafanomezantsoa S et al (2021) (−)-6-epi-artemisinin, a natural stereoisomer of (+)-artemisinin in the opposite enantiomeric series, from the endemic Madagascar plant Saldinia proboscidea, an atypical source. Molecules 26:5540. https://doi.org/10.3390/molecules26185540

Ren P, Shang W, Yin W et al (2022) A multi-targeting drug design strategy for identifying potent anti-SARS-CoV-2 inhibitors. Acta Pharmacol Sin 43:483–493. https://doi.org/10.1038/s41401-021-00668-7

Roger T, Pierre-Marie M, Igor VK, Patrick VD (2015) Phytochemical screening and antibacterial activity of medicinal plants used to treat typhoid fever in Bamboutos division, West Cameroon. J App Pharm Sci 5:034–049. https://doi.org/10.7324/JAPS.2015.50606

Salima D, Radia D, Nesrine H, Karima M (2022) Ethnobotanical study on the use of medicinal plants with antiviral interest, case of sars-cov-19, in the region of Seraidi (Annaba, north-east Algeria). Plant Archives 22:184–192. https://doi.org/10.51470/PLANTARCHIVES.2022.v22.no2.033

Sampangi-Ramaiah MH, Vishwakarma R, Shaanker RU (2020) Molecular docking analysis of selected natural products from plants for inhibition of SARS-CoV-2 main protease. Curr Sci 118:1087

Saraiva VB, de Silva LS, Ferreira-DaSilva CT et al (2011) Impairment of the plasmodium falciparum erythrocytic cycle induced by angiotensin peptides. PLoS One 6:e17174. https://doi.org/10.1371/journal.pone.0017174

Schwarz S, Sauter D, Wang K et al (2014) Kaempferol derivatives as antiviral drugs against the 3a channel protein of coronavirus. Planta Med 80:177–182. https://doi.org/10.1055/s-0033-1360277

Seanego CT, Ndip RN (2012) Identification and antibacterial evaluation of bioactive compounds from Garcinia kola (Heckel) seeds. Molecules 17:6569–6584. https://doi.org/10.3390/molecules17066569

Shaji DD (2020) Computational identification of drug Lead compounds for COVID-19 from moringa oleifera. https://doi.org/10.26434/chemrxiv.12535913.v1

Sharma V, Chaudhary U (2015) An overview on indigenous knowledge of Achyranthes aspera. J Crit Rev 2:13. https://doi.org/10.31838/jcr.02.01.03

Sharma A, Ahmad Farouk I, Lal SK (2021) COVID-19: a review on the novel coronavirus disease evolution, transmission, detection, control and prevention. Viruses 13:202. https://doi.org/10.3390/v13020202

Singh S, Florez H (2020) Bioinformatic study to discover natural molecules with activity against COVID-19. F1000Res 9:1203. https://doi.org/10.12688/f1000research.26731.1

Sinha SK, Shakya A, Prasad SK et al (2021) An in-silico evaluation of different Saikosaponins for their potency against SARS-CoV-2 using NSP15 and fusion spike glycoprotein as targets. J Biomol Struct Dyn 39:3244–3255. https://doi.org/10.1080/07391102.2020.1762741

Sisakht M, Mahmoodzadeh A, Darabian M (2021) Plant-derived chemicals as potential inhibitors of SARS-CoV-2 main protease (6LU7), a virtual screening study. Phytother Res 35:3262–3274. https://doi.org/10.1002/ptr.7041

Sommerstein R, Kochen MM, Messerli FH, Gräni C (2020) Coronavirus disease 2019 (COVID-19): do angiotensin-converting enzyme inhibitors/angiotensin receptor blockers have a biphasic effect? J Am Heart Assoc 9:e016509. https://doi.org/10.1161/JAHA.120.016509

Stewart JA, Lazartigues E, Lucchesi PA (2008) The angiotensin converting enzyme 2/ang-(1-7) axis in the heart. Circ Res 103:1197–1199. https://doi.org/10.1161/CIRCRESAHA.108.189068

Su H, Yao S, Zhao W et al (2020) Discovery of baicalin and baicalein as novel, natural product inhibitors of SARS-CoV-2 3CL protease *in vitro*. Mol Biol

Sumitha A, Brindha Devi P, Hari S, Dhanasekaran R (2020) COVID-19—In silico structure prediction and molecular docking studies with doxycycline and quinine. Biomed Pharmacol J 13:1185–1193. https://doi.org/10.13005/bpj/1986

Talapko J, Škrlec I, Alebić T et al (2019) Malaria: the past and the present. Microorganisms 7:179. https://doi.org/10.3390/microorganisms7060179

Taoheed AA, Tolulope AA, Saidu AB et al (2017) Phytochemical Properties, proximate and mineral composition of Curcuma longa Linn. And Zingiber officinale Rosc.: a comparative study. J Sci Res Rep 13:1–7. https://doi.org/10.9734/JSRR/2017/32623

Tardif J-C, Bouabdallaoui N, L'Allier PL et al (2021) Colchicine for community-treated patients with COVID-19 (COLCORONA): a phase 3, randomised, double-blinded, adaptive, placebo-controlled, multicentre trial. Lancet Respir Med 9:924–932. https://doi.org/10.1016/S2213-2600(21)00222-8

Tegen D, Dessie K, Damtie D (2021) Candidate anti-COVID-19 medicinal plants from Ethiopia: a review of plants traditionally used to treat viral diseases. Evid Based Complement Alternat Med 2021:e6622410. https://doi.org/10.1155/2021/6622410

ter Ellen BM, Kumar ND, Bouma EM et al (2020) Resveratrol and pterostilbene potently inhibit SARS-CoV-2 replication in vitro. Microbiology

Thuy BTP, My TTA, Hai NTT et al (2020) Investigation into SARS-CoV-2 resistance of compounds in garlic essential oil. ACS Omega 5:8312–8320. https://doi.org/10.1021/acsomega.0c00772

Tijani KB, Alfa AA, Sezor AA (2019) Studies on phytochemical, nutraceutical profiles and potential medicinal values of Allium sativum Linn (Liliaceae) on bacterial meningitis. Int Neuropsych Dis J 13:1–15. https://doi.org/10.9734/indj/2019/v13i230105

Trivedi GN, Karlekar JT, Dhameliya HA, Panchal H (2020) A review on the novel coronavirus disease based on in-silico analysis of various drugs and target proteins. J Pure Appl Microbiol 14:849–860. https://doi.org/10.22207/JPAM.14.SPL1.22

Trotter RT, Logan MH (1986) Informant consensus: a new approach for identifying potentially effective medicinal plants. In: Plants and Indigenous medicine and diet, 1st edn. Routledge, New York, NY, p 22

Tsai Y-C, Lee C-L, Yen H-R et al (2020) Antiviral action of tryptanthrin isolated from Strobilanthes cusia leaf against human coronavirus NL63. Biomol Ther 10:366. https://doi.org/10.3390/biom10030366

Verma S, Abbas M, Verma S et al (2021) Impact of I/D polymorphism of angiotensin-converting enzyme 1 (ACE1) gene on the severity of COVID-19 patients. Infect Genet Evol 91:104801. https://doi.org/10.1016/j.meegid.2021.104801

Vroh BTA (2020) Diversity of plants used in traditional medicine against the main symptoms of COVID-19 in sub-Saharan Africa: review of the literature. Ethnobot Res Appl 20:1–14

Wen C-C, Kuo Y-H, Jan J-T et al (2007) Specific plant terpenoids and lignoids possess potent antiviral activities against severe acute respiratory syndrome coronavirus. J Med Chem 50: 4087–4095. https://doi.org/10.1021/jm070295s

WHO (2021) World malaria report 2021. World Health Organization, Geneva

World Health Organization (2020) Coronavirus disease 2019 (COVID-19): situation report, 51. World Health Organization, Geneva

Xuan TD, Tsuzuki E, Hiroyuki T et al (2004) Evaluation on phytotoxicity of neem (Azadirachta indica. A. Juss) to crops and weeds. Crop Prot 23:335–345. https://doi.org/10.1016/j.cropro.2003.09.004

Yan R, Zhang Y, Li Y et al (2020) Structural basis for the recognition of SARS-CoV-2 by full-length human ACE2. Science 367:1444–1448. https://doi.org/10.1126/science.abb2762

Yetein MH, Houessou LG, Lougbégnon TO et al (2013) Ethnobotanical study of medicinal plants used for the treatment of malaria in plateau of Allada, Benin (West Africa). J Ethnopharmacol 146:154–163. https://doi.org/10.1016/j.jep.2012.12.022

Zahedipour F, Hosseini SA, Sathyapalan T et al (2020) Potential effects of curcumin in the treatment of COVID-19 infection. Phytother Res 34:2911–2920. https://doi.org/10.1002/ptr.6738

Zhou Y, Hou Y, Shen J et al (2020a) Network-based drug repurposing for novel coronavirus 2019-nCoV/SARS-CoV-2. Cell Discov 6:1–18. https://doi.org/10.1038/s41421-020-0153-3

Zhou P, Yang X-L, Wang X-G et al (2020b) A pneumonia outbreak associated with a new coronavirus of probable bat origin. Nature 579:270–273. https://doi.org/10.1038/s41586-020-2012-7

Zhou Y, Gilmore K, Ramirez S et al (2021) In vitro efficacy of artemisinin-based treatments against SARS-CoV-2. Sci Rep 11:14571. https://doi.org/10.1038/s41598-021-93361-y

# Chapter 20
# Exploring the Potential Antiviral Properties of *Nigella sativa* L. Against SARS-CoV-2: Mechanisms and Prospects

**Mohammed Dalli, Salah-eddine Azizi, Ali Azghar, Abderrazak Saddari, Elmostapha Benaissa, Yassine Ben Lahlou, Mostafa Elouennass, and Adil Maleb**

M. Dalli (✉)
Laboratory of Microbiology, Faculty of Medicine and Pharmacy, University Mohammed First, Oujda, Morocco

Laboratory of Bioresources, Biotechnology, Ethnopharmacology and Health, Faculty of Sciences, University Mohammed the First, Oujda, Morocco
e-mail: m.dalli@ump.ac.ma

S.-e. Azizi
Laboratory of Microbiology, Faculty of Medicine and Pharmacy, University Mohammed First, Oujda, Morocco

A. Azghar · A. Maleb
Laboratory of Microbiology, Faculty of Medicine and Pharmacy, University Mohammed First, Oujda, Morocco

Laboratory of Bioresources, Biotechnology, Ethnopharmacology and Health, Faculty of Sciences, University Mohammed the First, Oujda, Morocco

Laboratory of Microbiology, Mohammed VI University Hospital, Oujda, Morocco

A. Saddari
Laboratory of Microbiology, Faculty of Medicine and Pharmacy, University Mohammed First, Oujda, Morocco

Laboratory of Microbiology, Mohammed VI University Hospital, Oujda, Morocco

E. Benaissa · Y. Ben Lahlou · M. Elouennass
Department of Bacteriology, Mohammed V Teaching Military Hospital, Rabat, Morocco

Epidemiology and Bacterial Resistance Research Team/BIO-INOVA Center, Faculty of Medicine and Pharmacy, University Mohammed V, Rabat, Morocco



J.-T. Chen (ed.), *Ethnopharmacology and Drug Discovery for COVID-19: Anti-SARS-CoV-2 Agents from Herbal Medicines and Natural Products*,
https://doi.org/10.1007/978-981-99-3664-9_20

## 20.1 Introduction

The utilization of medicinal and aromatic plants is widespread among low-income communities, particularly in developing countries, as they are easily accessible and affordable (UNIDO 2006). Black cumin or *Nigella sativa* L., a plant belonging to the Ranunculaceae family, has been widely used to treat a variety of medical conditions throughout the world. Moroccan locals claim to have used this specific plant to treat a number of ailments, including cardiovascular disorders, stomachache, queasy, and arthritis (Fakchich and Elachouri 2021). *Nigella sativa* has been shown to possess a diverse range of properties, including antioxidant (Burits and Bucar 2000; Dalli et al. 2021a, b, c), antimicrobial (Dalli et al. 2021a, b, c), anti-inflammatory (Ali 2015), immunomodulatory (Salem 2005), antidiabetic (Dalli et al. 2021a, b, c, 2022), and low toxicity (Kehili et al. 2018). The existence of various bioactive substances, such as phenolic acids, flavonoids, alkaloids, and terpenoids, has been noted in *N. sativa* through several phytochemical investigations (Bourgou et al. 2010; Malhotra 2012).

Coronaviruses (CoV), a group of viruses, have been discovered to be capable of infecting livestock and inducing a range of disorders in the intestines, respiratory system, and other parts of the body, varying in severity. These viruses are responsible for a variety of illnesses in diverse wildlife species such as birds, bats, mice, giraffes, and whales. In humans, they cause both mild ailments like the common cold and more severe respiratory infections like pneumonia, resulting in substantial economic losses (Chan et al. 2013; Hasoksuz et al. 2007). Four of the seven coronavirus strains that can infect humans (HCoV OC43, HCoV 229E, HCoV HKU1, and HCoV NL63) typically cause moderate cold symptoms in persons with healthy immune systems. MERS-CoV, SARS-CoV, and SARS-CoV-2 are the three remaining coronaviruses, are all of zoonotic origin, and can all result in serious respiratory illnesses and even death. Chinese authorities informed the WHO about cases of pneumonia in China on December 31, 2019, which were later determined to be SARS-CoV-2 virus (Hasöksüz et al. 2020). The illnesses quickly spread over the world in March 2020, and the WHO proclaimed COVID a new pandemic (Salzberger et al. 2021). Similar to SARS-CoV and MERS, bats are found to be the natural reservoir for SARS-CoV-2 (Song et al. 2019); however, the difference lies in the intermediary, which is pangolins for MERS and civet cats for SARS-CoV (Prompetchara et al. 2020). Furthermore, 2.34% is the rate mortality of SARS-CoV-2, which is lower than the rates for SARS-CoV and MERS, which are 9.56% and 35.37%, respectively (Ganesh et al. 2021; Xie and Chen 2020). SARS-CoV-2 typically replicates in the upper and lower respiratory passages. Droplets and aerosols from infected asymptomatic and symptomatic patients are the main means of transmission with an incubation period of 5.7 days (Salzberger et al. 2021).

The ongoing COVID-19 pandemic has shed light on the importance of medicinal and aromatic plants as a tank of biomolecules that could help in the treatment and prevention of viral infections by SARS-CoV-2 which brings back to the surface the exploration of *N. sativa* and its derivates that are known to be endowed with high antiviral properties.

## 20.2 Epidemiology of COVID-19

The WHO has reported roughly 756,581,850 confirmed cases of COVID-19, including 6,844,267 deaths, according to recent numbers released on February 17th, 2023. While calculating the number of persons who have received vaccinations, it was discovered that 13,195,832,385 vaccine doses have been given globally. Corresponding to recent WHO statistics, Europe registered the highest number of confirmed cases, with a total of 272,634,146 cases since the start of the pandemic, with France as an example reporting 38,475,606 cases, followed by Western Pacific countries with 200,792,487 confirmed cases, then North and South America with a cumulative total of 189,648,697 cases, dominated by the USA with 101,496,168 confirmed cases. However, Southeast Asia has a total of 60,762,712 cases, followed by the Eastern Mediterranean countries with 23,251,360 confirmed cases. Finally, the African continent has the lowest number of confirmed cases, with 9,491,684 recorded. Regarding the number of deaths, the North and South America region has the highest number so far, with 2,923,918 deaths, followed by Europe with 2,189,540 deaths, the Southeast Asia region with 803,789 deaths, the Western Pacific countries with 402,289 deaths, then the Eastern Mediterranean countries with 349,434 deaths, and finally the African continent with 175,284 deaths. In Morocco, from 3 January 2020 to February 17, 2023, the number of COVID-19 recorded has reached more than 1,272,353 cases since the start of the pandemic with 16,296 deaths, while the total administered vaccine doses on the Moroccan territory were 55,379,860 (WHO 2023).

## 20.3 Molecular Basis of SARS-CoV-2 Pathogenesis

The new coronavirus, sometimes referred to as SARS-CoV-2, is a highly contagious virus that is causing the continuing COVID-19 epidemic worldwide. The virus belongs to the Coronaviridae family and exactly to the Betacoronavirus genus and is characterized by positive-sense single-stranded RNA (Zhu et al. 2020). A 30 kilobase genome makes up the SARS-CoV-2 and encodes 27 proteins, including the spike protein (S), which has 1273 amino acids and is composed of three similar chains, the envelope protein (E), the membrane protein (M), and the nucleocapsid protein (N) (Bailey-Elkin et al. 2017; Seyed Hosseini et al. 2020). Also, it was indicated the presence of hemagglutinin-esterase on the virus surface, which could help with virus entry (Boopathi et al. 2020). The S protein mediates viral entry into host cells by binding to the host cell receptor ACE2, which is expressed on the surface of human cells, particularly in the respiratory tract (Alnefaie and Albogami 2020; Boopathi et al. 2020) (Fig. 20.1). After SARS-CoV-2 enters host cells, the viral RNA genome is released into the cytoplasm, where it serves as a template for the translation of viral proteins. The viral RNA-dependent RNA polymerase (RdRp) and other viral enzymes also replicate and transcript the viral RNA genome in the cytoplasm. The replicated and transcribed viral RNA can be packaged into new virus

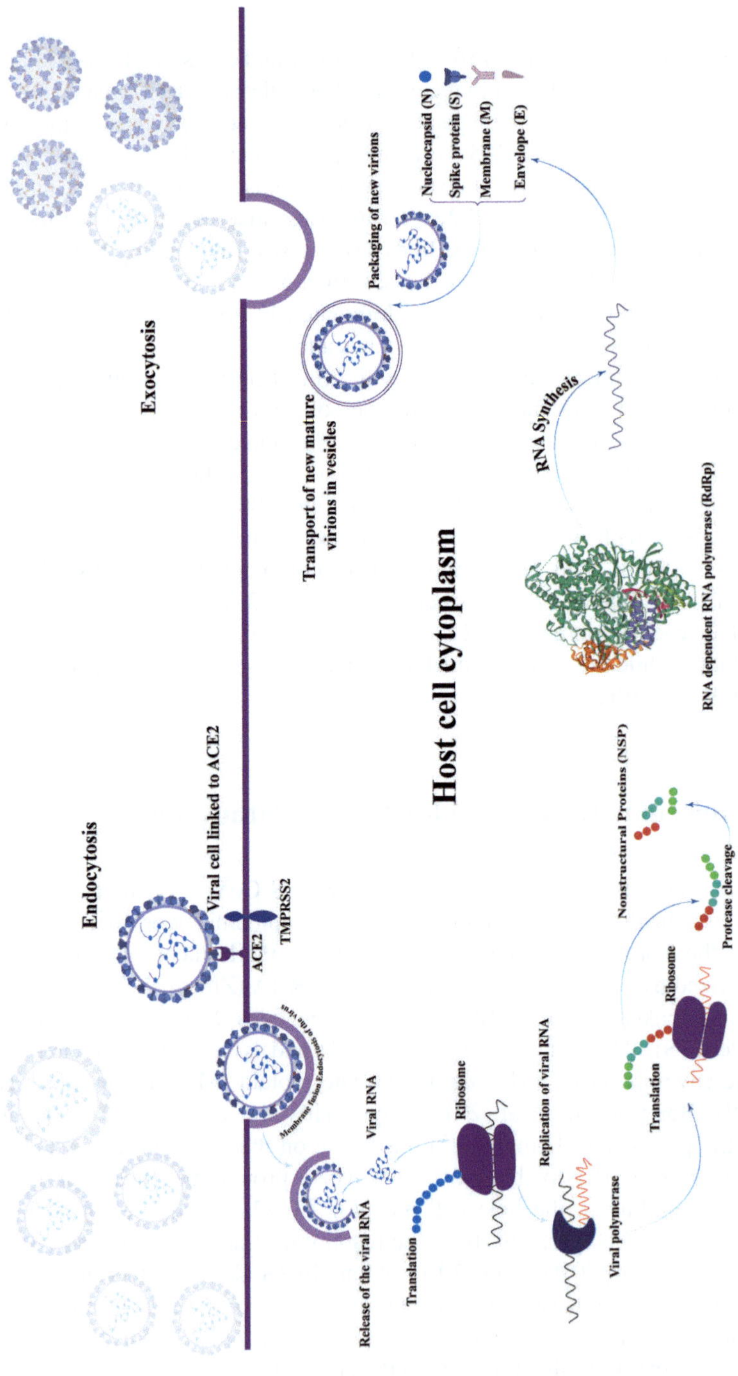


**Fig. 20.1** SARS-CoV-2 structure and pathogenesis mechanism *Nigella sativa* bioactive compounds

particles, which can then infect neighboring cells and spread the infection figure. One of the challenges of SARS-CoV-2 is its ability to mutate and evolve, leading to the emergence of new variants of the virus that may have increased transmissibility and virulence, as well as potential resistance to current treatments and vaccines. The viral RdRp is error-prone meaning that it can make mistakes when copying the viral RNA, leading to the accumulation of mutations in the viral genome over time (Alnefaie and Albogami 2020) (Fig. 20.1). Non-structural proteins (NSPs), which aid in viral replication and immune system evasion, are also encoded by the SARS-CoV-2 genome. The RNA helicases, proteases, and RNA processing enzymes that make up the NSPs are essential for viral replication and translation (Boopathi et al. 2020). A trimeric transmembrane protein called S protein is essential for viral entrance into host cells. The S protein is made up of two functional subunits, S1 and S2, the former of which is in charge of receptor binding and the latter of which is in charge of membrane fusion. The S1 subunit's receptor-binding domain (RBD) interacts with host cells' ACE2 receptors to permit virus entry. Increased binding affinity to the ACE2 receptor brought on by mutations in the RBD region may boost the virus's virulence and transmissibility (Boopathi et al. 2020) (Fig. 20.1). Moreover, the presence of TMPRSS2 was found to be overexpressed in alveolar cells and also was reported to increase the binding between ACE2 and viral S proteins which facilitates viral entrance into host cells (Alnefaie and Albogami 2020). The deregulation of the host immune response, which results in the release of proinflammatory cytokines and chemokines, is one of the characteristics of SARS-CoV-2 infection. In some situations, this immunological reaction can result in multiple organ failures, severe respiratory distress, and even death. The ability of SARS-CoV-2 to inhibit interferon signaling and elude identification by innate immune cells is one of the factors contributing to the virus's ability to circumvent the host immune response and cause severe disease. The viral NSP1 protein has been shown to inhibit host translation and immune responses by binding to the host 40S ribosome and blocking the production of antiviral proteins (Yuan et al. 2021) (Fig. 20.1).

Two proteases that are essential for the replication and maturation of the virus are encoded for by SARS-CoV-2: the main protease (Mpro) and the papain-like protease (PLpro). Structure and function of SARS-CoV-2 proteases Mpro and PLpro are both cysteine proteases that are essential for the maturation of the viral polyproteins that are translated from the viral RNA (Hilgenfeld 2014). Mpro, also known as 3CLpro, cleaves the polyproteins at 11 specific sites, while PLpro cleaves at three specific sites. The cleavage of the polyproteins by these proteases releases the individual proteins necessary for the assembly of new viral particles. Mpro is a homodimeric protease that consists of two identical chains of approximately 33 kDa. The active site is situated in a cleft site between domains I and II in each of the three domains that make up the chain. The peptide link between the glutamine residue and the subsequent residue is broken when Mpro detects a specific sequence in the polyproteins that contains a glutamine residue at the P1 location. The structure of Mpro has been extensively studied, and several inhibitors have been developed that target its active site (Hilgenfeld 2014; Jin et al. 2020). PLpro is a multidomain protease that consists of three domains: an N-terminal ubiquitin-like domain (Ubl), a protease domain, and a C-terminal domain. The protease domain contains the active site, which recognizes a specific sequence in the polyproteins, containing a leucine residue at the P1 position. PLpro cleaves the peptide bond between the leucine and the next residue. The Ubl domain of PLpro is involved in the recognition of ubiquitin and interferes with the host immune response by cleaving and removing ubiquitin from host proteins which protects the viral protein from proteolysis (Hilgenfeld 2014). Because of their significance in the SARS-CoV-2 life cycle, Mpro and PLpro are desirable targets for the creation of COVID-19 antiviral medications (Bailey-Elkin et al. 2017; Hilgenfeld 2014). The newly synthesized viral proteins in the cytoplasmic environment were inserted in the endoplasmic reticulum (ER) and then transferred intermediate compartment between the endoplasmic reticulum and the Golgi apparatus (ERGIC). Moreover, the N protein duplicated genomes make nucleocapsids in the cytoplasm, which then aggregate inside the ERGIC membrane to create new virions. Via the process of exocytosis, fresh virions are exported from infected cells to the cell membrane in smooth-walled vesicles before being secreted to infect more cells (Boopathi et al. 2020) (Fig. 20.1). The ongoing COVID-19 pandemic has highlighted the importance of molecular biology in the diagnosis, treatment, and prevention of viral infections which brings back to the surface the exploration of new alternatives that are endowed with high antiviral properties.

## 20.4 *Nigella sativa* Bioactive Compounds

The seeds and aerial sections of *Nigella sativa* contain bioactive substances from several natural sources, according to phytochemical analysis performed in different studies. The investigation of black cumin volatile compounds composition has demonstrated the presence of thymoquinone, thymol, carvacrol, β-cymene, α-phellandrene (Dalli et al. 2021a, b, c; Kabir et al. 2020). Among the discovered phenolic acids and flavonoids, the presence of rutin, nigelflavonoside B, flavone, apigenin, catechin, quercetin, vanillic acid, p-coumaric acid, ferulic acid, and gallic acid was also noted (Dalli et al. 2021a, b, c; Malhotra 2012; Parveen et al. 2020). Additionally, various bioactive alkaloid compounds including nigellimine, nigellimine N-oxide, nigellicine, and magnoflorine were identified and characterized using different spectral techniques such as NMR (Malik and Zaman 1992; Malik et al. 1995; Atta-ur-Rahman 1985). Additionally, saponins were also identified in *Nigella sativa* extracts such as hederagenine; nigelloside; kaempferol 3-O-rutinoside, flaccidoside (Parveen et al. 2020; Taşkin et al. 2005). Crude fiber, minerals including salt, copper, zinc, phosphorus, and calcium, as well as vitamins like thiamine, niacin, and folic acid were discovered to be present in black cumin seeds (Takruri and Dameh 1998). The GC-MS analysis of the black cumin seeds showed that several different kinds of fatty acids were present. While lauric acid, myristic acid, linolenic acid, and eicosadienoic acid were found in smaller levels and ranged in concentrations from 0.5% to 3.4%, linoleic acid, oleic acid, and palmitic acid were the main fatty acids (Fig. 20.2) (Dalli et al. 2021a, b, c; Nickavar et al. 2003).

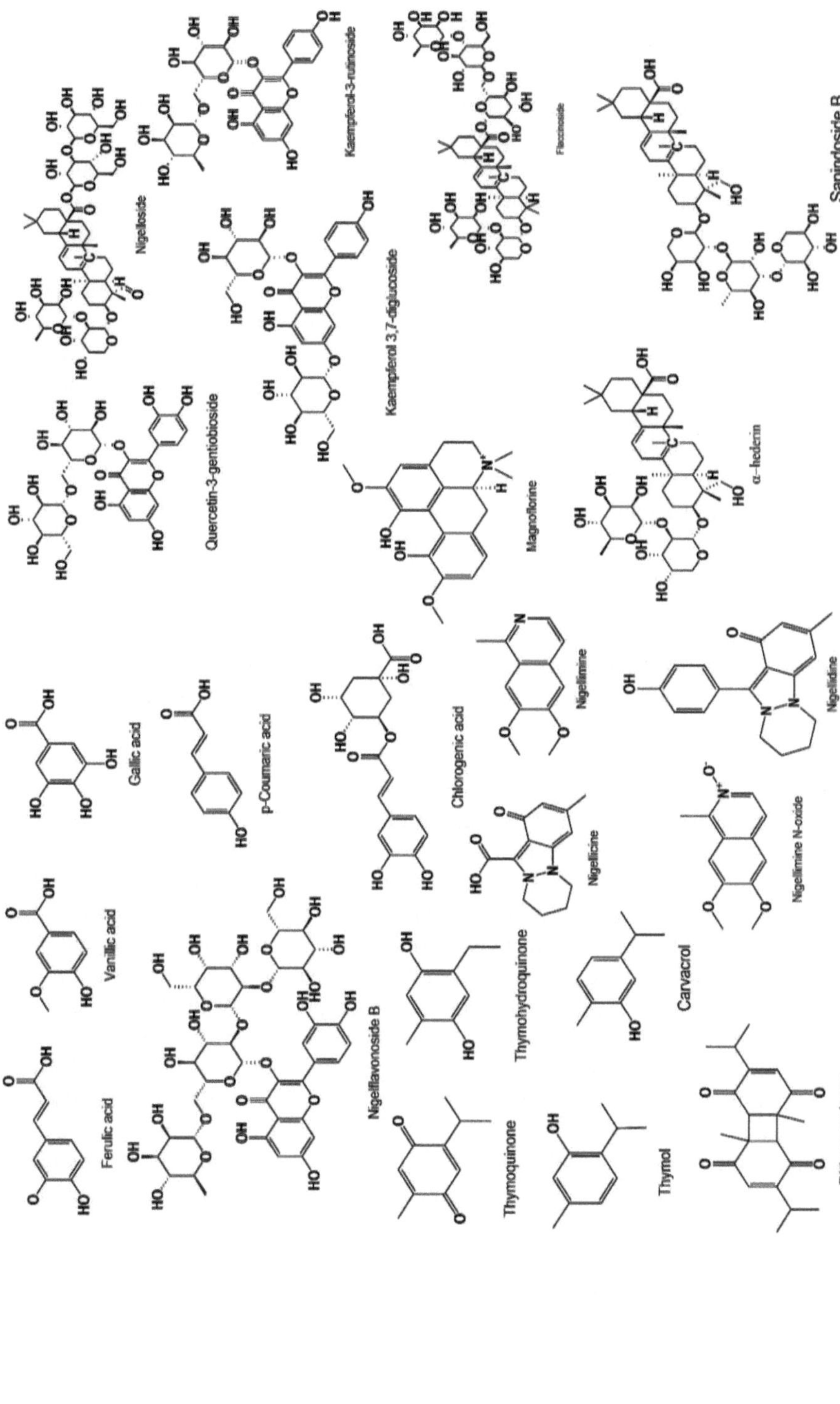


**Fig. 20.2** Principal chemical compounds found in *Nigella sativa* L.

## 20.5 Anti-Inflammatory and Immunomodulatory Activity

In earlier reports, Eo and TQ the major compound identified in NS seeds were reported to reduce inflammation which was comparable to the data collected using indomethacin in the carrageenan model. Actually, when administered intraperitoneally at a dosage of 1.55 mL/Kg, Eo suppressed inflammation by 96.26%. Conversely, at a concentration of 5 mg/mL, TQ exhibited a greater anti-inflammatory effect by suppressing it by 104.88% (Mutabagani and El-Mahdy 1997). Similarly, it was mentioned that TQ was endowed with a great capacity to inhibit cyclooxygenase and 5-lipoxygenase pathways, the same study has demonstrated the ability of NS aqueous extract to attenuate the inflammation and to exert an important analgesic effect which was comparable to aspirin (Al-Ghamdi 2001) (Fig. 20.3). Numerous studies evaluate the NS effect on rats injected with carrageenan which is a chemical compound that can induce inflammation involving phospholipase A2 and by increasing inflammation mediators such as cytokines, serotonin, histamine, and prostaglandins. The findings showed that NSO balm stick application helped reduce the Carrageenan-induced paw edema inflammation in rats which is very obvious in the reduction of oedema. In the same context, the application of a 10% NSO balm stick gave the highest anti-inflammatory effect with an inhibition of 60.64%, also an inhibition of TNF-α, and a reduction of leucocytes (Dwita et al. 2019) (Fig. 20.3). When used in a formaldehyde-induced arthritis model, *N. sativa* has a potent anti-inflammatory effect that is comparable to aspirin which could be attributed to NSO's ability to hinder the generation of eicosanoids (Pise and Padwal 2017). Regarding the cytokines profile, it was observed that IL-1A increased after the application of fresh and stored NS oil. Also, it was observed that stored oil was able to decrease IL-1B, while fresh oil decreased IL-6. Conversely, it was reported that TQ was not able to reduce any cytokine (Bordoni et al. 2019) (Fig. 20.3). Recently it was reported that macrophages increase their phagocytosis ability after exposure to 50 and 100 μg/mL of TQ which is considered a principal mechanism allowing the engulfing of various types of antigens by breaking them down into fragments and presenting these antigen peptides on the cell membrane. This presentation enables the adaptive immune system to identify and target the antigen for the attack. This process is mainly linked to an increase and the expression of the human leukocyte antigen DR isotype and the production of interferon-gamma. These findings suggest that TQ could enhance different properties of macrophages (Alkhattabi et al. 2022). TQ administered orally to carrageenan-mice models over approximately 14 days showed outcomes that were consistent with earlier studies at doses of 50 and 100 mg/Kg. Furthermore, proinflammatory parameters were reduced due to a decrease in MDA and an increase in antioxidant indicators (Hijazy et al. 2022) (Fig. 20.3).

Patients infected with SARS-CoV-2 may experience changes in various health markers. Elevated levels of C-reactive protein, IL-6, serum ferritin, and erythrocyte sedimentation rate may be among these modifications. In addition, the numbers of $CD^{4+}$ and $CD^{8+}$ lymphocytes, monocytes, and platelets may drop, while IL-7, IL-10, TNF-α, and other indicators may be overproduced. Moreover, those who are

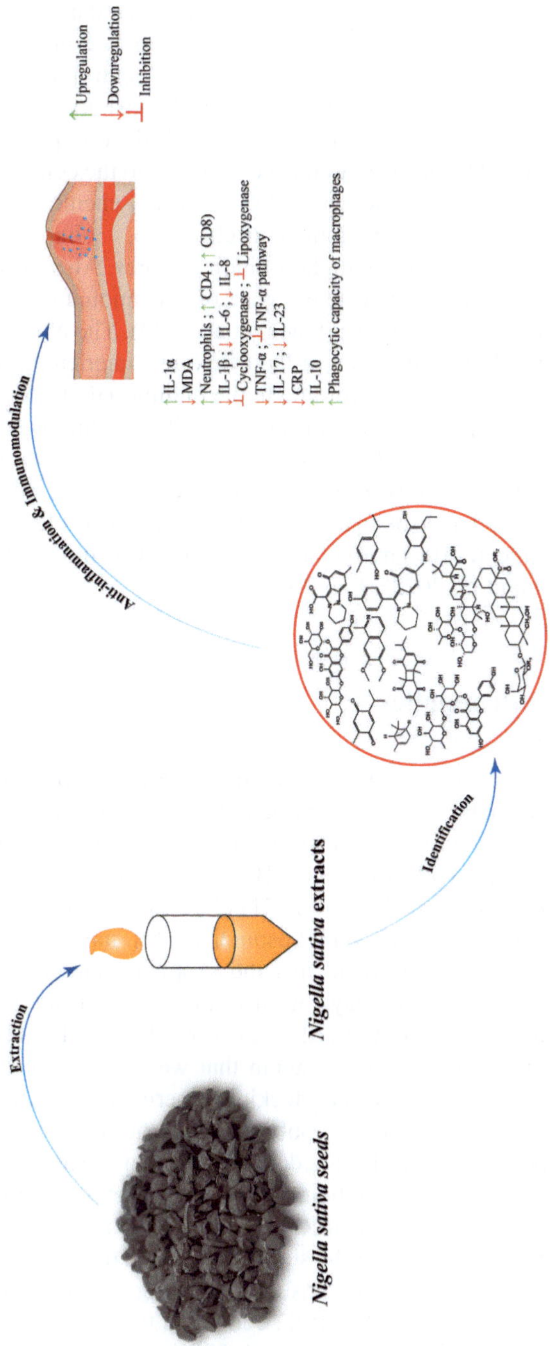


**Fig. 20.3** An overview of the anti-inflammatory and immunomodulatory effects of *Nigella sativa* and its derivatives

infected may have an increase in neutrophils (Rahman 2020). Aside from its anti-inflammatory properties, consuming 2 grams of NS seeds daily for 3 months has been shown to enhance antioxidant activity and cell-mediated immunity in patients with β-thalassemia. This is evidenced by an increase in neutrophil count, as well as an increase in CD4 and CD8 counts (Fig. 20.3) (El-Shanshory et al. 2019). Moreover, the detected amount of IL-1B, IL-6, IL-8, and TNF-α dropped in various NS fractions, although IL-10 levels increased in comparison to the control. According to a molecular docking analysis of the major iNOS pathway chemicals, nigelloside C, rosmarinic acid, quercetin O-rhamnosyl-triglucoside, γ-linolenic acid, and nigellidine-4-O-sulfite exhibited the highest binding scores. Also, it was discovered that roasting had a considerable negative impact on the amount of certain bioactive chemicals with high binding affinities (Salem et al. 2023). The *Nigella sativa* EtOH extract therapy reduced the expression of the cytokine's interleukin in the study on lupus mice produced by pristane, according to the findings (IL-17, IL-6, and IL-23). Moreover, a decline in the expression of regulatory T cells that was nearly normal was observed (Hikmah et al. 2022). The supplementation with *Nigella sativa* showed an important decrease of proinflammatory cytokines in mice like TNF-α, CRP, and IL-6 in infected mice by malaria. It was also noted, an elevation of the anti-inflammatory cytokine IL-10, antioxidant makers in mice model (Gholamnezhad et al. 2021; Ojueromi et al. 2022) (Fig. 20.3).

## 20.6 Antiviral Properties

Many investigations were established to learn more about the effects of this panacea in order to better understand the antiviral activity of *Nigella sativa* and its derivatives. By inhibiting the angiotensin-converting enzyme 2 (ACE2), which stops viral proteins from interacting with host cells, TQ has a significant antiviral potential against COVID-19 in silico (Kadil et al. 2021) (Fig. 20.4). When black cumin seed oil was tested on patients who were hospitalized for having mild COVID-19 symptoms, Koshak et al. 2021 found that the supplementation with 500 mg/Kg reduces the average number of days needed to recover from the viral attack (10.7 days), compared to the control (12.5 days) (Koshak et al. 2021). Four of the nine bioactive compounds from black cumin that were investigated, according to theoretical findings based on molecular docking experiments, displayed high interaction with RNA-dependent RNA polymerase. For instance, compared to Remdesivir, the positive control, α-hederin showed a greater binding affinity (RdRp). The affinities of TQ, nigellicine, and nigellidine were lower than those found in α-hederin but close to those of Remdesivir (Mir et al. 2022) (Fig. 20.4). The inhibition of TMPRSS2 was among the strategies exerted to alleviate SARS-CoV-2 by inhibiting the truncation of S proteins thus blocking its endocytosis. Carvacrol demonstrated high affinity with TMPRSS2 compared to Comstat used as a control while thymol had a weak affinity (Yadav et al. 2021). Kaempferol, nigellidine, and dithymoquinone were some of the compounds with high affinity with COVID-19

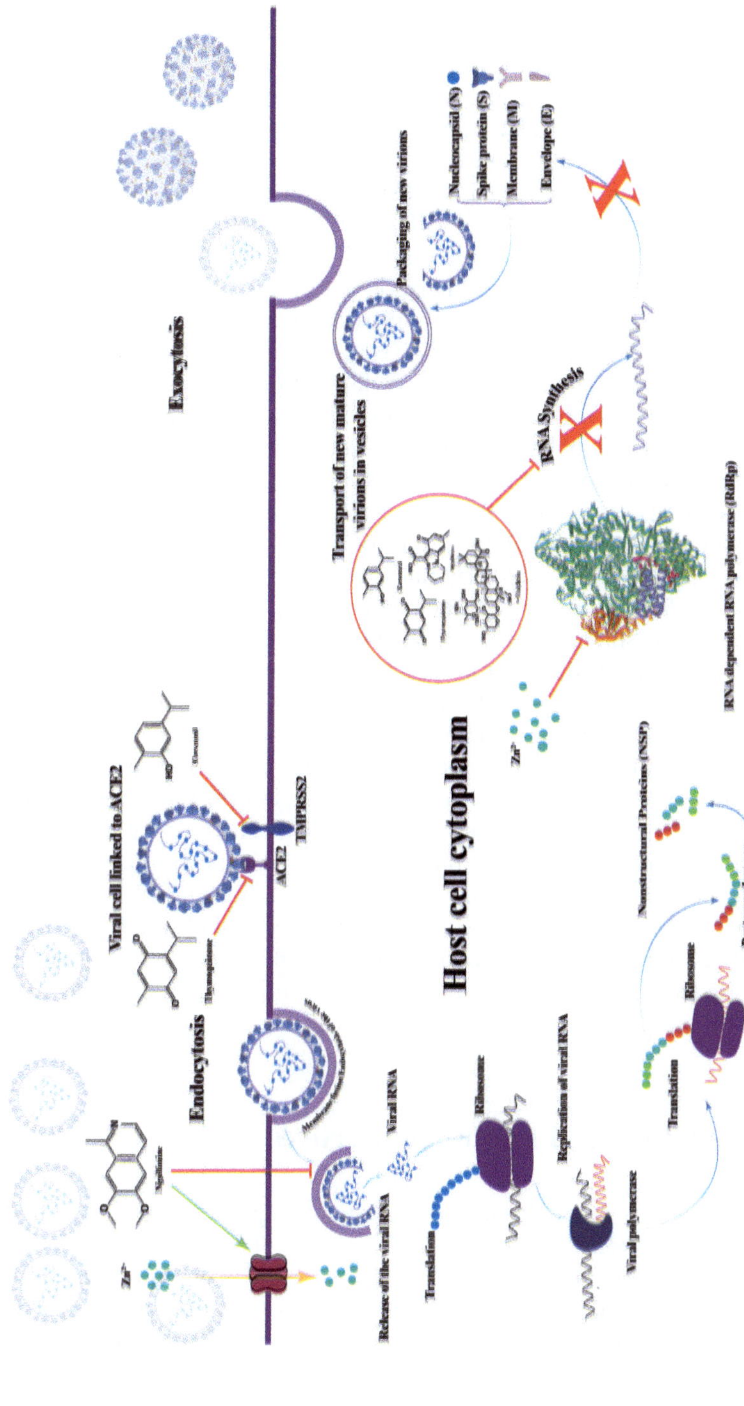


**Fig. 20.4** The antiviral properties of *Nigella sativa* bioactive compounds

proteases (Ahmad et al. 2021). Likewise, *Nigella sativa* biocompounds (TQ, DTQ, nigellicine, and others) were discovered to have antiviral capabilities in addition to their antioxidant and anti-inflammatory effects. The various compounds were also found to have a high interaction with different amino acids (Baig and Srinivasan 2022; Sherwani et al. 2022) (Fig. 20.4). Similar to this, it was observed that hydroxychloroquine (HCQ) and black cumin seeds co-administration decreased toxicity and enhanced HCQ's antiviral effect against COVID-19. A number of theories contend that the alkaloid nigellimine can prevent SARS-CoV-2 from entering its host by inhibiting ACE2 (Khan 2021). In addition, it was shown that nigellimine might have similar ionophore actions to encourage zinc uptake and entry into pneumocytes, assisting in the stimulation of the immune system against SARS-CoV-2 (Rahman 2020) (Fig. 20.4). In silico analysis performed by Khan et al. revealed that TQ, a phytochemical derived from NS essential oil, could be a promising candidate to combat SARS-CoV-2 due to its interaction with the main protein active site (C145, H41), which changes the viral protein's catalytic activity which encourages the application of this molecule for clinical applications, which was checked according to the drug likeliness guidelines (Khan et al. 2024). For more accuracy, the in vitro antiviral of the various *NS* derivates was also verified. In fact, the methanolic extract induced inhibition of SARS-CoV-2 main protease by a percentage of 31.5%. Meanwhile, TQ inhibited the main protease by 63.21%. The $IC_{50}$ value recorded was 10.26 μM. The finding reported in this study confirmed the results obtained in silico. Conversely, a low selectivity index supports the anti-SARS-CoV-2 action (Abdallah et al. 2022) (Fig. 20.4). Recently, it was indicated that the supplementation with NS and Vitamine D at different dosages during 14 days to 120 patients diagnosed to be COVID-19 positive. During the supplementation period, it was noted that the three treated groups by NS or by NS+ VitD played an important role in reducing the severity of caught, diarrhea, fatigue, and pharyngitis compared to the control group. Also, supplementation with the mentioned mixture has reduced the clearance time from the viral infection and reduced the severity and progression of the virus (Said et al. 2022). In counterpart, other studies have noted that the supplementation of 262 subjects with whole seeds or the powder of NS showed no significant changes in symptoms and inflammatory markers during 10 days of treatment which suggest that the *Nigella sativa* action requires more time to exert its antiviral effect (Bin Abdulrahman et al. 2022).

## 20.7 Conclusion and Future Perspectives

Available evidence suggests that due to its antiviral properties, *Nigella sativa* and its derivatives may be beneficial against SARS-CoV-2. These substances have been demonstrated to affect host immunological responses as well as viral entrance, replication, and transcription. To confirm these results and establish the ideal dosage and treatment plan, additional research is required. Future studies should focus on finding the precise active ingredients in *N. sativa* that are responsible for its antiviral activities and thoroughly examining their mechanisms of action (in vitro, in vivo). *N. sativa* must also undergo clinical trials to determine its efficacy and safety as a potential treatment for SARS-CoV-2 and other viral diseases. Overall, Nigella sativa represents a viable route for the creation of fresh antiviral treatments and calls for additional research.

**Acknowledgments** The authors of this chapter would like to express their gratitude to the laboratory technician Mr. Ahmed Khalaf.

## References

Abdallah HM, El-Halawany AM, Darwish KM, Algandaby MM, Mohamed GA, Ibrahim SRM, Koshak AE, Elhady SS, Fadil SA, Alqarni AA, Abdel-Naim AB, Elfaky MA (2022) Bio-guided Isolation of SARS-CoV-2 Main protease inhibitors from medicinal plants: in vitro assay and molecular dynamics. Plants (Basel) 11(15):1914. https://doi.org/10.3390/plants11151914

Ahmad S, Abbasi HW, Shahid S, Gul S, Abbasi SW (2021) Molecular docking, simulation and MM-PBSA studies of nigella sativa compounds: a computational quest to identify potential natural antiviral for COVID-19 treatment. J Biomol Struct Dyn 39(12):4225–4233. https://doi.org/10.1080/07391102.2020.1775129

Al-Ghamdi MS (2001) The anti-inflammatory, analgesic and antipyretic activity of Nigella sativa. J Ethnopharmacol 76(1):45–48. https://doi.org/10.1016/S0378-8741(01)00216-1

Ali BH (2015) Pharmacological and toxicological properties of Nigella sativa. Phytother Res 17(4): 299–305. https://doi.org/10.1002/ptr.1309

Alkhattabi NA, Hussein SA, Tarbiah NI, Alzahri RY, Khalifa R (2022) Thymoquinone effect on monocyte-derived macrophages, cell-surface molecule expression, and phagocytosis. Nutrients 14(24):1–13. https://doi.org/10.3390/nu14245240

Alnefaie A, Albogami S (2020) Current approaches used in treating COVID-19 from a molecular mechanisms and immune response perspective. Saudi Pharm J 28(11):1333–1352. https://doi.org/10.1016/j.jsps.2020.08.024

Atta-ur-Rahman SM (1985) Isolation and structure determination of nigellicine, a novel alkaloid from the seeds of nigella sativa. Tetrahedron Lett 26:2759–2762

Baig A, Srinivasan H (2022) SARS-CoV-2 inhibitors from nigella sativa. In Applied biochemistry and biotechnology. 194(3). Springer, Cham, pp 1051. https://doi.org/10.1007/s12010-021-03790-8

Bailey-Elkin BA, Knaap RCM, Kikkert M, Mark BL (2017) Structure and function of viral deubiquitinating enzymes. J Mol Biol 429(22):3441–3470. https://doi.org/10.1016/j.jmb.2017.06.010

Bin Abdulrahman KA, Bamosa AO, Bukhari AI, Siddiqui IA, Arafa MA, Mohsin AA, Althageel MF, Aljuaeed MO, Aldeailej IM, Alrajeh AI, Aldosari KM, Hawsawi NA, Zawbaee KI, Alsurayea SM (2022) The effect of short treatment with Nigella sativa on symptoms, the cluster of differentiation (CD) profile, and inflammatory markers in mild COVID-19 patients: a randomized, double-blind controlled trial. Int J Environ Res Public Health 19(18). https://doi.org/10.3390/ijerph191811798

Boopathi S, Poma AB, Kolandaivel P (2020) Novel 2019 coronavirus structure, mechanism of action, antiviral drug promises and rule out against its treatment. J Biomol Struct Dyn 39:1–10. https://doi.org/10.1080/07391102.2020.1758788

Bordoni L, Fedeli D, Nasuti C, Maggi F, Papa F, Wabitsch M, De Caterina R, Gabbianelli R (2019) Antioxidant and anti-inflammatory properties of Nigella sativa oil in human pre-adipocytes. Antioxidants 8(2):1–12. https://doi.org/10.3390/antiox8020051

Bourgou S, Pichette A, Marzouk B, Legault J (2010) Bioactivities of black cumin essential oil and its main terpenes from Tunisia. S Afr J Bot 76(2):210–216. https://doi.org/10.1016/j.sajb.2009.10.009

Burits M, Bucar F (2000) Antioxidant activity of Nigella sativa essential oil. Phytother Res 14(5): 323–328. https://doi.org/10.1002/1099-1573(200008)14:5<323::AID-PTR621>3.0.CO;2-Q

Chan JF-W, To KKW, Tse H, Jin D-Y, Yuen K-Y (2013) Interspecies transmission and emergence of novel viruses: lessons from bats and birds. Trends Microbiol 21(10):544–555. https://doi.org/10.1016/j.tim.2013.05.005

Dalli M, Azizi S, Benouda H, Azghar A, Tahri M, Boufalja B, Maleb AG, N. (2021a) Molecular composition and antibacterial effect of five essential oils extracted from Nigella sativa L seeds against multidrug-resistant bacteria: a comparative study. Evid Based Complement Alternat Med 2021:6643765. https://doi.org/10.1155/2021/6643765

Dalli M, Azizi S, Kandsi F, Gseyra N (2021b) Evaluation of the in vitro antioxidant activity of different extracts of Nigella sativa L. seeds, and the quantification of their bioactive compounds. Mater Today Proc 45(P 8):7259–7263. https://doi.org/10.1016/j.matpr.2020.12.743

Dalli M, Daoudi NE, Azizi S, Benouda H, Bnouham M, Gseyra N (2021c) Chemical composition analysis using HPLC-UV/GC-MS and inhibitory activity of different Nigella sativa fractions on pancreatic α-amylase and intestinal glucose absorption. Bio Med Res Int 2021:9979419

Dalli M, Daoudi NE, Abrigach F, Azizi S, Bnouham M, Kim B, Gseyra N (2022) In vitro α-amylase and hemoglobin glycation inhibitory potential of Nigella sativa essential oil, and molecular docking studies of its principal components. Front Pharmacol 13:1036129. https://doi.org/10.3389/fphar.2022.1036129

Dwita LP, Yati K, Gantini SN (2019) The anti-inflammatory activity of nigella sativa balm sticks. Sci Pharm 87(1). https://doi.org/10.3390/scipharm87010003

El-Shanshory MH, Aboonq NM, Fakhreldin MS, Attia AR, Arafa M, Walaa Mariah RA, Baghdadi H, Ayat M, Zolaly M, Nabo MMH, Almaramhy HH, El-Sawy SA, Zidan M, Elshazley M, Alharbi R, Moustafa S, Abu-el Naga M, El Sayed SM (2019) Nigella sativa improves anemia, enhances immunity and relieves iron overload-induced oxidative stress as a novel promising treatment in children having beta-thalassemia major. J Herb Med 16:100245. https://doi.org/10.1016/j.hermed.2018.11.001

Fakchich J, Elachouri M (2021) An overview on ethnobotanico-pharmacological studies carried out in Morocco, from 1991 to 2015: systematic review (part 1). J Ethnopharmacol 267(part 1): 113200. https://doi.org/10.1016/j.jep.2020.113200

Ganesh B, Rajakumar T, Malathi M, Manikandan N, Nagaraj J, Santhakumar A, Elangovan A, Malik YS (2021) Epidemiology and pathobiology of SARS-CoV-2 (COVID-19) in comparison with SARS, MERS: an updated overview of current knowledge and future perspectives. Clin Epidemiology Glob Health 10:100694. https://doi.org/10.1016/j.cegh.2020.100694

Gholamnezhad Z, Boskabady MH, Hosseini M (2021) The effect of chronic supplementation of Nigella sativa on splenocytes response in rats following treadmill exercise. Drug Chem Toxicol 44(5):487–492. https://doi.org/10.1080/01480545.2019.1617301

Hasoksuz M, Alekseev K, Vlasova A, Zhang X, Spiro D, Halpin R, Wang S, Ghedin E, Saif LJ (2007) Biologic, antigenic, and full-length genomic characterization of a bovine-like coronavirus isolated from a giraffe. J Virol 81(10):4981–4990. https://doi.org/10.1128/JVI.02361-06

Hasöksüz M, Kiliç S, Saraç F (2020) Coronaviruses and SARS-COV-2. Turkish J Med Sci 50 (SI-1):549–556. https://doi.org/10.3906/sag-2004-127

Hijazy HHA, Dahran N, Althagafi HA, Alharthi F, Habotta OA, Oyouni AAA, Algahtani M, Theyab A, Al-Amer O, Lokman MS, Alsharif KF, Albrakati A, Amin HK, Dawood SM, Kassab RB, Ellethy RA (2022) Thymoquinone counteracts oxidative and inflammatory machinery in carrageenan-induced murine paw edema model. Environ Sci Pollut Res 30:16597. https://doi.org/10.1007/s11356-022-23343-8

Hikmah Z, Endaryanto A, Ugrasena IDG, Rahaju AS, Arifin S (2022) Nigella sativa L. as immunomodulator and preventive effect on renal tissue damage of lupus mice induced by pristane. Heliyon 8(4):e09242. https://doi.org/10.1016/j.heliyon.2022.e09242

Hilgenfeld R (2014) From SARS to MERS: crystallographic studies on coronaviral proteases enable antiviral drug design. FEBS J 281(18):4085–4096. https://doi.org/10.1111/febs.12936

Jin Z, Du X, Xu Y, Deng Y, Liu M, Zhao Y, Zhang B, Li X, Zhang L, Peng C, Duan Y, Yu J, Wang L, Yang K, Liu F, Jiang R, Yang X, You T, Liu X, Guddat LW, Xu W, Xiao G, Qin C, Shi Z, Jiang H, Rao Z, Yang H (2020) Structure of Mpro from SARS-CoV-2 and discovery of its inhibitors. Nature 582(7811):289–293. https://doi.org/10.1038/s41586-020-2223-y

Kabir Y, Akasaka-Hashimoto Y, Kubota K, Komai M (2020) Volatile compounds of black cumin (Nigella sativa L.) seeds cultivated in Bangladesh and India. Heliyon 6(10):e05343. https://doi.org/10.1016/j.heliyon.2020.e05343

Kadil Y, Mouhcine M, Filali H (2021) In silico investigation of the SARS CoV2 protease with thymoquinone, the major constituent of nigella sativa. Curr Drug Discov Technol 18(4): 570–573. https://doi.org/10.2174/1570163817666200712164406

Kehili N, Saka S, Aouacheri O (2018) L'effet phytoprotecteur de la nigelle (Nigella sativa) contre la toxicité induite par le cadmium chez les rats. Phytothérapie 16(4):194–203

Khan SA (2021) Combating COVID-19 : the role of drug repurposing and medicinal plants. J Infect Public Health 14(4):495–503. https://doi.org/10.1016/j.jiph.2020.10.012

Khan MT, Ali A, Wei X, Nadeem T, Muhammad S, Al-Sehemi AG, Wei D (2024) Inhibitory effect of thymoquinone from Nigella sativa against SARS-CoV-2 main protease. An in-silico study. Braz J Biol 84:1–7. https://doi.org/10.1590/1519-6984.25066

Koshak AE, Koshak EA, Mobeireek AF, Badawi MA, Wali SO, Malibary HM, Atwah AF, Alhamdan MM, Almalki RA, Madani TA (2021) Complementary therapies in medicine Nigella sativa for the treatment of COVID-19 : an open-label randomized controlled clinical trial. Complement Ther Med 61:102769. https://doi.org/10.1016/j.ctim.2021.102769

Malhotra SK (2012) Nigella. In: Handbook of herbs and spices, vol 2, 2nd edn. Woodhead Publishing, Oxford, pp 391–416. https://doi.org/10.1533/9780857095688.391

Atta-ur-Rahman, Malik S, Zaman K (1992) Nigellimine: a new isoquinoline alkaloid from the seeds of nigella sativa. J Nat Prod 55(5):676–678. https://doi.org/10.1021/np50083a020

Atta-ur-Rahman, Malik S, Hasan SS, Choudhary MI, Ni CZ, Clardy J (1995) Nigellidine—a new indazole alkaloid from the seeds of Nigella sativa. Tetrahedron Lett 36(12):1993–1996. https://doi.org/10.1016/0040-4039(95)00210-4

Mir SA, Firoz A, Alaidarous M, Alshehri B, Bin Dukhyil AA, Banawas S, Alsagaby SA, Alturaiki W, Bhat GA, Kashoo F, Abdel-Hadi AM (2022) Identification of SARS-CoV-2 RNA-dependent RNA polymerase inhibitors from the major phytochemicals of Nigella sativa: an in silico approach. Saudi J Biol Sci 29(1):394–401. https://doi.org/10.1016/j.sjbs.2021.09.002

Mutabagani A, El-Mahdy SAM (1997) A study of the anti-inflammatory activity of Nigella sativa L. and thymoquinone in rats. Saudi Pharma J 5(2–3):110–113

Nickavar B, Mojab F, Javidnia K, Roodgar Amoli MA (2003) Chemical composition of the fixed and volatile oils of Nigella sativa L. from Iran. Z Naturforsch C J Biosci 58(9–10):629–631. https://doi.org/10.1515/znc-2003-9-1004

Ojueromi OO, Oboh G, Ademosun AO (2022) Effect of black seeds (Nigella sativa) on inflammatory and immunomodulatory markers in plasmodium berghei-infected mice. J Food Biochem 46(11):1–27. https://doi.org/10.1111/jfbc.14300

Parveen A, Farooq MA, Kyunn WW (2020) A new oleanane type saponin from the aerial parts of nigella sativa with anti-oxidant and anti-diabetic potential. Molecules 25(9):1–13. https://doi.org/10.3390/molecules25092171

Pise HN, Padwal SL (2017) Evaluation of anti-inflammatory activity of nigella sativa: an experimental study. Natl J Physiol Pharm Pharmacol 7(7):707–711. https://doi.org/10.5455/njppp.2017.7.0204705032017

Prompetchara E, Ketloy C, Palaga T (2020) Immune responses in COVID-19 and potential vaccines: lessons learned from SARS and MERS epidemic. Asian Pac J Allergy Immunol 38(1):1–9. https://doi.org/10.12932/AP-200220-0772

Rahman MT (2020) Potential benefits of combination of Nigella sativa and Zn supplements to treat COVID-19. J Herb Med 23:100382. https://doi.org/10.1016/j.hermed.2020.100382

Said SA, Abdulbaset A, El-Kholy AA, Besckales O, Sabri NA (2022) The effect of Ni gella sativa and vitamin D3 supplementation on the clinical outcome in COVID-19 patients: a randomized controlled clinical trial. Front Pharmacol 13(November):1–14. https://doi.org/10.3389/fphar.2022.1011522

Salem ML (2005) Immunomodulatory and therapeutic properties of the Nigella sativa L. seed. Int Immunopharmacol 5(13–14):1749–1770. https://doi.org/10.1016/j.intimp.2005.06.008

Salem MA, El-Shiekh RA, Aborehab NM, Al-Karmalawy AA, Ezzat SM, Alseekh S, Fernie AR (2023) Metabolomics driven analysis of Nigella sativa seeds identifies the impact of roasting on the chemical composition and immunomodulatory activity. Food Chem 398(August 2022):133906. https://doi.org/10.1016/j.foodchem.2022.133906

Salzberger B, Buder F, Lampl B, Ehrenstein B, Hitzenbichler F, Holzmann T, Schmidt B, Hanses F (2021) Epidemiology of SARS-CoV-2. Infection 49(2):233–239. https://doi.org/10.1007/s15010-020-01531-3

Seyed Hosseini E, Riahi Kashani N, Nikzad H, Azadbakht J, Hassani Bafrani H, Haddad Kashani H (2020) The novel coronavirus disease-2019 (COVID-19): mechanism of action, detection and recent therapeutic strategies. Virology 551:1–9. https://doi.org/10.1016/j.virol.2020.08.011

Sherwani S, Rajendrasozhan S, Khan MWA, Saleem M, Khan M, Khan S, Raafat M, Othman Alqahtani F (2022) Pharmacological profile of Nigella sativa seeds in combating COVID-19 through in-vitro and molecular docking studies. PRO 10(7):1346. https://doi.org/10.3390/pr10071346

Song Z, Xu Y, Bao L, Zhang L, Yu P, Qu Y, Zhu H, Zhao W, Han Y, Qin C (2019) From SARS to MERS, thrusting coronaviruses into the spotlight. Viruses 11(1):59. https://doi.org/10.3390/v11010059

Takruri HRH, Dameh MAF (1998) Study of the nutritional value of black cumin seeds (Nigella sativa L). J Sci Food Agric 76(3):404–410. https://doi.org/10.1002/(SICI)1097-0010(199803)76:3<404::AID-JSFA964>3.0.CO;2-L

Taşkin MK, Alankuş Çalişkan Ö, Anil H, Abou-Gazar H, Khan IA, Bedir E (2005) Triterpene saponins from Nigella sativa L. Turk J Chem 29(5):561–569

UNIDO (2006) Compendium of medicinal and aromatic plants. https://www.unido.org/sites/default/files/2009-10/Compendium_of_medicinal_and_aromatic_plants_volume_2_0.pdf

WHO (2023) WHO coronavirus (COVID-19) dashboard|WHO coronavirus (COVID-19) dashboard with vaccination data; https://covid19.who.int/

Xie M, Chen Q (2020) Insight into 2019 novel coronavirus—an updated interim review and lessons from SARS-CoV and MERS-CoV. Int J Infect Dis 94:119–124. https://doi.org/10.1016/j.ijid.2020.03.071

Yadav PK, Jaiswal A, Singh RK (2021) In silico study on spice-derived antiviral phytochemicals against SARS-CoV-2 TMPRSS2 target. J Biomol Struct Dyn 1–11:11874. https://doi.org/10.1080/07391102.2021.1965658

Yuan S, Balaji S, Lomakin IB, Xiong Y (2021) Coronavirus Nsp 1: immune response suppression and protein expression inhibition. Front Microbiol 12:752214. https://doi.org/10.3389/fmicb.2021.752214

Zhu N, Zhang D, Wang W, Li X, Yang B, Song J, Zhao X, Huang B, Shi W, Lu R, Niu P, Zhan F, Ma X, Wang D, Xu W, Wu G, Gao GF, Tan W (2020) A novel coronavirus from patients with pneumonia in China, 2019. N Engl J Med 382(8):727–733. https://doi.org/10.1056/nejmoa2001017

MIX
Papier aus verantwortungsvollen Quellen
Paper from responsible sources
FSC® C105338

If you have any concerns about our products,
you can contact us on
**ProductSafety@springernature.com**

In case Publisher is established outside the EU,
the EU authorized representative is:
**Springer Nature Customer Service Center GmbH**
**Europaplatz 3, 69115 Heidelberg, Germany**

Printed by Libri Plureos GmbH
in Hamburg, Germany